Lecture Notes in Networks and Systems

Volume 386

The series "Lecture Notes in Networks and Systems" publishes the latest developments in Networks and Systems—quickly, informally and with high quality. Original research reported in proceedings and post-proceedings represents the core of LNNS.

Volumes published in LNNS embrace all aspects and subfields of, as well as new challenges in, Networks and Systems.

The series contains proceedings and edited volumes in systems and networks, spanning the areas of Cyber-Physical Systems, Autonomous Systems, Sensor Networks, Control Systems, Energy Systems, Automotive Systems, Biological Systems, Vehicular Networking and Connected Vehicles, Aerospace Systems, Automation, Manufacturing, Smart Grids, Nonlinear Systems, Power Systems, Robotics, Social Systems, Economic Systems and other. Of particular value to both the contributors and the readership are the short publication timeframe and the world-wide distribution and exposure which enable both a wide and rapid dissemination of research output.

The series covers the theory, applications, and perspectives on the state of the art and future developments relevant to systems and networks, decision making, control, complex processes and related areas, as embedded in the fields of interdisciplinary and applied sciences, engineering, computer science, physics, economics, social, and life sciences, as well as the paradigms and methodologies behind them.

Indexed by SCOPUS, INSPEC, WTI Frankfurt eG, zbMATH, SCImago.

All books published in the series are submitted for consideration in Web of Science.

For proposals from Asia please contact Aninda Bose (aninda.bose@springer.com).

More information about this series at https://link.springer.com/bookseries/15179

Liviu Moldovan · Adrian Gligor
Editors

The 15th International Conference Interdisciplinarity in Engineering

Conference Proceedings

Springer

Editors
Liviu Moldovan
Faculty of Engineering and Information Technology "George Emil Palade" University of Medicine, Pharmacy, Science, and Technology of Targu Mures
Târgu Mureş, Romania

Adrian Gligor
Faculty of Engineering and Information Technology "George Emil Palade" University of Medicine, Pharmacy, Science, and Technology of Targu Mures
Târgu Mureş, Romania

ISSN 2367-3370 ISSN 2367-3389 (electronic)
Lecture Notes in Networks and Systems
ISBN 978-3-030-93816-1 ISBN 978-3-030-93817-8 (eBook)
https://doi.org/10.1007/978-3-030-93817-8

This Springer imprint is published by the registered company Springer Nature Switzerland AG
The registered company address is: Gewerbestrasse 11, 6330 Cham, Switzerland

Foreword

These proceedings contain research papers that were accepted for presentation at the 15th International Conference Inter-Eng 2021 "Interdisciplinarity in Engineering", which was held on 7–8 October 2021, in the city of Târgu Mureş, Romania. The conference slogan is "Innovative aspects of Industry 4.0 concepts aimed at consolidating the digital future of manufacturing in companies". This proceedings book contains a rich experience of the academic and research institutions and the industry on diverse themes related to advances and innovation in technology.

The International Conference Inter-Eng "Interdisciplinarity in Engineering" is a leading international professional and scientific framework dedicated to engineers and scientists, mainly focused on disseminating research results, scientific contributions and recent technological developments, as well as current practices in engineering. This yearly established conference represents an important traditional scientific event, taking place at Faculty of Engineering and Information Technology, "George Emil Palade" University of Medicine, Pharmacy Science and Technology of Târgu Mureş, Romania.

Inter-Eng marks this year eighteen years of existence, the first edition taking place in 2003. Meanwhile, the conference prestige grew, attracting an important number of regional and international institutions in becoming conference supporters and partners. So this year's edition is organised in cooperation with the Romanian Academy of Technical Sciences, McMaster University from Canada and Zagazig University from Egypt. The conference partners from Romania are Romanian General Association of Engineers in Romania, Mureş County Council and Târgu Mureş City Hall.

From another perspective, the international visibility of the conference results is an outstanding increase, the conference proceedings in the previous seven editions Inter-Eng 2013, up to Inter-Eng 2021 being published as dedicated issues in Elsevier's Procedia Technology Journal, Procedia Engineering Journal and Procedia Manufacturing Journal, as well as MDPI's Proceedings and made available in open access on Elsevier's Science Direct for researchers worldwide. These volumes are indexed in the Clarivate Analytics Conference Proceedings Citations Index.

Inter-Eng conference has a new structure regarding the organisation frequency that became annual, but also the organisation of the sessions by including modern technologies of distance communication within online virtual panel, which facilitates mobility and interaction of participants with reduced costs and high efficiency.

Inter-Eng conference starts from the observation that today in full twenty-first century, in the era of high technology, without new approaches in research one cannot speak of a harmonious society. The theme of the conference proposing a new approach related to "Industry 4.0" the development of a new generation of smart factories grounded on the manufacturing and assembly process digitalization is related to advanced manufacturing technology, lean manufacturing, sustainable manufacturing, additive manufacturing, manufacturing tools and equipment.

The conference slogan in this year is "Innovative aspects of Industry 4.0 concepts aimed at consolidating the digital future of manufacturing in companies".

Industry is a central pillar of the European economy, and the EU production sector represents more than 95 million jobs. Nowadays, the challenge is to ensure that all industrial sectors make the best use of new technologies and to manage their transition to higher value products and processes, commonly known as "Industry 4.0".

Currently, manufacturing production is determined by international competition and the requirement to rapidly adapt production to market needs. But manufacturing companies are facing a huge challenge in implementing Industry 4.0 solutions. The concept of the industry is not only limited to the manufacturing system, but also comprises the supply system and the sales system of the enterprise, its entire value chain which constitutes a globalised system of links of all the enterprises including all their functions and services.

These demands can only be met through the radical advances of current manufacturing technology supported by Industry 4.0 which is integrating business and production processes, as well as integrating suppliers and customers into the company's value chain.

Through vertical networking of the smart factory's cyber physical systems, changes in stocks or demand or even faults in equipment are quickly addressed. Both production and maintenance management of the factory can be organised automatically and independently of each other. With real-time virtualisation of everything in the factory, parts and equipment can be located anywhere, and the workflows are self-organised for optimum efficiency.

Horizontal integration brings the efficiency of the network to other outside organisations, such as subcontractors, suppliers, logistics service providers, distribution points and customers. As products become more integrated with the Internet of Things, through embedded electronics and communication technologies, the link with the manufacturer is maintained throughout its life cycle.

Exponential technologies represent one of the major characteristics of Industry 4.0 serving as a catalyst for improvements in the manufacturing process. These technologies are evolving and enabling change at an accelerating pace. Innovation through exponential technologies can help manufacturers develop faster, be more flexible and unlock new forms of value.

Governed by vertical and horizontal integration and enabled by cyber physical systems, as well as by the Internet of Things, the organisation of a factory of the future will be more flexible, changeable, decentralised and not as deterministic as today's organisations.

Meeting these advances is different, because at present there are only a few "islands" of the Industry 4.0 concept.

Delays of manufacturing companies in adapting to the transformations produced by digitization will lead to low competitiveness, with adverse effects for their future. The better the new technologies of "Industry 4.0" are understood and applied in practice, the more these companies will gain a competitive advantage.

This year, the Inter-Eng conference has distinguished keynote speakers:

- Prof. Dr. Dan Centea, McMaster University, Canada
- Prof. Dr. Habil. Laszlo Kovacs, University of Miskolc, Hungary
- Prof. Dr. Abdelazim M. Negm, University of Zagazig, Egypt
- Prof. Dr. Ahmed Hussein Ibrahim, University of Zagazig, Egypt.

They are researchers with outstanding results in their field of activity, giving the scientific weight to the conference. On behalf of the scientific committee, we thank them for attending the Inter-Eng 2021 conference.

During this edition of the conference, 74 papers distributed in five sessions are listed. It is the remarkable international participation with papers from nine countries on four continents such as Romania, Hungary, Portugal, Spain, Canada, Egypt, Morocco, Saudi Arabia or Turkey.

The Inter-Eng 2021 conference submissions have been anonymously reviewed by two independent reviewers, to ensure the final standard of the accepted submissions. On behalf of the scientific committee of Inter-Eng 2021, we thank all reviewers for their hard work.

We are especially grateful to the authors who submitted their papers to this conference and to the presenters who provided the substance of the meeting.

This proceedings book contains a rich experience of the academic and research institutions and the industry on diverse themes related to advances and innovation in technology. We do hope that researchers, knowledge workers and innovators both in academia and industry will find it a valuable reference material.

These conference proceedings are made available with the professional support from:

Răzvan Cazacu
Mircea Dulău
Lucian-Ioan Dulău
Stelian-Emilian Oltean
Liviu-Dorin Pop
Technical Editing.

The Inter-Eng conference becomes a tradition, and we are optimistic that it will be maintained in future, contributing to the development of emerging technologies in various fields.

Last but not least, we hope that everybody had a good time in Târgu Mureş, and we invite participants for the next year's edition of the Inter-Eng conference.

October 2021

Liviu Moldovan
President of Scientific Committee

Adrian Gligor
Executive Director

Reference

https://inter-eng.umfst.ro/2021/.

Contents

Power Systems, Energy Efficiency and Renewable Technologies

Advanced Manufacturing Technologies and Materials

Zn-Al Anticorrosive Coating Adapted to Obtain Protected Steel Wires

Marius Tintelecan[1(✉)], Dana-Adriana Iluțiu-Varvara[1], Oscar Rodriguez Alabanda[2], Ioana-Monica Sas-Boca[1], and Gustavo Aristides Santana Martinez[3]

[1] Technical University of Cluj Napoca, B-dul. Muncii, no. 103-105, Cluj Napoca, Romania
marius.tintelecan@ipm.utcluj.ro

[2] University of Córdoba, Edificio Paraninfo, Primera Planta. Campus de Rabanales, 14071 Córdoba, Spain

[3] University of São Paulo, Engineering School of Lorena, Lorena 12602-810, Brazil

Abstract. The present work describes and quantifies the process of adding Al in thermally deposited Zn on the surface of a S235 steel wire in order to increase the anticorrosive efficiency of this protective layer. Zn crystallizes in an hexagonal system that will inevitably deform mostly by twinning. A first fundamental question is to determine which is the percentage of Al, which must be found integrated in the Zn matrix, necessary for the loss of protective layer registered when the wire is processed by drawing to the lower reduction possible. Zn-Al alloy specimens with an increasing percentage of Al have been processed by pressing. The pressing process was conducted under certain conditions of temperature and speed, simulating to a certain extent its deformation by drawing, being affected the surface of the coated steel wire. In other words, the first fundamental question of the work must be reformulated: what is the percentage of Al, which must be found in the Zn-Al alloy, in order to "forget" its deformation by twinning and to adopt its deformation by sliding? An optimal chemical composition of the Zn-Al alloy have to be determined considering that, after the deposition of "Hot Dip" on the surface of a steel wire it will be processed by drawing. It has been also analyzed the way of depositing Al in the protective layer as well as the resistance in the salt fog of the steel wires protected with this alloy. In each technical stage (subsequent deposition and drawing process) a practical comparison has been made between the variants object of this study: pure Zn (protective variant 1) and the variant: Zn-Al alloy (protective variant 2).

Keywords: Zn · Al · Deposition · Steel wiredrawing · Anticorrosiveness

1 The Technical Process

Obtaining anti-corrosion coated steel wire in any protection variant required the following steps:

L. Moldovan and A. Gligor (Eds.): Inter-Eng 2021, LNNS 386, pp. 3–12, 2022.
https://doi.org/10.1007/978-3-030-93817-8_1

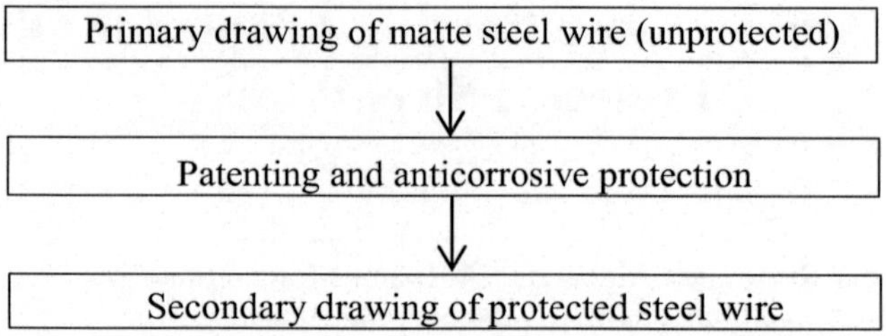

The primary drawing of the matte steel wire was performed starting from a diameter of 6.0 mm to a diameter of 2.0 mm, in the dry version using partial reductions of 20%.

The patenting process was carried out in the unrolled wire, using an tunnel type oven for heating it in the austenitic field, and a sudden immersion in baths of molten Pb for sorbitization. The patenting process was made after the construction of the optimal structure (by austenitization and sorbitization) of steel wire, pickling the surface of steel wire in hydrochloric acid baths followed by fluxing the surface of the wire (in a bath containing a mixture of $ZnCl_2$ salts and NH_4Cl) and by drying its surface also in the tunnel oven. Anticorrosive protection was the technical operation by which either a layer of pure Zn (protective variant 1) or a layer of Zn-Al alloy (protective variant 2) was deposited on the surface of the steel wire. This "Hot Dip" process basically involved immersion of the wire in a bath in which the protective material is melted (so liquid) the melt having a certain temperature, and maintaining the immersion during a certain time. This experimental deposition bathtub of the Zn-Al alloy was mounted on the surface of the galvanizing bathtub itself, being partially immersed in molten zinc. The sketch of this experimental bathtub is shown in the Fig. 1:

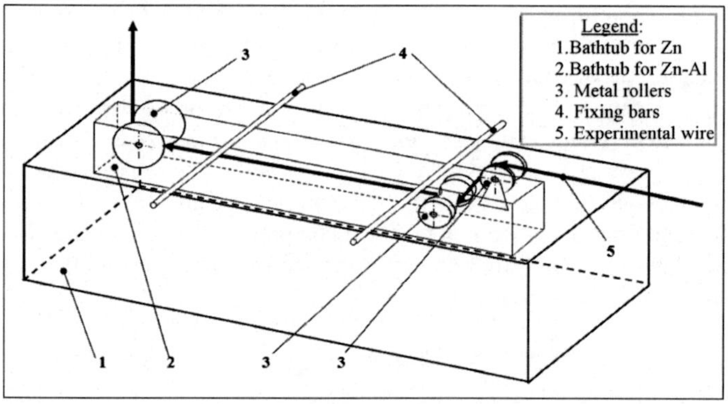

Fig. 1. Sketch of the Zn-Al alloy deposition bathtub.

The temperature range used was from 430 °C to 450 °C and the removal speed (wiping speed) of the steel wire from the bath with molten Zn-Al alloy is identical to that of regular galvanizing. The temperature of the contents of the test bathub must to be extremely precisely known since being semi-submerged (partially submerged) in the galvanizing bathtub, the steel wire adopts the temperature of the outer melt. The fixed part of the test bathtub was made of ARMCO iron sheet with a thickness of 4.0 mm and the movable part (rollers) with a low carbon content steel.

A second drawing sequence of the protected steel wire was performed starting from an intermediate diameter of 2.0 mm to a final diameter of 1.1 mm, working in a dry version and using dies whose deformation angle 2α was 14° and appliying partial reductions of 20%. The drawing series used was: Ø 2.0 mm → Ø 1.8 mm → Ø 1.6 mm → Ø 1.42 mm → Ø 1.26 mm → Ø 1.1 mm [1].

2 The Chemical Composition of the Protective Layer Deduction Pressing

2.1 Analysis of the Zn-Al Equilibrium Diagram

It has been assumed that the cause lies precisely in the crystallization of the Zn-Al protective layer. The crystallization mode is determined by the chemical composition [2]. In deducing the exact chemical composition, we started from the analysis of the Zn-Al equilibrium diagram that is shown in the Fig. 2.

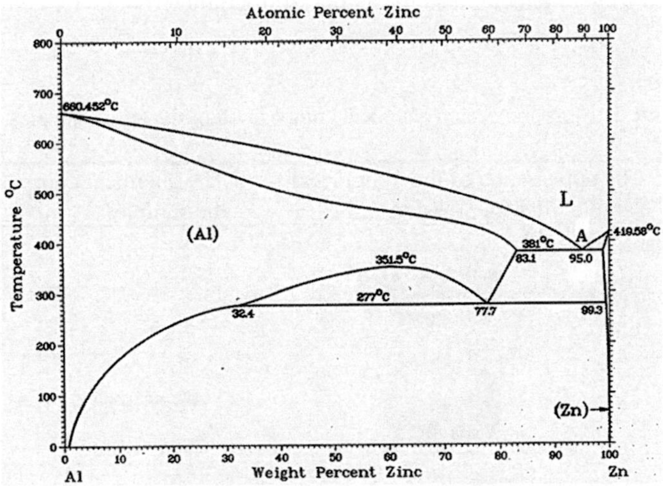

Fig. 2. Zn-Al equilibrium diagram

The really interesting fact in this diagram, from the point of view of the anticorrosive protection of the steel wire but also of the wire drawing manufacturer of that type of wire, is the eutectic, whose chemical composition was 95% Zn and 5% Al. Being Zn in the majority, it is obvious that the structure of the whole resulting ensemble is in hexagonal system. In the case of deformation of the steel wire by wire drawing, when the wire is protected against corrosion in this way, it was observed the almost or total scraping of the protective layer of Zn-Al alloy caused by the pressure and friction inherent to its passage through the die.

2.2 Pressing Process

It has been considered that the cause of such problems in the wiredrawing process behavior lies precisely in the crystallization way of this protective layer of Zn-Al. For the purpose of this study, different samples with various chemical compositions were prepared, somewhat adjacent to the eutectic A (see Fig. 2), but which no longer crystallized in a hexagonal system. It was considered that, to determine the behavior of the specimens in which the percentage of Zn and Al (initially 95% Zn and 5% Al according to the eutectic A) varied slightly in the vicinity of the initial point A, it is sufficient to evaluate the behavior of a sample of this material when it has undergone compression processing. The following samples of Zn-Al alloy were therefore realized: sample 1–96% Zn and 4% Al, sample 2–95.5% Zn and 4.5% Al, sample 3–95.0% Zn and 5.0% Al, sample 4–94.5% Zn and 5.5% Al and sample 5–94% Zn and 6% Al [3]. We underline the following:

1. Firstly, cylindrical samples were made (their diameter being 10 mm and their height 10 mm).
2. The pressing force was identical for all experiments.

All samples were obtained through casting from liquid (from melted alloy) in metal casting molds.

Table 1. Appearance and chemical composition of the tested samples.

Sample number	The appearance of the cylindrical sample after its pressing	The chemical composition of the sample
1.		96%Zn+4%Al
2.		95.5%Zn+4.5%Al
3.		95.0%Zn+5.0%Al
4.		94.5%Zn+5.5%Al
5.		94.0%Zn+6.0%Al

After solidification the samples were processed on a mechanical press, which performed the pressing at a speed of 0.1 m/s.

The behavior resulting from their pressing process are specified in Table1.

For the documentation of the presumption of the crystallographic structure and for the demonstration of the new variant of crystallization, the diffractograms of pure Zn and Zn-Al were made (see Fig. 3 and Fig. 4) respectively.

2.3 Resulting Diffractograms

The diffractogram of pure Zn is shown in the Fig. 3 and can be compared with the diffractogram of the Zn-Al alloy with 94.0% Zn and 6.0% Al (Fig. 4):

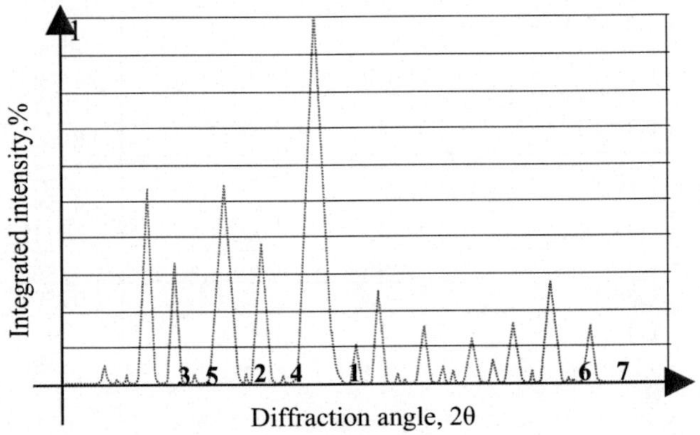

(a)

Peak number	Diffraction angle 2θ [°]	Diffracted radiation intensity	Integrated intensity [%]	Latice parameter [Å]
1	70,1	3978	100	1,342978
2	43,1	2142	54	2,09982
3	36,3	2118,5	53	2,476027
4	54,35	1516	38	1,688775
5	39,5	1320	33	2,282499
6	127,4	1131	28	0,859461
7	138,75	625	16	0,823357

(b)

Fig. 3. a) The pure Zn diffractogram; b) explanations of the pure Zn diffractogram.

The arrangement of the peaks "betrays" the adoption for the Zn-Al alloy, in a different crystallization manner than that of pure Zn, which is obviously different in terms of the behavior of the two variants of anticorrosive protection processed by wire drawing. Analyzing the diffractograms of the two crystallization ways, the highlighted characteristics lead to the conclusion of a hexagonal crystallization system, for the case of pure Zn, in contradiction with the cubic crystallization system shown by the Zn-Al alloy.

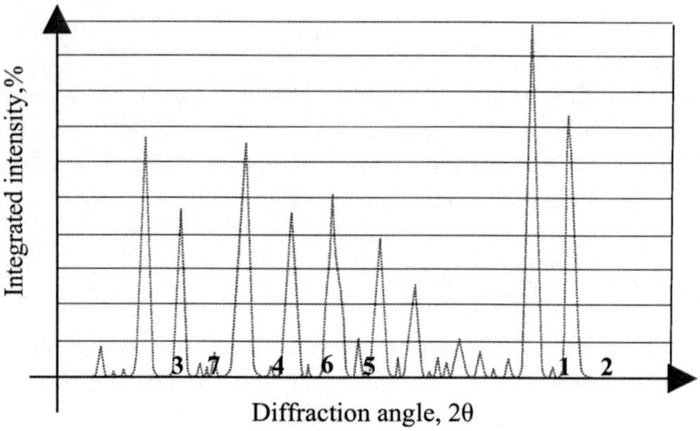

(a)

Peak number	Diffraction angle 2θ [°]	Diffracted radiation intensity	Integrated intensity [%]	Latice parameter [Å]
1	127,2	3980	100	0,860205
2	138,7	2865,8	72	0,812568
3	36	2666	67	2,49597
4	43	2626,8	66	2,104472
5	70	2030	51	1,344651
6	64,1	1870,6	47	1,695987
7	38,75	1831	46	2,324932

(b)

Fig. 4. a)- Zn-Al alloy diffractogram; b)- explanations of the Zn-Al alloy diffractogram.

3 Aluminum Distribution in the Zn-Al Alloy Layer, Deposited on the Surface of a Steel Wire

Following with chemical measurements of the thickness of the deposited layers, an average of 85 μm was obtained for the case of pure Zn coating and about 40 μm for the protective coating with Zn-Al alloy.

Also, photomicrographs of the protective layer of both protective variants were made on the Ø 2.00 mm steel wire [4]. The area visualized in the photomicrographs is shown in the Fig. 5 and it was made by embedding the samples in epoxy resin and allowed visualize a portion of the cross section of the steel wire, after applying the specified protective variant. The preparation of the samples and the visual results are shown in the Figs. 5, 6, 7 and 8

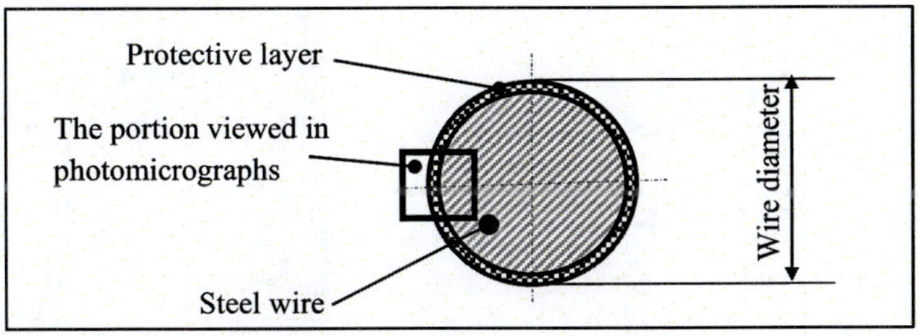

Fig. 5. The cross-sectional portion of the protective coated steel wire, viewed in photomicrographs.

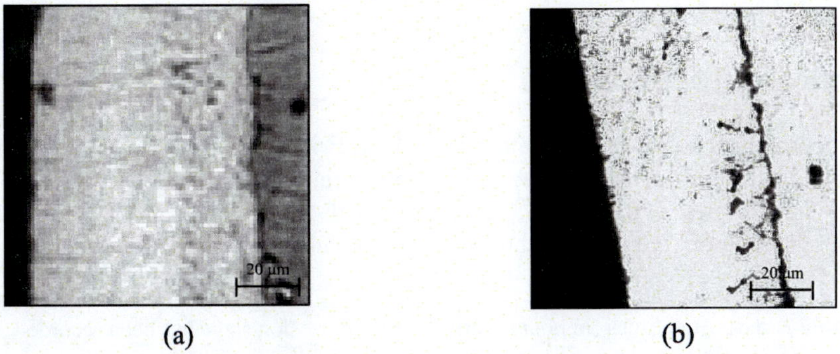

Fig. 6. Appearance of the protective layer of pure Zn (at a magnification of 2500X): a) respectively Zn-Al alloy; b) deposited on the surface of a steel wire

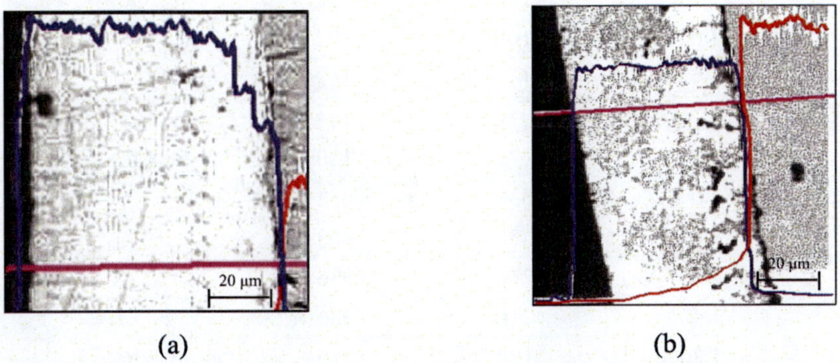

Fig. 7. The compositional profile of Zn (blue) and Fe (red), respectively, on the analysis line (magenta) passing through the deposited protective layer and on a steel wire with a) Zn pure, b) alloy Zn-Al

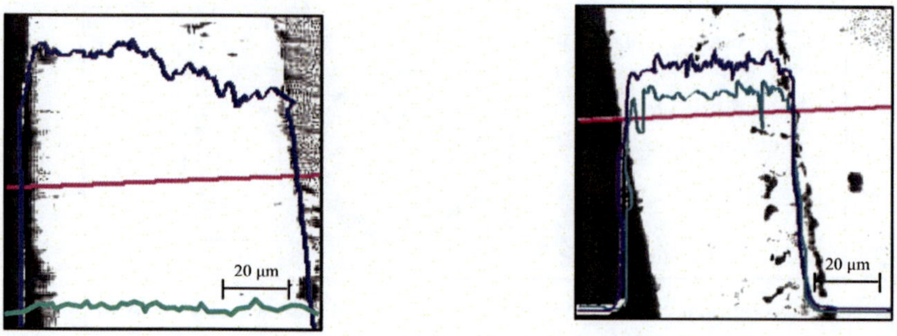

(a) (b)

Fig. 8. The compositional profile of Zn (blue) and Al (green), respectively, on the analysis line (magenta) passing through the deposited protective layer and on a steel wire with a) Zn pure, b) alloy Zn-Al [5, 6]

4 Anticorrosive Resistance

The anticorrosive resistance was determined in an Erichsen salt fog installation at the Universitá degli Studi di Trento, Facoltá di Ingegneria dei Materiali. The duration was monitored as follows: 1. the appearance of "red rust", 2. the extension of "red rust", 3. the generalization of "red rust" for hours [7]. The results are specified in the Table 2.

Table 2. Exposure times [in hours] until a certain corrosion stage has been recorded

Type of corrosion protection	Test wire diameter [mm]	Duration until "red rust" appears	Duration until the extension of "red rust"	Duration until the generalization of "red rust"
pure Zn [protection variant 1]	2.0	216	336	432
	1.8	192	216	336
	1.6	168	192	216
	1.42	144	168	192
	1.26	72	144	168
	1.1	72	120	144
Zn-Al alloy (94.0% Zn and 6.0% Al) [protection variant 2]	2.0	336	432	–
	1.8	336	408	–
	1.6	216	336	–
	1.42	216	336	432
	1.26	192	216	336
	1.1	168	192	216

Transposed to graphs, the data deduced for corrosion are shown in the Fig. 9:

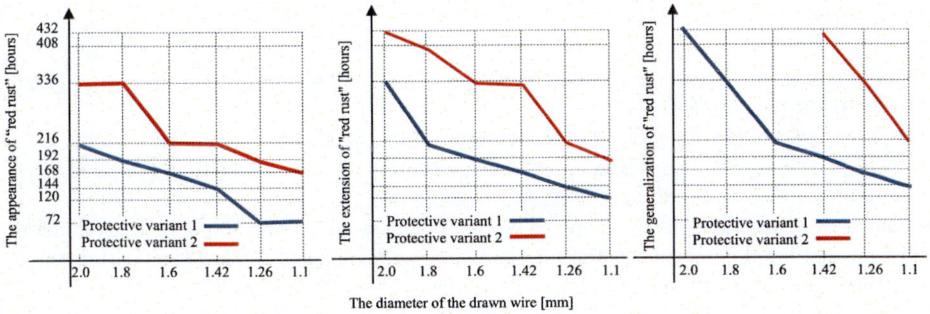

Fig. 9. The stage of corrosion of the steel wire surface during its wire drawing process

5 Conclusions

It has been proven that was adopted the cubic character of the crystallization system of protective variant 2 (with Zn-Al alloy having 94.0% Zn and 6.0% Al) compared to the hexagonal crystallization system adopted by the variant 1 (pure Zn) trough diffractograms and pressing behavior [8, 9].

The distribution of Zn and Al were observed both in protective variant 2 (with Zn-Al alloy) and in variant 1 (with pure Zn) in the protective layer deposited on the surface of a steel wire with Ø 2.0 mm.

The steel wires protected in one of the specified variants were drawn with the drawing series Ø 2.0 mm → Ø 1.8 mm → Ø 1.6 mm → Ø 1.42 mm → Ø 1.26 mm → Ø 1.1 mm and the corrosion was controlled (according to the German standard ASTM 117) in an Erichsen salt fog installation at the Università degli Studi di Trento, Facoltá di Ingegneria dei Materiali. It was quantified based on observations made after drawing process and corrosion, considering three different time periods in which certain different stages of corrosion were visible. The recorded time periods (in hours) corresponded to: stage I: appearance of "red rust", stage II: extension of "red rust" and stage III: generalization of "red rust". These recorded durations for each specified diameter practically denoted two things:

- The corrosion duration regresses during both the drawing process, in both protective variants. Regardless of the diameter of the steel wire, the protective variant 2 (with Zn-Al alloy) is always much superior to the protective variant 1 (with pure Zn) [10] (see Table 2)
- Protective variant 2 (with Zn-Al alloy) also proves to be better than protective variant 1, even at a diameter of the steel wire not still deformed by drawing (see the values corresponding to the wire of 2.0 mm in the Table 2)

So, for steel wires either undeformed nor drawn (in a specified technical version), the protective variant 2 (with Zn-Al alloy) confers a corrosion protection clearly superior to the protective variant 1 (with pure Zn) [11].

In the future work, authors will study adhesion of the coating to the steel, processability, and its formability.

References

1. Wright, N.R.: Wire Technology. Process Engineering and Metallurgy, Elsevier, ISBN: 978-0-12-382092-1 (2011)
2. Pop, M.: Plastic Deformations. ISBN: 978-606-543-509-4 (2014)
3. Yexin, J., Guangling, L.: Development and application of Zn-5% Al-RE alloy coated steel wire. In: Steel Wire Products (2001)
4. Pandis, K.P., Papaioannou, S., Siaperas, V., Terzopoulos, A., Stathopoulosa, V.N.: Evaluation of Zn- and Fe- rich organic coatings for corrosion protection and condensation performance on waste heat recovery surfaces. Int. J. Thermofluids **3–4**, 100025 (2020)
5. Wei, W.A.N.G., Zhi, L.I., Shen, W.J., Yin, F.C., Ya, L.I.U.: Phase equilibria of Zn-Al-Ti ternary system at 450 and 600 °C. In: Transactions of Nonferrous Metals Society of China, vol. 30, no. 4, pp. 1005–1016 (2020)
6. Lervik, A., et al.: Atomic structure of solute clusters in Al–Zn–Mg alloys. In: Science Direct Acta Materialia, vol. 205, Elsevier (2021)
7. Zhang, L., Ma, A., Jiang, J., Song, D., Chen, J., Yang, D.: Anti-corrosion performance of waterborne Zn-rich coating with modified silicon-based vehicle and lamellar Zn (Al) pigments. Prog. Nat. Sci. Mater. Int. **22**(4), 326–333 (2012)
8. Zhang, Z., Yua, J., He, D.: Effect of contact solid solution treatment on peak aging of Al-Zn-Mg-Cu alloys. J. Mater. Res. Technol. **9**(3), 6940–6943 (2020)
9. Zha, Z., Tang, J., Haq Tariq, N., Wang, J., Cui, X., Xiong, T.: Microstructure and corrosion behavior of cold-sprayed Zn-Al composite coating. Coatings **10**(10), 931 (2020)
10. Tailor, S., Modi, A., Modi, S.C.: Synthesis, microstructural, corrosion and antimicrobial properties of Zn and Zn–Al coatings. Surf. Eng. **35**(8), 736–742 (2019)
11. Sugimaru, S., Tanaka, S., Hikita, N., Ohba, H., Yoshie, A., Nishida, S.: Zinc alloy coated steel wire with high corrosion resistance. In: Nippon Steel Technical Report, no. 96, pp. 34–38 (2007)

Bacillus sp. R2: Promising Marine Bacterium with Chitinolytic/Agarovorant Activity and Multiple Enzymes Productivity

Ben Amar Cheba[1,2(✉)]

[1] Biology Department, College of Science, Jouf University, P.O. Box: 2014, Sakaka, Saudi Arabia

[2] Department of Biotechnology, Faculty of Nature and Life Sciences, University of Sciences and Technology of Oran -Mohamed Boudiaf (USTOMB), BP 1505, Al Mnaouar, 31000 Oran, Oran, Algeria

Abstract. Seventy-two bacterial isolates from soil and aquatic sources were tested for their ability to produce chitinase. After successive rounds of primary and secondary screening, an agar-degrading marine bacterium from the Red Sea (Hurghada- Egypt), designated R2 attracted our attention not by its higher chitinolytic agarovorant activity but also by its multiple enzyme production (protease, gelatinase, lipase, esterase, amylase, cellulase, pectinase, dextranase, alginase, arabinase, peroxidase, manganese peroxidase ....) and its moderately halophilic/halotolerant physiology. This bacterial isolate was selected and identified using conventional methods as well as the 16S rRNA technique. The morphological, physiological, and biochemical characterization showed that the strain was Gram-positive or Gram- variable, endospore-forming rods, motile by a single polar flagellum, positive for oxidase, catalase, peroxidase, and urease, negative for nitrate reductase, indole, MR, and VP., therefore, it was assigned to the genus *Bacillus.* Moreover, the isolate was identified molecularly and submitted in the Gen Bank sequence database as Bacillus sp. R2 with a given accession number DQ 923,161. the higher chitinolytic agarovorant activity and multi enzymatic productivity of *Bacillus sp.* R2 makes it a suitable candidate for marine enzymes production in addition to its ecological role in marine biogeocycles.

Keywords: *Bacillus sp.* R2 · Identification · Chitinase · Hydrolytic enzymes · Marine biotechnology

1 Introduction

Enzymes are highly efficient biocatalysts produced by living organisms, they accelerate biochemical reactions rates faster by millions of times faster than any chemical reaction [1]. Enzymes from microbial origin were the most preferable source due to their availability, diversity, and stability than plant and animal enzymes, in addition to lower cost, high production rate in a short time and space [2, 3]. moreover, microbial enzymes have gained more attention globally for their widespread applications in various industrial

L. Moldovan and A. Gligor (Eds.): Inter-Eng 2021, LNNS 386, pp. 13–24, 2022.
https://doi.org/10.1007/978-3-030-93817-8_2

sectors ranging from chemicals, food, feed, fuel, textile, detergent, paper, cosmetics, and pharmaceuticals to agricultural, environmental, and medical fields [4–7].

Many hydrolytic enzymes from fungal and bacterial sources were already produced and used in various commercial processes. chitinases and agarases are glycosyl hydrolases that attracted the attention of many researchers due to their versatile biotechnological uses and produced from various sources [8, 9]. The present work was designed to isolate chitinase and agarase potent bacterium and screen its ability for multiple enzyme productivity, moreover, identifying the isolate conventionally and molecularly for further biotechnological valorizations.

2 Material and Methods

2.1 Sampling and Chitinolytic and/or Agarolytic Bacteria Isolation

Different samples of soils, water, sands, and sediments samples from Mediterranean and red seas and Mariout lake, Nile River (Egypt) were collected, and 0.1 ml were spread directly, whereas soil, sewage, and wastewaters samples were diluted in sterilized seawater and plated on chitinase detection agar (CHDA) plates that contained seawater, 1% (v/v) colloidal chitin, 1.5% (w/v) agar, pH 7. Whereas agarolytic bacteria were screened on agarase detection agar (ADA) plates containing 1.5% (w/v) Agar, 0.02% (w/v) yeast extract dissolved in seawater pH: 7 and incubated at 30OC for few days (48−120 h).

2.2 Chitinolytic and Agarolytic Bacteria Detection

Chitinolytic bacteria can be detected by visualizing the clear zone formed on (CHDA) plates. while the agarolytic activity was assessed by liquefaction or shallow depression appearing around the colonies or with Gran's test [10] when the (ADA) plates were flooded with Lughole's iodine and kept at 4 °C for 1 h. Agarolysis was seen as clear yellow haloes formed around the colonies in contrast to the purple-brown background. This indicated that agarase diffused out from the colonies and reducing compounds were released during agar degradation. All colonies showing clear zone or formed depression, liquefaction, or pits on (CHDA) or (ADA) plates were picked up and purified by successive streaking on the same screening media.

2.3 Potent Strain Conventional Identification

Screenings resulted in potent isolate were subjected to conventional and molecular identifications. The conventional one was carried out according to the morphological, physiological, and biochemical tests described in Bergey's manual of systematic Bacteriology [11, 12]. Morphological characteristics, such as colonial characteristics, pigmentation, luminescence, and swarming were performed on plates of nutrient agar, marine LB agar, seawater agar, and chitin agar. Cell morphology, spore formation was studied by staining (simple and differential) and observed microscopically. Physiological tests were performed at different temperatures and different pHs and different sodium chloride concentrations as well as Biochemical tests were performed as described before. [13, 14,

15] : These tests included oxidation-fermentation (O-F) test, oxidase, catalase, nitrate reduction, methyl red (MR), Vogues-Proskauer (VP), indole, urease, citrate, carbohydrates or sugars fermentation, gelatinase, hemolysin, and utilization of some substrates. Moreover, plasmid miniprep of the potent strain was done according to the method of Zaghloul et al. (1985) [16] with minor modification.

2.4 Potent Strain Molecular Identification

DNA Extraction, Purification, and Amplification. The DNA was extracted according to Sambrook et al. 1989 [17] their purity and concentration were measured at 260 and 280 nm. The conserved 16S rDNA gene was amplified by polymerase chain reaction (PCR) using universal primers designed to amplify the full length (1,500 bp) of the 16S rDNA gene according to the Escherichia coli genomic DNA sequence. The forward primer was 5'- AGAGTTTGATCMTGGCTCAG-3' and the reverse primer was 5'- TACGYTACCTTGTTACGACTT-3'. Amplification of the entire 16S rDNA gene was performed by PCR using the thermocycler (Progene – Techne "Cambridge" LTD – UK). Hundred ng of purified genomic DNA was used in 50 μl reaction mixtures containing 30 p moles of each primer and 2 units of Taq DNA polymerase. The thermocycler was programmed as follow: an initial denaturation of 5 min at 94 °C followed by 30 cycles of 1 min at 94 °C (denaturation), 1 min at 55 °C (primer annealing), and 1.5 min at 72 °C (extension) plus one additional cycle for final elongation at 72 °C for 5 min. The single DNA band of approximately 1.5 Kb (PCR product) was detected by agarose gel electrophoresis.

PCR Product Purification, DNA Sequencing, and Phylogenetic Analysis. The PCR product was purified to remove incorporated nucleotides and excess primers using (QIA gen PCR Purification Kit). The pure PCR product (1.5 Kb) was subjected to automated sequencing by ABI PRISM sequencer based on the dideoxy chain termination method developed by Sanger et al. (1977) [18]. The obtained nucleotide sequence of the 16S rDNA of the isolate R2 was aligned with various sequences of Bacillus members using the Blast search database [19] The sequence has been deposited in the GenBank sequence database and the phylogenetic analysis of the isolate R2 16S rDNA sequence and their relationship with Bacillus group was evaluated using the distant matrix method [20] followed by neighbor-joining analysis [21].

2.5 Potent Strain Multiple Enzymes Production And Assays

After growing the strain in seawater + 0.05% yeast extract supplemented separately with 0.5% of tested substrate (colloidal chitin, agar, cellulose, xylan, starch, blue dextran, pectin, alginate, Arabic gum, lactose, casein, and Tween 80), the flasks incubated at 37 0 C with 180 rpm agitation rate. The quantitative assays of cellulase, pectinase, amylase, dextranase, and alginase were carried out with the DNSA method as described below for agarase and chitinase. While xylanase activity was assayed according to Bailey et al. (1992) [22] using(sigma) oat spelt xylan. Proteolytic activity (neutral and alkaline) was measured according to the method of Cliffe and Law (1982) [23] using Hide Powder

Azure (HPA). One unit of proteolytic activity was defined as the amount of enzyme that developed a change of absorbance (0.1) against the control at 595 nm at 37 °C. Lipolytic or esterolytic activity was determined according to the method described earlier using P- nitro phenyl palmitate as substrate [24]. One unit of lipolytic or esterolytic activity was defined as the amount of enzyme that released one micromole of P- nitro phenyl from the substrate P- nitro phenyl palmitate per min. Furthermore, Peroxidase activity was assayed according to [25] using 2,20 -azino-bis (3-ethyl benzthiazoline-6-sulfonic acid) (ABTS) as a substrate, whereas manganese peroxidase was assessed according to [26, 27].

2.6 Analytical Methods

Electron Microscopy

Potent isolate cell and spore morphologies, size, and flagellation were studied by using JEOL scanning and transmission electron microscopy (scanning electron microscopy – SEM –and transmission electron microscopy –TEM), the true dimensions of the cell were calculated according to the following equation: True length (μm) = L (mm) × 103/M. Where L (mm) = Length on the photograph in millimeter. M = Magnification.

Chitinase, Agarase, and Protein Assays

Chitinase and agarase activities were analyzed according to the Miller method [28] by estimating the released reducing ends of sugar using N-acetyl - D-glucosamine (NAG) and galactose as standards, respectively. One unit of chitinase and agarase activities were defined as the amount of enzyme required to release 1 μ mol of NAG and galactose per minute respectively during these conditions. as described in the Bradford method (1976) [29] Soluble proteins were assessed using bovine serum albumin as a reference for standard curve preparation.

3 Results and Discussion

3.1 Potent Strain Conventional Identification

Seventy-two bacterial isolates from soil and aquatic sources were tested for their ability to produce chitinase. After successive rounds of primary and secondary screening, a marine bacterium from the Red Sea (Hurghada- Egypt), designated R2 attracted our attention by its higher chitinolytic agarovorant activity (Fig. 1). Consequently, it was selected for further identification. The morphological, cultural, as well as physiological, and biochemical characteristics summarized in Table 1: showed that the strain R2 was Gram-positive or variable, strictly aerobic, endospore-forming Bacilli, motile by a single polar unsheathed flagellum (Fig. 2). Cells were straight rods approximately 1 μm in diameter and 3–4 μm in height (Fig. 2). The endospores are oval or ellipsoidal, (1.25–1.66 × 2.95–3.95 μm), lie centrally or sub terminally in slightly swollen sporangia (Fig. 3). The isolate colonies were smooth, circular to slightly irregular, slightly raised, opaque, cream to pale yellow and 2–4 mm in diameter after one day of growth at 30 °C on MLB agar medium.

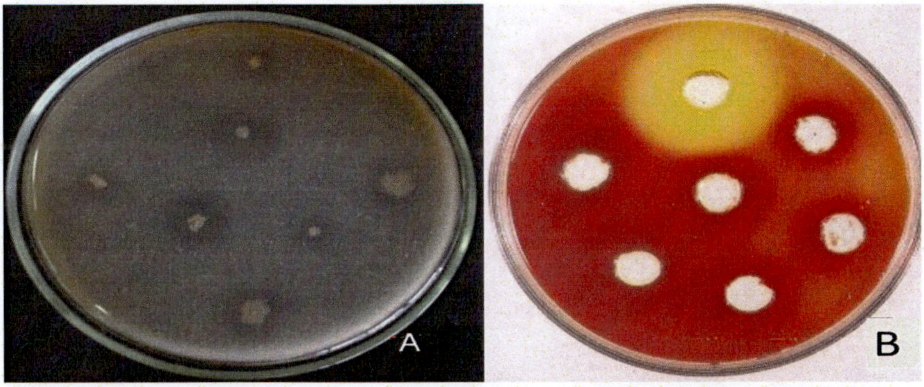

Fig. 1. A: Chitinase screening on (CHDA) plates, B: *Bacillus sp.* R2 agarase activity detection through well cut diffusion assay on agar plates.

Fig. 2. A: Transmission electron photomicrographs of the R2 isolate (negative form) cells have single polar flagellum (15000X), B: Ultra-thin section showing the rod-shaped morphology of the cells during division (40000X).

The growth of isolate R2 was between 4 and 42 °C while the optimal growth lied between 30–37 °C Table 1. Good growth occurred at pH 5–10 and was limited at pH 4 and 11. No growth was observed in the absence of NaCl, while good growth was noted in the presence of NaCl 2–18%. The strain tolerated weakly NaCl concentration 18–30%. A similar range was tolerated by closer Gram-positive or variable marine bacilli strains, such as *B. aquimaris* isolated from the yellow sea [30] which tolerated up to 18% NaCl.

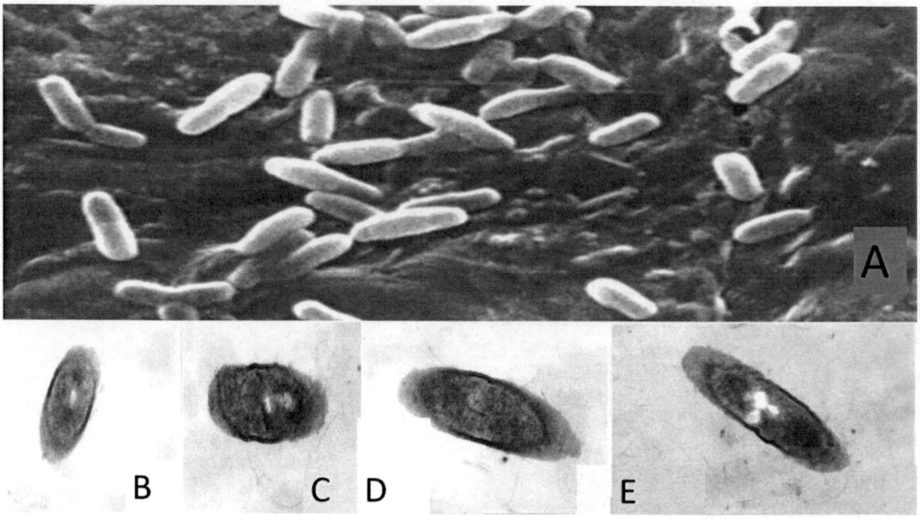

Fig. 3. Electron photomicrographs of the isolate R2 (sporulated form). Panel A; SEM micrograph showing the paracentral/subterminal spore position. Panels B, C and D and E; ultra-thin Sects. (50,000 X) showing the oval and ellipsoidal shapes of the spore, respectively.

Moreover, *Jeotgalibacillus alimentarius* grew at 19% NaCl [31]. The isolate R2 was aerobic and gave a positive reaction for oxidase, catalase, peroxidase, and urease. The isolate was negative for nitrate reductase, indole, MR, and VP. The isolate produced several enzymes such as agarase, chitinase, protease, lipase, amylase, gelatinase, dextranase, and β-galactosidase (Table 2). The isolate R2 fermented the following sugars to produce acid from glucose, galactose, mannose, xylose, glycerol, lactose and maltose, sucrose, arabinose, sorbose; however, fructose was not fermented (Table 1). Additionally, the plasmid profile of the isolate revealed the absence of any plasmids. The conventional identification, based on the criteria described in Bergey's manual, of systematic bacteriology [11, 12] assigned the isolate R2 to the genus Bacillus..

Table 1. Bacillus *sp. R2* morphological, physiological, and biochemical characteristics.

Characteristics	Results	Characteristics	Results
Gram-staining	V or +	(18–30)%	(vw)
Cell morphology	Straight rods	Growth in MacConkey	+
Cell size	(1.65 × 4) μm	Hemolysin	+
Cell flagellation	Single polar	Protease	+
Cell arrangement	Mono	Gelatinase	+
Sporangia:		Lipase	+
Spore shape	Ellipsoidal/oval	Amylase	+
Spore size	(1.45 × 3.35) μm	Dextranase	+
Spore position	Central/subterminal	Agarase	+
Sporangia swollen	Slightly	Chitinase	+
Motility	+	Hydrolysis of:	
Swarming	+	Casein	+
Colony color	Cream to pale yellow	Gelatin	+
Pigmentation	–	Tween 20	+
Luminescence	–	Tween 80	+
Salt requirement	+	Starch	+
Aerobe/anaerobe	Aerobe	Agar	+
O/F test	±	Agarose	+
Oxidase	+	Filter paper	–
Catalase	+	Cellulose	±
Peroxidase	+	CMC	+
Nitrate reductase $N^{O_3^-} \rightarrow N^{O_2^-}$	–	Chitosan	+
Methyl red (MR)	–	Alginate	+
Voges-Proskauer (VP)	–	Pectin	±
Indole from tryptophan	–	Arabic gum	+
Urease	+	Dextran blue	+
Citrate	+	Acid from:	
Growth at:		Glucose	+
(4, 20, 30, 37, 42) °C	+	Galactose	(w)
(50, 55, 60) °C	–	Fructose	(Alk)
Growth at pH:		Mannose	+

(*continued*)

Table 1. (*continued*)

Characteristics	Results	Characteristics	Results
4	(vw)	Sorbose	–
(5, 6, 7, 8, 9, 10)	+	Arabinose	–
11	W	Xylose	+
Growth in NaCl:		Lactose	+
0%	–	Maltose	+
(2–3)%	+	Sucrose	–
(3–12)%	+	Glycerol	+
(12–18)%	±	Presence of plasmids	–

(+) Positive, (−) Negative: (w) Weak positive, (v) Variable, (vw) very weak positive, Alk; alkaline reaction.

3.2 Potent Strain Molecular Identification

The resulting data indicated that the isolate R2 understudy belonged to Bacillus. sp (probability 97%). This confirmed the conventional identification and as evident from the taxonomy report that the isolate R2 exhibited 98% identity with Bacterium JL-74 [32] and 97% similarity with *Bacillus sp.* CNJ9O4 PLO4, Bacillus. sp. CNJ815 PLO4, Bacillus. sp. T5-12, *Bacillus. sp.* PO1, Bacillus holothurians, Marine *Bacillus* NRRLB–14851, and *B. barbaricus*. It was noticed that all the above strains were Gram-positive or variable moderately halophilic, halotolerant marine bacilli. Part of the nucleotide sequence (561 bp) was submitted to the GenBank database, with the accession number (DQ923161). A phylogenetic tree (Dendrogram) was generated using the Bio-Edit program and neighbor-joining algorithm. Figure 4 showed that the isolate R2 was phylogenetically related to members of the Bacillus group and formed a coherent cluster with *B. halodurans* and *B. clausi*. We must point out that the taxonomic position of our strain was very far from the terrestrial Bacilli such *as B.subtilus, B.cereus*…and very closer to the gram variable marine Bacilli with moderately halophilic or halotolerant properties, Interestingly, many recently isolated positive or gram variable bacilli with halophilic or halotolerant properties have been isolated from marine sources (Yellow sea of China and Korea, Red sea (this study), Died sea (2) and from salt lakes of China [33] and Mongolia [34] and have been identified as members of related new genera such as *Virgibacillus, Lentibacillus, Tenuibacillus, Cerasibacillus, Halobacillus* [35] *and Salinibacillus* [36] rather than members of the genus Bacillus. All these facts indicate the importance of the polyphasic approach for the determination of the exact taxonomic position of this special group of Bacilli which our strain R2 was a newly isolated strain from the red sea belongs to this group. Accordingly, the future study aimed to determine the exact taxonomic status of the strain R2 using a combination of phenotypic properties, chemotaxonomy since the strain was gram variable, and phylogenetic analysis based on full 16S rDNA gene (1500 bp) not partial sequence and genomic DNA-DNA relatedness and based on these data collectively the novel strain R2 may be placed at least in novel distinct specie.

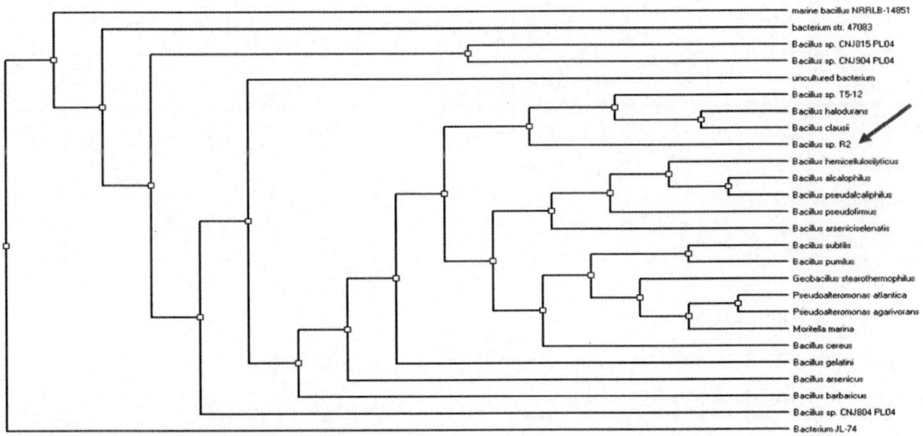

Fig. 4. Dendrogram (based on 16S DNA sequence comparisons) showing the taxonomic position and the phylogenetic relationship of Bacillus sp. R2 with several Bacillus species. Scale bar: 0.1 substitution pair nucleotide position.

3.3 Potent Strain Multiple Enzymes Productivity

Besides the agarovorant and chitinolytic activities, the results presented in Table 2 revealed that Bacillus sp. R2 produces multiple enzymes include cellulase, xylanase, pectinase, amylase, dextranase, arabinase, alginase, and peroxidase, in addition to other hydrolytic enzymes such as protease, gelatinase, lipase, and lactase. Bacillus sp. R2 multiple enzymes productivity was not surprising for the Bacillus genus that secretes numerous enzymes degrading various substrates, enabling them to survive in a continuously changing environment. Furthermore, Bacillus has become the major microbial cell factory for many industrial products [37, 38], including enzymes [39], heterologous proteins, amino acids, vitamins, and antibiotics [40].

Table 2. The enzymes produced by Bacillus sp. R2.

Enzyme	Production	Enzyme activity (U/ml)
Agarase	+++	32
Chitinase	+++	34.5
Cellulase	+	11.32
Xylanase	+	0.36
Dextranase	+	29
Amylase	++	22
Lipase	+++	198.86
Protease (Neutral)	++	10.2
Protease (Alkaline)	+	6.25

(*continued*)

Table 2. (*continued*)

Enzyme	Production	Enzyme activity (U/ml)
Gelatinase	++	ND
Urease	++	ND
Pectinase	+	19.24
Alginase or alginolytic act	++	21.12
Arabinase	+	12.5
Lactase (B-galactosidase)	+	45.6
Peroxidase	+	16 (a)
Manganese peroxidase	+	3.8 (b)

ND: not determined, (a) and (b) enzymes activities were estimated by Khelil et al. 2015, 2016 [41, 42]

4 Conclusion

Bacillus genus and its enzymes continue to show great potential for practical applications in various biotechnological fields. In these concepts *Bacillus sp.* R2 multiple enzymes productivity enables it to become a potential microbial cell factory for many hydrolytic enzymes which will be exploitable in food, feed, fuel, and fertilizers biotechnologies. in addition, the various substrates degradability by *Bacillus sp.* R2 explained its ecological role in marine organic matter mineralization.

Acknowledgment. Thanks, are due to the Algerian ministry of higher education and scientific research (AMHESR) for their financial support to this work.

References

1. Gurung, N., Ray, S., Bose, S., Rai, V.: A broader view: microbial enzymes and their relevance in industries, medicine, and beyond. BioMed Res. Int. (2013)
2. Anbu, P., Gopinath, S.C.B., Chaulagain, B.P., Tang, T.-H., Citartan, M.: Microbial enzymes and their applications in industries and medicine 2014. BioMed. Res. Int. **2015**, 3 (2015). Article ID 816419
3. Gopinath, S.C.B., Anbu, P., Lakshmipriya, T., Hilda, A.: Strategies to characterize fungal lipases for applications in medicine and dairy industry. BioMed. Res. Int. **2013**, 10 (2013). Article ID 154549
4. Banerjee, G., Ray, A.K.: Impact of microbial proteases on biotechnological industries. Biotechnol. Genet. Eng. Rev. **33**(2), 119–143 (2017)
5. Nigam, P.: Microbial enzymes with special characteristics for biotechnological applications. Biomolecules **3**(3), 597–611 (2013)
6. Raveendran, S., et al.: Applications of microbial enzymes in the food industry. Food Technol. Biotechnol. **56**(1), 16–30 (2018)
7. Danilova, I., Sharipova, M.: The practical potential of bacilli and their enzymes for industrial production. Front. Microbiol. **11** (2020)

8. Dukariya, G., Kumar, A.: Distribution and biotechnological applications of Chitinase: a review. Internat. J. Biochem. Biophys. **8**, 17–29 (2020)
9. Jahromi, S.T., Barzkar, N.: Future direction in marine bacterial agarases for industrial applications. Appl. Microbiol. Biotechnol. **102**, 6847–6863 (2018). https://doi.org/10.1007/s00253-018-9156-5
10. Gran, H.H.: Studien über meeresbakterien. II über die hydrolyse des agar agars durch ein neues enzyme, die Gelase. Bergen Mus. Aarbog. **2**, 1–16 (1902)
11. MacLeod: Bergey's Manual of Systematic Bacteriology, vol. 1. Williams and Wilkins, London (1968) (1984)
12. Garty, G.M.: Bergey's Manual of Systematic Bacteriology. 2nd edn., vol. 2. The Proteobacteria, Springer, Verlag (2005)
13. Mac Faddin, J.F.: Biochemical Tests for Identification of Medical Bacteria. 3rd Lippincott Williams and Wilkins 227, East Washington (2000)
14. Koneman, E.W., Allen, S.D., Janda, W.M., Schrechenberger, P.C., Washington, C., Winn, J.R.: Color Atlas and Textbook of Diagnostic Microbiology, 5th edn. (1997)
15. Rosovitz, M.J., Voskuil, M.I., Chambliss, G.H.: Bacillus, chapter 31 in Topley and Wilson's-Microbiology and microbial infections 9th edition, edited by Collier, L., Ballows, Sussman, M.A. (eds.), vol. 2. Systematic Bacteriologies (1998)
16. Zaghlool, T.I., Kawamura, F., Doi, R.H.: Translational coupling in Bacillus subtilis of heterologous Bacillus subtilis – Escherichia coli gene fusion. J. Bacteriol. **164**, 550–555 (1985)
17. Sambrook, J., Firtsch, E.F., Maniatis, T.: Molecular Cloning a Laboratory Manual Cold Spring Harbor Laboratory. cold spring Harbor, New York (1989)
18. Sanger, F., Nicklen, S., Coulson, A.R.: DNA sequencing with chain-terminating inhibitor. Proc. Natl. Aca. Sci. USA **74**, 5463–5467 (1977)
19. Altschul, S.F., Thomas, L.M., Schäffer, A.A., Zang, J., Zhang, Z., Miller, W., Lipman, D.J.: Gapped BLAST and PSI-BLAST: a new generation of protein database search programs. Nucleic Acids Res. **25**, 3389–3402 (1997)
20. Jukes, T.H., Contor, C.R.: Evolution of protein molecules. In: Munro, H.N. (ed.) Mammalian Protein Metabolism, vol. 3, pp. 21–132. Academic Press Inc., New York (1969)
21. Saitou, N., Nei, M.: The neighbor-joining method – a new method for reconstructing phylogenetic trees. Mol. Biol. Evol. **4**, 406–425 (1987)
22. Bailey, M.J., Biely, P., Poutanen, K.: Interlaboratory testing of methods for assay of xylanase activity. J. Biotechnol. **23**(3), 257–270 (1992)
23. Cliffe, A.J., Law, B.A.: A new method for the detection of microbial proteolytic enzyme. J. Dairy. Res. **49**, 209–219 (1986)
24. Vorderwuelbecke, T., Kieslich, K., Erdmann, H.: Comparison of lipases by different assays. Enz. Microb. Technol. **14**, 631–639 (1992)
25. Sigma-Aldrich, Enzymatic assay of peroxidase (EC 1.11.1.7) 2,20 -Azino-bis (3-Ethylbenzthiazoline-6-Sulfonic Acid as a Substrate (1996). https://www.sigmaaldrich.com/content/dam/sigma-aldrich/docs/Sigma/Enzyme_Assay/p6782enz.pdf
26. Boer, C.G., Obici, L., de Souza, C.G.M.: Decolorization of synthetic dyes by solid-state cultures of Lentinula (Lentinus) edodes producing manganese peroxidase as the main ligninolytic enzyme. Bioresour. Technol. **94**, 107–112 (2004)
27. Wariishi, H., Valli, K., Gold, M.H.: Manganese (II) oxidation by manganese peroxidase from the basidiomycete Phanerochaete chrysosporium. J. Biol. Chem. **267**, 23688–23695 (1992)
28. Miller, G.R.: Use of Dinitro salicylic Acid reagent for determination of reducing sugar. Anal. Chem. **31**(3), 426–428 (1959)
29. Bradford, M.M.: A rapid and sensitive method for the quantitation of microgram quantities of protein utilizing the principle of protein-dye binding. Anal. Biochem. **72**, 248–254 (1976)

30. Yoon, J.H., Kim, I.G., Kang, K.H., Oh, T.K., Park, Y.K.: Bacillus marisflavi sp. nov. and Bacillus aquimaris sp.nov., isolated from seawater of a tidal flat of the yellow sea in Korea. Inter. J. Syst. Evol. Microbiol. **53**, 1297–1303 (2003)
31. Yoon, J.H., Weiss, N., Lee, K.C., Lee, I.S., Kang, K.H., Park, Y.H.: Jeotgalibacillus alimentaris gen. nov. sp.nov. a novel bacterium isolated from jeotal with L-lysine in the cell wall, and reclassification of Bacillus marinus Rüger 1983 as Marinibacillus marinus gen.nov., comb. nov. Inter. J. Syst. Evol. Microb. **51**, 2087–2093 (2000)
32. Du, H., Jiao, N., Hu, Y., Zeng, Y.: Diversity and distribution of pigmented heterotrophic bacteria in marine environments. FEMS. Microbiol. Ecol. **57**(1), 92–105 (2006)
33. Lim, J.-M., et al.: Bacillus salarius sp. nov., a halophilic spore-forming bacterium isolated from a Salt Lake in China. Int. Syst. Evol. Microbiol. **56**(2), 373–377 (2006)
34. Carrasco, I.J., et al.: Gracilibacillus orientalis sp. nov., a novel moderately halophilic bacterium isolated from a Salt Lake in Inner Mongolia. Int. Syst. Evol. Microbiol. **56**(3), 599–604 (2006)
35. Amoozegar, M.A., Malekzadeh, F., Malik, K.A., Schumann, P., Sproer, A.: Halobacillus karajensis sp. nov., a novel moderately halophilic. Int. Syst. Evol. Microbiol. **53**(4), 1059–1063 (2003)
36. Ren, P.G., Zhou, P.J.: Salinibacillus aidingensis gen. nov., sp. nov. and Salinibacillus kushneri sp. nov., a moderately halophilic bacterium isolated from a neutral saline lake in Xin–Jiang, China, Syst. Evol. Microbiol. **55**(2), 949–953 (2005)
37. Schallmey, M., Singh, A., Ward, O.P.: Developments in the use of Bacillus species for industrial production. Can J Microbiol. **50**, 1–17 (2004)
38. Liu, Y., Li, J., Du, G., Chen, J., Liu, L.: Metabolic engineering of Bacillus subtilis fueled by systems biology: recent advances and future directions. Biotechnol Adv. **35**, 20–30 (2017)
39. van Dijl, J., Hecker, M.: Bacillus subtilis: from soil bacterium to super-secreting cell factory. Microb Cell Fact. **12**, 3 (2013)
40. Cherukuri, P.K., Songkiatisak, P., Ding, F., Jault, J.M., Xu, X.N.: Antibiotic drug nanocarriers for probing of multidrug ABC membrane transporter of Bacillus subtilis. ACS Omega **5**, 1625–1633 (2020)
41. Khelil, O., Choubane, S., Cheba, B.A.: Polyphenol's content of spent coffee grounds subjected to physicochemical pretreatments influences lignocellulolytic enzymes production by Bacillus sp. R2. Bioresour. Technol. **211**, 769–773 (2016).
42. Khelil, O., Choubane, S., Cheba, B.A.: Co-production of cellulases and manganese peroxidases by Bacillus sp. R2 and Bacillus cereus 11778 on waste newspaper: application in dyes decolorization. Procedia Technol. **19**, 980–987 (2015)

Calculation Program for an Elastic Coupling with Bolts and Non-metallic Elements from Natural Rubber

Marilena Ghiţescu[1(✉)], Ion-Marius Ghiţescu[1], Sorin Vlase[1], and Arina Modrea[2]

[1] Department of Mechanical Engineering, Transilvania University of Brasov, 500036 Brasov, Romania
{marius.ghitescu,svlase}@unitbv.ro
[2] University of Medicine, Pharmacy, Science and Technology of Târgu Mures, Tîrgu-Mureş, Romania

Abstract. The paper presents a design program and verification of an elastic coupling with bolts and non-metallic elements. The elastic coupling are composing from semi-couplings, intermediary disk, metallic plates, cylindrical pin of centered of a in-termediary disk to the one semicoupling, bolts, screws of fixing of metallic plates to the intermediary disc, Grower washer, nuts and nonmetallic elements. The program allows, starting from the establishment of the input data (engine power, input shaft speed, input shaft diameter - known or obtained from the calculation) to establish the operating conditions of the transmission of which the coupling is part, and further the dimensions of the wedges, the semicoupling, the outer diam-eter of the coupling, the arrangement of the bolts the dimensions of the half-couplings, the choice of materials for the coupling components and to perform verification calculations for the main components (bolt, non-metallic element) through which the movement leading to the driven semicoupling.

Keywords: Design program · Elastic coupling · Nonmetallic element · Bolt

1 Establishing of Entry Data of Program

The paper presents a calculation program of a new elastic coupling with bolts and non-metallic elements that allows the design of these coupling, sizing components, checking feathers, bolts and non-metallic elements and drawing diagrams using analytical relationships that are introduced in the program.

The calculation program was made in Delphi software, in the Delphi programming language. For computer modeling of elastic couplings with non-metallic elements, made, original software packages, designed in Delphi, were used.

A computer with the following characteristics was used for the computer modeling of the prototype:

- Processor: Pentium II - MMX; Base Memory: 640 K.

L. Moldovan and A. Gligor (Eds.): Inter-Eng 2021, LNNS 386, pp. 25–37, 2022.
https://doi.org/10.1007/978-3-030-93817-8_3

- Memory: 640 RAM.
- Hard disk: C - 5 GB; D - 1 GB.
- Disk drive: A - 1.44 Mbytes.
- CD-ROM drive: E.
- Monitor: CTX.

Computer modeling of elastic couplings with non-metallic elements involved the determination of general analytical relationships of functional modeling. Thus, a calculation program was designed to allow both the computerized constructive design and the graphical visualization of some characteristic functional diagrams.

The elastic coupling transmits the motion in both directions, regardless of the direction of rotation chosen at the input shaft. This new elastic coupling has in their componence, compared to the classic solutions of elastic couplings with bolts, one disk between driving and driven semicouplings which has milled four holes arranged symmetrically on its circumference, metallic plates mounting on the disc, the non-metallic elements that are mounted of bolts. Two metal plates are mounted, fixed by screws, left to right of each channel. The non-metallic element has various construc-tive shapes and may be realized from different qualities rubber. The non-metallic element is mounted on the bolts fixed by the driving semicoupling and is mounted lightly pressed, left - right, in the channels of two plates.

The following input data were chosen: engine power *P* [kW], engine speed *n* [rpm] and shaft diameter *d* [mm] - considered known, on which the coupling is mounted (see Fig. 1), or the diameter of the shaft *d* [mm] can be calculated if it is not known [1].

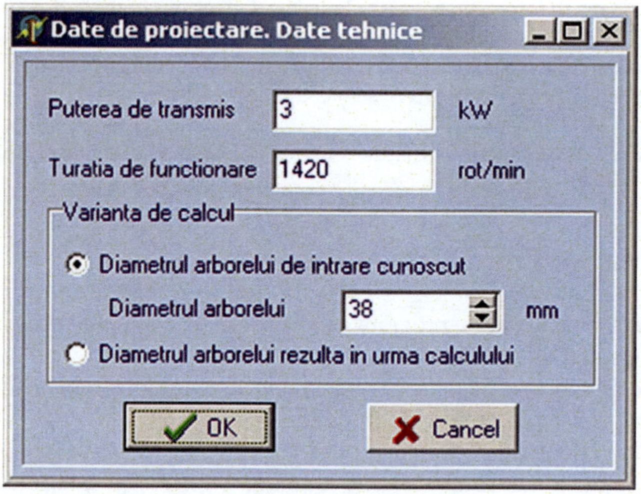

Fig. 1. Introduction of entry data.

From window "Design dates. Functionary conditions", Fig. 2, is selected the load character for the motor machine (electric motor), the chosen load character being with non-uniform operation with shocks, average inertial masses. For the safety coef-ficient corresponding to the load character chosen for the electric motor, is adopted a value, $K_s = 2, 4$, in the recommended range [2–6].

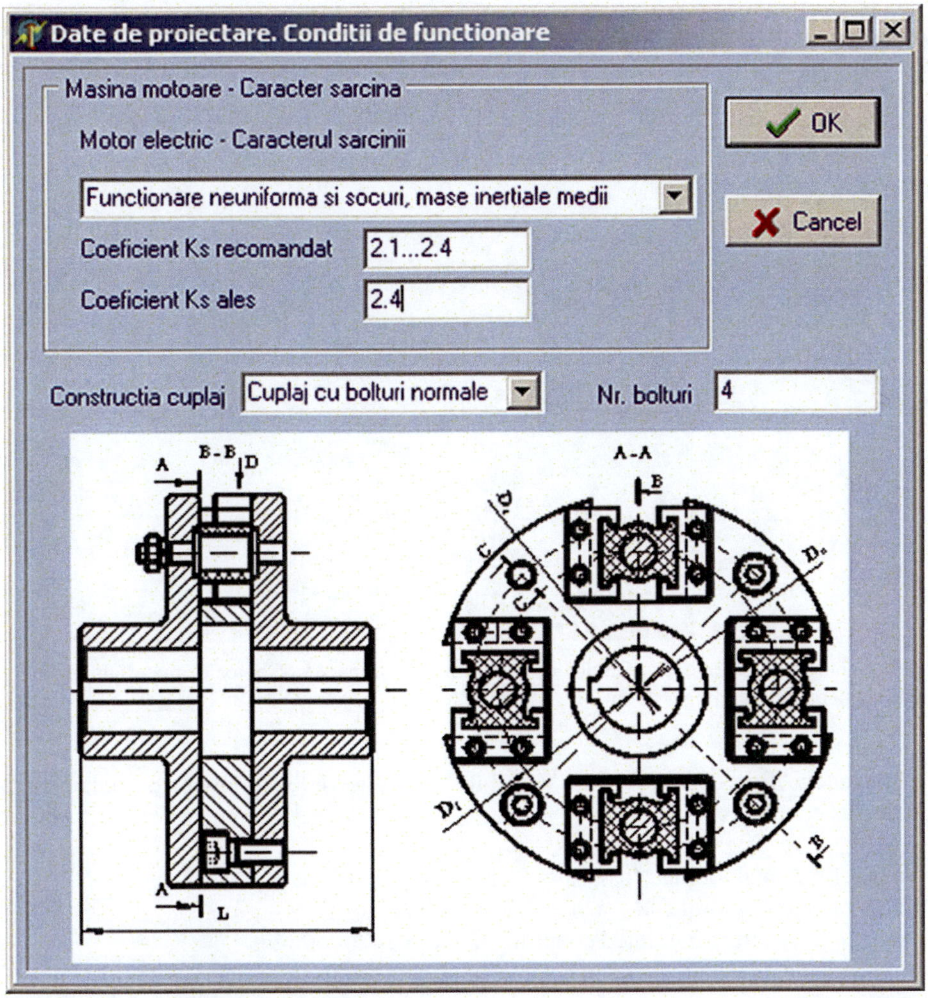

Fig. 2. Window "Design dates. Functionary conditions" - Choice of the constructive variant of the coupling and of the safety coefficient Ks.

For the selected coupling variant, in the "Design dates. Functionary conditions" displays the corresponding drawing of the coupling and the number of bolts (z = 4 bolts) in the coupling component.

From window "Design dates. Choice of materials", Fig. 3, are chosen the materials from which the intermediate disc are made the metal plates, the bolts, the feathers, the semi-couplings and the non-metallic elements. For the intermediate disc, plates, bolts, semi-couplings was adopted brand OLC 45 improvement steel (equivalent to C45) for a good resistance of these main elements of the coupling component. For non-metallic elements was adopted natural rubber NR. The feathers that are mount-ed on the shafts are made of general purpose steel for construction brand OL60.

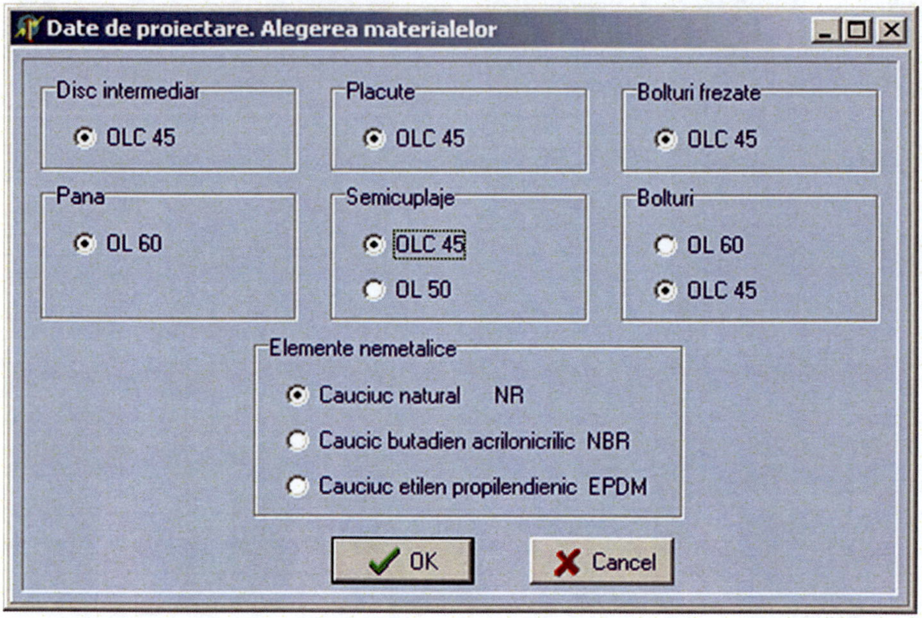

Fig. 3. Window "Design dates. Choice of materials Functionary" - Choice of materials for different elements of coupling.

In window "Entry dates. Results" (see Fig. 4) are shown the following partial results of calculus coupling:

- Value of nominal torque $M_{tn:}Mt_n = 20176\,\text{Nmm}$;
- Value of calculus torque $M_{tc:}Mt_c = 48423\,\text{Nmm}$;
- speed to the shaft on mounted coupling, n : $n = 1420\,\text{rot/min}$;
- diameter of entry shaft, d – known value/value resulted from calculus, with possibility of standardization a shown value for shaft diameter: $d = 38\,\text{mm}$;
- standardization length of end shaft – in function of shaft series (long/short), $larb = 80\,\text{mm}$ – for short shaft;
- feather dimensions $\left(\text{b}_{\text{feather}}, h_{feather}\right)$ in function of shaft diameter d: $b_{feather} = bpana = 10\,\text{mm}$, $h_{feather} = hpana = 8\,\text{mm}$;
- recommended lengths in STAS for feather, $l_{feather}$, real length of resulted feather from calculus with possibility of adoption of one standardized length in function of length of semicoupling: $l_{feather} = lpanaST = 50\,\text{mm}$;
- establish of value of disposer diameter of bolts, D_1, in recommended interval: $D_1 = 124\,\text{mm}$;
- bolt diameter, d_b: $d_b = 16\,\text{mm}$;

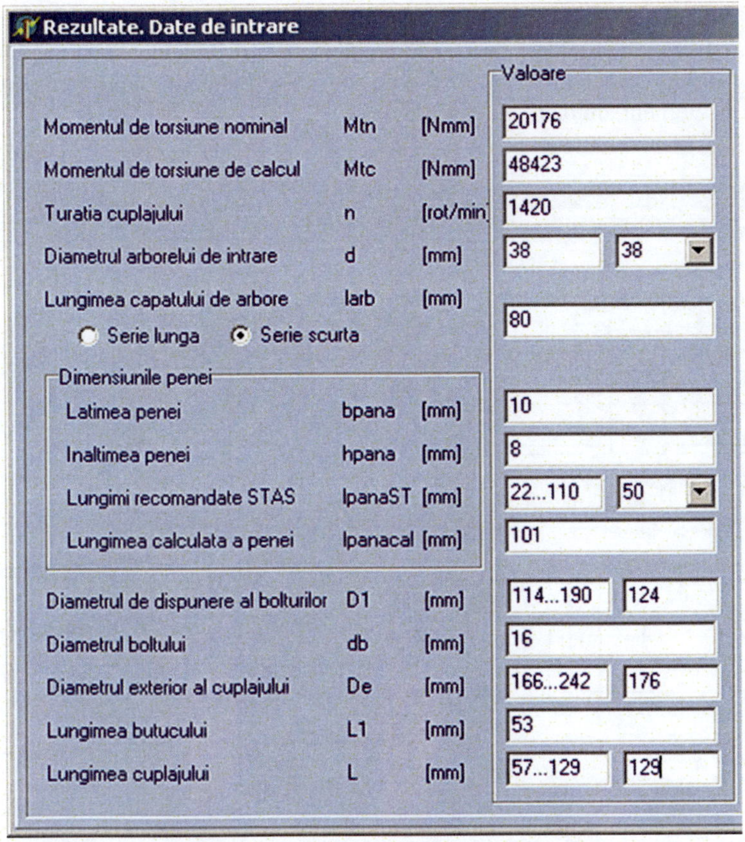

Fig. 4. Windows "*Results. Input data*".

- exterior diameter of coupling, D_e: $D_e = 176\,\text{mm}$;
- hub length, L_1: $L_1 = 53\,\text{mm}$;
- coupling length, L, with possibility of choice of value in recommended interval: $L = 129\,\text{mm}$.

Also is possible set of values for disposer diameter D_1 of bolts, exterior diameter D_e, of coupling, length feather $l_{feather} = l_{panaST}$ and coupling length L.

2 Coupling Check Calculation

In this section will be chosen in turn a calculation scheme of the coupling corresponding to the constructive variant that is calculated and a constructive form of non-metallic element, and through the program will be calculated and verified each shape of the non-metallic element, will determine the torques capable of being transmitted to the coupling, as well as the calculation of the bending bolt check, the relative rota-tion angle between the half-couplings, the static rigidity of the coupling.

From Fig. 5 is choosing the calculus scheme for first constructive variant of cou-pling (the scheme from left side, variant I), and for non-metallic element is choosing form 1. Depending on the selected design of the non-metallic element, the dimensions of the non-metallic element are displayed immediately.

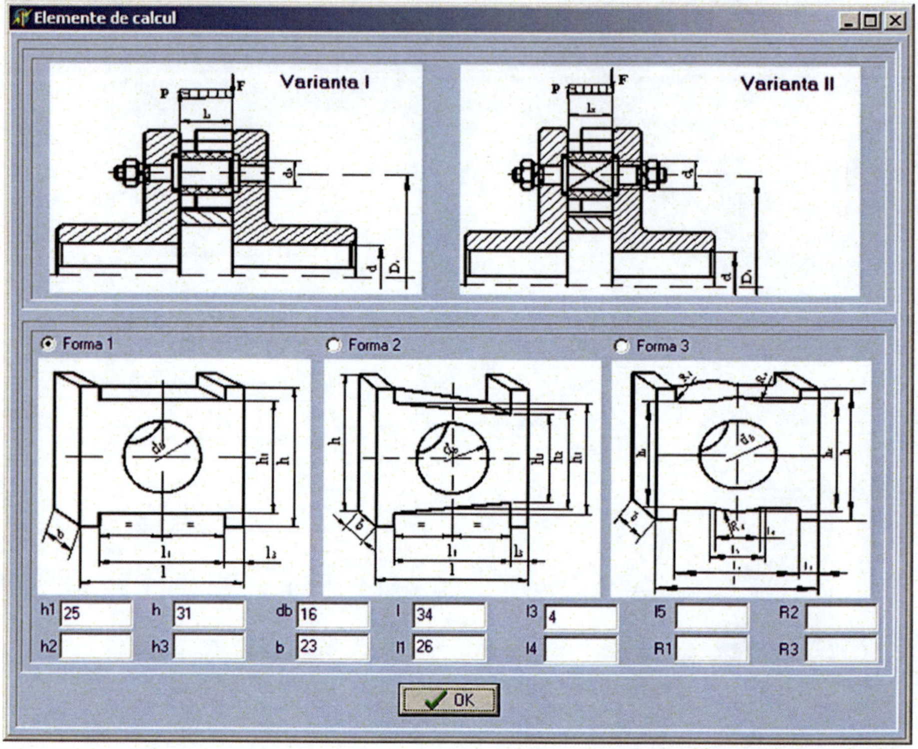

Fig. 5. Select of form 1 of non-metallic element and dimensions of form 1.

The final results of program calculus may be show in Fig. 6. The final results contain information about: value of the force $F1$ that loads a bolt, value of the effective traction stress σ_t of the nonmetallic element, value of the effective crushing stress σ_s of the nonmetallic element, values of the capable torques of being transmitted to the coupling from the condition of resistance to crushing and traction of the non-metallic element, determination of the ratio of capable moments K (as well as the calculation of the bending bolt check - the effective bending tension), the relative rotation angle θ between the half-couplings, the static rigidity kr of the coupling.

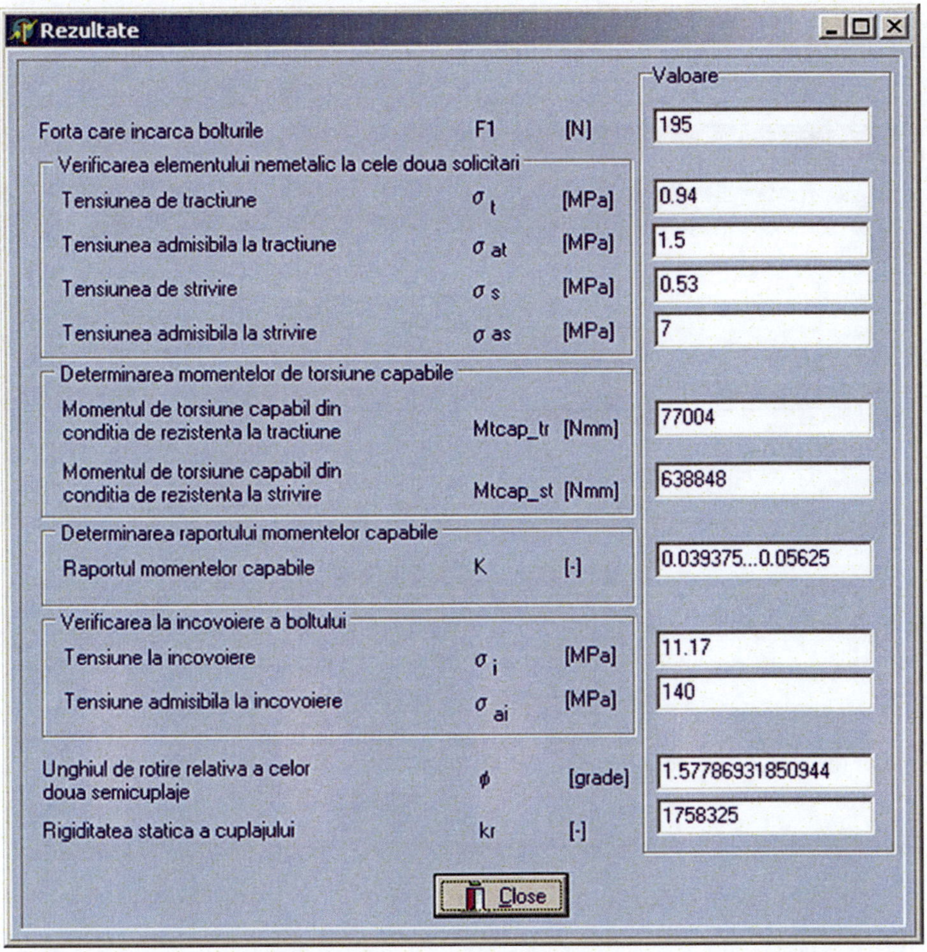

Fig. 6. Results of check calculus for form 1 of non-metallic element from natural rubber NR.

Figure 6 shows results of check calculus for form 1 of non-metallic element the following results of check calculus are obtained:

- the effective tensile stress of the non-metallic element is $\sigma_t = 0.94$ MPa ,the admisible tensile stress is $\sigma_{at} = 1.5$ MPa, $\sigma_t = 0.94$ MPa $\langle \sigma_{at} = 1.5$ MPa, which means that the non-metallic element resists tensile;
- the effective crushing stress of the non-metallic element is $\sigma_s = 0.53$ MPa, the admisible crushing stress is $\sigma_{as} = 7$ MPa, $\sigma_s = 0.53$ MPa $\langle \sigma_{as} = 7$ MPa, which means that the non-metallic element resists crushing;
- the torque capable of being transmitted by form 1 of non-metallic element is $M_{tcapt} =$ 77004 Nmm from resistance condition to traction.

Figure 7 presenting the diagrams plotted for ratio of capable torque in function of ratio (h_1/d_b).

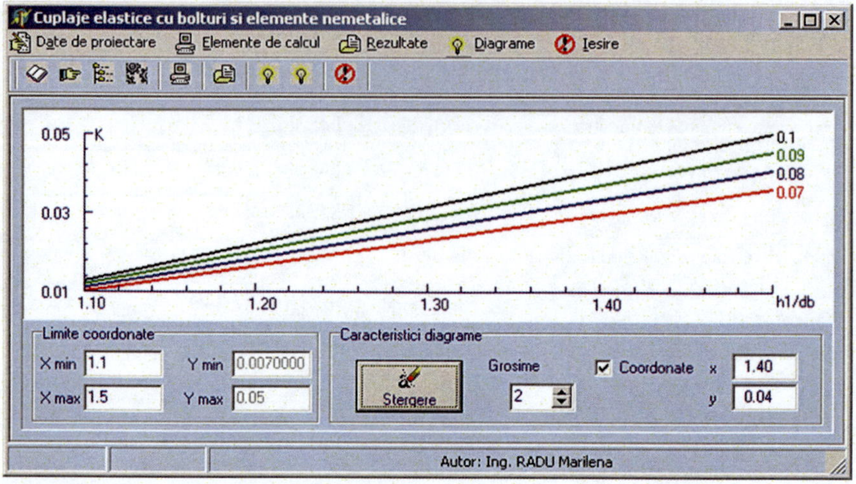

Fig. 7. Ratio of capable torques in function of (h_1/d_b) for form 1 of non-metallic element from NR

Figure 8 presenting the the diagram calculus torque in function of relative rotation angle of those two semicouplings $M_{tc} = M_{tc}(\Delta\varphi)$.

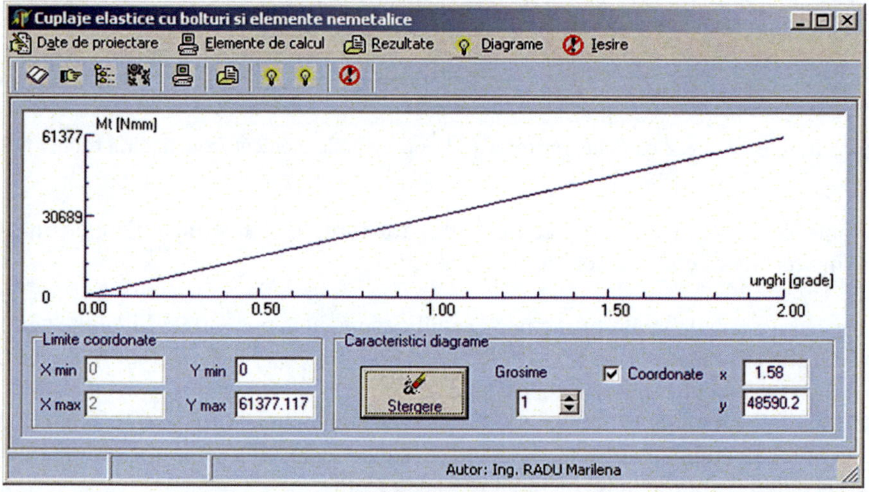

Fig. 8. Calculus torque in function of relative rotation angle of those two semi-couplings $M_{tc} = M_{tc}(\Delta\varphi)$– form 1 of non-metallic element from NR

Further on it will maintain the same value of entry dates (P, n, d) and functionary conditions - $K_s = 2, 4$, the same constructive variant of coupling, the same materials, respective dimensions of bolts, disposer diameter of those and exterior diameter of coupling. From window "***Calculus Elements***" is choosing form 2 for nonmetallic elements (see Fig. 9), for making the verification calculations, the results being presents in Fig. 10 for form 2.

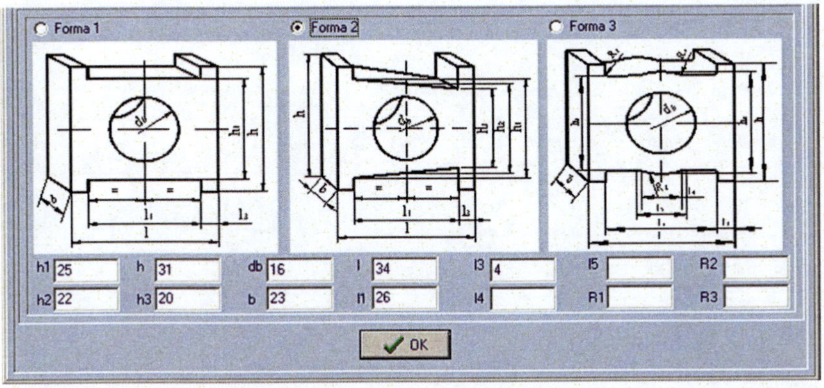

Fig. 9. Select of form 2 and dimensions of form 2 of non-metallic element, realized from natural rubber NR

Further on it will maintain the same value of entry dates ($P = 3$ kw, $n = 1420$ rot/min, $d = 38$ mm) and functionary conditions - $K_s = 2{,}4$, the same constructive variant of coupling, the same materials for elements from componence of coupling, respective dimensions of bolts, disposer diameter of those and exterior diameter of coupling.

From window "Calculus Elements" is choosing respective form 3 (see Fig. 11) and the dimensions of this shape are displayed, and the program continues to perform the verification calculations, the results being presents in Fig. 12 for form 3.

Figure 10 shows that for form 1 the following results are obtained:

- the effective tensile stress of the non-metallic element is $\sigma_t = 1.41$ MPa, the admisible tensile stress is $\sigma_{at} = 1.5$ MPa, $\sigma_t = 1.41$ MPa $\langle\ \sigma_{at} = 1.5$ MPa, which means that the non-metallic element resists tensile;
- the effective crushing stress of the non-metallic element is $\sigma_s = 0.53$ MPa, the admissible crushing stress is $\sigma_{as} = 7$ MPa, $\sigma_s = 0.53$ MPa $\langle\ \sigma_{as} = 7$ MPa, which means that the non-metallic element resists crushing;
- the torque capable of being transmitted by form 1 is $M_{tcapt} = 51336$ Nmm;
- the ratio of the capable torques K = 0.02625… 0.0375;
- the effective bending tension of the bolt is $\sigma_i = 11.17$ MPa, the admissible bending tension is $\sigma_{ai} = 140$ MPa, $\sigma_i = 11.17$ MPa $\langle\ \sigma_{ai} = 140$ MPa, which means that the bolt resists to bending;
- the relative rotation angle of the two half-couplings is $\varphi = 1.57786931850944°$;
- the static rigidity of the coupling determined theoretically is $kr = 1758325$.

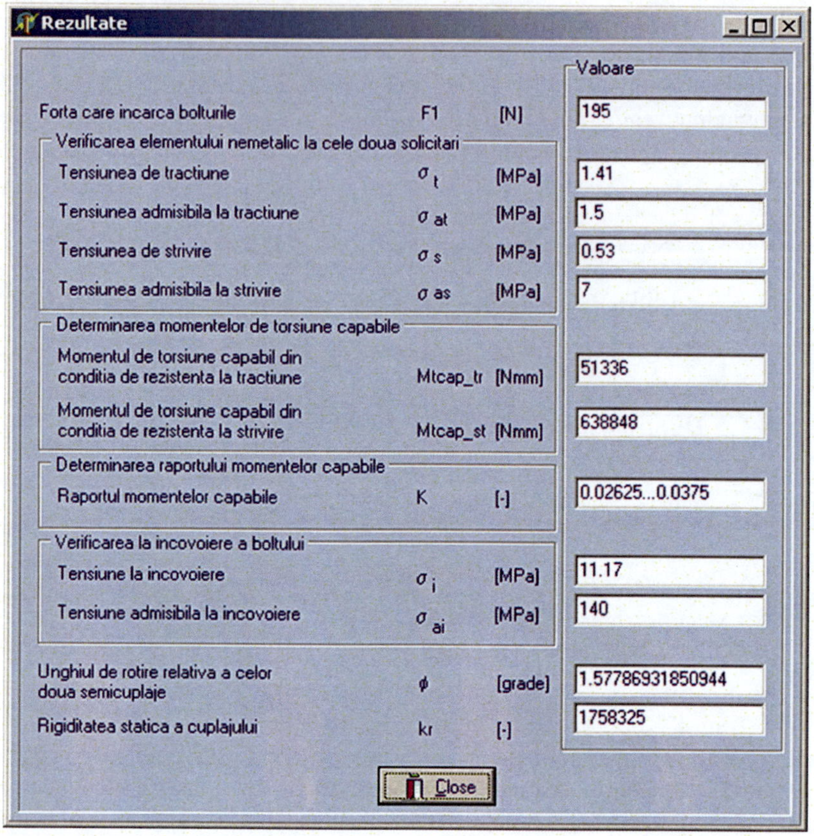

Fig. 10. Results of check calculus for form 2 of non-metallic element from natural rubber NR

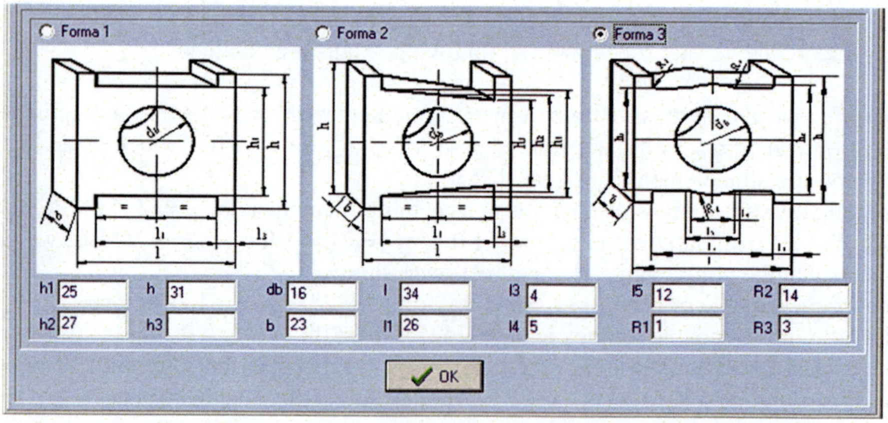

Fig. 11. Select of form 3 of non-metallic element, realized from natural rubber NR

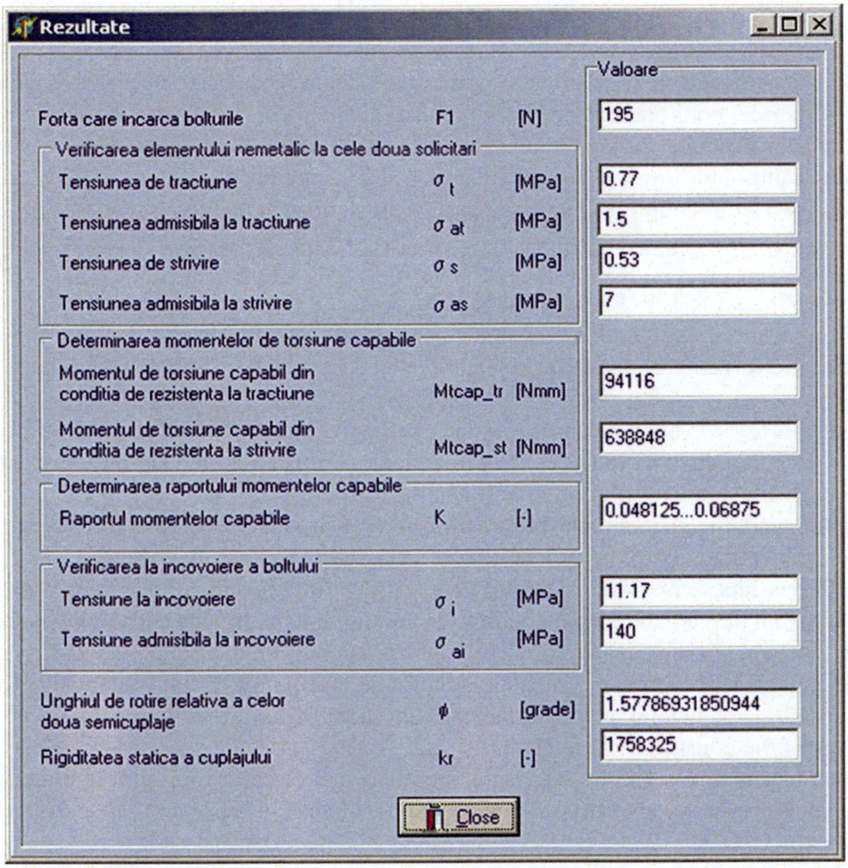

Fig. 12. Results of check calculus for form 3 of non-metallic element from natural rubber NR

Figure 12 shows that for form 3 of noon-metallic element the following results of check calculation are obtained:

- the effective tensile stress of the non-metallic element is $\sigma_t = 0.77\,\text{MPa} \langle \sigma_{at} = 1.5\,\text{MPa}$, which means that the non-metallic element resists tensile;
- the effective crushing stress of the non-metallic element is $\sigma_s = 0.53\,\text{MPa} \langle \sigma_{as} = 7\,\text{MPa}$, which means that the non-metallic element resists crushing;
- the torque capable of being transmitted by form 1 is $M_{tcapt} = 94116\,\text{Nmm}$;
- the ratio of the capable torques K = 0.048125… 0.06875;
- the effective bending tension of the bolt is $\sigma_i = 11.17\,\text{MPa} \langle \sigma_{ai} = 140\,\text{MPa}$;
- the relative rotation angle of the two half-couplings is $\varphi = 1.57786931850944°$;
- the static rigidity of the coupling determined theoretically is $kr = 1758325$.

3 Conclusion

In paper was presented an original program by informatized design of elastic coupling with cylindrical bolts and intermediary nonmetallic elements, (the nonmetallic element having 3 constructive variants, realized from natural rubber), which permits constructive designing and drawing the calculus torque in function of relative rotation angle of those two semi-couplings $M_{tc} = M_{tc}(\Delta\varphi)$, in static functionary conditions, *for different values of geometric parameters and* ratio of capable torques $K_{F1} = K_{F1}\left(\frac{h_2}{d_b}\right)$, $K_{F2(3)} = K_{F2(3)}\left(\frac{h_2}{d_b}\right)$ in function of $\left(\frac{h_1}{d_b}\right)$, $\left(\frac{h_2}{d_b}\right)$.

From achieved calculations for this constructive solution of coupling with those 3 constructive forms of nonmetallic element realized from natural rubber result that:

- The bolts resist to bending, and the value of effective stress of bending is not influence by the constructive form for nonmetallic element, this stress having the same value for each constructive solution of nonmetallic element.
- The nonmetallic element resist to traction and crush, the crush solicitation being the principal solicitation of non-metallic element.
- From resistance condition to traction and crush of nonmetallic element, it is establishing that the capable torque obtain from condition to resistance to crushing is some big that value of the capable torque obtain from condition to traction.
- Value of effective stress of crushing is the same indifferent by the constructive form choose for nonmetallic element, the crushing area being the same regardless of the shape of the non-metallic element.
- Value of effective stress by traction is modified (increase for form 2 of non-metallic element, decrease for form 3 of non-metallic element) to rapport of value displayed for form 1.
- Value of capable torque from resistance condition to traction is modified that value displayed for form 1, increase for form 3, and decrease for form 2 of non-metallic element.
- The relative rotation angle between the semi-couplings has the same value for each shape of non-metallic element made of natural rubber, regardless of the constructive shape of the non-metallic element, because in its calculation relation appear the same terms (length of non-metallic element l, bolt diameter d_b, height of non-metallic element h, number of bolts z, width of non-metallic element b, disposer diameter D_1 of bolts, modulus of elasticity of the material of the non-metallic element E, calculus torque M_{tc}) which have the same value for each term regardless of the form chosen.
- The rigidity of the coupling is constant, it has the same value for each form of non-metallic element made of natural rubber, regardless of the constructive form of the non-metallic element, since in its calculation relation the same terms appear (length of non-metallic element l, bolt diameter d_b, height of non-metallic element h, number of bolts z, the width of non-metallic element b, disposer diameter D_1 of bolts, the material of the non-metallic element E) whose values do not change.

Designing program is a program with a very big generality and which permits of the designer further developments.

The study method propose from intermediary program is interactive from the facility offer of operator, on the road process by functional modeling, to choose between more work hypotheses.

References

1. Radu, M.: Doctoral thesis. Theoretical and experimental studies as concerns couplings with nonmetallic elastic elements. Transilvania University of Brasov (2005)
2. Jula, A., Lateş, M.: Machine parts. Transilvania University of Brasov Publishing House (2004)
3. Jula, A., et al.: Machine Parts and Mechanical Transmissions. Transilvania University of Brasov Publishing House (2005)
4. Moldovean, G., et al.: Designing the Shape of Straight Shafts. Lux Libris Publishing House, Brasov (1998)
5. Collection of norms and extracts from standards for the design of machine components, vol. I and II. University of Brasov Publishing House (1984)
6. Săvescu, D., Budală, A., Ghiţescu, M.: Machine Parts Mechanical Transmissions used in Industrial Technical Constructions. Lux Libris Publishing House, Brasov (2013)
7. Ghitescu, M., Ghitescu, I.M., Borza, P.N., Vlase, S.: New optimized solution for a flexible coupling with bolts used in the mechanical transmissions. Symmetry-Basel **13**(2), 171 (2021). https://doi.org/10.3390/sym13020171
8. Modrea, A., Vlase, S., Calin, M.R., et al.: The influence of dimensional and structural shifts of the elastic constant values in cylinder Fiber composites. J. Optoelectron. Adv. **15**(3–4), 278–283 (2013)
9. Vlase, S., Nastac, C., Marin, M., et al.: A method for the study of the vibration of mechanical bars systems with symmetries. Acta Technica Napocensis. Ser.- Appl. Math. Mech. Eng. **60**(4), 539–544 (2017)
10. Marin, M., Ellahi, R., Vlase, S., et al.: On the decay of exponential type for the solutions in a dipolar elastic body. J. Taibah Univ. Sci. **14**(1), 534–540 (2020)
11. Vlase, S., Marin, M., Scutaru, M.L., et al.: Coupled transverse and torsional vibrations in a mechanical system with two identical beams. AIP Adv. **7**(6), 065301 (2017)
12. Vlase, S., Marin, M., Oechsner, A.: Considerations of the transverse vibration of a mechanical system with two identical bars. Proc. Inst. Mech. Eng. Part L-J. Mater.-Des. Appl. **233**(7), 1318–1323 (2019)

Legal Requirements Versus Customer Requirements in Machine Cab Design

Lates Daniel(✉)

George Emil Palade University of Medicine, Pharmacy, Science, and Tehnology of Targu Mures, Târgu Mureș, Romania
daniel.lates@umfst.ro

Abstract. This paper deals with the problem of legal requirements related to the requirements formulated by customers regarding the equipment cabs. In order for a cab to be used on roads, it must meet certain requirements that correspond to the directives in force. Customers would like to have the most spacious cab, the widest possible field of vision, ergonomics, while the legislation requires first of all to respect the safety requirements of the passenger and the proper use of the equipment from which it is made. Some of these aspects will be presented in the paper and as a methodology it will start from the legal requirements that are in force.

Keywords: Cab · Homologation · Customer requirements · Legal requirements · ROPS testing · FOPS testing · Field of view · Windows

1 Legislative Framework

The cab is used on any motor vehicle, on wheels or tracks which has at least two axles, the function of which consists essentially in its traction force and which is specially designed for towing, pushing, carrying or operating certain machines, machinery or trailers intended for use in agricultural, forestry, construction, etc. [1].

Approval of a cab means the procedure whereby the approval authority certifies that a type of vehicle, system, component or separate technical unit complies with the applicable administrative stipulations and technical requirements [2].

Approval is required to ensure that the cab meets all technical and safety requirements so as not to endanger the somatic integrity of the driver. The tractor cab is particularly important given that it is the area of the vehicle in which the driver of the vehicle operates, consequently, the component of the vehicle that has exclusive direct contact with it, and in case of accident or failure can damage it. The equipment cab must meet a series of requirements and norms according to ISO Standards, EU Directives and Regulations, but also international ones [3].

Approval authorities shall ensure that manufacturers applying for type-approval comply with their obligations under the Regulation. The homologation authorities homologate only those cabs that comply with the requirements of Regulation 2013–167, art.6. [3].

L. Moldovan and A. Gligor (Eds.): Inter-Eng 2021, LNNS 386, pp. 38–46, 2022.
https://doi.org/10.1007/978-3-030-93817-8_4

2 Customer Requirements

Customers would like to have the most spacious cab, with air conditioning and audio systems as high as possible, field of view as large as possible, the possibility of using outdoor video cameras, ergonomics on the seats, operator seat and passenger with regulation and relaxation possibilities, while the legislation requires first of all to respect the safety requirements of the passenger and those in traffic as well as the part of the compliant use of the equipment of which it is part.

Customers are of the opinion or have reason to assume that the placing on the market or putting into service of the vehicle, system, component or separate technical unit is not in accordance with the regulations or delegated or implementing acts adopted in on that basis, it shall immediately take the necessary corrective action to restore the conformity of the vehicle, system, component or separate technical unit or to withdraw or recall it, as appropriate [3].

The manufacturer's intervention in this process is to inform without delay the approval authority that granted the approval, providing, in particular, details of the non-compliance and the corrective measures taken [4].

3 Cab Design

3.1 Making Design by Manufacturers

Manufacturers shall ensure that vehicles are designed, constructed and assembled in such a way as to minimize the risk of injury to the occupants of the vehicle and others in the vicinity of the vehicle.

- Manufacturers shall ensure compliance of vehicles, systems, components and separate technical units with the applicable requirements set out in the Regulation, including requirements relating to:
- integrity of the vehicle structure;
- systems that provide the driver with visibility and information on the condition of the vehicle and the surrounding area, including windows, mirrors and driver information systems;
- vehicle lighting systems;
- vehicle occupant protection systems, including interior fittings, head restraints, seat belts, doors;
- heating systems;
- devices to prevent unauthorized use;
- vehicle identification systems;
- masses and dimensions;
- rear protection devices;
- lateral protection, etc.

3.2 Field of Vision

These requirements ensure market competitiveness, by constantly pursuing that all components, assemblies and subassemblies of the product are of high performance, instead the machine cab requires a wider field of view (Fig. 1 and Fig. 2) to facilitate when handling the machine, therefore, additional windows are added between the bonnet and the extremities of the tractor to allow the driver of the machine to see exactly where he is stepping on the wheel, in order to avoid certain undesirable accidents during use.

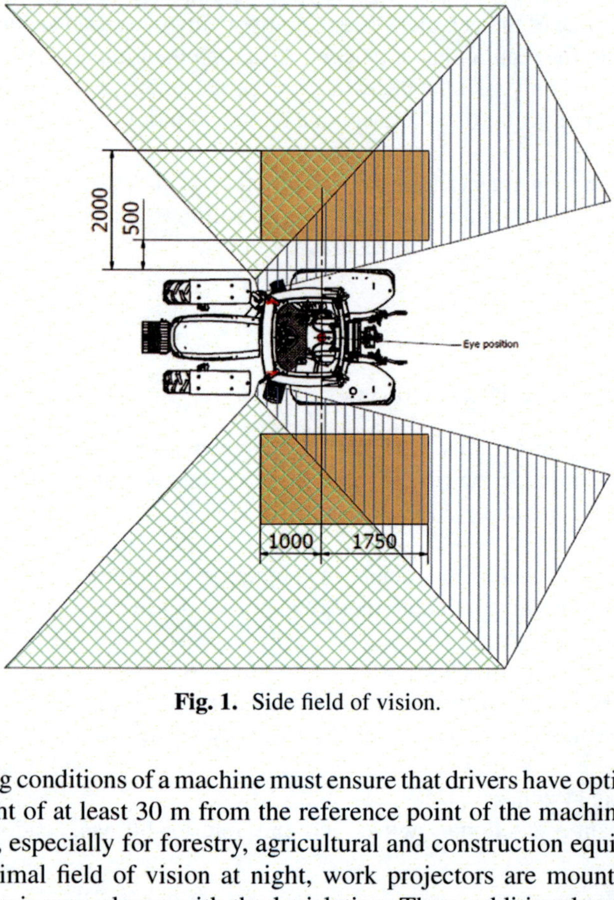

Fig. 1. Side field of vision.

The lighting conditions of a machine must ensure that drivers have optimum visibility during the night of at least 30 m from the reference point of the machine both in front and in the rear, especially for forestry, agricultural and construction equipment.

For an optimal field of vision at night, work projectors are mounted to ensure a lighting surface in accordance with the legislation. These additional projectors can be mounted with a 360° orientation on the inside of the ceiling to ensure extended visibility.

The lighting conditions also include the signal lights corresponding to the road traffic: headlights with road light, with passing light, direction indicator lights, etc.

When choosing the type of cab, the rules concerning the glass elements of the cab must also be taken into account. The glass of the windscreen, rear window and side windows must be a safety glass supported (Fig. 3) by a protective membrane, taking into account the percentage of glass components of the agricultural tractor cabs, the fractions of which in the event of an impact, be of very small dimensions, which do not cause

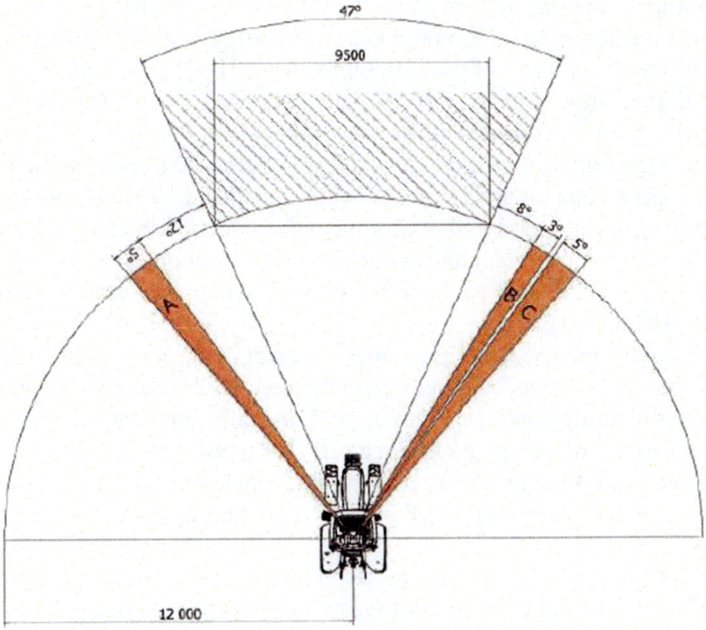

Fig. 2. Field of vision.

injury to the driver, if the membrane is pierced by objects and fails to support the broken glass [3].

Fig. 3. Broken secure windshield.

In order to be approved, the cab must meet a number of conditions in accordance with Directive 2006/42/EC of the European Parliament and of the Council of 17 May 2006 on technical equipment. Looking at things more explicitly, the legislation in force provides certain regulations and standards directly on the product in question, regulations and rules much stricter and more explicit in the Delegated Regulation (EU) No. 1322/2014 Of the Commission From 19 September 2014 [5], as well as Commission Implementing Regulation (Eu) 2015/504 From 11 March 2015, [6]. This Regulation lays down detailed technical requirements and test procedures for the design, construction and assembly of the vehicle for the approval of agricultural and forestry vehicles and their systems, components and separate technical units, detailed modalities and requirements for type-approval, virtual tests and conformity of production, technical specifications regarding access to vehicle repair and maintenance information, as well as performance standards and technical service evaluation criteria in accordance with Regulation (EU) No 182/2011. 167/2013 [2].

The technical equipment sector is an important part of the construction industry. The cost of a number of accidents caused directly by the use of technical equipment can be reduced by the safe design and construction of technical equipment, its proper handling and maintenance. In order to be considered safe, the equipment must comply with the safety conditions and regulations imposed by the legislation in force. These rules are particularly important, given that it is about the safety of the operator and implicitly his life [3].

In view of the nature of the risks involved in the use of the technical equipment provided for in this Directive, procedures must be established to assess compliance with the essential health and safety requirements of the operator. These procedures must be developed in the light of the degree of inherent danger involved. Manufacturers should bear full responsibility for certifying the conformity of their technical equipment with the provisions of this Directive. However, for certain types of technical equipment, with a higher risk factor, a stricter certification procedure is recommended.

According to Directive 42 of 2006, a process of risk assessment and reduction must be carried out. The manufacturer or his authorized representative must:

- to establish the limits of the technical equipment, which includes the anticipated destination and any possible inappropriate use;
- to identify the dangers that may be generated by the equipment and the dangerous situations associated with them;
- to estimate the risks, taking into account the severity or possible injuries or damage to health and the probability of their occurrence;
- assess the risks in order to establish the need to reduce the risk in accordance with the objective of this Directive;
- eliminate hazards or reduce the risks associated with hazards by applying protective measures [7].

3.3 Performing Virtual Tests

The technical service must provide a test report on the results of the virtual test. The test report must be clear, consistent with the correspondence report and the validation report and must include at least the following elements: creation of a virtual prototype, input data and simulation results in terms of the technical requirements imposed.

3.4 Requirements Applicable to Rops Protection Structures (Roll-Over-Protective-Structure - Against Rolling Over)

Protective structure in the event of an overturning means the structure provided on a tractor with the essential purpose of avoiding or limiting the risks to which the driver is exposed if the tractor overturns during normal use. The ROPS tests (Fig. 4) have the role of ensuring the observance of safety norms regarding the deformability of the cab, this must not have a deformability that enters the free space of the driver and neither its elements or subassemblies which can injure the person by deforming or detaching them when the cab impacts the ground. In this sense, virtual and physical tests are performed that simulate the impact of the cab with the ground from different positions (front, rear, side and vertical) to test the resistance of weldable elements, but also prefabricated elements [9].

Fig. 4. The four phases of the ROPS test [8].

The ROPS test (Fig. 5). is not a "dynamic" test by sending a car into the wall. We use a hydraulic cylinder to push slowly cab beams in four directions (rear push, side push, rear flattening, front flattening). By regulating manually the pressure inside the hydraulic cylinder through the piloting system (third party manufacturer), and knowing its section area, the test operator is able to control the force/displacement and energy injected into the cab to follow the homologation standard [8].

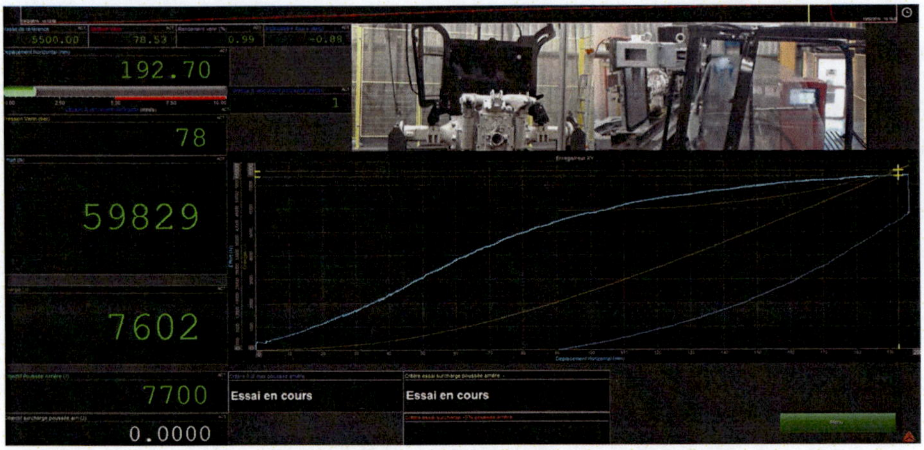

Fig. 5. ROPS cab testing [8]

3.5 Requirements for FOPS Protection Structures (Falling-Objects-Structure - Resistance in Case of Heavy Objects Falling from Above)

Assembly which protects, in the event of a fall of objects, the area above the head of an operator in the driving position. The protection structure may be manufactured by the tractor manufacturer or by an independent company. In both cases, the test is valid only for the tractor model which has been tested.

The protection structure must be tested for each tractor model on which it is to be mounted [3]. The protective structure under test shall include at least all components that transfer the load from the impact area with the falling object used for the test to the safety area. The protective structure under test must be rigidly attached to the test bench at its normal points of attachment (Fig. 6).

Fig. 6. FOPS cab testing [10].

4 Conclusions

The legislation is the one that says its last word, while the clients are the only ones who can come up with proposals and suggestions that are analyzed by the specialist and then regulated by law in terms of security and operator comfort. The regulations established at European and international level want to regulate a process of studying the risk and reducing it.

Machinery manufacturers must manufacture cabs that comply with the regulations in force and meet customer requirements, but these requirements must not be above the law.

The legislatively regulated visual field wants the cabs to have the best possible overview both front and side and rear so that the operator of the equipment can observe any road danger.

ROPS and FOPS tests add to the operator's safety in case of overturning or falling objects, and not to endanger his life. The ergonomics of the cab also depend on the cab manufacturer and his experience, who will consult with customers so that the stress is as low as possible and the handling of the machine is a pleasure.

References

1. Directive 2011/77/EU of the European Parliament and of the Council of 27 September 2011 amending Directive 2006/116/EC on the term of protection of copyright and certain related rights - art.1 alin.1 (2011)
2. Regulation (EU): No. 167/2013 of the European Parliament and of the Council on the approval and market surveillance of vehicles, 5 February 2013

3. Cengher, A.M.: Technical Documentation for Tractor Cab Approval, Diploma Project (coordinator Lates Daniel). University of Medicine, Pharmacy, Science and Technology Târgu Mureş (2019)
4. Regulation (EU): No. 167/2013 of the European Parliament and of the Council art.8 Obligations of producers, 5 February 2013
5. Delegated Regulation (EU): No. 1322/2014 of the Commission, supplementing and amending Regulation (EU) no. Regulation (EC) No 167/2013 of the European Parliament and of the Council as regards vehicle construction and general requirements for type-approval, 19 September 2014
6. Commission Implementing Regulation (EU): 2015/504 of 11 March 2015 implementing Regulation (EU) No 182/2011 Regulation (EC) No 167/2013 of the European Parliament and of the Council on administrative requirements for the approval and market surveillance of agricultural and forestry vehicles Text with EEA relevance (2015)
7. Directive 2006/42/EC of the European Parliament and of the Council on technical equipment and amending Directive 95/16/EC (recast), 17 May 2006
8. https://www.naro.go.jp/english/laboratory/iam/research/safetyevaluationandstandardization/index.html
9. https://encryptedtbn0.gstatic.com/images?q=tbn:ANd9GcR26v5usmRx7GTxsro3eNu2fwZ_ESTqWXYZlA&usqp=CAU
10. https://pronar.pl/en/produkt/tests-of-falling-object-protective-structures-fops/
11. https://dewesoft.com/case-studies/rops-testing-tractor-cabin-structural-safety-test

Value Stream Map Importance in the Field of Electrostatic Powder Painting

Jozsef Boer[1] and Petruta Blaga[2(✉)]

[1] SC Allcolors Serv SRL, Vidrasau, Parc Industrial Mures 1/G/5, 47612 Mures, Romania
calitate@allcolors.ro

[2] "George Emil Palade" University of Medicine, Pharmacy, Science and Technology of Targu Mureş, 38 Gheorghe Marinescu, 540139 Tirgu Mures, Mures, Romania
petruta.blaga@umfst.ro

Abstract. Many companies struggle to find the right solution to grow their business. The main purpose of a business is to make a profit that is to get money by the investment and the work done. One useful instrument to do it is a Value Stream Map (VSM). A value stream always begins and ends with a customer. Based on this, we had created a tested and implemented VSM to increase in profits which can be achieved the field of electrostatic powder painting business. Regardless of the results obtained by using a VSM, it is very important to identify those points and those factors that determine the results. Management must consider the results obtained for the decisions on the decision to reduce the lead-time and parallel with this the cost of the process.

Keywords: Electrostatic field painting · Pretreatment · Value stream map · Efficiency · Waste · Work-in-Progress (WIP)

1 Introduction

Electrostatic field painting with powder paint is a process by which a layer of decorative powder paint is applied to a metallic surface to color it and provide protection against the action of external factors.

This process employs charged particles to paint a work piece more efficiently.

The raw pieces go down a conveyor belt towards a paint booth, or paint tank, where it is sprayed with, electrostatically charged paint particles. Before this, all raw pieces are passing a pretreatment step to clean and passivate the surface which is going to be painted. After, the raw pieces are coated; it continues the conveyor belt to an oven, where the paint is cured. The benefits to the process of electrostatic coating are the ability to recover the little over-spray and having the process automated which will cut costs. To be more efficiently, the whole quantity of the raw products should be going true the entire process on the shorter time.

Saving time, we save money, so the process is more efficiently [1].

L. Moldovan and A. Gligor (Eds.): Inter-Eng 2021, LNNS 386, pp. 47–55, 2022.
https://doi.org/10.1007/978-3-030-93817-8_5

2 Theoretical Foundations

The paper aims to present a solution which can generate profit in the field of electrostatic powder painting business [2–4].

The value-stream mapping [5] identifying and reducing "the waste" in value streams, thereby increasing the efficiency of a given value stream [6–8].

The "waste" removal increasing productivity by creating leaner operations which in turn make waste and quality problems. The used methods are often used in Lean environments to analyze and design flows at the system level [9].

Commonly accepted types of waste are:

- Processes which are creating too much of a good or service that damages production flow, quality [10], productivity
- Steps when the goods are not being transported or worked on
- Unnecessary inventory, excess stock
- Poor layout and communication and unnecessary motion in the process
- Double-handling and excessive movement
- Unnecessary movements, excess energy using
- Resources and costs required to correct the defects.

3 The VSM Build Up Process

Schematically, the electrostatic field painting with powder paint process takes place in 20 steps and it the looks as follows (Fig. 1):

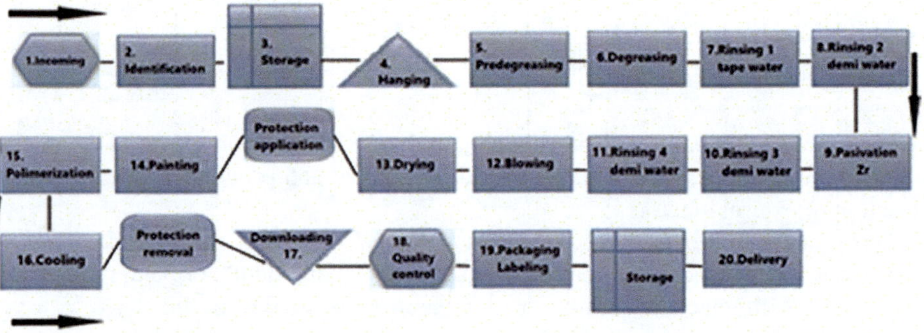

Fig. 1. The process flow diagram in the electrostatic field powder paint.

After a sampling quality control at the workshop for the reception of raw products [11], they are placed on the painting line by hanging them on the conveyor. They pass through the pre-treatment tunnel where a chemical attack by sprayers [12, 13] made in several steps:

1. Coarse washing with water

- Degreasing using a strongly alkaline solution
- Rinse with water
- Rinse with demineralized water
- Passivation with a nanoceramic multimetal solution
- Rinse with recirculated demineralized water
- Rinse with fresh demineralized water.

The products enter the drying oven to dry and completely remove the water residues.

The painting step is the next step done in a special booth, ready to apply the powder paint on the surface of the products by means of automatic guns and manual guns based on the electrostatic field created between the ends of the guns and the body of the paint product. The powder passes on the surface of the product and adheres to it thanks to the electronic loading of the paint powder.

The next stage is the polymerization of the paint layer on the surface of the product takes place in the polymerization oven. The products are cooled and after quality control they are sent for packaging and delivered to the customer.

The essence of the implementation consists in reducing all the unproductive times and movements known as waste, without creating dysfunctions and discount from the painting process and the quality of the painted product.

a.
b.
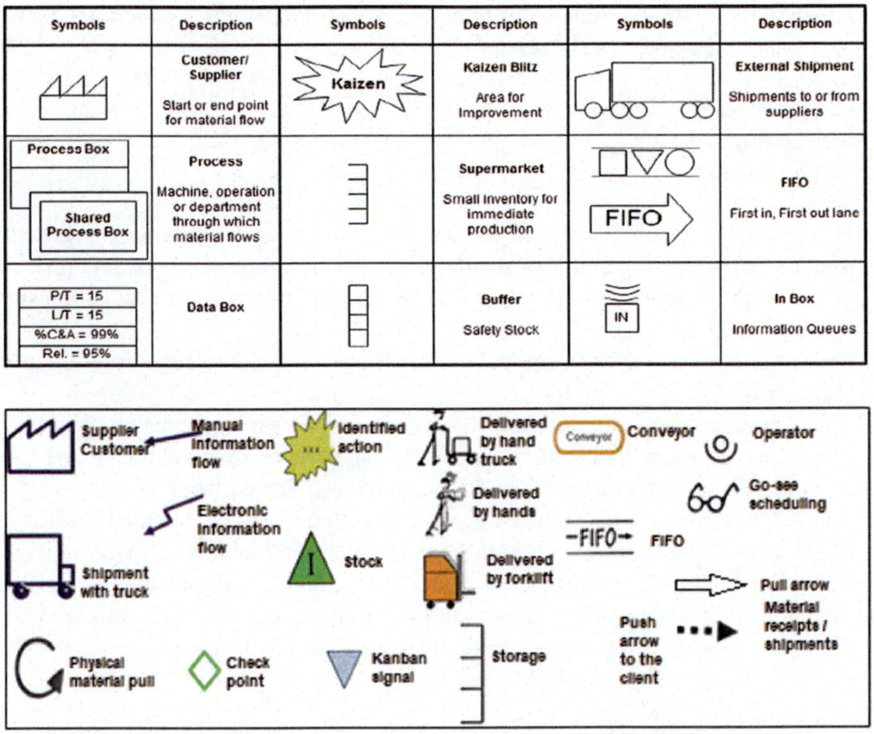

Fig. 2. a, b. Common symbols to build up a value stream map.

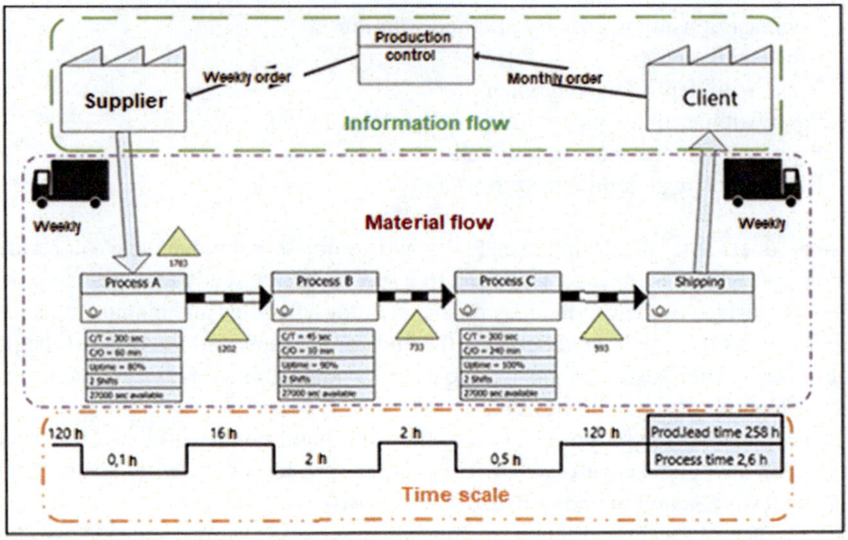

Fig. 3. VSM raw representation.

The value stream map (VSM) gives the possibility to do this [14].

Are three main flows which helps to build up the map helping to split up the process into steps and see where are the waste:

- Information flow
- Material flows
- Lead time ladder.

Using simple graphical instruments, we can build up the VSM based on those three main flows. The mapping employs standard symbols to represent items and processes. The knowledge of these symbols is essential to correctly interpret the production system problems (Fig. 2).

Knowing the process steps and using the symbols we have a raw representation of the VSM: (Fig. 3).

Using these instruments, the activities from the process becomes easily separated into the value stream, which is the focus of one type of attention and the "waste" steps, another type. Identifying them, the waste would be eliminated from the process.

The VSM manager using sheets, pens, and colored pencils, together with 6 team members, is creating the map. One of these members is Lean expert who is working together with a production supervisor, operator, a quality engineer, a logistical leader and maintenances responsible. Before starting, is necessary to clarify the aim of the mapping. It is done in a representative day, not peak or leave.

As a result, the VSM it looks like in Fig. 4.

Baseline Metrics

Based on the internal management report the administrator of the company reviewed the financial balance and determined the baseline metrics and set the next goals [15]:

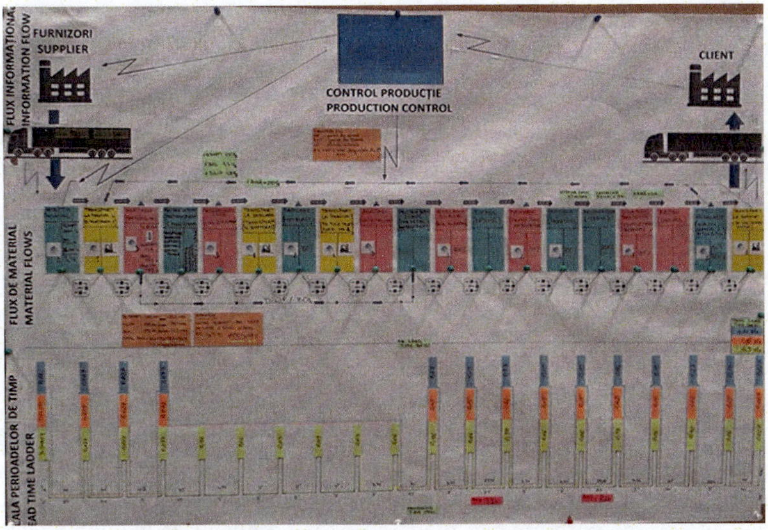

Fig. 4. The value stream map after implementation.

- Reduce lead-time from 9 to 5 days (−44,44%)
- Reduce the price from 5 to 4,6 euro/m^2 (−8%)
- Maintain profit margin.

VSM Data – Takt Time

Takt time is a tool used to design work and it measures the average time interval between the start of production of one unit and the start of production of the next unit when items are produced sequentially.

Current volumes were 1000 m^2/day painted products but after the new goals setting the total painted surface increase to 1500 m^2/day. The company's employees work 2 shifts of 8 h with a lunch break of 20 min and 2 breaks of 5 min break/shift. So, total work time is 7.5 h, means 450 min/shift. Takt time currently painting 1000 m^2/day/shift is 0.45 min/m2 and for 2 shifts is 0.90 min/m^2 (Table 1).

Table 1. Key metrics of the daily volumes.

Key metrics	Current values	Target values
Volumes (m^2/day)	1000	1500
Shifts/day	2	2
Effective working time (min)	450	450
Takt time/day (min/m^2)	0,90	0,60

Takt time, the rate at which the company would need to be created to meet the goals, would be: 900 min/1500 m^2, respectively 0.60 min/m^2.

Other Key Metrics
Other key metrics that have been taken into account are presented in Table 2.

Table 2. Other daily key metrics.

Key metrics	Current values
Employees	56
Labor cost/hour (euro)	560
Painting cost euro/m^2	15
Rework rate (of total order) %	2

Current State VSM
Reviewing the map, we are checking the resources as Work-In-Process (WIP) and Lead-Time. These match the current state metrics of the company. If wouldn't match, the actual current state is useless and must be corrected before processing.

The term Work-In-Progress (WIP) is a production and supply-chain management term describing partially finished goods awaiting completion. These costs are subsequently transferred to the finished goods account and eventually to the cost of sales [16].

Key Observations
Process step "Hanging" has less resources than needed to keep up with demand. Discussions with the line operators show that many times the person from the "Downloading" process is helping out at "Hanging" process and is contributing to the low FPY at "Hanging". The person at "Downloading" does not have the proper training to be working at "Hanging". Process" Painting" has more resources than needed.

FPY (First Pass Yield) is a measure of quality in a process that reflects the percentage of product made correctly without any rework or corrective activity.

Current Profit Margin
The calculation of the profit margin assumes no lost painted material for rework.

In the working conditions with a cost of 10 euro/hour/employee, with a takt time of 0.84 min/m^2 and a material cost of 5 euro/m^2, the total cost for painted m^2 becomes 13.4 euro. Under these conditions, the marginal profit is 11.94 euros (Table 3).

Current state – Takt time 0.84 min/m^2
Since the company goal is to achieve 1500 m^2/day the Future State needs to take into consideration these increased volumes: 1500 m^2/day make at takt time of 0.60 min.

Future Profit Margin
With the same conditions of infrastructure and human resources, the company tends to lower production costs and obtain a more competitive price. This assumes no lost material for scrap or rework. The goal is to obtain the same marginal profit with a reduction of the costs of the painted surface from 13.4 to 10.2 euro/m2 without losing from the marginal profit (Table 4).

Table 3. Current profit margin.

Key metrics	Current values
Labor cost/employee (euro/hour)	10
Takt time (min/m^2)	0,84
Labor cost (euro/m^2)	8,4
Material cost (euro/m^2)	5
Total cost (euro/m^2)	13,4
Profit margin (euro)	11,94

Table 4. Future profit margin.

Key metrics	Target values
Labor cost/employee (euro/hour)	10
Takt time/m^2 (min)	0,56
Labor cost/m^2 (euro)	5,6
Material cost /m^2 (euro)	4,6
Total cost /m^2 (euro)	10,2
Profit margin	7,84

We notice that the result is not as expected: the marginal profit is reduced. It means that despite the constructive intention to reduce production costs, management cannot make the decision to pursue this action as the marginal profit decreases, as such in the long run the business will suffer.

4 Conclusion

This potential Future State is not meet the goals of 1500 m^2/per day and a 5 days lead-time. It will not meet the product price reduction to 4.6 euro/m^2 while maintaining the profit margin.

The type of projects selected need to be realistic. An experienced operations and supply chain project manager needs to facilitate project selection so that the Future State is achievable just with the right management parameters settings.

Before any projects are started their expected results need to be plugged into a Future State map to confirm that company goals will be met.

By using this methodology, it was desired to obtain results that would strengthen the fairness of the decision to reduce the lead-time and parallel with this the costs of the process creating the decision to make couple implementations on the "waste" removing as are in the process, logistic or human resources.

We started from a thematic qualitative empirical research by observing the results obtained from the process during the production. Empirical research was followed by applied research aimed at finding a solution or tool to reduce production costs [17].

After data collection and information processing, the conclusion was that at this moment isn't the right time and way to decrease the price of the painting process in the condition of the 5 days processing time aim.

During the process observation, has been found couple not useful elements which are creating leaner operations which in turn make waste and quality problems.

The main identified "waste" elements during the monitoring are:

- Steps when the goods are not being transported or worked on
- Unnecessary inventory, excess stock
- Poor layout and communication and unnecessary motion in the process
- Double-handling and excessive movement
- Unnecessary movements, excess energy using
- Resources and costs required to correct the defects.

Monitoring reported in the paper is carried out for the entire painting process, started from the raw material incoming, hanging of the products on the pre-treatment line to the last, packing and final control after the painting and polymerization of the powder paint from the surface of the products, finalized with the storage and delivery steps.

Every step was monitored individually; these are parts of the whole painting process, so the obtained results helped to take the right decision at the level of the management regarding the proposed goals to be achieved.

Based on the results has be decided to continue with a deeper involvement and create a new VSM using the results after the last implementations.

In this way we will have later a new VSM, subject of a new upcoming research.

Even if using a developed and technologically advanced system of painting in electrostatic field is not sufficient to purchase and connecting the performant devices to the painting and pre-treatment system, but must be carried out the surveillance, monitoring, namely to achieve continuous implementations to reduce costs of production and to cope with increasing competition in the market in the field, without forgetting the other important elements that make up the final price of the transaction cost of painting in an electrostatic field [18].

In this way we will have later a new VSM, subject of a new upcoming research.

References

1. Boer, J., Blaga, P.: Streamlining the work process by reducing procedural times in the field of electrostatic powder painting. MDPI J. Proc. **63**(1), 71 (2020)
2. Kramer, H.: Safety aspects of electrostatic paint, powder and flock spraying processes. J. Electrostat. **30**, 77–92 (1993)
3. Boer, J., Blaga, P.: Factors that generate nonconformities in the electrostatic powder pain-ting. Procedia Technol. **19**, 1083–1088 (2015)
4. Karaoglan, A.D., Ozden, E.: Electrostatic powder coating process optimisation by implementing design of experiments. Trans. Inst. Met. Finish. **99**(1), 46–52 (2021)

5. Bungau, C., Gherghea, I.C., Prichici, M.: Value stream mapping analysis, efficiency methods of operational management. In: Abrudan, I. (ed) Twenty Years After: How Management Theory Works, Proceedings, pp. 188–198. Todesco Publishing House, Cluj-Napoca (2010)
6. Galushko, O.S.: Value stream map and methodics of mapping. Actual Prob. Econ. **108**, 96–104 (2010)
7. Schillig, R., Stock, T., Müller, E.: Energy value-stream mapping a method to visualize waste of time and energy. In: Umeda, S., Nakano, M., Mizuyama, H., Hibino, H., Kiritsis, D., von Cieminski, G. (eds.) APMS 2015. IAICT, vol. 459, pp. 609–616. Springer, Cham (2015). https://doi.org/10.1007/978-3-319-22756-6_74
8. Salwin, M., Jacyna-Golda, I., Banka, M., Varanchuk, D., Gavina, A.: Using value stream mapping to eliminate waste: a case study of a steel pipe manufacturer. Energies **14**(12), 3527 (2021)
9. Harris, G.A., Donatelli, A.: Value stream mapping as a lean management tool. In: Engineering Management: It's About People, Proceedings, pp. 245–250. AMER Soc Engineering Management, MO (2001)
10. Pop, L.D., Nagy, E.: Improving product quality by implementing ISO/TS 16949. Procedia Technol. **19**, 1004–1011 (2015)
11. Pop, L.D.: Study on creating a simplified model of quality management system in a SME from the Central Region of Romania. Procedia Technol. **22**, 1084–1091 (2016)
12. Boer, J., Blaga, P.: The influence of the specific indicators of the chemical treatment on the production costs of painting in electrostatic field. Procedia Manuf. **32**, 325–330 (2019)
13. Boer, J., Blaga, P.: Optimizing production costs by redesigning the treatment process of the industrial waste water. Procedia Technol. **22**, 419–424 (2016)
14. Kuhlang, P., Edtmayr, T., Sunk, A., Muehlbradt, T.: Enhancing work system design and improvement by further developments of value stream mapping. In: 2014 IEEE International Conference On Industrial Engineering and Engineering Management (IEEM), pp. 464–469. IEEE, New York (2014)
15. Pei, X.B., Pei, Z.J.: Considering the cost analysis of value stream in the lean improvement. In: Kong, T., Wang, C. (eds.) Proceedings of the 2016 2nd International Conference on Education Technology, Management and Humanities Science, vol. 50, pp. 623–626. Atlantis Press, Paris (2016)
16. Boer, J., Blaga, P.: Production cost optimization in industrial wastewater treatment. Procedia Econ. Fin. **15**, 1463–1469 (2014)
17. Boer, J., Blaga, P.: Innovative method to reduce process costs in the field of electrostatic powder painting. Procedia Manuf. **46**, 44–48 (2020)
18. Boer, J., Blaga, P.: Making production more efficient using analysis and continuous improvement methods. MDPI J. Proc. **63**(1), 73 (2020)

Low-Velocity Transverse Impact Investigations of CFRP Composite Laminated Plates - Simplified Static Simulations Versus Dynamic Experimental Tests

Marius Nicolae Baba[1(✉)] and Florin Dogaru[2]

[1] Transilvania University of Braşov, Eroilor Blvd no. 29, 500036 Braşov, Romania
mariusbaba@unitbv.ro

[2] Consaro Engineering SRL, Mihail Kogalniceanu Blvd no. 23, 500090 Braşov, Romania

Abstract. Nonlinear contact and large displacements static simulations were conducted using Ansys FE software to assess the response of CFRP composite laminated plates tested to low-velocity impact. The scope is to present a solution of enough accuracy and a method of analysis equivalent to a transient dynamic model but more straightforward. The experimental tests were performed upon plate specimens with rectangular geometry dimensions of $150 \times 100 \times 2.5$ mm^3, made of 8 unidirectional laminae (carbon fiber/epoxy vinyl ester resin), stacked in a [0/-45/45/90]s layup configuration. Results are presented in terms of contact force versus central plate deflections, covering a range of impactor velocities between 0.25 m/s to 3 m/s. Within this interval, the predicted numerical response agrees well with the experimental data.

Keywords: Low-Velocity Transverse Impact (LVTI) · Carbon Fiber Reinforced Plastics (CFRP) · Finite Element Analysis (FEA) · Laminated composite

1 Introduction

Although considerable efforts have been made over the years to improve the design, analysis, and prevention of CFRP laminated composites against damages due to low-velocity impact, their use in large-scale industrial applications, however, remains quite limited. The polymeric composite materials reinforced with carbon fibers are already used conventionally in aerospace and automotive industries and soon appear to become more involved in a broader range of engineering applications requiring enhanced mechanical properties for reduced overall weight [1–3]. Indeed, the potential of these materials to sustain the barely visible impact damages (i.e., BVID) is still a particular topic of concern for many researchers in the field of design for manufacturing laminated composite structures. At its core, the low-velocity impact damage might affect the composite material's outer layers, the inner layers, or both the outer and inner layers [4–7].

A large amount of computational and experimental work has been carried out over the years toward developing an accurate prediction of the low-velocity transverse impact

L. Moldovan and A. Gligor (Eds.): Inter-Eng 2021, LNNS 386, pp. 56–63, 2022.
https://doi.org/10.1007/978-3-030-93817-8_6

of CFRP laminated plates. Comprehensive reviews are available in the literature [8–12] and will not be treated here. We note, however, that in pre-sizing design phases, the use of complex finite element models, intended to accurately represent the heterogeneous nature of these materials as well as their complex intra- and inter-ply damage mechanisms, is often limited due to extremely low computational efficiency and numerical integration issues [13]. Moreover, during the design workflow of CFRP laminated composites prone to low-velocity impact damages, experimental testing is often unlikely to provide all the design information over the material, loading and geometric parameters adopted in current structural engineering practices. Therefore, simplified and well-documented finite element models must be available to verify and supplement the existing input data.

Based on the results obtained through the use of a simplified nonlinear finite element model, the purpose of this paper is to determine the level of accuracy between the impact response acquired through experimental tests at different thresholds of impactor velocity and the predicted responses obtained through static FE numerical simulations. In addition, the results of a quasi-static analytical assessment based on the impact energy conservation proposed by the authors in a previously published paper [14] are considered.

2 Materials and Methods

2.1 Experimental Set-up

Experimental investigations were done on plate specimens of rectangular geometry, made of epoxy vinyl ester matrix (Derakane 470–30), reinforced with carbon fibers.

The composite base plates utilized to cut off the rectangular specimens were manufactured manually at laboratory scale by brushing on the resin (a fraction of 30% volumetric ratio of carbon fiber was estimated). The composite plate thickness of 2.5 mm was obtained by stacking eight unidirectional laminae and following a symmetrical layup configuration with [0/-45/45/90]s. The tests were carried out using a tower test device (see Fig. 1), designed to adjust the impact velocity to any particular value ranging from 0.25 m/s to 3 m/s by changing the height of a projectile of 1,9 kg weight.

The impacted specimens were leaned against a perforated steel plate support with a rectangular cut-out of 125×75 mm^2. An intermediate wood plate of 6 mm thickness was attached upon, as schematically represented in Fig. 1(a), to avoid the damages of specimens at the contact with the edges of the support. The specimen was clamped to the supporting steel plate for each test through four tighten screws placed laterally, close to the specimen edges. Rubber-tip clamps were used to avoid local damages due to excessive tightening and to reduce the vibration effects. At the beginning of each test, a hemispherically nosed projectile of 16 mm diameter, made of high hardness alloy steel and vertically aligned on two guiding bars, is raised to the required drop height and then released to fall onto the specimen at its center. A rigidly connected accelerometer to the rear side of the projectile is placed to measure the acceleration and the resulting contact force during the impact. Besides this accelerometer, the involved hardware acquisition system also includes a signal amplifier, a data acquisition board, and a PC station. For each test, after the first impact event, the projectile was manually caught to avoid multiple impacts.

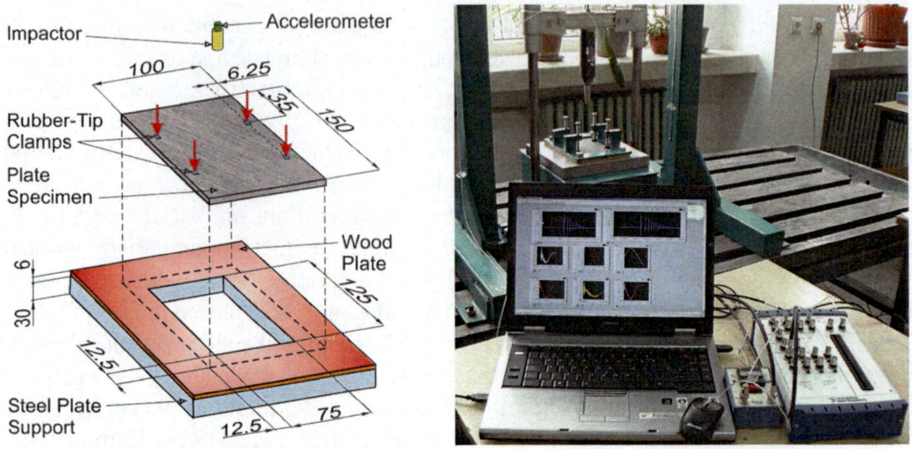

a) Specimen fixtures

b) Lab testing assembly

Fig. 1. The tower test device and specimen fixtures for low-velocity impact tests

2.2 FE Simulation

The commercial FE software Ansys was employed to simulate the three-dimensional low-velocity impact tests of CFRP laminated composite plates. Since the inertia effects are pretty minor, they were disregarded so that a straightforward static nonlinear analysis was performed implicitly, which provides a more detailed stress distribution [15]. The primary source of nonlinear behavior is assumed to occur due to the contact area change with the magnitude of the applied load by using SOLID187 tetrahedron elements to model the projectile impactor. The effects of large displacements were also considered with the aid of higher-order 3D elements composed of 20 nodes named SOLID186 in Ansys FE software, used hereafter to model the laminated composite plate. Each node has three degrees of freedom: i.e., translations along the nodal x, y, and z directions. This kind of solid element exhibits a quadratic displacement response, and thus, it appears to be more accurate in estimating the deflections than any other three-dimensional element with linear shape functions.

On the other hand, since the underlying plate elements have mid-side nodes at the contact interface due to quadratic shape function, specific elements satisfying this condition have had to be used to represent and compute both the contact and sliding between 3-D surfaces, including the projectile and the laminated composite plate under investigation. Here, it is worth mentioning that in Ansys FE software, these kinds of elements are CONTA174 and TARGE170. The pair of potential contact surfaces are usually referred to as either contact surface or target surface. The CONTA174 elements model the contact and sliding response between the target surface and the deformable plate surface, defined by these particular elements. Moreover, the Coulomb friction, as well as shear stress friction, are allowed for CONTA174 elements.

The augmented Lagrangian method was utilized as a contact algorithm in the simulation. In such a way, the contact stresses are enlarged during equilibrium iterations leading to a value of final penetration lower than the allowable tolerance. Typically, the augmented Lagrangian method generates better conditioning relative to the penalty method, while it is less sensitive to the contact stiffness magnitude [16].

The laminated plate was simply supported upon the edges, and the load was applied through the projectile impactor with the maximum value acquired throughout the experimental investigations. Due to the symmetry, only a quarter of the model was considered to minimize the size of the FE model and hence, the computational time. The edges and symmetry restraints, both for the plate and the projectile, are schematically represented in Fig. 2.

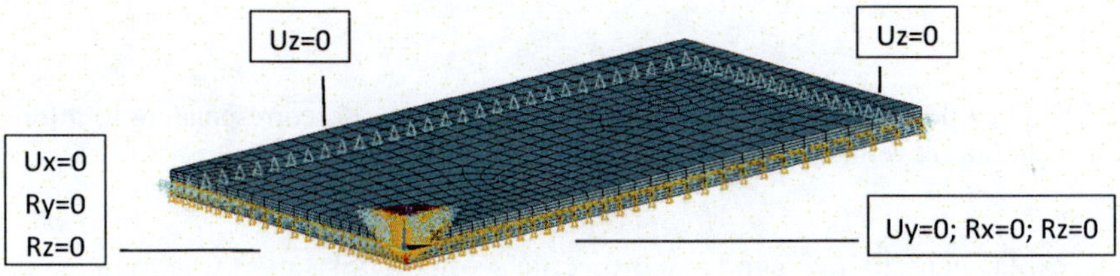

Fig. 2. The quarter FE-model of projectile and plate in contact

Plane orthotropic and linear elastic material properties (see Table 1) were assigned to each unidirectional composite layer with reference to a global predefined coordinate system. The fiber direction of 0° layer is directed along the higher side of the specimen.

Table 1. The elastic properties of unidirectional composite lamina

E_1 [GPa]	E_2 [GPa]	E_3 [GPa]	υ_{12} [-]	G_{12} [GPa]	G_{23} [GPa]
55	4	4	0.3	1.35	1.35

The mesh density in the contact region was found to influence the solution accuracy directly. Thus, a comparative study was performed by employing different element sizes (1, 1.5 and 3 mm) in the contact area. The results obtained for an element size of 1.5 mm provided satisfactory convergence both in stresses and strains.

3 Results and discussion

In the current study, FEA-based predictions and experimental tests, under the aforementioned conditions and assumptions, were made for nine particular values of the projectile's initial velocities (i.e., 0.25 m/s, 0.35 m/s, 0.5 m/s, 0.75 m/s, 1 m/s, 1.5 m/s, 2 m/s, 2.5 m/s and 3 m/s). Figure 3 shows the deflection fringe plot of the laminated plate under question, acquired by FEA at step number 17, corresponding to an impact velocity of 1 m/s (i.e., an applied force of 760 N). It can be observed that a maximum displacement of 2.78 mm is reached at the contact area. For the sake of illustrating its level, the central plate deflection obtained for the maximum applied contact force of

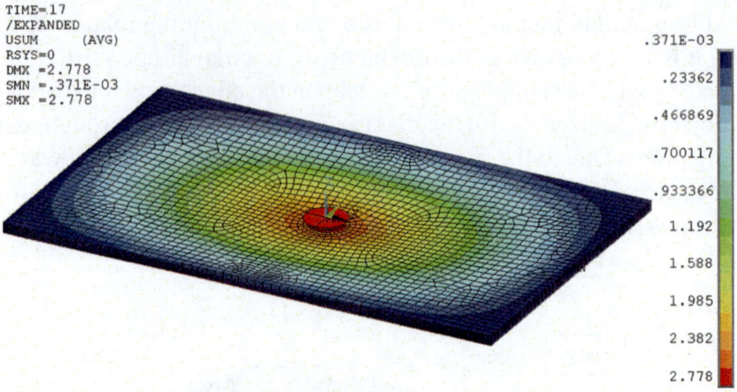

Fig. 3. The plate deflection acquired by FEA at time step no. 17, corresponding to a force of 760 N (i.e., an impact velocity of 1 m/s)

3750 N (experimentally averaged at a projectile's velocity of 3 m/s) was found equal to 7.35 mm (not represented graphically here, but referenced later, in Fig. 6).

Photographs of damages that occurred upon the plate specimens impacted at a projectile's velocity of 3 m/s are portrayed in Fig. 4. A slight indentation is visible on the impacted side, while upon the non-impacted side, complex interactions of matrix cracks, delaminations, and fibers breakage can be observed, although they are not much apparent. No damages were visible for the projectile's velocities with initial values lower than 3 m/s (corresponding to an average indentation in the contact area equal to 0.4 mm). A specific threshold value of 0.3 mm to 0.5 mm, pointing out the permanent indentation for BVID (i.e., substantial damages/failures that occur in the underlying layers with only a minor surface indent detectable by visual inspection on their external surfaces) is outlined by Bouvet and Rivallant [2]. Thus, for the laminate under investigation, the projectile's velocity of 3 m/s establishes the impact energy corresponding to the level of BVID (i.e., 8.6 J).

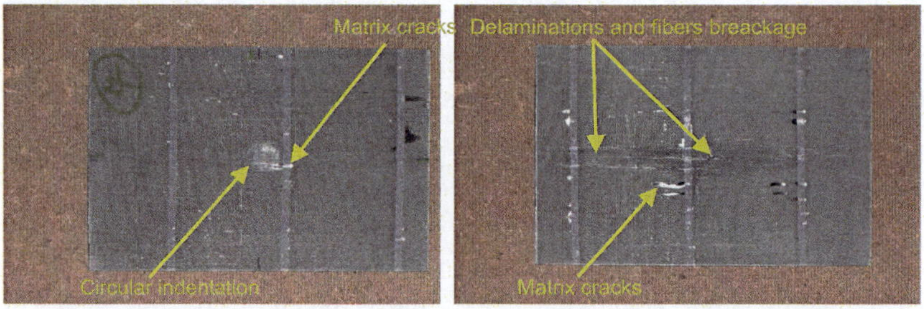

a) Face side (Impacted) b) Back side (Non-impacted)

Fig. 4. The damages occurred upon the plate specimen impacted at a projectile's velocity of 3 m/s (corresponding to BVID level)

Figure 5 depicts a time-varying deflection event obtained by both FEA, experimental test and analytical, corresponding to a projectile's velocity of 1 m/s. For this velocity, the prediction of maximum displacement based on FE-model and analytical are about the same and consistent with the experimental result. However, the estimated deflection is 8% higher than the empirically determined level of 2.5 mm. This difference might be the result of the peculiar failure behavior of CFRP laminates under low-velocity impact.

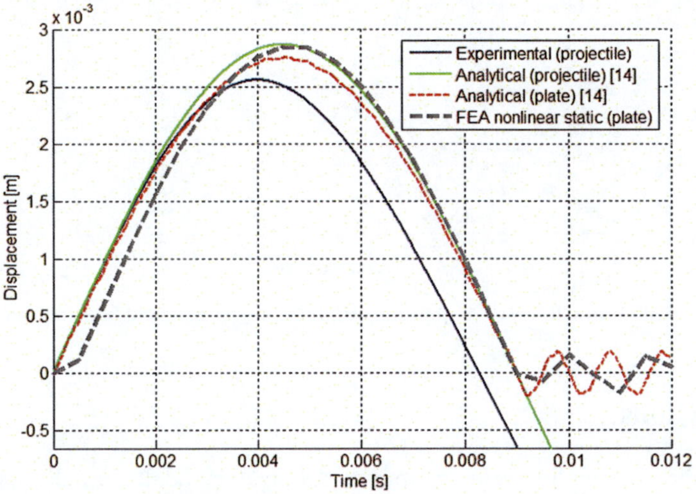

Fig. 5. The time-varying deflection event corresponding to a projectile's velocity of 1 m/s (experimental, FEA and analytical [14])

An envelope plot of contact force versus central plate deflections obtained for all the particular values of the projectile's initial velocities, as underlined at the beginning of this section, is represented in Fig. 6. It can be seen that the analytical solution developed in reference [14] is accurate only for small displacements, generally speaking, for a maximum transversal displacement lower than the thickness of the plate. Nevertheless, the mean value of empirically recorded contact force falls somewhere between the linear analytical solution and the nonlinear numerical solution. It is also worth noting that the results are pretty scattered for projectile's velocities higher than 1 m/s. Such a response behavior may be expected for this loading level since the intra- and interlaminar damages are introduced in the laminated plate, thus determining a considerable reduction of stiffness as well as the resultant contact force. Moreover, as already mentioned in reference [14], for values of the initial projectile's velocity up to about 3 m/s, the maximum values of contact force and deflection are reached at the same time, which means that within this velocity range, the impact response may be assumed as quasi-static.

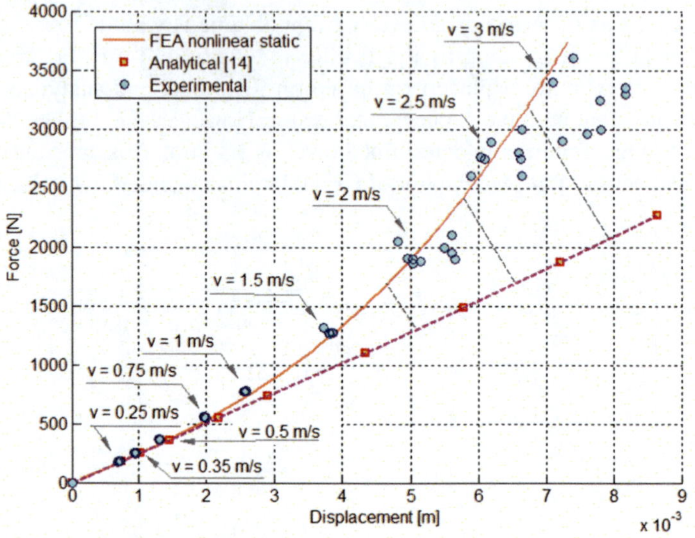

Fig. 6. Force - displacement envelope plot for different projectile's velocities between 0.25 m/s and 3 m/s

4 Conclusions

Employing the current low-velocity impact testing methodologies, the energy that induced the level of BVID equal to 8.6 J was found for a CFRP laminated composite plate made of 8 unidirectional laminae (carbon fiber/epoxy vinyl ester resin), stacked in a [0/-45/45/90]s layup configuration of 2.5 mm thickness.

Although the testing results are pretty scattered for the projectile's velocities higher than 1 m/s, the mean value of empirically recorded contact force falls somewhere between the linear analytical solution and the nonlinear static FEA solution. Such a response behavior may be expected since, beyond this loading level, matrix cracks, fibers breakage and interlaminar damages are introduced in the laminated plate, thus determining a considerable reduction in stiffness and resultant contact force. However, the analytical solution based on classical plate theory (see reference [14]) gives accurate results only for the initial projectile's velocities lower than 1 m/s (corresponding to a maximum displacement that is smaller relative to the thickness of the laminate plate under analysis).

References

1. Lopresto, V., Caprino, G.: Damage Mechanisms and Energy Absorption in Composite Laminates Under Low Velocity Impact Loads. In: Abrate, S., Castanié, B., Rajapakse, Y. (eds.) Dynamic Failure of Composite and Sandwich Structures. Solid Mechanics and Its Applications, vol. 192, pp. 209–289. Springer, Dordrecht (2013). https://doi.org/10.1007/978-94-007-5329-7_6
2. Bouvet, C., Rivallant, S.: Damage tolerance of composite structures under low-velocity impact. In: Dynamic Deformation, Damage and Fracture in Composite Materials and Structures, pp. 7–33. Woodhead Publishing (2016)

3. Abrate, S.: The dynamics of impact on composite structure, impact response and dynamic failure of composites and laminate materials, Part 2. In: Kim, J.K., Yu, T.X. (eds.) Trans Tech Publications, Switzerland (1998)
4. Kersys, A., Kersiene, N., Ziliukas, A.: Experimental research of the impact response of E-Glass/Epoxy and Carbon/Epoxy composite systems. Mater. Sci. **16**(4), 1392–1320 (2010)
5. Bogenfeld, R., Kreikemeier, J., Wille, T.: Review and benchmark study on the analysis of low-velocity impact on composite laminates. Eng. Fail. Anal. **86**, 72–99 (2018)
6. Salvetti, M., Gilioli, A., Sbarufatti, C., Manes, A., Giglio, M.: Analytical model of the dynamic behaviour of CFRP plates subjected to low-velocity impacts. Compos. B Eng. **142**, 47–55 (2018)
7. Austin, A., Priyadarsini, R.S.: Low-velocity impact behaviour of composite laminates – a review. Emerg. Trends Eng. Sci. Technol. Soc. Energy Environ. 83–90 (2018)
8. Pandian, A., Sultan, M.T.H., Marimuthu, U., Shah, A.U.M.: Low Velocity Impact Studies on Fibre-Reinforced Polymer Composites and Their Hybrids – Review. Encyclopedia of Renewable and Sustainable Materials. Elsevier, Amsterdam (2020)
9. Kaware, K., Kotambkar, M.: Low velocity impact response and influence of parameters to improve the damage resistance of composite structures/materials: a critical review. Int. J. Crashworthiness 1–25 (2021)
10. Qiu, A., Fu, K., Lin, W., Zhao, C., Tang, Y.: Modelling low-speed drop-weight impact on composite laminates. Mater. Des. **60**, 520–531 (2014)
11. Prentzias, V., Tsamasphyros, G.J.: Simulation of low velocity impact on CFRP aerospace structures: simplified approaches, numerical and experimental results. Appl. Compos. Mater. **26**(3), 835–856 (2019)
12. Li, X., Ma, D., Liu, H., Tan, W., Gong, X., Zhang, C., Li, Y.: Assessment of failure criteria and damage evolution methods for composite laminates under low-velocity impact. Compos. Struct. **207**, 727–739 (2019)
13. Alfano, G., Crisfield, M.: Finite element interface models for the delamination analysis of laminated composites: mechanical and computational issues. Int. J. Numer. Methods Eng. **50**(7), 1701–1736 (2001)
14. Baba, M.N., Dogaru, F., Guiman, M.V.: Low velocity impact response of laminate rectangular plates made of carbon fiber reinforced plastics. In: 13th International Conference Interdisciplinarity in Engineering (INTER-ENG 2019) Proceedings, pp. 95–102. Procedia Manufacturing, vol. 46 (2020)
15. Constantin, N., Găvan, M., Sandu, M., Sorohan, Ş, Anghel, V.: Damage assessment through impact force history recording. Key Eng. Mater. **347**, 665–670 (2007)
16. ANSYS homepage. https://www.ansys.com/resource-center. Accessed 01 May 2021

An Assessment of the Metallic Iron Content from Steel Mill Scale – Essential Factor for Sustainability and Circular Economy

Dana-Adriana Iluțiu-Varvara(✉), Marius Tintelecan, Claudiu Aciu, Carmen Maria Mârza, and Ioana Monica Sas-Boca

Technical University of Cluj-Napoca, 28 Memorandumului Street, 400114 Cluj-Napoca, Romania

Abstract. The aim of this paper is the assessment of the total and metallic iron contents from steel mill scale coming from the technological flow of the steel rolling and from the Crişeni landfill (Sălaj County, Romania). The ten steel mill scale samples were collected from the technological flow of the steel rolling and the ten steel mill scale samples were collected from the Crişeni landfill. The determination of the total and metallic iron in the steel mill scale samples was achieved by using a spectrometer based on X-ray fluorescence. There were analyzed ten samples from the technological flow of the steel rolling and ten samples from the landfill. The metallic iron content from the ten samples coming from the technological flow of the steel rolling, varies from 64.4 (sample 8) to 72.7 (sample 7). The metallic iron content from the ten samples coming from the landfill, varies from 58.7 (sample 4) to 63.2 (sample 9). The results obtained show that the steel mill scale is a significant source of the metallic iron. The reuse of the metallic iron content, as a raw materials or auxiliary materials, would contribute to sustainability and circular economy in the iron and steel industry.

Keywords: Metallic iron · Steel mill scale · Metallurgical wastes management · Sustainability · Reuse · Closed - loop system · Circular economy

1 Introduction

Mill scale is considered a by-product of the steelmaking, which comes from the rolling mill in the steel hot rolling process. Mill scale it is a valuable metallurgical raw material for iron making, steelmaking, and construction industries because it contains valuable metallic fractions. The annual quantity of the oily sludge and mill scales, generated in Europe, it is approximately 500,000 tones/yr. From this quantity, more than 30% is not valorized. Due to this fact, significant quantities of valuable metallic minerals are lost forever. In the world, the quantity of the steelmaking by-products, such as dust and mill scale represent approximately 5 million tons [1–9].

The main characteristic of the circular economy is the redesign of the industrial processes, so that the materials constantly circulate in a "closed-loop system", which assures that the waste generation is minimized [10].

L. Moldovan and A. Gligor (Eds.): Inter-Eng 2021, LNNS 386, pp. 64–70, 2022.
https://doi.org/10.1007/978-3-030-93817-8_7

A circular economy enables economic growth in a sustainable manner and at the same time encourages environmental protection and social prosperity.

The European Commission has estimated that manufacturing sector from the European Union (EU) would gain an additional 600 billion Euros each year, if the transition to a circular economy would be accomplish. At first, the circular economy concept was based on the 3R principle (reduce, reuse, recycle), while more recently it was expanded to the 6R principle (reuse, recycle, redesign, remanufacture, reduce, recover) [11–13].

Sustainable development in the steel industry, involves [14, 15]:

- recovery and reuse the metallic iron contents from the wastes;
- recovery and reuse the valuable components from the wastes;
- conserving the natural resources such as iron ore, coal, dolomite, magnesite etc.;
- minimizing the quantity of wastes landfilled;
- increasing the degree of the metallurgical wastes recycling;
- minimizing the quantity of hazardous wastes;
- minimizing the emissions.

The purpose of the paper is the assessment of the total and metallic iron contents from steel mill scale, in order to improve the management of metallurgical wastes, for sustainability and circular economy in the steel industry.

The objectives of the paper are:

- the assessment of the total iron contents from steel mill scale (from landfill Crişeni);
- the assessment of the metallic iron contents from steel mill scale (from landfill Crişeni);
- the assessment of the total iron contents from steel mill scale (coming from the technological flow of the steel rolling);
- the assessment of the metallic iron contents from steel mill scale (coming from the technological flow of the steel rolling);
- improving the metallurgical wastes management from landfill;
- improving the metallurgical wastes management, coming from the technological flow of the steel rolling;
- reuse of the iron from this waste (as a raw material or as an auxiliary material) for closed - loop system and for sustainability and circular economy in the iron and steel industry.

2 Materials and Methods

The ten mill scale samples were collected from the technological flow of the steel rolling (metallurgical plant, Sălaj County, Romania) and the ten steel mill scale samples were collected from the Crişeni landfill. The mill scale landfilled to the landfill, comes from the different cooling and cleaning operations taking place in the rolling process. The steel mill scale samples were collected from ten points to the edge of the landfill.

There were analyzed ten samples from the technological flow of the steel rolling and ten samples from the landfill. The determination of the total and metallic iron in the steel mill scale samples was achieved by using a spectrometer (Niton type) based on (XRF) X-ray fluorescence. The chemical composition of the metallic iron from steel mill scale samples was determined in compliance with the methodology described in the references [15–17].

3 Results and Discussions

In the Fig. 1 are presented the changes in the percentage concentrations of the total and metallic iron in the steel mill scale samples from landfill.

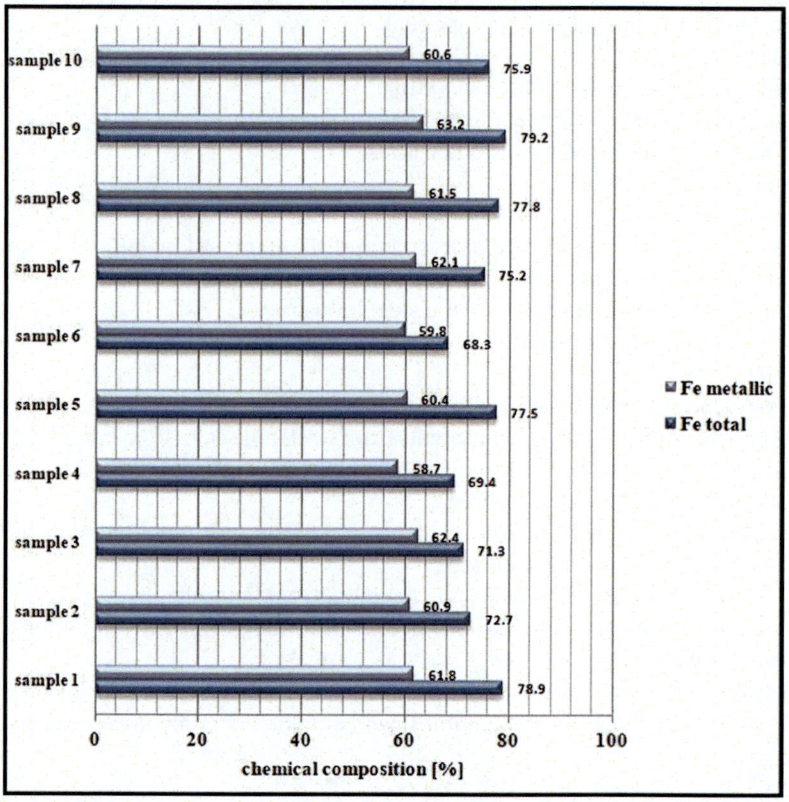

Fig. 1. Changes in the level of total and metallic iron of the steel mill scale from landfill.

From the analysis of the data presented in the Fig. 1 it results that:

- the concentrations of the total and metallic iron in all ten steel mill scale samples, are significant;
- the percentage concentration of the total and metallic iron, from the ten steel mill scale samples varies from one sample to another;
- the ten steel mill scale samples have variable concentrations of the total iron, between 68.3 (sample 6) to 79.2 (sample 9);

- the metallic iron content from the ten steel mill scale samples, varies from 58.7 (sample 4) to 63.2 (sample 9);
- the reuse of the iron content from this waste, as a source of raw material or as a source of auxiliary material, in the technological process from which it comes, would contribute towards a closed - loop system for the steelmaking industry.

In the Fig. 2 are presented the changes in the percentage concentrations of the total and metallic iron in the steel mill scale samples coming from the technological flow of the steel rolling.

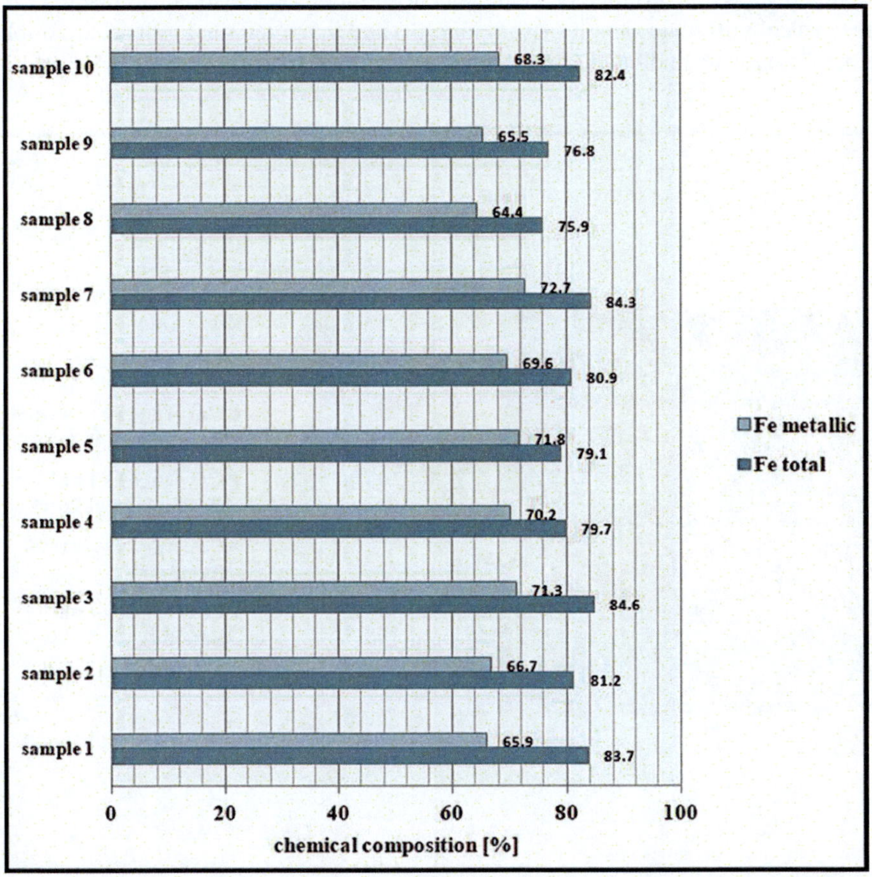

Fig. 2. Changes in the level of total and metallic iron of the steel mill scale coming from the technological flow of the steel rolling.

From the analysis of the data presented in the Fig. 2 it results that:

- the concentrations of the total and metallic iron in all ten steel mill scale samples, coming from the technological flow of the steel rolling, are significant;

- the percentage concentration of the total and metallic iron, from the ten steel mill scale samples varies from one sample to another;
- the ten steel mill scale samples have variable concentrations of the total iron, between 75.9 (sample 8) to 84.6 (sample 3);
- the metallic iron content from the ten steel mill scale samples, coming from the technological flow of the steel rolling, varies from 64.4 (sample 8) to 72.7 (sample 7);
- the reuse of the metallic iron content from this waste, as a raw material or as an auxiliary material (to the steelmaking in the EAF), would contribute to sustainability and circular economy in the iron and steel industry.

Figure 3 shows the average percentage concentrations of the total and metallic iron in the ten steel mill scale samples (from landfill) and in the ten steel mill scale samples (coming from the technological flow of the steel rolling).

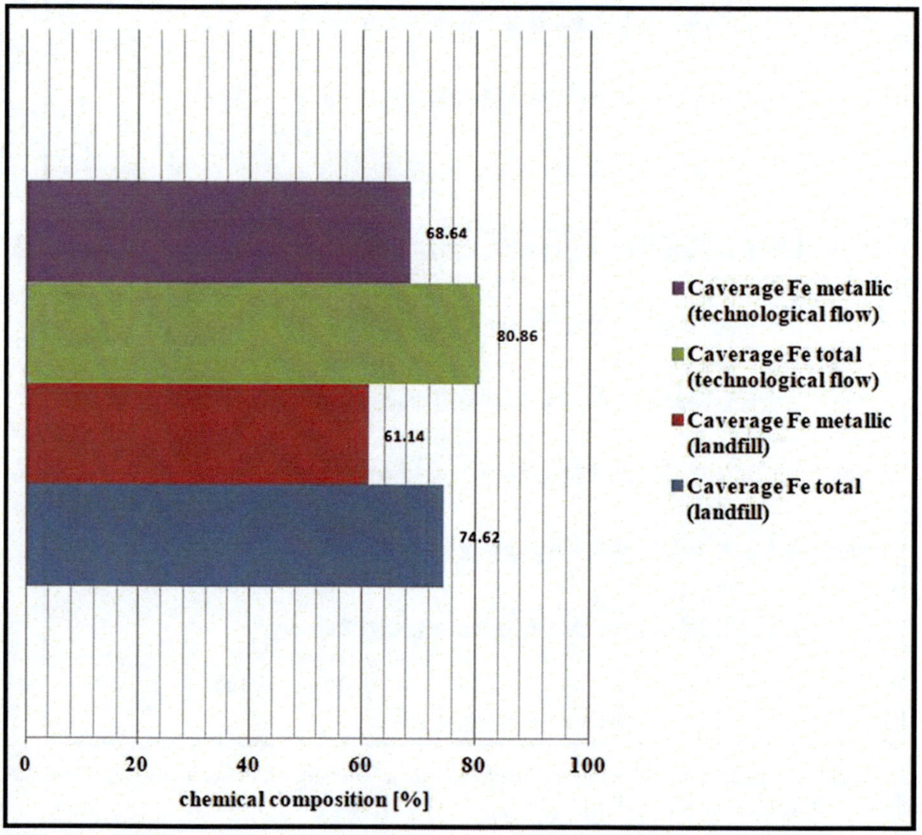

Fig. 3. Average concentrations of the total and metallic iron in the steel mill scale samples (from landfill) and in steel mill scale samples (coming from the technological flow of the steel rolling).

From the analysis of the data presented in the Fig. 3 it results that:

- the average concentrations of the total and metallic iron, in the steel mill scale samples, coming from the technological flow of the steel rolling and from the landfill, are significant;
- the landfill Crişeni is a significant source of iron;
- the reuse of iron, as a raw material, in the steelmaking process or in the other technological process, may be considered a positive economic factor would contribute to sustainability and circular economy.

4 Conclusions

For sustainability and circular economy in the steel industry, in a first stage, it is necessary, to assess the total and metallic iron from landfill wastes and from technological flow.

From the analysis of the obtained results, regarding the assessment of the metallic iron content from steel mill scale, it results that:

- the steel mill scale is a significant source of the metallic iron;
- the metallic iron content from the ten samples coming from the technological flow of the steel rolling, varies from 64.4 (sample 8) to 72.7 (sample 7);
- the metallic iron content from the ten samples coming from the landfill, varies from 58.7 (sample 4) to 63.2 (sample 9).

Without an assessment of the total and metallic iron content from steel mill scale, followed by reuse in the steel industry or in the other technological processes, this resource is lost forever. The reuse of the metallic iron from this waste, as a raw material or as an auxiliary material to the steelmaking in the electric arc furnace, would contribute to sustainability and circular economy in the iron and steel industry.

Acknowledgements. This paper is written within the TUCN Internal Research Project Competition 2016 "Research concerning the characterization of the oily mill scale in order to identify an optimum method for reduction of the quantities of hazardous wastes landfilled", Internal competition for Research/Development/Innovation – Project 16362/07.07.2016, C.I. type 1.1 - T4, Technical University of Cluj-Napoca (2016). The Internal Research Project Competition is funded by the Technical University of Cluj-Napoca in order to support the internal accredited research structures.

References

1. Saberifar, S., Jafari, F., Kardi, H., Jafarzadeh, M.A., Mousavi, S.A.: Recycling evaluation of mill scale in electric arc furnace. J. Adv. Mater. Process. **2**(3), 73–78 (2014)
2. Cartwright, D., Clayton, J.: Recycling oily mill scale and dust by injection into the EAF. Steel Times Int. **24**, 42–43 (2000)
3. Murthy, Y., Agarwal, A., Pandey, A.: Characterization of mill scale for potential application in construction industry. Indian J. Eng. **14**(35), 71–76 (2017)

4. Iluţiu-Varvara, D.A., Tintelecan, M., Aciu, C., Sas-Boca, I.M.: Reuse of the steel mill scale for sustainable industrial applications. In: Proceedings, vol. 63, no. 14, pp. 1–4 (2020)
5. Houbart, M.: PLD-Erection of a Demonstrative De-Oiling Plant for Recycling Oily Steelmaking Sludge and Mill Scales. LIFE11 ENV/LU/000855; Layman Report: Luxembourg (2017)
6. Bienvenu, Y., Rodrigues, S.: Manufacture of metal powders from pulverulent waste. ENSMP; Centre des Matériaux, CNRS UMR 7633: Corbeil-Essonnes, France (2007)
7. Iluţiu-Varvara, D.A., Aciu, C., Pică, E.M., Sava, C.: Research on the chemical characterization of the oily mill scale for natural resources conservation. Procedia Eng. **181**, 439–443 (2017)
8. Iluţiu-Varvara, D.A., Aciu, C., Mârza, C.M., Sas-Boca, I.M., Tintelecan, M.: Assessment of recycling potential of the oily mill scale in the steelmaking industry. Procedia Manuf. **22**, 228–232 (2018)
9. Iluţiu-Varvara, D.A., Aciu, C., Tintelecan, M., Sas-Boca, I.M.: Assessment of recycling potential of the steel mill scale in the composition of mortars for sustainable manufacturing. Procedia Manuf. **46**, 131–135 (2020)
10. Fischer, A., Pascucci, S.: Institutional incentives in circular economy transition: the case of material use in the Dutch textile industry. J. Clean. Prod. **155**, 17–32 (2017)
11. Jawahir, I.S., Bradley, R.: Technological elements of circular economy and the principles of 6R-based closed-loop material flow in sustainable manufacturing. Procedia Cirp **40**, 103–108 (2016)
12. Korhonen, J., Honkasalo, A., Seppala, J.: Circular economy: the concept and its limitations. Ecol. Econ. **143**, 37–46 (2018)
13. Grdic, Z.S., Nizic, M.K., Rudan, E.: Circular economy concept in the context of economic development in EU countries. Sustainability **12**(7), 3060 (2020)
14. Iluţiu-Varvara, D.A., Mârza, C.M., Sas-Boca, I.M., Ceclan, V.A.: The assessment and reduction of carbon oxides emissions at electric arc furnaces - essential factors for sustainable development. Procedia Technol. **19**, 402–409 (2015)
15. Iluţiu-Varvara, D.A., Mârza, C.M., Domniţa, F.V., Sas-Boca, I.M., Tintelecan, M.: An assessment of the metallic iron content from metallurgical wastes - essential factor for sustainable development in the steelmaking industry. Procedia Eng. **181**, 357–362 (2017)
16. Xu, Z., Hwang, J., Greenlund, R., Huang, X., Luo, J., Anschuetz, S.: Quantitative determination of metallic iron content in steel-making slag. J. Miner. Mater. Charact. Eng. **2**, 65–70 (2003)
17. Iluţiu-Varvara, D.A., Mârza, C.M., Brânduşan, L., Aciu, C., Balog, A., Cobîrzan, N.: Assessment of the metallic iron content from steelmaking slags in order to conserve natural resources. Procedia Technol. **12**, 615–620 (2014)

Research on Quick-Closing Systems for Classroom Doors

Liviu Dorin Pop[1](✉) and Majlath Sándor Dániel[2]

[1] "George Emil Palade" University of Medicine, Pharmacy, Science, and Technology of Târgu Mureş, 540142 Târgu Mureş, România
liviu.pop@umfst.ro

[2] S.C. PLASMATERM S.A., Str. Budiului, Nr. 66/A, 540390 Târgu Mureş, România

Abstract. This paper presents the research on the realization/design of an accessory (prototype) of the anti-panic type for rapid closing of classroom doors which has the purpose of blocking/locking the mortise lock in the door by pressing a button. Generally, classroom doors are locked/bolted from the outside by means of a key, the accessory described in the paper is intended to increase the security of these doors by quick locking from the inside without the use of a key. Locking is achieved by pressing a red button built into the housing of the accessory. During the installation process of the accessory, the first step is to connect it to the existing door die by means of a square rod that serves as a rotating axle, thus transmitting a rotational moment produced in-side the locking body towards the cog in the door die in order to lock it. The CAD model of the fixture was created using the SOLIDWORKS 2015 software.

Keywords: Accessory · Lockset · Mortise · Rotation · Block

1 Introduction

School security encompasses all measures taken to combat threats to people and property in education environments [1]. One term connected to school security is school safety, which is defined as the sheltering of students from violence, as well as exposure to harmful elements such as drugs and gang activity [2].

This paper presents the research on the realization/design of an "(Prototype) anti-panic accessory, for quick lock of classroom doors", which aims to block the mortise lock in the door. The locking occurs by pressing a red button built into the accessory housing. The role of this accessory is to increase the security of the classroom doors, which are generally locked from the outside by use of a key.

L. Moldovan and A. Gligor (Eds.): Inter-Eng 2021, LNNS 386, pp. 71–80, 2022.
https://doi.org/10.1007/978-3-030-93817-8_8

In the process of accessory installation, firstly the connection is made with the existing mortise lock in the door, by using of a square rod (spindle) that serves as a rotating axis, thus transmitting the moment of rotation produced inside the locking body to the cam, with the aim of lock it. After the connection the "Locking body assembly" is mounted on the door with four rounded head wood screws with Philips's drive style; subsequently, the protective housing is added, by screwing four conical screws with flat head. This housing contains the shutter-release button and the transparent indicator that allows you to track the status of the accessory: Locked or Open.

The quick lockset accessory consists of screws, pins, spacers, springs but also components that are made of metallic materials by casting and cold rolled (sheets), plastics (by plastic injection and 3D printing). The CAD model of the lockset was made using SOLIDWORKS 2015 [3–5].

2 CAD Designing of the Lockset Accessory

2.1 Description

The role of this mechanism is to turn the linear motion into a rotational motion. Due to the unique design of the lockset accessory, the shaft rotation can be:

- counterclockwise (see Fig. 1a);
- clockwise (see Fig. 1b).

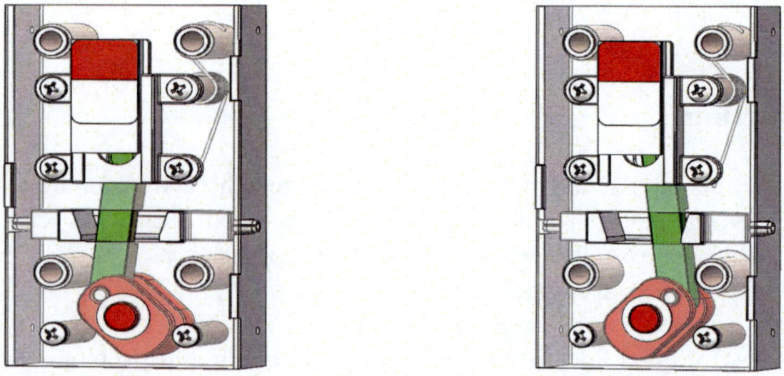

a b

Fig. 1. (a) Counterclockwise rotation; (b) Clockwise rotation

The camshaft achieves an angular rotation of 90° (±3°). The lockset accessory is only compatible with "Mortise Cylinder" locks (see Fig. 2a, b).

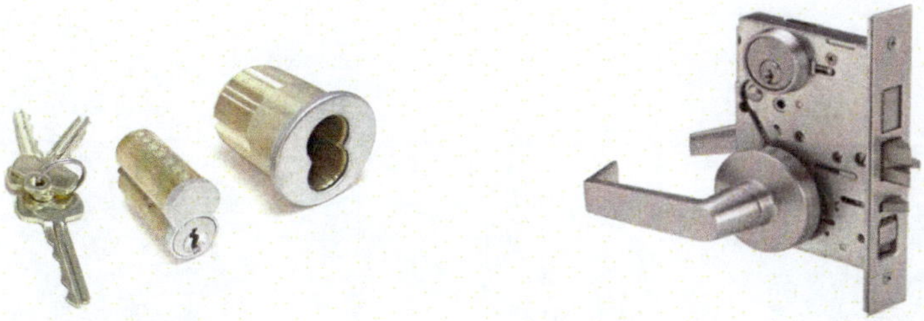

a b

Fig. 2. (a) Mortise cylinder; (b) Mortise cylinder lockset

The accessory can be mounted mainly on wooden doors. If the door handle is fitted with a collar, its outer size may not exceed 3 in./76.2 mm (see Fig. 3a, b).

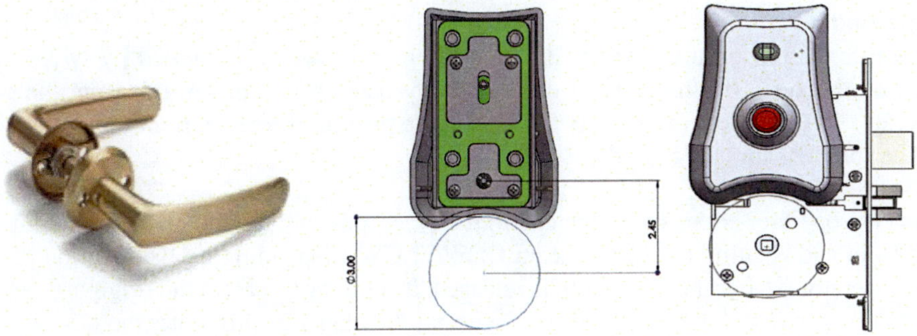

a b

Fig. 3. (a) Door handle with collar; (b) Overall size for handle collar Ø 3″

The shell has a clear plastic lens; thanks to it, you can see in which stage the indicator is:

- Locked/Closed (red label);
- Unlocked/Open (white label).

When the door is locked by the device; to reset it, the cam shaft must be rotated in the opposite direction. This can be done by applying a simple movement of the doorknob, and at the same time ensuring the simultaneous withdrawal of the dice locks. When the device is reset, locking from the outside can be done with a key.

2.2 Positions of the Device

a. Initial position.
The device is in "ready – to – use" position. The cam operator is tensioned by a torsion spring, but it's locked in by the blocker (see Fig. 4a).

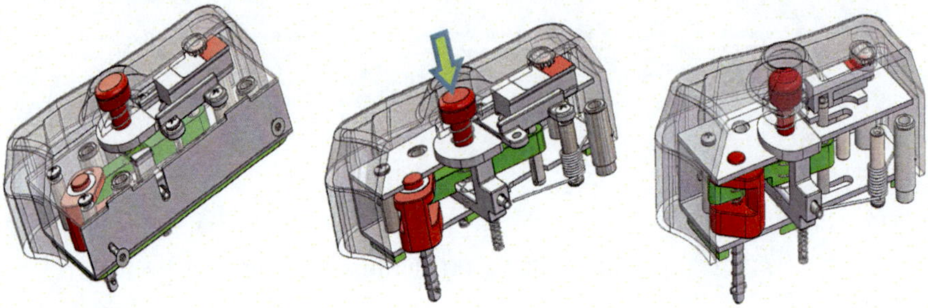

a b c

Fig. 4. (a) Initial position; (b) Second position; (c) Third position

b. Second position.
When the button on the cover is pressed; the blocker which is guided by two spring pins on the sides (in the slots of the lock body base) slides down, releasing the cam operator. The cam operator starts to transfer the pushing force from the torsion spring to the spindle cam (see Fig. 4b).

c. Third position.
The cam operator slides from one end to the other, pushing the indicator from white to red label and rotating the spindle cam CCW or CW (depending by the position of the cam operator). After the red button is released, it should go back to the original position (it can be pressed, but it can't interact with the blocker). In this state, the device is in closed position (Fig. 4c).

3 The Execution of a Chosen Part

3.1 Details About the Chosen Part and It's Material

From the entire set of parts of the quick-closing device, we chose a bended sheet metal part, for which we have collected data of the cold pressing processing technology (see Fig. 5a, b).

For the beginning we chose a material [6] from which the sheet metal part will be made and collected data regarding the analysis of the chemical composition (Table 1).

The chosen material was: C10E (DIN 1,1121/AISI 1010) is a cementing steel with low strength in the core [7, 8].

Secondly, we have realized a material cutting analysis in terms of material savings. The general terms, cutting means the placement on a semi-finished product of products

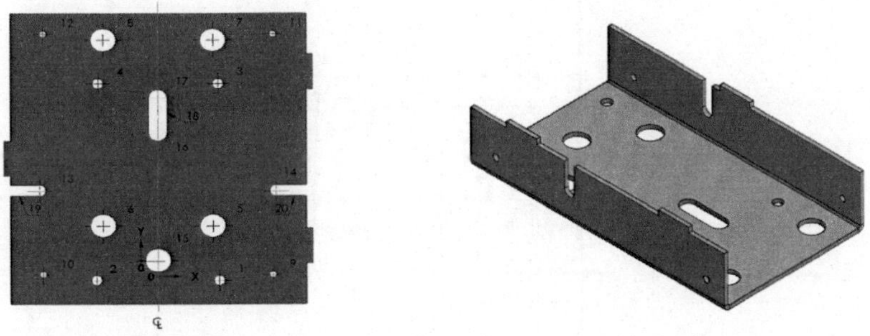

a b

Fig. 5. (a) Flat sheet metal part; (b) Bended sheet metal part

Table 1. Chemical composition of C10E material (1.1121)

C%	Si %	Mn %	P % (max.)	S % (max.)
0,07–0,13	0,4	0,3–0,6	0,035	0,035
±0,02	+0,03	±0,04	+0,005	+0,005

with certain technological forms, in order to separate them, so as to result in a minimum amount of waste, because depending on the configuration of the part and the way the piece is placed, the amount of waste may vary.

There were three variants from which we have chosen the most productive one (see Fig. 6).

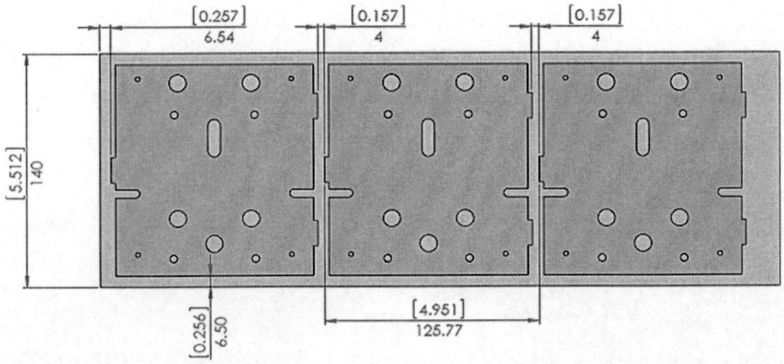

Fig. 6. Straight cutting of the material (Sheet metal strips 1400 mm × 140 mm × 2 mm)

Following the calculation of the cutting force (force for pushing the part through the cutting plate) the mechanical work during cutting and punching; and finally, according to the calculations, we chose a mechanical press with a capacity of 84 tons [9].

The calculation of the center of pressure was performed by the analytical method according to the figure below [10] (see Fig. 7).

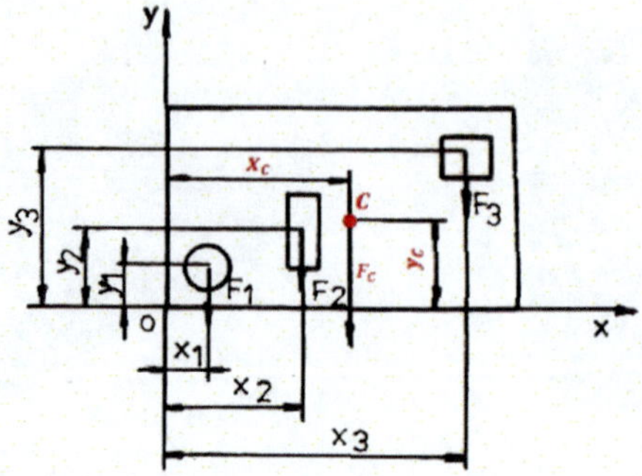

Fig. 7. Scheme for determining the analytical calculation of the pressure center

The position of the center of pressure "C" is determined by the Xc and Yc coordinates by the following relations:

$$X_C = \frac{F_1 \cdot x_1 + F_2 \cdot x_2 + \ldots + F_n \cdot x_n}{F_1 + F_2 + \ldots + F_n} = \frac{L_1 \cdot x_1 + L_2 \cdot x_2 + \ldots + L_n \cdot x_n}{L_1 + L_2 + \ldots + L_n} \quad (1)$$

$$Y_C = \frac{F_1 \cdot y_1 + F_2 \cdot y_2 + \ldots + F_n \cdot y_n}{F_1 + F_2 + \ldots + F_n} = \frac{L_1 \cdot y_1 + L_2 \cdot y_2 + \ldots + L_n \cdot y_n}{L_1 + L_2 + \ldots + L_n} \quad (2)$$

3.2 Designing of a Simultaneous Punching and Cutting Die

We made constructive calculations for the design of the components in terms of resistance to both punches and the cutting plate.

Next, we have performed compression calculations and buckling verifications, followed even by applying tolerances [11] to all active dimensions for both cutting and punching.

Finally, the design of each component of the die (Fig. 8) was done using the SOLIDWORKS 2015 3D software.

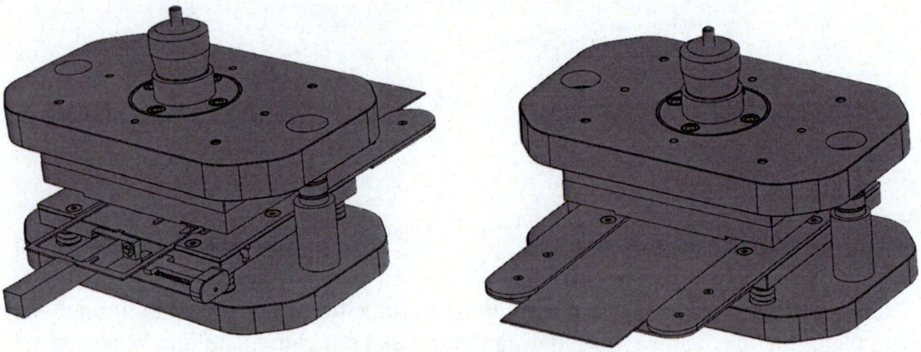

Fig. 8. Simultaneous punching and cutting die - 3D

3.3 Design of a U-Shaped Bending Mold

Changing the shape of a sheet plate was done by flat bending around an axis with a given radius of curvature. The deformation process of the material is performed differently in the layers of the bent area (Fig. 9).

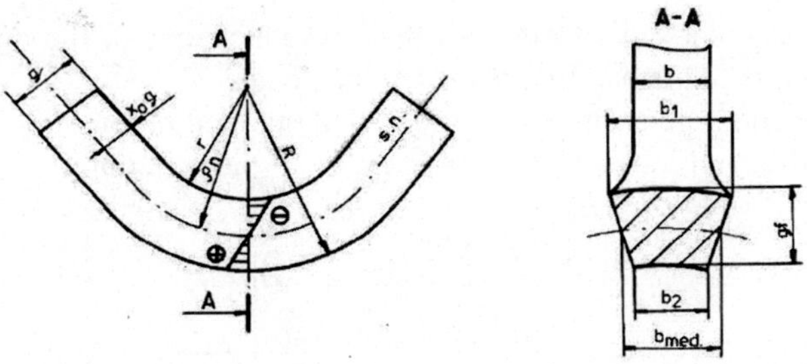

Fig. 9. The bending deformation processes [12]

When bending the parts from narrow semi-finished products, there is an accentuated deformation of the cross section. This consists in reducing the thickness of the semi-finished product in the bent portion, widening it towards inside the part and narrowing it towards outside, with the formation of curves in the cross section. Therefore, during bending, the rectangular section of the blank transforms into a deformed trapezoid. The stress state is flat, and the strain state is spatial.

When bending parts from wide blanks, the cross section of the blank deforms very little. For this reason, the deformation state is flat, and the stress state is spatial.

Next, we made determination of the dimensions of the semi-finished products required for bending (Fig. 10).

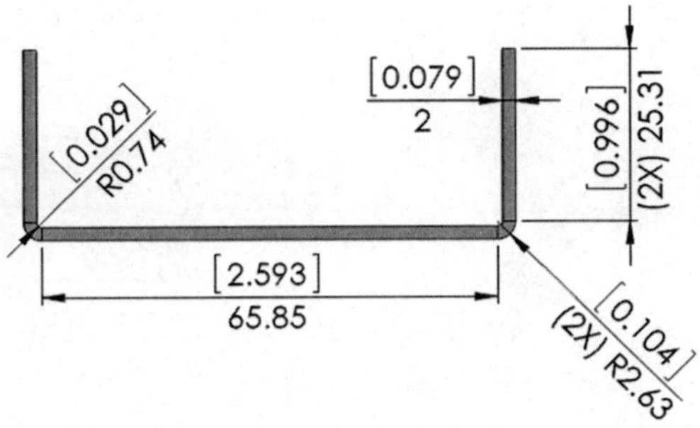

Fig. 10. Bended piece

The formula for calculating the length of the blank in case of bending is [13]:

$$L = l_1 + l_2 + \ldots + l_n + \frac{\pi}{2} \cdot (r_1 + x_1 \cdot g) + \frac{\pi}{2} \cdot (r_1 + x_2 \cdot g) + \ldots$$

$$+\frac{\pi}{2}\cdot(r_{n-1}+x_{n-1}\cdot g) \tag{3}$$

$$L=25{,}31+65{,}85+25{,}31+2\cdot\left[\frac{\pi}{2}\cdot(2{,}63+0{,}37\cdot 2)\right]=116{,}47\cdot 10{,}587$$

$$\Rightarrow \mathbf{L=127{,}05\,mm}$$

In the next was made calculations for the active elements of the mold for U-bent parts, the dimensions of the bending plate and the punch.

When the bending operation is performed without calibration, the active elements must be corrected with the size of the elastic return of the material.

$$tg\ \beta=0{,}75\cdot\frac{l}{k\cdot g}\cdot\frac{\sigma_c}{E} \tag{4}$$

$$tg\ \beta=0{,}75\cdot\frac{l}{k\cdot g}\cdot\frac{\sigma_c}{E}=0{,}75\cdot\frac{18{,}25}{0{,}63\cdot 2}\cdot\frac{20{,}45\,\mathrm{daN/mm^2}}{21000\,\mathrm{daN/mm^2}}=0{,}75\cdot 14{,}484\cdot 9{,}738$$

$$\Rightarrow \mathbf{tg}\ \boldsymbol{\beta}=\mathbf{105{,}783}\Rightarrow\mathbf{3{,}537}$$

Finally, the design of each component of the U-shaped bending mold was done using the SOLIDWORKS 2015 3D software (Fig. 11).

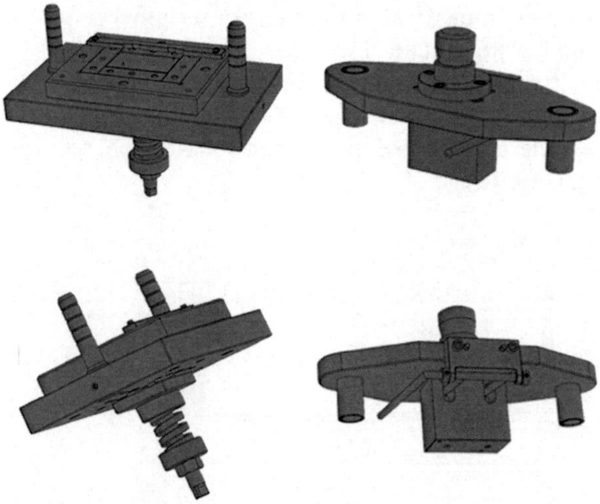

Fig. 11. U-shaped bending mold

4 Conclusion

In the introductory part, we developed data on the locking accessory itself, more precisely on its role, compactness, dimensions, and functionality. Next, was de-signed the cold pressing processing technology of the chosen part, where we did research on the metal part, on the conditions of shape, on appearance, and delivery condition.

The analysis of the material includes physical, thermal, chemical, and mechanical properties [14]. It was calculated the forces required for perforation and cutting, the center of pressure, constructive calculations, strength, compression and buckling check for the active elements, plus dimensional tolerances.

After the calculations mentioned above, all the components belonging to the "Simultaneous perforation and cutting die" were designed, as well as the general and detailed drawings.

The data we collected for the design of a U-shaped bending mold for cold plastic deformation processing, includes calculations relating to the active elements of the mold and to the bending force, respectively to the mechanical work produced during bending.

Therefore, an analysis was performed to "Optimizing the cost of the material, for each component of the quick release accessory".

In this assessment, the component of the quick lock accessory was renewed, by replacing and adding other new components, which are made of other materials, in order to give it new functions to remove restrictions and find an optimal solution to reduce the cost of construction.

Emotional and physical safety in school are related to academic performance and safe schools promotes the protection of students from violence, exposure to weapons or threats [15]. In this sense, this research presents an accessory of the anti-panic type for rapid closing of classroom doors which has the purpose of locking the mortise lock in the door by pressing a button.

References

1. Enews, Goodwin University: The growing importance of school safety and security training. https://www.goodwin.edu/enews/growing-importance-of-school-safety-and-security/
2. Boccanfuso, B., Kuhfeld, M.: Multiple Responses, Promising Results: Evidence-Based, Nonpunitive Alternatives to Zero Tolerance. Child Trends, Washington, DC (2011)
3. Stirbu, C.: The Designer's SolidWorks Friend, Tehnopress Publishing House (2007)
4. Lombard, M.: Solidworks 2013 Bible. The Comprehensive Tutorial Resource. Wiley edition (2013)
5. Fulkerson, F.: Solidworks Basics. A Project Based Approach. Industrial Press, Inc. (2015)
6. Socaciu, T.: Elements of materials science and engineering. "Petru Maior" University Press, Tg. Mures (2011)
7. SR EN 10084:2008 standard
8. http://www.steelnumber.com/en/steel_composition_eu.php?name_id=222. Accessed March 2020
9. Braha, V., Nagit, Gh.: Stamping technologies – design guide. Tehnica - Info Press, Chisinau (2002)
10. Socaciu, T., Pop, L.: Cold-pressing technology design guide. "Petru Maior" University Press, Tg. Mureş (2014)
11. Tero, M.: Tolerances and dimensional control. "Petru Maior" University Tg. Mureş (2015)
12. Socaciu, T., Pop, L.: Cold-pressing equipment and devices. "Petru Maior" University Press, Tg. Mures (2013)

13. Socaciu, T., Pop, L.: Cold-pressing technology. "Petru Maior" University Press, Tg. Mureş (2014)
14. Boer, J., Blaga, P.: Reducing production costs by monitoring the roughness of raw product surface. Procedia Manuf. **22**, 202–208 (2018)
15. Gottfredson, G., Gottfredson, D.: What schools do to prevent problem behavior and promote safe environments. J. Educ. Psychol. Consult. **12**(4), 313–344 (2001)

Innovative Technologies, Design and Materials in Civil Engineering

Comparative Analysis Between Two Constructive Solutions of a Steel Tied-Arch Road Bridge

Ştefan Guţiu, Cătălin Moga, Mircea Suciu, and Alexandra-Denisa Danciu(✉)

Technical University From Cluj-Napoca, Cluj-Napoca, Romania
{stefan.gutiu,mircea.suciu,alexandra.danciu}@cfdp.utcluj.ro,
catalin.moga@dst.utcluj.ro

Abstract. Steel arches used as the road bridge superstructures are economically competitive solutions for the field of medium spans, offering at the same time architecturally appreciated solutions. In the urban areas, where considerations related to the aesthetic aspects and integration of the artwork with the architecture of the buildings in the vicinity of the site are important, there is a tendency to adopt the trough arch road bridges, the arches being carefully analyzed in terms of aesthetic and architectural aspect offered. Within a Feasibility Study for the construction of a new bridge over the Someş River in Cluj-Napoca, several constructive variants were analyzed: bridge on concrete beams with several spans, bridge on beams with composite steel-concrete structure and steel bridge on arches. The paper presents two variants of the bridge on the arches, with two and four carriage-lanes, including some aspects related to the design of the super-structure.

Keywords: Trough road bridges · Open vs. closed tied-arched · Eurocodes EN SR 1993 · Arch buckling resistances

1 Introduction

The road bridges designed with the resistance structure on steel arches, are economically competitive for the field of medium spans, compared to other structural types such as trusses, offering at the same time architecturally appreciated solutions.

The field of optimal spans for road bridges on arches largely overlaps with that of bridges on trusses, but often factors links as the duration of execution, consumption of manufacture operations and consideration of architecture aspects, situates the constructive solutions on arches on a more advantageous place compared to those on trusses.

Also, for a large part of the field of medium spans, bridges on steel arches can be compared economically with the bridges of concrete or composite steel-concrete structures, the latter being used mainly for bridges with more lanes of traffic with more spans, as is the case of highways. In urban areas, where the aesthetic, compatibility and integration of the work of art with the architecture of the buildings in the vicinity of

L. Moldovan and A. Gligor (Eds.): Inter-Eng 2021, LNNS 386, pp. 83–92, 2022.
https://doi.org/10.1007/978-3-030-93817-8_9

the site are important, there is a tendency to adopt the solution with steel arches being carefully analyzed in terms of aesthetic and architectural aspect.

It should be mentioned that arches with vertical hangers are preferred in terms of visual aspect, having in view from the parallelism of the hangers from any eye angle of observation, compared to the solutions with inclined hangers, more structurally efficient.

In a Feasibility Study for the construction of a new bridge over the Someş River in Cluj-Napoca municipality, several constructive variants were analyzed: bridge on concrete beams with more spans, bridge on beams with composite steel-concrete structure and metal bridge on arches [1].

In the end, the bridge solution on the arches was chosen, but for this solution two variants were analyzed: the bridge with two lanes of traffic and the bridge with four lanes of traffic. The agreed variant was a tied-arches bridge of 46.0 m span and four carriage-lane with console elements for pedestrian sidewalks and cycle paths.

It should be noted that the bridge's decking had to be in a relatively small depth, driven by the connection of the carriage-lanes to the adjacent street roads and the assurances of the safety space resulting from the hydraulic calculation.

The paper presents the two variants of the bridge on the arches, including some aspects of a technical and structural nature, respectively aspects related to the design of the main elements of the structures, the circular arches with vertical hangers made of semirigid bars.

2 Technical Solutions Analyzed for the Superstructure

In the Feasibility Study conducted with Cluj Municipality for the Bridge over the Someş River in Cluj-Napoca, two constructive solutions have been developed for the bridge with a single span of 46.0 m, as it follows [1]:

1. Bridge with two carriage-lanes and pedestrian and cyclist sidewalks.
2. Bridge with four carriage-lanes and pedestrian and cyclist sidewalks.

2.1 Two-Carriage Lanes Bridge

Figure 1 [1] shows the plan view of the site with the solution found for solving the traffic flows, pedestrian, and bike paths on one-lane bridge and on the adjacent zones.

The superstructure of the bridge is on steel arches with semi-rigid hangers, a closed type bridge, where the arches are equipped with an upper bracing.

The elevation of the bridge is presented in Fig. 2 [1].

In Fig. 3 the bridge transversal cross-section and the bridge' rendering is presented.

The bridge superstructure consists of a composite steel-concrete deck, suspended on two steel box arches, with a variable depth of the cross-section in the vertical plane.

The span of the bridge structure is of 46.00 m, the distance between the axis of the longitudinal tie-girders and between of the arches is of 9.00 m, and the total width of the deck is of 18,00 m.

The carriage deck width is of 7.80 m consisting of two 3.50 m lanes and 2 × 0.40 m safe spaces, and laterally on the 4.50 m cantilevers, the 1.50 m pedestrian sidewalks and 2.00 m cycle paths are arranged.

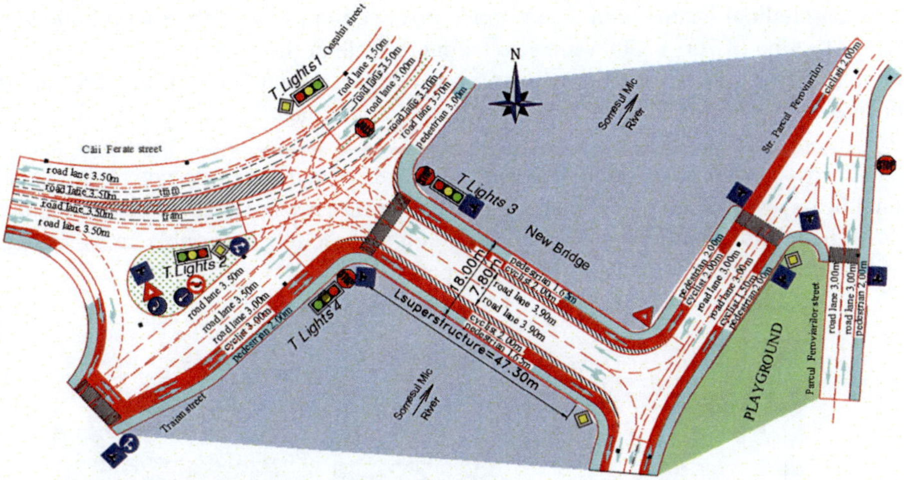

Fig. 1. Plan view of the Bridge with two carriage-lanes [1].

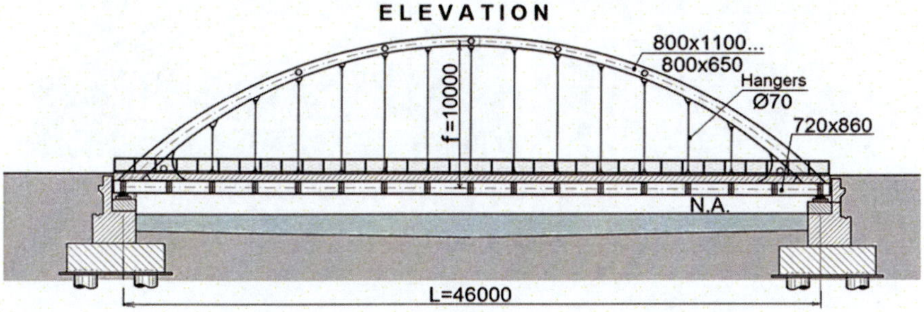

Fig. 2. Elevation of the Bridge with two carriage-lanes.

The main materials used for the resistance structure of the bridge structure are steel grade S355 M/N – for steel structure, S460 ML – for hangers; concrete Class C30/37 – for reinforce slab; reinforcement steel grade S500.

The arch rise measured between the axis of the longitudinal beams and the axis of the crown arch is of 10,00 m. The arches are made of 3 sections, and the site mounting joints are made by welding, resulting in a complete and watertight box section, so that the corrosion felt on the inside is practically negligible, by the lack of aeration and of a wetness.

For the horizontal plane buckling stability (in a transverse direction) connecting elements made of circular pipes, located outside the traffic gauge, are provided. The arches take over the loading from the deck by vertical round steel hangers located in the right of the cross-girders, at a distance of 2.85 m.

The deck beams consist of the following elements:

- two longitudinal beams with a semi-open box section (lower flange with hollows), with a section of 700 × 800 mm, also taking the role of the arch tie;
- double T cross-girder section, with variable depth between 800 mm to 900 mm, to ensure a cross-sectional slope of 2% for the water drainage.

The cross-girders are located at a distance of 2,85 m and are connected to the longitudinal beams with SIRP and a welded plate that ensure a semi-fixed connection at the ends of them. Side cantilevers – the construction with double T variable cross-section supports of the path for sidewalks and cycle paths.

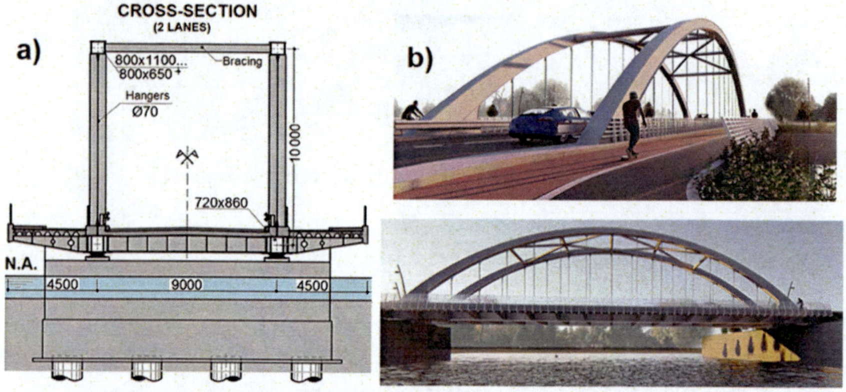

Fig. 3. Bridge with two carriage-lanes: a) Cross-section; b) Rendering.

2.2 Four-Carriage Lanes Bridge

Figure 4 [1] shows the plan view of sites for solving flows of traffic, pedestrian and bike paths on four carriage-lane bridge and on the adjacent zones.

The superstructure of the bridge is on steel arches with semi-rigid hangers, an open type bridge, the arches not being equipped with a superior brace, respective a free-standing arch.

The elevation of the bridge is presented in Fig. 5.

In Fig. 6 [1] the bridge cross-section and the rendering are presented.

In the four-carriage lane bridge variant, the distance between the axis of the longitudinal tie-beams and between the axis of the arches is of 16.50 m and the total width of the bridge deck is 26.00 m. The deck has a carriage way width of 14.40 m, consisting of four 3.50 m carriage-lanes with 2 × 0.20 m safety spaces and laterally on the 4.50 m cantilevers, are arranged pedestrian sidewalks of 1.50 m and cycle paths of 2.00 m.

The arches are designed as box with a variable section of 1200 × 1600 mm at the supports to 1200 × 850 mm the crown. The box cross-section is stiffened inside with longitudinal stiffening and transverse diaphragms between the walls. The arch rise measured between the axis of the longitudinal beams and the axis of the crown arch shall

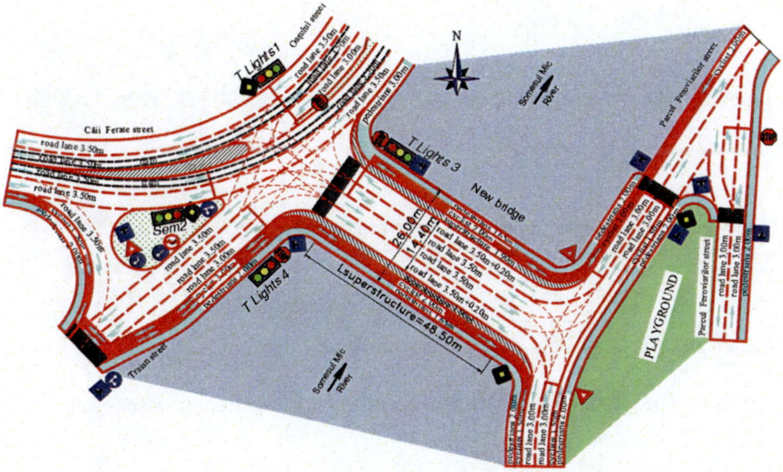

Fig. 4. Plan view of the Bridge with four carriage-lanes [1].

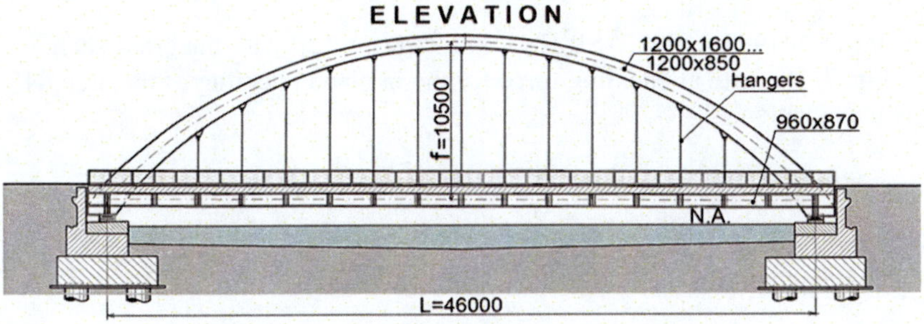

Fig. 5. Elevation of the Bridge with four carriage-lanes.

be of 10,50 m and the vertical hangers are located in the right of the cross-girders at a distance of 2.85 m.

The deck beams consist of the following elements:

- two longitudinal tie-beams with a semi-open box section (lower flange with hollows), with a section of 960 × 870 mm
- current double-T-section cross-girders with variable depth to ensure the 2% cross-slope between 870 mm to 1020 mm.

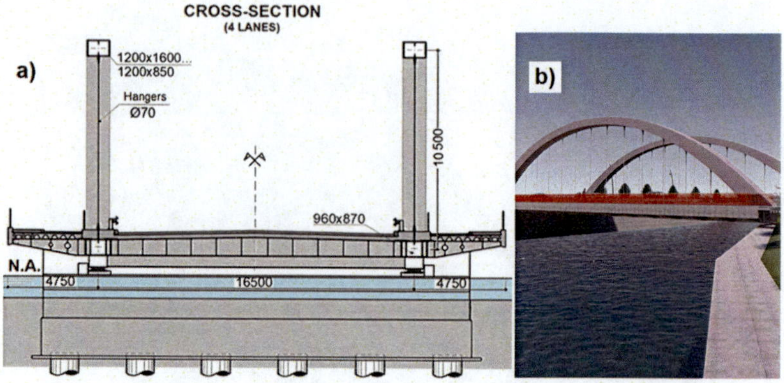

Fig. 6. Bridge with four carriage-lanes: a) Cross-section; b) Rendering

3 Design Aspects. Critical Buckling Force in the Arch

3.1 Critical Force for the In-Plane Arch Buckling for Both Type of Bridges

According to EN 1993-2: 2005 (SR EN 1993-2:2007), [2], with the cross-section axes given in Fig. 7, the critical buckling force for the in plane buckling of the arch, $N_{cr.y}$ is given by the equation:

$$N_{cr.y} = \frac{\pi^2 EI_y}{(\beta \overline{s})^2} \tag{1}$$

The critical length for the in-plane buckling of the arch:

$$L_{cr.y} = \beta \overline{s} \tag{2}$$

where: $\overline{s}$ – half length of the arch, EI_y – the flexural stiffness in the plane of the arch, β – the coefficient of the buckling length in plane of the arch (Fig. 8).

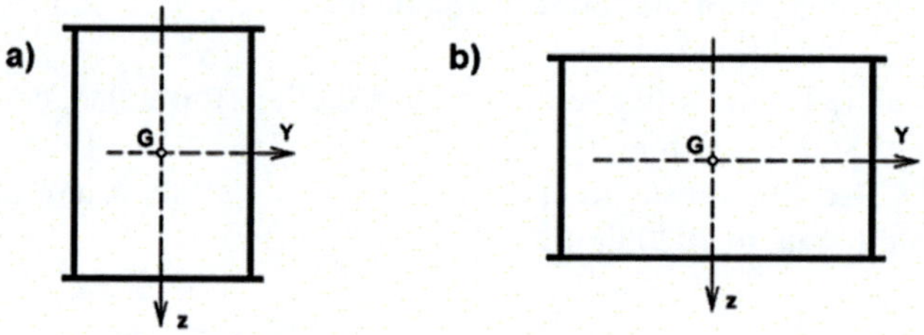

Fig. 7. Arch cross-sections: a) cross-section for two carriage-lanes; b) cross-section for four carriage-lanes.

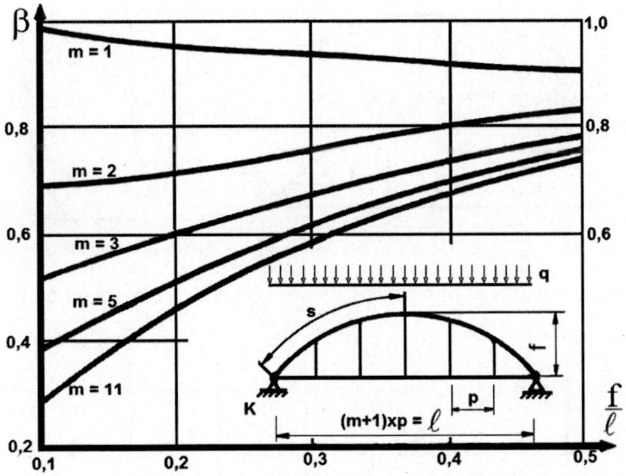

Fig. 8. Buckling factor β for arches with vertical hangers and tie-girder [2].

3.2 Critical Force for the Out of Plane Buckling of the Arch

Critical force for the out of plane buckling of arches with overhead bracing and final frames. Case of two- carriage lanes bridge.

For the out of plane of the arch, according to EC3-2 - Annex D [2], the critical buckling force, in the case of arch systems fitted with upper bracing and end frames (portal frames), shall be determined with the equation:

$$N_{cr.z} = \frac{\pi^2 EI_z}{(\beta h)^2} \tag{3}$$

Critical length in the out of plane of the arch: $L_{cr.z} = \beta \cdot h$

The geometrical buckling characteristics of the final portals for arches are given in Fig. 9 and the value of h_r may be taken as the mean of all hanger's length multiplied by $1/sin\alpha_k$.

The coefficient of buckling length β for the out of plane buckling of arches with overhead bracing and final frames are taken in accordance with EC3-2 - Annex D, [2], and is given in Fig. 10.

Critical force for the out of plane buckling of arches without overhead bracing. Case of four- carriage lanes bridge (free-standing arches).

The critical buckling force in the out of plane of the arches without overhead bracing is given by the equation:

$$N_{cr.z} = \frac{\pi^2 EI_z}{(\beta l)^2} \tag{4}$$

where: l – the arch span, EI_z – the out of plane flexural stiffness, β – the out of plane buckling coefficient.

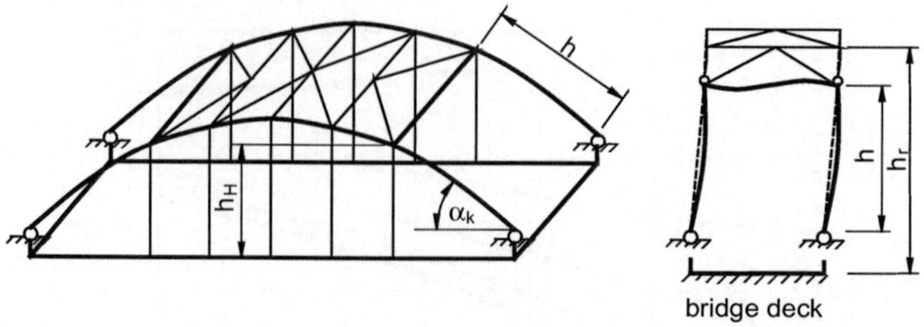

Fig. 9. Geometrical buckling characteristics of the final portals for arches [2].

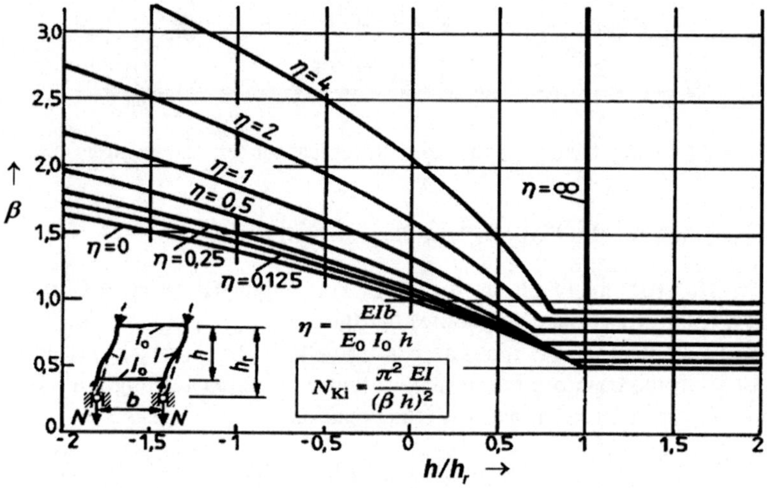

Fig. 10. Diagrams of buckling length factor β - final portals [2].

In the out of plane of unbracing arches according to EC3-2, [2], the coefficient of buckling length is calculated using Eq. (5):

$$\beta = \beta_1 \beta_2 \tag{5}$$

The coefficients β_1 and β_2 shall be taken in accordance with Table 1 and Table 2, taken from [2].

For the evaluation of the coefficient β_2, EC3-2 does not specify how the rate $\frac{q_H}{q}$ is assessed or evaluated. For the evaluation of this rate, the following equation can be used, [3]:

$$q_H = \frac{q}{1 + \frac{EI_G}{EI_{ARCH}}} \tag{6}$$

The papers [4] and [5] were also used in the elaboration of the resistance calculations.

Table 1. β_1 factor.

Value of β_1 factor					
f/l	0.05	0.10	0.20	0.30	0.40
I_z - constant	0.50	0.54	0.65	0.82	1.07
$I_z = \frac{I_{z.0}}{\cos\alpha_B}$	0.50	0.52	0.59	0.71	0.86

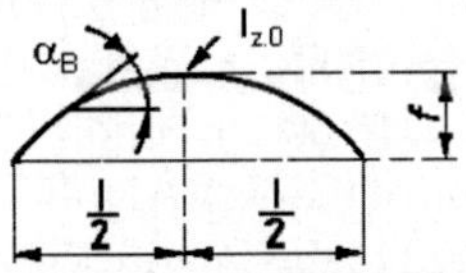

Table 2. β_2 factor.

	Loading mode	β_2	Comments
1	Conservative (deck is fixed to the top of the arch)	1	q – total load
2	By hangers	$1 - 0.35\frac{q_H}{q}$	q_H– part of the load transmitted by the hangers
3	By posts	$1 + 0.45\frac{q_{st}}{q}$	q_{st}– part of the load transmitted by the posts

4 Conclusions

Road bridges with a resistance structure made of steel tied arches are economically competitive for the field of medium spans and offers architecturally successful solutions. The optimal spans for road bridges on arches largely overlaps with that of truss bridges and bridges having composite steel-concrete structure, but often factors related to the duration of execution, the volume of manufacture, maintenance operations, and especially architectural considerations, place constructive solutions on tied-arches on a more advantageous place compared to other solutions.

In a Feasibility Study for the construction of a new bridge over the Someş River in Cluj-Napoca, Romania, several constructive variants were analyzed, and, in the end, the bridge solution based on the tied arches was chosen.

For this solution two variants were analyzed: the bridge with two carriage-lanes of and the bridge with four lanes of traffic. The agreed variant was a tied-arches bridge of 46.0 m span and four carriage-lane with console elements for pedestrian sidewalks and cycle paths.

The stability check of steel arches is an important technical problem, and in Euronorms SR EN 1993-2:2007. Eurocode 3: Design of steel structures. Part 2: Steel Bridges (EC3-2), are presented equations and diagrams with which critical buckling forces can be assessed for the in-plane arch and for the out of plane arch.

As regards the consumption of the main materials at the 4 carriage-lane bridge in comparison to the 2 carriage-lanes bridge have resulted as follows:

– Steel S355 bridge superstructure: S_(4-Lanes) = 1.65·S_(2-Lanes)

– Concrete C 30/37 in slab, sidewalks: C_(4-Lanes) = 1.45·C_(2-Lanes)
– Reinforcement steel BST 500: R_(4-Lanes) = 1.40·S_(2-Lanes).

In the case of the open bridges (without upper bracing), one of the service limit state (SLS) verification condition consists to limit the horizontal arch deflection to a value of $\delta\_H = f/1500$, a condition from which a much larger cross-section in the horizontal direction of the arch (rigidity about to z-z axis) can results, or a stronger end cross-girder compared to the intermediate ones.

Both solutions were applied in case of the open bridge structure.

References

1. Feasibility study elaborated by S.C. Miranda Project SRL and XC PROJECT SRL, Cluj-Napoca
2. SR EN 1993-2/2007. Eurocode 3: Steel structures. Part 2: Steel bridges
3. Romeijn, A.: Steel Bridges, Course CT 5125, TU Delft (2006)
4. Moga, P., Gutiu, St.: Poduri metalice. Conformarea suprastructurilor, UTPress, Cluj-Napoca (2020)
5. Moga, P., Gutiu, St., Danciu A.D., Moga C.: Poduri metalice. Ghid de proiectare, UTPress, Cluj-Napoca (2020)

Behavior Analysis of One–Component Waterproofing Mortars by Mechanical and NMR Investigations

Daniel Cadar[1(✉)], Daniela Lucia Manea[1], Dumitrita Moldovan[1], Elena Jumate[1], and Radu Fechete[1,2]

[1] Technical University of Cluj-Napoca, 28 Memorandumului Street, 400114 Cluj-Napoca, Romania
daniel.cadar@ccm.utcluj.ro
[2] Faculty of Physics, Babeş-Bolyai University, 1 M. Kogălniceanu Street, 400084 Cluj-Napoca, Romania

Abstract. Waterproofing mortars are used for many applications in the field of construction. The behavior of waterproofing mortars, after application on surface, is different due to the use of aggregates and polymers from different sources, even if it satisfies the waterproofing needs of the substrate and meets quality standards. Measurements performed at 3, 7 and 28 days by mechanical tests combined with Nuclear Magnetic Resonance (NMR) relaxometry highlight the physical and mechanical properties of the mortar in relation to the presence of hydration water and water in small, medium and large pores. During the 28 days of hydration, the size and homogeneity of the pores change radically so that the mortar becomes more resistant to compression and flexural tensile.

Keywords: One-component · NMR · Waterproofing

1 Introduction

On the market of building materials there is an important category of mortars that provide protection against water, this is represented by waterproofing mortars. The recipe of such a mortar is similar to that of a classic mortar, the difference is the type of polymer used and the method of application on the designated surface. Depending on the application method, waterproofing mortars can be classified as one-component or two-component.

This paper presents the characterization of one-component mortars by traditional methods complemented by NMR investigations. The samples are prepared on site by adding water according to the statements of each producer.

They were chosen from three one-component waterproofing mortars, currently sold on the Romanian market and which are used for protection against moisture of surfaces where water comes from outside to inside (positive pressure), applying only on mineral substrates (concrete, masonry bricks or cement-based plasters). Typical examples of use of waterproofing mortars are: drinking water tanks, foundation walls, wet areas, elevator shafts, underground passages, irrigation canals.

L. Moldovan and A. Gligor (Eds.): Inter-Eng 2021, LNNS 386, pp. 93–99, 2022.
https://doi.org/10.1007/978-3-030-93817-8_10

Through the investigations carried out for this study, three objectives were pursued. The first objective was to analyze the evolution of the mechanical strengths of the mortars during 28 days in relation to the amount of hydration water of the aggregates. The second objective was to characterize the distribution of water in the pores by observing the changes of the distribution of transverse relaxation times (T_2). The third objective was to extract from the homogeneity of the pores the best receipt to obtain the waterproofing effect of the mortar.

2 Materials and Methods

2.1 Materials

The materials used are one-component waterproofing mortars. These are based on Portland cement, fine aggregates, additives and polymers in the form of redispersible powder. Each of the components of the waterproofing mortar has a very important role in the workability required during application or in ensuring the protection against water.

Unlike a classic mortar, the waterproofing mortar is applied in thin layers, of maximum 2 mm, so that the size of the aggregates used must be less than 0.5 mm. The chosen additives are stabilizers or emulsifiers of mortars and are in the form of antifoaming or surfactant agents. In all of three types of investigated waterproofing mortars, the type of polymer used was based on latex, in an elastomeric or thermoplastic redispersible form. The role of the polymer is i) to prevent the formation of cracks due to the agglomerated structure of calcium silicate hydrates and calcium hydroxide bound together by the weakest van der Waals forces [1], ii) to have adhesion and iii) to form a matrix structure that stops water infiltrations.

Table 1 presents the three waterproofing mortar recipes analyzed in this paper. An equal amount of powder and water were used to perform the tests, according to the specifications of each mortar.

Table 1. Type of mortars.

Type of mortar	Powder [kg]	Water [l]	Water/Powder ratio
C65	1.8	0.446	0.248
P88	1.8	0.396	0.22
AQ	1.8	0.594	0.33

2.2 Methods

The time evolution of the waterproofing mortar's physical properties was investigated using modern ^{1}H NMR relaxometry methods to get the distributions of transverse relaxation times which were associated with various pores [2].

NMR measurements were performed with the low-field NMR Bruker Minispec spectrometer working at the proton frequency of 19.69 MHz. The transverse relaxation time (T_2) was measured by the CPMG (Carr-Purcell-Meiboom-Gill) pulse sequence [3–5] with 3000 echoes and 0.07 ms echo time. To enhance the signal-to-noise ratio, the number of scans was set at 1024. For the determination of the T_2 transverse relaxation time distributions, the acquired CPMG decays were analyzed using a Laplace inversion algorithm. This technique was developed for the characterization of relaxation in porous materials [6–10].

The mechanical strengths were determined on prismatic specimens (40 × 40 × 160 mm) using a hydraulic press, according to [11]. This were determined at 3, 7 and 28 days. The NMR measurements were performed on the same samples (or part of samples) subjected to flexural tensile and compressive strengths measurements, procedures which leads to the sample's failure [2].

3 Results and Discussions

The distribution of relaxation times T_2 for waterproofing mortars is presented in Fig. 1. At a first analysis it can be seen that all three types of waterproofing mortars have the distribution composed by 4 peaks. The amplitude, shape and linewidth of each peak provide information about the position and amount of water chemically bound to the

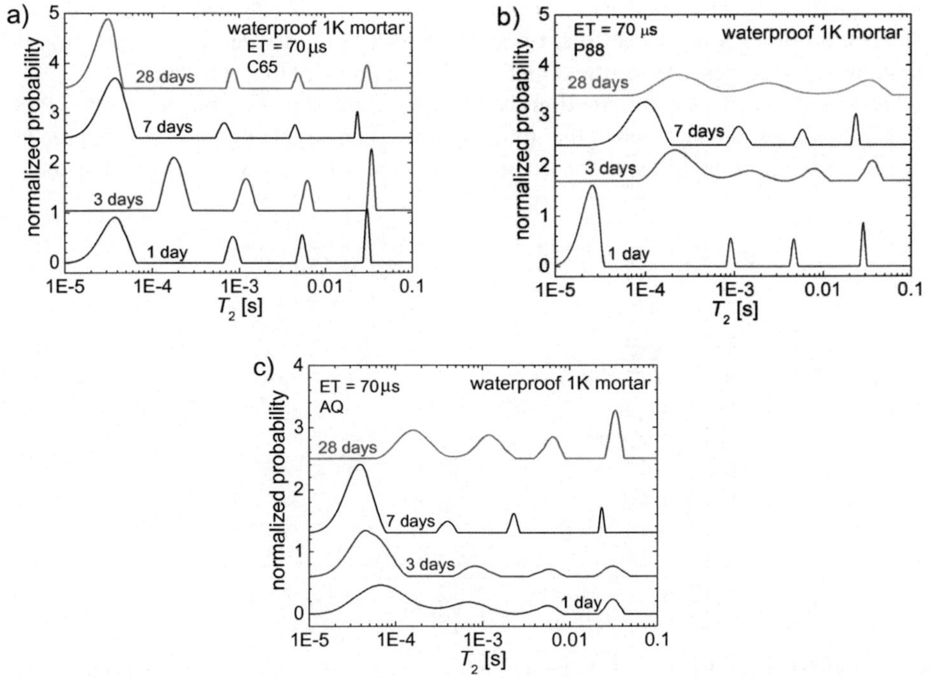

Fig. 1. Normalized probability of transverse relaxation time at 1, 3, 7 and 28 days after preparation for samples of waterproof 1K mortars: (a) C65; (b) P88 and (c) AQ.

aggregates and about the amount of water in the small, medium and large pores as well as their homogeneity and mobility. Although the general behavior is similar, the values of the relaxation times for each mortar show differences that are due to the amount of water used for preparation but also to the own recipe.

In Fig. 1a mortar C65 at 1 day after preparation has the highest amount of ^{1}H located as hydration water of the aggregates. At 3 days after preparation the distribution of T_2 relaxation times changes significantly, which can be associated with changes in the distribution of water in the pores. At 7 days after preparation the amounts of water are much smaller in small, medium and large pores, they change very little from day 7 to day 28.

Figure 1b presents the mortar P88 which, at first day after preparation, has quite rigid components, with homogeneous and well-defined pores. At 3 days, the distribution of relaxation times shifts to higher values, both for those associated to the hydration water and for pores. At 7 days, the mortar presents an increased rigidity, observed from the decreases of the T_2-values associated to the hydration water and to the pores. Large pores are more homogeneous than medium and small pores. At 28 days, only 3 peaks are observed. Probably the small pores were merged with medium pores, phenomena observed from the wide measured distributions.

In Fig. 1c it is observed that at the first day after preparation, the AQ mortar show wide peaks, meaning a high heterogeneity for water used for the hydration of mortar's components and to fill the pores. At 3 days, the increase in homogeneity is observed by decreasing the width of the peaks, which leads also to an increase in stiffness. At 7 days after preparation the homogeneity and rigidity are increasing. At 28 days the distributions of relaxation times are very different compared to those measured previously, this is interpreted by a decrease in stiffness and increase in porosity of the mortar.

Figure 2 compares the distributions of the measured relaxation times for C65, P88 and AQ mortars at 28 days, when theoretically it is assumed that the mortar is hardened and with maximum qualities to ensure protection against water, so that no subsequent

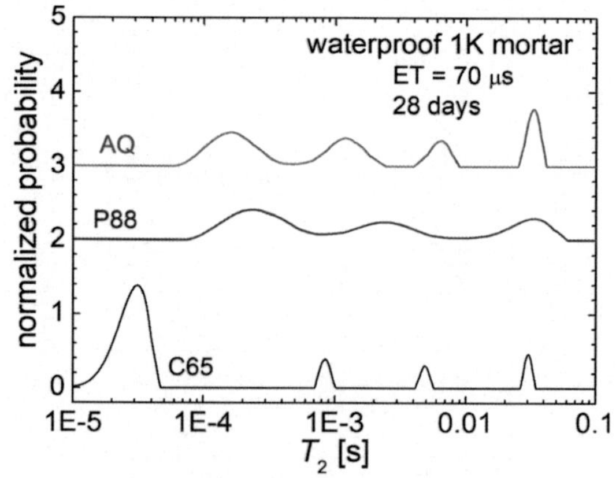

Fig. 2. Normalized probability of transverse relaxation time at 28 days after preparation for samples of waterproof 1K mortars.

evolution is expected. Even if the position and linewidth of the peaks is different for each mortar, the amounts of water in the integral areas are relatively the same.

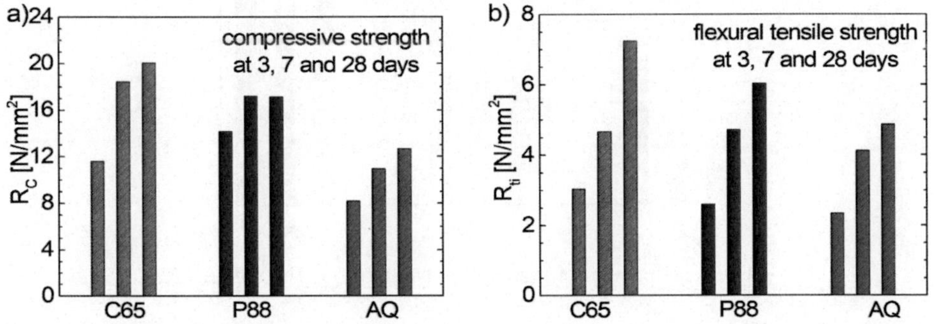

Fig. 3. a) Compressive strength measured for samples of waterproof 1K mortars at 3, 7 and 28 days after preparation; b) Flexural tensile strength measured for samples of waterproof 1K mortars at 3, 7 and 28 days after preparation.

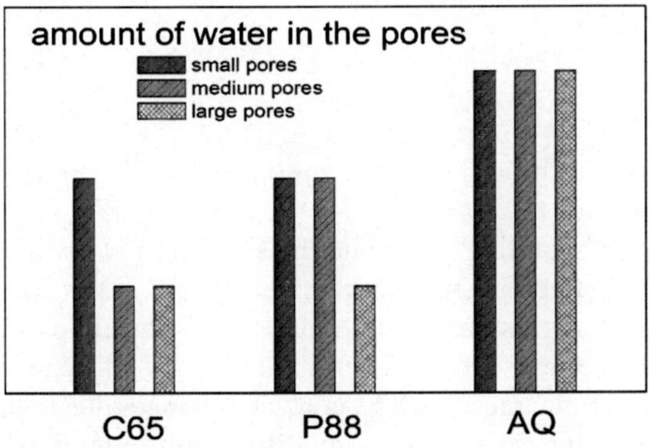

Fig. 4. Amount of water in the pores for samples of waterproof 1K mortars.

Compressive strength and flexural tensile strength were investigated starting with day 3 after preparation. The increasing evolution of the resistance to both compressive (Fig. 3a) and flexural tensile (Fig. 3b) can be observed at the three measured intervals. The mechanical properties of the mortars change less in the interval of 7–28 days compared to the changes measured for the interval 0–7 days. The most relevant mechanical parameter is the compressive strength. By comparison, at 28 days, the mortar C65 has the best compressive strength, followed by the mortar P88 and then the mortar AQ.

The evolution of the amount of water in the pores was measured by the method of NMR relaxometry at intervals of 1, 3, 7 and 28 days. The results show how the amount of water decreases the most in the first 3 days. In the interval of 3–28 days the amount of water is too small to change its proportion considerably.

In Fig. 4 is presented a comparison, at 28 days after the preparation of the samples. It is observed that there is a larger amount of water in the large, medium and small pores in the AQ mortar, followed by the P88 mortar and the C65 mortar.

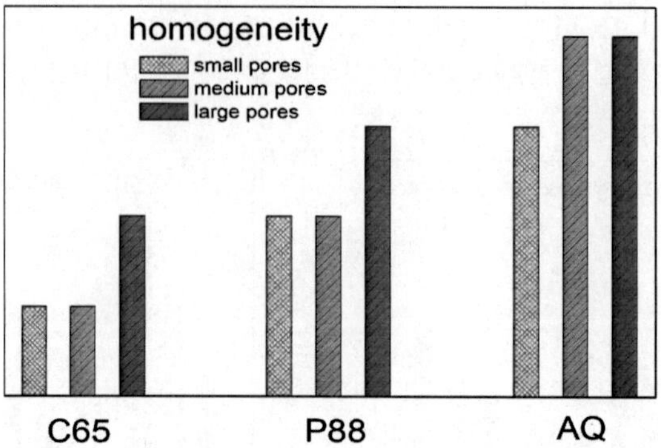

Fig. 5. Homogenity for samples of waterproof 1K mortars.

The diversity of homogeneity in each type of investigated mortar within 28 days of preparation can be seen in Fig. 5. The most important information provided by this measurement is that the homogeneity at 28 days is less than that measured in the first day.

4 Conclusions

Carrying out investigations on waterproofing mortars and interpreting the results led to the following conclusions:

Both, the amount of hydrating water and the amount of water in the small, medium and large pores have an effect on the compressive strength of the mortar. As the amount of water decreases the compressive strength increases inversely proportionally.

The distribution of T_2 transverse relaxation times showed a change in pore size and homogeneity in all four-time intervals. The pore size changes due to the amount of water that decreases over time and the homogeneity differs quite a lot at the same interval for each mortar.

The lack of homogeneity at the end of the hardening period of the mortar may suggest that it has the desired waterproofing effect, due to the hydrating effect of the cement and the formation of the sealing matrix. For a more advanced analysis of this sealing effect, it is necessary to carry out investigations using SEM (Scanning Electron Microscope), this being one of the future objectives of the research team.

References

1. Ohama, Y.: Handbook of Polymer-Modified Concrete and Mortars, pp. 13–19. Noyes Publications, New Jersey (1995)
2. Jumate, E., Manea, D.L., Moldovan, D., Fechete, R.: The effects of hydrophobic redispersibele powder polymer in portland cement based mortars. Procedia Eng. **181**, 316–323 (2017)
3. Borgia, G.C., Brown, R.J.S., Fantazzinit, P.: Uniform-penalty inversion of multiexponential decay data. J. Magn. Reson. **132**, 65–77 (1998)

4. Callaghan, P.T.: Translational Dynamics & Magnetic Resonance – Principles of Pulse Gradient Spin Echo NMR. Oxford University Press, New York (2011)
5. Venkataramanan, L., Song, Y.Q., Hürlimann, M.D.: Solving Fredholm integrals of the first kind with tensor product structure in 2 and 2.5 dimensions. IEEE Trans. Signal Process. **50**, 1017–1026 (2002)
6. Jumate, E., Moldovan, D., Fechete, R., Manea, D.: NMR relaxometry study of plaster mortar with polymer additives. AIP Conf Proc. **1565**, 112–116 (2013)
7. Jumate, E., Moldovan, D.C., Fechete, R., Manea, D.L.: The porosity and exchange processes in Portland cement pastes by 1D and 2D NMR relaxometry. CE-PhD, pp. 421–427, Cluj-Napoca, Romania (2012)
8. Fechete, R., Moldovan, D., Demco, D.E., Blümich, B.: Laplace inversions applied to multicomponent T_2-T_2 exchange experiments. Diffus. Fundam. **10**, 14.1–14.3 (2009)
9. Hürlimann, M.D., Flaum, M., Venkataramanan, L., Flaum, C., Freedman, R., Hirasaki, G.J.: Diffusion-relaxation distribution functions of sedimentary rocks in different saturation states. Magn. Reson. Imag. **21**, 305–310 (2003)
10. Jumate, E., Moldovan, D., Manea, D.L., Demco, D.E., Fechete, R.: The effects of cellulose ethers and limestone fillers in portland cement-based mortars by ^{1}H NMR relaxometry. Appl. Magn. Reson. **47**(12), 1353–1373 (2017). https://doi.org/10.1007/s00723-016-0844-y
11. SR EN 1015-11:2002/A1:2007, Methods of test for mortar for masonry. Part 11: Determination of flexural and compressive strength of hardened mortar

Characterization of Lightweight Concrete with Chopped Plastic Bottles

Sabina Scripca(✉), Gabriel Bejan, Marinela Barbuta, and Liliana Bejan

Faculty of Civil Engineering and Building Services, "Gheorghe Asachi" Technical University of Iasi, 1, Prof. Dimitrie Mangeron Blvd., 700050 Iasi, Romania

Abstract. In the building industry, concrete is the most used construction materials due to its high performances and advantages related to costs, durability and security in service life. Considering the requirements related to the environment, the concrete has a high potential to be ecological (green concrete) because in its mix different types of waste can be incorporated without affecting its properties or, in some cases, special properties can be obtained. As green concrete the lightweight concrete obtained by using wastes in the mix is also studied in the last decades. The article presents the experimental results obtained on concrete prepared by using chopped PET as aggregate substitution, in dosages of 50%, 70% and 90% and fly ash used as addition to the cement. The density, compressive strength, flexural strength and split tensile strength of concrete were experimentally determined. For comparison of mechanical characteristics, a control mix of concrete without chopped plastic waste was used. The influence of this kind of wastes on mechanical strengths is discussed.

Keywords: Fly ash · Chopped plastic bottle · Lightweight concrete · Structural concrete · Mechanical strengths

1 Introduction

In the building industry, concrete is the most used construction materials due to its high performances and advantages related to costs, durability and security in service life. The lightweight concrete is an important product due to its advantages such as: it presents reduced density that contributes to obtain smaller sizes of elements cross-section, structures and foundations, it has a better thermal conductivity, a better resistance to fire than the traditional concrete, it is prepared similar to traditional concrete, etc. [1–3].

Lightweight concretes are of many types: compact lightweight concrete, macroporous lightweight concrete, aerated lightweight concrete, foamed lightweight concrete, etc. [4].

The using domain of lightweight concretes is extended from masonry units for partition walls, thermal insulation panels, floors or roofs screeds to the use for cast in situ external and internal walls, structural concrete elements such as reinforced beams and slabs, lintels, scaffolds etc. [5].

L. Moldovan and A. Gligor (Eds.): Inter-Eng 2021, LNNS 386, pp. 100–109, 2022.
https://doi.org/10.1007/978-3-030-93817-8_11

Lightweight concrete has the density between 800 kg/m^3 and 2000 kg/m^3, obtained with different types of aggregates (lightweight aggregates or wastes of different kind). Materials used as aggregates for obtaining lightweight concrete can be of various types: natural (scoria, expanded clay, tuff, perlite, etc.) or by using wastes which are replacing the aggregates, such as recycled glass, polystyrene granules, chopped plastic bottles, vegetable wastes etc. [6–11]. The lightweight with recycled aggregates has reduced values of mechanical strength which can be improved by using other types as wastes, such as: rice husk, banana leaves ash, bamboo ash, bagasse ash, steel slag, glass powder, etc. [12–16].

The use of different types of waste in producing lightweight concrete is a necessity that responds to the new requirements related to the environment pollution and sustainability of building materials and constructions.

In the paper the physical-mechanical characteristics of lightweight concrete prepared with fly ash and chopped plastic bottles as aggregates replacement are analysed.

2 Experimental Program

2.1 Materials

The experimental lightweight concrete was prepared with cement, river aggregates in three sorts, wastes type fly ash, chopped plastic bottles (PET), water and superplasticizer. A control mix in which was added only 10% fly ash from the cement dosage was used as control mix for comparing the results.

The cement used for experiments was type CEM II/A-LL 42,5 R in a dosage 324 kg/m^3.

Aggregates were used in three sorts: sort I (0–4 mm), was a mix of natural sand with crushed gravel and sort II and III were natural river gravel of 4–8 mm and 8–16 mm, respectively. The quantities used in the control mix, noted BCM, were: 803 kg/m^3 of sand; 384 kg/m^3 of sort II and 559 kg/m^3 of sort III. The apparent specific mass of aggregate was evaluated to 2.7 kg/m^3. In the concrete with PET wastes substitutions of aggregate sort I, the dosages of all types of substitutions were 50%, 70% and 90% from aggregate quantity, measured in volume. The mixes with chopped plastic bottles (PET) were noted, B.C.PET 1, B.C.PET 2 and B.C.PET 3.

The fly ash used in the study was from Electric Power Plant Holboca Iasi, that was also used in previous researches [17]. The physical properties of fly ash are: grey colour, the particles are spherical with diameters between 0.01 μm and 400 μm, specific area of 480–520 m^2/kg and density of 2400–2550 kg/m^3. In the chemical composition there are SiO_2 (18.3%), CO_2 (17.15%), Al_2O_3 (13.9%).

The wastes of plastic bottles were obtained by cutting the bottles in pieces with sizes between 1–4 mm. The experimentally estimated unit weight of chopped PET was 433 kg/m^3.

The mixing water was in a dosage of 172 l/m^3 and as super-plasticizer admixture was used Master Glenium SKY 617 from BASF in a dosage of 1% from the cement dosage.

2.2 Samples for Experiment

The concrete with chopped plastic bottles (noted B.C.PET 1, B.C.PET 2, B.C.PET 3) was prepared by mixing all components in the following order: gravel, sand, chopped PET, cement, water and superplasticizer. After the concrete mixing the samples were poured: cubes of 150 mm size for determining compressive strength (f_c), [18] and density [19] and prism of 100 × 100 × 500 mm for determining flexural strength (f_{ti}) [20] and split tensile strength (f_{td}) [21] of experimental concretes.

3 Testing Results and Discussions

3.1 The Density of Experimental Concretes

Experimental values of density determined at 28 days, are presented in Fig. 1.

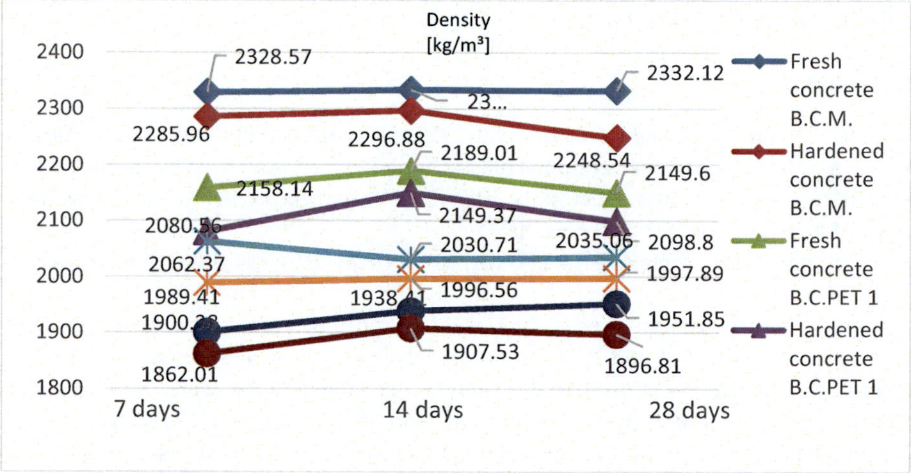

Fig. 1. Bulk density for experimental concrete with chopped plastic waste.

Bulk density of concrete with fly ash and chopped plastic waste in wet and dry state is smaller than that of control concrete (BCM). In the case of density of fresh concrete the values are between 2149.6 kg/m^3 (B.C.PET 1) and 1951.85 kg/m^3 (B.C.PET 3), that means a decrease of 7.82% in comparison with control mix for BCPET 1 and of 16.31% for B.C.PET 3. In the case of hardened concrete, the density vary between 2098.8 kg/m^3 (B.C.PET 1) and 1896.81 kg/m^3 (B.C.PET 3), that means a decrease of 6.7% in comparison with control mix for B.C. PET 1 and 15.65% for B.C. PET3.

The mix B.C. PET 1 has the density bigger than 2000 kg/m^3 and cannot be considered as lightweight concrete. The other concretes BCPET 2 and BCPET3 have the density smaller than 2000 kg/m^3. Lightweight concrete BCPET2 has the highest density and also the highest compressive strength.

Concrete B.C.PET 2 and B.C.PET 3 can be classified in D2,0, in function of the volumic mass [23]. The strength class of BCPET2 is LC20/22 and BCPET 3 is LC 12/13. The concrete BCPET 2 can be used as lightweight resistant concrete and and as thermal concrete. The concrete BCPET 3 can be used as thermal concrete.

The dosage of chopped plastic as replacement of aggregate sort 0–4 mm influenced the bulk density of concrete: with the increasing of chopped plastic dosage, the density is decreasing.

3.2 Compressive Strength of Concrete with PET Wastes

The values of compressive strength determined at 28 days are presented in Fig. 2.

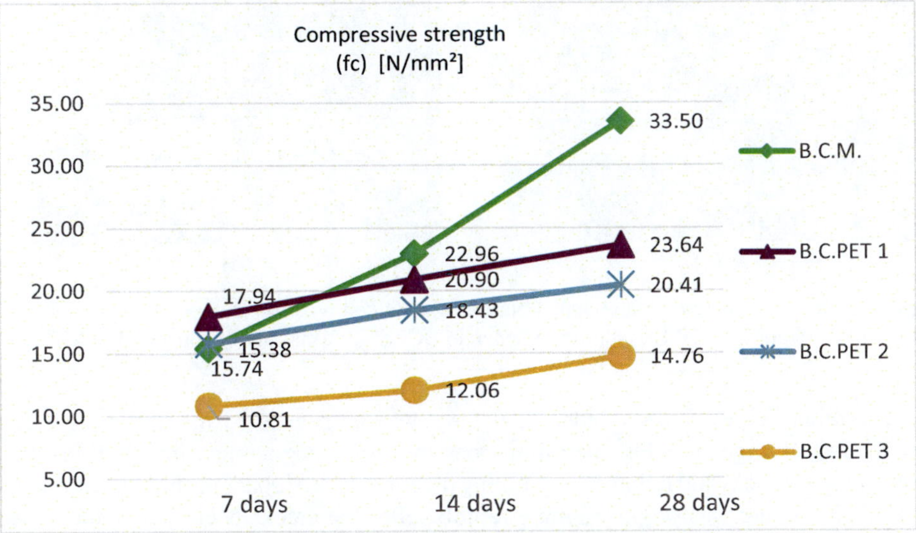

Fig. 2. Compressive strength of experimental concretes.

Compressive strength of concrete with fly ash and chopped plastic waste presents an evolution similar to that of control mix, from 7 days to 28 days and the values are smaller than that of control concrete (BCM).

At 28 days the values of fc are between 23.64 N/mm^2 (B.C.PET 1) and 14.76 N/mm^2 (B.C.PET 3), values smaller than that of control mix with percentages of 29.5% and 56%, respectively. Concrete type B.C.PET 1 and B.C.PET 2 presents values that define them as structural concrete [22]. The dosage of chopped plastic as replacement of sort 0–4 mm influences compressive strength: fc is decreasing with the increasing of chopped plastic waste.

The failure in axial compression of samples of concrete with chopped plastic is presented in Fig. 3.

Fig. 3. Failure of concrete with chopped plastic in axial compression: a) testing the sample; b) failure of concrete BCPET 1; c) failure of concrete BCPET 2; d) failure of concrete BCPET 3.

The failure type of cubes under axial force was influenced by the dosage of PET. For reduced dosages of PET, the failure was in stages, beginning with small cracks, developing of cracks and failure by separating parallel plans of concrete. For high dosages of PET the failure was produced more rapidly, after the cracks occurred. In the areas where PET waste was not uniformly distributed, less cracks occurred.

3.3 Flexural Strength of Concrete with PET Wastes

The values of flexural strengths are analyzed in Fig. 4.

The evolution diagram of flexural strength from 7 days to 28 days is with a small increasing in comparison with that of control mix, BCM. All values of concretes with chopped plastic are smaller than that of control concrete with percentages between 43% for B.C.PET 1 and 49.9% for B.C.PET 3.

The dosage of chopped plastic as replacement of sort 0–4 mm influences flexural strength: f_{ti} is decreasing with the increasing of plastic waste.

The failure in flexural strength test of samples of concrete with chopped plastic is presented in Fig. 5.

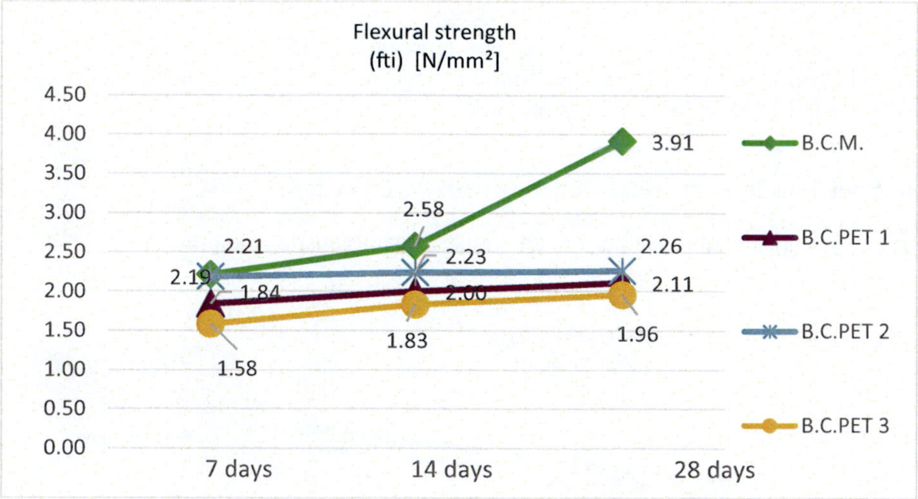

Fig. 4. Flexural strength of experimental lightweight concrete.

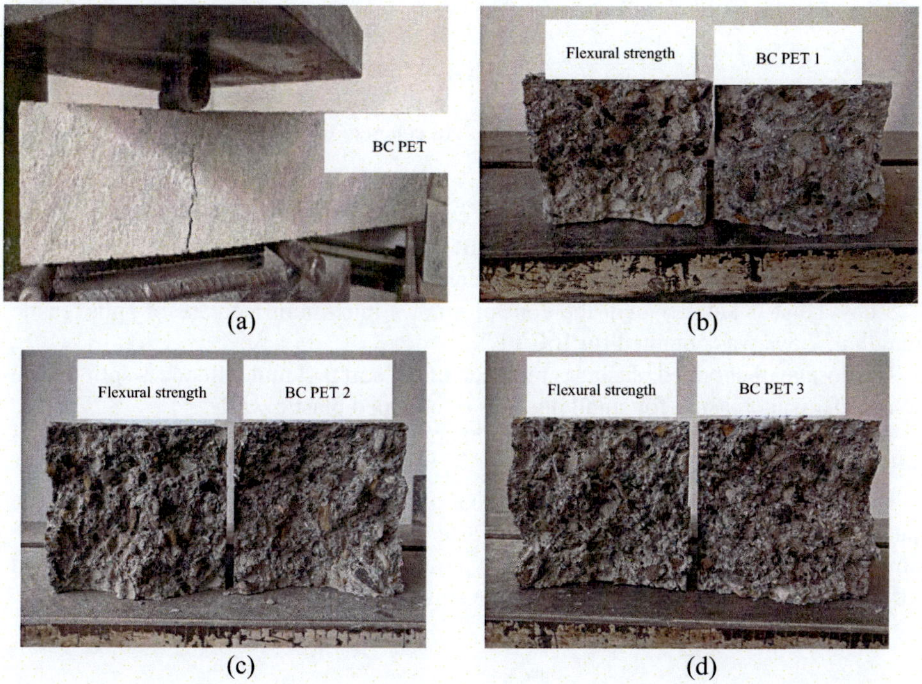

Fig. 5. Failure of concrete with chopped plastic in flexural test: a) testing the sample b) failure of concrete BCPET 1; c) failure of concrete BCPET 2; d) failure of concrete BCPET 3.

The failure of samples subjected to flexure was similar to that of the traditional concrete with only one plane of failure. The aggregates of PET waste were not broken, they were pulled-out from the matrix for all dosages of PET. A good adhesion between plastic waste (PET) and matrix was observed.

3.4 Split Tensile Strength of Concrete with PET Wastes

In Fig. 6 are presented the experimental values of split tensile strengths.

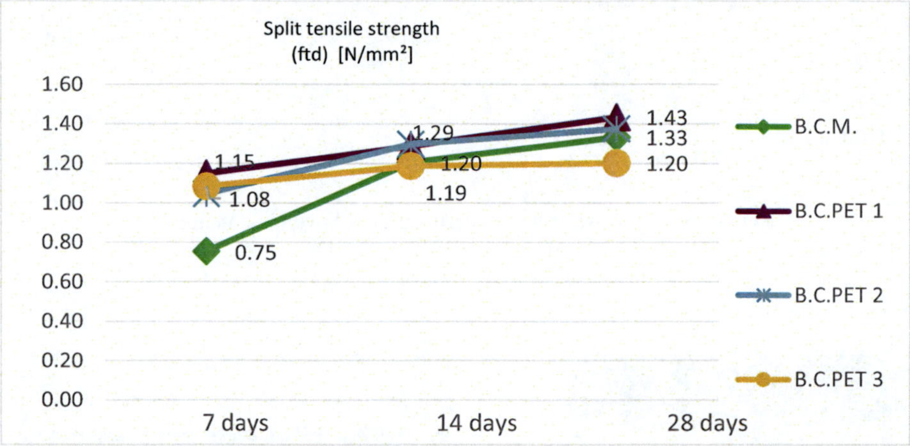

Fig. 6. Split tensile strength of concrete with wastes.

The evolution of split tensile strength of concretes with chopped plastic from 7 days to 28 days is similar to that of control mix, BCM. All values of concretes with chopped plastic are bigger than that of control concrete, exception is the value of B.C.PET 3 at 28 days, that is smaller than the control mix. A maximum increase of split tensile strength of 7.5% was obtained for B.C.PET 1.

The dosage of chopped plastic as replacement of sort 0–4 mm influences split tensile strength: f_{td} is increasing for small dosages of chopped plastic waste.

The failure in split test of samples of concrete with chopped plastic is presented in Fig. 7.

The failure in split test was similar to that of the traditional concrete with only one plane of failure. The aggregates of PET waste were not broken; they were pulled-out from the matrix for all dosages of PET. A good adhesion between plastic waste and matrix was observed, which can explain the small increasing of the strength.

Fig. 7. Failure of concrete with chopped plastic in split tensile test: a) testing the sample b) failure of concrete BCPET 1; c) failure of concrete BCPET 2; d) failure of concrete BCPET 3.

4 Conclusions

In the experimental program, chopped plastic wastes (PET) were used for preparing concrete with substitution of sort 0–4 mm in volume from 50% to 90% from the weight of the sand. The dosage of chopped plastic waste is influencing the mechanical characteristics of concretes. The bulk density, compressive strength and flexural strength are decreasing with the increasing of waste dosage. In the case of density, only over dosages of 60% of replacement of sand with chopped plastic bottle the lightweight concrete can be obtained. For reduced dosages of PET, the lightweight concrete can be used as resistant and as thermal concrete. For high dosages of PET the lightweight concrete can be used only for thermal protection. In the case of split tensile strength small dosages of chopped plastic are increasing the split strength value and only for higher dosages the split tensile strength is decreasing. An optimum dosage of chopped plastic waste in the case of replacing sand is around 70% from the aggregate volume.

The lightweight concrete obtained by using chopped plastic bottle wastes as replacement of sand is one of the solutions for introducing wastes in concrete mix. Also the other sorts of aggregates can be replaced with chopped plastic bottle, separately or in combination. The use of wastes of different types as replacement of aggregates contributes to the consumption of wastes that polluted the environment and to the rational use of natural raw resources. A better understanding of these new types of construction materials is necessary for an adequate and profitable use.

References

1. Ahmad, M.R., Chena, B., Farasat, S., Shahab, A.: Investigate the influence of expanded clay aggregate and silica fume on the properties of lightweight concrete. Constr. Build. Mater. **220**, 253–266 (2019)
2. Youm, K.S., Moon, J., Cho, J.Y., Kim, J.J.: Experimental study on strength and durability of lightweight aggregate concrete containing silica fume. Constr. Build. Mater. **114**, 517–527 (2016)
3. Timu, A., Bejan, G., Sosoi, G., Barbuta, M., Rotaru, A.: Mechanical characteristics of lightweight concretes obtained by aggregate replacement. Bull. Transilvania Univ. Braşov **10**(59), 181–187 (2018)
4. Tang, Q., Ma, Z., Wu, H., Wang, W.: The utilization of eco-friendly recycled powder from concrete and brick waste in new concrete: A critical review. Cem. Concr. Compos. **114**, 103807 (2020)
5. Bolden, J., Abu-Lebdeh, T., Fini, E.: Utilization of recycled and waste materials in various construction applications. Am. J. Environ. Sci. **9**(1), 14–24 (2013). https://doi.org/10.3844/ajessp.2013.14.24
6. Olofinnade, O., Chandra, S., Chakraborty, P.: Recycling of high impact polystyrene and low-density polyethylene plastic wastes in lightweight based concrete for sustainable construction. Mater. Today Proc. **38**(Part 5), 2151–2156 (2021)
7. Islam, M.J., Meherier, M.S., Islam, A.R.: Effects of waste PET as coarse aggregate on the fresh and harden properties of concrete. Constr. Build. Mater. **125**, 946–951 (2016)
8. Stefan, I., Barbuta, M., Budescu, M., Petru, M., Banu, O.M., Taranu, N.: Particularities regarding the mechanical behavior of some types of sustainable concrete mixes with waste materials. Roman. J. Mater. **48**(2), 236–244 (2018)
9. Cadere, C., Barbuta, M., Rosca, B., Serbanoiu, A.A., Burlacu, A., Oancea, I.: Engineering properties of concrete with polystyrene granules. 11th International Conference Interdisciplinarity in Engineering, INTER-ENG 2017, 5–6 October 2017, Tirgu-Mures, Romania. Proc. Manuf. **22**, 288–293 (2018)
10. Kanning, R.C., Portella, K.F., Bragança, M.O., Bonato, M.M., dos Santos, J.C.: Banana leaves ashes as pozzolan for concrete and mortar of Portland cement. Constr. Build. Mater. **54**, 460–465 (2014)
11. He, Z., Chang, J., Liu, C., Du, S., Huang, M.A.N., Chen, D.: Compressive strengths of concrete containing rice husk ash without processing. Roman. J. Mater. **48**(4), 499–506 (2018)
12. Khan, R., Jabbar, A., Ahmad, I., Khan, W., Khan, A.N., Mirza, J.: Reduction in environmental problems using rice-husk ash in concrete. Constr. Build. Mater. **30**, 360–365 (2012)
13. Van, V.T.A., Rößler, C., Bui, D.D., Ludwig, H.M.: Mesoporous structure and pozzolanic reactivity of rice husk ash in cementitious system. Constr. Build. Mater. **43**, 208–216 (2013)
14. Dwivedi, V.N., Singh, N.P., Das, S.S., Singh, N.B.: A new pozzolanic material for cement industry: bamboo leaf ash. Int. J. Phys. Sci. **1**, 106–111 (2006)
15. Akram, T., Memon, S., Obaid, H.: Production of low cost self-compacting concrete using bagasse ash. Constr. Build. Mater. **23**, 703–712 (2009). https://doi.org/10.1016/j.conbuildmat.2008.02.012
16. Xin, Y., Tao, Z., Song, T.Y., Pan, Z.: Performance of concrete made with steel slag and waste glass. Constr. Build. Mater. **114**, 737–746 (2016)
17. SR EN 12390-3: 2011. Testing Hardened Concrete: Part 3. Compressive Strength of Test Specimens. Romanian Standard Association, Bucharest, Romania (2011)
18. SR EN 12390-7:2005. Testing Hardened Concrete: Part 2. Density of Hardened Concrete. Romanian Standard Association, Bucharest, Romania (2005)

19. SR EN 12390-4:2010. Testing Hardened Concrete: Part 5. Flexural Strength of Test Specimens. Romanian Standard Association, Bucharest, Romania (2010)
20. SR EN 12390-4:2010. Testing Hardened Concrete: Part 7. Split Tensile Strength of Test Specimens. Romanian Standard Association, Bucharest, Romania (2010)
21. Norm for producing and execution of concrete, reinforced concrete and prestressed concrete works, NE 012/1-2010 (2010)
22. C155-2012 Norm regarding the production of lightweight concrete

Comparative Analysis Between the Hanger Arrangement in an 80 m Network Arch Bridge with Circular Hollow Cross-Sections

Alexandra Danciu(✉), Ștefan Guțiu, Cătălin Moga, and Maria Bucerzan

Technical University from Cluj-Napoca, Cluj-Napoca, Romania
{Alexandra.Danciu,Stefan.Gutiu}@cfdp.utcluj.ro,
Catalin.Moga@dst.utcluj.ro

Abstract. The hanger arrangement in network arch bridges is fundamental to the behavior of the bridge. As by definition, a network arch is a tied-arch bridge with inclined hangers that cross each-other at least twice. In the present paper, a comparative analysis of the hanger arrangement and the stresses distribution for a road bridge with a span of 80 m was conducted. Bridge spans in most parts of Romania are small and medium, up to 100 m, therefore there is a need to investigate and come with solutions to spans in the range of 50 to 100 m. The present study was conducted for a span of 80 m, with the bridge superstructure made of steel using circular hollow cross-sections for both the arch and the tie. The permanent and live loads were applied to the bi-dimensional structure.

Keywords: Trough road bridges · Network arch bridge · Hanger arrangement · Hollow circular cross-section

1 Introduction

Nielsen (1930), in his Ph.D. thesis [1], developed arches with sloping hangers. The V-configuration used reduces the bending in the arch, because a load on the left side of the span activates the sloping hangers to the right, resulting in an even load on the arch. Nielsen also applied for a patent in 1926, for a structure where the hangers crossed once. Tveit (1959), while working on his Master thesis [2] and then his PhD came to think of a configuration where hangers crossed each other at least twice, with the objective in mind of the best possible load distribution over the arch. The world first network arch was built in 1964, Steinkjer Bridge in Norway, with a span of 79.75 m and a rise of 12 m and an f/L ration of 0.15. Since than a great number of such structures have been erected around the world, for all bridge types: pedestrians, railway bridges and road bridges.

As already state, Network Arch Bridges (NAB) are arch bridges that have hangers that cross each other at least twice. The arches are made of steel, while, for distances between the arches of up to 20 m, Tveit [3] considers the tie should be made of concrete. Pipinato [4] published a study of a multiple span network arch bridge on a lightweight superstructure, where the arches are made of steel, using a circular hollow cross-section

L. Moldovan and A. Gligor (Eds.): Inter-Eng 2021, LNNS 386, pp. 110–119, 2022.
https://doi.org/10.1007/978-3-030-93817-8_12

and the tie is a steel-concrete composite cross-section The structure proposed by Pipinato [4] has a span of 130 m. A similar structure is the Bridge over the river Debra, a steel network arch with a span of 110 m with two inclined steel arches that link together in the keystone area [5]. In 2015 another study [6] investigating the influence of different hanger arrangements for a road bridge with a span of 100 m and an arch rise of 17 m was con-ducted, for circular hollow steel arches, inclined inward with 15 degrees after the tie beam-axis.

In the present study, based on the network arches with hollow circular cross-sections mentioned, we proposed and investigated the behavior of a similar structure, but for a smaller span, 80 m. The network arch proposed has vertical arches connected through wind bracings.

2 Structural Consideration

One of the major research interests when discussing NAB is the hanger arrangement. The available studies show that the number of hangers and their arrangement have a decisive influence over the bending moment distribution along the arch.

Teich in his PhD [7] devised and investigated five different hanger arrangements for bridge spans 100 m or larger and for 24, 36, 48 and 60 hangers. Based on his investigation, the best solutions are given by hanger arrangements 2 and 4. Hanger arrangement 2 is also known as the constant angle increase and 4 is the radial arrangement.

2.1 Constant Angle Increase (Incr)

For the 80 m structure investigated in the present paper, a variant of the constant angle increase of the hangers was investigated for 20 hangers, Fig. 1, 22 hangers, Fig. 2 and 24 hangers, Fig. 3.

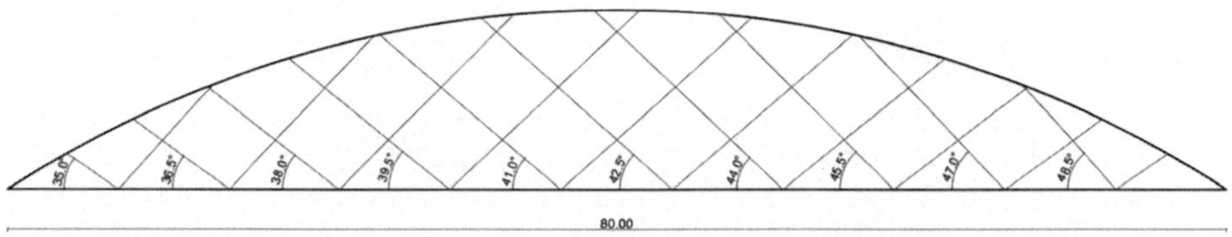

Fig. 1. Hanger arrangement for 20 hangers, constant angle increase with 1.5°.

Fig. 2. Hanger arrangement for 22 hangers, constant angle increase with 1.5°.

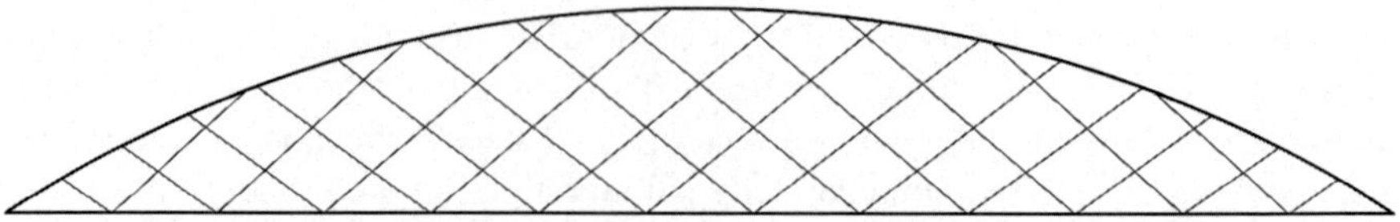

Fig. 3. Hanger arrangement for 24 hangers, constant angle increase with 1.5°.

The method to construct the hanger arrangement starts from dividing the length of the tie in equal parts. The starting hanger is inclined with an angle of 35°, then the slope increase is of 1.5° so that the angle variation is in the range 35°–48.5° for 20 hangers, 35°–50° for 22 hangers and 35°–51.5° for 24 hangers. Teich recommendation for 100–250 m span range and 24+ hangers in terms of the hangers' slope is between 35°–45°.

2.2 Radial Arrangement (RA)

The radial arrangement was proposed by Brunn and Schanack [8] in 2013 and has at its center the fact that if uniformly distributed loads act in the radial direction on the arch, then the bending in the arch about the horizontal axis is minimized. Following this idea, the hanger arrangement in Fig. 4 was devised by Brunn and Schanack [8].

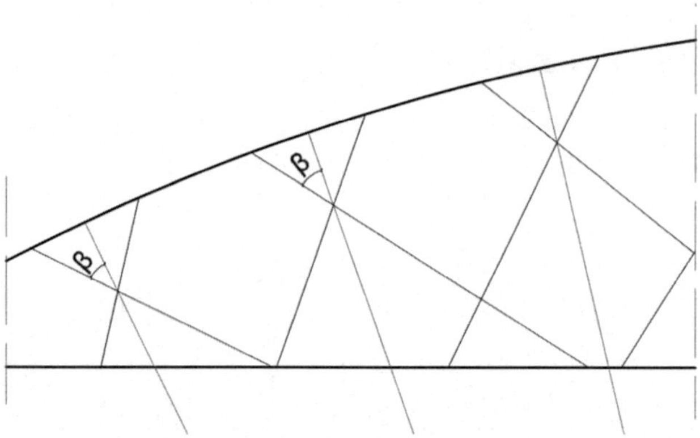

Fig. 4. The radial arrangement.

The hanger upper nodes are spaced equidistantly along the arch, while the hanger intersection lies on the radii of the arch that is part of a circle. The only variable is the angle of intersection between the hangers and the radii to the arch. The angle of intersection is the same for all hangers. The best results have been found for angles in the range 38° to 45°. The radial arrangement for the present study, for 20 hangers, is shown in Fig. 5, the value chose for β is of 38°, for 24 hangers is shown in Fig. 6 and for 24 hangers in Fig. 7.

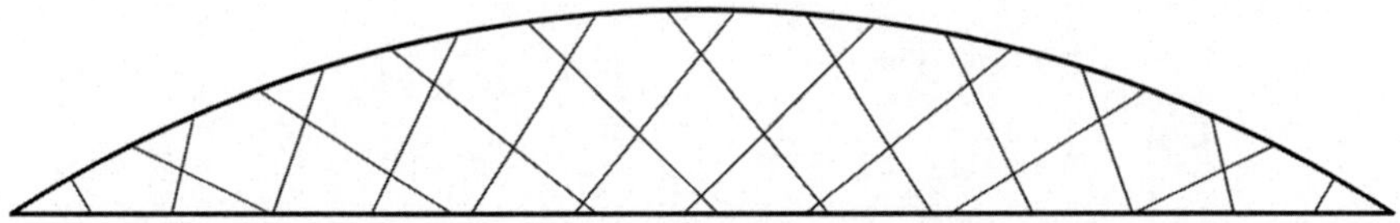

Fig. 5. Radial arrangement for 20 hangers, the angle between the hangers and the radii of 38°.

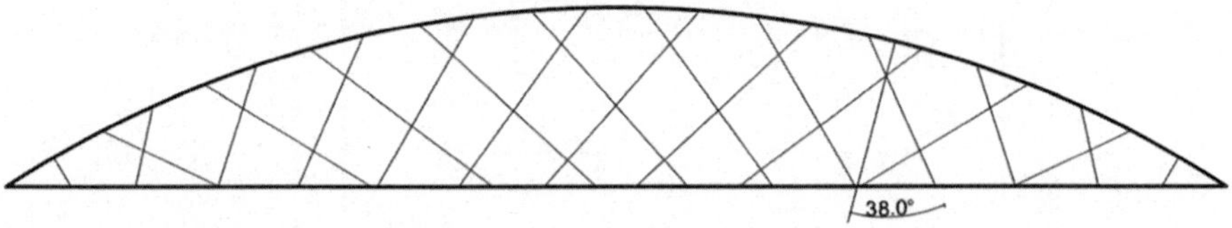

Fig. 6. Radial arrangement for 22 hangers, the angle between the hangers and the radii of 38°.

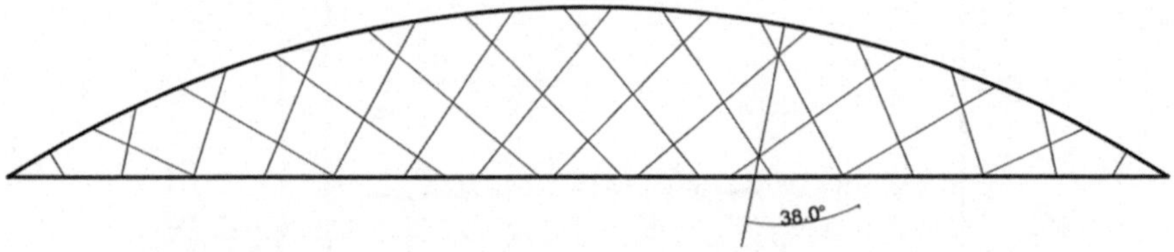

Fig. 7. Radial arrangement for 24 hangers, the angle between the hangers and the radii of 38°.

2.3 Network Arch Design

The dimensions of the bridge cross-section were chosen considering the Romanian requirements [9] for road bridges located on a class III road, with an upper bracing. The total distance between the arch axis is of 10 m, the two sidewalks are placed outside the arches.

The bridge cross-section elements are presented in Table 1 and the bridge cross-section is presented in Fig. 8. The geometrical arrangement of the lower bracing and the upper bracing is shown in Fig. 9. It is worth mentioning that the exact location of the upper wind bracing varies with the type of hanger arrangement proposed.

The analysis was undertaken considering the dead loads and the LM1 convoy. As stated in Eurocode 1 [10], the tandem system (TS) can be considered for bridges with a span larger that 10 m, as having a single axle with the load value equal to the TS load. The behavior of the superstructure in terms of stresses was investigated for the TS modeled as a single axle and as a moving vehicle with two axles. In the analysis the TS load was modeled as a moving load, while the distributed load from the LM1 convoy, UDL, was considered as loading the entire length and half of the length of the bridge, Fig. 10. The ratio between the distributed live load and the dead load is UDL/g = 0.42.

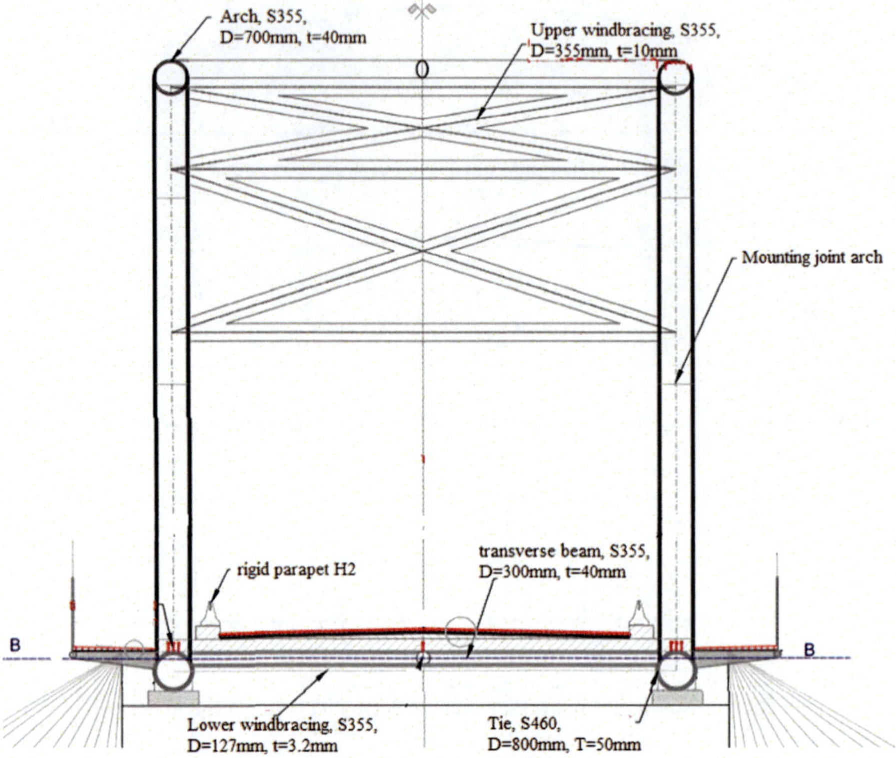

Fig. 8. Bridge cross-section.

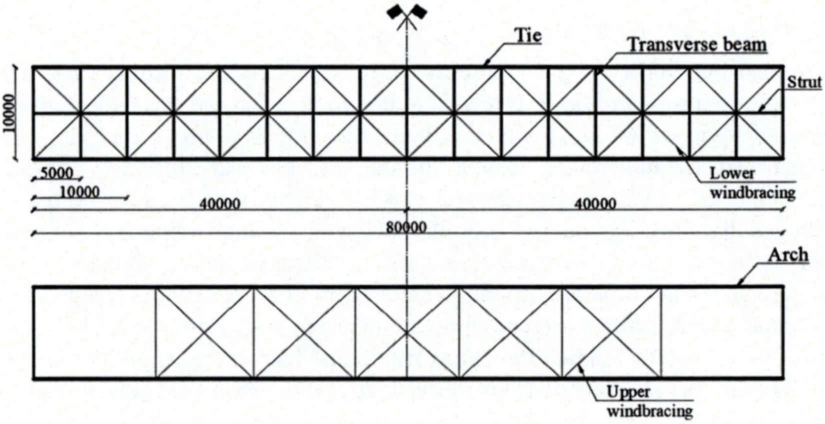

Fig. 9. Geometrical arrangement for the lower and upper wind-bracings of the bridge.

The comparative analysis of the behavior of the network arch was investigated for 20, 22 and 24 hangers, for two different hanger arrangements as stated above.

Table 1. Bridge cross-sections.

Element	Cross-section	Element	Cross-section
Arch S355	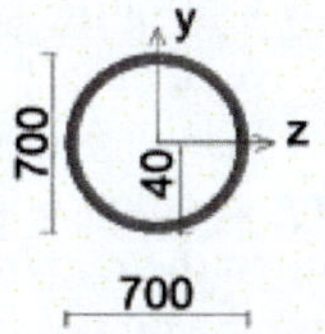	Transverse beam S355	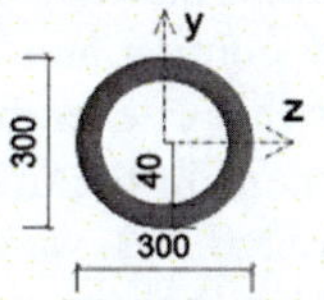
Tie S460	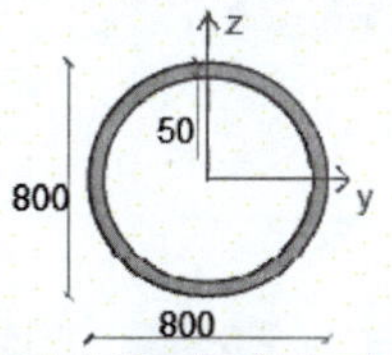	Strut S355	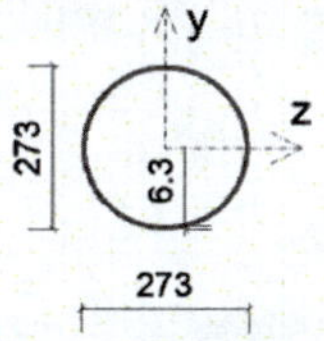
Hanger S355	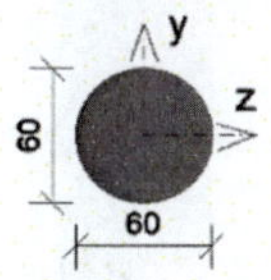	Upper bracing S355	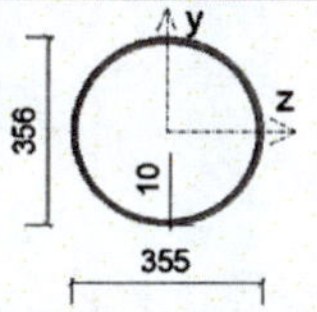

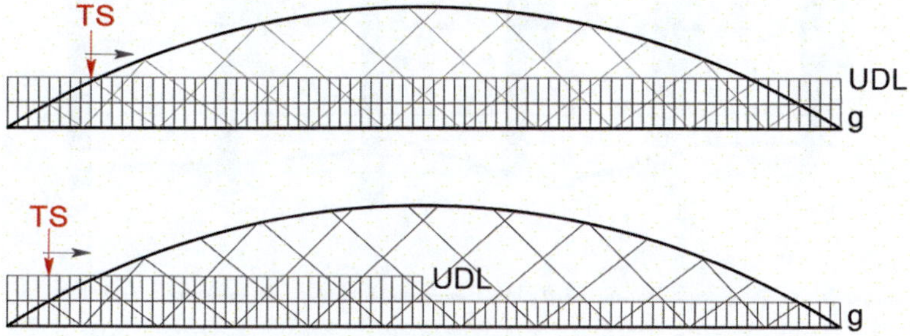

Fig. 10. Loading cases.

3 Stress Distribution

The bending moment and axial forces in the arch for the hanger arrangements under consideration are presented in Fig. 11 for the TS modeled as a vehicle with one axel. The investigation was also carried out for TS modeled as a vehicle with two axels, but the variation in the value of the unitary stresses is very small.

In Fig. 11 the compression force in the arch is represented as a column, while the bending moment as a line. The blue column represents the maximum compression force in the arch, while the entire span is loaded with the dead load and the convoy, TS+UDL. The best distribution of the compression force in the arch is obtained for the Radial Arrangement (RA) with 20 and 22 hangers, while for 24 hangers a slight increase in the value of the corresponding bending moment is observed. For the Constant angle Increase Arrangement (Incr) in Fig. 11, both the compression in the arch and the bending moment are larger than in the case of RA. For the case of only half of the structure loaded with UDL (yellow column and line in Fig. 11), the improvement in the stress distribution in the RA vs Incr is even better. When analyzing the behavior in terms of the highest bending moment in the arch with the corresponding compression force (orange and green representations), we can notice a small increase in the maximum value of M and N in the RA, but the compression force in the arch, for the highest value of the bending moment is over 20% smaller than the maximum compression force.

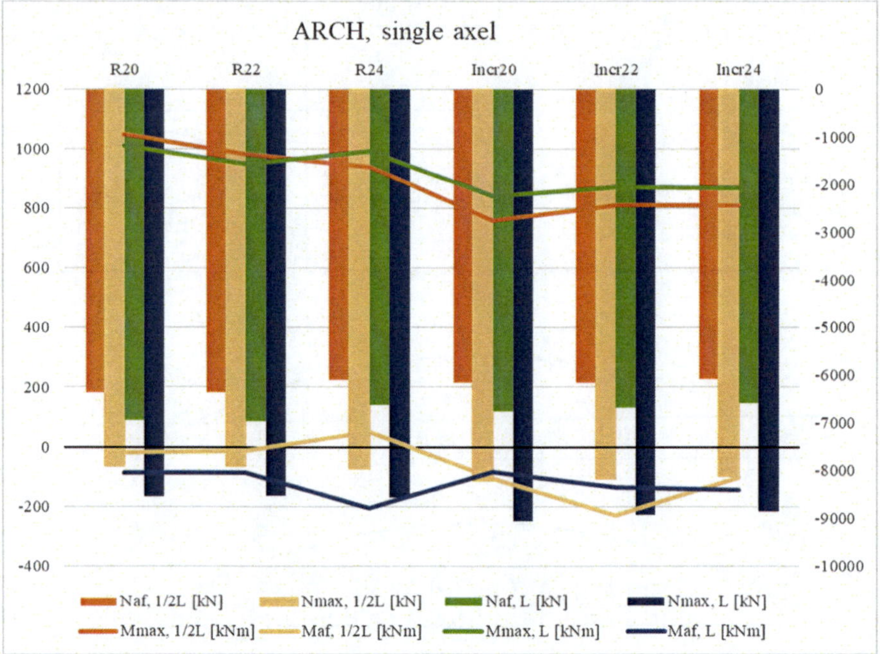

Fig. 11. Bending moment and compression force in the ARCH.

In Fig. 12 the bending moment and the axial force in the Tie is represented. For the maximum axial force in the tie, the best bending moment distribution can also be seen in the RA, with a minimum of M for 24 hangers, that is expected as the higher the number of hangers, the smaller is the value of the bending in the tie. On the other hand, the maximum bending moment sees an increase in the RA for a smaller number of hangers. In Incr arrangement, for the maximum value of the axial force, the bending moments are also increased with 30 to 50%.

In case of NAB a very important aspect is the hanger behavior in terms of hanger relaxation. For the medium span investigated, both RA and Incr with 24 hangers lead to hanger relaxation in at least one of the loading cases. As expected, the worst behavior is to be seen in case of half the span loaded by UDL force from the convoy. For 22 hangers, only the Incr arrangements leads to hanger relaxation for half the span loaded by the convoy. The best behavior for the investigated span is observed for RA with 20 and 22 hangers, where no relaxation and no bending occur in the hangers. The lowest value of the tension in the hangers occur for 22 hangers, the value of the axial force is 8% smaller than for 20 hangers and 38.5% smaller that the case of Incr with 20 hangers, the only other configuration where no relaxation occurs. The variation of the tension force vs. the number of relaxed hangers occurring in each configuration is presented in Fig. 13.

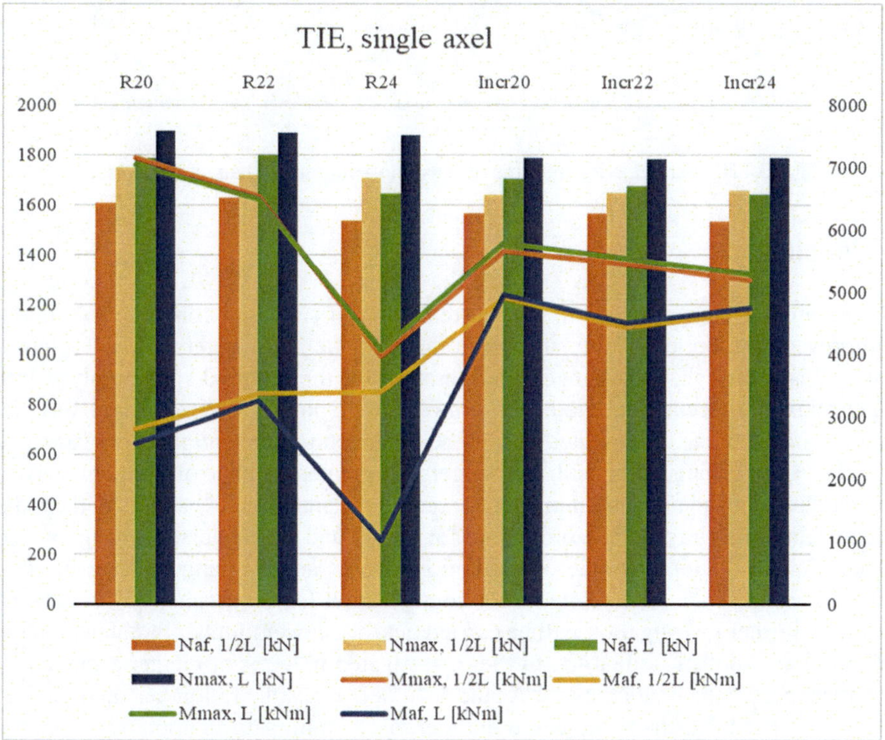

Fig. 12. Bending moment and axial force in the TIE.

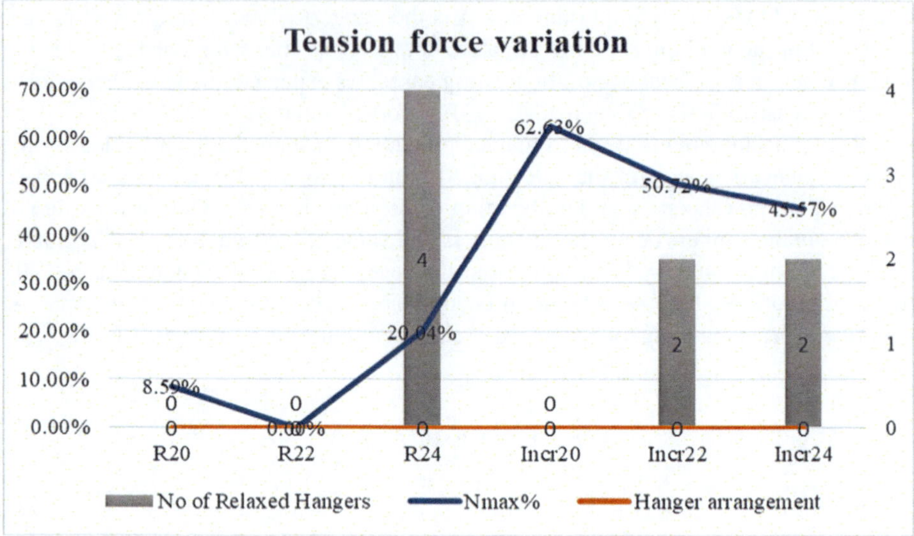

Fig. 13. Hanger behavior in terms of tension variation and hanger relaxation.

4 Conclusions

Network arch bridges can lead to a great reduction of steel and concrete used due to an improved distribution of the stresses in the arch, tie and hangers. The problem to be solved is to find the best hanger configuration for the required span so no or little hanger relaxation occurs and the bending moments to have a lower constant variation along the arch. For the investigated span of 80 m, the optimum configuration was found to be the radial arrangement with 22 hangers. A greater number of hangers leads to hanger relaxation and higher compression forces in the arch. For 24 hangers in the radial arrangement, 4 hangers are relaxed, while for the 20 and 22 hangers, no hangers were relaxed. For the Increases angle configuration, for 22 and 24 hangers, 2 of the total number of hangers are relaxed, while for configuration Incr20 no hangers are relaxed. Therefore, further investigation will be carried out for a lower number of hangers in this configuration. Another issue to be further investigated in terms of hanger arrangement is what happens with smaller spans. As the span decreases, the ration between the arch rise and the span needs to be increased in order to accommodate an upper wind bracing and the free passage gauge.

References

1. Nielsen, O.F.: Foranderlige Systemer med anvendelse på buer med skraatstillede Hængestenger (Discontinuous systems used on arches with inclined hangers), Ph.D. Thesis, Gad Copenhagen (in Danish) (1930)
2. Tveit, P.: "Bogebruer med skrå krysstilte hengestenger", (Arch bridges with inclined intersecting hangers), Ph.D. thesis presented at the Tech. Univ. of Norway. 64 pages, 78 drawings, in Norwegian (1959)

3. Tveit, P.: How to design economical network arches. IOP Conf. Ser. Mater. Sci. Eng. **471**, 052078 (2019). https://doi.org/10.1088/1757-899X/471/5/052078
4. Pipinato, A.: Structural analysis and design of a multispan network arch bridge. Bridge Eng. **169**(1), 54–66 (2016). https://doi.org/10.1680/jbren.14.00013
5. Mato, F., Cornejo, M.O., Sánchez, J.: Design and construction of composite tubular arches with network suspension system: recent undertakings and trends. J. Civ. Eng. Architect. **5**(3), 191–214 (2011)
6. Vlad, M., Kollo, G., Marusceac, V.: A modern approach to tied-arch bridge analysis and design. In: Acta Tehnica Corviniensis – Bulletin of Engineering Tome VIII, Fascicule 4, ISSN: 2067–3809 (2015)
7. Teich, S.: Beitrag zur Optimierung von Ntezwerkbogenbrucken (Contribution to Optimizing Network Arch Bridges), PhD Thesis, Technischen Universitat Dresden, Germany (in German) (2012)
8. Brunn B., Schanack F.: Calculation of a double track railway network arch bridge applying the European standards, Diploma Thesis, Dresden University of Technology (2013)
9. STAS 2924-91. Poduri de sosea. Gabarite, Institutul Roman de Standardizare (in Romanian)
10. SR EN 1991-2: Actions on structures – Part 2: Traffic loads on bridges

Paper Ash Used as Substitute of Cement, in Cement Mixtures

Maria Loredana Ţinţişan(✉), Adrian-Cristian Siomin, Anamaria Zaharie, and Daniela Lucia Manea

Faculty of Construction, Technical University of Cluj Napoca, Memorandumului Street, No. 28, 400114 Cluj-Napoca, Romania
daniela.manea@ccm.utcluj.ro

Abstract. The construction industry together with the construction materials industry are some of the most important consumers of natural resources. This industry has an important effect on the unsustainable development trend of the global economy. A sustainable approach to mortars, precisely from the perspective of the fact that it is an intensely used material in construction works, brings a contribution to sustainable development efforts in the construction industry.

Paper on the other hand is a material that is used in bulk. Paper recycling is relatively easy, but since it is not a material that can be recycled indefinitely, large amounts of waste are produced during the process. This article addresses the issue of using waste from cellulose particles processing, as a substitute for cement in mortars.

The importance of reducing the amount of cement used annually in construction works is supported by the fact that cement production contributes eight percent to annual CO_2 emissions. Green cement is a must in order to achieve the objectives of sustainable development in construction.

Following the experimental program, it is found that the replacement of an percentage of cement with paper ash brings many advantages.

Keywords: Green cement · Paper ash · Mortars · Tensile strength

1 Introduction

The feasibility of replacing a percentage of cement in the composition of concrete and mortar with cellulose waste is supported in two directions.

First of all, cement production is one of the processes that generate the most carbon dioxide. The process of obtaining Portland cement involves large quantities of fossil fuels and the mass exploitation of natural stone deposits. Concrete industry is the industry that consumes the largest amount of natural raw material and is responsible for 7% of greenhouse gas emissions into the atmosphere.

To reduce the impact that the cement industry has on the environment, it is necessary to develop alternative binders. Previous research shows that the ash resulting from burning of various materials (straw, plastic) has physicochemical properties similar to those of cement [1, 2]. Paper and cardboard do not have a recycling rate of 100% from raw

L. Moldovan and A. Gligor (Eds.): Inter-Eng 2021, LNNS 386, pp. 120–128, 2022.
https://doi.org/10.1007/978-3-030-93817-8_13

material. The process of making new paper consists of breaking cellulose fibers and fitting them into a fabric pattern that forms the sheet of paper. This process is valid both for the extraction of cellulose fibers from wood and from used cardboard and paper [3]. For this reason, paper or cardboard can only have between 3 and 8 recycling cycles, unlike steel, for example, which can be recycled indefinitely, without its chemical components suffering from repeated process. During the paper waste processing process, a so-called "mill sludge" is produced [4]. This is manufacturing waste, which is an economic and environmental threat. All over the world, huge quantities are generated as a result of the paper recycling process. India accounts for 0.7% of total waste stored in nature, and England produces no less than 1.5 million tons a year. Currently, the management of this waste is its scattering on agricultural land, in order to fertilize the land, or incineration, to obtain green energy, the resulting ash being also scattered on agricultural fields [5]. Therefore, "Mill Sludge" is a by-product, obtained from the process of removing inks from recycled paper, as well as the process of bringing the recycled paper to the pulp state. It contains organic matter, cellulose fibers, but also inorganic compounds such as calcium carbonate and glue, mixed with water. Currently, it is found that this material has a very good energy potential, being useful as an alternative fuel for production of Portland cement. Studies show that the ash resulting from burning paper waste in furnaces at a temperature of 700 °C for 2 h, contains silica, metakaolin, and CaO_2 (calcium oxide-quicklime) [6], which contribute chemically to Portland cement ingredients. Studies carried follow the behavior of ash as a partial substitute for cement in concrete mixtures. Because ash brings a higher intake of CaO_2, it is expected to increase the compressive strength [7].

2 Materials and Experimental Study

2.1 Characteristics of the Materials Used

The binder used is a Portland cement without additives, type CEM I 52.5R. The physical and mechanical characteristics were determined in the laboratory, as follows: the standard compressive strength is 52.0 N/mm^2 according to SR EN 196-1: 2016, the initial setting time is 45 min. according to SR EN 196-3: 2017, and the specific density is 2.94 g/cm^3, according to SR EN 196-4: 2008.

The aggregate used is a river aggregate. The sorts used for the recipes that entered in the experimental program are 0.00–0.25 mm, 0.25–0.5 mm, 0.5–1 mm, 1–2 mm, and 2–4 mm, according to the particle size distribution curve (see Fig. 1). The aggregate was sorted on fine sorts in the laboratory, was chosen, and used according to SR EN 13139: 2003/C91: 2009.

The raw paper waste was processed to obtain ash. For studied samples, only office paper waste was used. The ash resulting from the combustion was screened to separate the particles that have a size less than or equal to the size of the cement particles, 0.025 mm. The difference in density between cement powder and ash powder, presented in Table 1, was analyzed in order to establish the influence of low density RPA on water-cement ratio.

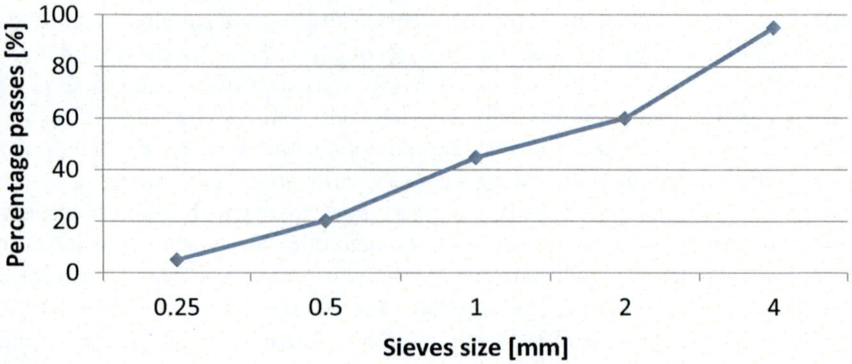

Fig. 1. Ganulometric curve of the aggregate used in the recipes included in the experimental program. Percentage passes on sieves with square mesh size of 0.25, 0.5, 1, 2 and 4 mm are analyzed.

Table 1. Measured density for blinders.

Material	Density [g/cm^3]
Cement CEM I 52,5R	2,94
Recycled office paper ash	2,44

2.2 Tests Performed on Fresh Mortar

The basic recipe that is the subject of this work is a CS IV plaster mortar, in which cement was replaced as a percentage of mass with RPA (recycled paper ash). To obtain a constant consistency, the variable is water volume added to each recipe.

The spread, process by which the consistency of fresh mortar is physically measured, follows constant values of 18–18.5 cm, for all the studied recipes. This characteristic was measured according to SR EN 1015-3: 2001/A2: 2007 using the spreading table. The variation of the specific density of the fresh mortar is also followed in order to determine the changes brought about by the difference in density of the cement compared to that of the ash. An important indicator regarding the behavior of fresh mortar in application is closely monitored. The segregation is checked for each recipe in order to comply with the limits established in the regulations.

2.3 Tests Performed on Hardened Mortar

For each of the 5 studied recipes, 9 parallelepiped specimens with dimensions of 4 × 4 × 16 cm and 2 cubic samples with a side of 10 cm were cast. Rectangular samples will be tested for bending stress, and the resulting ends are tested for compression stress, at fixed time intervals according to SR EN 196-1: 2020, namely 3, 7 and 28 days. The samples were kept for testing at the established intervals, under normal conditions, according to SR EN 1015-2: 2001/A1: 2007. The short name of the studied recipes are presented in the Table 2.

Table 2. Name of the studied recipes

Short name	Composition
CS IV	Standard plaster mortar
CS IV+5%	Standard CS IV mortar, with 5% cement replacement with recycled paper ash
CS IV+10%	Standard CS IV mortar, with 10% cement replacement with recycled paper ash
CS IV+15%	Standard CS IV mortar, with 15% cement replacement with recycled paper ash
CS IV+20%	Standard CS IV mortar, with 20% cement replacement with recycled paper ash

Results of the bending tensile strength tests obtained for the 5 recipes are presented numerically in Table 3, representing the average of the 3 test pieces cast for each recipe.

Table 3. Bending tensile strength.

Sample	Avr. load [N]	Avr. tensile strength [N/mm^3]	Sample age
CS IV	1906.67	4.47	3 days
CS IV+5%	1730.00	4.05	
CS IV+10%	1803.33	4.23	
CS IV+15%	1695.00	3.97	
CS IV+20%	1343.33	3.15	
CS IV	2226.67	5.22	7 days
CS IV+5%	2231.67	5.23	
CS IV+10%	2270.00	5.32	
CS IV+15%	1495.00	3.50	
CS IV+20%	1830.00	4.29	
CS IV	2593.33	6.08	28 days
CS IV+5%	2780.00	6.52	
CS IV+10%	2710.00	6.35	
CS IV+15%	2190.00	5.13	
CS IV+20%			

The results of the compressive strength tests obtained for the 5 recipes are presented numerically in Table 4, representing the arithmetic mean of the 6 molded specimens for each recipe (the prism halves resulting from the bending test).

3 Discussions and Results

The numerical results obtained from the tests of the 5 recipes under study, from Table 2 and Table 3, are transformed into graphs that highlight the behavior of the material and

Table 4. Compression strength.

Sample	Avr. load [KN]	Avr. compressive strength [N/mm^2]	Sample age
CS IV	55.82	34.85	3 days
CS IV+5%	44.84	28.11	
CS IV+10%	41.83	26.08	
CS IV+15%	45.68	28.55	
CS IV+20%	29.36	18.35	
CS IV	58.71	36.69	7 days
CS IV+5%	53.84	33.64	
CS IV+10%	50.44	31.52	
CS IV+15%	48.12	30.07	
CS IV+20%	30.67	20.50	
CS IV	62.57	39.11	28 days
CS IV+5%	59.40	37.13	
CS IV+10%	66.80	41.74	
CS IV+15%	60.73	49.12	
CS IV+20%			

the evolution of resistance depending on the percentage of RPA addition in the basic recipe (see Fig. 2 and Fig. 4).

In terms of bending strength at 28 days, the best results are for the mixture with 5% RPA substitute. The average values have an increase of about 7%, while the recipe CS IV+10% has an increase of only 4.3% compared to the basic recipe (see Fig. 3). The CS VI+20% recipe ends up equaling the tensile strength at 28 days with the basic recipe. In conclusion, reducing the amount of cement by up to 20%, being replaced by RPA, does not lead to significant changes in bending strength. We can appreciate the fact that the addition of ash determines a ductile behavior of the material.

The compressive strength is analyzed for the same samples as named before. It can be observed the same growth trend up to CSVI+10%, after which the compressive strength decreases by 8% for CSVI+15% recipe. The relevant difference appears in the CVI+20% recipe where the results decrease considerably, by 30%.

Even if the additive material does not behave as well in compression as in stretching, we can say that the recipe CS IV+10% is the one that reaches values comparatively equal to those in the basic recipe, which makes this recipe the best choice. Analyzing the case of the CS VI+15% recipe the decrease of tensile strengths does not exceed 15%, and the compressive strength does not decrease more than 8%, so it falls within the limits accepted in regulations, we can consider the feasibility of improving this mix to obtain better results (Fig. 5).

Based on these results, the CS VI+10% recipe is considered the most feasible choice. Thus, the specific strengths of the mortar are not affected, while the material economy,

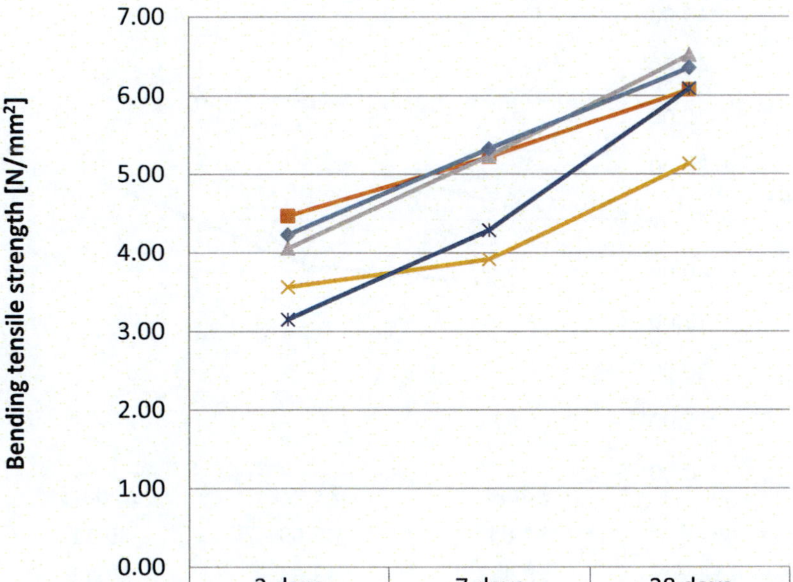

	3 days	7 days	28 days
CS4	4.47	5.22	6.08
CS4+5%	4.05	5.23	6.52
CS4+ 10%	4.23	5.32	6.35
CS4+15%	3.56	3.91	5.13
CS4+20%	3.15	4.29	6.09

Fig. 2. Bending tensile strength results measured in N/mm^2 for the 5 recipes, at 3, 7 and 28 days after casting.

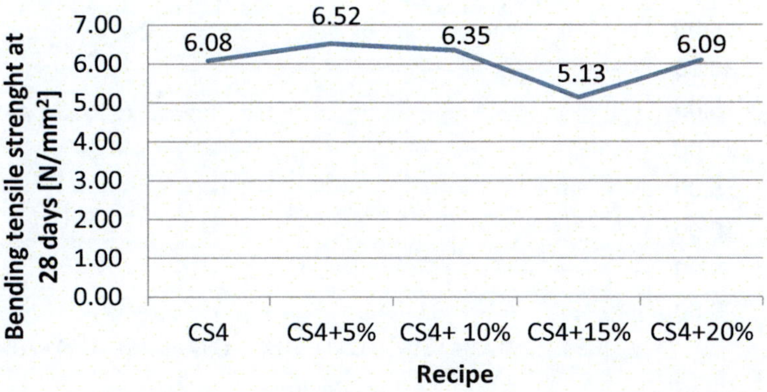

Fig. 3. Variation of the tensile strength by bending measured in N/mm^2 for the 5 recipes, 28 days after casting.

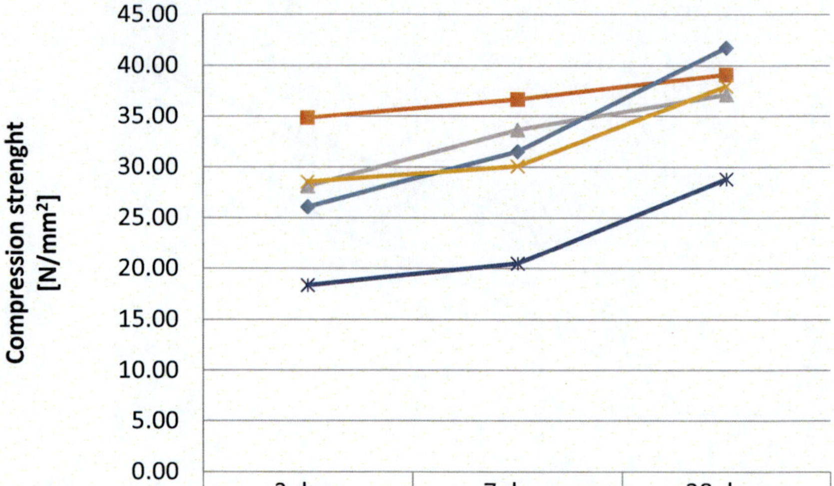

	3 days	7 days	28 days
CS4	34.85	36.69	39.11
CS4+5%	28.11	33.64	37.13
CS4+ 10%	26.08	31.52	41.74
CS4+15%	28.55	30.07	37.95
CS4+20%	18.35	20.50	28.81

Fig. 4. Results of the measured compressive strength in N/mm^2 for 5 recipes, 3, 7 and 28 days after casting.

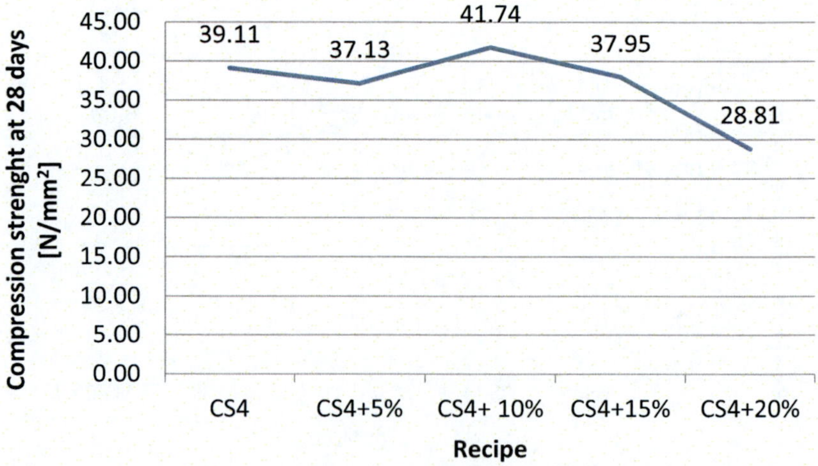

Fig. 5. Variation of measured compressive strength in N/mm^2 for 5 recipes at 28 days after casting.

but also the possibility of recycling a waste are exploited sustainably. The importance of the figures is accentuated by the following fact: globally 2,500 million T of Portland cement are produced annually. Europe produces only 5% of this quantity. Turkey and the USA produce another 5%. It follows from this calculation that replacing 10% of the cement with RPA would reduce the amount of cement produced by Europe, Turkey and USA.

4 Conclusions

Paper ash is a waste obtained by processing recycled paper, so it is a material that has zero costs. an economic plus being the fact that in the combustion process is obtained caloric energy and biogas, which can be transformed into alternative energy sources. The ash, which until now was also a waste, is collected and used at no additional cost, in order to bring a notable benefit to the cement both from an economic point of view and to the protection of the environment. Due to ongoing studies, it can be stated that the entire amount of ash, even particles larger than 0.025 mm that are used in this study, can be used in mixtures. So, over 90% of the amount of ash resulting from the combustion of cellulosic waste can be used as such, without the need for further processing, such as grinding.

The personal contribution regarding the study of this material is fueled by the need to find environmentally friendly solutions. The study of the current situation brings with it the awareness of the problems that humanity is facing. The solution that is supported by the present experimental study, comes to meet the need to reduce the amount of cement used worldwide, but also to find a feasible way to use a waste in construction works. My real job is to find a recipe that incorporates as much waste as possible without compromising the physical properties of the new hybrid material.

As previously mentioned, the study continues with the introduction of the coarse part of the ash, consisting of particles in the form of flakes with dimensions between 0.025 and 4 mm. At the same time, the hydrophobic behavior of these mortars containing RPA is studied.

Acknowledgements. This work was supported by the Post-Doctoral Program "Entrepreneurial competences and excellence research in doctoral and postdoctoral programs - ANTREDOC", project co-funded from the European Social Fund POCU/380/6/13/123927.

References

1. Bolden, J., Abu-Lebdeh, T., Fini, E.: Utilization of recycled and waste materials in various construction applications. Am. J. Environ. Sci. **9**, 14–24 (2013)
2. Tamanna, K., Raman, S.N., Jamil, M., Hamid, R.: Utilization of wood waste ash in construction technology: a review. Construct. Build. Mater. **237**, 117654 (2020)
3. Čabalová, I., Kačík, F., Geffert, A., Kačíková, D.: The effects of paper recycling and its environmental impact. Environ. Manage. Pract. **17**, 329–350 (2011)
4. Fava, G., Ruello, M.L., Corinaldesi, V.: Paper mill sludge ash as supplementary cementitious material. J. Mater. Civ. Eng. **23**(6), 772–776 (2011)

5. Tay, J.H., Show, K.Y.: Resource recovery of sludge as a building and construction material – a future trend in sludge management. Water Sci. Technol. **36**, 259–266 (1997)
6. Gluth, G.J., Lehmann, C., Rübner, K., Kühne, H.C.: Reaction products and strength development of wastepaper sludge ash and the influence of alkalis. Cem. Concr. Compos. **45**, 82–88 (2014)
7. Ahmad, S., Malik, M.I., Wani, M.B., Ahmad, R.: Study of concrete involving use of waste paper sludge ash as partial replacement of cement. IOSR J. Eng. **3**, 2278–8719 (2013)
8. Ahmadi, B., Al-Khaja, W.: Utilization of paper waste sludge in the building construction industry. Resour. Conserv. Recycl. **32**(2), 105–113 (2001)

Recovery of Used Paints in the Field of Plaster Mortars

Adrian-Cristian Siomin(✉), Maria Loredana Ţinţişan, Anamaria Zaharie, and Daniela Lucia Manea

Faculty of Civil Engineering, Technical University of Cluj-Napoca, Cluj-Napoca, Romania
Daniela.Manea@ccm.utcluj.ro

Abstract. Waste from the use of construction materials represents the largest amount of waste generated in the world. Used paints could enter an extensive recycling process, and will be reused for other purposes, namely in the preparation of plaster mortars.

Industrial production and the construction of residential buildings leave a significant amount of waste, and burying them underground is a negative factor on the environment, resulting waste that can be recycled and reused.

The obligation to implement a waste management, with clear measures regarding the recycling methods by type of material will lead to the reduction of raw materials used in the manufacture of construction materials.

Just as there is the circuit of water in nature and the circuit of municipal household waste, we must consider a circuit of construction materials, from their manufacture, their labeling, to their reuse in the construction of buildings.

Human, as a component of society, must know and be responsible for the fact that the resources of this planet are limited and that methods must be found to use waste through efficient management, while respecting the environment.

To reduce the negative effects on the environment caused by construction and demolition waste, we emphasized the importance of protecting the environment and developed an experimental program through which used paint is used as a substitute for water in the production of plaster mortar.

Keywords: Recycling · Paint · Plaster mortars

1 Introduction

1.1 Generalities

Taking into account the industrial growth of all developing countries, especially the construction materials production sector, it is necessary to focus on waste management, respectively the recycling of materials and their reuse in the field of construction.

Through an integrated and customized waste management on various types of materials, the cost of materials could be reduced, so that in addition to reducing costs it will be possible to limit raw materials, using materials recovered from waste from new constructions and rehabilitation of old ones.

L. Moldovan and A. Gligor (Eds.): Inter-Eng 2021, LNNS 386, pp. 129–138, 2022.
https://doi.org/10.1007/978-3-030-93817-8_14

Just as there is the circuit of water in nature and the circuit of municipal household waste, we must consider a circuit of construction materials, from their manufacture, their labeling, to their reuse in the construction of buildings.

Industrial production and the construction of residential buildings cause a significant amount of waste, and burying them underground is a negative factor on the environment, resulting in a waste of waste that can be recycled and reused.

Waste from the use of construction materials represents the largest amount of waste generated in the world. They consist of various types of materials: wood, glass, plaster, concrete, plastic, brick, ceramic wall and floor tiles, solvents, asbestos, **used paints**, which could go into an extensive recycling process and be reused for other purposes.

The objective of the work is to obtain a new construction material, in which used paint is a basic component in the preparation of plaster mortars, designed as apparent, decorative mortars.

2 Experimental Program

2.1 Component Materials and Their Characteristics

The experimental program involves the development of plaster mortar recipes with reference to CS IV (cement-based mortar, sand and water) in which the used paint is a substitute for water in different percentages.

Portland cement, respectively the clinkerized hydraulic binder obtained by grinding the cement clinker, having in addition the gypsum to regulate the setting time is the essential component in order to obtain the new mortar.

In the experimental program, for the preparation of plaster mortar recipes was used Portland cement without addition, type CEM I 52.5 R, in accordance with SR EN 197-1: 2011 Cement Part 1: Composition, specifications and conformity criteria of common cements [1]. The composition of EMC type cement I 52.5 R is as follows: clinker 95–100% and limestone 0–5%. Also, this type of cement has the following specifications: initial setting time min. 45 min, stability/expansion max. 3 mm, initial compressive strength min. 35 N/mm^2, standard compressive strength max. 57 N/mm^2, calcination loss max. 5%, in accordance with SR EN 413-1: 2011 Cement for masonry. Part 1: Composition, specifications and conformity criteria [2].

The real density of this cement according to the measurements performed in the laboratory of the Faculty of Constructions within the Technical University of Cluj-Napoca is $\rho = 2.94$ g/cm^3, by the pycnometer method (equipment used for this determination: burette and pycnometer).

Aggregate (sand) from the alluvial layer of a river, in compliance with SR EN 13,139: 2003/C91: 2009 Aggregates for mortars [3].

For the preparation of the plaster mortars, the granulometric composition of the aggregate is established, respectively the establishment of the sand sorts, with the help of the set of sieves/sieves with square meshes of: 0.25 mm; 0.50 mm; 1 mm; 2 mm; 4 mm (See Fig. 1).

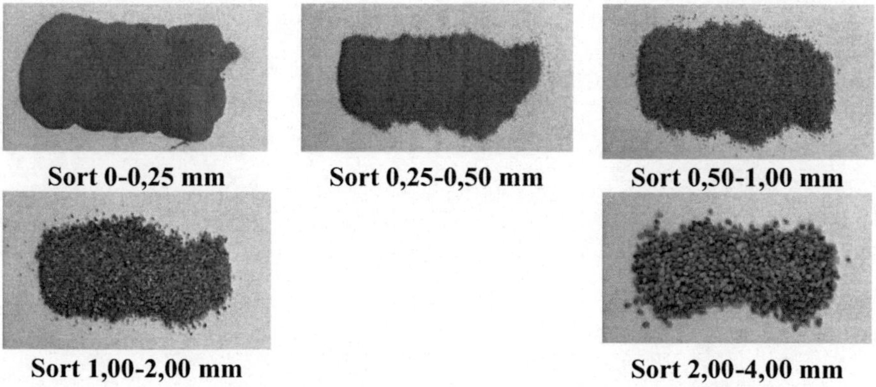

Fig. 1. Sand sorts: (0–0.25), (0.25–0.50), (0.50–1.00), (1.00–2.00), (2.00–4.00) mm.

For the preparation of plaster mortars, **water** complying with SR EN 1008: 2003 Concrete preparation water is used. Specifications for the sampling, testing and assessment of the suitability for use of water, including water recovered from concrete industry processes, as concrete preparation water [4].

The used water-based paint, red in color and recovered from a construction site, it was proposed through the experimental program to replace in the composition of the plaster mortar a certain percentage of the amount of water used in the preparation of this mortar, respectively 5%, 10%, 15% and 20% of the amount of water.

The density of used paint according to the measurements performed in the laboratory of the Faculty of Civil Engineering within the Technical University of Cluj-Napoca is $\rho = 1.04$ g/cm^3, by the pycnometer method (equipment used for this determination: burette and pycnometer).

3 Results

3.1 Research Methodology

In order to determine the characteristics and perform the determinations, the requirements of the harmonized European standard SR EN 1015 1 $\div$ 21/2020 on Test methods for mortars, which include how to determine the particle size distribution, sampling, determination of fresh mortar consistency, determination of bending strength, were observed. of hardened mortar, determination of compressive strength [5].

In order to carry out the experimental program, a recipe was used for a CS IV plaster mortar, with the following composition for 7 L: Portland cement CEM I 52.5R - 2.695 kg, aggregate on sorts 6.96 kg (0.35 kg of sort 0–0.25, 1.05 kg of sort 0.25–0.50, 1.71 of sort 0.50–1.00, 1.05 kg of sort 1.00–2.00, 2.80 kg of sort 2.00–4.00) and water 1.50 L, respecting a consistency of 10 cm (See Table 1).

Table 1. Composition recipes used in the preparation of plaster mortar.

Recipe	Portland Cement CEM I 52,5R [kg]	Aggregate by sorts [kg]	Water [l]	Red used paint [l]
CSIV	2.695	6.96	1.500	–
CSIV 5P			1.554	0.072
CSIV 10P			1.470	0.144
CSIV 15P			1.400	0.225
CSIV 20P			1.320	0.300

The density of fresh mortar according to the measurements performed in the faculty laboratory is $\rho = 2274$ kg/m^3, measured with a 1 L vessel.

where:

- CSIV - plaster mortar;
- CSIV 5P - plaster mortar in which water was replaced with 5% used paint;
- CSIV 10P - plaster mortar in which water was replaced with 10% used paint;
- CSIV 15P - plaster mortar in which water was replaced with 15% used paint;
- CSIV 20P - plaster mortar in which water was replaced with 20% used paint.

The experimental program proposes the gradual replacement of water with different percentages of paint. Table 1 contains the coding for the mortars in this paper.

For the recipe of the **CSIV 5P** plaster mortar, 5% water was replaced, respecting the mortar consistency of 10 cm. Thus, the recipe CS IV 5P has the following composition: Portland cement CEM I 52.5R - 2.695 kg, aggregate on sorts 6.96 kg (0.35 kg from sort 0–0.25, 1.05 kg from sort 0.25–0.50, 1.71 kg from sort 0.50 -1.00, 1.05 kg of sort 1.00–2.00, 2.80 kg of sort 2.00–4.00), water 1.554 L and red used paint 0.072 L (See Table 1).

In the fresh state, for all 5 mortars determinations on the density were made, and the results are presented in the table below (See Table 2).

For the recipe of the **CSIV 10P** plaster mortar, 10% water was replaced, respecting the mortar consistency of 10 cm. Thus, the recipe CS IV 10P has the following composition: Portland cement CEM I 52.5R - 2.695 kg, aggregate on sorts 6.96 kg (0.35 kg of sort 0–0.25, 1.05 kg of sort 0.25–0.50, 1.71 of sort 0.50–1.00, 1.05 kg of sort 1.00–2.00, 2.80 kg of sort 2.00–4.00), water 1.470 L and red used paint 0.144 L (See Table 1).

For the recipe of the **CSIV 15P** plaster mortar, 15% water was replaced, respecting the mortar consistency of 10 cm. Thus, the recipe CS IV 15P has the following composition: Portland cement CEM I 52.5R - 2.695 kg, aggregate on sorts 6.96 kg (0.35 kg of sort 0–0.25, 1.05 kg of sort 0.25–0.50, 1.71 of sort 0.50- 1.00, 1.05 kg of sort 1.00–2.00, 2.80 kg of sort 2.00–4.00), water 1,400 L and red used paint 0.225 L (See Table 1).

For the recipe of the **CSIV 20P** plaster mortar, 20% water was replaced, respecting the mortar consistency of 10 cm. Thus, the recipe CS IV 20P has the following composition: Portland cement CEM I 52.5R - 2.695 kg, aggregate on sorts 6.96 kg (0.35 kg from sort

Table 2. Density of plaster mortars.

Recipe	Apparent density ρ_a [kg/m^3]
CSIV	2274,00
CSIV 5P	1981,30
CSIV 10P	2021,00
CSIV 15P	1906,00
CSIV 20P	1931,00

0–0.25, 1.05 kg from sort 0.25–0.50, 1.71 from sort 0.50- 1.00, 1.05 kg of sort 1.00–2.00, 2.80 kg of sort 2.00–4.00), water 1,320 L and red used paint 0.30 L (See Table 1).

3.2 Bending Strength

The fresh mortar was poured into molds so as to obtain 9 specimens measuring 4 × 4x16cm, being stripped, stored and tested for bending according to the standard mentioned in Sect. 3.1 at intervals of 3 days, 7 days and 28 days, using the equipment of the laboratory of the Faculty of Constructions (device Fruhling - Michaelis), and the results are summarized in Table 3 and presented in Fig. 2 (See Fig. 2, Fig. 3 and Table 3).

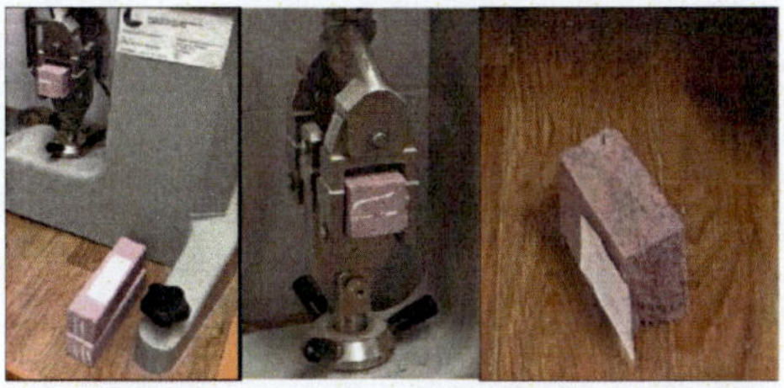

Fig. 2. Test specimens to determine the bending strength

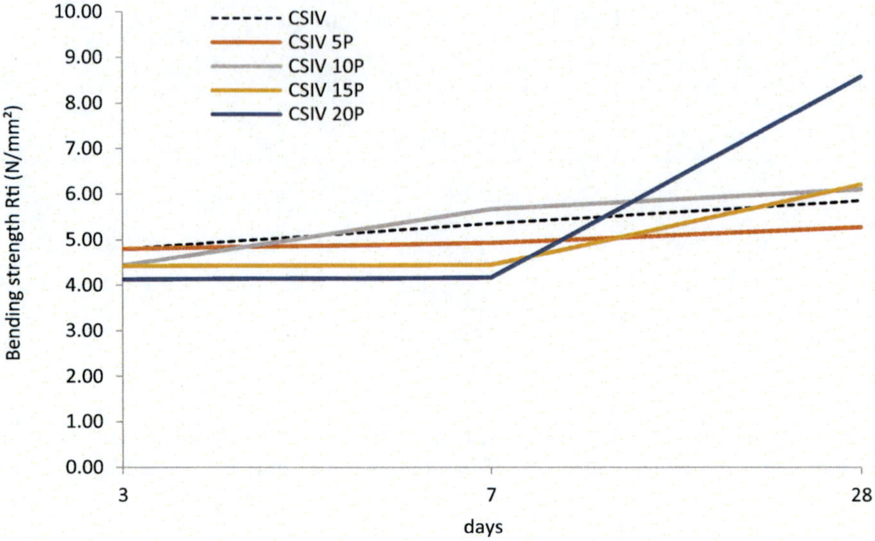

Fig. 3. Bending strength Rti [N/mm^2]

Table 3. Bending strength results.

Sample	Average bending strength [Rti] [N/mm^2]		
	3 days	7 days	28 days
CSIV	4,80	5,36	5,86
CSIV 5P	4,80	4,93	5,28
CSIV 10P	4,43	5,67	6,11
CSIV 15P	4,42	4,44	6,21
CSIV 20P	4,13	4,16	8,58

3.3 Compressive Strength

The halves/ends of 4 × 4 × 16 cm test specimens resulting from bending were tested for compression according to the standard referred to in Sect. 3.1 at intervals of 3 days, 7 days and 28 days, using the equipment of the laboratory of the Faculty of Constructions (Hydraulic press), and the results are summarized in Table 4 and presented in Fig. 5 (See Fig. 4, Fig. 5 and Table 4).

Fig. 4. Test specimens to determine the compressive strength

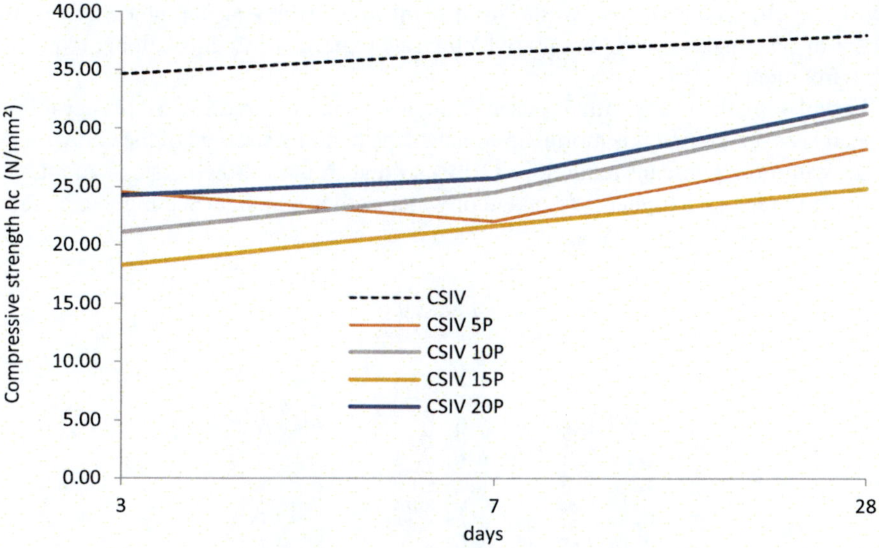

Fig. 5. Compressive strength Rc [N/mm^2]

Table 4. Compression strength results

Sample	Average compressive strength [Rc] [N/mm^2]		
	3 days	7 days	28 days
CSIV	34,65	36,52	38,03
CSIV 5P	24,56	22,07	28,29
CSIV 10P	21,10	24,53	31,31
CSIV 15P	18,29	21,66	24,29
CSIV 20P	24,23	25,76	32,00

4 Interpretation of Results

4.1 Contributions

The experimental program was carried out in the faculty laboratory, using the equipment and identifying the properties of the component materials (cement, sand, paint), but also the 5 samples of mortar both fresh and hardened.

The personal contribution of the experimental program is the replacement of a percentage of the amount of water with used paint, respectively 5%, 10%, 15% and 20% for the preparation of plaster mortars.

The results presented in Chapter 3 highlight the beneficial contribution made by the idea of recycling construction waste, used paint, thus achieving the objectives of the concept of sustainable development, reducing raw materials and thus helping to protect the environment.

By replacing the water with 5% used paint, according to the CSIV 5P plaster mortar recipe, it is observed that this mortar approaches the bending strength of the CSIV recipe mortar, with no significant reductions (10%) so that in the following steps an attempt was made to replace a higher percentage of water with used paint (See Fig. 6 and Table 3).

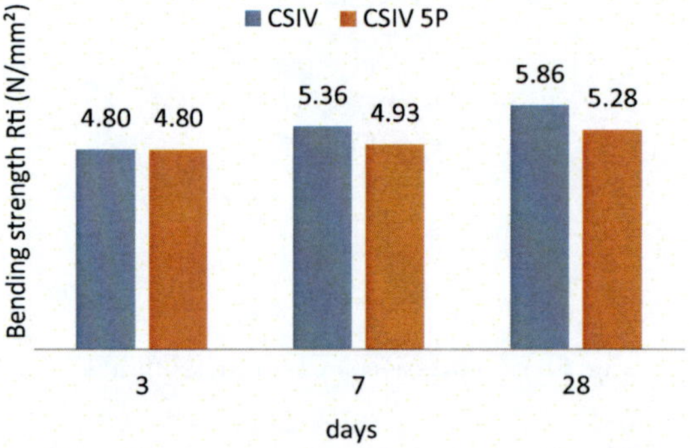

Fig. 6. Comparison of bending strength values CSIV - CSIV 5P [N/mm^2]

Replacing water with a percentage of 10% used paint, according to the CSIV 10P plaster mortar recipe, causes changes in the mortar structure and you can see the increase in bending strength compared to the mortar in the CSIV recipe by 0.25 N/mm^2 (See Fig. 7 and Table 3).

Also, by replacing the water with 15% used paint, according to the CSIV 15P plaster mortar recipe, you can see the increase in bending strength compared to the CSIV recipe mortar by 0.35 N/mm^2 (See Fig. 8 and Table 3), at 28 days, but not every 3 and 7 days.

By replacing the water with 20% used paint, according to the CSIV 20P plaster mortar recipe, you can see the increase in bending strength compared to the CSIV recipe by 2.72 N/mm^2 (See Fig. 9 and Table 3), at 28 days, although at 3 and 7 days there is a slight decrease. Given the purpose and scope of use of the material as a plaster mortar, the values at 28 days and later are important.

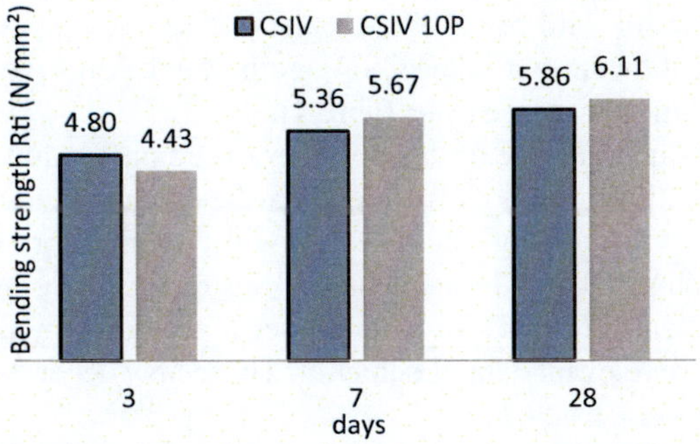

Fig. 7. Comparison of bending strength values CSIV - CSIV 10P [N/mm^2]

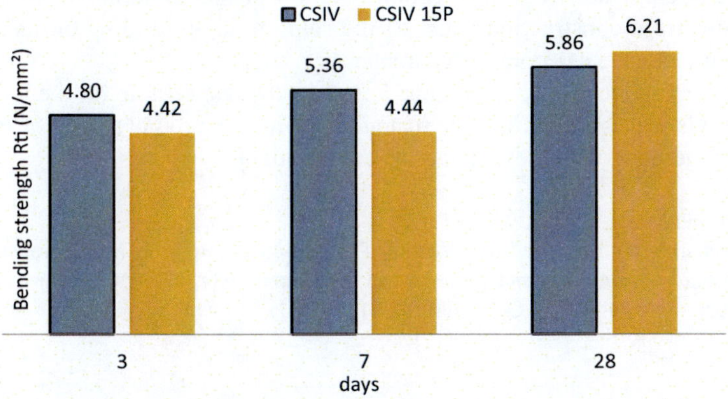

Fig. 8. Comparison of bending strength values CSIV - CSIV 15P [N/mm^2]

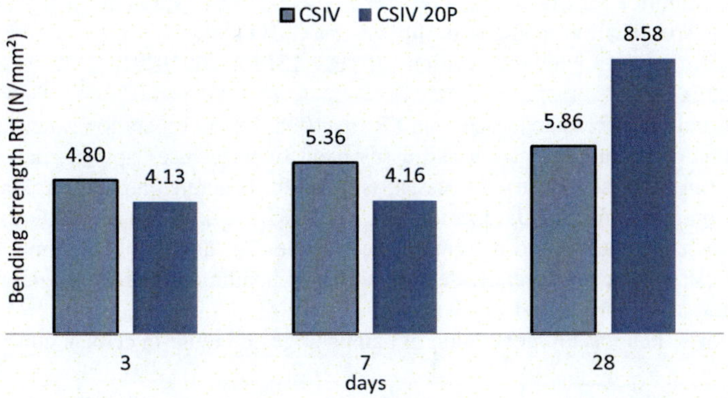

Fig. 9. Comparison of bending strength values CSIV - CSIV 20P [N/mm^2]

By replacing the water with 5%, 10%, 15% and 20% used paint, a decrease in the compressive strength of the mortar in the CSIV recipe can be observed, especially in the case of the CSIV 15P mortar recipe (See Table 4).

The results highlighted the role of this polymer - used paint equivalent to the role of a fluidizing additive, influencing both fresh and hardened characteristics.

The results are very promising, the research will continue through another experimental program through which the cement will be replaced by used paint, and the evaluation of mechanical properties to extend over a period of more than 28 days, so that there are even more relevant data of behavior. These new decorative plaster mortars, over time.

An investigation by SEM electron microscopy will be even more edifying explaining how the structure of the mortar is modified by using used paint, at the same time intervals as in previous studies.

The application of the concept of sustainable development, based on the idea of waste management, leads us to the creation of a new plaster material, to which we can use used paint from construction sites. At the same time, depending on its color, the plaster mortar can have a decorative character.

Man, as a component of society, must know and be responsible for the fact that the resources of this planet are limited and that methods must be found to use waste through efficient management, while respecting the environment.

Acknowledgements.
This work was supported by the Post-Doctoral Programme "Entrepreneurial competences and excellence research in doctoral and postdoctoral programs - ANTREDOC", project co-funded from the European Social Fund POCU/380/6/13/123927.

References

1. European Committee for Standardization: SR EN 197-1: 2011 Cement Part 1: Composition, specifications and conformity criteria for common cements (2011)
2. European Committee for Standardization: SR EN 413-1: 2011 Cement for masonry. Part 1: Composition, specifications and conformity criteria (2011)
3. European Committee for Standardization: SR EN 13139: 2003/C91: 2009 Aggregates for mortars (2009)
4. European Committee for Standardization: SR EN 1008: 2003 Preparation water for concrete. Specifications for sampling, testing and assessment of water use capacity, including water recovered from concrete industry processes, as water for concrete preparation (2003)
5. European Committee for Standardization: SR EN 1015 1 ÷ 21/2020 on Methods for testing mortars, which include how to determine the particle size distribution, sampling, determination of fresh mortar consistency, determination of bending strength of hardened mortar, determination of compressive strength (2020)
6. Feng, E., Sun, J. Feng, L.: Regeneration of paint sludge and reuse in cement concrete (2018)

Influence of Iron Trioxide Addition on Alkali-Activated Fly Ash-Based Geopolymer Paste

Brăduț Alexandru Ionescu[1,2](✉), Mihail Chira[1], Adrian-Victor Lăzărescu[1,2], and Carmen Florean[1]

[1] NIRD URBAN-INCERC Cluj-Napoca Branch, 117 Calea Florești, 400524 Cluj-Napoca, Romania
bradut.ionescu@incerc-cluj.ro

[2] Faculty of Civil Engineering, Technical University of Cluj-Napoca, 15 Constantin Daicoviciu Street, 400020 Cluj-Napoca, Romania

Abstract. This paper presents the study conducted in order to study the influence of the use of iron trioxide as an addition to an alkaline activated geopolymer paste on its physical and mechanical performance. The experimental test followed the variation of the flexural and compressive strength of alkali-activated geopolymer pastes at the age of 7 days, which takes place after the introduction as an addition to the preparation of 1%, 5% and 10% (mass percent) iron trioxide powder and compared to the control sample performed without addition. The experimental results showed that the flexural strength increased by 7.2% for a 1% addition of iron trioxide to fly ash, by 30.7% and by 33.7% for cases of addition of iron trioxide 5%, respectively 10% to fly ash. Test regarding the compressive strength revealed an increase of 21.7% for a 1% addition of iron trioxide to fly ash, by 15.3% and 14.8% respectively for cases of addition of iron trioxide 5% and 10% respectively.

Keywords: Geopolymer paste · Alkali-activation · Iron trioxide powder

1 Introduction

In the last years, worldwide, as a result of increasing pollution, there has been the issue of environmental protection, preservation of non-renewable natural resources and the identification of minimally polluting or non-polluting technologies in all areas of activity.

In line with current development strategies, sustainable development and implementation of the concepts of the Circular Economy, the construction industry must align itself with innovative technologies to achieve alternative, environmentally friendly construction materials. This is urgently needed because the construction industry is the fastest growing in the world, with current world statistics indicating that 260 million tons of cement are needed annually [1]. On the other hand, the cement industry is responsible for producing CO_2 emissions because, for the production of one ton of Portland cement, about one ton of CO_2 is released into the atmosphere [2]. An alternative to traditional

L. Moldovan and A. Gligor (Eds.): Inter-Eng 2021, LNNS 386, pp. 139–149, 2022.
https://doi.org/10.1007/978-3-030-93817-8_15

concrete is represented by geopolymer ecological concretes based on alkaline activated mineral binders.

The term of geopolymer was originally introduced by Davidovits [3], representing a wide range of inorganic materials. According to Davidovits [4], nine different classes of geopolymers are defined, geopolymer concrete, with aluminosilicate structure, of particular interest because it could be used to completely replace traditional concrete. In 1999, Palomo identified the possibility of activating pozzolanic materials such as blast furnace slag and power plant ash "using alkaline liquids to form a binder and completely replace the use of Portland cement in concrete production" [5].

The geopolymer is an inorganic polymer composed of two main constituents, namely source materials of aluminosilicates and alkaline activating solutions [6]. In general, sources of aluminosilicates containing high amounts of silicon, alumina and calcium oxide are used in the preparation of geopolymers. The combination of sodium silicate (Na_2SiO_3) and sodium hydroxide (NaOH) has been widely used as an alkaline activator in liquid form [7]. Currently, experimental research has shown that the technology of making geopolymer materials offers the possibility of using industrial waste as raw materials, for example, fly ash, thus responding to the concepts of the Circular Economy. Thus, reports from the literature indicated the obtaining of geopolymer materials which, among other types of cementitious materials fall within the research interest worldwide [8].

The relative types and quantities of coal incombustible matter determine the chemical composition of the fly ash. The chemical composition is composed mainly of oxides of silicon (SiO_2), aluminum (Al_2O_3), iron (Fe_2O_3) and calcium (CaO), while magnesium, potassium, sodium, titanium and sulfur are present in a smaller amount. The major influence on the chemical composition of flying ash comes from the type of coal. However, recent studies have shown that precursors with a higher iron content than those commonly found in fly ash can be activated in an alkaline environment [9–11] with applications in engineering.

Although the presence of iron trioxide (Fe_2O3) could play important roles in the structure and properties of geopolymers, the replacement of aluminum trioxide with iron trioxide has not yet been fully studied.

The compression strength of geopolymers generally increases with increasing specific concentration of alkaline activators [12, 13]. A higher concentration gives rise to a stronger ion pair formation and ensures a more complete and faster poly-condensation process of the particle interface [14] improving the dissolution of silicon and aluminum-containing materials in the presence of activators [15]. However, too high a concentration could lead to an increase in the coagulated structure [16], resulting in lower workability with rapid hardening behavior [17]. According to the literature, the most used NaOH concentrations are 6, 8 and 10 M. The optimum alkaline concentration can also vary depending on a large number of conditions and factors, such as the specific properties of the raw materials, the alkaline activator/raw material ratio, the Na_2SiO_3/NaOH ratio, the curing temperature, or even the age of testing.

Research on geopolymer paste and mortars with heat treatment (between 30 and 90 °C) has shown an increase in the chemical reaction, leading to an increase in compression strength from an early age [18]. Other researchers [19] claim that too high a heat treatment temperature (for example above 90 °C) would lead to a geopolymer with a porous structure due to rapid water loss, causing a negative effect on mechanical properties of the geopolymer.

The optimum temperature of the heat treatment proved to be around 60 to 75 °C, which could accordingly improve the geopolymerization process and an adequate microstructural development [20]. The testing age of geopolymers, according to the literature was 3, 7, 14, 28, 56 and 90 days. However, most reports showed that geopolymer samples gain their strength at the age of 7 days, due to heat treatment [21].

The aim of this paper is to present the study on the influence of the use of iron trioxide as an addition to an alkaline activated geopolymer paste on its physical and mechanical performance. The experimental test followed the variation of the tensile strength by bending and the compression strength of alkaline activated geopolymer pastes at the age of 7 days, which takes place after the introduction as an addition to the preparation of 1%, 5% and 10% (mass percent) trioxide iron powder compared to the control sample performed without addition.

2 Materials and Methods

2.1 Raw Materials

For the design and production of the alkali-activated geopolymer paste samples, the following raw materials were used: fly ash (FA), iron trioxide Fe_2O_3, sodium hydroxide solution NaOH(6 M), sodium silicate solution (Na_2SiO_3) 35% (Fig. 1).

The ash that was used in this study was obtained from Rovinari power-plant, Romania and its chemical composition was established through X-ray fluorescence analysis (XRF analysis) (Table 1). Iron trioxide Fe_2O_3 in powder form was used as an additive. After dry mixing the fly ash with iron trioxide, an alkaline activating solution was added to make the geopolymer paste. A mixture consisting of 35% sodium Na_2SiO_3 silicate solution and a sodium silicate solution (NaOH, 6 M) was chosen as alkaline activator, with mass ratio Na_2SiO_3: NaOH = 2:1. The sodium hydroxide solution was prepared dissolving the NaOH pearls, 99% purity; into water in order to obtain the desired concentration of the solution.

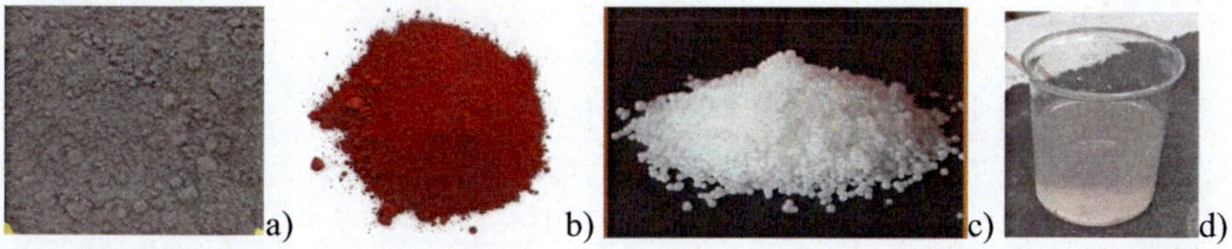

Fig.1. Raw materials used in the production of the alkali-activated geopolymer samples: (a) fly ash; (b) Iron trioxide; (c) NaOH; (d) Na_2SiO_3 solution.

Table 1. Chemical composition of the fly ash used in the production of the alkali-activated geopolymer samples.

Rovinari fly ash										
Oxides	SiO_2	Al_2O_3	Fe_2O_3	CaO	MgO	SO_3	Na_2O	K_2O	P_2O_5	TiO_2
Wt (%)	46.94	23.83	10.08	10.72	2.63	0.45	0.62	1.65	0.25	0.92

2.2 Mix-Design, Molding and Curing Treatment

Preliminary results obtained on geopolymer paste, using the same type of fly ash, led to an alkaline activator/fly ash ratio of 0.78, because it has shown good workability of the alkali-activated fly ash-based geopolymer paste. In order to study the effect of Fe_2O_3 addition on the mechanical properties of the material, the following mix-design ratios were produced:

- R21: fly ash + solution [35% Na_2SiO_3 + NaOH (6M)], with a Na_2SiO_3/NaOH ratio = 2.0;
- R22: fly ash + 1% Fe_2O_3 + solution [35% Na_2SiO_3 + NaOH (6M)], with a Na_2SiO_3/NaOH ratio = 2.0;
- R23: fly ash + 5% Fe_2O_3 + solution [35% Na_2SiO_3 + NaOH (6M)], with a Na_2SiO_3/NaOH ratio = 2.0;
- R24: fly ash + 10% Fe_2O_3 + solution [35% Na_2SiO_3 + NaOH (6M)], with a Na_2SiO_3/NaOH ratio = 2.0.

The quantities of raw materials used to produce the alkali-activated geopolymer samples are presented in Table 2.

Table 2. Mix-design ratio of the alkali-activated geopolymer samples.

Mixt.	Fly ash		Fe_2O_3		Fly ash + Fe_2O_3	Alkaline activator			Na_2SiO_3/NaOH	Alkaline activator solution /(FA + Fe_2O_3)
						Na_2SiO_3	NaOH	Total		
	[%]	[g]	[%]	[g]	[g]	[g]	[g]	[g]	–	–
R21	100	267	0	0	267.00	138.7	69.3	208.0	2.0	0.78
R22	100	267	1	2.67	269.67	140.0	70.0	210.0		
R23	100	267	5	13.35	280.35	145.6	72.8	218.4		
R24	100	267	10	26.70	293.70	152.5	76.3	228.8		

After the preparation of the alkaline activator by combining the sodium silicate solution with the sodium hydroxide solution 24 h prior to mixing, the fly ash and/or the Fe_2O_3 addition and the activator were mixed together until a homogeneous mixture was obtained. The samples were then placed in 40 mm × 40 mm × 160 mm molds and heat cured at 70 °C for 24 h (Fig. 2). During the heat treatment a film was put on top of the molds in order to prevent the quick release of the water from the mixtures. After demolding, the samples were kept in the climatic chamber at (20 ± 1) °C and (60 ± 5)% humidity until the test were conducted (Fig. 3).

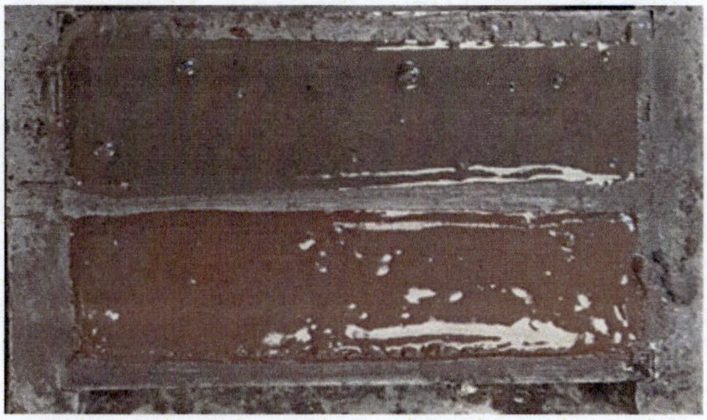

Fig. 2. Alkali-activated geopolymer fresh-state samples: without Fe_2O_3 addition (above); with 1% Fe_2O_3 addition (below).

Fig. 3. Alkali-activated geopolymer samples after demolding: (a) after heat treatment and (b) prior to the 7 days testing.

2.3 Testing Methods

The apparent density in the fresh state was evaluated according to EN 12,350-1, as the ratio between mass and volume immediately after the mixing finished. The apparent density in hardened state was evaluated according to EN 12,390-7, right after the demolding of the samples and at the age of 7 days.

To obtain the result for the mechanical properties of the samples, for all the alkali-activated geopolymer pastes, minimum three samples were tested to determinate the average value.

The flexural strength (Fig. 4 and Fig. 5) and the compressive strength of the samples were determined according to EN196-1, the standard used for ordinary Portland cement mortars.

Fig. 4. Alkali-activated geopolymer samples during flexural strength test.

Fig. 5. Alkali-activated geopolymer samples after flexural strength test.

The percentage increases of the flexural and compressive strength of the samples with the addition of iron trioxide were calculated by comparing them to the control sample (R21), without the addition of iron trioxide.

3 Results and Discussions

Iron-containing fly ash takes time to react with the alkaline solution in the geopolymer system to form the iron silicate binder gel. This mainly happens due to the high atomic mass and large atomic diameter of the iron contained in silica and alumina [22]. Cannio [23] identified ferro silicate in addition to the sodium aluminum silicate gel in the geopolymer, which contributes to the durability of the geopolymer. Thus, the longer curing time led to an increase in the compressive strength to which iron oxide contributed.

3.1 Apparent Density

The results regarding the apparent density in the fresh and hardened state of the alkali-activated geopolymer samples are presented in Table 3 and Fig. 6.

Table 3. Apparent density of the alkali-activated geopolymer samples.

Mixture	Fly ash	Fe_2O_3	$Na_2SiO_3/NaOH$	$\rho_{fresh\ state}$	ρ_{2days}	ρ_{7days}
–	[%]		–	[kg/m^3]		
R21	100	0	2.0	1580	1380	1290
R22	100	1		1620	1390	1330
R23	100	5		1640	1450	1350
R24	100	10		1670	1510	1380

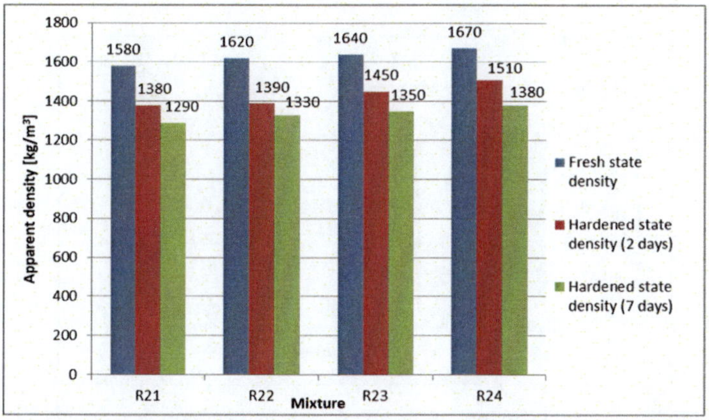

Fig. 6. Apparent density of the alkali-activated geopolymer samples.

In terms of apparent density in fresh state. it is observed that, in relation to the composition considered as control, R21, by the addition of iron trioxide (Fe_2O_3) in percentages of 1, 5 and 10% (compositions R22, R23, R24) to the fly ash, the apparent density in the fresh state increases on as the percentage of Fe_2O_3 added increases. Thus, there were increases of 2.2% for a 1% addition of iron trioxide to fly ash, 3.3% and 5.4% for cases of addition of 5% iron trioxide, respectively 10% to fly ash.

The same behavior was observed for both the 2 and 7 days apparent density in hardened state, with an increase of 0.2% for an addition of 1% iron trioxide to fly ash, by 5.2% and 8.7% for cases of addition of iron trioxide 5% and 10%, respectively (2 days) and 3.1% for an addition of 1% of iron trioxide to fly ash, by 4.5% and 6.7% respectively for cases of addition of iron trioxide 5% and 10%, respectively (7 days).

3.2 Mechanical Properties of the Alkali-Activated Geopolymer Samples

The results regarding the mechanical properties of the alkali-activated geopolymer samples are presented in Table 4 and Fig. 7 (R_{ti} - flexural strength) and Fig. 8 (R_c - compressive strength).

Table 4. Mechanical properties of the alkali-activated geopolymer samples.

Mixture	Fly ash	Fe_2O_3	$Na_2SiO_3/NaOH$	R_{ti}	R_c
–	[%]		–	[N/mm^2]	[N/mm^2]
R21	100	0	2.0	2.401	18.19
R22	100	1		2.574	22.13
R23	100	5		3.190	21.57
R24	100	10		3.475	21.39

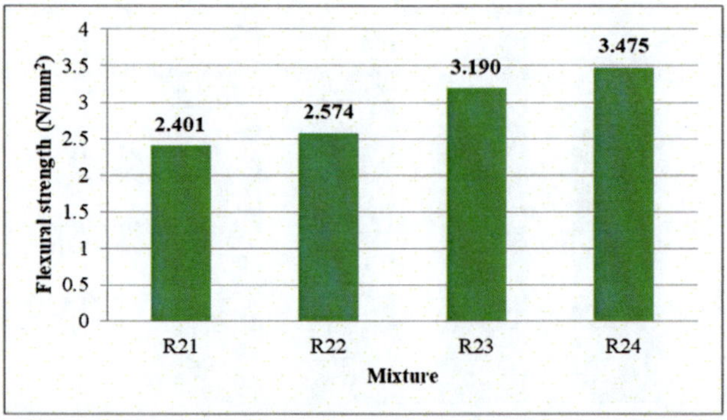

Fig. 7. Flexural strength of the alkali-activated geopolymer samples.

From Table 4 and Fig. 7 it is observed that the flexural strength of the samples increased linearly with the increase of the addition of iron trioxide. Thus, there were increases of 7.2% for an addition of 1% iron trioxide to fly ash, by 32.9% and 44.7% for cases of addition of iron trioxide 5% and 10%, respectively.

The compressive strength of the samples at 7 days increased by the introduction of the addition of iron trioxide. Thus, there were increases by 21.7% for an addition of 1% of iron trioxide to fly ash, by 18.6% and by 17.6% for cases of addition of iron trioxide 5% and 10%, respectively.

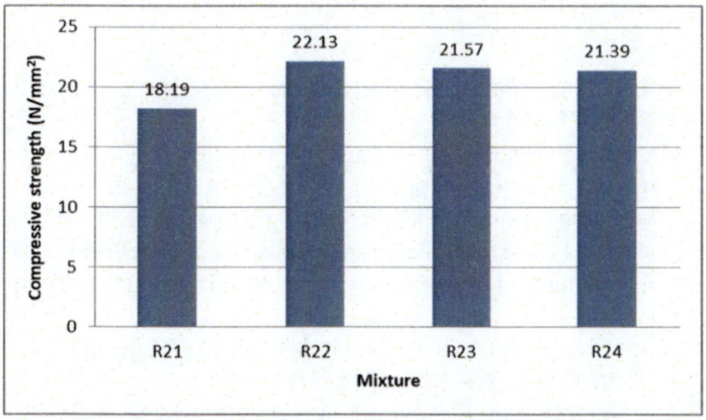

Fig. 8. Compressive strength of the alkali-activated geopolymer samples.

4 Conclusions

Studies have shown that precursors with a higher than usual iron content found in fly ash can be activated in an alkaline environment, a conclusion that is consistent with the literature. This paper aimed to investigate the implications of the presence of iron in the geopolymer structure on preliminary alkali-activated fly ash-based geopolymer materials.

By the addition of iron trioxide in proportions of 1%, 5% and 10% to alkaline activated geopolymer pastes based on fly ash, it was observed that the mechanical properties of the material increased, both in terms of flexural strength and also, compressive strength.

Flexural strength of the alkali-activated geopolymer samples increased with the addition of 1%, 5% and 10% (mass percent) iron trioxide powder compared to the control sample without addition. Thus, there were increases of 7.2% (Mixture R22) for a 1% addition of iron trioxide to fly ash, by 32.9% (Mixture R23) and by 44.7% (Mixture R24) for cases of addition of iron trioxide 5% and 10% respectively to fly ash.

The best compressive strength at 7 days was obtained by adding 1% iron trioxide to fly ash. The mechanical strength for this mixture R22 increased by 21.7% compared to the control sample R21, without the addition of iron trioxide. For mixtures R23 (5% Fe_2O_3 addition) and R24 (10% Fe_2O_3 addition) the compression strengths at the age of 7 days increased by 18.6% and 17.6% respectively compared to the R21 control sample, without the addition of iron trioxide.

Further studies will focus on studying the influence of Fe_2O_3 addition in the production of alkali-activated fly ash-based geopolymer paste by varying several parameters that can affect the mechanical properties of the samples (i.e. raw material ratios in the mix-design) and study the possibility of transition to geopolymer mortars, by adding aggregates into the mixtures.

Acknowledgements. This paper was financially supported by the Project "Entrepreneurial competences and excellence research in doctoral and postdoctoral programs - ANTREDOC", project co-funded by the European Social Fund financing agreement no. 56437/24.07.2019.

References

1. Abdul Aleem, M.I., Arumairaj, P.D.: Geopolymer concrete - a review. Int. J. Eng. Sci. Emerg. Technol. **1**(2), 118–122 (2012)
2. Davidovits, J.: High-alkali cements for 21st century concretes. In: Concrete Technology, Past, Present and Future, ACI SP, vol. 144, pp. 383–397 (1994)
3. Davidovits, J.: Geopolymer, green chemistry and sustainable development solutions. In: Proceedings of the World Congress Geopolymer 2005. Geopolymer Institute (2005)
4. Davidovits, J.: Geopolymers: inorganic polymeric new materials. J. Therm. Anal. **37**(8), 1633–1656 (1991)
5. De Silva, P., Sagoe-Crenstil, K., Sirivivatnanon, V.: Kinetics of geopoly-merisation: role of Al_2O_3 and SiO_2. Cem. Concr. Res. **37**(4), 512–518 (2007)
6. Zainal, F.F., Fazill, M.F., Hussin, K., Rahmat, A., Abdullah, M.M.A., Wazien, W.: Effect of geopolymer coating on mild steel. Solid State Phenom. **273**, 175–180 (2018)
7. Davidovits, J.: Properties of geopolymer cements. In: Proceedings of 1st International Conference on Alkaline Cements and Concrete, pp. 131–150. VIPOL Stock Comp., Kiev (1994)
8. Toader, T.P., Mircea, A.C.: Self-healing concrete mix-design based on engineered cementitious composites principles. In: Multidisciplinary Digital Publishing Institute Proceedings, vol. 63, no. 1, p. 5 (2020)
9. De Barros, S., De Souza, J.R., Gomes, K.C., Sampaio, E.M., Barbosa, N.P., Torres, S.M.: Adhesion of geopolymer bonded joints considering surface treatments. J. Adhes. **88**(4–6), 364–375 (2012)
10. Rego, S.R., Gomes, K.C., Rosas, M., Torres, S.M., De Barros, S.: Application of geopolymeric adhesives in ceramic systems subjected to cyclic temperature environments. J. Adhes. **90**, 120–133 (2014)
11. Komnitsas, K., Zaharaki, D.: Geopolymerisation: a review and prospects for the minerals industry. Mater. Eng. **20**, 1261–1277 (2007)
12. Hardjito, D., Fung, S.: Fly ash-based geopolymer mortar incorporating bottom ash. Mod. Appl. Sci. **4**(1), 44–52 (2010)
13. Xu, H., van Deventer, J.S.J.: The Geopolymerisation of alumino-silicate minerals. Int. J. Miner. Process. **59**(3), 247–266 (2000)
14. Raijiwala, D., Patil, H.: Geopolymer concrete: a green concrete. In: 2nd International Conference on Chemical, Biological and Environmental Engineering (ICBEE 2010), pp. 202–206. IEEE, Cairo (2010)
15. Mishra, A., Choudhary, D., Jain, N., Kumar, M., Sharda, N., Dutta, D.: Effect of concentration of alkaline liquid and curing time on strength and water absorption of geopolymer concrete. ARPN J. Eng. Appl. Sci. **3**(1), 14–18 (2008)
16. Alonso, S., Palomo, A.: Alkaline activation of metakaolin and calcium hydroxide mixtures: influence of temperature, activator concentration and solids ratio. Mater. Lett. **47**(1), 55–62 (2001)
17. Memon, F., Nuruddin, M.F., Khan, S.H., Shafiq, N.R.: Effect of sodium hydroxide concentration on fresh properties and compressive strength of self-compacting geopolymer concrete. J. Eng. Sci. Technol. **8**(1), 44–56 (2013)
18. Van Jaarsveld, J., Van Deventer, J., Lukey, G.: The effect of composition and temperature on the properties of fly ash-and kaolinite-based geopolymers. Chem. Eng. J. **89**(1), 63–73 (2002)
19. Rovnaník, P.: Effect of curing temperature on the development of hard structure of metakaolin-based geopolymer. Constr. Build. Mater. **24**(7), 1176–1183 (2010)
20. Chindaprasirt, P., Chareerat, T., Hatanaka, S., Cao, T.: High-strength geopolymer using fine high-calcium fly ash. J. Mater. Civ. Eng. **23**(3), 264–270 (2010)

21. Lăzărescu, A., Szilagyi, H., Ioani, A., Baeră, C.: Parameters affecting the mechanical properties of fly ash-based geopolymer binders-experimental results. In: IOP Conference Series: Materials Science and Engineering, vol. 374, p. 012035 (2018)
22. Zaillani, W.W.A., Abdullah, M.M.A., Arshad, M.F., Burduhos-Negris, D.D., Tahir, M.F.M.: Effect of Iron Oxide (Fe_2O_3) on the properties of fly ash based geopolymer. In: IOP Conference Series: Materials Science and Engineering, vol. 877, p. 012017 (2020)
23. Cannio, M., Cyriaque, R., Myllyam, L., Rosa, R., Kamseu, E., Chinje, U., Leonelli, C.: Microstructure and engineering properties of Fe_2O_3(FeO)-Al_2O_3-SiO_2 based geopolymer composites. J. Clean. Prod. **199**, 849–859 (2018)

Cementitious Composite Materials with Self-healing Properties Using Integral Waterproofing Admixtures by Mass Crystallization

Tudor Panfil Toader[1,2(✉)], Carmen Dico[1], and Călin Mircea[2]

[1] NIRD URBAN-INCERC Cluj-Napoca Branch, 117 Calea Florești, Cluj-Napoca, Romania
tudor.toader@incerc-cluj.ro

[2] Faculty of Civil Engineering, Technical University of Cluj-Napoca, 15 Constantin Daicoviciu Street, 400020 Cluj-Napoca, Romania

Abstract. Cementitious composite materials are the most used material in the realization of transport infrastructures. In order to increase the service life of these structures, it is necessary to design cementitious composite materials with special properties, such as the self-healing capacity, which allows the occurrence of a self-closing phenomenon after cracking. This paper shows the performance recorded for two cementitious composites with self-healing capacity induced by the use of an integral waterproofing admixture by mass crystallization. The experimental results obtained indicate a good self-healing capacity, quantified by a degree of healing of at least 57% after 192 h of conditioning, respectively, total closure of cracks after 336 h of conditioning. This work contributes to the increase of knowledge in the field of cementitious materials with self-healing capacity, indicating a possibility of obtaining this effect through the use of a waterproofing admixture by mass crystallization, simultaneously with the presentation of the possibilities of use of industrial waste such as fly ash and limestone slurry.

Keywords: Self-healing · Waterproofing by mass crystallization · Durability

1 Context

The life of building elements produced using cementitious composite materials is influenced by the external factors to which they are exposed and by current mechanical stresses or accidental overloads in operation, to which they are subjected. As a result of this combination of factors, the building elements made of cementitious compounds end up cracking, thus allowing the corrosive agents to penetrate the reinforcement, which corrodes it. Following the cracking of the cover represented by the cementitious composite, access routes (microcracks/cracks) are created towards the reinforcement. Through these cracks/microcracks, the corrosive agents from the outside come into contact with the reinforcement producing corrosion, which, over time, leads to the loss of the bearing capacity of the reinforcement, respectively of the structural elements because the reinforcement has the role of taking over the bending efforts that occur in the elements.

L. Moldovan and A. Gligor (Eds.): Inter-Eng 2021, LNNS 386, pp. 150–165, 2022.
https://doi.org/10.1007/978-3-030-93817-8_16

The use of self-healing cementitious materials improves the durability performance and extends life cycle of the elements, without requiring costly and time-consuming repair of cracks. Although cracking cannot be avoided over the life of a structure, the ability of certain materials to heal these cracks makes them extremely valuable materials for the construction industry.

This phenomenon is called "self-healing (SH)". Autogenously crack healing phenomena can occur largely within cement-based materials through four mechanisms, including (1) calcite formation (calcium carbonate, $CaCO_3$), (2) continuous hydration of the cement matrix upon contact with moisture, (3) swelling of the cementitious matrix, and (4) particle sedimentation in the crack [1, 2]. SH results in the recovery of certain mechanical or durability properties of cement paste. Over the past two decades, numerous research studies have been conducted to establish the various characteristics of natural healing in cement-based materials [3, 4]. Although the concepts and mechanisms of autogenic and autonomous healing have been defined and are known, from a design and application perspective, it is necessary to evaluate the effectiveness of various self-healing technologies based on the intention of practical application [1, 5–11]. SH intrinsic has limited applications due to its limited crack closure capability (~200–300 μm) and the fact that the phenomenon occurs conditionally [12–14].

From the literature it is known that the self-healing mechanism can be initiated in: (1) concrete with mineral additions, (2) concrete with magnesium oxide addition, (3) improved concrete with crystalline mixtures, (4) reinforced concrete with high performance fibers improved with crystalline mixtures, (5) concrete with pre-placed microcapsules containing polymeric healing agent and (6) concrete with encapsulated bacteria [1, 15].

The general approach to research methodology indicates that methods of quantitative assessment of the self-healing capabilities of various technologies are needed. In particular, since cracks have a direct impact on durability performance rather than mechanical properties, it is very important to evaluate the recovery of durability performance through self-healing. In general, crack healing was evaluated using non-destructive testing (NDT) or microstructure analysis [16]. NDT, which has been performed using radiation testing [17], acoustic emission [18], ultrasonic testing [19, 20] and image analysis [21], is a relatively simple process, but has a low degree of reliability and limited applicability in the direct evaluation of mechanical or durability performance. Alternatively, water permeability tests were also used in the direct evaluation of self-healing durability perforation recovery [22] and, concurrently with the water absorption test [23], accelerated chloride diffusion tests [24, 25].

One of the methods of inducing self-healing ability is the addition of discontinuous and randomly dispersed fibers that can control the formation and limit the growth of cracks inside the matrix due to the bridging effect. They also provide sufficient support and time for any type of SH mechanisms to emerge later.

Among these synthetic fibers, polypropylene (PP) fibers are commonly used for reinforcing cement-based materials and have attracted the attention of researchers due to their relatively low cost and weight, being inert in a high pH environment, resistant to corrosion and cracking from shrinkage, allow easy mass dispersion, have high melting point and chemical stability [1, 26–28]. Studies have shown that the inclusion of fibers results in improved post-cracking behavior as well as less brittle concrete behavior.

Also, for the purpose of inducing self-healing capacity, the researchers investigated the effectiveness of introducing specific crystalline waterproofing admixtures (CWA) into concrete, in varying proportions, depending on the water-binder ratio. Since the durability of concrete is one of the most important properties, both financially and in terms of sustainability, reducing water penetration as much as possible should be a priority. A mix of the concrete, properly engineered, produced, and performed properly, with a low ratio of water-binding agent, resulting in a final product with low permeability and high durability [29, 30], but according to the literature it is recommended to be used in the design and preparation of the concrete has a water-cement ratio greater, because the high content of water in the mixture promotes the self-healing [22]. Azarsa et al. specifies that each type of CWA produces different crystals responsible for reducing the penetration of water and the occurrence of the self-healing phenomenon of concrete [31].

With regard to self-healing, the opening of cracks that can heal completely varies, depending on the mode of induction of this property, from 0.1 mm [31–33] to 0.3 mm or even 0.4 mm [31, 34]. As for other properties, CWA has been reported to improve resistance to freeze-thaw cycles, reduce chloride ion penetration [35], improve sulfate attack resistance and do not significantly affect the compressive strength of concrete [36].

The aim of the research activity carried out and presented in this paper is to evaluate the self-healing capacity of two cementitious composite mixtures produced using a waterproofing admixture of crystallization in mass, compared to a control mixture (produced without using this type of admixture).

2 Materials and Methods

The materials used for the preparation of the cementitious composite mixtures were: Portland cement EN 197-1-CEM I 42.5 R, produced at Aleşd cement factory, Bihor County, Romania; fly ash from the thermal power plant resulting from the burning of coal to obtain electricity at Mintia thermal power plant, Hunedoara, Romania; washed river aggregates (0–4) mm granular class; limestone slurry, resulting from the cutting of marble rock taken as a by-product of a marble processing plant in Cluj-Napoca, Romania, superplasticizing admixture type Master Glenium 51 (BASF), integral waterproofing admixture by crystallization (CWA), PVA fibers, 8 mm and water. The preparation of the mixtures was carried out in the laboratory NIRD URBAN-INCERC Cluj-Napoca Branch.

The research methodology consisted in following the following steps:

1. compositional design (according to Table 1) of two cement compositions with self-curing capacity (T1 and T2) and a control composition (TM);
2. the maturation of the specimens (reaching the age of 28 days after casting) was carried out by keeping in the first 24 h after casting, in metal patterns, at a constant temperature (20 ± 2) °C and relative humidity min. 90%, and subsequently, after

Table 1. Mix-design ratios for TM, T1, T2.

Mixtures	CEM [2]	FA [3]	Sand	Water	PVA	LS [4]	CWA[5]	Superplasticizer
TM	1,00	1,12	0,82	0,56	0,04	0,24	–	0,02
T1	1,00	1,11	0,82	0,56	0,04	0,24	0,01	0,02
T2	1,00	1,10	0,82	0,56	0,04	0,24	0,02	0,02

values reported to cement quantity, [2] CEM-cement; [3] FA-fly-ash, [4] LS-limestone slurry, [5] CWA – crystalize waterproof admixture

demolding, by immersing water with temperature (20 ± 1) °C, until the age of 28 days;

3. upon reaching the age of 28 days after casting, the tests for determination of compressive strength (R_c) and bending strength (R_{ti}) were performed, for which the results indicated in Table 2, expressed in N/mm^2, were obtained.

Table 2. Mix-design ratios for TM, T1, T2.

Mixtures	R_c	R_{ti}
TM	57.6	15.8
T1	58.2	16.5
T2	56.6	16.3

According to the mean value of the bending tensile strength, the cracking force required to induce the controlled cracking was determined, representing (87 ± 1) % of the mean bending tensile strength, respectively, P = 6000 [N].

4. evaluation of self-healing capacity by tracking the degree of closure of cracks. For this purpose, through microscopic evaluation, crack openings were measured, initially and subsequently, during exposure to conditioning. Conditioning in order to induce the self-curing effect consisted in exposure to alternative wet-dry cycles, a cycle consisting of immersion 16 h in water (20 ± 1) °C and 8 h kept in dry environment at (23 ± 2) °C.

For a better evaluation, the surface of the prisms were divided into several evaluation areas, depending on the cracking mode, these areas being kept throughout the evaluation process.

The mean healing degree ($GV^m{}_t$) was calculated as a percentage reduction in the mean crack opening at time t, compared to the mean crack opening initially recorded at the time of the crack, Eq. (1):

$$GV_t^m = (D_t - D_0) * 100/D_0(\%) \quad (1)$$

The degree of healing of the maximum crack opening ($GV^{M}{}_{t}$) was calculated as a percentage reduction of the maximum crack opening at time t, compared to the maximum crack opening initially recorded at the time of the crack, Eq. (2):

$$GV_t^M = (D_t - D_0) * 100/D_0(\%) \quad (2)$$

The mean moment healing degree ($GV^{Mm}{}_{t}$) was calculated as a reduction in the mean opening of the crack at time t, compared to the mean opening of the crack recorded in the previous stage, (t − 1), in relation to the length of time spent on conditioning, Eq. (3):

$$GV_t^{Mm} = (D_t - D_{t-1})/(t - (t-1))\ (\mu m/h) \quad (3)$$

Where: D_0 - mean crack opening, measured immediately after the crack (mm); D_t - mean crack opening measured after t hours of conditioning (mm); D_{t-1} - the average opening of the crack, measured in the previous step (mm); t – conditioning time (24 h; 96 h; 192 h; 336 h; 480 h).

3 Results and Discussions

Evaluation of the self-healing capacity for the projected mixtures T1, T2 and TM following microscopic crack analysis and to evaluate the opening of cracks at set time intervals, namely: at the time of cracking at 24 h, 92 h, 192 h, 336 h and 480 h, the surface of the prism was divided into 9 zones for the composition T1, 20 zones for the composition T2 and 16 zones for the composition TM.

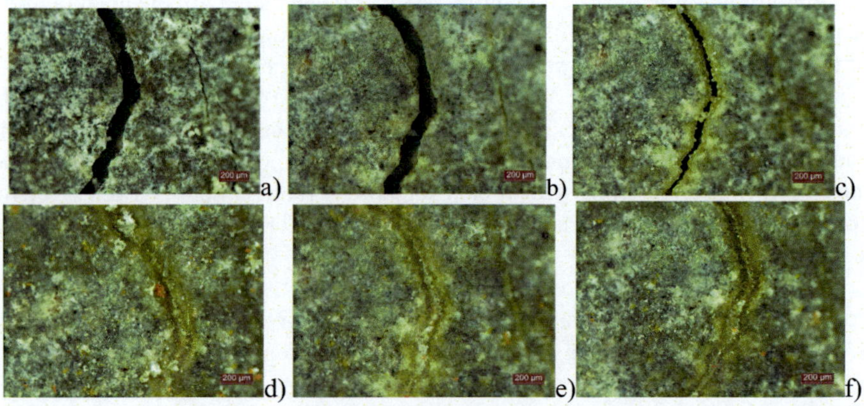

Fig. 1. Example of evolution of self-curing phenomenon for T1 composition: a) induced crack appearance; b) crack appearance after 24 h conditioning for self-curing; c) crack appearance after 96 h self-healing conditioning; d) self-healing crack appearance after 192 h self-healing conditioning; e) self-healing crack appearance after 336 h self-healing conditioning; f) self-healing crack appearance after 480 h self-healing conditioning.

After the initial analysis of the cracks present on the sample composition T1, T2 and TM, they were subjected to wet-dry cycles (16 h wet – 8 h dry), during which the evolution of the cracks was also followed at the predetermined time intervals (24 h, 96 h, 192 h, 336 h, 480 h).

The evolution of the self-curing process of a crack (composition T1) during 480 h of exposure to conditioning in order to induce the self-curing effect is exemplified is shown in Fig. 1.

To induce the cracking state for the mixtures T1, T2 and TM a cracking force of 6000 N was applied, representing (86–88%) of the maximum cracking force.

Experimental results for mixture T1 show that:

- following the cracking, depending on the area assessed, it is observed that cracks with variable openings were obtained;
- the maximum initial opening of the cracks falls within the range (0.139–0.033) mm and decreases as the conditioning period passes, Fig. 2 and Fig. 3;
- the average initial opening of the cracks falls within the range (0.1049–0.031) mm and decreases as the conditioning period passes, Fig. 2 and Fig. 3;
- the healing degree of the maximum opening of the cracks, ($GV^{M}{}_{t}$), increases as the conditioning period passes, reaching 100%, after 96 h in the case of zones 4, 6 and 8, after 192 h in the case of zones 5 and 7, after 336 h in the case of zone 9, after 480 h in the case of zone 1, remaining open with a maximum opening of 0.063 mm, Fig. 4;
- the average degree of healing of the identified cracks, ($GV^{m}{}_{t}$), increases as the conditioning period passes, reaching 100%, after 96 h in the case of Zones 4, 6, 8, after 192 h in the case of Zones 5 and 7, after 336 h in the case of Zone 9, after 480 h in the case of Zone 1, remaining open, with the final average healing degree of 91.42%, Fig. 5;

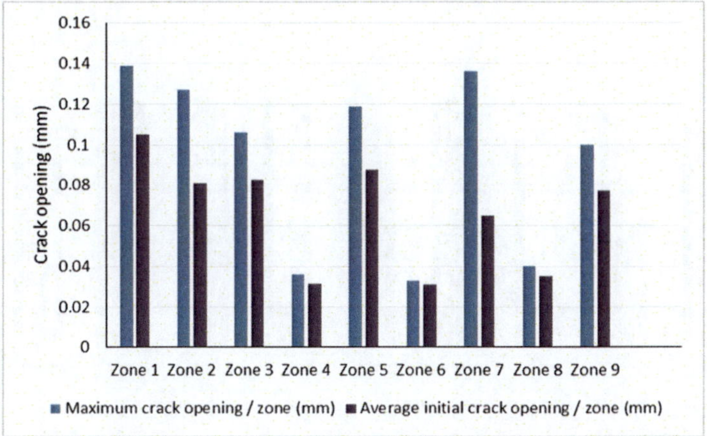

Fig. 2. Maximum opening and average opening of cracks, values measured immediately after cracking.

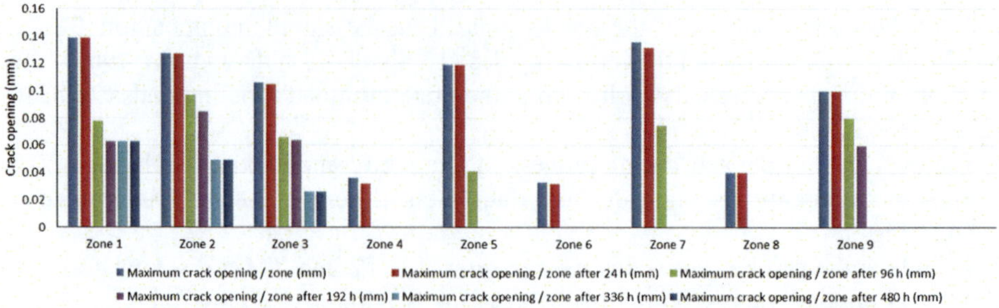

Fig. 3. Evolution of maximum crack opening during conditioning.

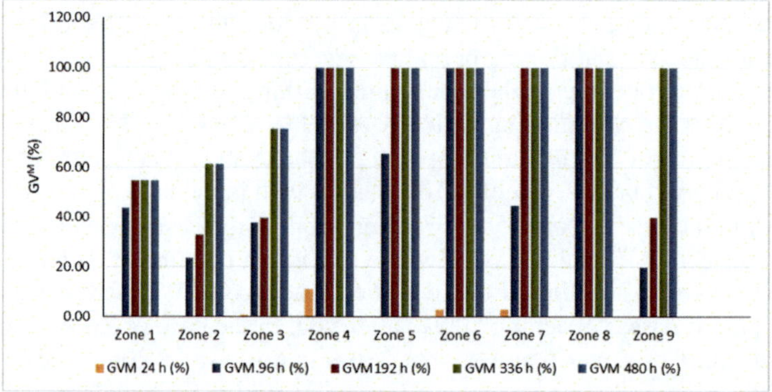

Fig. 4. Evolution of the healing degree calculated for the maximum opening of the cracks, during T1 conditioning

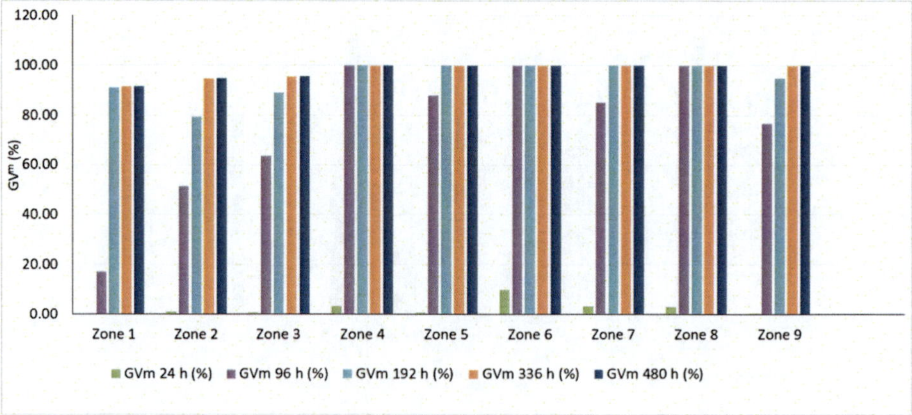

Fig. 5. Evolution of the healing degree calculated for the average opening of the cracks, during T1 conditioning.

– The average moment healing rate ($GV^{Mm}{}_t$) is a measure of the closing speed of the crack between two conditioning time intervals and indicates that in the first hours of conditioning (maximum 96 h) the closing speed is higher, in contrast to the subsequent development of the self-healing phenomenon that occurs at a lower speed, Fig. 6. Exception to this finding shall be made only in the case of the indicator calculated for zone 1. This kinetically delaying behavior of the crack closing process was attributed to the large opening of this crack characterized by an average initial opening of 104.9 μm.

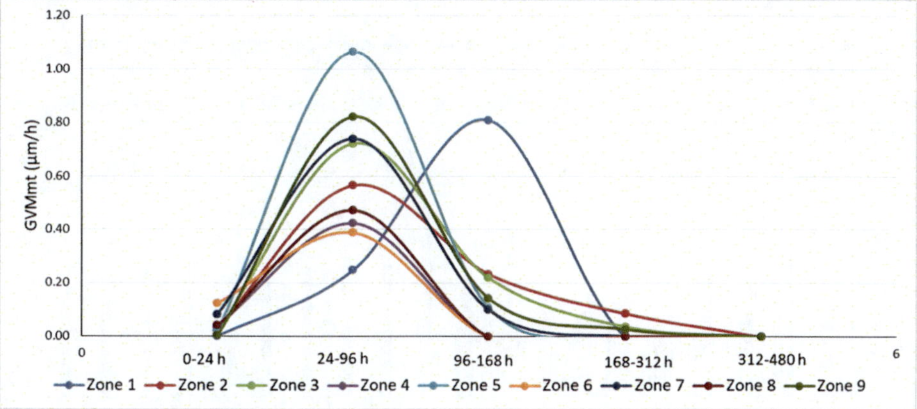

Fig. 6. Evolution of the kinetic indicator of the healing process, average degree of moment healing ($GV^{Mm}{}_t$) T1

Experimental results for T2 mixture show that:

– following the cracking, depending on the area assessed, it is observed that cracks with variable openings were obtained;
– the maximum initial opening of the cracks falls within the range (0.085–0.040) mm and decreases as the conditioning period passes, Fig. 7 and Fig. 8;

– The healing degree of the maximum crack opening, ($GV^{M}{}_t$), increases as the conditioning period passes, up to 100%, after 96 h in the case of Zones 16, 18, 19, 20, after 192 h in the case of Zone 1, 3, 8, 9, 10, 11, 12, 13, 14 17, after 336 h in the case Zone 2, after 480 h in the case of Zone 4. Only cracks in Zone 5 remained open with an average opening of 0.048 mm, cracks in Zone 6 with a maximum opening of 0.060 mm, Zone 7 with a maximum opening of 0.051 mm and Zone 17 with a maximum opening of 0.079 mm, see Fig. 9.

– The average degree of healing of the identified cracks, ($GV^{m}{}_t$), increases as the conditioning period passes, reaching 100%, after 96 h in the case of Zones 16, 18, 19,

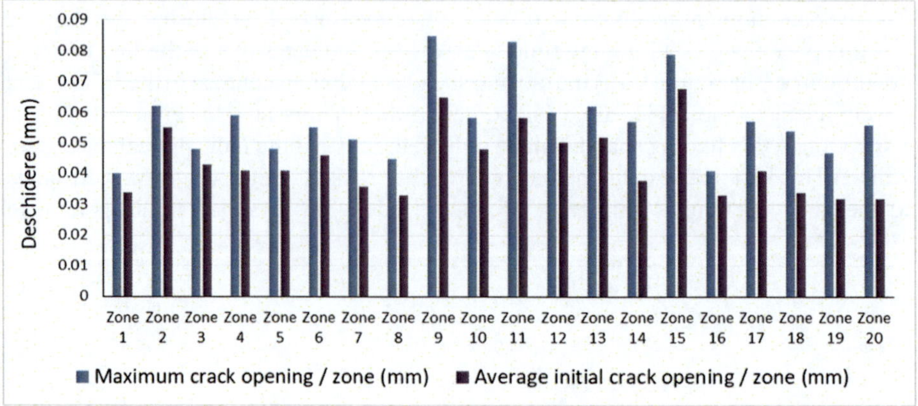

Fig. 7. Maximum opening and average opening of cracks, values measured immediately after cracking.

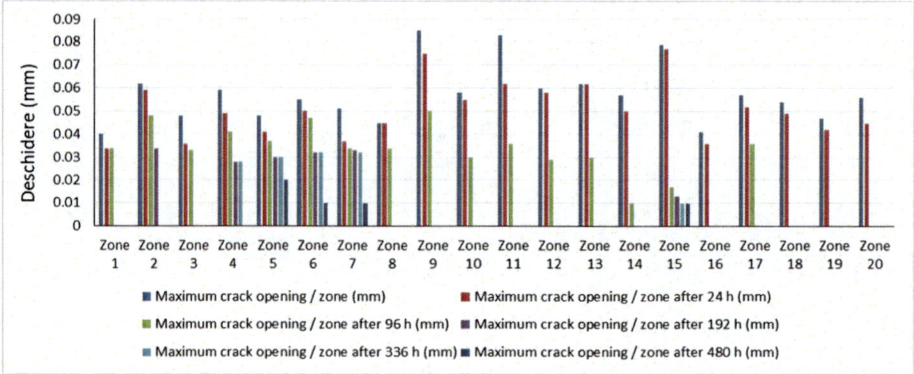

Fig. 8. Evolution of maximum crack opening during conditioning.

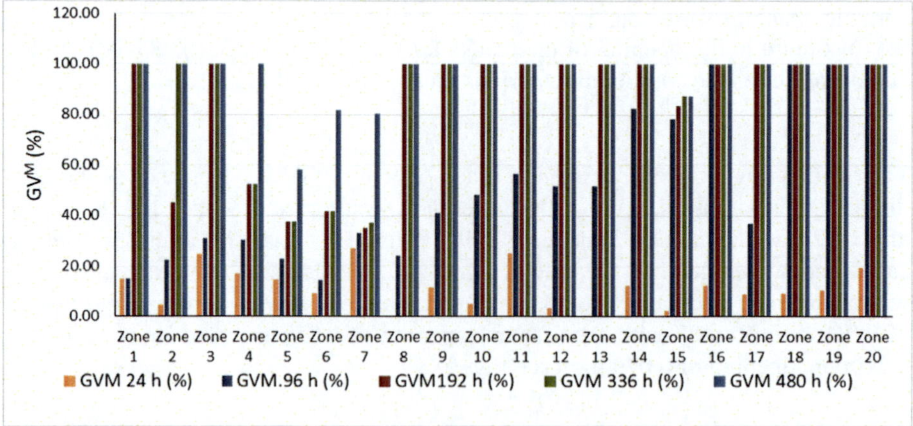

Fig. 9. Evolution of the healing degree calculated for the maximum opening of the cracks, during T2 conditioning.

20, after 192 h in the case of Zones 1, 3, 8, 9, 10, 11, 12, 13, 14 17, after 336 h in the case of Zone 2, after 480 h in the case of Zone 4. Cracks in Zone 5 remained open with an average 97.56%, Zone 6 with 80.43%, Zone 7 with 82.22%, and Zone 17 with 98.52%, Fig. 10.

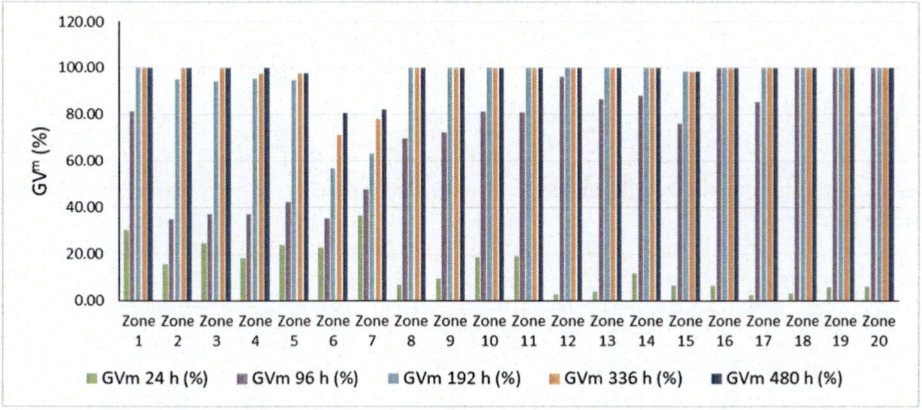

Fig. 10. Evolution of the healing degree calculated for the average opening of the cracks, during T2 conditioning.

- The average moment healing rate ($GV^{Mm}{}_t$) is a measure of the closing speed of the crack between two conditioning time intervals and indicates that in the first hours of conditioning (maximum 96 h) the closing speed is higher, in contrast to the subsequent development of the self-healing phenomenon that occurs at a lower speed, Fig. 11. Exception to this finding shall be made only in the case of the indicator calculated for Zones 5, 6, 7 and 15. This kinetically delaying behavior of the crack closing process was attributed to the large opening of this crack characterized by an average initial opening of between 48 and 79 μm.

Experimental results for TM mixture show that:

- as a result of cracking, according to the assessed area it is observed that cracks with variable openings have been obtained;
- the maximum initial opening of the cracks falls within the range (0.109–0.033) mm and decreases as the conditioning period passes, Fig. 12 and Fig. 13;
- the average initial opening of the cracks falls within the range (0.830–0.029) mm and decreases as the conditioning period passes, Fig. 12 and Fig. 13.

- The healing degree of the maximum crack opening, ($GV^{M}{}_t$), increases as the conditioning period passes, up to 100%, after 192 h in the case of Zones 8, 9, 10, 15, after

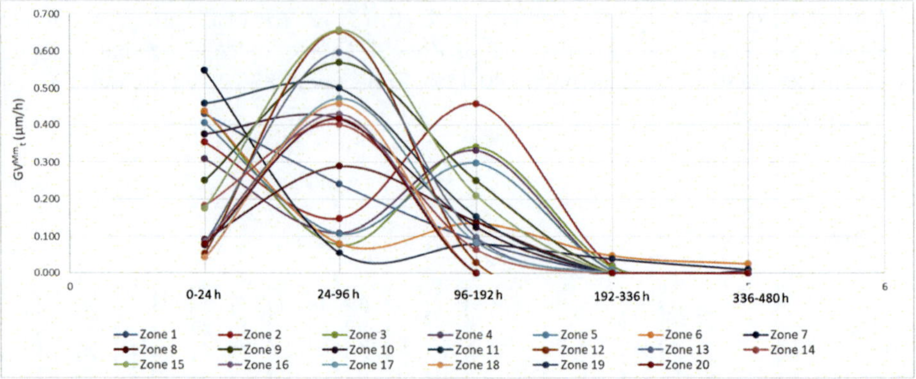

Fig. 11. Evolution of the kinetic indicator of the healing process, average degree of moment healing ($GV^{Mm}{}_{t}$) T2.

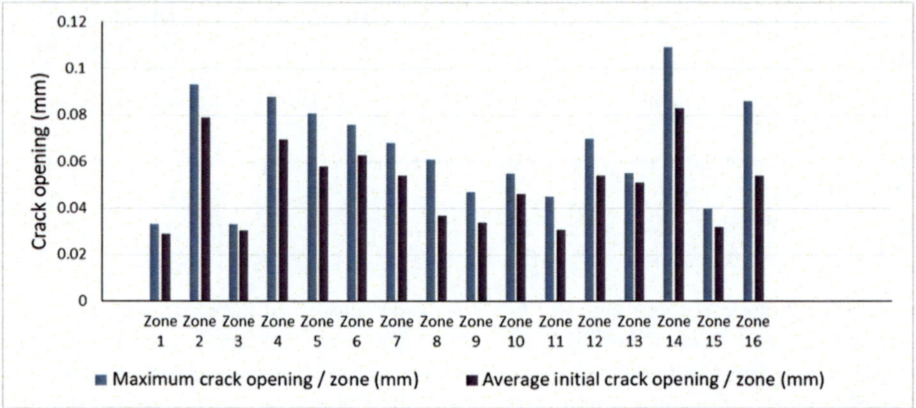

Fig. 12. Maximum opening and average opening of cracks, values measured immediately after cracking.

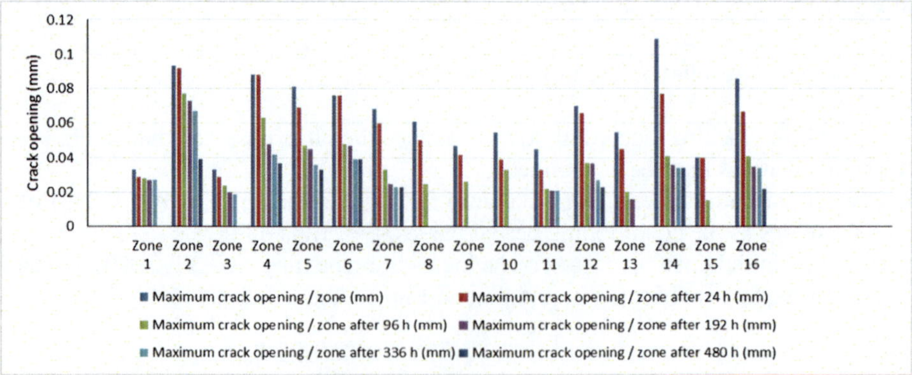

Fig. 13. Evolution of maximum crack opening during conditioning.

336 h in the case Zone 13, after 480 h in the case of Zones 1, 3 and 11. Cracks in Zones 2, 4, 5, 6, 7, 12, 14 and 16 remained open with an average opening between 0.068 and 0.109 mm, see Fig. 14.

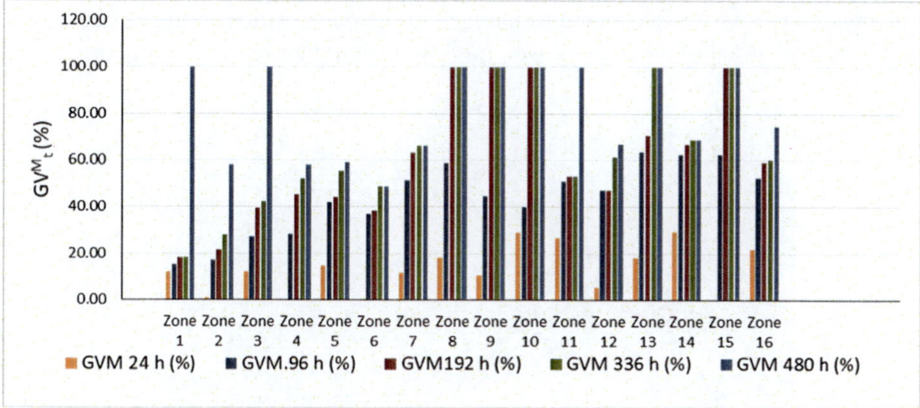

Fig. 14. Evolution of the healing degree calculated for the maximum opening of the cracks, during TM conditioning.

– The average degree of healing of the identified cracks, (GV^m_t), increases as the conditioning period passes, reaching 100%, after 192 h in the case of Zones 8, 9, 10 and 15, after 336 h in the case of Zone 13, after 480 h in the case of Zones 1, 3 and 11. Cracks that remained open had an average between 77.78% and 98.15%, Fig. 15.

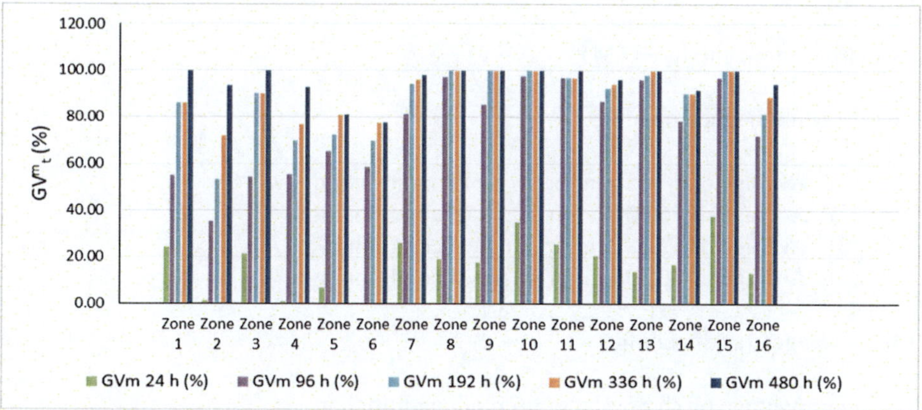

Fig. 15. Evolution of the healing degree calculated for the average opening of the cracks, during TM conditioning.

- The average moment healing rate ($GV^{Mm}{}_{t}$) is a measure of the closing speed of the crack between two conditioning time intervals and indicates that in the first hours of conditioning (maximum 96 h) the closing speed is higher, in contrast to the subsequent development of the self-healing phenomenon that occurs at a lower speed, Fig. 16. Exception to this finding shall be made only in the case of the indicator calculated for Zones 2, 4, 5, 6, 7, 12,14 and 16. This kinetically delaying behavior of the process of closing the cracks was attributed to the lack in the composition of the cementitious composite material of the integral waterproofing admixture by crystallization.

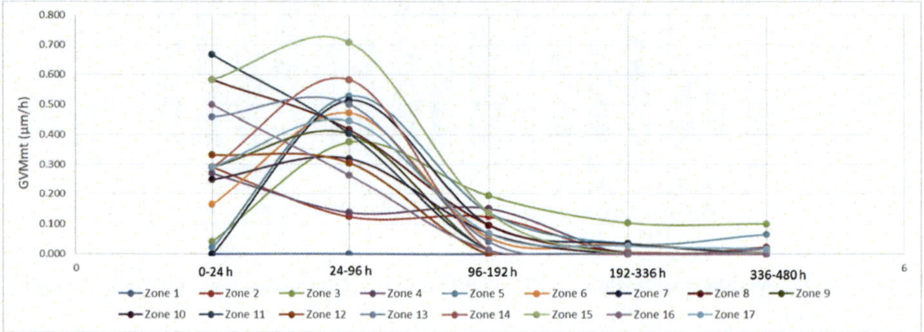

Fig. 16. Evolution of the kinetic indicator of the healing process, average degree of moment healing ($GV^{Mm}{}_{t}$) TM

4 Conclusions

The experimental results obtained indicate a good self-healing capacity, quantified by the following healing degrees:

1. Control sample mixture (TM):

   a. Maximum crack opening:

      (1) Average of 40.55% after 96 h of conditioning;
      (2) Average of 56.80% after 192 h of conditioning;
      (3) Average of 62.18% after 336 h of conditioning;
      (4) Average of 76.63% after 480 h of conditioning;

   b. Average crack opening:

      (1) Average of 78.85% after 96 h of conditioning;
      (2) Average of 87.20% after 192 h of conditioning;
      (3) Average of 90.69% after 336 h of conditioning;

        (4) Average of 95.63% after 96 h of conditioning;

2. Mixture T1:

    a. Maximum crack opening:

        (1) Average of 59.52% after 96 h of conditioning;
        (2) Average of 74.15% after 192 h of conditioning;
        (3) Average of 87.95% after 336 h of conditioning;
        (4) Average of 87.95% after 480 h of conditioning;

    b. Average crack opening:

        (1) Average of 75.78% after 96 h of conditioning;
        (2) Average of 94.92% after 192 h of conditioning;
        (3) Average of 97.99% after 336 h of conditioning;
        (4) Average of 97.99% after 96 h of conditioning;

3. Mixture T2:

    a. Maximum crack opening:

        (1) Average of 52.09% after 96 h of conditioning;
        (2) Average of 84.79% after 192 h of conditioning;
        (3) Average of 87.82% after 336 h of conditioning;
        (4) Average of 97.94% after 480 h of conditioning;

    b. Average crack opening:

        (1) Average of 72.61% after 96 h of conditioning;
        (2) Average of 94.91% after 192 h of conditioning;
        (3) Average of 97.15% after 336 h of conditioning;
        (4) Average of 97.94% after 96 h of conditioning;

Results obtained on the cementitious composite samples using integral waterproofing admixture by mass crystallization show the effectiveness of using this type of admixture in producing the self-healing effect of the cementitious composites and also speeding the process, thus obtaining very good results.

This work contributes to the increase of knowledge in the field of cement materials with self-healing capacity, indicating a possibility of obtaining this effect through the use of a waterproofing additive by mass crystallization, simultaneously with the presentation of the possibilities of use of industrial waste such as fly ash and limestone slurry.

Acknowledgment. This paper was financially supported by the Project "Entrepreneurial competences and excellence research in doctoral and postdoctoral programs - ANTREDOC", project co-funded by the European Social Fund financing agreement no. 56437/24.07.2019.

References

1. Garg, M., Azarsa, P., Gupta, R.: Self-healing potential and post-cracking tensile behavior of polypropylene fiber-reinforced cementitious composites. J. Compos. Sci. **5**, 122 (2021)
2. Van Tittelboom, K., De Belie, N.: Self-healing in cementitious materials. Materials **6**(6), 2182–2217 (2013)
3. Sahmaran, M., Keskin, S.B., Ozerkan, G., Yaman, I.O.: Self-healing of mechanically-loaded self-consolidating concretes with high volumes of fly ash. Cement Concr. Compos. **30**(10), 872–879 (2008)
4. Yang, Y., Lepech, M.D., Yang, E.-H., Li, V.C.: Autogenous healing of engineered cementitious composites under wet–dry cycles. Cem. Concr. Res. **39**(5), 382–390 (2009)
5. Sahmaran, M., Li, M., Li, V.C.: Transport properties of engineered cementitious composites under chloride exposure. ACI Mater. J. **104**(6), 604–611 (2011)
6. Granger, S., Loukili, A., Pijaudier-Cabot, G., Chanvillard, G.: Experimental characterization of the self-healing of cracks in an ultra-high-performance cementitious material: Mechanical tests and acoustic emission analysis. Cem. Concr. Res. **37**(4), 519–527 (2007)
7. Neville, A.: Autogenous healing—a concrete miracle? Concr. Int. **24**, 76–82 (2002)
8. Baeră, C., Szilagyi, H., Matei, C., Hegyi, A., Lăzărescu, A., Mircea, A.C.: Optimizing approach on fibre engineered cementitious materials with self-healing capacity (SH-FECM) by the use of slurry lime (SL) addition. In: MATEC Web of Conerences, vol. 289, p. 01001 (2019)
9. Szilagyi, H., Baeră, C., Hegyi, A., Lăzărescu, A.: Romanian resources of waste and industrial by-products as additions for cementitious mixtures. In: International Multidisciplinary Scientific GeoConference: SGEM, vol. 18, no. 6.3, pp. 325–332 (2018)
10. Florean, C., Szilagyi, H., Hegyi, A.: Environment and pollution management of pollution volatile organic compounds in Cluj-Napoca. Present Environment and Sustainable Development, pp. 207–218 (2016)
11. Nicula, L.M., Corbu, O., Iliescu, M., Sandu, A.V., Hegyi, A.: Study on the durability of road concrete with blast furnace slag affected by the corrosion initiated by Chloride. Adv. Civ. Eng. **2021**, 8851005 (2021)
12. Ramm, W., Biscoping, M.: Autogenous healing and reinforcement corrosion of water-penetrated separation cracks in reinforced concrete. Nucl. Eng. Des. **179**(2), 191–200 (1998)
13. Aldea, C.-M., Song, W.-J., Popovics, J.S., Shah, S.P.: Extent of healing of cracked normal strength concrete. J. Mater. Civ. Eng. **12**, 92–96 (2000)
14. Homma, D., Mihashi, H., Nishiwaki, T.: Self-Healing capability of fibre reinforced cementitious composites. J. Adv. Concr. Technol. **7**(2), 217–228 (2009)
15. Hearn, N., Morley, C.T.: Self-sealing property of concrete—experimental evidence. Mater. Struct. **30**, 404–411 (1997)
16. Ferrara, L., et al.: Experimental characterization of the self-healing capacity of cement based materials and its effects on the material performance: a state of the art report by COST Action SARCOS WG2. Constr. Build. Mater. **167**, 115–142 (2018)
17. Snoeck, D., Steuperaert, S., Van Tittelboom, K., Dubruel, P., De Belie, N.: Visualization of water penetration in cementitious materials with superabsorbent polymers by means of neutron radiography. Cem. Concr. Res. **42**, 1113–1121 (2012)
18. Van Tittelboom, K., De Belie, N., Lehmann, F., Grosse, C.U.: Acoustic emission analysis for the quantification of autonomous crack healing in concrete. Constr. Build. Mater. **28**, 333–341 (2012)
19. Liu, S., Bundur, Z.B., Zhu, J., Ferron, R.D.: Evaluation of self-healing of internal cracks in biomimetic mortar using coda wave interferometry. Cem. Concr. Res. **83**, 70–78 (2016)

20. In, C.-W., Holland, R.B., Kim, J.-Y., Kurtis, K.E., Kahn, L.F., Jacobs, L.J.: Monitoring and evaluation of self-healing in concrete using diffuse ultrasound. NDT E Int. **57**, 36–44 (2013)
21. Feiteira, J., Tsangouri, E., Gruyaert, E., Lors, C., Louis, G., De Belie, N.: Monitoring crack movement in polymer-based self-healing concrete through digital image correlation, acoustic emission analysis and SEM in-situ loading. Mater. Des. **115**, 238–246 (2017)
22. Roig-Flores, M., Pirritano, F., Serna, P., Ferrara, L.: Effect of crystalline admixtures on the self-healing capability of early-age concrete studied by means of permeability and crack closing tests. Constr. Build. Mater. **114**, 447–457 (2016)
23. Anglani, G., et al.: Sealing efficiency of cement-based materials containing extruded cementitious capsules. Constr. Build. Mater. **251**, 119039 (2020)
24. Şahmaran, M.: Effect of flexure induced transverse crack and self-healing on chloride diffusivity of reinforced mortar. J. Mater. Sci. **42**(22), 9131–9136 (2007). https://doi.org/10.1007/s10853-007-1932-z
25. Van Belleghem, B., Heede, P.V.D., Van Tittelboom, K., De Belie, N.: Quantification of the service life extension and environmental benefit of chloride exposed self-healing concrete. Materials **10**(1), 5 (2016)
26. Banthia, N., Gupta, R.: Influence of polypropylene fiber geometry on plastic shrinkage cracking in concrete. Cem. Concr. Res. **36**, 1263–1267 (2006)
27. Hao, Y., Cheng, L., Hao, H., Shahin, M.A.: Enhancing fiber/matrix bonding in polypropylene fiber reinforced cementitious composites by microbially induced calcite precipitation pre-treatment. Cem. Concr. Compos. **88**, 1–7 (2018)
28. Afroughsabet, V., Biolzi, L., Monteiro, P.J.M.: The effect of steel and polypropylene fibers on the chloride diffusivity and drying shrinkage of high-strength concrete. Compos. B Eng. **139**, 84–96 (2018)
29. Gojević, A., Ducman, V., Netinger Grubeša, I.N., Baričević, A., Pečur, I.B.: The effect of crystalline waterproofing admixtures on the self-healing and permeability of concrete. Materials **14**, 1860 (2021)
30. Sideris, K.K., Chatzopoulos, A., Tassos, C., Manita, P.: Durability of concretes prepared with crystalline admixtures. In: MATEC Web Conferences, vol. 289, p. 09003 (2019)
31. Azarsa, P., Gupta, R., Biparva, A.: Inventive microstructural and durability investigation of cementitious composites involving crystalline waterproofing admixtures and portland limestone cement. Materials **13**, 1425 (2020)
32. Žáková, H., Pazderka, J., Reiterman, P.: Textile reinforced concrete in combination with improved self-healing ability caused by crystalline admixture. Materials **13**, 5787 (2020)
33. Cuenca, E., Mezzena, A., Ferrara, L.: Synergy between crystalline admixtures and nano-constituents in enhancing autogenous healing capacity of cementitious composites under cracking and healing cycles in aggressive waters. Constr. Build. Mater. **266**, 121447 (2021)
34. Cuenca, E., Rigamonti, S., Brac, E.G., Ferrara, L.: Crystalline admixture as healing promoter in concrete exposed to chloride-rich environments: experimental study. J. Mater. Civ. Eng. **33**, 04020491 (2021)
35. Al-Rashed, R., Jabari, M.: Dual-crystallization waterproofing technology for topical treatment of concrete. Case Stud. Constr. Mater. **13**, 00408 (2020)
36. García-Vera, V.E., Tenza-Abril, A.J., Saval, J.M., Lanzón, M.: Influence of crystalline admixtures on the short-term behaviour of mortars exposed to sulphuric acid. Materials **12**, 82 (2019)

Influence TiO_2 Nanoparticles Addition on the Physico-Mechanical Performances of Micro-concrete

Elvira Grebenișan[1,2](✉), Andreea Hegyi[1], Adrian-Victor Lăzărescu[1,2], Henriette Szilagyi[1], and Carmen Florean[1]

[1] NIRD URBAN-INCERC Cluj-Napoca Branch, 117 Calea Florești, 400524 Cluj-Napoca, Romania
elvira.grebenisan@incerc-cluj.ro

[2] Technical University of Cluj-Napoca, 28 Memorandumului Street, 400114 Cluj-Napoca, Romania

Abstract. Currently, worldwide, research on the production of cementitious composites with self-cleaning properties (using the photocatalytic character of TiO_2) is an area of real interest. The aim of this paper was to present a synthesis of the results of experimental research on the influence of the addition of TiO_2 nanoparticles on the physical-mechanical properties of cement composites based on white cement micro-concrete. Both the results of research reported to date and experimental ones have shown that the properties of concrete are positively influenced, as long as the amount of nanoparticles is not in excess. Research has indicated an increase in mechanical resistances, more pronounced in the first 7 days. The increase in tensile bending strength as a result of the addition of TiO_2 nanoparticles, experimentally recorded, is a maximum of 7% for testing at 7 days of age and a maximum of 4.5% for testing at 28 days of age, respectively. The increase in compressive strength due to the addition of TiO_2 nanoparticles, experimentally recorded, is a maximum of 3.6% for the 7-days tests and a maximum of 2% for the 28-days tests, respectively. In the case of an excess of TiO_2 nanoparticles, or their insufficiency, the effect on the properties of the micro-concrete is the opposite.

Keywords: Micro-concrete · Self-cleaning · TiO_2 nanoparticles

1 Introduction

At present, worldwide, the general research directions are aiming to identify the possibilities for the most sustainable use of building materials and to identify new opportunities for improving performance and durability, simultaneously with as little impact as possible on the environment. Cementitious composites produced with the addition of TiO_2 nanoparticles come to meet these directions by their specificity in exploiting the photocatalytic property of TiO_2 nanoparticles, thus obtaining a high-performance, durable material with self-cleaning capacity and increased resistance to the development of microorganisms on its surface, for use in the construction field.

L. Moldovan and A. Gligor (Eds.): Inter-Eng 2021, LNNS 386, pp. 166–181, 2022.
https://doi.org/10.1007/978-3-030-93817-8_17

With regard to the influence of TiO_2 nanoparticles on the hardened-state micro-concrete, by adding or replacing a part of cement with different amounts of TiO_2 nanoparticles, the mechanical properties of concrete are improved both due to the smaller CH crystal sizes and the formation of a larger and better organized amount of C-H-S gel [1]. It also increases freeze-thaw resistance, abrasion resistance and resistance to the action of chemical agents. The excess addition of TiO_2 nanoparticles decreases the performance of the compound relative to the control sample. This phenomenon is strictly conditioned by the ratios in which the raw materials are used, because if the amount of TiO_2 nanoparticles is added in excess, the performance of the composite is influenced in a decreasing way [1].

There are some controversies regarding the mechanical strength of this type of cementitious composite. Some research shows that they are not negatively influenced up to a maximum of 6% addition of TiO_2 nanoparticles [2], while others reduce this threshold to 5%, 3% [3] or even 1% [4].

A large number of experimental research has indicated that compressive strength increases with the increase of nanoparticles regardless of age of maturation, with 1% being optimal, but there are also reports which show that compressive strength decreases with the increasing percentage of TiO_2 [5].

On the other hand, haste in the maturation process of the micro-concrete was constantly noted. Compared to the control sample (without nano-TiO_2 content), the composite material with TiO_2 nanoparticles addition showed an increase in compressive strength recorded at 7 days of age and a smaller increase in compressive strength between 7 and 28 days [5].

According to the literature, the compressive strength is directly influenced by the content of TiO_2 nanoparticles and increases, up to a certain concentration threshold, with the increase in the content of TiO_2 nanoparticles. Thus, research has shown increases in compressive strength for: 2% nano-TiO_2, at the age of 28 days [6]; 0.5%, 1%, 1.5%, and 2% TiO_2, (water/cement ratio, w/b = 0.4), 1% being the optimal percentage [7]; 1%, 3%, and 5% TiO_2 (w/b = 0.42) at 28 days of age [8]; 1% TiO_2 [9]; 0.5%, 1%, 1.5% and 2% TiO_2, at the age of 7, 28 and 90 days, 1% optimal (when hardening in water), 2% optimal (when hardening in lime water) [10]; 1%, 2%, 3%, 4% and 5% TiO_2, (w/b = 0.4), at 28 days of age, 4% being the optimal percentage [11, 12]. According to other researchers, the compressive strength decreases with the increase in the amount of nanoparticles, respectively, with the introduction of 1%, 2%, 3%, 4% and 5% TiO_2 nanoparticles [13].

In terms of bending tensile strength, the use of TiO_2 nanoparticles of 1% and 3% relative to the amount of cement increased bending tensile strength, while 5% decreased it, with 1% being optimal [8]. Other research has shown that the bending resistance increases with the increase in the amount of nanoparticles upon the introduction of 1%, 2%, 3%, 4%, and 5% TiO_2 (w/b = 0.38) at 7, 14, 28 and 90 days of age, with 4% being the optimal percentage [14, 15]. Increased bending resistances are also reported by other authors, at the introduction of 0.5%, 1%, 1.5% and 2% TiO_2 (w/b = 0.4) at water hardening 1% being the optimal percentage of TiO_2 nanoparticles addition, and at lime water hardening, 2% being the optimal percentage [16, 17]. Li et al. and Nazari and Riahi also confirm the increase in bending resistance, at the introduction of 1% and

3% TiO_2 (w/b = 0.42), at the age of 28 days, 1% being the optimal percentage [18], respectively 1%, 2%, 3%, 4% and 5% TiO_2 (w/b = 0.4) at the age of 2, 7 and 28 days [11, 12].

In terms of water absorption, this parameter decreases with increasing the content of TiO_2 nanoparticles addition for: 2% nano-TiO_2, at the age of 28 days [6]; 1%, 2%, 3%, 4% 5% TiO_2 (w/b = 0,38), at the age of 7, 14, 28, and 90 days (4% optimal) [14, 15]; 0.5%, 1%, 1.5%, 2% TiO_2, (w/b = 0,4), and at the age of 7, 28 and 90 days, about 0.5% considered optimal [16]; 0.5%, 1%, 1.5%, 2% TiO_2 (w/b = 0,4), and at the age of 28 and 90 days, [19, 20]; 1%, 2%, 3%, 4% 5% TiO_2 (w/b = 0,4), and at the age of 7 and 28 days 4% is the percentage of the optimal [11, 12]. There are also authors whose studies show increased water absorption with increased content of TiO_2 nanoparticles, as follows: 1%, 2%, 3%, 4% and 5% TiO_2, (w/b = 0.4), at the age of 2 days [19, 20] or at the introduction of 0.5%, 1%, 1.5%, and 2% TiO_2, (w/b = 0.4), at the age of 7 days [19, 20] (Table 1).

Table 1. Water absorption according various references.

Author	TiO_2 (NT) (%)	Material type	w/b (%)	Age	Compressive strength	Bending resistances	Water absorption
Salemi et al. [6]	2	Concrete	ns	28 days	Increases	–	Decreases
Nazari et al. [7]	0.5, 1, 1.5	Concrete	0,4	–	Increases, 1% optimal	–	–
Zhang and Li [8]	1, 3, 5	Concrete	0,42	28 days	Increases with increasing of NT content	1% and 3% NT increases, 5% decreases, 1% optimal	–
Li et al. [9]	1	Concrete	–	–	Increases	–	–
Soleymani [10]	0.5, 1, 1.5, 2	Concrete	0,4	7 days 28 days 90 days	Increases, 1% optimal (when hardening in water), 2% optimal (when hardening in lime water)	–	–
Nazari and Riahi [11, 12]	1, 2, 3, 4, 5	Concrete	0,4	2 days	Increases, 4% optimal	Increases	Increases
				7 days 28 days			Decreases, 4% optimal

(*continued*)

Table 1. (*continued*)

Author	TiO_2 (NT) (%)	Material type	w/b (%)	Age	Compressive strength	Bending resistances	Water absorption
Behfarnia et al. [13]	1, 2, 3, 4, 5	Concrete	–	28 days	Decreases	–	Decreases, 4% optimal
Jalal et al. [14, 15]	1, 2, 3, 4, 5	Concrete	0,38	7 days 14 days 28 days 90 days	–	Increases, 4% optimal	Decreases, 4% optimal
Nazari [16] and Soleymani [17]	0.5, 1, 1.5, 2	Concrete	0,4	7 days 28 days 90 days	–	Increases, 1% optimal	Decreases, 0.5% optimal
Soleymani [19, 20],	0.5, 1, 1.5, 2	Concrete	0,4	7 days 28 days	–	–	Increases
				90 days			Decreases, 0.5% optimal

In all cases, regardless of conditioning temperature and age, mechanical strength of the cementitious composites increased to a content of 2% TiO2, after which they decrease [21]. Also, the decrease in conditioning temperature negatively influences mechanical resistances for all cases 0–5% TiO_2 [21].

In terms of porosity, some studies show that it decreases with increasing content of TiO_2 nanoparticles, at the introduction of TiO_2 nanoparticles by up to 10% [21], others show its increase at the introduction of more than 5% TiO_2 [22].

Reduced density and porosity, increased mechanical variation due to the rapid formation of hydration products, implicitly influences durability [21, 23–27].

In conclusion, based on the results of the researches presented in the literature, the influence of the introduction of nano-TiO_2 in cementitious binders causes changes in the physical-mechanical performance, but an optimal content of TiO_2 nanoparticles cannot be accurately assessed to ensure a general improvement in composite performance.

The aim of this paper is to present research carried out for the analysis of the influence of nano-TiO_2 addition into the mass of cementitious composites of micro-concrete has on their physical and mechanical performances.

2 Materials and Methods

2.1 Raw Materials

The materials used in the preparation of the nano-TiO_2 addition micro-concrete were: white Portland cement CEM I 52.5 R, Degussa P25 TiO_2-nanoparticles, aggregates granular class 0–4 mm and 4–8 mm, 6 mm and 19 mm PVA fibres and water.

2.2 Preparation and Conditioning

Five micro-concrete with nano-TiO_2 addition mixtures have been prepared with a percentage content of TiO_2 nanoparticles of 0% (control sample), 2%, 3%, 4% and 5%, with the following ratios (Table 2).

After all the constituent materials have been pre-conditioned, water and sand were mixed together for 30 s at the speed of 140 ± 5 rpm. A dry pre-mixing of cement with TiO_2 nanoparticles followed, after which a mechanical mixing with the water and sand was started. In the composition thus obtained, the two types of fibers were added during mixing. After the addition of the fibers, mixing followed for 30 s at a speed of 285 ± 10 rpm. A break of 60 s was initiated, followed by a mechanical mixing for 60 s at a speed of 285 ± 10 rpm.

On the newly obtained mixture, the fresh-state density has been measured, after which the mixture was poured into prismatic, 40 × 40 × 160 mm, metallic molds in order to assess the following parameters: apparent density in hardened state, mechanical strength (flexural and compressive strength) and water absorption (Fig. 1a). 24 × 85 × 130 mm samples were also prepared in order to assess the density of the material and water absorption at different time intervals, porosity, mechanical properties (flexural strength after freeze–thaw cycles and flexural strength after thermal shock), mortar adhesion and white degree (Fig. 1b).

The samples thus obtained were conditioned for 24 h in molds, at 90% humidity and 20 °C temperature, in the dark. After the 24 h the samples were demolded and immersed completely in water for 27 days, at the temperature of 20 °C, also in a dark environment.

Table 2. Micro-concrete with nano-TiO2 addition mixtures.

Raw materials	White Portland cement (%)	Fibres (19 mm) (%)	Fibres (6 mm) (%)	Aggre-gates granular (0–4 mm) (%)	Aggre-gates granular (4–8 mm) (%)	Nano-TiO_2 particles (%)	Admixture (%)	Water/cement ratio (%)
Control sample (0% TiO_2)	100	0.18	0.05	133.7	133.7	0	1	0.45
Composites with 2% TiO_2	100	0.18	0.05	133.2	133.2	2	1	0.45
Composites with 3% TiO_2	100	0.18	0.05	132.8	132.8	3	1	0.45
Composites with 4% TiO_2	100	0.18	0.05	132.5	132.5	4	1	0.45
Composites with 5% TiO_2	100	0.18	0.05	132.2	132.2	5	1	0.45

Until testing, the samples were stored in laboratory conditions, in the absence of light. Laboratory equipment was used to test the samples: press for resistance determination,

a)  b)

Fig. 1. Images with five micro-concrete with nano-TiO_2: a) before testing; b) after testing

Pull-off apparatus for adhesion determination, portable leukometer type WSB-1 for white degree determination, as well as heat-regulating oven.

2.3 Testing Methods

The measurements of physical and mechanical properties carried out, followed the tests in order to assess the following parameters: bulk density in the fresh state (EN 12,350-6) [28], the apparent density in hardened state (EN 12,350-7) [29], water absorption (EN 14,617-1) [30], porosity (EN 1936) [31], flexural strength (EN 12,390-5) [32], compressive strength (EN 12,390-3) [33], bending strength (EN 14,617-2) [34], freeze-thaw (EN 14,617-5) [35], thermal shock resistance (EN 14,617-6) [36] adhesion of mortar paste (EN 1015-12) [37] and the degree of white (parameter measured using a portable leukometer type WSB-1).

3 Results and Discussions

3.1 Apparent Density in Fresh and Hardened State

The results regarding both the apparent density in fresh state (measured after mixing stopped) and the apparent density in hardened state (measured at 28 days) are shown in Fig. 1. It is observed that an increasing or decreasing trend in the values recorded in the case of the introduction into the cementitious mixture of TiO_2 nanoparticles cannot be identified. This behavior cannot be motivated otherwise than by the heterogeneity of the degree of dispersion of TiO_2 nanoparticles in the cement mass. The experimental results however eloquently indicate an increase in the fresh apparent density of the composites into which the photosensitive nanoparticles were introduced, an increase due to their distribution in the pores of the cementitious matrix, thus resulting in a more dense material (Fig. 2a). In the case of the hardened state density assessment for matured samples 28 days after casting, a slight decrease in the density of composites with 2% and 4% TiO_2 content, and an increase in composites with 3% and 5% TiO_2 content is observed compared to the control sample, which could indicate, in correlation with the specifications in the literature [22], the difficulty of achieving a homogeneous distribution of nanoparticles the existence of agglomeration areas of nanoparticles, simultaneously with areas characterized by increased porosity (Fig. 2b).

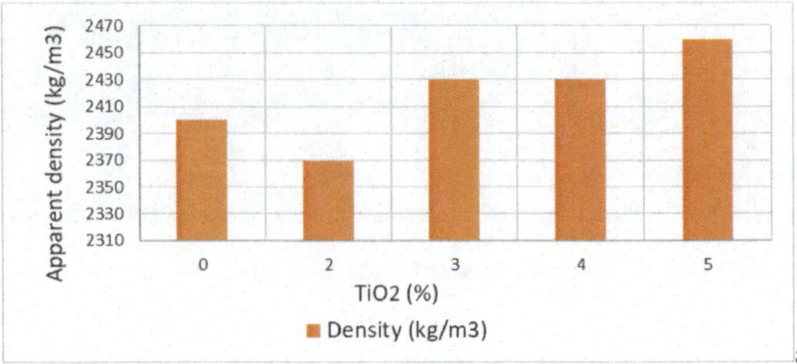

a)

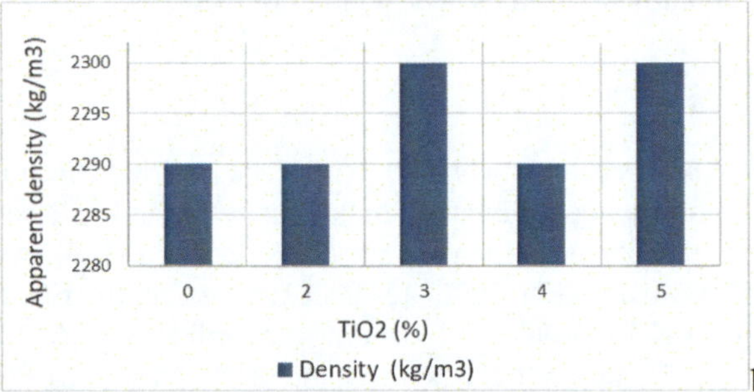

b)

Fig. 2. Cementitious composites with nano-TiO_2 addition apparent density in: a) fresh state and b) hardened state.

3.2 Water absorption

The results of water absorption for saturated samples are shown in Fig. 3. The results on the amount of water absorbed according to the duration of immersion (1 h, 8 h, 24 h, 48 h and 72 h respectively) are shown in Fig. 4.

In terms of the water absorption of the samples with nano-TiO_2 addition, there may be a tendency for the development of the values shown depending on the amount of TiO_2 nanoparticles that are introduced into the mass of the cementitious binder, whichever is the lower, resulting in a matrix with a content of 2% TiO_2 nanoparticles, and the maximum value in the matrix with a content of 5% TiO_2 nanoparticles (Fig. 2). This behavior cannot be motivated otherwise than by the heterogeneity of the degree of homogeneity of the composites, respectively, by the distribution of TiO_2 nanoparticles in the cement mass and by the degree of pore filling with these nanoparticles.

Regarding the amount of water absorbed according to the duration of immersion (1 h, 8 h, 24 h, 48 h and 72 h respectively) of the cementitious samples containing TiO_2 nanoparticles, this has an increasing trend. Thus, it increases, with increasing immersion time, for all mixtures, regardless of the percentage of TiO_2 nanoparticles introduced into the binder (Fig. 3).

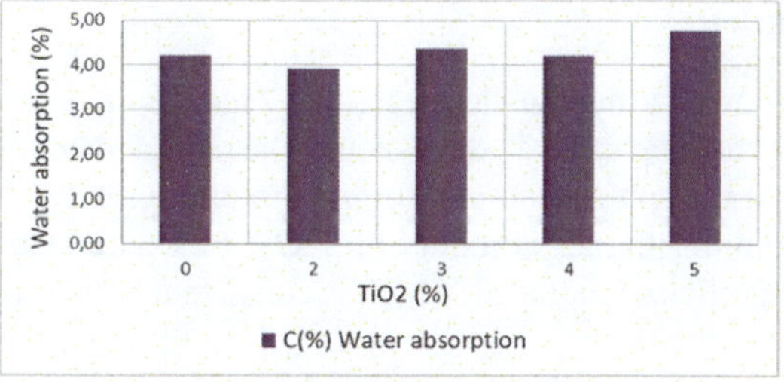

Fig. 3. Water absorbtion.

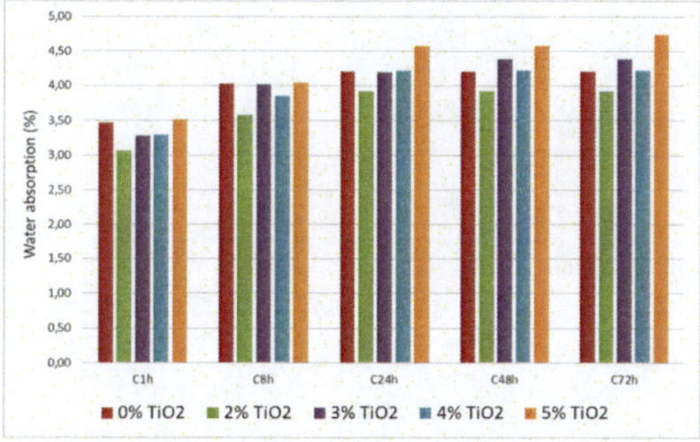

Fig. 4. Absorbed water based on immersion time.

3.3 Porosity

The porosity results are shown in Fig. 5. The sample containing 2% TiO_2 nanoparticles was found to have a lower open porosity compared to the control sample.

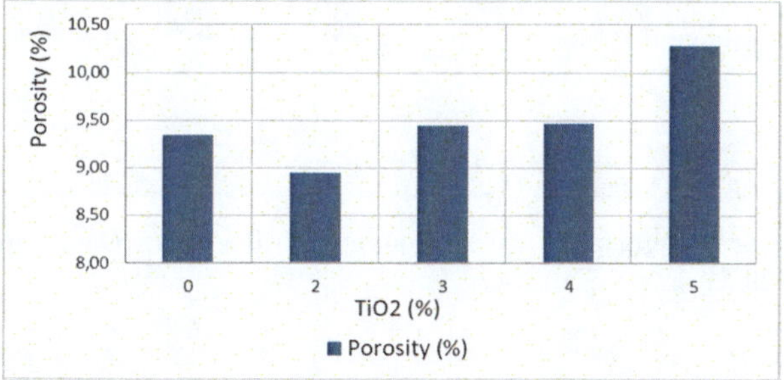

Fig. 5. Micro-concrete with nano-TiO_2 addition porosity.

For samples with a percentage content of 3% TiO_2, 4% TiO_2 and 5% TiO_2, they have a higher open porosity relative to the control sample, value increasing with the increase in the content of TiO_2 nanoparticles.

3.4 Bending strength

The results regarding the bending tensile strength at 7 and 28 days are shown in Fig. 6. It was observed that both at 7 days and 28 days of age, the recorded bending tensile strength increases, compared to the control sample, for the mixture with 2% TiO_2 nanoparticles. For composites containing higher amounts of TiO_2, this parameter decreased as the amount of TiO_2 nanoparticles increased, which is consistent with existing specifications in the literature [21].

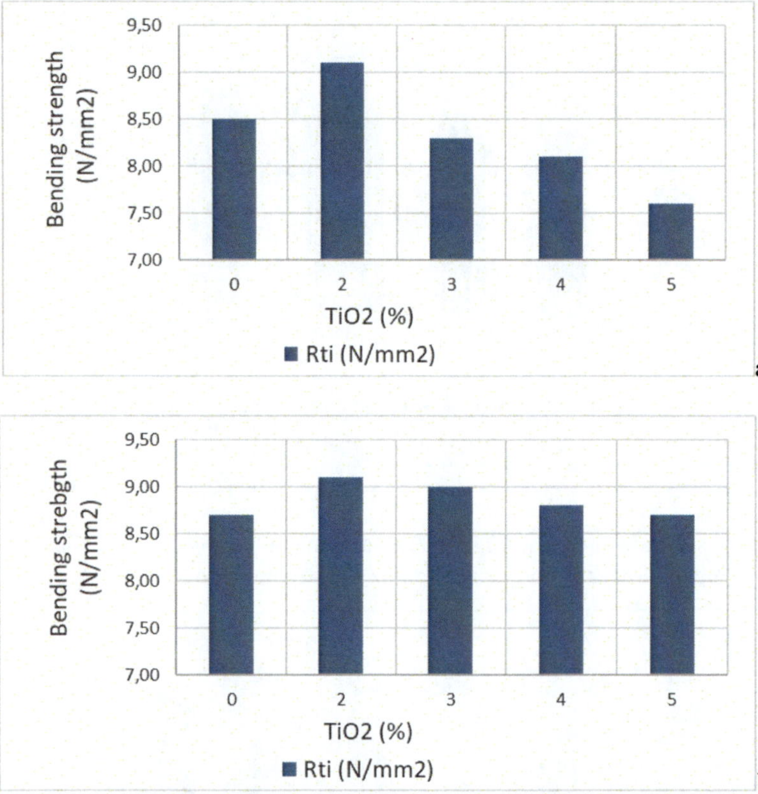

a)

b)

Fig. 6. Bending tensile strength of the cementitious composites with TiO_2 nanoparticles addition at: a) 7 days and b) 28 days.

3.5 Compressive Strength

The compressive strength results at 7 and 28 days are shown in Fig. 6. With regard to the compressive strength of mixtures with TiO_2 nanoparticles addition, the 7-day test shows an increasing trend in the parameter followed by the increase in the amount of nanoparticles to the percentage of 3% TiO_2, after which the value of compressive strength decreases with the increase in the percentage of nanoparticles (Fig. 7a). This increase in compressive strength can be considered a sign of the acceleration of the hardening process, a conclusion that is in correlation with some specifications in the literature [5]. The 28-day test also shows an increase in the value of compressive strength with the increase in the percentage of nanoparticles, but this time up to the percentage of 4% TiO_2, after which the compressive strength decreases (Fig. 7b).

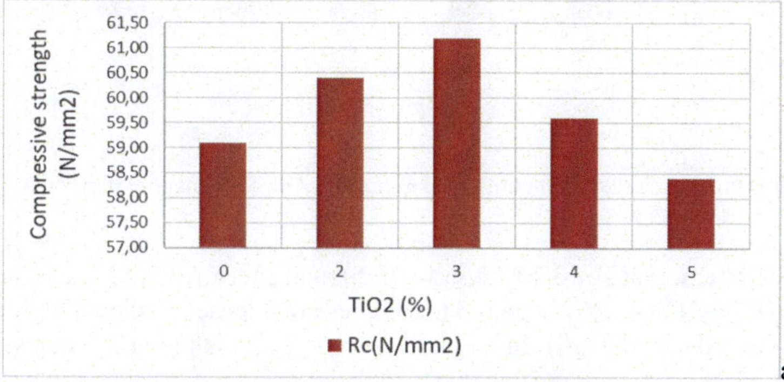

a)

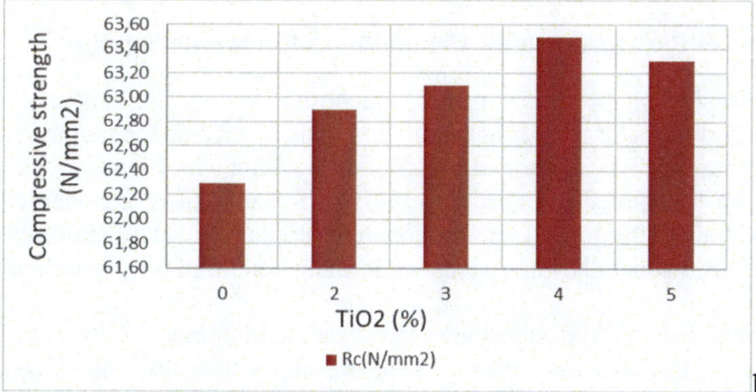

b)

Fig. 7. Compressive strength of the cementitious composites with TiO_2 nanoparticles addition at: a) 7 days and b) 28 days.

3.6 Flexural Strength and Influence of the Environment Conditions

The flexural strength results are shown in Fig. 8. Although a trend in the evolution of the values recorded in the introduction of different percentages of TiO_2 nanoparticles cannot be identified, the highest value was obtained in the sample with a content of 2% TiO_2 nanoparticles and the lowest value in the sample with 3% TiO_2 nanoparticles.

On the other hand, it was found that the exposure of cementitious composites, matured, to 25 freeze-thaw cycles induces a decrease in their performance by 2.5–11% compared to samples not exposed to these environmental conditions (Fig. 9a and Fig. 10).

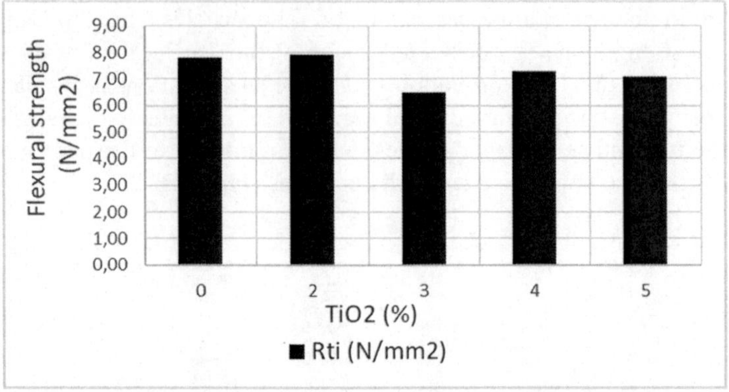

Fig. 8. Flexural strength of cementitious composites with TiO_2 nanoparticles addition.

The action of external stress in the form of thermal shock resulted in a reduction in flexural strength by 1.2–2.8% compared to samples not exposed to these environmental conditions (Fig. 9b and Fig. 10). In the case of a sample containing 4% nanoparticles, the percentage loss is likely to be high due to the inhomogeneity of the sample (Fig. 10).

3.7 Adhesion of the Mortar to the Cementitious Composite Support

As for the adhesion of the mortar to the cementitious composite support with TiO_2 nanoparticles addition, it increases as the amount of nano-TiO_2 in the samples increases, up to the percentage of 3% of nanoparticles introduced, after which it decreases (Fig. 11). However, it can be pointed out that, regardless of the amount of nanoparticles introduced into the binder, the adhesion to the concrete support has values above the limit of 0.5 N/mm^2, a limit generally imposed as a minimum condition for plastering/finishing materials.

However, the fact that the adhesion to the substrate of composites with 4% and 5% nano-TiO_2, respectively, is lower than the control composition (0% nano-TiO_2), may be an indicator of the maximum amount of nanoparticles that can be introduced into the cement mass, so that this performance is not negatively influenced.

3.8 White Degree

The results regarding the evaluation of the degree of whiteness are shown in Fig. 12. As expected, the degree of white increases continuously as the amount of nano-TiO_2 in the cementitious binder increases, this is a very well-known effect and reported in the literature [38], being called "chalk-effect". A more obvious white degree increase is observed for the 2% nano-TiO_2 samples, followed by a slower increase for the 3% and 4% nano-TiO_2 samples, as evidenced by the graphic representation (Fig. 11).

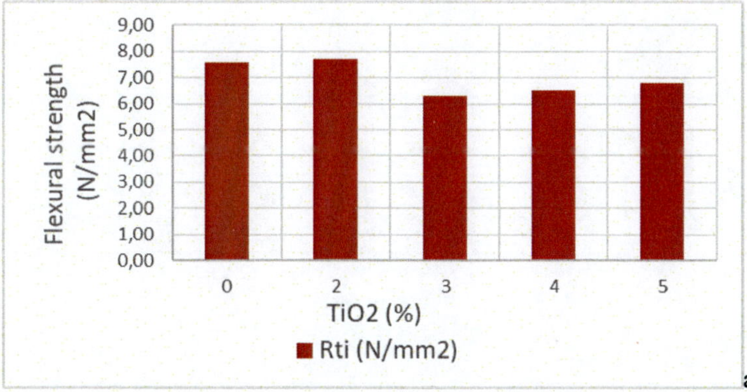

a)

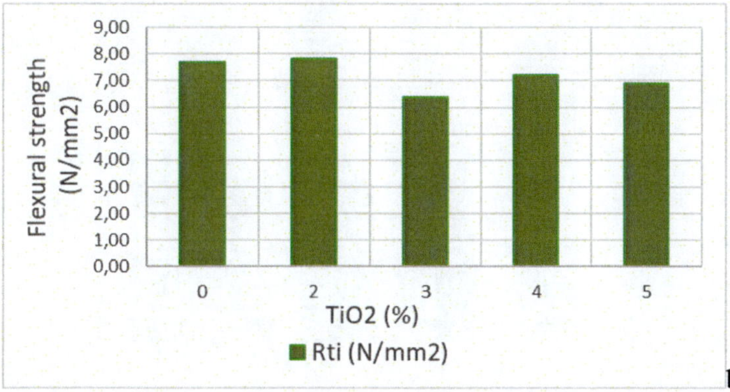

b)

Fig. 9. Flexural strength of the cementitious composites with TiO_2 nanoparticles addtion after: a) freeze-thaw cycles and b) after thermal shock.

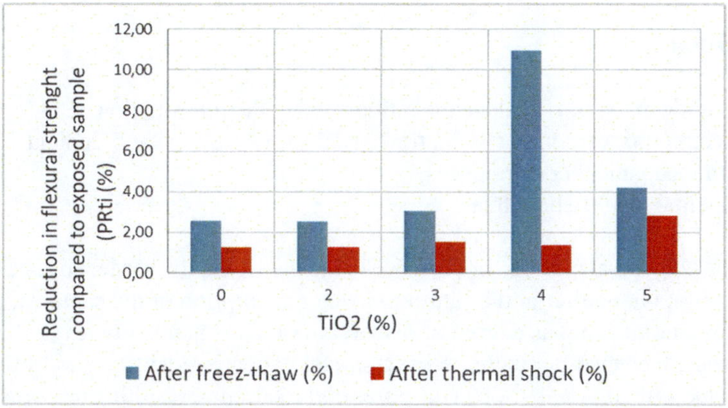

Fig. 10. Reduce in flexural strength of TiO_2-containing samples exposed to freeze-thaw and thermal shock compared to unexposed samples

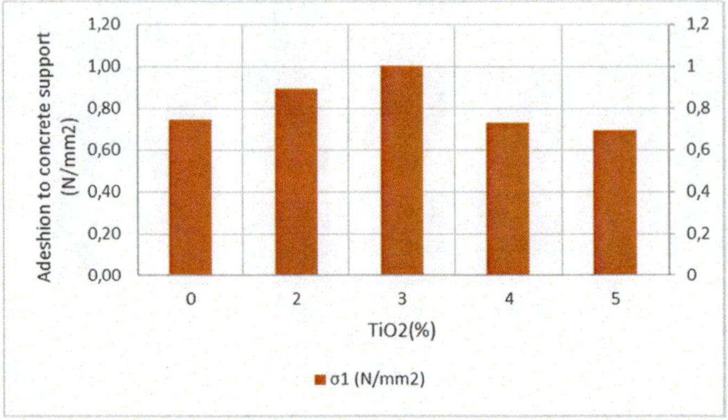

Fig. 11. Adhesion of mortar to concrete support based on white Portland cement containing TiO_2 nanoparticles

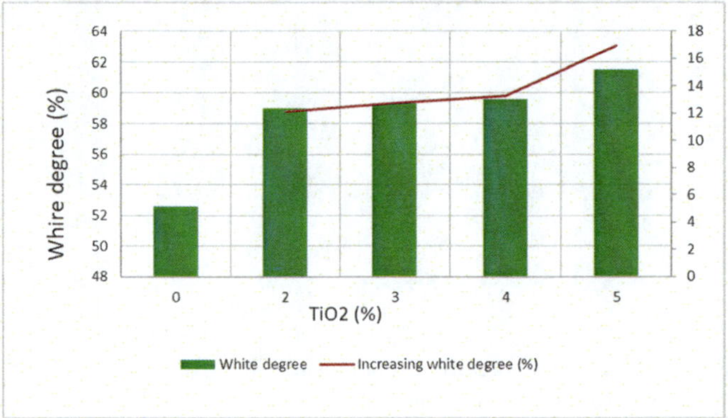

Fig. 12. White degree evaluation of cementitious composites with TiO_2 nanoparticles addition.

4 Conclusions

The aim of this work was to analyze the influence that the introduction of TiO_2 nanoparticles in a cementitious composite matrix based on white Portland cement has on its physical and mechanical performances.

Experimental results show that:

- There was no general increase or decrease in the density in the hardened state (28 days after casting), but neither in the saturation water absorption of the samples, probably due to the inhomogeneous degree of dispersion of TiO_2 nanoparticles in the cementitious mass. However, an increase in fresh-state apparent density was observed, an increase due to their distribution in the pores of the samples, resulting in a densification of the material;
- The amount of water absorbed according to the immersion duration (1 h, 8 h, 24 h, 48 h and 72 h respectively) of the samples containing TiO_2 nanoparticles has an

increasing trend. Thus, it increases with increasing immersion time, in the case of all compositions, regardless of the percentage of TiO_2 nanoparticles introduced into the mixtures;

- Mechanical properties increase at the introduction of 2% TiO_2 nanoparticles (for bending tensile strength at 7 and 28 days) and at the introduction of up to 3% and 4% in the cement matrix (for compressive strength at 7 and 28 days respectively), after which they decrease. Thus the induction of more than 4% nanoparticles of TiO_2 is not motivated;
- The flexural strength under the conditions of exposure of the samples to certain environmental factors is reduced by 2.5–11% in the case of samples exposed to freeze–thaw cycles and by 1.2–2.8% compared to samples not exposed to these conditions;
- The adhesion of the mortar to the support samples with TiO_2 increases as the amount of nano-TiO_2 in the samples increased, up to the percentage of 3% of nanoparticles introduced, after which it decreases.

Acknowledgements. This paper was financially supported by the Project "Entrepreneurial competences and excellence research in doctoral and postdoctoral programs - ANTREDOC", project co-funded by the European Social Fund financing agreement no. 56437/24.07.2019

References

1. Ma, B., Li, H., Mei, J., Li, X., Chen, F.: Effects of nano-TiO_2 on the toughness and durability of cement-based material. Adv. Mater. Sci. Eng. **2015**, 583106 (2015)
2. Zhang, S.M.-H., Tanadi, D., Li, W.: Effect of photocatalyst TiO_2 on workability, strength, and self-cleaning efficiency of mortars for applications in tropical environment. In: 35th Conference on Our World in Concrete & Structures, Singapore (2010)
3. Janus, M., Zając, K.: Concretes with photocatalytic activity, high performance concrete technology and applications. INTECH (2016).
4. Sorathiya, J.V., Shah, S.G., Kacha, S.M.: Effect on Addition of nano "titanium dioxide" (TiO_2) on compressive strength of cementitious concrete. In: International Conference on Research and Innovations in Science, Engineering & Technology, vol. 1. Birla Vishvakarma Mahavidyalaya, Gujarat (2017)
5. Rashad, A.M.: A synopsis about the effect of nano-titanium dioxide on some properties of cementitious materials - a short guide for civil engineer. Rev. Adv. Mater. Sci. **40**, 72–88 (2015)
6. Salemi, N., Behfamia, K., Zaree, S.A.: Effect of nanoparticles on frost durability of concrete. Asian J. Civ. Eng. (BHRC) **15**(3), 411–420 (2014)
7. Nazari, A., Riahi, S., Riahi, S., Shamekhi, S.F., Khademno, A.: Benefits of Fe_2O_3 nanoparticles in concrete mixing matrix. J. Am. Sci. **6**(4), 102–106 (2016)
8. Zhang, M., Li, H.: Pore structure and chloride permeability of concrete containing nano-particles for pavement. Constr. Build. Mater. **25**(2), 608–616 (2011)
9. Li, H., Xiao, H., Guan, X., Wang, Z., Yu, L.: Chloride diffusion in concrete containing nano-TiO_2 under coupled effect of scouring. Compos.: Part B **56**, 698–706 (2014)
10. Soleymani, F.: The filler effects TiO_2 nanoparticles on increasing compressive strength of limestone aggregate-based concrete. J. Am. Sci. **8**, 734–737 (2012)

11. Nazari, A., Riahi, S.: The effect of TiO_2 nanoparticles on water permeability and thermal and mechanical properties of high strength self-compacting concrete. Mater. Sci. Eng. A **528**(2), 756–763 (2010)
12. Nazari, A., Riahi, S.: RETRACTED: splitting tensile strength of concrete using ground granulated blast furnace slag and SiO_2 nanoparticles as binder. Energy Build. **43**(4), 864–872 (2011)
13. Keivan, A., Keivan, A., Behfarnia, K.: The effects of TiO_2 and ZnO nanoparticles on physical and mechanical properties of normal concrete. Asian J. Civ. Eng. (BHRC) **14**(4), 517–531 (2013)
14. Jalal, M., Fathi, M., Farzad, M.: Effects of fly ash and TiO_2 nanoparticles on rheological, mechanical, microstructural and thermal properties of high strength self compacting concrete. Mech. Mater. **61**, 11–27 (2013)
15. Jajal, M., Ramezaianpour, A.A., Pool, M.K.: Effects of titanium dioxide nanopowder on rheological properties of self compactingconcrete. J. Am. Sci. **8**(4), 285–288 (2012)
16. Nazari, A.: The effects of curing medium on flexural strength and water permeability of concrete incorporating TiO_2 nanoparticles. Mater. Struct. **44**(4), 773–786 (2011)
17. Soleymani, F.: Assessments of the effects of limewater on water permeability of TiO_2 nanoparticles binary blended limestone aggregate-based concrete. J. Am. Sci. **7**(11), 7–12 (2011)
18. Jayapalan, A.R., Lee, B.Y., Kurtis, K.E.: Efect of nano-sized titanium dioxide on early age hydration of portland cement. Nanotechnol. Constr. **3**, 267–273 (2009)
19. Kaykha, M.M., Soleymani, F.: The filler effects of TiO_2 nanoparticles in concrete. J. Am. Sci. **7**(12), 158–161 (2011)
20. Soleymani, F.: Assessments of the effects of limewater on water permeability of TiO_2 nanoparticles binary blended palm oil clinker aggregate-based concrete. J. Am. Sci. **8**(5), 698–702 (2012)
21. Chen, J., Kou, S., Poon, C.: Hydration and properties of nano-TiO_2 blended cement composites. Cement Concr. Compos. **34**(5), 642–649 (2012)
22. Essawy, A.A., El Aleem, A.A.: Physico-mechanical properties, potent adsorptive and photocatalytic efficacies of sulfate resisting cement blends containing micro silica and nano-TiO_2. Constr. Build. Mater. **52**, 1–8 (2014)
23. Ma, B., Li, H., Mei, J., Ouyang, P.: Effect of nano-TiO_2 addition on the hydration and hardening process of sulphoaluminate cement. J. Wuhan Univ. Technol.-Mater. Sci. Edn. **30**, 768–773 (2015)
24. de Mendiburu, F.: agricolae: Statistical Procedures for Agricultural Research. R package version 1.3-3 (2020). https://CRAN.R-project.org/package=agricolae
25. Firmino, H.C., et al.: Antifungal activity of TiO_2-CeO_2 nanofibers against Candida fungi. Mat. Lett. **283**, 128709 (2021)
26. Šebesta, M., Nemček, L., Urík, M., Kolenčík, M., Bujdoš, M., Hagarová, I., Matúš, P.: Distribution of TiO_2 nanoparticles in acidic and alkaline soil and their accumulation by Aspergillus niger. Agronomy **10**(11), 1833 (2020)
27. Burduhos Nergis, D.D., Vizureanu, P., Corbu, O.: Synthesis and characteristics of local fly ash based geopolymers mixed with natural aggregates. Rev. Chim. **70**(4), 1262–1267 (2019)
28. SR EN 12350-6: 2019 - Tests on fresh concrete. Part 6: Density
29. SR EN 12390-7: 2019 - Tests on reinforced concrete. Part 7: Density of reinforced concrete
30. SR EN 14617-1:2013 - Agglomerated stone. Test methods. Part 1: Determination of apparent density and ab-sorption of water
31. SR EN 1936:2007 - Methods of testing natural stone. Determination of actual and apparent density and total and open porosity
32. SR EN 12390-5: 2019 - Test on reinforced concrete. Part 5: bending strength of specimens

33. SR EN 12390-3:2019 - Încercare pe beton întărit. Partea 3: Rezistenţa la compresiune a epruvetelor
34. SR EN 14617-2:2016 - Agglomerated stone. Test methods. Part 2: Determination of bending strength (tensile strength)
35. SR EN 14617-5:2012 - Agglomerated stone. Test methods. Part 5: Determination of frost-thaw resistance
36. SR EN 14617-6:2012 - Agglomerated stone. Test methods. Part 6: Determination of thermal shock resistance
37. SR EN 1015-12:2016 - Methods of testing mortars for masonry. Part 12: determination of adhesion of reinforced plastering and gluing mortars on supports
38. Cassar, L.: Nanotechnology and photocatalysis in cementitous materials. In: NICOM'2, pp. 1–7 (2015)

The Use of Ceramic Waste in the Construction Materials Industry Based on the Concept of Sustainable Development

Anamaria Zaharie(✉), Maria Loredana Țințișan, Adrian-Cristian Siomin, Daniela Lucia Manea, and Monica Luminița Pleșa

Faculty of Civil Engineering, Technical University of Cluj-Napoca, Cluj-Napoca, Romania
{anamaria.zaharie,daniela.manea}@ccm.utcluj.ro

Abstract. Sustainable development involves: community responsibility, environmental protection, prosperity and economic efficiency. Protecting the environment and ensuring the long-term prosperity of society means that any product must be manufactured, consumed and transported in a sustainable way. Obtaining new, environmentally friendly materials is currently the main concern of researchers in the field. At the same time, obtaining sustainable materials by replacing components with construction and demolition waste can reduce the consumption of naturally non-renewable resources. The cost of recycling and recovery of construction and demolition waste is mainly given by the cost of storing this type of waste. Massive waste but with great potential for recovery is ceramic waste. In order to reduce the large waste storage resulting from construction and demolition, this paper proposes the construction of new types of plaster mortars by phased and percentage replacement of the aggregate component. The study involves the execution of a series of recipes for CS IV type plaster mortars in which the aggregate was replaced in proportion of 10%, 25%, 35% and 50% with ceramic waste. The present study aims to identify the physical-mechanical behavior over time at 3, 7 and 28 days, respectively, of mortars with ceramic waste, both fresh and hardened. By using ceramic waste as a partial substitute for aggregates in the proposed recipes, it is desired to reduce the consumption of natural resources and their reuse and recovery from the construction industry.

Keywords: Sustainable development · Ceramic wastes · Plastering mortars

1 Introduction

Globally, the generation of construction and demolition waste as well as waste from the processing of various construction materials is a major problem in environmental pollution.

Depending on the source of raw materials, ceramic waste can be classified into two categories: waste generated by ceramic factories that use only red paste for the manufacturing of their products (brick, ceramic blocks and roof tiles) and burnt waste produced from ceramic tiles, such as tiles and sanitary objects [1].

L. Moldovan and A. Gligor (Eds.): Inter-Eng 2021, LNNS 386, pp. 182–191, 2022.
https://doi.org/10.1007/978-3-030-93817-8_18

Recent studies have estimated that about 30% of daily production in the ceramic industry is lost [2]. At present, this waste is used to make landfills, but this reuse does not meet the necessary storage requirements for this waste, which has a high degree of occupancy and environmental pollution.

The process of burning ceramic products results in pollutants of the type: total dust, flue gases: nitrogen oxides, sulphur oxides, carbon dioxide, carbon monoxide, heavy metals, vapour-saturated air. The amount of greenhouse gas emissions can be reduced by recycling existing ceramic products and revaluating them to obtain construction materials with low environmental impact [3].

At present, there are various studies on the production of concrete with aggregates from brick waste, which replace natural aggregates in concrete recipes. In these experiments, their durability, microscopic analysis and their mechanical strength are studied [4–6]. Another method of reusing ceramic waste is to make bricks from a mixture of discarded pottery, crushed toiletries and other ingredients [7].

2 Materials and Methods

2.1 Materials

In order to obtain an ecological mortar the materials used in the experimental program were Portland cement, aggregates, ceramic wastes and water. To obtain optimal results, the aggregates were replaced with crushed ceramic wastes in different proportions as following: 10%, 25%, 35% and 50%.

Cement. Portland cement is a construction material in the form of a fine powder resulting from the grinding processes of the cement clinker, having the role of a water-based binder in mortar and concrete recipes.

The type of cement used is Portland cement EN 197-1: 2011 CEM I 52,5R without addition in which the composition of the clinker is in proportion of 95 ÷ 100%. *The following* Table 1 *gives the physical properties of the used cement.*

Fine Aggregate. The aggregate used in the experimental program is sand. River sand with a grain size between 0÷4 mm, washed, dried and subsequently sorted, whose foreign matter content is 0%. The sand was screened as follows: (0–0.25) mm, (0.25–0.50) mm, (0.50–1.00) mm, (1.00–2.00) mm and (2.00–4.00) mm.

Ceramic Waste. The experimental program proposed the use of solid brick ceramic waste. This type of waste results from construction and demolition works. Solid brick is a raw, coloured ceramic product with a porous structure, of parallelepiped shape, whose raw material is clay, resulting from the firing process at temperatures between (900–1000) °C.

In order to be able to be used, the brick waste was subjected to the breaking process with the help of the compactor and subsequently screened sorts. (See Fig. 1). For the production of ceramic sorts in industrial quantities, a crushing equipment for the exploitation of stone waste can be used, which can produce 150 tons/hour.

The resulting sorts are similar in size to the aggregates used in the CSIV mortar recipe, referred to as standard from now on. Each sort of sand was replaced in proportion of 10%, 25%. 35% and 50% with solid brick waste (see Fig. 2).

Table 1. Physical properties of Cement*.

Physical properties	Test result
Fineness: specific surface, Blaine method [cm^2/g]	5568,8
Setting time, Vicat's method	
Initial [hrs: min.]	1:45
Final [hrs: min.]	3:20
Compressive strength of standard mortar	
3 days [N/mm^2]	34.82
7 days [N/mm^2]	36.71
28 days [N/mm^2]	39.11

*Physical test was carried out in the Laboratory of Building Materials - Technical University of Cluj-Napoca.

Fig. 1. Device of grinding ceramic waste.

Water. In order to make the mortar-type mixture, it is important that the used water is clear, does not contain impurities, and does not have an odor to avoid the existence of certain substances that can prevent the hardening of the mortar.

In the experimental study, drinking water from the Laboratory of the Faculty of Constructions from Cluj-Napoca was used.

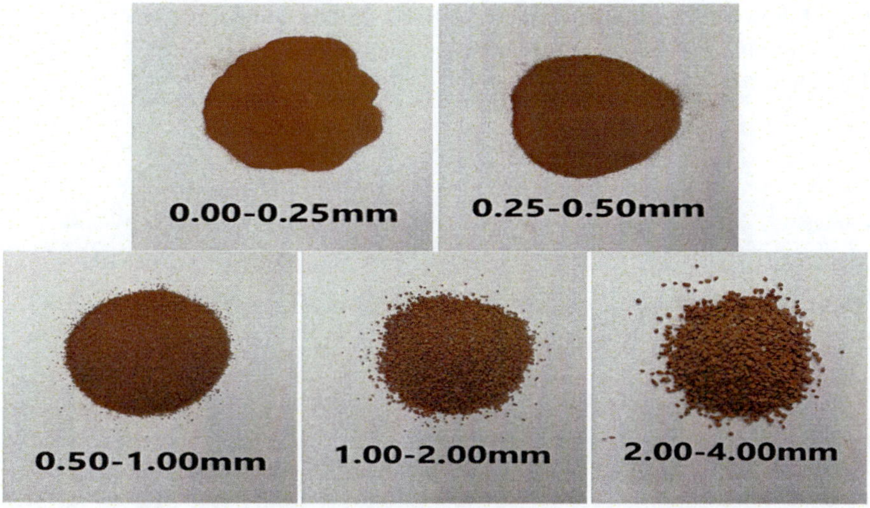

Fig. 2. Representation of ceramic waste after the sorting stage.

2.2 Composition of Ceramic Waste Mortars

Ecological mortars are the mortars in which the raw materials are fully or partially replaced by recyclable waste, for the purpose of environmental sustainability and sustainable development in civil engineering, while preserving the properties of the material aimed at.

The basic recipe used in the experimental program is the recipe for CS IV type plaster mortar, the quantities for 1 m3 being presented in Table 2.

Table 2. Plaster mortar standard.

Plaster Mortar S		
Materials	Tip of materials	Amount
Cement	Portland Cement 52,5R	385 kg
Fine agg.	(0÷0.25) mm	77,5 kg
	(0.25÷0.50) mm	232,5 kg
	(0.50÷1) mm	378,5 kg
	(1÷2) mm	232,5 kg
	(2÷4) mm	620 kg
Water	Water	267 l
w/c		0.69

In the experimental part, the authors aimed to produce several types of ecological plastering mortars, where the sand was replaced in various proportions with ceramic waste, as follows: 10% ceramic waste (CW 10%), 25% ceramic waste (CW 25%), 35% ceramic waste (CW 25%) and 50% ceramic waste (CW 50%).

The Table 3 below presents four variants and the recipes used.

Table 3. Ceramic waste mortars compositions.

Mortar types	CW 10%	CW 25%	CW 35%	CW 50%
Cement [kg/m^3]	1	1	1	1
Fine agg. [kg]				
0÷0.25 mm	0.9	0.75	0.65	0.50
0.25÷0.50 mm	0.9	0.75	0.65	0.50
0.50÷1 mm	0.9	0.75	0.65	0.50
1÷2 mm	0.9	0.75	0.65	0.50
2÷4 mm	0.9	0.75	0.65	0.50
Crushed bricks [kg]				
0÷0.25 mm	0.1	0.25	0.35	0.50
0.25÷0.50 mm	0.1	0.25	0.35	0.50
0.50÷1 mm	0.1	0.25	0.35	0.50
1÷2 mm	0.1	0.25	0.35	0.50
2÷4 mm	0.1	0.25	0.35	0.50
Ration w/c	0.59	0.65	0.67	0.75

All mortar recipes were cast in two layers, each layer being compacted with a mechanical vibrating machine, in steel moulds with the size of 40 × 40 × 160 mm. The recipes were prepared in the Laboratory of Building Materials, they were kept in wet air boxes up to 7 days and moved in a room with the humidity (65 ± 5)% at a temperature of (20 ± 4)°C, up to the test age.

2.3 Methods

The determination of the physical and mechanical characteristics were performed according to the current standards regarding the testing of mortars SR EN 1015 1 ÷ 21/2020.

A number of three specimens were subjected to the tests for each method performed, taking into account their final arithmetic value.

Apparent Density. The determination of the apparent density of the mortars in fresh state, respectively in hardened state was made according to the European standard SR EN 1015-6: 2001/A1: 2007 [8] respectively SR EN 1015-10: 2002/A1: 2007 [8, 9].

Consistency of the Mortar. The determination of the fresh mortar consistency is made with the spreading mass according to SR EN 1015-3: 2001 [10] (see Fig. 3).

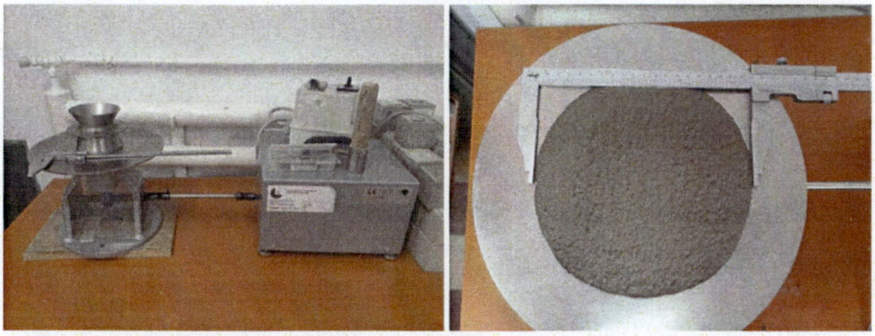

Fig. 3. The mass of spreading. Consistency of standard mortar.

Determination of Flexural Strength. The flexural strength test was determined according to SR EN 1015-11:2020, performed in time at 3, 7 and 28 days on prism sized 40 × 40 × 160 mm [11].

Determination of Compressive Strength. This test is performed on prism rests obtained after the flexural strength. The test was performed in time at 3, 7 and 28 days with the hydraulic press. The testing was performed according to the European standard SR EN 1015-11: 2002 [11].

3 Results and Discussion

3.1 Apparent Density

On the ecological plaster mortars there was determined the apparent density in fresh state and the apparent density of the hardened mortars at 3, 7 and 28 days, respectively, according to the following tables: Table 4 and Table 5.

Table 4. Apparent density of fresh mortars.

Mortar types	S	CW 10%	CW 25%	CW 35%	CW 50%
Aparent density [kg/m^3]	2274	2695	2153	1961	1948

The apparent density of fresh mortars is between 1948-2695 Kg/m^3. (See Table 4) The mortar with the highest value of bulk density is CW10% mortar. All values obtained are in accordance with the current regulations. The average values of the apparent density of fresh mortar show a proportional decrease with the increase in the amount of ceramic waste.

Table 5. Apparent density of hardened mortars (from 3 to 28 days).

Mortar types	Mortar age					
	3 days		7 days		28 days	
	Weight [kg]	Apparent density [kg/m^3]	Weight [kg]	Apparent density [kg/m^3]	Weight [kg]	Apparent density [kg/m^3]
S	0.575	2250	0.563	2200	0.567	2210
CW 10%	0.566	2210	0.564	2200	0.550	2140
CW 25%	0.520	2030	0.530	2070	0.524	2040
CW 35 %	0.512	2000	0.504	1970	0.511	1990
CW 50%	0.497	1940	0.498	1950	0.479	1870

The apparent density of hardened mortars is between 1940-2250 Kg/m^3. (see Table 5). The mortar with the highest value of apparent density is the S-type mortar, a mortar that does not contain brick waste. According to the table above, it can be seen that the more brick waste is added to the mortar recipes, the lower the value of the apparent density (at 3, 7 and 28 days).

3.2 Consistency of Fresh Ecological Mortars

The representation of the average values of the consistency of the fresh mortar determined with the spreading mass, are found in Table 6 [10].

Table 6. Mortar consistency values.

Mortar type	S	CW 10%	CW 25%	CW 35%	CW 50%
Average value of the consistency [mm]	182.5	200.0	186.0	179.0	174.0

The mortar with the highest value of consistency is the mortar in which the aggregate has been replaced in proportion of 10%. As for the average values obtained for the other types of mortars, they decrease in value as the percentages of brick waste in the proposed recipes increase.

3.3 Bending Tensile Strength

The results are summarized in Table 7 and shown in Fig. 4 (see Fig. 4. and Table 7).

According to the table above, it can be seen that the bending tensile strengths increase over time for all mortars studied in the experimental program. All mortars with brick waste have higher values than the standard mortar.

Table 7. Tensile strength of plaster mortars.

	Flexural strength [N/mm²]				
Mortar type	S	CW 10%	CW 25%	CW 35%	CW 50%
Mortar age					
3 days	4.47	4.88	4.26	4.22	4.09
7 days	5.22	5.28	4.96	4.34	4.28
28 days	6.08	6.60	6.72	6.74	7.63

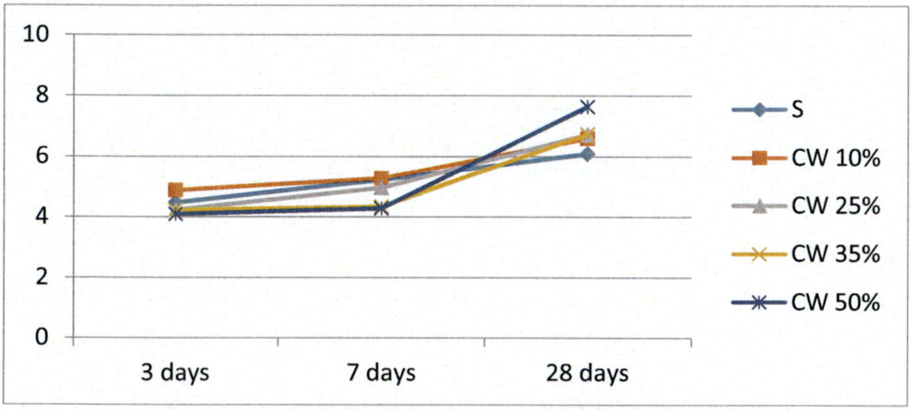

Fig. 4. Graphic representation of Flexural strength [N/mm^2].

3.4 Compressive Strength

The results are summarized in Table 8 and shown in Fig. 6 (See Fig. 6 and Table 8). The values of the compressive strength are above the average values according to the current standards, namely 6 N/mm^2. The figure below shows the breaking of the mortar samples in which the waste replaced the sand in proportion of 25% (Fig. 5).

Table 8. Compressive strength of plaster mortars.

	Compressive strength [N/mm²]				
Mortar type	S	CW 10%	CW 25%	CW 35%	CW 50%
Mortar age					
3 days	34.86	28.56	22.69	22.68	22.63
7 days	36.71	35.24	28.48	29.45	25.70
28 days	39.11	44.41	42.28	39.57	33.96

Fig. 5. Tests to determine the compressive strength.

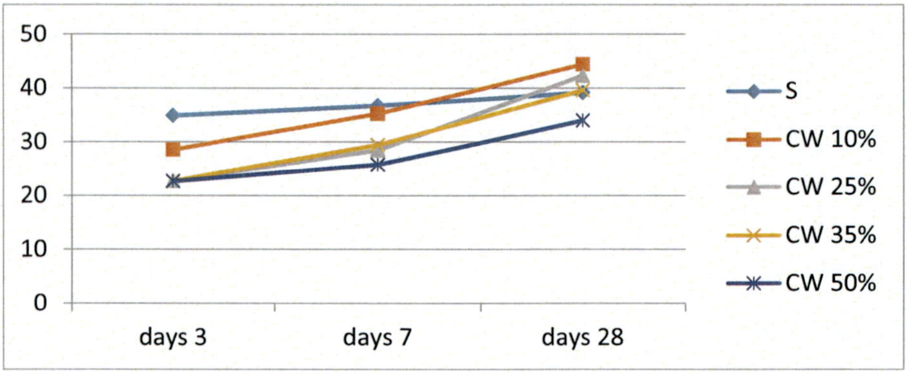

Fig. 6. Compressive strength [N/mm^2].

Table 8 shows the average values obtained in the experimental program. The experimental study was carried out in time, to highlight the influence of brick waste, in terms of the evolution of mechanical properties. It was observed that using brick waste in different proportions does not negatively influence the results, so that the CW type mortar 10%, has the highest values of compressive strength, higher than standard mortar.

CW type mortars 25%, CW 35% all have higher values than the standard mortar at 28 days.

It was highlighted that the 50% CW type mortar has lower values of compressive strength, compared to the standard mortar, at each test (3, 7 and 28 days), but falls within the limits imposed by standards.

4 Conclusion

Waste management aims to save natural resources by reusing recyclable waste. The issue of waste and the decrease in the amount of resources, especially non-renewable ones, are being discussed more and more. The waste recovery process will eliminate pollutant gas emissions. Ceramic waste can be used to make new construction products.

The present study analyzes the realization of some plaster mortars in which the aggregates were replaced with brick waste in different proportions.

According to the results obtained, the use of brick waste as a sand substitute in different proportions (10%, 25%, 35%) does not negatively influence the physical and mechanical properties of plaster mortars, except for the 50% variant.

The average values obtained in terms of mechanical strength (bending tensile strength as well as compressive strength) are above the average values according to the current standards. In some cases these resistances exceed even the values obtained on the reference mortar.

Following the results obtained, it can be seen that the brick waste can be used as a partial replacement of the aggregate. By revaluating and reusing them in the composition of mortars, the negative impact of this waste on the environment will be significantly reduced.

The present study is a first step in following the behaviour of these types of mortars, and in the future to study by SEM the influence of brick waste in the process of hardening of the mortar and the way of working of the component materials. It will be also studied the process of hydration of Portland cement taking in considering also the influence of ceramic waste in the composition of mortars.

References

1. Pacheco-Torgal, F., Jalali, S.: Reusing ceramic wastes in concrete. Constr. Build. Mater. **24**(5), 832–838 (2010)
2. Senthamarai, R.M., Devadas Manoharan, P.: Concrete with ceramic waste aggregate. Cem. Concr. Compos. **27**(9–10), 910–913 (2005)
3. Production of ceramics (CER). https://circabc.europa.eu. Accessed 05 Apr 2021
4. Medina, C., Frías, M., Sánchez de Rojas, M.I.: Microstructure and properties of recycled concretes using ceramic sanitary ware industry waste as coarse aggregate. Constr. Build. Mater. **31**, 112–118 (2012)
5. Halicka, A., Ogrodnik, P., Bartosz, Z.: Using ceramic sanitary ware waste as concrete aggregate. Constr. Build. Mater. **48**, 295–305 (2013)
6. Ray, S., Haque, M., Soumic, S.A., Mita, A.F., Rahman, M.M., Tanmoy, B.B.: Use of ceramic wastes as aggregates in concrete production: a review. J. Build. Eng. **43**, 102567 (2021)
7. Transforming waste into high-end brick design. https://www.stonecycling.com/our-projects. Accessed 05 Mar 2021
8. SR EN 1015-6:2001/A1:2007 Methods of testing masonry mortar. Part 6: Determination of apparent density of fresh mortar
9. SR EN 1015-10:2002/A1:2007 Methods of testing masonry mortar. Part 10: Determination of apparent density of hardened mortar
10. SR EN 1015-3:2001 Methods of testing masonry mortar. Part 3: Determination of consistency of fresh mortar (with spreading table)
11. SR EN 1015-11:2020 Methods of testing masonry mortar. Part 11: Determination of bending and compressive strength of hardened mortar

Self-compacting Concrete with Recycled Aggregates

Vlad Constantin Panaite(✉) and Marinela Barbuta

Faculty of Civil Engineering and Building Service, "Gheorghe Asachi" Tehnical Univerity of Iasi, 1, Prof. Dimitrie Mangeron Bld., 70050 Iasi, Romania

Abstract. The article presents an experimental study on self-compacting concrete prepared by using different percentages of recycled aggregates, in order to obtain a sustainable and environmentally friendly self-compacting concrete. For this purpose, ten different compositions of self-compacting concrete were made. The main recipe used for comparing results has been discovered from trial and error using the recommendations from the Eurocode. After the recipe was obtained the replacement of cement, limestone and fine aggregates had been made with the following recycled materials: fly ash, blast furnace slag, crumb rubber and brick dust in different percent and replacing different materials as presented in Table 1. The influence of the type and dosage of recycled aggregates is discussed. The main objective of the article was to analyze the properties of self-compacting concrete prepared with these recycled materials. In order to have the possibility to certify the difference between the ten recipes no changes had been made from one recipe to another except variations of water and super plasticizer quantity for obtaining SCC due to the different absorption by the recycled material. The w/c ratio varies from 0.38 to 0.63 and the superplasticizer varied from 0.8 to 1.3 % of the cement.

Keywords: Self-compacting concrete · Recycled aggregates · Environmentally friendly

1 Introduction

Self-compacting concrete is a special type of concrete which possess the ability to self-compact under its own weight without any vibration. SCC offers high compressive strength and low permeability. It is highly flowable, non-segregation concrete, it can spread all the space of the formwork. This reduces the cost, saves time and improves quality of concrete. Since cement is the most expensive component of concrete, reducing cement content is an economical solution. The high percentage extraction of natural aggregates for the production of concrete harms the environment, compromising the sustainable development of the construction industry. In the meantime, the volume of waste materials is increasing. This waste can and should be recycled and reused in the concrete production. Therefore, utilizing waste concrete recycling materials in engineering will contribute to protecting the environment and capturing the residual value of waste concrete [1–3]. The influence of fly ash on the properties of fresh concrete is improving

L. Moldovan and A. Gligor (Eds.): Inter-Eng 2021, LNNS 386, pp. 192–201, 2022.
https://doi.org/10.1007/978-3-030-93817-8_19

workability. In combination with other additions fly ash concrete can present improved properties (higher durability, a lower shrinkage) [7, 8]. As SCC requires high cement content which leads to increase in cost and temperature rise during heat of hydration, additives or pozzolanic material such as fly ash, limestone powder or slag can generally be used as partial replacement of cement to reduce the cost and heat of hydration. The superplastizer is necessary for producing a highly fluid concrete mix, while powder materials or viscosity agent are required to maintain sufficient stability/cohesion of the mix, hence reducing bleeding, segregation and settlement [16].

Used tires is another waste that is used in concrete [9, 10]. The mechanical properties of rubberized concrete are smaller compared to ordinary concrete. But these can be improved by adding fly ash [11]. The possibility to design self-compacting rubberized concrete appears particularly attractive because this new material might join the characteristics of SCC (high flowability, high mechanical strength, low porosity, etc.) with the tough behavior of the rubber phase, thus leading to a building material with more versatile performances [14].

Recycled aggregates are used for obtaining concrete and it was demonstrated in various research that workability is not affected. Compressive strength can be the same as in the case of structural concrete or even higher by reducing the ratio water/cement [12].

The reduction of the water-cement ratio results in a decrease in porosity and refinement of capillary pores in matrix [15].

2 Experiment Description

2.1 Materials and Method

For the experimental tests a mix of concrete was used as control with the following components: cement: 380 kg/m^3, aggregate sort 0–4 mm: 918 kg/m^3, sort 4–8 mm: 306 kg/m^3, and 8–16 mm: 540 kg/m^3 and limestone 120 kg/m^3. The water was 145 L/m^3 and the super-plasticizer (Sika VascoCrete 20HE Gold) 0.8 % of the amount of cement. This mix was obtained by trying several recipes and the one that gave the best results in fresh state was chosen. The other nine compositions were prepared by replacing the components of the control mix with different types and different proportions of wastes (Fig. 1), as is presented in Table 1. The quantities were chosen as a first step and will be further investigated. The trials for finding the closest recipe for SCC began February 2021 and finished March 2021 with all samples poured and ready to get hardened for in April to get the 28 day strength that is presented in the graphics below. The fresh concrete was sampled and tested for filling ability, passing ability and segregation resistance. Slump flow, T50 cm spread, and V-funnel flow tests were conducted to measure the filling ability of concrete. The results of fresh properties of SCC are presented in Table 2.

The mechanical properties determined where: compressive strength, flexural strength and split tensile strength with the results revealed in the graphics below concluding that when the amount of recycled material increases the strength decreases except for the recipe when fly ash replaced limestone.

The cement used for obtaining of all the mixes was type ExtraDur 52 CEM I,5R, according to Romanian Standard [17].

Fig. 1. The materials used in the preparation of the 10 mixes.

Table 1. Mix proportion for SCC mixes.

Mix	C_0	C_1	C_2	C_3	C_4	C_5	C_6	C_7	C_8	C_9
Cement (kg/m^3)	380	380	380	380	380	380	380	380	380	380
Limestone (m^3)	0.044	0	0.044	0.044	0.04	0.044	0.044	0.044	0.044	0.044
Limestone (kg/m^3)	120	0	120	120	120	120	120	120	120	120
Sand (m^3)	0.57	0.57	0.57	0.57	0.57	0.51	0.51	0.34	0.46	0.34
Sand (kg/m^3)	918	918	918	918	918	826.2	826.2	550.8	734.4	550.8
Sort1 4–8 mm (kg/m^3)	306	306	306	306	306	306	306	306	306	306
Sort2 8–16 mm (kg/m^3)	540	540	540	540	540	540	540	540	540	540
Water (l/m^3)	145	145	150	153	145	194	145	238	172	239
SP (%)	0.8	0.8	0.8	0.85	0.8	1	0.8	1.1	1	1.3
Fly ash (kg/m^3)	0	120	38	57	0	0	0	0	0	0
Blast furnace slag (kg/m^3)	0	0	0	0	0	0	0	367.2	0	0
Crumb rubber (kg/m^3)	0	0	0	0	12	91.8	9.18	0	0	0
Brick dust (kg/m^3)	0	0	0	0	0	0	0	0	183.6	367.2
W/c ratio	0.38	0.38	0.394	0.4	0.38	0.51	0.38	0.62	0.45	0.63

In Table 1 listed mixes are as following:

- C_0 - control mix
- C_1 - mix with complete replacement of the limestone filler with fly ash C_2 - mix with replacement of 10% of cement with fly ash.
- C_3 - mix with replacement of 15% of cement with fly ash.
- C_4 - mix with replacement of 10% of limestone filler with crumb rubber
- C_5 - mix with replacement of 10% of sand with fly crumb rubber
- C_6 - mix with 10% replacement of the amount of 10% sand removed from the recipe with crumb rubber
- C_7- mix with replacement of 40% of sand with blast furnace slag
- C_8 - mix with replacement of 20% of sand with brick dust
- C_9 - mix with replacement of 40% of sand with brick dust.

2.2 Samples

The samples were poured in cubes of 150 mm in size for determining the compressive strength, and prisms of 100 × 100 × 500 mm in dimension for determining the flexural strength and split tensile strength (NE 012/2-2010). After 24 h, the specimens were removed from the formwork and kept in the laboratory in the water until testing. The temperature of water in which the cubes were submerged was maintained at 25 ± 2 C. The specimens were cured for 28 days until testing for determining the compressive strength (fc), flexural strength (fti) and split tensile strength (ftd) according to standard on minimum three samples [18–20].

3 Testing Result and Discussions

3.1 Fresh Properties of SCC

The fresh concrete was sampled and tested for filling ability, passing ability and segregation resistance as shown in Fig. 2. Slump flow, T50 cm spread, and V-funnel flow tests were conducted to measure the filling ability of concrete. The results of fresh properties of SCC are presented in Table 2.

(a) (b) (c)

Fig. 2. Tests performed on fresh concrete: (a) Slump flow, (b) V-funnel, (c) L-box.

Table 2. Fresh properties of SCC.

Mix	Slump flow 550–899 mm	T500 2–5 s	V-funnel 8–27 s	L-box 0.8–1.0 cm
C_0	670	3.7	9.4	0.8
C_1	660	3.5	8.2	0.8
C_2	660	3	9	0.8
C_3	680	2.9	9	0.8
C_4	640	3.8	9	0.8
C_5	450	–	–	–
C_6	550	3.2	9.2	0.8
C_7	400	–	–	–
C_8	570	4.5	14	0.8
C_9	540	–	–	–

The workability tests revealed that the viscosity of the SCC varied from 680 at C3 to 400 mm at C7 (pictured in Fig. 3). The minimum value of slump has to be 550 mm and the maximum value 899 mm for a fresh SCC. Consequently, three of the mixtures (C5, C7, C9) do not behave like a self-compacting concrete, requiring a generous surplus of water and additive. Thereby materials such as rubber, blast furnace slag and brick dust cannot be used in self-compacting concrete without additional costs.

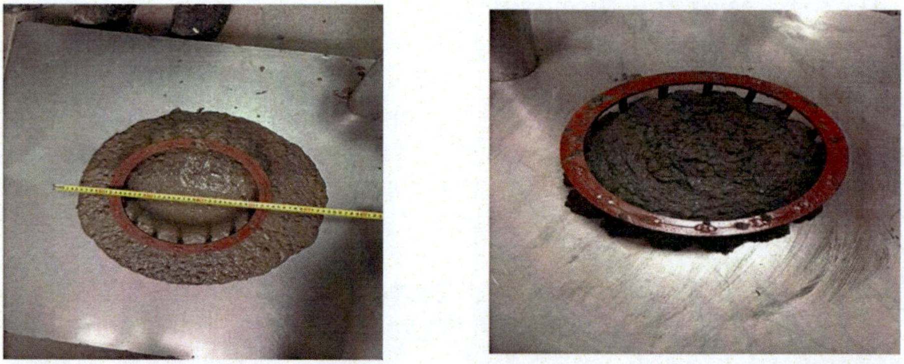

Fig. 3. Sample results on fresh concrete: (a) Slump flow of C9, (b) Slump flow of C7.

3.2 Mechanical Properties of SCC

Compressive Strength. The compressive strength determined on cubes at 28 days of the different mixes is shown in Fig. 4.

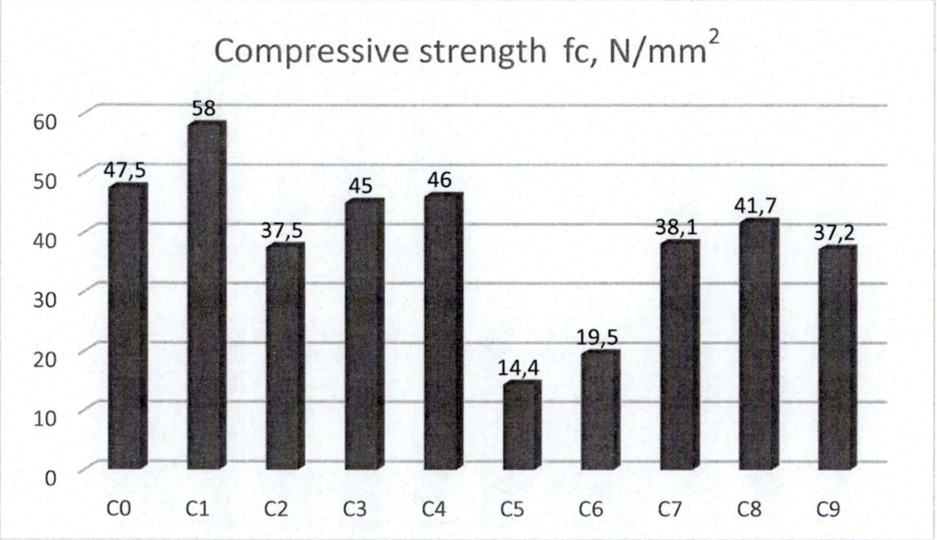

Fig. 4. Compressive strength comparison.

The compressive strength of experimental mixes of self-compacting concrete varied from 58 N/mm^2 for mix C1 to a minimum value of 14.4 N/mm^2 for mix C5. Fly ash improved the compressive strength, the value of fc for C1 is with 22.1% bigger than that of the control mix. All other values of fc were smaller than that of the control mix. Except the values of concretes with crumb rubber (C5 and C6), all other values of fc were near the value of control mix and define the concrete as structural concrete (NE 012/2-2010). Figure 5 depicts the testing procedure.

(a)

(n)

Fig. 5. Compressive strength: (a) Testing setup, (b) The failure of cubes in axial compression.

Flexural Strength. The flexural strength of the concrete specimens was determined after 28 days. Figure 6 illustrates the flexural strength of all mixes.

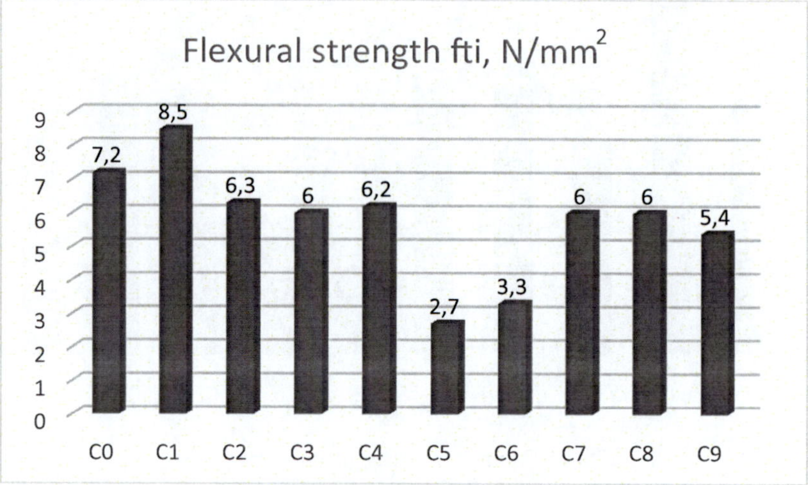

Fig. 6. Flexural strength comparison.

The flexural strength of experimental mixes of self-compacted concrete varied from 8.5 N/mm^2 for mix C1 to a minimum value of 2.7 N/mm^2 for mix C5. Fly ash improved the flexural strength, the values of fti for C1 is bigger with 15,3% than that of the control mix. All other values of fti were smaller than that of the control mix. Except the values of concretes with crumb rubber (C5, C6), all other values of fti were near the value of control mix. Figure 7 depicts the flexural testing procedure.

(a) (b)

Fig. 7. Flexural strength: (a) Testing setup, (b) The surface of failure.

Split Tensile Strength. The split tensile strength of concrete was tested in the laboratory conditions at the age 28 days and the results are given in Fig. 8.

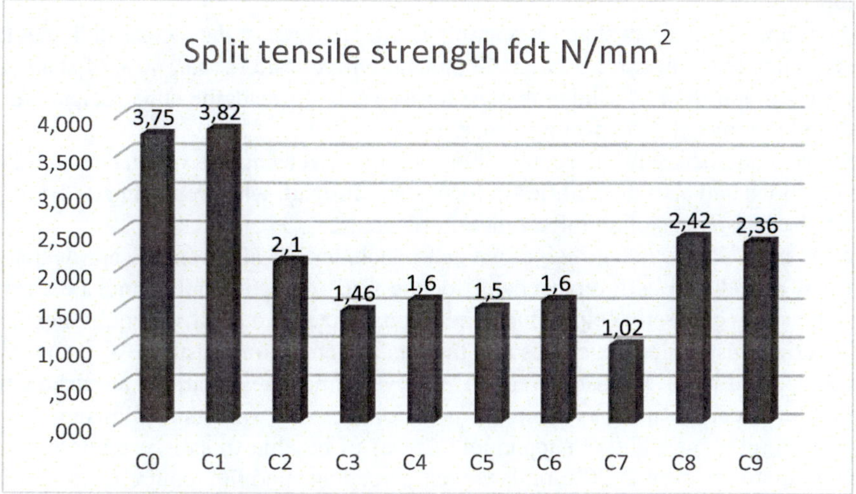

Fig. 8. Split tensile strength comparison.

The split tensile strength of experimental mixes of self-compacted concrete varied from 3.82 N/mm^2 for mix C1 to a minimum value of 1.02 N/mm^2 for mix C7. Fly ash improved the split tensile strength, the value of ftd for C1 is with 1.8% bigger than that of the control mix. All other values of ftd were smaller than that of the control mix. Figure 9 depicts the split testing procedure.

(a) (b)

Fig. 9. Split tensile test: (a) Testing setup, (b) The type of failure.

4 Conclusions

In the experimental program, four types of wastes (fly ash, blast furnace slag, crumb rubber and brick dust) were used in self-compacting concrete. From test results, we can conclude that:

The properties in fresh state for obtaining SCC for three of the recipes do not obtain the recommended values for SCC: 10% of sand with crumb rubber, 40% of sand with blast furnace slab, 40% of sand with blast furnace slag. Instead, the other recipes can be used for obtaining self-compacting concrete.

Fly ash can successfully replace limestone by increasing the resistance of all the three functions and also the abilities in fresh state. Instead, when replacing a part of the cement, the resistances of all three functions decrease.

In the blast furnace slag recipe, we can conclude that all strengths are decreased when it is introduced as a replacement for the fine aggregate and cannot be used in self-compacting concrete recipes due to additional water and additive requirements.

The brick dust recipe concludes that there is a decrease in resistance as the amount of sand replaced by brick dust increases. However, the values of all the tests show that the brick dust can be used as a replacement for the sand with the following impediment: it requires an additive supplement (20%) to be able to satisfy a minimum of fresh requirements even at 20% replacement percentage and the requirements of fresh self-compacting concrete are not satisfied when the replacement percentage is 40%.

When crumb rubber replaces sand, we see a decrease in strength and a worse behavior in fresh state. Following the studies, the fly ash will be used as a limestone replacement in the SCC recipe.

References

1. Figueiras, H., Nunes, S., Coutinho, J.S., Figueiras, J.: Combined effect of two sustainable technologies: self-compacting concrete (SCC) and controlled permeability formwork (CPF). Construct. Build. Mater. **23**, 2518–2526 (2009)
2. Kou, S.C., Poon, C.S.: Properties of self-compacting concrete prepared with coarse and fine recycled concrete aggregates. Cem. Concr. Coposites **31**(9), 622–627 (2009)
3. Sun, C., Chen, Q., Xiao, J., Liu, W.: Utilization of waste concrete recycling materials in self-compacting concrete. Resour. Conserv. Recycl. **161**, 104930 (2020)
4. Assie, S., Escadeillas, G., Waller, V.: Estimates of self-compacting concrete 'potential' durability. Constr. Build. Mater. **21**(10), 1909–1917 (2007)
5. Unal, O., Topcu, I.B., Uygunoglu, T.: Use of marble dust in self-compacting concrete. In: Proceedings of V Symposium MERSEM0 on Marble and Natural Stone, Afyon, Turkey, pp. 413–420 (2006)
6. Felekoglu, B., Tosun, K., Baradan, B., Altun, A., Uyulgan, B.: The effect of fly ash and limestone fillers on the viscosity and compressive strength of self-compacting repair mortars. Construct. Build. Mater. **36**(9), 1719–1726 (2006)
7. Dinakar, P., Babu, K.G., Santhanam, M.: Durability properties of high volume fly ash self-compacting concrete. Cem. Concr. Compos. **30**, 880–886 (2008)
8. Leung, H.Y., Kim, J., Nadeem, A., Jayaprakash, J., Anwar, M.P.: Sorptivity of self-compacting concrete containing fly ash and silica fume. Constr. Build. Mater. **113**, 369–375 (2016)

9. Khaloo, A.R., Dehestani, M., Rahmatabadi, P.: Mechanical properties of concrete containing a high volume of tire-rubber particles. Waste Manag. **28**, 2472–2482 (2008)
10. Yilmaz, A., Nurhayat, D.: Possibility of using waste tire rubber and fly ash with Portland cement as construction materials. Waste Manag. **29**, 1541–1546 (2009)
11. Abaza, O.A., Hussein, Z.S.: Flexural behavior of steel fiber-reinforced rubberized concrete. J. Mater. Civil Eng. **28**(1), 04015076 (2015)
12. Thomas, C., Sosa, I., Setién, J., Polanco, J.A., Cimentada, A.I.: Evaluation of the fatigue behaviour of recycled aggregate concrete. J. Cleaner Prod. **65**, 397–405 (2014)
13. Yazdi, M.A., Yang, J., Yihui, L., Su, H.: A review on application of waste tire in concrete. Int. J. Civil Environ. Struct. Constr. Archit. Eng. **9**(12), 1648–1653 (2015)
14. Bignozzi, M.C., Sandrolini, F.: Tyre rubber waste recycling in self-compacting concrete. Cem. Concr. Res. **36**, 735–739 (2005)
15. Sharobim, K., Mohammedin, H., Mohamed, A., Fattouh, M.: Effect of using different supplementary cementitious materials in high strength self compacting concrete. Am. J. Eng. Res. **6**(10), 212–220 (2017)
16. Narashimhan, H.S.: Behavior of self-compacting concrete using using pozzolanic materials. J. Mech. Civil Eng. (IOSR-JMCE) **15**(1), 57–64 (2018)
17. Romanian Standards Association (ASRO), Cement, Part 1: Composition, specifications and conformity criteria for common cements, SR EN 197-1:2011, Romanian Standard Association, Bucharest, Romania) (2002)
18. Romanian Standards Association (ASRO), Testing hardened concrete. Part 3: Compressive strength of test specimens, SR EN 12390-3:2011, Romanian Standard Association, Bucharest, Romania (2005)
19. Romanian Standards Association (ASRO), Testing hardened concrete. Part 5: Flexural strength of test specimens, SR EN 12390-5:2009, Romanian Standard Association, Bucharest (2005)
20. Romanian Standards Association (ASRO), Testing hardened concrete. Part 6: Split tensile strength of test specimens, SR EN 12390-6:2010, Romanian Standard Association, Bucharest, Romania (2005)

Aspects Regarding Reinforced Concrete Pillars Strengthening Methods

Stanca Simona(✉)

Tehnical University of Cluj Napoca, Cluj-Napoca, Romania
simona.stanca@ccm.utcluj.ro

Abstract. The aim of the present study lies in analysing techniques of rehabilitation for reinforced concrete pillars. The paper presents an analysis regarding the aspect and efficiency of the rehabilitation techniques aiming at improving and strengthening a structure, either individually or as a whole.

Regarding the case study presented in the paper, a correct intervention strategy is assessed in the condition in which individual deficiencies of structural and non-structural elements are identified, as well as their mixed effect upon the building seismic behaviour mechanism and overall defects regarding structural strength, deformability, redundancy, and regularity [1].

The role of any intervention analyzed and proposed, dependent upon de character and signification of damages, lies in amplifying the life span of the element itself and of the exciting structure.

The proposed and taken action/solution should be flexible, so as to provide a balance between the conservation of older elements and the adaptation to present-day standards.

The modern strategy adopted, and the action/intervention program derived therefrom has the essential aim of obtaining innovative responses, in line with new concepts.

Keywords: Rehabilitation · Strengthening · Fibre reinforced polymers · Recommendations

1 Introduction

As time passes, various physical, chemical, and biological alterations can occur in the structure of a building and they can act simultaneously, sometimes linked in causal succession, the specific evolution being related to the physical and chemical properties of the materials and environmental factors.

Functionally, the rehabilitation of a structural member or of a building refers to bringing it back to an active state, by building, some of its abilities damaged during service for various reasons.

In the case of constructions, irrespective of the structural type adopted, interventions in damages require repairing, rehabilitation and strengthening works.

L. Moldovan and A. Gligor (Eds.): Inter-Eng 2021, LNNS 386, pp. 202–210, 2022.
https://doi.org/10.1007/978-3-030-93817-8_20

Strengthening reinforced concrete elements requires a complex knowledge. It includes the use of conventional cement-based materials, as well as new materials that include advanced composites [2].

Research undertaken over time has led to an improvement in the level of knowledge, many intervention methods used in the past have been revised and developed on the basis of new requirements, and new methods have been proposed, often based on new materials [3, 4].

Each recommendation/solution that involves consolidation methods is analyzed with a focus on its contribution/performance, relative advantages, and disadvantages in terms of application details [5].

The local upgrading techniques comprise measures applied to specific structural elements of a building, in order to enhance their mechanical characteristics. Traditional techniques make use of conventional materials: concrete and structural steel, whilst novel ones employ more innovative materials like: fiber reinforced polymers (FRP), textile reinforced mortars (TRM) etc. [6].

2 Recommended Intervention Measures for Reinforced Concrete Pillars

2.1 Reinforced Concrete Pillars Have a Wide Range of Use and are Present in Almost All Types of Current and Special Reinforced Concrete Structures

Recommendations/interventions are based upon the results of the building qualitative evaluation report that points out the nature and size of the damages in the building structural elements.

Recommended intervention measures for reinforced concrete pillars are presented below, (see Fig. 1) [7]:

2.2 Using the Rebuild Own Software [8]

The Rebuild software was developed as an application with Visual Studio Code, a source code editor developed by Microsoft for Windows.

The software accepts, through the modules developed, to access a set of provisions required to design intervention works for the structural rehabilitation of a building.

Considering the wide variety of the constructions and deficiencies found, the application exhibits a large number of solutions, with large applicability related to efficiency, erection conditions and economy.

Following structure diagnosis, defects are revealed, their severity and the need to intervene and the kinds of interventions to be applied.

With the help of Rebuild's own program, following the analysis of the building presented in the case study, several measures were recommended to strengthen the reinforced concrete pillars.

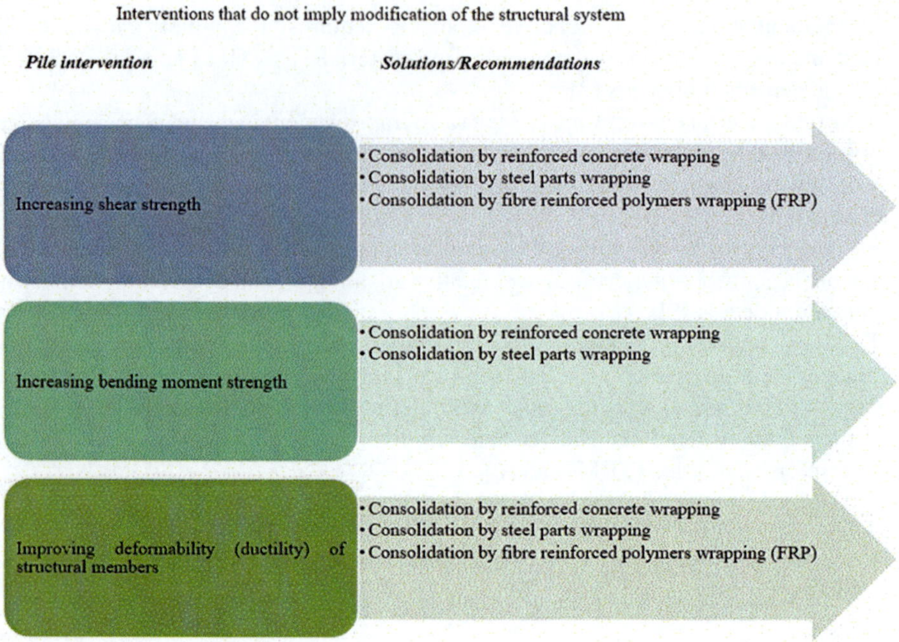

Fig. 1. Recommended intervention measures for reinforced concrete pillars.

3 Case Study

3.1 Generalities About the Building [8]

The building under investigation has four levels, a ground floor and three upper floors.

The superstructure consists of reinforced concrete frames: the marginal columns present a cross section of 30 cm × 40 cm, the central columns have the size 35 cm × 40 cm, the monolith reinforced concrete floor has a slab of 15 cm thick. The exterior dividing walls are made of 30 cm thick brick masonry, while the interior dividing walls have brick masonry, 25 cm thick at the level of the staircase and corridor, and 12.5 cm in the rest of the building. The roof is a traffic-free terrace form. The foundations are insulated under the columns, of dimensions 200 cm × 160 cm under the marginal columns and 220 cm × 180 cm under the central columns.

Geometry of structural members and structure (see Fig. 2):

3.2 Recommended Intervention Measures, Strengthening Measures for Reinforced Concrete Pillars [8]

The solutions suggested are destined to bring back the structure/structural members to their optimal operational constants, according to the provisions in force.

Through the Rebuild software, two types of strengthening measures for the reinforced concrete pillars were recommended, out of which the optimal functional and structural variant being further adopted:

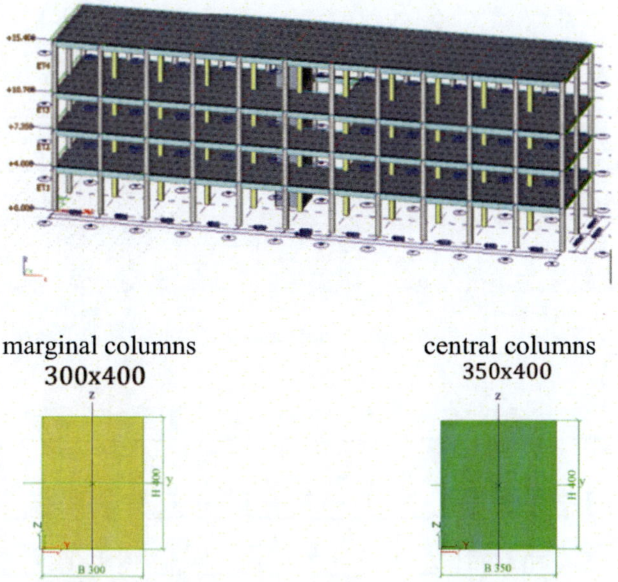

Fig. 2. Geometry of structural members and structure.

- Covering pillars with reinforced concrete [8];
- Covering reinforced concrete pillars with fibre reinforced polymers (FRP) [8].

3.3 Similitudes/Comparisons Between the Two Methods of Strengthening Reinforced Concrete Pillars

Covering pillars with reinforced concrete leads to increased shear force strength, increased concrete deformability through confining and preventing damages due to bending in the compressed area of the concrete [9, 10]. Covering consists in increasing the cross section of the member by coating it with a reinforced concrete sleeve. The purpose of covering reinforced concrete pillars with concrete lies in increasing the initial bearing capacity of the member under investigation [11].

Fibre reinforced polymer composites are used in a larger and larger range for rehabilitate structures or structural members, especially when conventional consolidation solutions present various forms of deficits. The role of covering reinforced concrete pillars with fibre reinforced polymers (FRP) consists in increasing the bearing capacity of the member, enlarging ductility, and dampening or stopping the damaging speed during building servicing [12].

Axial load N diagram, axis 2 longitudinal frame, in the two initial/strengthened models (see Fig. 3a, b, c; Fig. 4):

It is noticed a redistribution of efforts from the existing situation to the proposed one, the axial effort for analyzed models is about 15–30% lower in the consolidated version, depending on the adopted solution.

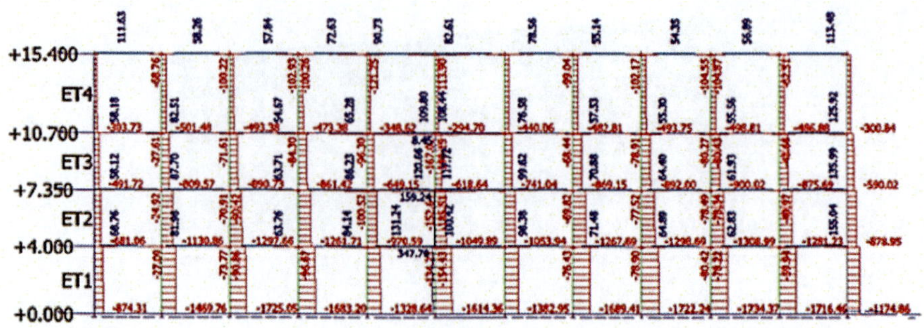

a. Initial model

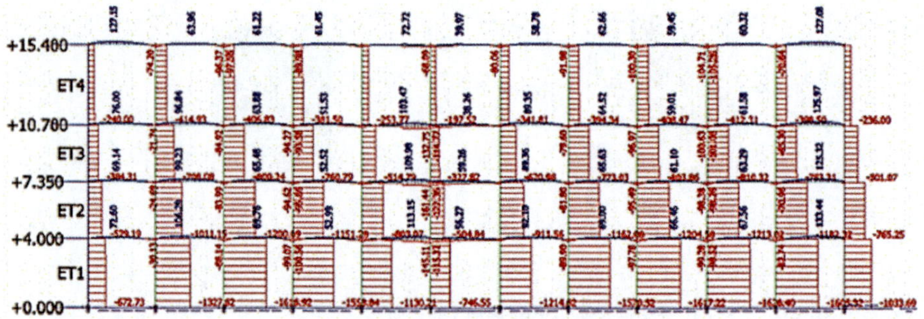

b. Covering pillars with reinforced concrete

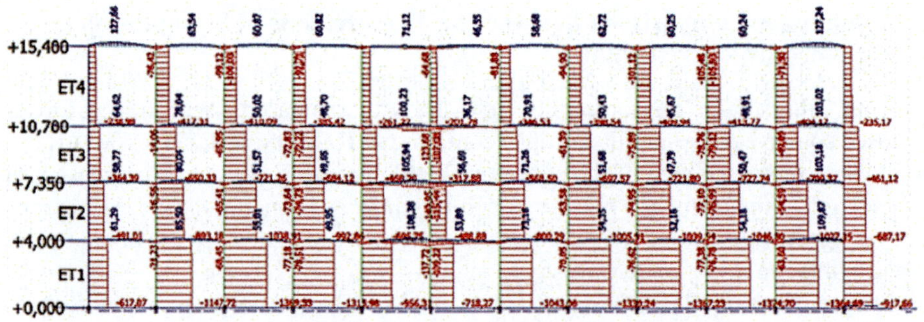

c. Covering reinforced concrete pillars with fibre reinforced polymers (FRP).

Fig. 3. Axial load N diagram

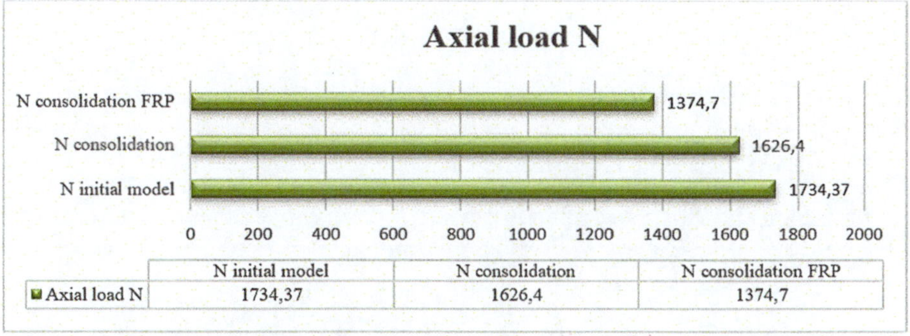

Fig. 4. Variation of the *axial load N* for the models in analysis [KN]

Normalised axial load νd, calculation and verification, in the initial/strengthened models (see Fig. 5):

$$\nu d = N/bxhxfcd, \nu adm = 0,45$$

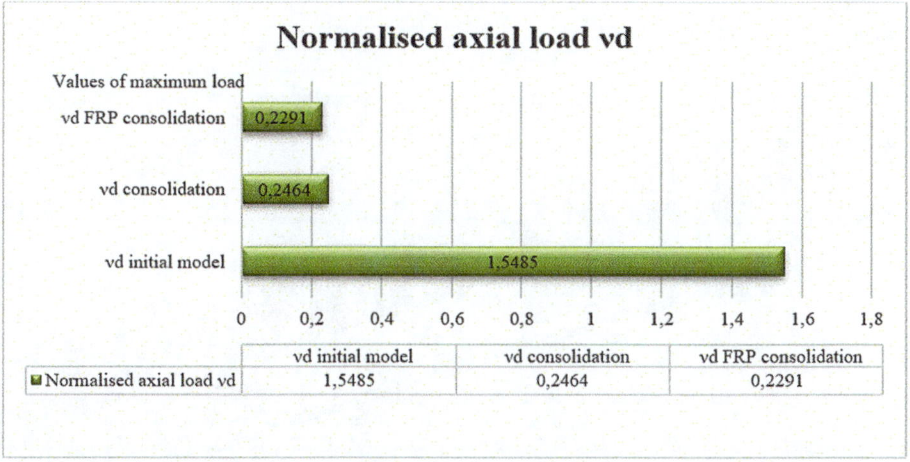

Fig. 5. Normalised axial load νd.

The normalised axial load νd for the analyzed models, depending on the adopted solution, meets the condition $\nu_{adm} = 0,45$.

Bearing capacity of the pillars in the two initial/strengthened models (see Fig. 6, Fig. 7, Fig. 8):

Analyzing the figures above, it can be observed a significant increase in the load-bearing capacity at the axial force of the reinforced concrete pillars by their consolidation, regardless of the consolidation method adopted.

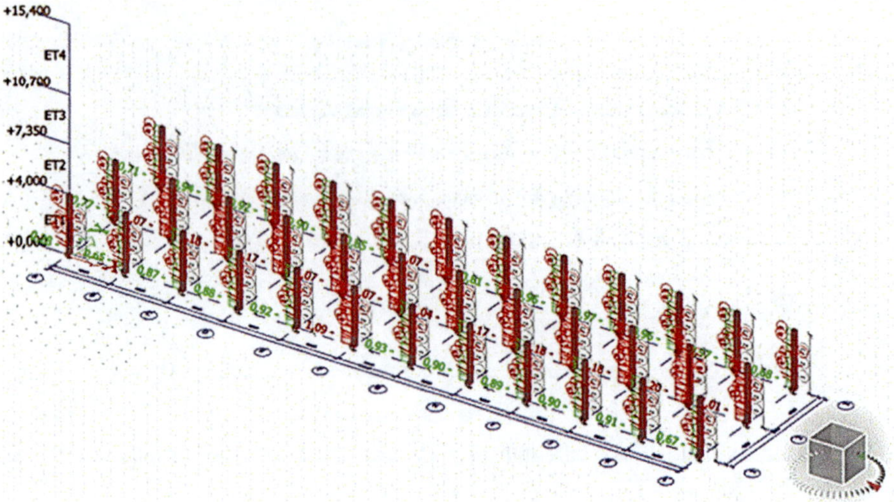

Fig. 6. Bearing capacity of the pillars - Initial model.

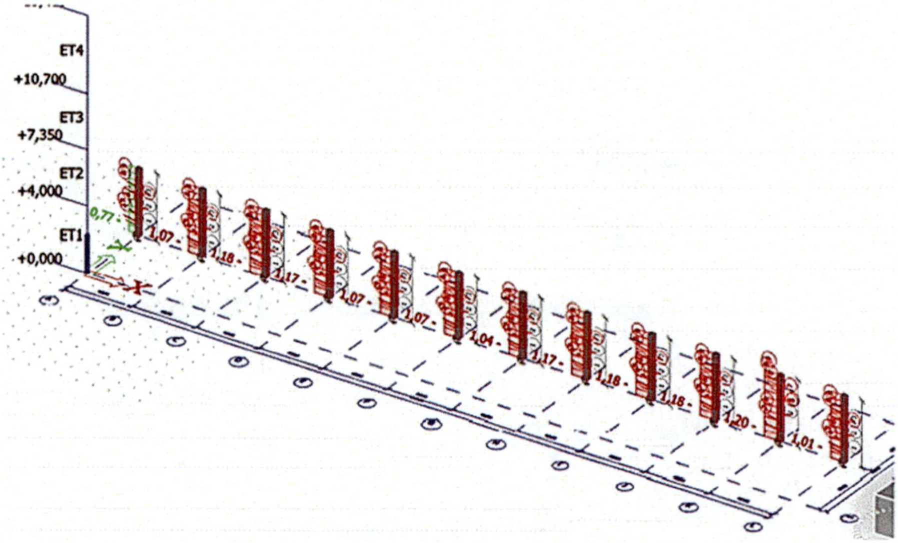

Fig. 7. Bearing capacity of the pillars - Covering pillars with reinforced concrete.

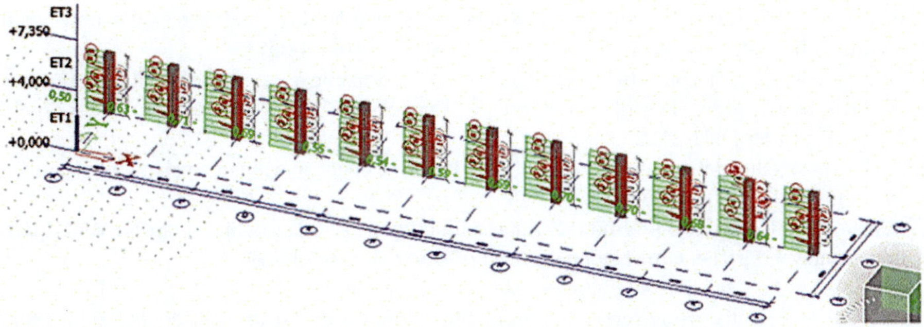

Fig. 8. Bearing capacity of the pillars - Covering reinforced concrete pillars with fibre reinforced polymers (FRP).

4 Conclusions

The recommended intervention methods have the purpose of increasing the strength of the members to cutting forces, to the bending moment and/or axial force, improving deformability (ductility), extending stiffness, and enlarging the post-elastic deformation capacity.

The results of this study highlight the significant increase of the bearing capacity to the axial force of the reinforced concrete pillars through their strengthening. It is noticed that stress is redistributed from the existing to the suggested situation, the axial load being lower in the consolidated variant, out of the two cases shown.

The solutions outlined in this paper approach classical and modern formulae, agreed upon by experts, which made use of conventional materials such as concrete, steel, polymer composites, and which re-establish or improve the structural performance in the building members that require interventions.

The rehabilitation of reinforced concrete pillars using modern intervention techniques, raises different opinions, but we must take into consideration that rehabilitation of the building elements is motivated it is motivated by a functional and structural interest.

References

1. Stanca, S.: Interventions upon brick masonry walls – constructive solutions. Bul. Inst. Polit. Iași **66–70**(3), 61–71 (2020). ISSN 1224-3884 (p), ISSN 2068-4762 (e)
2. Heiza, K., Nabil, A., Meleka, N., Tayel, M.: State-of-the art review: strengthening of reinforced concrete structures. In: Different Strengthening Techniques, Conference: Sixth International Conference on NANO-Technology in Construction (NTC 2014), Cairo, Egypt, vol. 6 (2014)
3. Mahmoud, M.H., Afefy, H.M., Kassem, N.M., Fawzy, T.M.: Strengthening of defected beam–column joints using CFRP. Mater. Sci. Med. J. Adv. Res. **5**(1), 67–77(2014). https://doi.org/10.1016/j.jare.2012.11.007.
4. Pohoryles, D., Melo, J., Rossetto, T., Varum, H.: Seismic retrofit schemes with FRP for deficient RC beam-column joints: state-of-the-art review. J. Compos. Constr. **23**(4) (2019). https://doi.org/10.1061/(ASCE)CC.1943-5614.0000950.

5. Engindeniz, M., Kahn, L., Zureick, A.: Repair and strengthening of reinforced concrete beam-column joints: state of the art. ACI Struct. J. **102**(2), 1–14 (2005)
6. Gkournelos, P.D., Triantafillou, T.C., Bournas, D.A.: Seismic upgrading of existing reinforced concrete buildings: a state-of-the-art review. Eng. Struct. **240**, 1 (2021). https://doi.org/10.1016/j.engstruct.2021.112273
7. Cod de proiectare seismică – partea a III-a – Prevederi pentru evaluarea seismică a clădirilor existente, Indicativ P100-3/2008
8. Stanca, S.E.: Contribuţii privind reabilitarea funcţională şi structurală a clădirilor vechi şi/sau dezafectate, UTPRES Cluj-Napoca (2021). ISBN 978-606-737-499-5
9. https://www.isomat.ro/products/materiale-pentru-reparatii/
10. INCERC, Normativ privind consolidarea cu fibre a elementelor structurale din beton, Institutul Naţional de Cercetare - Dezvoltare în Construcţii şi Economia Construcţiilor, Bucureşti (2005)
11. Mircea Crişan, Curs de Restaurare Structurală, UAUIM (2012)
12. https://rou.sika.com/ro/solutii-pentru-constructii/repara/consolidare-structurala.html

Management and Education for Sustainable Manufacturing

Comparative Quantitative Analysis of Air Quality Indicators and Macroeconomic Indicators for EU and Non-EU Member Countries

Ioan-Bogdan Bacoș and Manuela Rozalia Gabor(✉)

"George Emil Palade" University of Medicine, Pharmacy, Sciences and Technology of Târgu Mureș, Târgu Mureș, Romania
bacos.ioan-bogdan.20@stud.umfst.ro, manuela.gabor@umfst.ro

Abstract. Air quality in Europe is gradually moving forward. In any case, between 2008 and 2018, a significant proportion of the urban population within the EU-28 was exposed to concentrations of certain discuss pollutants over the EU limit or target values. The numbers of individuals exposed were indeed higher in relation to the more rigid World Health Organization (WHO) air quality guideline values set for the assurance of human wellbeing. The purpose of this paper is to analyse with statistical methods the corelations and causal relationship between macroeconomic indicators and air quality index. Data used in the study was the most important pollution indicators (PM2.5, PM10 and CO_2) to determine the impact of economic, industrial, and social development on air quality. Our results highlight the presence of significant correlations between macroeconomic indicators and air pollution indices.

Keywords: Air quality · Index · Macroeconomics · Statistical analysis · Tourism impact · Competitiveness

1 Introduction

The social and environmental problems imposed on contemporary society in recent times are proving to be a great challenge for all economies. Environmental protection becomes an essential prerequisite for achieving sustainable competitive advantage and an integral part of the proactive management of any country industry [1, 2]. Tourism is mostly considered in logical writing an industry with a critical impact on the financial, social and useful structure of rural areas and as a basic calculation in the revitalization and expansion of the provincial economy [3].

The term of environmental pollution became a key point to have in mind by different types of health organizations, or important category of people, there are several authors as: [4–7]; that integrates the term environmental quality into the quality of natural attractions, green countries or destinations thus resulting in qualitative countries from the point of view of the environment [8, 9]. The emergence of the term air quality comes with

L. Moldovan and A. Gligor (Eds.): Inter-Eng 2021, LNNS 386, pp. 213–223, 2022.
https://doi.org/10.1007/978-3-030-93817-8_21

the technological improves of medicine and the awareness of health problems related to massive pollution [10]. Although the term air quality is often analysed by public health specialists and environmental analysts, existing studies show little attention to the influence of these quality indicators in the macro-economic field of each country industry [11]. Of the existing studies for the analysis of competitiveness between countries, only a small part presents and analyses an interdependent relationship between air quality and the strategic/competitive positioning of destinations with lower air pollution indices, or so-called environmental competitiveness [12–16].

In 2019, according to the World Tourism Organization, the number of international tourist arrivals worldwide reached 1.5 billion, two years before it was predicted to do so [28]. That year also marked the seventh consecutive year in which the increase in tourism exports (+5%) exceeded the increase in exports of goods (+4%). Given this rapid pace of growth, the forecast that international arrivals will reach 1.8 billion by 2030 may be conservative [8]. Even worldwide the travellers generally did not regard air pollution as a concern while making travel decisions, in some regions/cities or countries, tourists started to realise the impact of the environment issues after their visit [17–19].

This represents huge potential for the tourism sector and the growth of global economies [20]. Thus, emerging economies contribute large proportions of travellers to this global trend and most destinations are becoming increasingly desirable for tourists, thus challenging both competitiveness and increasing the quality of tourist products [20, 21].

Since 2007, the World Economic Forum (WEF) has started formulating indices that analyse the competitiveness of the tourism industry (TTCI). This system of indicators reports and measures the competitiveness of the major tourist destinations around the world, with the objective of assessing the factors and policies that make it an attractive destination for international tourism. By hiring tourism leaders, the World Economic Forum aims to conduct an in-depth analysis of the competitiveness of this sector. Published biennially, the Travel and Tourism Competitiveness Report and Index compares the competitiveness of the tourism sector of 140 global economies and measures the set of factors and policies that enable the sustainable development of the "Travel & Tourism (T&T)" sector, which in turn contributes to a country's economic development and tourism competitiveness [22].

Air pollutants stand out in nature through their different action on the human body and the environment. Thus, in nature, pollutants are classified according to different characteristics, such as: chemical composition, persistence in the environment, ability to be transported, reactions, impact on human health and the environment [23–26].

According to National Centre for Environmental Health, European Environment Agency and World Health Organization the most common air pollutants are:

- Pollutants based on gas (e.g., SO_2, CO, NO_2, ozone or volatile organic compounds)
- Particulate matter (e.g., PM2,5 and PM10, coarse and fine particulate matter, the most recently discovered air pollutants) [27].
- Toxic chemicals (Persistent Organic Pollutants) (e.g. insecticides, pesticides, dioxins).
- Toxic heavy metals (e.g., lead, mercury)

The purpose of this paper is to present and analyse the corelations between macroeconomic indicators and air quality index. These results will be the research hypothesis for a complex study regarding the *air quality and its influence on the tourism industry.*

The main objective is to analyse the qualitative, quantitative, alternative qualitative and numeric quantitate indicators according to the data that were found in this field. *The second objective* is to analyse the causal relationship between indicators using statistical methods. For data analysis the statistical software SPSS 23.0 (licenced) by was used.

The *macroeconomic indicators* from the study are: gross domestic product (GDP), tourists arrivals, human development index (HDI), percentage of tourism industry in GDP, country surface, EU accession year, EU/non-EU country.

The *air pollution* and *environment indicators* from the study are: Environmental Sustainability Index, Consumption of renewable resources, Fine particle emissions 10, Fine particle emissions 2.5, Carbon Dioxide Emissions.

The following statistical tools and methods was applied:

- The *descriptive statistics* according to the type of variables from the study, respectively *categorial* (absolute and relative frequencies) or *continuous* (mean $\pm$ standard deviation, minimum, maximum) separately for EU and non-EU countries;
- The *parametrical bivariate correlations* (*Pearson coefficients*) was used to analyse the direction, intensity and statistical significance of association between macroeconomics indicators and air quality/ environmental sustainability indicators;
- The *chi square bivariate test* to analyse if there are statistical significant differences between the European countries according to the *EU accession year* regarding the level of *Environmental Sustainability Index* and the level of *percentage of tourism industry in GDP;*
- *The Oneway ANOVA* to analyse the differences between members or non-members of European Union regarding the average values of air quality indicators.

Given the complexity of PhD research and the various dimensions of tourism, on the one hand, and the high interest in air quality standardization, we will address an interpretation and analysis of data both qualitatively and quantitatively. The quantitative approach will verify the research hypotheses, in order to confirm or not the existing relationship between air quality and tourism.

2 Methodology/Description of the Individual Database

The data were gathered from the databases of the World Economic Forum and European Commission namely: Eurostat: name of the countries, year of accession in European Union, surface; World Economic Forum: gross domestic product; Human Development Report: Human development index; Tourism Highlights: Tourist arrivals, Percentage of tourism industry in GDP; Travel & Tourism Competitiveness Reports: environmental Sustainability Index European Environmental Indicators: Fin particle emissions 10 and 2.5; Pollution of European Areas with Carbon Dioxide: Carbon Dioxide Emissions; European Progress Index: Consumption of renewable resources. The database contains 90 statistical observations, respectively data recorded in 2008, 2013 and 2015 and the following variables found in Table 1

Table 1. Variables, variable type and unit of measure

No. crt	Variable name	Variable type	Unit of measure
1	Countries	Qualitative	–
2	EU/non-EU members	Alternative Quality	–
3	EU Accession Year	Numeric Quantity	Year
4	Surface	Numeric Quantity	Km^2
5	Gross domestic product	Numeric Quantity	Mil euro
6	Tourist arrivals	Numeric Quantity	Mil people
7	Environmental Sustainability Index	Qualitative	Score 0–100
8	Consumption of renewable resources	Numeric Quantity	Percent
9	Human Development Index	Qualitative	Score 0–1
10	Percentage of tourism industry in GDP	Quality (percentages show increase, maintenance or decrease)	Percent
11	Fine particle emissions 10	Numeric Quantity	$\mu g/m^3$
12	Fine particle emissions 2.5	Numeric Quantity	$\mu g/m^3$
13	Carbon Dioxide Emissions	Numeric Quantity	Mil tons

The descriptive statistics discipline points to acquire information on the definition of a statistical population, statistical variables, getting data on the wonder subject to research, organizing the information and data, presenting them within the form of statistical series, highlighting the population structure in connection to the observed variables, highlighting the advancement of a marvel in time or space as well as graphic representation of the information and data.

The calculation of the central trend indicators was used using The SPSS 23.0 software. For the qualitative variables in the study, frequency and graph tables of the structure were made. Analyzing the year of accession of countries to the EU zone, we see a more common frequency for 2004, the year in which they have 9 countries joined the European Union, according to the Table 2.

According to the comparative analysis between EU member countries and non-EU countries, present in Tables 3 and 4, we note that there are significant differences between the two groups. Thus, the average gross domestic product is 37% higher for EU member countries. The environmental sustainability index variable ranks the group of non-EU states above EU states, with a significant difference of 0.4 points on average. The Human Development Index for non-EU countries ranks above EU Member States with an average of 0.9219 for non-EU countries and 0.895 for EU states.

Air quality differs significantly between the two groups, so fine particulate emissions 10 have an average of 23.5 for EU states and 15 for non-EU states. Fine particulate emissions 2.5 averages 14.2 $\mu g/m^3$ for EU states and 8.1 $\mu g/m^3$ for non-EU states.

Table 2. Year of accession of countries to the EU

		Frequency	Percent	Valid percent	Cumulative percent
Valid	Not part	8	8.9	9.0	9.0
	1958	18	20.0	20.2	29.2
	1973	9	10.0	10.1	39.3
	1981	3	3.3	3.4	42.7
	1986	6	6.7	6.7	49.4
	1995	9	10.0	10.1	59.6
	2004	27	30.0	30.3	89.9
	2007	6	6.7	6.7	96.6
	2013	3	3.3	3.4	100.0
	Total	89	98.9	100.0	
Missing	System	1	1.1		
Total		90	100.0		

Table 3. Descriptive statistics of variables in EU Member States

	N	Minimum	Maximum	Mean	Std. deviation
Surface (km^2)	78	2586	633,186	159,358.19	164,666.322
Gross domestic product (bil euro)	78	17	89,721	2139.51	10,927.130
Tourist arrivals (mil)	78	923,713	86,917,700	15,346,446.47	19,899,042.422
Environmental sustainability index (0–100)	78	3.90	11.50	5.1962	0.93356
Consumption of renewable resources (%)	78	2.80	54.65	18.8244	11.38841
Human Development Index (0–1)	78	0.77	0.94	0.8695	0.04104
Percentage of tourism industry in GDP (%)	78	1	11	3.40	1.913
Fine particle emissions 10 ($\mu g/m^3$)	78	0.0	60.4	23.518	9.4688

(continued)

Table 3. (*continued*)

	N	Minimum	Maximum	Mean	Std. deviation
Fine particle emissions 2.5 (μg/m^3)	78	0.0	41.5	14.253	7.7540
Carbon dioxide emissions (million tonnes)	78	6.5	853.7	144.232	206.9977
UE_nonUE = 2 (FILTER)	78	1	1	1.00	0.000
Valid N (listwise)	78				

Table 4. Descriptive statistics for non-EU countries

	N	Minimum	Maximum	Mean	Std. deviation
Surface (km^2)	12	41,285	385,207	192,996.75	138,672.990
Gross domestic product (bil euro)	12	11	2421	797.26	884.710
Tourist arrivals (nil)	12	520,642	37,651,000	12,132,136.58	13,035,092.628
Environmental sustainability index (0–100)	12	4.80	6.00	5.4750	0.35707
Consumption of renewable resources (%)	12	0.00	76.69	36.4000	34.97961
Human Development Index (0–1)	12	0.85	0.95	0.9219	0.02847
Percentage of tourism industry in GDP (%)	12	2	9	3.74	2.171
Fine particle emissions 10 (μg/m^3)	12	0.0	20.8	15.042	5.6805
Fine particle emissions 2.5 (μg/m^3)	12	0.0	14.3	8.158	4.6586
Carbon dioxide emissions (million tonnes)	12	3.5	544.5	139.250	201.5217
UE_nonUE = 2 (FILTER)	12	1	1	1.00	0.000
Valid N (listwise)	12				

Finally, carbon dioxide emissions show an average difference of 5 million tones between EU member countries and non-EU countries.

According to Chi Square's analysis, on the Table 5 and 6 we found a direct link to the values recorded by each country on the environmental sustainability index and the year of accession to the European Union.

Table 5. Environmental sustainability index (0–100) * EU accession year

		Not part	EU accession year				Total
			1958–1973	1981–1986	1995–2004	2007–2003	
Environmental Sustainability Index (0–100)	3,90- 4,30	0	2	2	3	1	8
	4,40–4,60	0	0	2	1	3	6
	4,70–4,90	1	3	3	6	2	15
	5,00–5,20	1	4	1	13	1	20
	5,30–5,50	2	8	1	7	0	18
	5,60–5,80	2	10	0	3	0	15
	5,90–6,10	2	0	0	3	0	5
	8,20–11,5	0	0	0	0	2	2
Total		8	27	9	36	9	89

Table 6. Chi square environmental sustainability index test (0–100) * EU accession year

	Value	df	Asymptotic significance (2-sided)
Pearson Chi-Square	265.432a	176	0.000
Likelihood Ratio	171.698	176	0.578
Linear-by-Linear Association	0.001	1	0.979
N of Valid Cases	89		

Another direct link according to Tables 7 is between the percentage of the tourism industry and the year of EU accession. One country becoming an EU member opens its borders and opportunities for the development and practice of the tourism industry. Thus, we can see an increase in the percentage of the tourism industry in the GDP with accession to the EU area.

In order to achieve correlations, the following variables were chosen: Environmental Sustainability Index, Fine Particle Emissions 10 and 2.5, Carbon Dioxide Emissions, Consumption of Renewable Resources and Human Development Index (Table 8).

Between the variables Renewable resource consumption and fine particulate emissions 10 $\mu g/m^3$ there is a statistically significant INDIRECTED and intensity correlation ($0.2 < 0.259 < 0.4$) for 98.6% of cases (p-value = 0.014).

Table 7. Chi-square test of percentage of tourism industry in GDP (%) * Year of EU accession

	Value	df	Asymptotin Significance (2-sided)
Pearson Chi-Square	414.193a	328	0.001
Likelihood Ratio	227.059	328	1.000
Linear-by-Linear Association	0.208	1	0.648
N of Valid Cases	89		

Table 8. Parametrical bivariate correlations

		Environmental Sustainability Index (0-100)	Fine particle emissions 10 (μg/m3)	Fine particle emissions 2.5 (μg/m3)	Carbon dioxide emissions (million tonnes)	Consumption of renewable resources (%)	Human Development Index (0-1)
Environmental Sustainability Index (0-100)	Pearson Correlation	1	-.259*	-.179	-.116	.070	.149
	Sig. (2-tailed)		.014	.092	.276	.510	.161
	N	90	90	90	90	90	90
Fine particle emissions 10 (μg/m3)	Pearson Correlation		1	.736**	.156	-.254*	-.543**
	Sig. (2-tailed)			.000	.141	.016	.000
	N		90	90	90	90	90
Fine particle emissions 2.5 (μg/m3)	Pearson Correlation			1	.169	-.237*	-.315**
	Sig. (2-tailed)				.111	.024	.003
	N			90	90	90	90
Carbon dioxide emissions (million tonnes)	Pearson Correlation				1	-.268*	.092
	Sig. (2-tailed)					.011	.386
	N				90	90	90
Consumption of renewable resources (%)	Pearson Correlation					1	.103
	Sig. (2-tailed)						.335
	N					90	90
Human Development Index (0-1)	Pearson Correlation						1
	Sig. (2-tailed)						
	N						90

*. Correlation is significant at the 0.05 level (2-tailed).
**. Correlation is significant at the 0.01 level (2-tailed).

Also, if we relate the environmental sustainability index variables and fine particle emissions 10 $\mu g/m^3$; fine particle emissions 2.5 $\mu g/m^3$ and carbon dioxide emissions there is a statistically significant negative correlation and an ACCEPTABLE ASOQUALITY GRADE ($0.2 < 0.254;0.237;0.268 < 0.4$) statistically significant for approximately 98% of cases. The Human Development Index variable and fine particle emissions

10 μg/m^3 present a statistically significant INDIRECTED and intensity correlation (0.5 < 0.543 < 0.6) for 100% of cases (p-value = 0.00).

And finally, between the variables Human Development Index and fine particle emissions 2.5 μg/m^3 there is an negative correlation and medium intensity (0.5 < 0.543 < 0.6) statistically significant for 99.7% of cases (p-value = 0.03).

According to the analysis "oneway ANOVA" (see Table 9) it is noted that there are statistical differences between the variable of EU or non-EU member countries in terms of fine particle emission variables 10, 2.5 and carbon dioxide.

Table 9. One way anova.

		Sum of Squares	df	Mean Square	F	Sig.
Fine particle emissions 10 (μg/m3)	Between Groups	747.213	1	747.213	9.059	.003
	Within Groups	7258.644	88	82.485		
	Total	8005.857	89			
Fine particle emissions 2.5 (μg/m3)	Between Groups	386.252	1	386.252	6.982	.010
	Within Groups	4868.364	88	55.322		
	Total	5254.616	89			
Carbon dioxide emissions (million tones)	Between Groups	258.110	1	258.110	.006	.938
	Within Groups	3746021.863	88	42568.430		
	Total	3746279.973	89			

3 Conclusion

In conclusion the most important results found by analyzing the macroeconomic indicators with air quality by part or non-part of EU countries, includes the Human Development Index which ranks non-EU countries above EU Member States. Also, According to Chi Square's analysis we found a direct link to the values recorded by each country on the environmental sustainability index and the year of accession to the European Union.

If we relate the environmental sustainability index variables and fine particle emissions 10 μg/m^3; fine particle emissions 2.5 μg/m^3 and carbon dioxide emissions there is a statistically significant indirect correlation and an acceptable asoquality grade, statistically significant for approximately 98% of cases.

The future research related to this subject are to study, analyze and present existing research and different approaches of the authors based on the topic "Air quality and its influence on the tourism industry" [28], analyzing both the terms of air quality related to the tourism industry and health also publishes the effects of pollution on the health of the pawns of this industry, tourists. According to the Global Sustainable Tourism Council and Booking's Sustainable Travel Report in 2019, around 70% of travelers around the world say that would be more likely to book an accommodation, knowing that it is environmentally friendly and in a sustainable environment, regardless of whether the expenses are higher. Finally, the tourist will start to indirect look for qualitative destinations according also with the Air Quality indicators. This research fill a gap into the literature by corelate indicators as Human Development Index and Sustainably Environmental Index with others macroeconomic indicators, or Air Quality pollutants as: Pm10, Pm2.5 and CO2. As we can extract from the research results, a country which is both from a macroeconomic point of view and from the point of view of air quality, human development, promoting sustainable development, the more this country can rank superior to the others, developing a competitive advantage both on general market as well as in the tourism industry.

References

1. Pang, S.F.H., McKercher, B., Prideaux, B.: Climate change and tourism: an overview. Asia Pac. J. Tour. Res. **18**, 4–20 (2013)
2. Maxim, A., Stana, D., Grigoraş, M.: Alphabet of Ecology and Agrotourism–Cluj Napoca Risoprint, Chapter 3, pp. 75–88 (2010)
3. Pizam, A.: The Management of Quality Destination, vol. 33, pp. 79–88. Niedermann Druck, St. Gallen (1991)
4. Mihalic, T.: Environmental management of tourist destinations: a factor of tourism competitiveness. Tour. Manag. **21**, 65–78 (1999)
5. Dwyer, L., Kim, C.: Destination competitiveness: determinants and indicators. Curr. Issue Tour. **6**(5), 369–414 (2003). https://doi.org/10.1080/13683500308667962
6. Singal, S.P.: Air Quality Monitoring and Control Strategy, pp. 80–120. Alpha Science International, Oxford (2012)
7. Mieczkowski, Z.: Environmental Issues of Tourism and Recreation, pp. 109–132. University Press of America, London (1995)
8. Middleton, V.T.: Sustainable tourism: a marketing perspective. In: Stabler, M.J. (ed.) Tourism Sustainability, pp. 129–142. CAB International, Wallingford (1997)
9. Costa, S., et al.: Integrating health on air quality assessment – review report on health risks of two major European outdoor air pollutants: PM and NO2. J. Toxicol. Environ. Health Part B Crit. Rev. **17**(6), 305–330 (2014)
10. Celeste, E., et al.: The impact of air quality on tourism: a systematic literature review. J. Tourism Futures (2020)
11. Saenz-de-Miera, O., Rossello, J.: Tropospheric ozone, air pollution and tourism: a case study of Mallorca. J. Sustain. Tour. **21**, 1198–1230 (2013)
12. Rossello-Nadal, J.: How to evaluate the effects of climate change on tourism. Tour. Manag. **42**, 332–341 (2014)
13. Sajjad, F., Noreen, U., Zaman, K.: Climate change and air pollution jointly creating nightmare for tourism industry. Environ. Sci. Pollut. Res. **21**(21), 12403–12418 (2014)
14. Cucculelli, M., Goffi, G.: Does sustainability enhance tourism destination competitiveness? Evidence from Italian destinations of excellence. J. Cleaner Prod. 370–382 (2016)
15. World Tourism barometer - International Tourism Highlights (2019). https://www.e-unwto.org/doi/pdf/10.18111/9789284421152. Accessed 27 Apr 2020
16. Vallero, D.: Fundamentals of Air Pollution, 5th edn, Chapter 7 "Air Pollutant Hazards", pp. 197–214 (2014)
17. Anaman, K.A., Looi, C.N.: Economic impact of haze-related air pollution on the tourism industry in Brunei Darussalam. Econ. Anal. Policy **30**, 132–145 (2000)
18. Law, R., Cheung, C.: Air quality in Hong Kong: a study of the perception of international visitors. J. Sustain. Tour. **15**, 390–401 (2007)

19. Zhang, A., Zhong, L., Xu, Y., Wang, H., Dang, L.: Tourists' perception of haze pollution and the potential impacts on travel: reshaping the features of tourism seasonality in Beijing China. Sustainability **7**, 2397–2414 (2015)
20. Hamilton, J.M., Maddison, D.J., Tol, R.S.J.: Effects of climate change on international tourism. Clim. Res. **29**, 245–254 (2005)
21. Widjaja, Y.I., Khalifa, G.S.A., Abuelhassan, A.E.: The effect of destination reputation on the revisit intention to halal tourism destination of Jakarta. Int. J. Bus. Econ. Law **20**(5), 104–111 (2019)
22. World Economic Forum. The Travel & Tourism Competitiveness Report 2019 published by the World Economic Forum's Platform (2019). http://www3.weforum.org/docs/WEF_TTCR_2019.pdf. Accessed 25 Apr 2021
23. Arya, S.P.: Air Pollution Meteorology and Dispersion. Oxford University, New York (1999)
24. Chen, J., et al.: Characteristics of trace elements and lead isotope ratios in PM2.5 from four sites in Shanghai. J. Hazard. Mater. **156**, 36–43 (2008)
25. Cocheo, C., Zaratin, L.: Assessment of human exposure to air pollution. In: Encyclopedia of Environmental Health, Science Direct, pp. 230–237 (2011)
26. Mike, A.: Air Pollution Encyclopaedia of Biodiversity, 2nd edn. Science Direct, pp. 136–147 (2013)
27. Jillian, M.: Air Pollution: Everything You Need to Know Natural Resources Defence Council (2016). https://www.nrdc.org/stories/air-pollution-everything-you-need-know. Accessed 02 Apr 2020
28. Bacoş, I.B., Gabor, M.R.: Air quality indices - case study: environmental sustainability pillar and Romania's positioning in the european and global context. Acta Marisiensis. Seria Technologica **18**(35), 22–27 (2021)

Work Regulation and Time Management to Avoid Occupational Stress in Industry

Camelia Angelica Dâmbean(✉) and Manuela Rozalia Gabor

Engineering and Management Department, "George Emil Palade" University of Medicine, Pharmacy, Sciences and Technology of Târgu Mureş, Gh. Marinescu, 5400139 Tirgu Mureş, Mureş County, Romania

dambean.camelia-angelica.20@stud.umfst.ro, manuela.gabor@umfst.ro

Abstract. Nowadays, due to the changes in recent years, when a series of unexpected changes took place, such as the pandemic, they determined all people to adapt and adopt a rapid development if they wanted to stay in the labor market and not give bankruptcy. Business life has become complex, many have tried to work from home in the pandemic, where it was possible, in companies being more complicated, only TESA staff having this possibility. Time pressure and ubiquitous stressors can negatively affect productivity. In the context of our study we examined the productivity sector which is complex and which requires collaboration, and the work environment is very important for the good development of the activity in order not to have scraps. Working under time pressure and meeting the norm exposes the employee to high stress if he does not know how to manage or control the time allocated to work.

This study is part of a wider research that highlights the importance of work regulation and the ability of individuals to know how to properly manage their work time to avoid work stress. The study addresses the literature on work standardization and time management, as well as professional stress and the relationship between them.

The aims of the research are the following: highlighting the relationships between time management and professional stress, the relationship between stress and work regulation. The research results validate the above statements. Excessive stress has been found to lead to disease and lower performance and productivity.

Keywords: Time · Work regulation · Stress · Time management

1 Introduction

Business and production life has become more demanding and complex, with many developments and changes that must be made on the go so as not to be removed from the labor market, a change that forces employees and the management team to adopt more tasks in a short time [1].

Time pressure on employees affects their effectiveness, increases anxiety and decreases concentration and performance at work [2].

L. Moldovan and A. Gligor (Eds.): Inter-Eng 2021, LNNS 386, pp. 224–234, 2022.
https://doi.org/10.1007/978-3-030-93817-8_22

The present research tries to identify the stressors, to be known by the management team in order to reduce or eliminate them where possible, thus offering a new perspective on the approach of correct time management in order to eliminate professional stress.

1.1 Work Regulation

Work standardization is a key element being given for performing in a time interval the stages of a work task and the sequence in which it should take place.

The standardization of work includes three basic elements, namely: the work time which is an element that the product must meet to satisfy the customer, the available production time which is established by the law of the scope of duties corresponding to the position and which is valid for all employees.

The standardization of work applies both to employees with executive positions and to the rest of the staff, it must be established in such a way as to ensure a normal pace of work without tiring the employee mentally, physically and intellectually. The normalization of work consists in the time allowed for the execution of a task according to the technical and organizational conditions necessary for the performance of operations or works by the employee working with normal intensity, in optimal working conditions, conditions of technological work processes productive, time for breaks, time needed for lunch breaks or other needs.

The calculation of the production norm is established by an analytical method taking into account the duration of the productive time multiplied by the timed quantity/duration of an operation.

$$\text{Nprod} = (\text{Top} * q * k1 * k2)/\text{top}, \text{ where } k1 = (100\text{th})/100;\ k2 = (100 - b)/100$$

k1 - loss coefficient; k2 - waste coefficient, a - percentage of admissible losses; b - the percentage of admissible waste [3].

The most important labor norms are the technical norm of production and the norm of time. The technical norm of time consists in the time necessary for the execution of a work in certain technical-economic conditions and the use of all means of production, instead the technical norm of production consists in the actual number of parts or production units to be executed in a certain period set time. The technical rules must constantly evolve in order not to keep production in place and they must take into account the conditions in which the work is carried out and the progress made.

The rules are considered to be met only when the quantity of products on the route is made on time without errors, being executed correctly according to all technical requirements. If the work is not executed correctly then the norm is not considered to be fulfilled, the worker being forced to repair or repair the respective product. A technical time rule must be well established because if it is not accurate it can give rise to errors. In order to establish a norm, the actual working time must be taken into account, but also the times necessary to carry out the work for the preparation of the material, the tools, the service, supply, rest and necessities. There are also unnecessary times, EQ those that are not measured, that interrupt the production, leave the work motivated or not, interruptions for the organization of the activity that were not established from the beginning. According to Alexandrescu L, stress consists in a state translated by a specific syndrome corresponding to all the non-specific changes thus induced in a biological system [4]. As the pandemic coronavirus spread around the world at a fast pace that is almost impossible to control, national authorities have been forced to urgently adopt a series of measures to restrict the spread of the virus among the population and thus, in order to ensure safety and health. workers, the Ministry of Labor and Social Protection recommended measures to make labor relations more flexible, namely: establishing individualized work programs, with the consent or at the request of the employee concerned; temporary modification of the workplace at the employee's home or telework, regulated by Law no. 81/2018, which at art. 2 lit. a and art 3 [5].

Stress manifests itself as a feeling of anxiety due to the pressure from the internal and external environment that exceeds the individual's ability to fulfill activity, and to master the respective situation [6, 7]. Organizational stress is called work-related stress in relation to one's relationship with the individual, aspects highlighted by researchers such as Low & Lee, Ramachandra R [8, 9] as opposed to by the organizational one highlighted by Lazrus and others [6, 7].

1.2 Time and Time Management

Time is a precious resource that is not reversible and cannot be stored or bought, but only used usefully. In order to be able to use their time properly, individuals must allocate sufficient periods of time for reflection, for work, family, relaxation, friends or cognitions. A person who manages to control and manage his time properly will have benefits on several levels, they must be aware that time is irreversible and that the service must give him satisfaction for a job well done, there must be less frustration and more self-confidence to succeed and no stress around him.

Time is a relatively abstract concept created by the individual in order to measure the flow of moments. As Aristotle said, time is one of the unknown things we do not know, being ubiquitous, essential to life, but it remains an enigma by its nature. Time is a structure created by people, a tool that helps us to perceive more easily the world around us. As Eistein found, time is different in that its passage depends on how a person travels in space, how he perceives the flow of moments.

For proper time management there must be a clear delegation of responsibilities, a prioritization of activities according to their importance for both the individual and society. Each person has his own report of the perspective of time. A person who is too passive about time means that he allows others around him to manage his own time, instead an active person who knows how to plan his activities, manage his time efficiently and occupy his own time not letting others manipulate it as they wish. Time can become an obsession for some individuals, as people often lose their temper at work when they enter a time crisis. If employees manage to improve their time management at work, they will also be able to improve their performance by achieving their own goals and the institution making a less intense effort. Time management at work consists of a special ability, in the person's ability to schedule their activities possible and on hours so that he is not pressed by time. It is known that employees currently face a large number of tasks, but individuals must give priority to their activities, and those without priority must plan according to their priority, their importance, in order to have professional satisfaction. Time management is necessary to get rid of daily stress, people being forced to adapt their work and lifestyle at a faster pace, to cope with competition. As the researchers Diploye, Smith and Howel stated, if a person knows how to approach intelligently, planned the time needed to re-solve activities, it means that he knows how to properly manage his time at work, thus eliminating stress.

There are certain factors that can lead to poor time management, namely: missed or delayed deadlines and appointments at the last minute; lack of concentration and professionalism, low quality of work; unforeseen and unwanted stress, financial sanctions or bad reputation; tense interpersonal relationships at work; short-term time pressure to perform activities.

1.3 Stress

Stress was first defined by researcher H. Selye in 1984 as a state of the body in response to the action of strong stimuli that cause that state.

Mack said in 1998 that work-related stress was a major problem internationally [10]. Salopsky stated in 2004 that when the human body feels threatened to survive, it takes extreme measures.

Hans Selyefirst used this term in 1950 to describe a set of reactions of the body to an external action exerted on it by a wide range of causative, physical agents such as trauma, burns, chemicals, biological factors. Such as infections, pandemics, consisting in the appearance of various morphofunctional changes, endocrine such as the pituitary gland, and adrenal glands that do not function normally [28].

Stress is defined by Neveanu P. as a state of tension, tension and discomfort that can be caused by various pathogens, and affectogens with negative charge, frustration or repression of motivational states, or incompetence -the ability to solve problems [11].

Stress is a state perceived as negative being accompanied by mental, physical or social discomfort and which is the consequence of the fact that employees can not meet the expectations of employers and their demands. We talk about occupational stress when there are harmful emotional, physical and behavioral responses, when there are conflicts between the requirements of the employee and those of the employer.

Research by researchers such as Curry, Wakefield, Price & Mueller [12] shows that sources of stress and workload are a predictor of satisfaction.

A recent research [13] shows that there are significant correlations between anger and the scales of work involvement. Another more recent research conducted in 2000 by Mikkelsen, Ogaard & Lovrich [14] shows that occupational stress has significant negative correlations with job satisfaction and emotional organizational attachment.

Organizational factors that cause stress include mental overload, ambiguity, role conflict, job insecurity, inadequate working conditions, lack of autonomy. Work can become a stressor if there is too much work, too many tasks in too short a time, if it is a difficult job or too many unclear responsibilities and demands. Another stressor at work can be ergonomic when there are too loud noises and vibrations, inadequate lighting and temperature, toxic substances, tiring posture, unhealthy, lack of lunch break, inadequate pay, sexual harassment, discrimination or insult. The psychic and behavioral changes existing at the time of occupational stress can have long-term repercussions if not intervened in time. An employee's stressed body may experience defense reactions such as low creativity, mnemonic and attentional disorders, low concentration, and fatigue reactions manifested by fatigue or overwork, delaying decisions or making hasty decisions, repeating mistakes, insomnia. All these manifestations present at the mental level can be accompanied by manifestations at the emotional and behavioral level. Manifestations at the emotional level in the phase of exhaustion are manifested by anxiety, frustration, irritability, lack of motivation, depression. Socio-professional activity and people's health can be influenced by occupational stress, and the consequences of stress can become devastating for the individual [29].

Behavioral manifestations in the phase of exhaustion when there is ubiquitous stress are manifested as follows: imbalance, psychomotor agitation or lack of energy, imprisonment, non-communicative or quarreling, staying overtime or shortening the program, or absenteeism, substance abuse such as alcohol, drugs. All these emotional, mental and behavioral changes are accompanied by changes that occur physiologically when there are palpitations, gastrointestinal problems, low immunity, arrhythmias, common diseases, muscle and headaches, thrombosis, fibrillation, cancer, heart attack or stroke.

If the management team is empathetic and has an objective and a goal to achieve, then it continuously monitors the subordinates, not stressing them, but observing their needs and the stressors. If the management team adopts a program that is too rigid and strict, it can be a stressor, but if it allows the employee some freedom in accordance with the work norm, clearly explaining to the employees the objectives of the institution to be achieved and good interpersonal communication and a spirit. By the team then the professional environment will be friendlier, less stressful, having success. Occupational stress is part of professional life, being present in many professions and having many sources and effects on the mental and physical health of employees [15–20, 22].

2 Method

The objectives of the study were to observe if employees face excessive stress, if productivity performance is affected by stressors, if work regulation is a stressor and if employees know or know how to manage their working time.

The following hypothesis were considered:

1. There is a significant relationship between time management and stress
2. There is a significant relationship between work regulation and stress.

The study was performed on a number of 70 employees from a car construction company in Tg. Mureș.

The participation in the study was made by simple randomization between 1.02-30.04.2021. Correlational design was used, and the group of subjects was in the age range of 20–70 years, of both sexes, with differentiated qualifications and coming from different backgrounds.

The management team took measures, through specific methods of quality management, for the company's employees to be aware of the importance of satisfying customer requirements, to be involved in the production of products at ever higher quality standards. In the realization of this study we had all the support of the management team.

Of the 70 subjects who were part of the study were the managerial staff, the TESA functional staff and the executive staff, the workers. The data were collected at the institution's headquarters, by directly addressing the respondents and distributing the questionnaires to them by the author of the research, questionnaires that were completed by 70 respondents. The time of conducting the data collection consisted starting from the direct conversation with the employees regarding the establishment of the time of the survey, until it was carried out, the post-coding of the answers and the grouping of the data was 4 months.

The parameters included in the research were sex, age, place of origin, seniority and time management test and stress test, as well as the work standardization observation sheet with the important elements of our study.

Dependent variables: Gender, age, seniority in the institution, marital status, education.

Independent variables: Stress assessment test, time management measurement test, work norm observation sheet:

a. Stress measurement test that has 25 items with yes or no answer,
b. Time management questionnaire that has 20 items with a response scale from 1 to 5, 1 representing never, 2-occasionally, 3 as a rule, 4-often, 5-always
c. The observation sheet for the observance of the work norm with two answer options yes or no.

3 Result

The first addictive variable taken into account was the sex of the subjects, noting that the percentage of men is dominant, because the specialty of the institution is focused on work specific to males (Fig. 1).

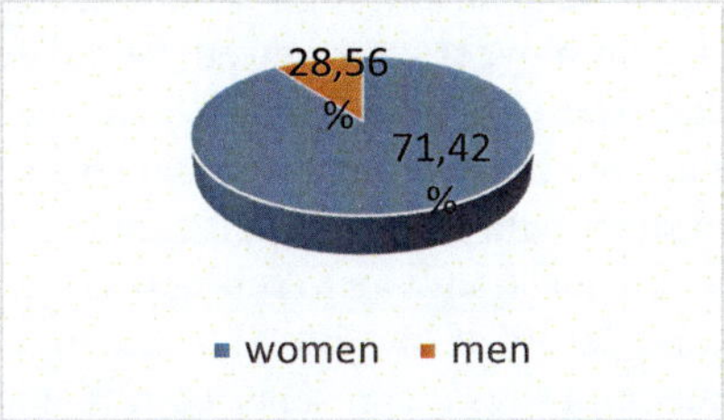

Fig. 1. Distribution of subjects by sex

We distributed a number of 70 questionnaires for stress N = 70, and the respondents were thus classified as 28,56.42% being women (N = 20) and 71.42% men (N = 50) (Fig. 2).

The employees who were in the age group with the highest share were in the 45–65 years group representing 48.28%, and the persons aged between 20–30 years represented 14.28%, with the age between 31–44 years old represented 20.44% and between 65–70 years old it represented 5.71%.

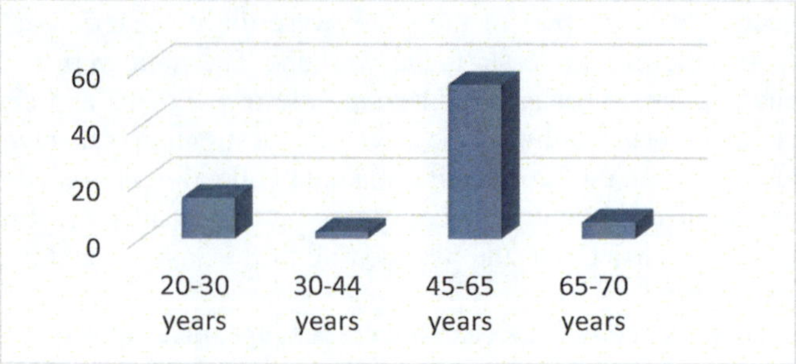

Fig. 2. Distribution of subjects by age

The third variable followed was the marital situation of the subjects, and the results also showed that 61.42% are married, 22.85% are single and 15.71% divorced (Fig. 3).

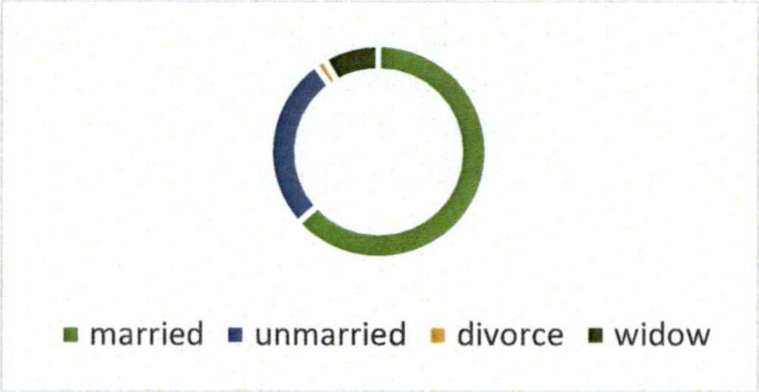

Fig. 3. Distribution of subjects by marital status

Regarding the level of schooling, 28.57% have a university degree, 17.14% have a post-secondary education, 31.42% have a high school education and 25.71% have a vocational school. Among the employees interviewed, 64.28% are from urban areas and 35.71% are from rural areas. The parameters included in the research were sex, age, place of origin, seniority and time management test and stress test, as well as the work standardization observation sheet with the important elements of our study.

To measure stress we used a questionnaire to self-assess the level of stress consisting of 25 items that highlight three levels of low, medium and high stress (Fig. 4).

Low stress is 28.59%, medium stress represents 41.42%, high level stress represents 18.57% of the interviewed population and 8.57% of employees have an exaggerated level of stress.

3.1 Data Analysis and Interpretation

We used the SPSS software to calculate the median, standard deviation and correlations of the 3 established parameters to demonstrate the established hypotheses.

A basic assumption underlying this model is that a prerequisite for organizational stress to be controlled is that there are substantial differences between it and time management.

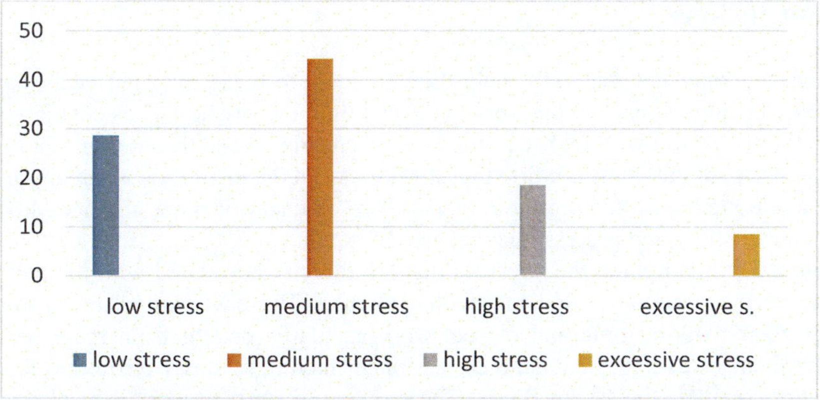

Fig. 4. Distribution of subjects according to the values obtained in the stress test

Another hypothesis is that there are significant differences between work regulation and stress, high stressors can be timing the various procedures necessary to calculate the work norm and check employees if they comply (Fig. 5).

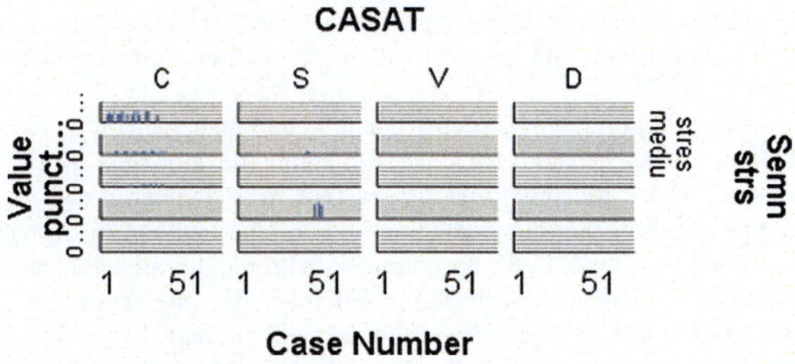

Fig. 5. Distribution of subjects according to marital status and comparison with stress

As can be seen, there is a significant close correlation at a significance threshold of $p < 0.05$ between:

1. Sex and seniority in the institution −0.376
2. Stress and seniority −0.345
3. Age and stress −0.361
4. time management and stress −0.274.

From the graph above we can see that in married and single people the stress level is higher, while in divorced and widowed people the stress level is low or non-existent.

From the table above we can see that there is a significant correlation at a significance threshold of $p < 0.01$ between: Seniority in the institution and stress 0.508; age and seniority 0.933.

4 Conclusions

Effective time management skills help the individual to make the right decisions that can have a positive impact on the work in the institution where he carries out his activity and implicitly of his own life, because it is more calculated and the stress is not high.

The results of this study confirm that stress exists in the institution where this experiment was done, consolidating the remarks of researchers of other authors who have written about stress and professional balance [23–25].

This study highlighted the stressors that are directly correlated with the ability to manage working time and compliance with work rules. Those who know how to properly manage their working time and respect work standards benefit from more personal time for themselves and their families, no longer having to stay overtime to fix their product or complete the assigned task. The problem of stress at work is recognized by many researchers and is highlighted in this study which emphasizes the importance of relaxation and calmness of the individual at work, which leads to better concentration, to make the right decisions leading to job satisfaction and productivity.

We can observe from the obtained results that we have a low individual stress if the individuals plan and manage their time correctly, being in the norm of work and not getting impatient when they are monitored. This study has many implications for the theoretical contribution, and the results of this study highlight that physiological and psychological stress causes a change in performance and job satisfaction.

This result is consistent with other studies by researchers such as Antoniou et al. [26], Stacciarini et al. [27].

For establishing and following a proper planning and management of activities and prioritizing them, in the current circumstances of a good development of work, family, friends and others. There are some principles of time management in everyday human life. Time can be a stress-generating factor, and if combined with professional dissatisfaction and frustration, it leads to the appearance of chronic diseases that can affect the production process. Time management begins with managing one's own time, imposing self-discipline, a skill that can be learned throughout life, and if this skill is learned it helps the person to be more efficient and productive. Mastering the skills of self-discipline leads to achieving the set goals, setting priorities, developing creativity and improving the quality of personal life and those around them. Among the methods of time management efficiency we mention self-discipline, keeping an hourly diary on the activities to be carried out daily, eliminating unnecessary activities that are not included in the job description, setting clear objectives and clear tasks, prioritizing tasks according to their importance or urgency, drawing up or drawing up an action plan, maintaining the person's consistency in activities, finding permanent motivations and maintaining them until the proposed objectives are reached, analyzing the problems that prevented the efficient development of time management.

In 2009, a European-level survey on health and safety at work was carried out, and the report focused on occupational safety and health management, examining how practices vary across Europe in terms of size unit and production sector [21].

For effective management, the management team must create a positive and empathetic emotional state capable of managing their own emotions, without giving in to stress. It is known that the inability of a manager to manage his own emotions can influence the emotions of subordinates, and thus can lead to decreased productivity, reducing its efficiency, which is reflected in the activity of the whole team. In the present situation it is observed that a relatively large number of employees are stressed, more than half of those interviewed, EQ 68.57%, taking into account the pandemic situation which is a pervasive factor and in addition to other stressors, but the management team manages the situation properly, being empathetic and having an emotional intelligence located at a high level, knowing how to manage stressful, problematic and under pressure situations, having employees who have been on sick leave in recent months or have worked from home some people, but the management team made decisions beneficial to the institution and advantageous for the company and employees.

References

1. Ozcelik, H., Langton, N., Adrich, H.: Doing well and doing good: the relationship between leadership practices that facilitate a positive emotional climate and organizational performance. J. Manag. Psychol. **23**(2) (2008)
2. Yılmaz, E., Altınok, V.: Okul yöneticilerinin yalnızlık ve yaşam doyum düzeylerinin incelenmesi. Kuram ve Uygulamada Eğitim Yönetimi **15**(59), 451–469 (2009)
3. Picos, C., Coman, Gh., et al.: Technical Standardization for Machining, vol. 1, 2. Bucharest Technical Publishing House (1979)
4. Alexandrescu, L.: Psychic stress. In: Prelipceanu et al. (eds.) Mental Health Treatise, vol. 1, Encyclopedic Ed. (2000)
5. Law no. 81/2018, regarding the regulation of the telework activity, published in the Official Gazette 296 of April 2 (2018)
6. Saygin, Y., Akdeniz, S., Deniz, M.E.: Loneliness and interpersonal problem solving as predictors of subjective well-being. Int. J. Inf. Educ. Technol. **5**(1), 32–35 (2015)
7. Lazrus, R.S.: Psychological stress in the workplace. Fifty Years of the Research and Theory of Rs Lazarus: An Analysis of Historical and Perennial Issues, 312 (2013)
8. Low, S.K., Kok, J.K., Lee, W.Y.: Perceived discrimination and psychological distress of Myanmar refugees in Malaysia. Int. J. Soc. Sci. Hum. **4**(3), 201–211 (2011)
9. Vijayalakshmi, P., Ramachandra, Reddemma, K., Math, S.B.: Perceived human rights violation in persons with mental illness: role of education. Int. J. Soc. Psychiatry **59**(4), 351–64 (2013)
10. Bellman, S., Forster, N., Still, N., Cooper, C.: Gender differences in the use of social support as a moderator of occupational stress. Stress and Health (2003)
11. Neveanu, P.P.: Dictionary of Psychology. Albatros Publishing House, Bucharest (1978)
12. Curry, J.P., Wakefield, D.S., Price, J.L., Mueller, C.W.: On the causal ordering of job satisfaction and organizational commitment. Acad. Manag. J. **29**(4) (1986)
13. Vîrgă, D., Sulea, C., Zaborilă, C., Bogathy, Z., Maricuţoiu, L.: Individual and Organizational indicators of employees' well-being: a predictive model. Lucrare prezentată la 14th European Congress of Work and Organizational Psychology, 13–16 mai, Santiago de Compostela, Spania (2009)
14. Mikkelsen, A., Ogaard, T., Lovrich, N.: Modeling the effects of organizational setting and individual coping style on employees subjective health job satisfaction and commitment. Public Adm. Q. **24**, 371–398 (2000)
15. Cooper, C.L., Robertson, I.: International Review of Industrial and Organizational Psychology, vol. 18. Wiley (2003)

16. Cooper, C.L., Payne, R. (eds.) Causes, Coping and Consequences of Stress at Work. Wiley (1988)
17. Dollard, M.F., Winefield, A.H.: Mental health: overemployment, unemployment, unemployment and healthy jobs. Aust. e-Journal Adv. Mental Health **1**, 170–195 (2002)
18. Giga, S.I., Cooper, C.L., Faragher, B.: The development of a frame work for a comprehensive approach to stress management interventions at work. Int. J. Stress. Manag. **10**, 280–296 (2003)
19. Greenhaus, J.H., Parasuraman, S.: Vocational and organizational behavior, 1985: a review. J. Vocat. Behav. **29**, 115–176 (1986)
20. Ivancevich, J.M., Matteson, M.T., Freedman, S.M., Phillips, J.S.: Worksite stress management interventions. Am. Psychol. **45**, 252–261 (1990). https://doi.org/10.1037/0003-066X.45.2.252
21. Management of occupational safety and health – An analysis of the findings of the European survey of enterprises on new and emerging risks, 2012, EU-OSHA (European Agency for Safety and Health at Work) (2012)
22. Greenhaus, J.H., Parasuraman, S.: Professional and organizational behavior: a review. J. Vocat. Behav. **29**, 115–176 (1985/1986)
23. Fetsch, R.J., Pergola, J.: Effective burnout prevention program. J. Extension [on-line serial] **29**(4) (1991, Winter)
24. Riggs, K., Beus, K.M.: Job satisfaction in Extension Journal of Extension [On-line serial] **31**(2) (1993, Summer). Available online
25. Fetsch, R.J., Kennington, M.S.: Balancing work and family in cooperative extension: history, effective programs, and future directions. J. Extension [On- line serial] **35**(1) (1997). Available online
26. Antoniou, A.S.G., Davidson, M.J., Cooper, C.L.: Occupational stress, job satisfaction and health state in male and female junior hospital doctors in Greece. J. Manag. Psychol. **8**(6), 592–621 (2003)
27. Stacciarini, J.M.R.: Occupational stress and constructive thinking: health and job satisfaction. J. Adv. Nurs. **46**(5), 480–487 (2004)
28. Selye, H.: Stress without Distress. Lippincott, New York (1974)
29. Pitariu, H.: Occupational stress in engineers: a contemporary reality. In: Psychology, n1, 2–4. Society of Science and Technology S.A. (1994)

The Effects of New Infrastructure on Traffic Dynamics. An Urban Simulated-Based Model

Mircea Rosca(✉), Eugen Rosca, Cristina Oprea, Florin Rusca, Oana Dinu, and Aura Rusca

University "Politehnica" of Bucharest, Splaiul Independentei 313, 060042 Bucharest, Romania
mircea.rosca@upb.ro

Abstract. Population growth, motorization, urbanization and population density changes lead to increased traffic congestion in cities. This article studies the impact of road infrastructure development and the new residential and office buildings on road traffic from the area that includes both the road access to a major highway in Romania and the access in the largest university campus in Bucharest. The study of road traffic is a complex problem, difficult to analyze by analytical methods and therefore it is suitable for the computer simulation. Thus, in order to evaluate the performance and the impact of designing new infrastructures, to explore new alternatives of traffic organization on the existing infrastructure, we have developed a road simulation model using AIMSUN. The values of traffic measures of effectiveness (delays, travel times, speeds, queue lengths, number of stops) obtained by modelling the proposed scenarios provide a clear picture of the impact that the development of a road overpass and of new buildings will have on the traffic in the studied area.

Keywords: Traffic congestion · Road infrastructure · Urban development · Microsimulation

1 Introduction

Road traffic congestion became one of the most irritating problems of the modern world due to the increasing of the motorization indexes, the expansion of the urban area, the growth of urban population and need of mobility [1].

While the urban road network development is extremely limited under physical and financial aspect [2], the solution for diminishing the congestion in large urban areas aims both actions for reducing the social mobility in terms of travel needs of persons and goods transfer) and actions for improving the use of existing infrastructure [3–5] by reducing spatial and temporal unevenness of traffic flows and by improving the traffic flow fluency due to the systematization and traffic regulation in conflict zones – intersections [6, 7].

The study of road traffic is a complex problem, difficult to analyze by analytical methods and therefore it is suitable for the computer simulation [8]. As a method of studying the activity of traffic systems, simulation tool presents the following advantages [9]:

L. Moldovan and A. Gligor (Eds.): Inter-Eng 2021, LNNS 386, pp. 235–246, 2022.
https://doi.org/10.1007/978-3-030-93817-8_23

- decision making without purchasing expensive tangible elements or interrupting system's current activities;
- compressing time and investigating phenomena within a convenient range;
- allows the analysis of a large number of alternatives by altering the initial conditions, having the advantage of returning to the response alternative in accordance with user requirements;
- the statistical and graphic indicators of congestion-generating points and also build-up, dissipation and duration of traffic congestion;
- exploring new ways of organizing traffic, of influences of the infrastructure investments;
- highlighting the links between the parameters of complex traffic systems and mutual influences.
- Depending on the level of detail of the analyzed system, the models characterizing the traffic flows are divided into three groups [10–12]:
- microscopic models (high degree of detail) - describe the spatio-temporal behaviour of each vehicle and the detailed interactions to which it takes part; For example, a vehicle lane change movement uses car-following models in relation to the current leading vehicle and then to the presumptive one, as well as to its presumed tracker from the desired lane;
- mesoscopic models - do not operate with individualization of vehicles or their trajectories but shape their behaviour in probabilistic terms; traffic is composed of small groups of entities whose actions and interactions are described in detail, based on the relationship between flow size, density and speed. For example, a vehicle lane change manoeuvre can be represented for individual vehicles as an instant event with decisions based on lane density rather than on the detailed interactions between vehicles.
- macroscopic models (low degree of detail) - analyzes the overall evolution of traffic flows, without taking into account the interaction between vehicles, but only the fundamental relationships between flow rate, density and speed. For example, lane change manoeuvres will not be represented.

According to Novacko et al. [13], driver and pedestrian behaviour significantly affect the safety and the flow of traffic at the microscopic and macroscopic level.

This article studies the impact of generated traffic as a direct result of a road infrastructure development and of the new residential and office buildings on an urban area that includes both the road access to a major highway in Romania and the access in the largest university campus in Bucharest [14]. For this purpose, we opted for road traffic modelling using the AIMSUN microsimulation software.

2 Methodology

The required data for running of dynamic traffic simulation in AIMSUN is included in scenario, experiment and replication. The scenario is composed of three types of data: network layout, traffic demand data and traffic control plans.

The traffic network is composed of a set of nodes (intersections) connected with a set of sections which may have different attributes (number of lanes, possible turning movements, speed limits, turning speed).

Traffic demand data are defined by vehicle types with their attributes, flows at the input sections (entrances to the network) and turn percentagesat intersections.

The input data required to define traffic control are the signal groups into which turning movements are grouped, the sequence of phases and, for each one, the signal groups that have right of way and the duration of each phase.

The traffic modelling process comprises a number of stages [15, 16]:

1. Collecting initial data and processing input data for the developed simulation model:

   - Traffic surveys were organized in two days, during the morning peak (7:30 am-9:30 am) and evening peak (4:30 pm-6:30 pm) for recording: the motorized traffic flows in the main intersections in the analyzed area, the motorized traffic flows and pedestrian flows at the access points in UPB Campus (UPB-S1 in Fig. 1)
   - Collecting and processing data related to under construction zones in the study area, which will contribute to new generated/attracted traffic flows. Also, data from public documents and reports (Area Town Planning, reports of real estate companies) and data collected on the field were used.

2. Development/calibration/validation of simulation models. Due to lack of devices needed for proper calibration of the microsimulation model, we have adjusted the model considering the following:

   - Checking/correcting the formalization errors of the traffic network imported from Open Street Map, involved a review of road sections characteristics - length, width, category, maximum speed, number of lanes, connections and allowed turns between intersections' sections etc.;
   - Using the microsimulator's default parameters;
   - The calibration process consisted in comparing the observed hourly flows and those obtained by experiments of the microsimulation model using the GEH indicator.

3. Analyzing the results and assessing the impact of urban development and road infrastructure on vehicle and pedestrian traffic in the study area (areas with new residential complexes, offices, opening of Ciurel overpass);

The simulation parameters are fixed values that describe the experiment (simulation time, warm-up period, statistics intervals) and some variable parameters used to calibrate the models (reaction times, lane changing zones etc.)

The performance indicators provided by AIMSUN software in order to analyze the traffic network are relative to the entire network, to each section, to turning movements at intersections, to sub-paths defined by the modeler. Some of the most important global and local traffic indicators [17] used are:

- Delay Time: average delay time per vehicle per kilometer. It is the difference between the expected travel time under ideal conditions and the observed travel time. It is calculated as the average for all vehicles in a certain time interval of statistics T;

- Flow: number of vehicles that exit the network during period T divided by T;
- Speed: average speed per vehicle;
- Mean Queue Length: average length of the queue in that section, expressed as the number of vehicles per lane.
- Number of stops: average number of stops per vehicle per km.

3 Study Area

University Politehnica of Bucharest (UPB) is the oldest and most prestigious school of engineers in Romania. With a large number of students (approximately 29,000 students), teaching staff, researchers and auxiliary staff working in the 15 faculties, the university occupies a significant place within the urban space where UPB Campus, with an area of approx. 120,000 m^2 (Fig. 1), is located.

At first, the university complex had a unitary character, and none of the arteries in the area crossed the courtyard of the institution. With the completion of the dormitory complex, the monolithic character of U.P.B. campus has been lost. After 1990, the growth of motorization degree, the development of residential buildings, commercial spaces and offices in the areas around UPB have contributed to the increase of road traffic flows.

The accessibility provided by the high capacity public transport services in the area - three metro stations (Politehnica, Grozăveşti and Petrache Poenaru), by two tram lines, mostly dedicated, was an additional factor in increasing the attractiveness in the area for

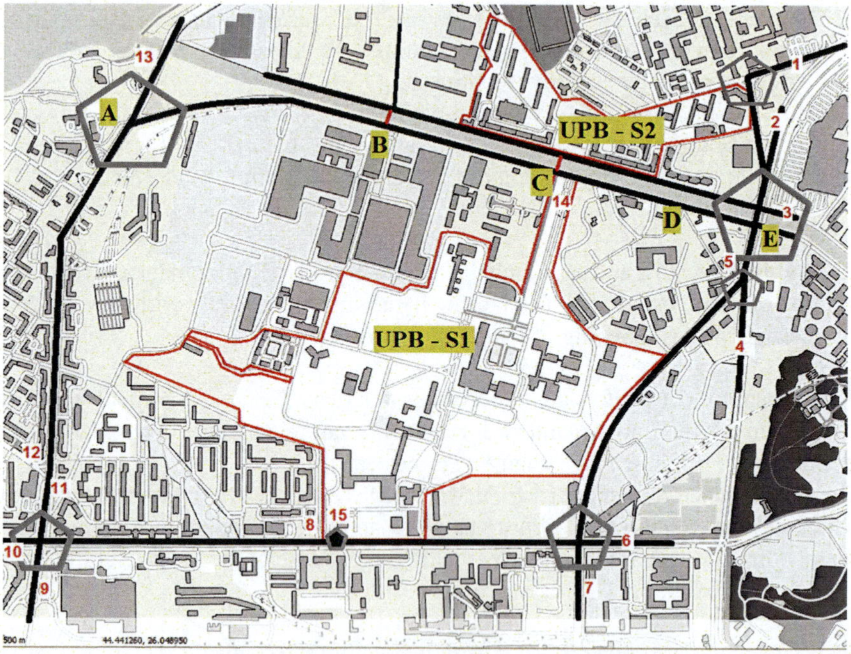

Fig. 1. Study area (UPB-S1 Educational/research/administration spaces; UPB-S2 Student dormitories spaces)

new real estate developments, leading to changes in urban functions and to restructuring the area where UPB Campus is located.

All these changes led to the saturation of high-capacity arteries that limits the area where UPB Campus is located: Iuliu Maniu Blvd., Splaiul Independenţei, Vasile Milea Blvd., Virtutea Blvd., symbolized with black line in Fig. 1. Furthermore, taking into consideration the "Ciurel Penetration" project [18], that includes an expressway sector and a set of passages having the role of connecting the A1 Bucharest - Pitesti highway with Splaiul Independentei, it is expected that the traffic indicators will change. As a consequence, the Splaiul Independenţei section, that split into two areas UPB Campus -with faculties and administrative spaces (denoted with UPB-S1 in Fig. 1) and the dormitory Regie Complex (denoted with UPB-S2), will become a busy artery that assures the takeover of the major traffic flows on West–East direction.

In Fig. 1 we have denoted with A-E the network nodes that will undergo changes in road infrastructure and/or traffic reorganization following the implementation of the project mentioned above [18], changes which will be detailed in the following sections.

The two areas UPB-S1 and UPB-S2 are bounded by a red line in Fig. 1 and separated by the Dambovita River and Splaiul Independenţei artery. The pedestrian crossing between the two areas is currently provided by a bridge over the river and two signalized crosswalks in node C.

4 Microsimulation Scenarios and Results

Road traffic microsimulation has been accomplished for the following scenarios:

- MS_S0 - "no project" scenario – the current pedestrian crossing is at the same level as the existing pedestrian bridge between UPB_S1 and UPB_S2 areas; without new urban development,
- MS_S1 - urban development scenario - Office/Residential Buildings and systematization of vehicle traffic - current pedestrian crossing at the same level as the existing pedestrian bridge between UPB_S1 and UPB_S2;
- MS_S2 - urban development scenario S1 and the building of a pedestrian overpass to ensure the connection between UPB_S1 and UPB_S2 without interruption of road traffic.

In urban development's scenarios MS_S1, MS_S2 were integrated the proposed solutions for road traffic systemization through the Splaiul Independenţei - Ciurel penetration - Bucharest – Pitesti highway project. Thus, the road infrastructure of nodes A and B (Fig. 2a) will be modified with overpass and roundabout (Fig. 2b).

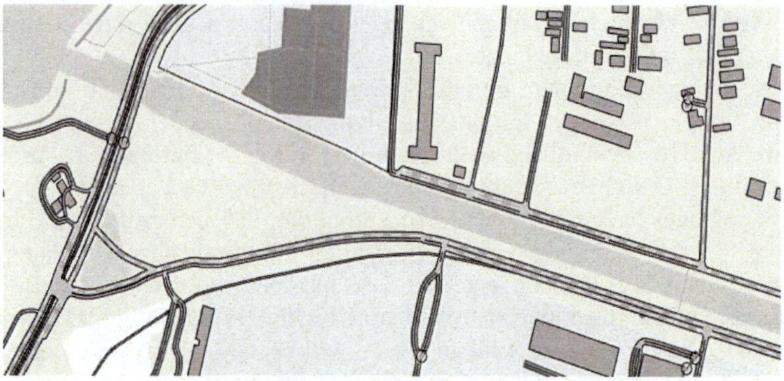

a) reference scenario MS_S0

b) road infrastructure development scenario – overpass and roundabout

Fig. 2. Road infrastructure and traffic systematization changes of node A (AIMSUN)

The regulation of the pedestrian circulation in the node C consists in the replacement of the level crossings (Fig. 3a and Fig. 3b) with a pedestrian overpass (Fig. 3c).

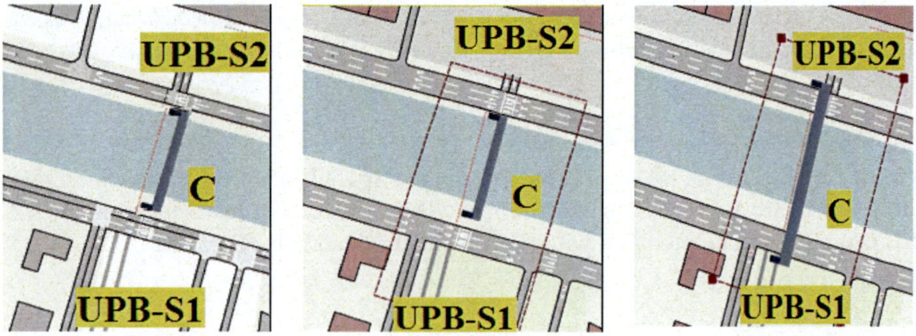

a) MS_S0 b) MS_S1 c) MS_S2

Fig. 3. Pedestrian crossing in node C for the simulated scenarios (AIMSUN)

Changes in traffic infrastructure and traffic regulation on Splaiul Independenţei (the arterial road from node A to node D) are summarized in Table 1.

Table 1. Road infrastructure and traffic regulation in the proposed scenarios. (SI - arterial road Splaiul Independenţei).

Changes on SI	MS_S0 Scenario	MS_S1, MS_S2 Scenarios
Roundabout node B	No	Yes
Traffic lanes on SI between node A and B/opposite	Two-way (one lane/two lanes)	One way (two lanes from A, two from overpass)
Between node B and C/opposite	Two-way (two/one lanes)	One way (four lanes)
Between node C and D/opposite	Two-way (two/one lanes)	One way (three lanes)
From Borneanu St. to node D	One way (two lanes)	Two-way (one/one lane)
Pedestrian crossing between UPB-S1 and UPB-S2 spaces in node C	Pedestrian crossing at level	Pedestrian overpass

The simulation experiments were realized in the following hypotheses:

- the warm-up time is 30 min, considered to be sufficient to limit the effects of the initial state and to reach the stationary regime;
- the simulation time is one hour in the peak interval 05:30 pm–06:30 pm, statistics intervals- 5 min
- in MS_S1 and MS_S2 scenarios the hourly flow rate was assumed to be q = 1500 PCU/h at the entrance to the Splaiul Independenţei Blvd from the Ciurel overpass, respecting the principle of moderation of the estimations;
- the vehicles flow generated by the development of office buildings is 10% of the number of employees, under the same principle of moderation;
- In all three analysis scenarios, the same traffic lights conditions at intersections were maintained (cycle length, phases, phase sequences).

The GEH statistic [19] used for traffic volume calibration compares expected or measured volumes with volume output from the microscopic simulation model and is calculated using the following formula:

$$GEH = \sqrt{\frac{2(q_m - q_c)^2}{q_m + q_c}} \tag{1}$$

where, q_m – output traffic volume from the simulation model (vehicles per hour);
q_c – real-world hourly traffic count (vehicles per hour);

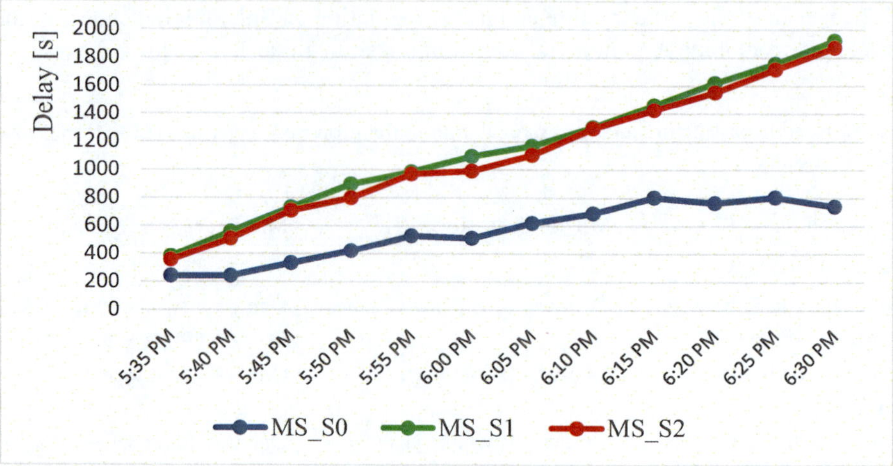

a) from B to E

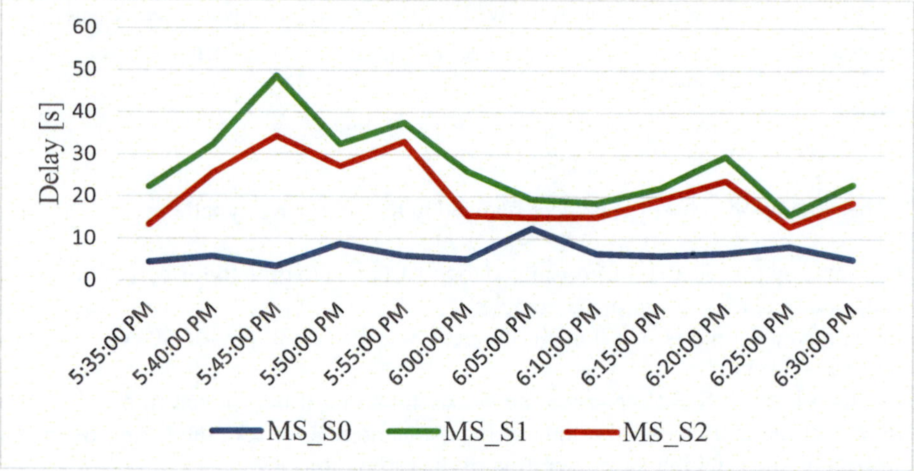

b) from E to B

Fig. 4. Average vehicle delays on SI section from B to E (a) and opposite direction (b)

Table 2 shows that all GEH values obtained on the road sections in the analyzed area (Virtuții Blvd.- Splaiul Independenței - Vasile Milea Blvd. - Orhideelor St.) are lower than the required limit value, 5.

By simulating the three scenarios MS_S0, MS_S1 and MS_S2 the time variation of the traffic indicators considered were obtained and compared for Splaiul Independenței (SI) section between node B and E. Thus, the differences between average vehicle delays on SI obtained from the three scenarios can be observed in Fig. 4. In the scenarios of urban development without (MS_S1) or with (MS_S2) the construction of the UPB-Regie pedestrian overpass, the delay times increase than in actual situation (MS_S0), although the traffic flow has increased only with 19%, respectively 36%.

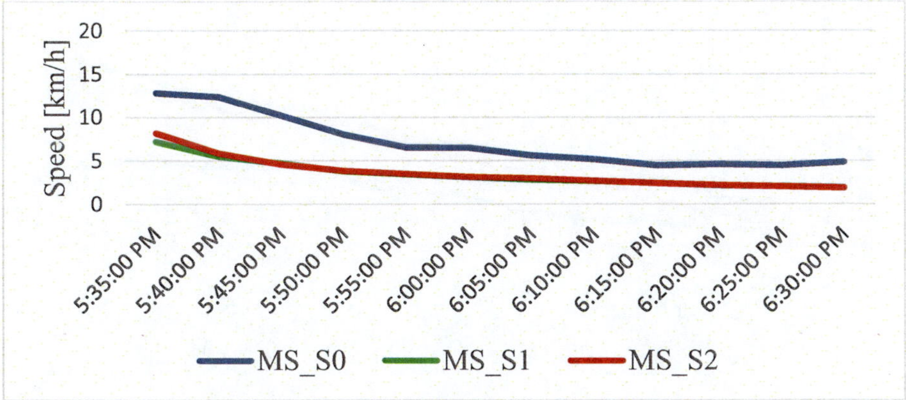

a) from B to E

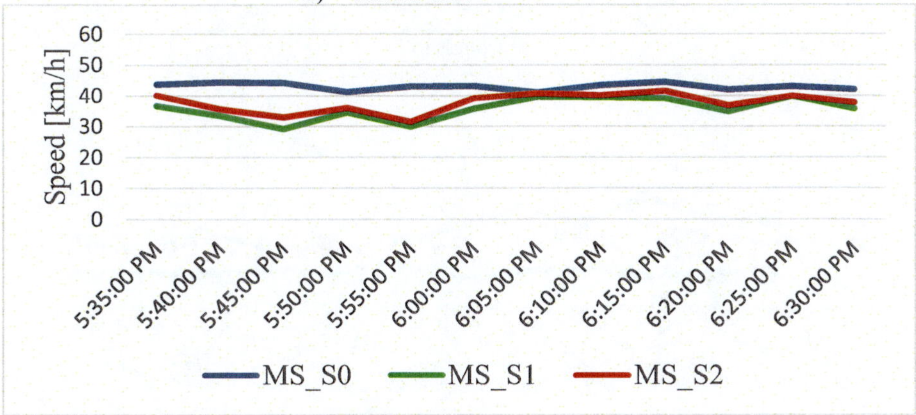

b) from E to B

Fig. 5. Average vehicle speeds on SI section from B to E (a) and opposite direction (b)

The differences between average vehicle speeds on SI obtained from the three scenarios are shown in Fig. 5. The vehicle speeds have drop by half for MS_S1 and MS_S2 scenarios (Fig. 6).

The traffic performance indicators on section between node B and E are summarized in Table 3.

The results show that in the scenarios of urban development, without (MS_S1) or with (MS_S2) the construction of the UPB-Regie pedestrian overpass, the delay times are much greater (107%, respectively 98%) than actual situation (MS_S0), although the traffic flow has increased only with 19%, respectively 36%. Also, the vehicle speeds decrease with 51% for MS_S1 scenario, respectively with 49% for MS_S2 scenario and the number of stops almost doubled.

The UPB-Regie pedestrian overpass (in node C) improves traffic performance indicators as a result of eliminating pedestrian crossings and the associated traffic lights. Also, this segregation of vehicle and pedestrian flows is a solution to increase pedestrian safety.

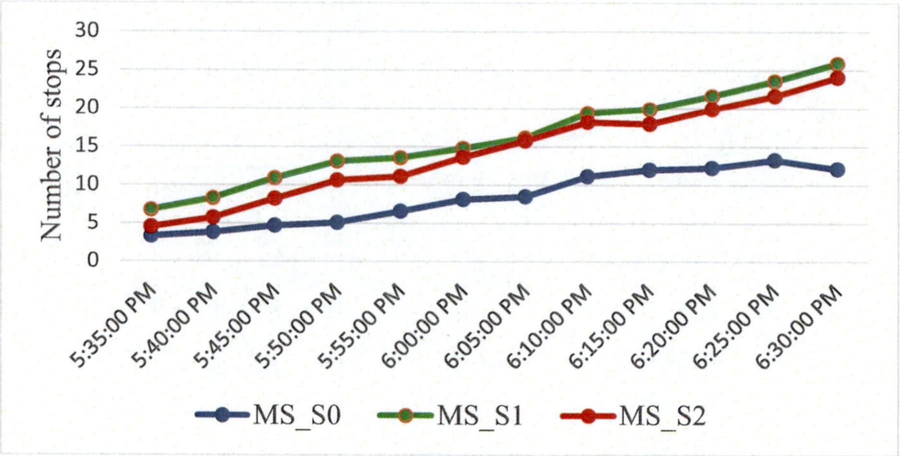

a) from B to E

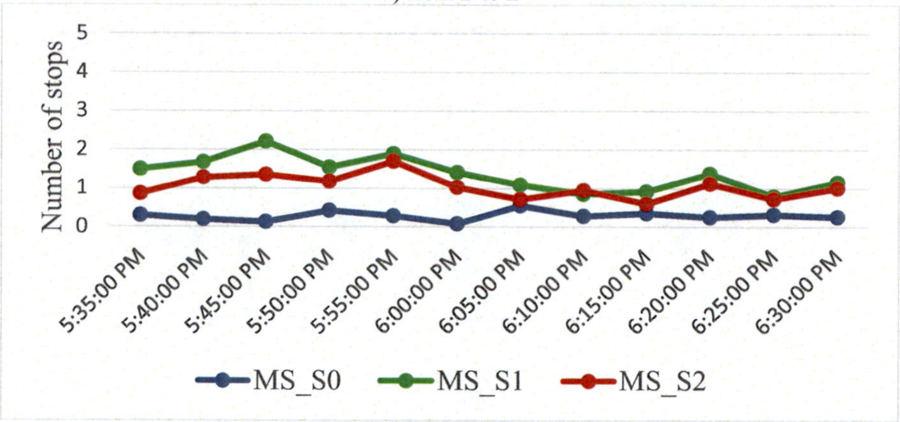

b)from E to B

Fig. 6. Average number of stops on SI section from B to E (a) and opposite direction (b)

5 Conclusion

Road traffic congestion is often the result of population and employment growth, socio-economic and cultural development, motorization, urbanization. Solutions to alleviate congestion in road traffic consist either in increasing capacity by improving infrastructure, building new arteries or widening existing ones, either in reducing social mobility and improving the use of existing infrastructure.

A new urban development has a direct influence on economic and social activities and implicitly on traffic flow and its associated risk. Thus, this research aims to identify the impact of road infrastructure development and the new residential and office buildings on road traffic from the area that includes both the road access to a major highway in Romania and the access in the largest university campus in Bucharest.

Table 2. GEH indicator values obtained on the road sections of the analyzed area (selection)

Road section	Simulated traffic volumes [PCU/h]	Measured traffic volumes [PCU/h]	GEH
Node A - Virtutii_South	1885	1745	3.29
Node A - Virtutii_East	731	740	0.33
Node A - Virtutii_North	1973	1958	0.34
Node B -IDM_West	729	853	4.41
Node B - IDM_East	735	801	2.38
Node B - IDM_South	221	225	0.27
Node C - SI	832	847	0.52
Node C - opposite SI	662	794	4.89
Node C - UPB_South	114	155	3.54
Node E - Grozavesti_West	874	774	3.48

Table 3. Comparison of the traffic indicators of simulated scenarios

	MS_S0	MS_S1	MS_S2
Delay Time [sec/km]	559	1156	1109
• in relation to MS_S0	–	107%	98%
• in relation to MS_S1	–		–4%
Flow [veh/h]	606	722	826
• in relation to MS_S0	–	19%	36%
• in relation to MS_S1	–	–	14%
Speed [km/h]	7.13	3.48	3.61
• in relation to MS_S0	–	–51%	–49%
• in relation to MS_S1			4%
Number of Stops	8	16	15
• in relation to MS_S0	–	93%	73%
• in relation to MS_S1	–	–	–10%

The traffic microsimulation model used to test the changes in road infrastructure and traffic regulation provides an image of the negative constraints measured by traffic performance indicators on population that transit the analyzed area. Therefore, the increase of road traffic induced by urban development and the construction of an road overpass leads to an increased on delays with 51% of the actual situation. In addition, building of an pedestrian overpass to ensure the connection between UPB Campus and Regie

Complex without interruption of road traffic will decrease delays with 4% and number of vehicle stops with 10% (Table 3).

Further research will focus on better calibration and validation of developed microsimulation model and on traffic risk estimation in order to generate a more robust model that will integrate adjacent intersections.

References

1. Rezaeinia, P., Palma, A., Lindsey, R.: Traffic congestion control: tradable permits vs road tolls, THEMA Working Papers 2021-09, Université de Cergy-Pontoise (2021)
2. Raicu, S., Costescu, D., Popa, M., Rosca, M.: Including negative externalities during transport infrastructure construction in assessment of investment projects. J. Eur. Transp. Res. Rev. **11**, 11–24 (2019). https://doi.org/10.1186/s12544-019-0361-9
3. Bădău, F., Abramović, B., Cormos, A.C., Iordache, V.: Management of urban and regional rail: case study Bucharest. LOGI – Sci. J. Transp. Logist. **11**(2), 120–131 (2020). https://doi.org/10.2478/logi-2020-0021
4. Macario, R.: Urban Public Transport. In: Finger, M., Holvad, T. (ed.) Regulating Transport in Europe, chapter 6, pp. 140–168. Edward Elgar Publishing (2013)
5. Zavada, J.B., Abramović, B., Sipus, D.: A strategic model of sustainable mobility in the city of Zagreb and its surrounding area. Int. J. Traffic Transp. Eng. (IJTTE) **7**(4), 430–442 (2017). https://doi.org/10.7708/ijtte.2017.7(4).03
6. Krasic, D., Novacko, L.: The impact of public transport network accessibility on trip generation model. J. PROMET-Traffic Transp. **27**(2), 165–172 (2015)
7. Gonzalo-Orden, H., Rojo, M., Perez-Acebo, H., et al.: Traffic calming measures and their effect on the variation of speed. In Efficient, Safe and Intelligent Transport: 12th Conference on Transport Engineering (CIT), pp. 349–356 (2016)
8. Tartaro, M.L., Toress, C., Wainer, G.: Defining models of urban traffic using the TSC tool. In: Proceedings of the Winter Simulation Conference, pp 1056–1063 (2001)
9. Transportation Research Board: Highway Capacity Manual the 6th Edition, The national academies of science engineering medicine (2016)
10. Hoogendoorn, S.P., Bovy, P.H.L.: State-of-the-art of vehicular traffic modeling. In: Proceedings of the Institution of Mechanical Engineers, Journal of Systems and Control Engineering. Part I, vol. 4, no. 215, pp. 283–304 (2001)
11. Akcelik, R.: Comparing lane based and lane-group based models of signalised intersection networks. Transp. Res. Procedia **15**, 208–219 (2016)
12. Kajalic, J., Celar, N., Stankovic, S., Kocic, A.: Vehicle platoon membership definition for unsaturated conditions. Int. J. Traffic Transp. Eng. **11**(2), 199–212 (2021)
13. Babojelić, K., Novačko, L.: Modelling of driver and pedestrian behaviour – a historical review. J. PROMET-Traffic Transp. **32**(5), 727–745 (2020)
14. Dinu, O.M., Stefanica, C.: Estimation of spatially diverted traffic on Bucuresti – Constanta A2 highway. In: Transportation and Land Use Interaction, pp. 139–149 (2008)
15. TSS-Transport Simulation Systems. AIMSUN Version 8.2 User's Manual (2017)
16. Costescu, D., Raicu, S., Rosca, M., Burciu, S., Rusca, F.: Using intersection conflict index in urban traffic risk evaluation. Procedia Technol. **22**, 319–326 (2016)
17. Barceló, J., Codina, E., Casas, J., Ferrer, J.L., García, D.: Microscopic traffic simulation: a tool for the design, análisis and evaluation of intelligent transport systems. J. Intell. Robot. Syst. **41**, 173–203 (2004)
18. Ciurel Penetration, Area Town Planning, Bucharest City Hall Homepage. https://www.pmb.ro/cautare/134. Accessed 20 Nov 2020
19. Nor Azlan, N.N., Md Rohani, M.: Overview of application of traffic simulation model. Matec Web Conf. **150**, 03006 (2018). https://doi.org/10.1051/matecconf/201815003006

Road Traffic Accidents - Safety Analysis of Urban Travel with Two-Wheel Vehicles

Cristina Oprea(✉), Anamaria Ilie, Mircea Roșca, Florin Ruscă, Sergiu Olteanu, and Isabela Ilie

Transport Faculty, Department Transport, Traffic and Logistics, University Politehnica of Bucharest, Splaiul Independentei, 313, 060042 Bucharest, Romania
cristina.oprea@upb.ro

Abstract. The number of deaths in road accidents in which are involved the users of two-wheel vehicles leads to the need of studying and identifying the causes and the solutions to reduce them. The specialists need to analyze the historical data of road accidents. According to European Transport Safety Council, pedestrian deaths in the EU decreased by 19%, powered two-wheeler (PTW), rider deaths by 20% and vehicle occupant deaths by 24% over the period 2010–2018. Cyclists were the only road user group that saw a stagnation. Romania is in the first three countries with the highest cyclist mortality. In Bucharest, the situation of accidents in which the users of two-wheeled vehicles are involved is not gratifying. The number of accidents from 2016 to 2020 according to the data obtained from General Police Directorate of Bucharest, is increasing. The municipality should speed up the construction of new bicycle lanes, starting with the arteries on which most accidents occurred.

Keywords: Vulnerable road participant safety · Two-wheel vehicles · Accident

1 Introduction

Road safety is currently an issue of global importance because both from studies on the causes of mortality worldwide and nationally [1–5], as well as from statistics conducted by many organizations (World Health Organization, Romanian Police, Eurostat), road accidents are one of the leading causes of mortality of young people.

Studying the factors that are involved in the occurrence of accidents, their causes and consequences, is a particularly important element in order to find effective measures to increase traffic safety.

The European Commission adopted in 2011 the strategy "Transport 2050: The major challenges, the key measures" which sets as a common goal of member countries "zero deaths" in road transport by 2050. Thus, the EU's goal is to halve the number of road deaths by 2020. Accidents in road transport are the most numerous and cause the most casualties worldwide (Fig. 1).

Eight EU Member States saw a record drop in the number of fatal accidents in 2019: Croatia, Finland, France, Germany, Greece, Latvia, Luxembourg and Sweden. However,

L. Moldovan and A. Gligor (Eds.): Inter-Eng 2021, LNNS 386, pp. 247–257, 2022.
https://doi.org/10.1007/978-3-030-93817-8_24

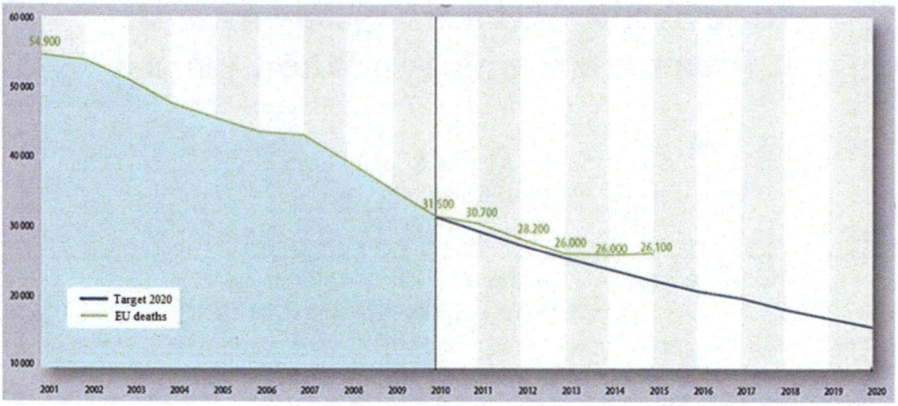

Fig. 1. Deaths in road accidents in the EU since 2001 (Source: [6])

progress has slowdown in most countries. Therefore, the EU target of halving the number of deaths from road accidents between 2010 and end of 2020 was not achieved, although perhaps that road fatalities dropped significantly in 2020 due to measures taken to combat the COVID'19 pandemic. This decrease was not sufficient to achieve the EU's goal.

Road collisions are the second important factor of death for people aged between 5 to 29 years and it is a third leading cause for the age group between 30–44 years [7].

The European Union (EU) faces many interconnected demographic and environmental challenges: climate change, the number of road deaths, increasing urbanization, population aging, increasing air pollution, increasing obesity, etc. There is a growing recognition at EU level that increasing the levels of active mobility, especially walking and cycling, can play an important role in overcoming many of these challenges. According to European Cyclist Federation [8], even the current levels of cycling in the EU produce benefits valued at around 150 billion euros per year.

In a recent study of the European Commission [9], the negative external costs of motorized road transport such as congestion, pollution and climate change are estimated at 800 billion euros per year.

In the period 2010–2018 pedestrian deaths in the EU decreased by 19%, powered two-wheeler (PTW) rider deaths by 20% and vehicle occupant deaths by 24% (Fig. 2).

In some countries, the reduction in the number of accidents involving cyclists is closely linked to the adopted road safety measures, as the deaths of bicycle users are reduced to the same extent as the deaths of motor vehicle users. Thus, according to the PIN Flash Report 38, only 9 EU countries have been able to adopt road safety measures so that the percentage of bicycle deaths decreases faster than deaths due to accidents between motor vehicles. An example is Lithuania, where the percentage of bicycle deaths has decreased with 11% annually - 4% faster than the reduce of the deaths of motor vehicle users.

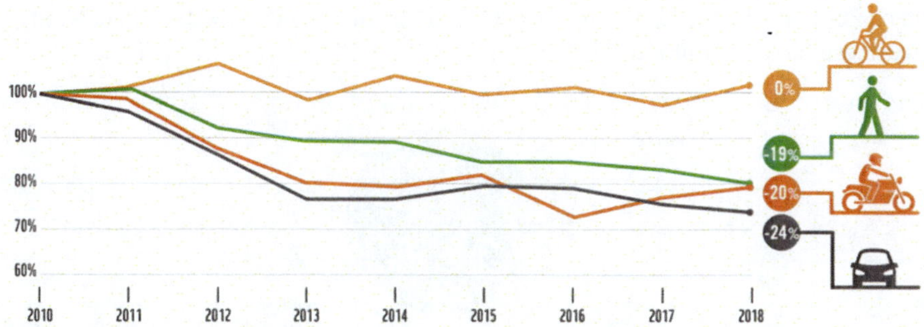

Fig. 2. Variation in the number of deaths of pedestrians, cyclists, PTW users reported by the police in 28 EU countries in the period 2010–2018. (Source: [10])

2 Literature Review

Hirasawa, M. and M. Asano developed in 2003 a traffic accident analysis system using GPS. The system provides the following information: road structure, road accessory facilities and weather information; it is used to analyze accident frequency, accident rate, and fatality rate [11].

Assegidew Yemane W/Yohannes and Amare Sewnet Minale developed a study to analyze traffic accidents using GIS to identify the hotspots of road accidents. They used digital maps in order to fix crash locations. They identified twenty-five hotspots using the Kernel Density estimation method [12].

Mohammed Taleb Obaidat, Thanaa M. Ramadan in 2012 developed a study to underline the most influential factors of accidents in dangerous locations of local urban roads and to correlate the characteristics of accidents with different factors, such as: geometric elements, pavement type, traffic speed, lighting conditions, existence of pedestrian facilities, traffic conditions etc. [13].

Jaikishan Damani and Perumal Vedagiri in 2021 realized a comprehensive review of the diverse studies that considered different sets of risk factors and have given surprising and even conflicting results of the safety of motorized two wheelers. They incorporated in their paper all the risk factors considered in previous research [14].

S Piantini et al. in 2016 analyzed 40 powered two-wheelers from the InSAFE database to other vehicle urban accidents. The results allowed them to identify possible countermeasures in terms of safety devices or re-design of vehicle sections to reduce the injuries in this subset of accidents [15].

H. Liers conducted a study based on the real-world accidents to describe the accident scenario involving motorcycles. The study gives a comprehensive overview about the German motorcycle accident scenario [16].

3 Safety Analysis of Urban Travel with Two-Wheel Vehicles

Traffic accidents have different causes, depending on the environment in which they occurred. The Department of Education and Training of New South Wales Australia found in a study that: - the human factor is the cause in 67% of accidents, - the vehicle is the cause in 4% of accidents and - road elements are the cause in 4% of accidents. In reality, these factors combine (Fig. 3).

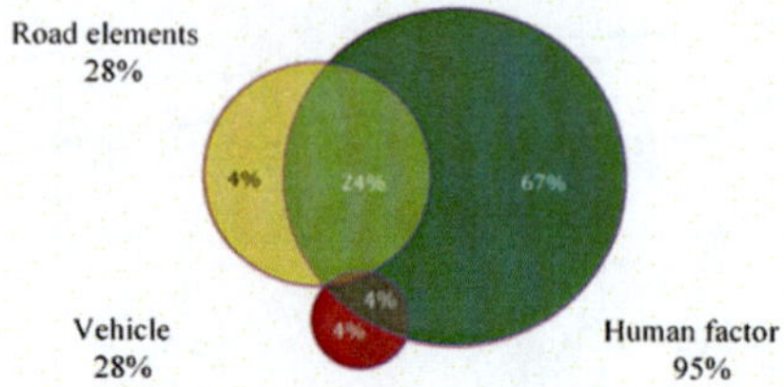

Fig. 3. The share of factors involved in accidents (Source: [17])

From 2010 to 2018, on the roads of European Union countries, approximately 51,500 pedestrians and 19,500 cyclists were killed, of which 5180 pedestrians and 2160 cyclists in 2018 alone. Thus, deaths among users of alternative travel (pedestrians, cyclists and scooters) represent 29% of all road deaths in the EU.

In the period 2010–2018, the number of deaths among cyclists decreased on average by only 0.4% each year in the EU, while among motor vehicle users there was an annual reduction in deaths of 3.1%.

Thus, deaths among cyclists fell in twelve EU countries, stagnated in two and rose in ten (Fig. 4).

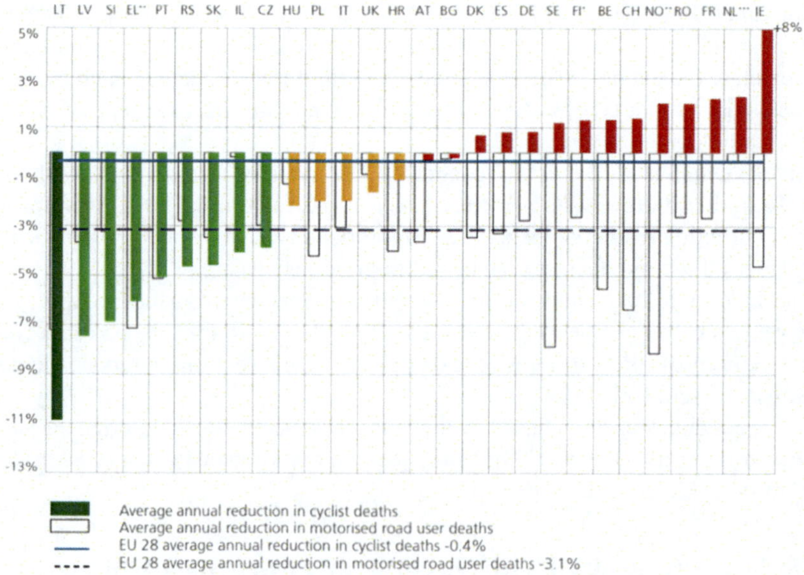

Fig. 4. Variation of the annual average death of bicycle users compared to the death of motor vehicle users (Source: [10])

In the EU, 83% of bicycle deaths were the result of a collision with a motor vehicle: 53% involved a car, 13% a heavy goods vehicle, 7% a van, 2% a bus or a coach, 2% a PTW and 6% another vehicle [18].

The percentage variations are very large between countries and come from both the differences related to the cycling tradition and the legislation.

If we take into account the road accidents in which cyclists are only injured in addition to those that result in their death, then it is considered that they represent 8% of the total accidents that occur on the roads of the European Union (Fig. 5).

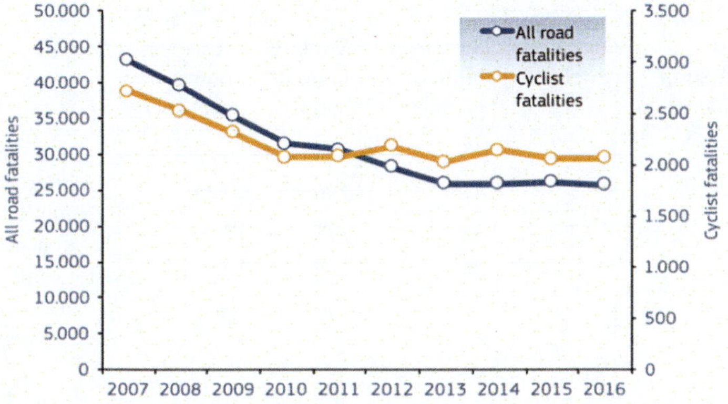

Fig. 5. Number of accidents involving cyclists in total road accidents at EU level (source: [19])

According to European Transport Safety Council, PIN Flash Report 38 [10], "the highest cyclist mortality is in the Netherlands with 12 cyclist deaths per million inhabitants, Romania with nine and Hungary with eight". The percentages of death can be observed in Fig. 6.

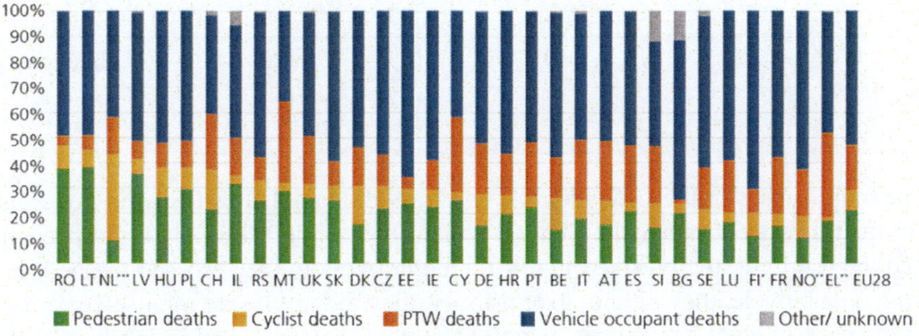

Fig. 6. Pedestrian, cyclist, power-two-wheeler (PTW) user and vehicle occupant deaths reported (Source: [10])

According to European Road Safety Observatory that used the statistics from 2010–2016, Romania is in the first three countries with the highest cyclist fatality rates per million population in the EU.

4 Safety Analysis of Urban Travel with Two-Wheel Vehicles in Bucharest

In Bucharest city, the situation of accidents in which the users of two-wheel vehicles are involved is not gratifying. According to the statistical data requested and obtained from the General Police Directorate of Bucharest, the number of accidents involving bicycle users (Table 1) and scooter users (Table 2) is increasing.

Table 1. Number of accidents involving bicycle users.

	Total number of accidents	Deceased people	Seriously injured people	Slightly injured people	Accidents due to cyclists	Cause
2016	329	1	43	285	177	- Non-compliance with traffic rules by cyclists - Non-prioritization by vehicle users - Insecurity when changing direction
2017	346	3	47	296	189	
2018	422	6	38	378	269	
2019	504	3	68	433	321	
2020	355	1	83	271	261	

Table 2. Number of accidents involving scooter users.

	Total number of accidents	Deceased people	Seriously injured people	Slightly injured people	Accidents due to scooter users	Cause
2016	2	0	0	2	2	- Speed - Insecurity when changing direction
2017	7	0	1	6	6	
2018	16	0	3	13	9	
2019	174	0	40	134	126	
2020	171	0	49	122	127	

The main causes of these accidents come from the fact that cyclists did not comply with traffic rules, according to the General Police Directorate: they had a high speed, they crossed the red light, they walked in a forbidden direction, they crossed the pedestrian crossing on a bicycle, they did not give priority, they did not insure themselves at turns.

In the case of electric scooters, which have often become used in Bucharest, there is also an alarming phenomenon. In 2019, the number of serious accidents with electric scooters began to increase. If in 2018 there was only one serious accident in which only one person was seriously injured, in 2019 the number of serious accidents reached 28, the main cause being non-compliance of speed limits.

Figure 7 shows the increasing trend of the number of accidents and the number of injured people.

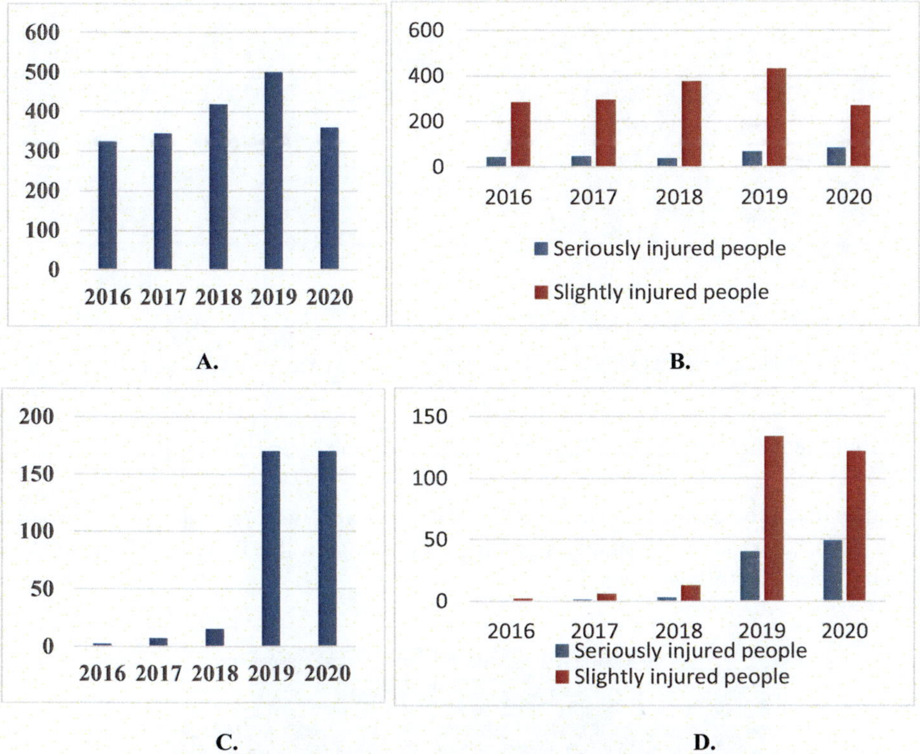

Fig. 7. Bicycle and scooter accidents in Bucharest during 2016–2020. **A**. Total number of bicycle accidents; **B**. Number of injured people in bicycle accidents. **C**. Total number of scooter accidents; **D**. Number of injured people in scooter accidents.

A survey realized in the National Strategy for Encouraging the Use of Bicycles, within a European project together with NGOs that promote cycling, shows that for 77% of cyclists, the main difficulty encountered in traffic is the infrastructure.

In fact, 45.8% of cyclists find it difficult to cross intersections. Therefore, in the case of intersections where many accidents occur (black spots), it is proposed to make arrangements that physically separate bicycle traffic from road traffic.

The areas in Bucharest with the most accidents involving cyclists (Fig. 8A), are:

- 2016: Calea Victoriei, Bd. Iuliu Maniu, Șos. Virturții;
- 2017: Calea Victoriei, Spl. Independenței, Șos. Colentina, Bd. I. Maniu, Șos. Olteniței
- 2018: Calea Victoriei, Bd. Unirii, Bd. Iuliu Maniu, Șos. Bucureștii-Noi, Șos. Mihai Bravu, Șos. Dorobanți, Șos. Pantelimon, Șos. Colentina;
- 2019: Calea Victoriei, Splaiul Independenței, Șos. Mihai Brav, Bd. Unirii, Șos. Pantelimon, Șos. Colentina, Calea Ferentari;
- 2020: Calea Victoriei, Bd. Unirii, Șos. Mihai Bravu, Bd. Timișoara.
- The arteries with the most accidents involving scooter users (Fig. 8B), are:
- 2019: Calea Victoriei, Splaiul Independenței;

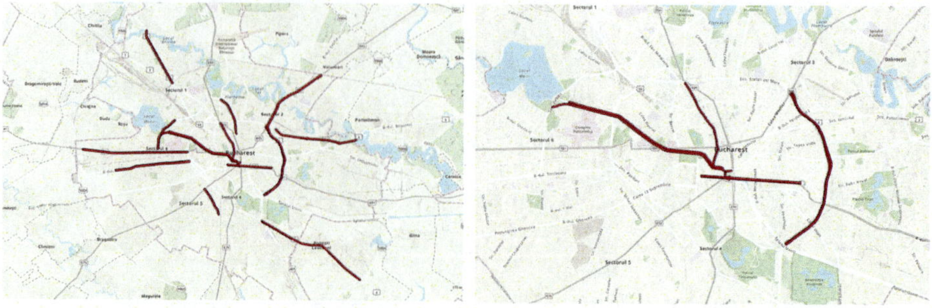

A. B.

Fig. 8. The road arteries on which most accidents occurred (ArcGIS) **A**. bicycle, **B.** scooter

- 2020: Calea Victoriei, Splaiul Independenței, Bd. Unirii, Șos. Mihai Bravu.

Currently, the routes that provide bicycle lanes in Bucharest and are appropriate in terms of road functionality/signaling, according to the BUCHAREST ROAD BRIGADE, are as follows (Fig. 9):

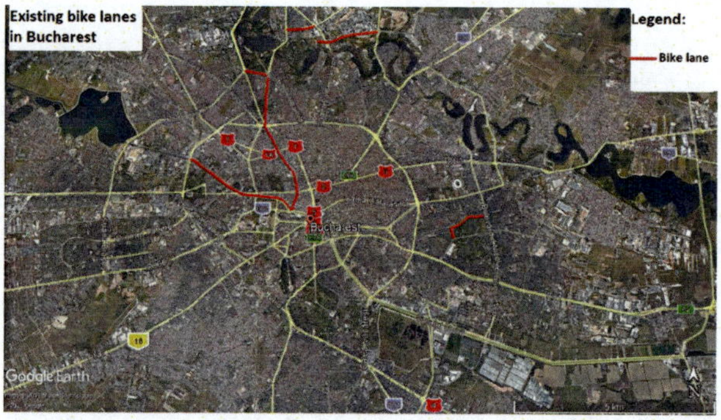

Fig. 9. Existing bicycle lanes in Bucharest

Calea Victoriei (both ways);
Splaiul Independenței (on the section between Calea Victoriei and Pod Basarab - both ways);
Bd. Aviatorilor (both ways);
Bd. Mareșal Constantin Prezan (both ways);
Șos. Pipera (on the section between Str. Nicolae G. Caramfil and Str. Aviației);
Str. Fabrica de Glucoză (both ways);
Câmpia Libertății (on the section Str. Baba Novac – Bd. Liviu Rebreanu);
Str. Baba Novac – Str. C-tin. Brâncuși (on the section Str. Câmpia Libertății – access to Al. I. Cuza park).

Considering the arteries on which the most bicycle and scooter accidents took place and taking into account the fact that in the Mobility Plan of the Bucharest municipality these arteries are included to be provided with bicycle lanes, they should be done with priority to reduce the risk of accidents (Fig. 10).

Fig. 10. The streets in Bucharest where most bicycle accidents took place

Figure 11 shows the map with the existing bike lanes and those proposed to be realized with priority.

Fig. 11. Existing bike lanes and those proposed to be realized with priority in Bucharest city

The construction of tracks for cyclists that delimit the bicycle traffic from the car one, contributes to the reduction of the risk of collisions with the vehicles. Also, applying lower speed limits is a useful measure. It is recommended that the bike lanes located on the road to be properly signalized and in accordance with the intersections they cross.

5 Conclusions

According to the statistics, in Romania the infrastructure dedicated to two-wheel vehicle users is almost non-existent, zero road education, zero awareness. Road accidents between users of two-wheel vehicles and motorized vehicles happen through common fault, but also through a lack of common commitment. The user of two-wheeled vehicles, when going out in traffic, must follow certain rules. On the other hand, local authorities and the police must provide the necessary infrastructure and context for things to go as smoothly as possible. If there were real infrastructure conditions in Bucharest, the number of accidents would decrease a lot.

We can say, therefore, that the lack of infrastructure for cyclists and the lack of education for the use of bicycles in traffic are the main causes for which cyclists are involved in road accidents.

In the absence of infrastructure dedicated to bicycles and with road traffic dominated by aggressive drivers, intolerant and sometimes untrained in traffic rules, cyclists and scooter users remain an endangered category of traffic participant.

References

1. European Commission: Mobility and Transport – Road Safety. https://ec.europa.eu/transport/road_safety/home_en. Accessed 6 May 2021
2. Spainhour, L.: Evaluation of traffic crash fatality, causes and effects, summary of final report (2005). https://www.fdot.gov/docs/default-source/safety/4-reports/bike-ped/FDOT_BD050.pdf. Accessed 06 May 2021
3. Archer, J., Vogel, K.: The Traffic Safety Problem in Urban Areas. Center for Traffic Research (2000)
4. Baron, W.: Accident analysis enables safety advances. ITS Int. **21**(2), 45–46 (2015)
5. Cordonescu, G.: Life has priority or the real costs of a car accident, Buletin AGIR (In Romanian). ISSN 2247-3548, vol. 4, pp. 70–75. AGIR, Bucharest (2014)
6. European commission, Mobility and transport - Road Safety, Statistics – accidents data (20170. https://ec.europa.eu/transport/road_safety/specialist/statistics_en. Accessed 10 May 2021
7. Anjuman, T., Hasanat-E-Rabbi, S., Siddiqui, C.K.A., Hoque, M.M.: Road traffic accident: a leading cause of the global burden of public health injuries and fatalities. In: Proceedings of the International Conference on Mechanical Engineering, ICME2007-AM-30, Dhaka, Bangladesh, pp. 29–31 (2007)
8. European Cyclist Federation: The benefits of cycling, Unlocking their potential for Europe. http://bit.ly/36L0zV0. Accessed 20 Apr 2021
9. European Commission: Sustainable transport. Internalization of transport external costs. http://bit.ly/2S1nDeq. Accessed 21 Apr 2021
10. European Transport Safety Council: How safe is walking and cycling in Europe? PIN Flash Report 38, January 2020. https://etsc.eu/wp-content/uploads/PIN-Flash-38_FINAL.pdf. Accessed 30 May 2021
11. Hirasawa, M., Asano, M.: Development of traffic accident analysis system using GIS. In: Proceedings of the Eastern Asia Society for Transport Studies, vol. 4, pp. 1193–1199 (2003)
12. Yohannes, A.Y., Minale, A.S.: Identifying the hot spot areas of road traffic accidents. Jordan J. Civ. Eng. **9**(3), 358–370 (2015). https://doi.org/10.14525/jjce.9.3.3077
13. Obaidat, M.T., Ramadan, T.R.: Traffic accidents at hazardous locations of urban roads. Jordan J. Civ. Eng. **6**(4), 436–447 (2012)
14. Damani, J., Vedagiri, P.: Safety of motorized two wheelers in mixed traffic conditions: literature review of risk factors. J. Traffic Transp. Eng. (Engl. Ed.) **8**(1), 35–56 (2021). https://doi.org/10.1016/j.jtte.2020.12.003
15. Piantini, S., Pierini, M., Delogu, M., Baldanzini, N., Franci, A., Mangini, M., Peris, A.: Injury analysis of powered two-wheeler versus other-vehicle urban accidents. In: IRCOBI Conference Proceedings, IRC-16-102, Malaga, Spain, pp. 840–853 (2016)

16. Liers, H., Hannawald, L., Prohn, M.: Analysis of the accident scenario of powered two-wheelers on the basis of real-world accidents (2014). https://bast.opus.hbz-nrw.de/opus45-bast/frontdoor/deliver/index/docId/658/file/03_Liers.pdf. Accessed 06 Jan 2021
17. Toadere, F.R.: Contribuţii la realizarea unor drumuri mai sigure, Ph.D thesis, Cluj Napoca, Romania (2015). (in Romanian)
18. Schepers, P., Stipdonk, H., Methorst, R., Olivier, J.: Bicycle fatalities: trends in crashes with and without motor vehicles in the Netherlands. Transp. Res. Part F: Traffic Psychol. Behav. **46**(B), 491–499 (2017)
19. European Road Safety Observatory: Traffic Safety. Basic Facts 2018 – Cyclists. https://ec.europa.eu/transport/road_safety/sites/default/files/pdf/statistics/dacota/bfs20xx_cyclists.pdf. Accessed 05 Feb 2021

Performance Evaluation of Dispatching Rules and Simulated Annealing in a Scheduling Problem from a Quality-Functionality Perspective

Diogo Almeida[1], Luís Pinto Ferreira[1,2](✉), José Carlos Sá[1,2], Manuel Lopes[1,2], Francisco José Gomes da Silva[1,2], and Mário Pereira[3]

[1] ISEP School of Engineering, Polytechnic of Porto, Porto, Portugal
lpf@isep.ipp.pt

[2] INEGI Instituto de Ciência e Inovação em Engenharia Mecânica e Engenharia Industrial, Porto, Portugal

[3] ESTG/CDRsp Polytechnic Institute of Leiria, Leiria, Portugal

Abstract. Production scheduling generates a direct impact on several aspects of manufacture, such as the number of delays in delivery to customers, total flow time, as well as the percentage of equipment used. It must, therefore, constitute a priority in production management, which should seek to implement scheduling techniques that will lead to positive results from the perspective of the quality of the solution. However, the methodology cannot overlook the functional aspect of the time which has elapsed until the solution is reached. This study is based on a real and specific module software improvement into a company devoted to the development of ERP software systems (Enterprise Resource Planning). It presents a solution for the production scheduling module focused on flow-shop operations, comprising a total of nine dispatching rules. An additional solution for scheduling is also proposed, which resorts to metaheuristic simulated annealing. Both solutions are compared to each other by using the quality-functionality binomial approach. These two environments are further contrasted with a third, where no effective solution for production scheduling exists. The environment which includes scheduling through dispatching rules was compared to the environment where no production scheduling was implemented. The results obtained from this analysis show an improvement of 13%. The simulated annealing solution presents an improvement of 3,6% when compared to a solution which uses dispatching rules. This improvement implies one extra minute in the calculation of the final solution.

Keywords: Production scheduling · Dispatching rules · Simulated annealing

1 Introduction

Production scheduling constitutes a complex problem that is of costly implementation and for which there are no methodologies to provide optimal solutions to most real

L. Moldovan and A. Gligor (Eds.): Inter-Eng 2021, LNNS 386, pp. 258–267, 2022.
https://doi.org/10.1007/978-3-030-93817-8_25

problems. Indeed, it is still an area which has been neglected by operational management; in many cases, there is no knowledge of the added value derived from the effective use of scheduling processes.

Over the past few decades, several problem-solving techniques have been published to address scheduling issues, thus responding to industrial contexts which have become more and more demanding. These solutions deal not only with the quality of scheduling sequences, but also with the functionality of these scheduling alternatives. It is precisely in the context of the quality-functionality binomial that the various techniques of scheduling production must be assessed; any added value ensuing from production scheduling cannot be hampered by the lengthy periods of time required to find a scheduling solution. The objective of this study is to compare the different solutions obtained: scheduling through dispatching rules; scheduling by means of metaheuristic simulated annealing (SA); and, finally, an environment where no systematized scheduling solution exists. The study was set in the context of a real case, where these solutions were implemented at an ERP (enterprise resource planning) software company.

The structure of this article comprises five sections: Sect. 1 contains the introduction; Sect. 2 consists of a literature review pertaining to production scheduling; Sect. 3 presents the methodology adopted to carry out this study, from the techniques used to a comparison of these; Sect. 4 deals with the results obtained and, finally, Sect. 5 discusses the conclusions of the work undertaken and presents a few points of departure for future research based on this study.

2 Review of Literature

Scheduling can be defined as a decision-making process which is used regularly both in industry and services. It deals with the allocation of resources to tasks within certain time periods, in order to optimize one or several performance measures [1, 2]. Scheduling has been implemented in various areas of activity, such as production, aviation and hospitals, amongst many others. Two fundamental restrictions can be identified in any scheduling problem: those associated to limitations in equipment capacity (machinery and others); and technological restrictions in the operating ranges presented by the various manufacturing orders (fixed operating sequences) [3].

The problems of scheduling production are essentially divided into two types: flow-shop problems and job-shop problems. Regarding flow-shop problems, each job must be executed on each machine, following exactly the same order. However, in the case of job-shop problems, the work can be processed on the machines without always following the same order. It is also important to distinguish deterministic scheduling from stochastic scheduling. When the conditions of the scheduling problem in question - such as processing times, setup times, job prioritization, etc. - are fully known, one is dealing with deterministic scheduling [4, 5]. On the other hand, stochastic scheduling does not assume that the values for the problem data have been established. Instead of these deterministic values, one considers the distributions which represent the behavior of these values.

One of the most common ways of representing the uncertainty inherent to a stochastic scheduling problem is by means of fuzzy models. There are numerous cases in which this uncertainty has been incorporated through fuzzy models: for example, in a case where the metaheuristic artificial bee colony was applied to solve a scheduling problem, using a fuzzy model for the processing times [6]; it can even be used in conjunction with the application of metaheuristic simulated annealing [7].

One of the main reasons for promoting research and interest in this area is the fact that this problem belongs to the NP-hard category; namely, it is impossible to obtain optimal solutions within an achievable period of time. Although the achievement of these optimal solutions is impossible, the use of heuristic methods has demonstrated that good solutions can be obtained; above all, this can be undertaken within the functional time periods required by real and industrial environments [8, 9]. This quality-functionality binomial is influenced by the heuristic chosen. However, the choice between using dispatching rules or a metaheuristic procedure is often related to whether one is faced with dynamic or stochastic environments (dispatching rules are well suited to dynamic environments) [8].

In this study, the choice resides in the management of the quality-functionality binomial. Typically, due to the fact that dispatching rules construct scheduling in an iterative manner, they generate better execution times and improved functionality, when compared to metaheuristic methods [10]. Nevertheless, ultimately, metaheuristic methods present better solutions overall [9].

With regard to metaheuristics, simulated annealing (SA) is one of the most widely adopted methods. This is chiefly due to the simplicity of implementation, the quality of its solutions and the ease with which it can be combined with other methods. The first SA algorithm records were addressed in [11] and were subsequently corroborated [12, 13]. The SA begins with an initial solution (usually random, but methods are sometimes applied to obtain an initial solution which is already considered to be a "good" one) and with the T value for temperature. Subsequently, one proceeds with several iterations, whereby better solutions are sought. Through changes in the solution considered, which are due to neighboring moves, there is a constant search for improved solutions. One of the main features of SA is that even a worse solution can be accepted with a certain probability. Namely, if the solution presented by each iteration is worse than the previous one, it can be accepted with a certain measure of probability.

The probability of accepting solutions of worse performance will decrease with T, making metaheuristics more selective during the research process. This means that, in the beginning, a solution which presents worse performance is highly likely to be accepted as a new solution, leading to further diversified research. However, during the course of additional research, this probability will drop, resulting in more intense and thorough research during the selection process [14]. It is during this stage that another binomial emerges, the study of which is of great interest in terms of meta-heuristics - the diversity-intensity binomial in the context of the research of solutions.

It should also be noted that, once the iterative process has been concluded, the current solution often does not correspond to the best solution found during the entire process. Therefore, and during the various iterations carried out, one must compare the result obtained for the current solution with the best result achieved so far, registering the solution which presents the best result.

In addition to simulated annealing, there are other very popular metaheuristics used to address the problems of production scheduling, which deserve to be mentioned here: amongst others, the genetic algorithm presented in the 70's, based on Darwin's theories regarding the evolution of the species [15]; the tabu search, inspired by the process of human memory [16]; or, still, that of the discrete artificial bee colony, which stems from natural phenomena, more specifically the foraging behavior of bee swarms [17]. In recent years, the combination of different metaheuristics has led to the appearance of hybrid methods, which have shown rather promising results.

3 Methodology of the Study

The starting point of this study resides in the implementation of a scheduling solution at an ERP. It uses two resolution techniques which focus on the two ends of the quality-function binomial: simple heuristic rules (associated to a low time period until the solution is obtained) and metaheuristic simulated annealing (associated with very positive results when compared to simpler techniques).

Regarding the dispatching rules, nine different rules were used. These are described in Table 1.

Table 1. Dispatching rules adopted.

Rule	Logic
Critical Ratio (CR)	Sort in ascending order according to quotient of subtraction of the current date from the delivery date by processing time to be executed
EDD	Sort the delivery dates in ascending order
SPT	Sort the processing times in ascending order
LPT	Sort the processing times in descending order
NOP	Sort in descending order the number of operations to be executed
Time interval (TP)	Sort in ascending order the gap between total processing time and the delivery date
FCFS	The first task to arrive is the first in the sequence
LCFS	The last task to arrive is the first in the sequence
Random	Sort randomly

After the implementation of each of the nine rules presented in Table 1, the sequences obtained were evaluated according to a total of eight different evaluation criteria: average delay; number of delays; maximum delay; total delay; total flow time; total machine downtime; average machine downtime; and percentage of machine use. Since one is dealing with ERP software, which reaches dozens of customers who present completely different needs, the greatest feature that this scheduling solution would have to present is that of flexibility/customization. One of the differentiating factors of this solution is the fact that it allows different weights to be allocated to each of the eight evaluation criteria adopted. Once the simple heuristic rules are implemented and a solution is obtained, the user can, for instance, choose to prioritize a drastic reduction in the number of delays in deliveries to customers, to the detriment of the other criteria; thus, the sequence obtained may be subjected to changes, which are more suitable to the prioritization of the criteria defined by the user. This flexibility/customization constitutes a relevant competitive advantage for the company selling the ERP.

This solution implemented on ERP software also possesses another path of resolution through metaheuristic simulated annealing. The use of eight distinct performance evaluation criteria, and the possibility of prioritizing certain criteria, are features which are carried over from the previously mentioned resolution procedure. This equality of conditions is a positive aspect, especially when proceeding with a comparison of the two techniques, which will be undertaken later.

This solution also incorporates a strategy that is interconnected with the logic of presenting the result of a given technique according to certain established evaluation criteria that use weightings. In the case of a production scheduling problem, the classic algorithm for simulated annealing accepts all the solutions which present a better makespan than the current customary solution. Within a certain probability, it also accepts solutions which present a worse makespan. In view of the fact that the result for a given sequence is not only dependent on the makespan from that sequence, but rather on the weighted assessment of three general performance criteria, then it makes perfect sense to consider this weighted result of the evaluation criteria as a comparison variable in simulated annealing rather than the makespan, thus enabling solutions that address the customers' needs.

The cooling procedure of the tool developed was the linear method (ß parameter); as criteria for downtime, one used the total number of iterations, the number of iterations without improvement and final temperature. The parameters considered can be seen in Table 2. The implemented solution and the respective tests were developed using the Visual Basic for Applications (VBA) programming language.

Table 2. Parameters considered in simulated annealing.

Parameters	Value
Iterations without improvement	3
Downtime temperature	1
N° of iterations	1000
Initial temperature	100
ß parameter	1

4 Analysis of Results

Once the two resolution techniques were fully implemented on the ERP, the following methodology was carried out in order to compare/evaluate performance:

1. Randomly generate values of processing times and delivery dates for a classic flow-shop scheduling problem by using a number of tasks, and machines selected at random;
2. Run the scheduling solution by means of dispatching rules, recording the result for each of the nine heuristic rules, as well as the best result;
3. Run the simulated annealing algorithm for the same initial problem data, recording the result;
4. Repeat the previous points 365 times (the objective is to simulate one year in an industrial environment, during which one scheduling action is carried out daily).

The result obtained for each sequence of tasks is the sum of the products between the weighting of each performance variable, and the value obtained for that performance variable (in the case of the criterion *machine use percentage*, and since one aims to maximize this value unlike the other criteria, one proceeds with subtraction and not addition). The results of the experiment undertaken (see Table 3) derive from a comparison of the three different environments:

- Environment without a solution for production scheduling (the use of the FCFS rule is considered in these cases. In the industrial context, companies which present no apparent solution for scheduling are using the FCFS rule in most cases);
- Environment with a solution, provided by dispatching rules (DR);
- Environment with a solution, provided by simulated annealing (SA).

Table 3. Results obtained.

Environment	Average	Rate	Percentage (%)	Time per solution
FCFS	4594	0	0	–
DR	3996	86	23,6	1 s
SA	3852	279	76,4	63 s
Total	–	365	100,0	–

In Table 3, the column designated as "Average" relates to the average value of the results obtained from the 365 iterations for each of the three environments considered. The "Rate" column indicates the number of times each environment presented the best result; that is, the one where the value obtained was the lowest of the three environments considered. Finally, the percentage of this rate in relation to the total of the 365 iterations was recorded in the "Percentage" column.

The experiment undertaken indicated that the implemented solution, the calculation of which was based on dispatching rules, presents an execution time (functionality) that is practically negligible, whereas the solution based on simulated annealing has a runtime of about 1 min.

In the component for the quality of the solution obtained (quality), one observed an improvement of 3.6% (the average changed from 3996 to 3852) in the results obtained through the solution which used SA, when compared with the solution using dispatching rules. It is also important to mention the improvement of 13% (a decrease in the average, from 4594 to 3996) obtained in the environment which implemented the solution using dispatching rules, when compared to an environment where there was no effective solution for production scheduling. These results are presented on the graph in Fig. 1.

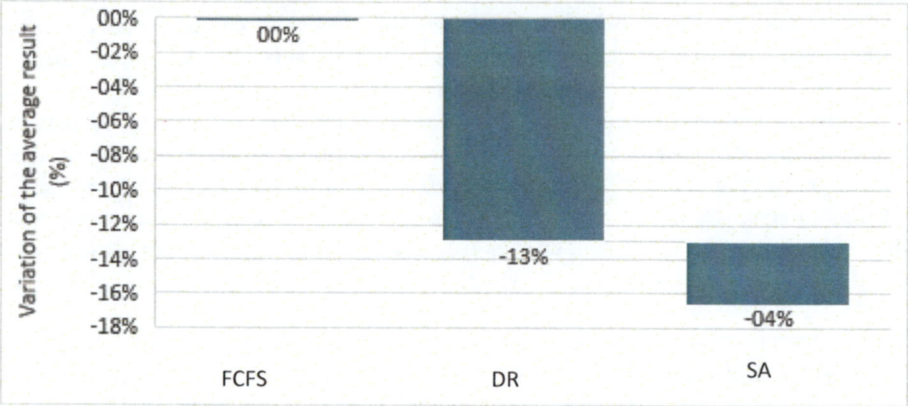

Fig. 1. Graph comparing the results obtained for the 3 environments considered.

One should further highlight that, of the 365 iterations performed, for the environment without a scheduling solution (the FCFS rule was used to simulate this environment), none of these presented the best solution.

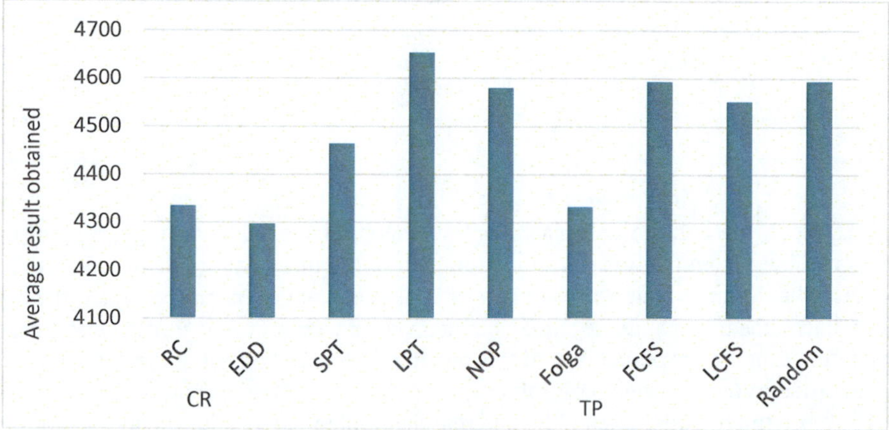

Fig. 2. Graph comparing the results obtained for the 9 dispatch rules considered.

In the specific case of dispatch rules, the experiment undertaken also allowed for the recording of the average results obtained for each of the nine rules used. Although the comparison of dispatch rules was not the main objective of this study, the experiment carried out enabled one to analyze the performance of these 9 rules. Figure 2 presents the average results obtained for each of the 9 rules implemented (one should consider that the lower the result, the better the performance of the dispatch rule).

5 Conclusions and Future Investigation

One of the conclusions to be drawn from this study is related to the profitability obtained by those companies which, since they do not have scheduling systems, have decided to implement a solution that uses dispatching rules. If one considers the significant improvement of 13% achieved in this context, as well as the practically negligible period of time that elapses until this solution is reached, the gains which ensue from the implementation of production scheduling procedures becomes clear, even when simple computational techniques are used.

Regarding a choice between the two techniques used, this is closely related to the characteristics of the context in which they are implemented. Indeed, when compared to the results which used dispatching rules, it is possible to observe a relevant improvement of 3.6% in the results obtained through SA. However, from the perspective of some companies, the additional 1 min required for the process of obtaining a scheduling solution may constitute a significant obstacle, especially when this minute must be multiplied by various shifts, various assembly lines and highly dynamic scheduling environments.

If one were to focus the analysis solely on the results obtained for solutions achieved through dispatch rules, it was the EDD, CR and time interval rules which presented the best results. In contrast, the random, LPT, NOP and FCFS rules displayed the worst results. Furthermore, one should point out the difference of 7.7%, between the value achieved for the rule presenting the best result (EDD) and the value obtained for the rule presenting the worst result (LPT). This significant difference in values shows that, even in environments which use simple rules to solve the problem of production scheduling, there should be careful analysis in the choice of rules as this can impact considerably on the results obtained.

This study is based on the comparison of the results obtained to address the problem of production scheduling through dispatch rules, as well as that of another independent path which uses metaheuristic simulated annealing. One of the conclusions that can be drawn was the virtually negligible weight in terms of functionality that a solution via dispatch rules presents (it only takes about 1 s to reach a solution which implements the set of 9 rules used). If one were to add to this the fact that the resolution of the scheduling problem through simulated annealing begins with an initial solution, there is clearly an opportunity here to study a third path of resolution: this connects the dispatch rules with metaheuristic simulated annealing, and considers the solution ensuing from the dispatch rules as the initial solution for metaheuristic simulated annealing. This use of simplistic methods to improve the initial solution considered for simulated annealing has already been implemented in several studies (consult, for example [14]). However, it would be pertinent to contextualize this method in the quality-functionality analysis followed in this study. Subsequently, it would be compared to an environment with no effective scheduling processes, as well as to environments which use dispatch rules and simulated annealing in a disconnected manner.

In this comparative analysis of methods under the quality-functionality binomial, another starting point which this article provides is the incorporation of other resolution methods. In view of the fact that the previously mentioned weight of the functionality of the dispatch rules is negligible, new dispatch rules may be used in addition to the 9 mentioned in this article to improve the quality of the solutions. Furthermore, there is also the possibility of studying other metaheuristics besides simulated annealing, for instance tabu search, discrete artificial bee colony or even hybrid methods, whose popularity has increased in recent years. The goal is to provide managers with varied solutions, which are suited to different levels of quality-functionality, so that they can choose the most adequate solutions for their shopfloor and its characteristics.

References

1. Leung, J.Y.-T.: Handbook of Scheduling: Algorithms, Models, and Performance Analysis. Chapman & Hall/CRC, Boca Raton (2004)
2. Pinedo, M.L.: Scheduling Theory, Algorithms, and Systems, 4th edn. Springer, Heidelberg (2012). https://doi.org/10.1007/978-1-4614-2361-4
3. Baker, K.R., Trietsch, D.: Principles of Sequencing and Scheduling. Wiley, Hoboken (2009)
4. Elgendy, A., Hussein, M., Elhakeem, A.: Optimizing dynamic flexible job shop scheduling problem based on genetic algorithm. Int. J. Current Eng. Technol. **7**(2), 368–373 (2017). http://inpressco.com/category/ijcet
5. Li, X., Gao, L., Zhang, C., Shao, X.: A review on integrated process planning and scheduling. Int. J. Manuf. Res. **5**(2), 161–180 (2010). https://doi.org/10.1504/IJMR.2010.031630
6. Gao, K.Z., Suganthan, P.N., Pan, Q.K., Chua, T.J., Chong, C.S., Cai, T.X.: An improved artificial bee colony algorithm for flexible job-shop scheduling problem with fuzzy processing time. Expert Syst. Appl. **65**, 52–67 (2016). https://doi.org/10.1016/j.eswa.2016.07.046
7. Zarandi, M.H.F., Hemmati, A., Davari, S., Turksen, I.B.: A simulated annealing algorithm for routing problems with fuzzy constrains. J. Intell. Fuzzy Syst. **26**(6), 2649–2660 (2014). https://doi.org/10.3233/IFS-130935
8. Đurasević, M., Jakobović, D.: A survey of dispatching rules for the dynamic unrelated machines environment". Expert Syst. Appl. **113**, 555–569 (2018). https://doi.org/10.1016/j.eswa.2018.06.053
9. Vlašić, I., Đurasević, M., Jakobović, D.: Improving genetic algorithm performance by population initialisation with dispatching rules. Comput. Ind. Eng. **137**, 106030 (2019). https://doi.org/10.1016/j.cie.2019.106030
10. Çağrı, S.E.L., Hamzadayi, A.: A simulated annealing approach based simulation-optimisation to the dynamic job-shop scheduling problem. Pamukkale Univ. J. Eng. Sci. **24**(4), 665–674 (2018). https://doi.org/10.5505/pajes.2017.47108
11. Metropolis, N., Rosenbluth, A.W., Rosenbluth, M.N., Teller, A.H., Teller, E.: Equation of state calculations by fast computing machines. Cit. J. Chem. Phys **21**(6), 1087 (1953). https://doi.org/10.1063/1.1699114
12. Talbi, E.G.: Metaheuristics: From Design to Implementation. Wiley, Lille (2009)
13. Boussaïd, I., Lepagnot, J., Siarry, P.: A survey on optimization metaheuristics. Inf. Sci. **237**, 82–117 (2013). https://doi.org/10.1016/j.ins.2013.02.041
14. Raaymakers, W.H.M., Hoogeveen, J.A.: Scheduling multipurpose batch process industries with no-wait restrictions by simulated annealing. Eur. J. Oper. Res. **126**(1), 131–151 (2000). https://doi.org/10.1016/S0377-2217(99)00285-4

15. Holland, J.H.: Adaptation in Natural and Artificial Systems, 2nd edn. University of Michigan Press, Ann Arbor (1975)
16. Glover, F.: Future paths for integer programming and links to artificial intelligence. Comput. Oper. Res. **13**(5), 533–549 (1986). https://doi.org/10.1016/0305-0548(86)90048-1
17. Karaboga, D.: An idea based on honey bee swarm for numerical optimization, Technical Report-TR06, Erciyes Univ. Eng. Fac. (2005)

A Lean Framework for Machining Budgeting Operations

Francisco J. G. Silva[1,2](✉), Vítor F. C. Sousa[1], José Carlos Sá[1,2], Matilde Tojal[1], Luís P. Ferreira[1,2], and Pedro Nogueira[1]

[1] ISEP - School of Engineering, Polytechnic of Porto, Porto, Portugal
{fgs,vcris,cvs}@isep.ipp.pt, luispintoferreira@eu.ipp.pt
[2] INEGI - Instituto de Ciência e Inovação em Engenharia Mecânica e Engenharia Industrial, Porto, Portugal

Abstract. Machining is widely used in many kinds of industries and outsourcing in this area is an increasing way to improve competitiveness. Thus, to get quotes for small, medium, or high series of parts is a common practice, which implies a strong effort from the possible suppliers to analyze the dimension, geometry, and specificities of their machining process. However, due to customization, the series of parts to be machined are increasingly reducing, but the time to analyze the costs involved in its production is the same. Moreover, budgeting is a boring and time-consuming task, because the level of creativity is reduced when the main concepts are already achieved, and the added value is relatively short, but requiring extreme attention to the details in order to get the best quote. This paper deals with a novel framework to easily create budgets, saving time and reducing the lead-time usually needed to provide a proposal to the customer. Thus, the lean principles were applied to budgeting tasks, reducing waste of time in this operation, and providing the companies with a real productive and effective tool to produce proposals in a very short period of time with the required accuracy. Tests carried out with the developed tool allowed for concluding that it presents deviations lower than 15%, value which can be improved through a fine tuning of some of the conditions presented.

Keywords: Lean · Machining · Budgeting · Lead-time · Outsourcing

1 Introduction

The job of preparing budgets is boring and time-consuming. However, purchases are usually carried out after market consultations, which involve a request for a quote by the customer, and the completion of the budget by those interested in making the supply. Automating this task as much as possible is to save time, one of the main missions of the Lean philosophy. The machining sector has been moving towards specialization, and it is increasingly common to resort to subcontracting these services by companies that only carry out the assembly of their final products [1]. Thus, this work had as main objective to elaborate a framework that allows, in a quick way, to elaborate budgets related to

L. Moldovan and A. Gligor (Eds.): Inter-Eng 2021, LNNS 386, pp. 268–279, 2022.
https://doi.org/10.1007/978-3-030-93817-8_26

machined parts, without the need for any highly qualified employee. Thus, the waste that this task normally introduces in the activities of companies dedicated to machining is eliminated, proving results whose can be reliable and accurate regarding the context.

After this preliminary introduction, this work is divided in six sections, as follows: Sect. 2 reports a literature review on the machining sector and the application of Lean concepts acting as theoretical support for the development, Sect. 3 specifies the methods and main settings need to develop the tool, Sect. 4 reports the results and corresponding deviations, Sect. 4 does a brief critical analysis of the results, trying to compare them with other results previously obtained, and Sect. 5 highlights the main conclusions.

2 Literature Review

The machining sector has a great impact on the global economy, having a significant weight in industries such as aerospace, aeronautics, automobile, railway and shipbuilding [2]. For this reason, machining has been the subject of numerous studies in various aspects, such as the optimization of manufacturing parameters [3–5], the extension of tool life [6–8], cutting forces analysis [9–11], or the application of Lean tools to optimize processes and operations [12–14]. Behind every order for machine parts, there is a quote request made to several companies, a budgeting process, a negotiation, and an award. Budgeting is a boring and time-consuming task because it involves a detailed analysis of its geometry, fastening, necessary tools, wear of these tools, machine and worker time, and all costs related to industrial activity, which include costs with cleanliness, security, facilities, etc. [15]. However, the machining time is of particular relevance, being around the same that many other costs are determined [16]. Automating this task is very important as it consumes time and resources. Indeed, this task is extremely important in responding to the customers' requests, but represents a necessary cost to the companies, which must be reduced as much as possible, improving the competitiveness. This is even more important as most of the companies dedicated to machining work under subcontract and are SMEs. Thus, resources are even more scarce, and they are requested by a high number of potential customers, but in many cases, they do not get the award. However, customers deserve a response, which is given through a detailed proposal.

The main focus of the Lean philosophy is to cut waste [17, 18]. If budgeting is carried out without automated procedures, it necessarily generates waste of time, which must be cut. Moreover, the proposal to the customer can be delayed due to manual procedures. In fact, several Lean tools have been used in order to optimize processes and operations in numerous companies linked to the metalworking and other kinds of industry, some of them automatizing processes and avoiding waste of time. Correia et al. [19] optimized production lines of electronic safety equipment, achieving a 10% increase in productivity using the Value Stream Mapping (VSM) and Lean Line Design (LLD) tools. Rosa et al. [20–22] expanded and optimized the productivity of Bowden cables production lines for automobiles by more than 40% using tools such as VSM, SMED and Line Balancing. In turn, Neves et al. [23] used tools such as the 5W2H (5 Whys + 2 Hows) and 5S, in a system supported by a PDCA (Plan-Do-Check-Act) cycle, which allowed a 10% increase in productivity in a textile industry that produced products with low added value, where this gain assumes even greater relevance. Martins et al. [24] applied the SMED (Single Minute Exchange of Die) methodology in a company that manufactures electrical power cables for automobiles, which had three electron beam welding equipment. The application of the methodology allowed to identify where the activities that generated excessive time consumption were found during the equipment setup, leading to the study of solutions capable of saving more than 50% of the time spent in these operations, significantly improving the product's competitiveness. Moreira et al. [25] used the OEE (Overall Equipment Effectiveness) tool to study the printing process in a printing company, finding the necessary solutions to increase this efficiency by 4%, with particular focus on the gains obtained through the reduction of non-conformities, which is another form of waste. Velmurugan et al. [12] improved the machining process flow and speed up the operations flow using time study methods and standardizing the work, achieving a better line balancing of the machining cells. Monteiro et al. [13] achieved gains of about 57% in setup times related to machining processes using SMED methodology. Thus, it seems that the literature is rich in practical examples of how Lean tools can be used to eliminate waste, increasing process efficiency. However, the literature is very scarce in terms of budgeting studies, which are very important to the life of large, medium, and small companies. In this work, it is intended to minimize the effort involved in the generation of estimates, creating an algorithm that allows to obtain reliable results in the estimation of parts obtained through machining, minimizing waste of time.

3 Methodology

A tool for the budgeting of machined parts was developed, calculating the overall production cost of each part based on the production and preparation times, as well as the amount of material needed for each of the parts. This enables the tool to provide accurate budgets at a considerably faster rate than the conventional budgeting process. The developed tool is also created taking into considerations the companies' resources (where this tool is applied), both in terms of workers, machines and even the type of orders that the company receives. The created model was validated in a manufacturing company, where it was found that one of the main causes for budget errors was high part variability (shape and required operations) that were produced in small series. This factor causes the budgeter to make a deep analysis of each of the produced part, thus increasing budgeting time.

The conventional budgeting process is comprised of four main steps, as can be observed in Fig. 1.

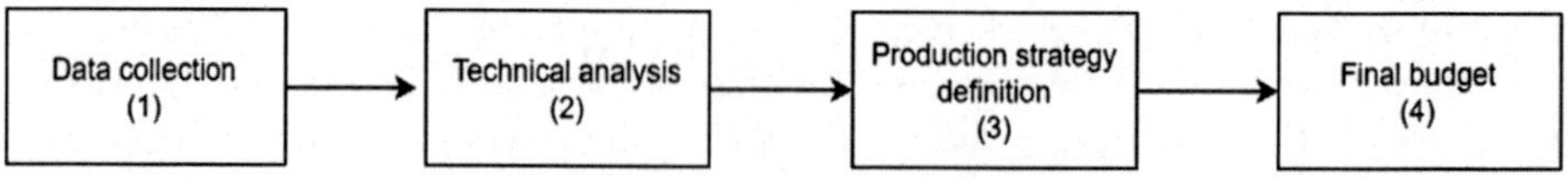

Fig. 1. Diagram of the conventional budgeting process, showing the four main steps.

These steps can be described as follows:

(1) Data collection – registering of the customer's particular requests for the part; drawing and improvement of part drawings;
(2) Technical analysis – evaluate part feasibility; check the tolerances for the part; verify available machines for part production;
(3) Production strategy definition – define the machining strategy; define the tools and devices that will be used; set the machining parameters;
(4) Final budget – considering the data provided from the definition of the production strategy, the overall cost of part production can be estimated, thus providing the customer with the budget.

This conventional budgeting process is prone to many errors, especially due to the quantity and variety of produced parts. Most of the provided budgets are calculated based on the amount of material that is needed to produce each part, because this method can provide accurate costs, as the amount of removed material is closely related to machining times. However, for the budgeting of complex parts, where additional operations are required or the part geometry is more complex, this method does not provide accurate budgets. Furthermore, for these complex parts, the data collection stage and the preparation for part production would be much more time consuming, when compared to simpler parts. Thus, there is here an opportunity to save time.

To try and solve the problems, a strategy for machining time calculation based on the final part's dimensions was adopted. This would relate the amount of required material, needed tools and machining times to produce the parts. These calculations were developed mainly for milling operations, due to the machines considered in this study. In addition to these calculations, a part complexity level list was created. These levels are dependent on the part's level of detail, geometry complexity and machining operations. These would influence the times estimated by the developed tool, including the drawing, part preparation, machine setup times and finishing times. The working principle of the developed budgeting tool can be observed in Fig. 2.

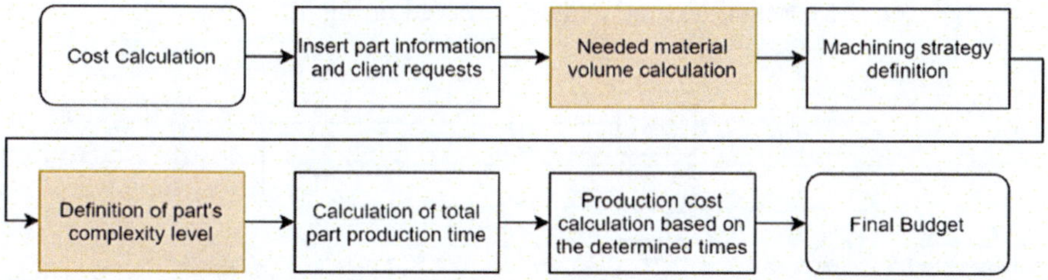

Fig. 2. Diagram of the working principle of the developed budgeting tool, in orange the more complex stages are marked.

As it can be observed in Fig. 2, the model determines the cost by considering inputs such as the client requests and part information (drawings), needed volume of material based on the part's dimensions and the machining strategy. Furthermore, a part complexity level is added, which will influence not only the machining times, but will also influence setup, preparation and finishing times. There are two stages marked in orange in Fig. 2, these stages are more complex than the others, as the material volume calculation and the definition of the part's complexity level is dependent on various factors. For example, the volume of material is determined based mainly on the final part's dimensions, with a higher margin being applied to bigger parts. Figure 3 shows the method for determining the required material volume, displaying the extra width of material considered for the calculation of this volume.

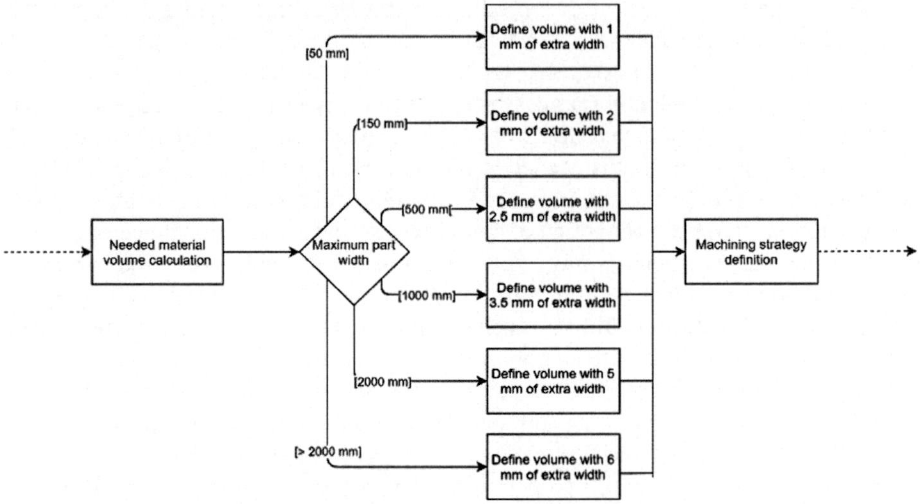

Fig. 3. Diagram of the method used for the required material volume calculation.

For the calculation of the material volume, the part's maximum width is considered, then attributing the appropriate extra width. The volume of the material is then used to estimate its cost, based on current market prices. Regarding the definition of the part's complexity level, the adopted method can be observed in Fig. 4.

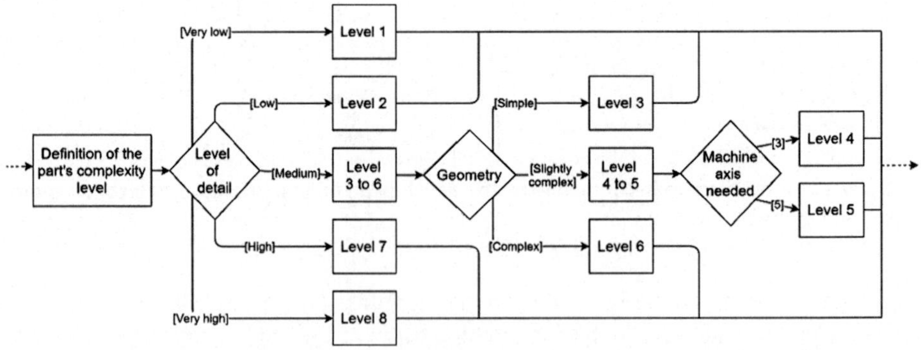

Fig. 4. Diagram of the method used for the definition of the different part complexity levels.

Regarding the calculation of the machining times for the part production, the developed tool mainly considers milling and drilling operations, as it was programmed for the machines considered in this study. Machining time equations were adopted for each of the operations that were considered, these are dependent on the machining parameters, tools, and part's dimensions. The considered operations were:

- Side-milling;
- Face-milling;
- End-milling,
- Drilling;
- Boring and threading.

In the following paragraphs the calculation method for each of the operations machining time that were mentioned are going to be presented, starting with side-milling. For the calculation of the machining time for this operation, the part's exterior perimeter (P_{ext}) is firstly calculated, considering its length and width values. Secondly, the estimated number of roughing passes (*R.P.*) is defined, by the Eq. (1), considering the part's thickness (t, in mm) and the radial depth of cut (a_e, in mm).

$$R.P. = \left(\frac{t}{a_e}\right). \tag{1}$$

Finishing operations are also considered, with the finishing passes (*F.P.*) being estimated according to Eq. (2), that considers the part's thickness and tool diameter ($\varnothing_{tool}$, in mm).

$$F.P. = \left(\frac{t}{0.5 \times \varnothing_{tool}}\right) \tag{2}$$

with these values obtained the operation time for side-milling operations ($T_{S.M}$, in minutes) can be calculated according to Eq. (3). This equation considers the determined values and the feed rate employed during the process.

$$T_{S.M.} = \frac{P_{ext} \times (No._{R.P.} + No._{F.P.})}{V_f} \tag{3}$$

The values for tool diameter and feed rate are determined based, primarily, on the part's thickness and material. Regarding the calculation of the face-milling operation time, as seen for the side-milling operations, the total machining time depends on the number of passes needed to face an area, the number of face-milling passes (*Fa.P.*) is determined by dividing the material's width by the chosen a_e value. Then, after obtaining the required number of facing passes, the total facing length (*F.L.*, in mm) is calculated, based on the part's length (l_p, in mm), the tool's diameter and the number of facing passes, as shown in Eq. (4).

$$F.L. = \left(l_p + \varnothing_{tool}\right) \times Fa.P \tag{4}$$

For the type of machines that were considered in this study, usually a value of 1 mm is adopted for a_p, however, in some cases, facing operations must be performed on both sides of the part. This requires the extraction of the part from the machine and its refixing. For the second facing operation the value for a_p is 5 mm. Thus the machining times for these operations are dividing into two, one regarding to the first face-milling time ($T1_{F.M.}$, in minutes), and second face-milling time ($T2_{F.M.}$, in minutes). These are calculated as seen in Eq. (5) and (6).

$$T1_{F.M.} = \frac{F.L.}{V_f} \tag{5}$$

$$T2_{F.M.} = \frac{F.L. \times \left(\frac{5}{ae}\right)}{V_f}. \tag{6}$$

If only one face-milling operation is required then only Eq. (5) is required to calculate the operation time, however, if the second step is required, in order to obtain the total operation time, the value obtained from Eq. (6) should be added to that obtained in Eq. (5).

For end-milling operations, the time is calculated based on the value of the radial depth of cut a_e, dependant on the tool's diameter (usually 40%). It is also required to know the distance that the tool moves during the process, dependant on the dimensions of the cavity that are being machined, after knowing the distance that the tool moves per increment (*E.l.*, in mm), this value is multiplied by the number of increments needed to machine the cavity (obtained by the a_p value). The obtained result is the total end-milling distance (*E.L.*, in mm). With this information and using the value of feed rate of the machining operation, the end-milling time ($T_{E.M.}$, in minutes) can be calculated, according to the Eq. (7).

$$T_{\cdot E.M.} = \frac{E.L.}{V_f}. \tag{7}$$

Generally, in cases where end-milling is required the finishing operations are performed in the cavities' interior wall, thus Eq. (7) can also be used to determine the machining times for finishing operations.

Regarding drilling operations, the machining time depends on the tool's diameter, the hole's depth (*D*, in mm), the feed per rotation value (*f*, in mm/rotation) and the rotational speed (*N*, in RPM) that are employed during the operation. Throughout the process, the tool performs a series of plunges, quickly retracting and then resuming drilling for a few millimetres, then retracting again, repeating this cycle until the operation is finished. The number of plunges (*Pl.*) that are required can be calculated according to the following Eq. (8).

$$Pl. = \frac{D}{\varnothing_{tool}}. \tag{8}$$

The value obtained from Eq. (8) should be rounded up, then, 1 more plunge should be added. After this value is determined, the drilling length (*D.L.*, in mm) is calculated, according to the following Eq. (9):

$$D.L. = \sum\nolimits_{n=1}^{Pl.} n \times \varnothing_{tool} \times 2 \tag{9}$$

with these values, the total drilling time ($T_{D.}$, in minutes) can be calculated, based on the drilling length, the feed per rotation and rotational speed values, as seen in Eq. (10).

$$T_{D.} = \frac{\left(\frac{D.L.}{f}\right)}{N}. \quad (10)$$

Finally, the calculation for the boring and threading operation times is quite simple. The total boring/threading time ($T_{Thr.}$, in minutes) is calculated based on the depth of the hole (D), the feed per rotation (f) and the rotational speed (N) values, being shown in the following Eq. (11):

$$T_{Thr.} = \frac{\left(2 \times \frac{D}{f}\right)}{N} \quad (11)$$

The developed budgeting tool also calculates and considers the preparation and other finishing operation times for the final budget. There are three main steps on the preparation process of each part: (1) CAD 2D/3D (computer-aided design) – for each part, a 2D technical drawing and a 3D drawing must be performed; (2) CAM (Computer Aided Manufacturing) – it is needed to create a CAM program for the machining of the various parts; and finally, the (3) Machine Setup – some parts require additional setup steps or even multiple extractions from the machine during the process.

Base times for the first step were determined based on the type of drawings received by SME companies. This CAD step is divided into two main steps, 2D technical drawings of the parts and 3D drawings. It was determined that the base time for each of these steps would be 5 and 15 min, respectively, adding $\boldsymbol{x}$ minutes of additional time for each detail added to the drawing (with $\boldsymbol{x}$ equal to the complexity level). For the CAM and Machine setup steps of the part preparation, it was determined that 10 min would be added for each step that requires part extraction from the machine.

Regarding finishing operations, an equation was adopted for the calculation of the times ($T_{finishing}$) associated with these. This Eq. (12) is based on the parts dimension (l_p -length and w_p – width), as well as the part's thickness (t) and the feed rate that was adopted (Vf).

$$T_{finishing} = \frac{(4 \times lp) + (4 \times wp) + (4 \times t)}{V_f} \quad (12)$$

After determining the part's complexity level, according to the method depicted in Fig. 3, the complexity levels will inflate the times for all the production stages, from preparation to finishing operations. In addition to the factors used to determine complexity level, the number of machining operations that are required also influence the part's level of complexity. The criteria used for the attribution of part complexity level can be observed in Table 1, as well as the inflation factors associated to each of the levels. These factors will be applied to each of the production steps, yielding the overall production time for the manufactured parts, enabling for total production cost determination.

Table 1. Part complexity level criteria and respective inflation factor, applied to the calculated operation times.

Complexity level	Part geometry	Number of threaded holes	Number of simple holes	Level of detail	Number of axis needed	Inflation factor
1	Simple	≤3	≤6	Very low	3	1,25
2	Simple	≤5	≤10	Low	3	1,2
3	Simple	≤7	≤14	Medium	3	1,2
4	Slightlycomplex	≤10	≤20	Medium	3	1,15
5	Slightlycomplex	≤15	≤30	Medium	5	1,1
6	Complex	≤20	≤40	Medium	5	1,08
7	Complex	≤30	≤60	High	5	1,08
8	Complex	≤35	≤70	Very High	5	1,08

4 Results and Discussion

Several tests were conducted for the validation of the developed budgeting tool, comparing the predicted machining times for each of the production steps, with the real registered productions times. A total of 11 parts of varying complexity level were tested, with at least one for each complexity level being produced. The budgeting tools uses these determined production times to calculate the overall machining cost of the parts, thus it was deemed appropriate to perform this time comparison. In Table 2, the average percentage deviation that was registered for each of the production steps can be observed, as well as the base times.

Table 2. Average percentage deviation of the estimated times compared to the real times for each part production step.

Production step	Average percentual deviation
CAD 2D/3D	+15%
CAM	+38%
Machining setup	−7%
Machining operations	+14%
Finishing operations	−71%

In terms of the determination of machining times, the tool can offer quite accurate prediction, exhibiting an average deviation value of 14%. This is quite satisfactory, as the machining times have a heavy impact on the part production cost. There were, however, some high deviation values, namely for finishing operations and the CAM step. This is due to the great variability of required operations from part to part. There was also registered higher deviation values for parts with higher complexity levels, however, this deviation was within acceptable range. Unfortunately, no terms of comparison have been found in the literature, because there are several studies about estimations of time in machine learning [26–28] and other simulation techniques, but budgeting seems to be a new issue in the field of machining investigation.

5 Conclusion

The developed tool can provide accurate and faster budgets, when compared to the conventional budgeting process, effectively reducing the time wasted in providing budgets to the customer. The following conclusions can be drawn from this work:

- The overall production cost is calculated based on the preparation, machining and finishing operation times for a certain part;
- Machining times influence greatly the process' cost, highlighting the importance of an accurate determination of these times;
- Calculation of the machining times are performed based on the operations that are carried out in the used machines;
- The creating of a part complexity level offers some level of standardization, reducing budgeting times while still providing accurate budgets;
- The tool contemplates the part's complexity level in its calculations, as this influences the overall production time of the parts;
- This tool was validated in a company setting, registering an average positive percentage deviation from the real machining times of 14%.

These results are quite satisfactory, with the tool exhibiting accurate prediction for the production times, reducing the budgeting time. There is room for improvement, mainly for the calculation of finishing operations. However, the tool is quite versatile and can be applied to multiple machine times, with the ability to be programmed to determined machining times for different operations. There is also the possibility of coupling novel time prediction methods and algorithms with this tool, thus improving its accuracy.

This tool will save a lot of time in the estimation of machining costs, allowing for a real cut in the waste of time spent in budgeting operations. Thus, despite none of the traditional tools of the Lean Manufacturing philosophy has been applied, the tool will undoubtedly contribute to save time and make these operations more competitive. Moreover, the principle can be conveyed to other processes just making a deep analysis of the processes and change the rules of the algorithm.

References

1. Ferreira, V., Silva, F.J.G., Martinho, R.P., Pimentel, C., Godina, R., Pinto, B.: A comprehensive supplier classification model for SME outsourcing. Procedia Manuf. **38**, 1461–1472 (2019)
2. Salem, A., Hegab, H., Kishawy, H.A.: An integrated approach for sustainable machining processes: assessment, performance analysis, and optimization. Sustain. Prod. Consum. **25**, 450–470 (2021)

3. Martinho, R.P., Silva, F.J.G., Martins, C., Lopes, H.: Comparative study of PVD and CVD cutting tools performance in milling of duplex stainless steel. Int. J. Adv. Manuf. Technol. **102**(5–8), 2423–2439 (2019). https://doi.org/10.1007/s00170-019-03351-8
4. Daniyan, I.A., Tlhabadira, I., Daramola, O.O., Mpofu, K.: Design and optimization of machining parameters for effective AISI P20 removal rate during milling operation. Procedia CIRP **84**, 861–866 (2019)
5. Gouveia, R.M., Silva, F.J.G., Reis, P., Baptista, A.P.M.: Machining duplex stainless steel: comparative study regarding end mill coated tools. Coatings **6**(4), 51–81 (2016)
6. Sousa, V.F.C., Silva, F.J.G.: Recent advances on coated milling tool technology—a comprehensive review. Coatings **10**, 235–261 (2020)
7. Daniyan, I., Mpofu, K., Ramatsetse, B., Gupta, M.: Review of life cycle models for enhancing machine tools sustainability: lessons, trends and future directions. Heliyon **7**(4), e06790 (2021)
8. Sousa, V.F.C., Silva, F.J.G., Pinto, G.F., Baptista, A., Alexandre, R.: Characteristics and wear mechanisms of TiAlN-based coatings for machining applications: a comprehensive review. Metals **11**, 260–309 (2021)
9. Wu, B., Yan, X., Luo, M., Gao, G.: Cutting force prediction for circular end milling process. Chin. J. Aeronaut. **26**(4), 1057–1063 (2013)
10. Brecher, C., Eckel, H.-M., Motschke, T., Fey, M., Epple, A.: Estimation of the virtual workpiece quality by the use of a spindle-integrated process force measurement. CIRP Ann. **68**(1), 381–384 (2019)
11. Sousa, V.F.C., et al.: Cutting forces assessment in CNC machining processes: a critical review. Sensors **20**, 4536–4561 (2020)
12. Velmurugan, V., Karthik, S., Thanikaikarasan, S.: Investigation and implementation of new methods in machine tool production using lean manufacturing system. Mater. Today: Proc. **33**(7), 3080–3084 (2020)
13. Monteiro, C., Ferreira, L.P., Fernandes, N.O., Sá, J.C., Ribeiro, M.T., Silva, F.J.G.: Improving the machining process of the metalworking industry using the lean tool SMED. Procedia Manuf. **41**, 555–562 (2019)
14. Kumar, S., Campilho, R.D.S.G., Silva, F.J.G.: Rethinking modular jigs' design regarding the optimization of machining times. Procedia Manuf. **38**, 876–883 (2019)
15. Mutilba, U., Sandá, A., Veja, I., Gomez-Acedo, E., Bengoetxea, I., Fabra, J.A.Y.: Traceability of on-machine tool measurement: uncertainty budget assessment on shop floor conditions. Measurement **135**, 180–188 (2019)
16. Liu, X., Zhou, Y., Tang, J.: A comprehensive adaptive approach to calculating the envelope surface of the digital models in CNC machining. J. Manuf. Process. **57**, 119–132 (2020)
17. Antosz, K., Stadnicka, D.: Lean philosophy implementation in SMEs – study results. Procedia Eng. **185**, 25–32 (2017)
18. Pena, R., Ferreira, L.P., Silva, F.J.G., Sá, J.C., Fernandes, N.O., Pereira, T.: Lean manufacturing applied to a wiring production process. Procedia Manuf. **51**, 1387–1394 (2020)
19. Correia, D., Silva, F.J.G., Gouveia, R.M., Pereira, T., Ferreira, L.P.: Improving manual assembly lines devoted to complex electronic devices by applying Lean tools. Procedia Manuf. **17**, 663–671 (2018)
20. Rosa, C., Silva, F.J.G., Ferreira, L.P.: Improving the quality and productivity of steel wire-rope assembly lines for the automotive industry. Procedia Manuf. **11**, 1035–1042 (2017)
21. Rosa, C., Silva, F.J.G., Ferreira, L.P., Pereira, T., Gouveia, R.: Establishing standard methods to improve the production rate of assembly lines used for low added-value products. Procedia Manuf. **17**, 555–562 (2018)
22. Rosa, C., Silva, F.J.G., Ferreira, L.P., Campilho, R.: SMED methodology: the reduction of setup times for Steel Wire-Rope assembly lines in the automotive industry. Procedia Manuf. **13**, 1034–1042 (2017)

23. Neves, P., Silva, F.J.G., Ferreira, L.P., Pereira, T., Gouveia, A., Pimentel, C.: Implementing lean tools in the manufacturing process of trimming products. Procedia Manuf. **17**, 696–704 (2018)
24. Martins, M., Godina, R., Pimentel, C., Silva, F.J.G., Matias, J.C.O.: Practical study of the application of SMED to electron-beam machining in automotive industry. Procedia Manuf. **17**, 647–654 (2018)
25. Moreira, A., Silva, F.J.G., Correia, A.I., Pereira, T., Ferreira, L.P., de Almeida, F.: Cost reduction and quality improvements in the printing industry. Procedia Manuf. **17**, 623–630 (2018)
26. Kim, M.S., Lee, B.S., Lee, H.S., Lee, S.H., Lee, J., Kim, W.: Robust estimation of outage costs in South Korea using a machine learning technique: Bayesian Tobit quantile regression. Applied Energy **278**, 115702 (2020)
27. Bodendorf, F., Franke, J.: A machine learning approach to estimate product costs in the early product design phase: a use case from the automotive industry. Procedia CIRP **100**, 643–648 (2021)
28. Yoo, S., Kang, N.: Explainable artificial intelligence for manufacturing cost estimation and machining feature visualization. Expert Syst. Appl. **183**, 115430 (2021)

Sustainability Strategy Methodology to Increase Brand Value

Michele Wong[1], Mihai Daniel Anitei[2], and Cristina Veres[3](✉)

[1] MLW Consultancy, Dubai, United Arab Emirates
michele@mlwconsultancy.com
[2] Technical University of Cluj-Napoca, Memorandumului Street 28, 400114 Cluj-Napoca, Romania
[3] George Emil Palade University of Medicine, Pharmacy, Science and Technology of Targu-Mures, Targu-Mures, Gh.Marinescu Street 28, 540142 Targu-Mures, Romania
cristina.veres@umfst.ro

Abstract. To increase brand value, companies should consider developing and continuously improving their sustainability strategies. With the spotlight on sustainability growing globally with focus on environment, social and governance (ESG), many are realizing the opportunity to utilize sustainability as a tool to attract investors and top talent, as well as increasing clients' fidelity. This work describes the methodology to build a sustainability strategy as a way to increase brand value and highlights the effects of sustainability reporting. The implementation of the presented methodology and the sustainability disclosure of Emirates NBD Bank in the United Arab Emirates, contributed to the improvement of a set of indicators and to the increase of their evaluated value with 230% from USD 1.78 billion in 2015 to USD 4.13 billion in 2020. The paper describes the importance of sustainability in the business environment, then details the professional approach for building a company sustainability strategy and report.

Keywords: ESG · Brand value · Rank · Strategy · Sustainability

1 Sustainability in the Business Environment

Sustainability is essential in business. Focus on a business' impact on Environmental, Social and Governance (ESG) practices has gained considerable traction in recent years and directly affects the organization's long-term operations and strategy. Studies have shown that ESG integration into a firm's valuation model improves its non-financial indicators, such as consumer satisfaction, market acceptance, lower cost of debt and the societal values it brings to its stakeholders [1].

The concept of sustainable development emerged in the 1960s with the rise of ecological concerns and the fear of resource scarcity [2–4]. Six decades later, the subject is more concrete than ever, gaining visibility, important efforts were made, and measures started being implemented.

The number of published articles in Science Direct - a world's leading source for scientific, technical, and medical research provides evidence of an increase of interest in Sustainability knowledge over the last decade (see Fig. 1).

L. Moldovan and A. Gligor (Eds.): Inter-Eng 2021, LNNS 386, pp. 280–288, 2022.
https://doi.org/10.1007/978-3-030-93817-8_27

Fig. 1. Published research related to sustainability on Science Direct [5]. *The number of articles in 2021 are reported as of 26th of June 2021.

Sustainability opens to business new opportunities as it improves and strengthens an organization's operations, practices and policies which enhances trust and integrity among its stakeholders. Some companies in the private sector began to misuse sustainability as a means for public relations and solely showcasing surface level practice of "doing good". The precarious economic environment reinforces the need for the existence of elements related to sustainability. One of the goals of sustainability is to ensure business continuity, regardless of market fluctuations, along with ensuring a balance between ESG and economic growth.

Implementing a sustainable business strategy entails assessment of corporate sustainability [6]. According to the Governance & Accounting Institute report in 2020, 90% of S&P 500 Index companies report on Sustainability [7].

Sustainability is a practical tool for enhancing a company's brand.

Investing in sustainability helps a company to create shared value across its core operations including:

- **Cost Savings** - efficient operations which lead to savings, such as using best practices in facility management for resources savings (i.e. water, transport, electricity);
- **Risk Mitigation** - remaining aware of the company's ESG risks can mitigate reputational loss such as unethical supply chains and labour rights;
- **Legal Requirements** - for companies with aspiration to grow globally, many countries around the world require sustainability disclosure to be publicly listed;
- **Government & Taxation Benefits** - Some governments have offered subsidies and incentives for companies to disclose on their ESG impact or their non-financial performance. In the European Green Deal approach New frameworks are planned to encourage more sustainable investment across the bloc and will enhance reporting

on how investment funds use ESG for portfolio construction [8], which will increase even more the importance of sustainability disclosure.
- **Attracting Investments** - conscious investors are more and more concerned about the transparency of a company's non-financial performance, thus publicly disclosed sustainability strategies and reports can help increase market value of an organization;
- **Increasing Customer Loyalty** - as sustainability is gaining more and more focus yearly, customers are starting to be aware of the impact their choices have on the long term.

As sustainability expectations increase globally, and demands for transparency rise from suppliers, partners, clients and investors, a company benefits from communicating its goals in the form of a sustainability strategy, followed by a public sustainability report.

Some of the companies choose to develop an internal department/function to work on the sustainability strategy and reporting, others may involve external experts to build up strategies for a determined period. Outsourcing is a good option when the company is lacking resources, people, competences, time and/or knowledge.

Sustainability strategy is closely linked to Brand Value. There is a growing interest in the value and management of intangible assets, specifically in brands since gather business environment becomes more complex than ever [9]. But we should also consider that brands are valuable assets for national economies, too.

2 Methodology to Build a Company Sustainability Strategy and Report

This paper presents the methodology of developing a sustainability strategy and report, the non-financial disclosure of a company. This methodology is one of the best practices by sustainability practitioners to develop reports that align to the Global Reporting Initiative (GRI) and the Sustainable Accounting Standards Board (SASB). 51% of S&P 500 Index reporting companies use GRI and 14% align with SASB [7].

The methodology to develop a sustainability strategy has 4 main steps:

1. Sustainability Strategy - Gap Analysis;
2. Sustainability Strategy - Materiality Analysis;
3. Sustainability Strategy - Final Framework;
4. Sustainability Report - on ESG Targets and Goals.

2.1 Step 1. Sustainability Strategy - Gap Analysis

Prior to building the company's personal strategy, a gap analysis is conducted by identifying and reviewing peer companies, top competitors, similar-sized companies in the industry as well as top sustainability performers within the same industry.

A peer review is conducted in the analysis with benchmark criteria:

- **Key terms** - what sustainability terminology is being used (i.e. social impact, ESG, Corporate Social Responsibility, sustainability, etc.);

- **Web content** - how are they disclosing publicly on the web;
- **Reports** - how is the company reporting and which international standards are they in alignment with, i.e. GRI, SASB, TFCD (Task Force on Climate-related Financial Disclosures), CDP (Carbon Disclosure Project);
- **Strategy/approach** - what are their sustainability focus areas;
- **Governance** - how is sustainability governed within the organization (i.e. board level, leadership team, etc.);
- **Engagement & materiality** - how does the company engage with their stakeholders;
- **Priorities & targets** - what are their priority sustainability topics and how do they report on it.

The gap analysis process provides an overview of their standing in sustainability among their peers and competitors. It also identifies what priority topics they can focus on to further benchmark in the future. Further, the process provides direction on operations and disclosure and identifies the material (significant) topics relevant to the company and its industry.

The gap analysis process documentation is usually not disclosed publicly by companies as they are documents that are solely used to assist in decision making on developing the sustainability strategy.

2.2 Step 2. Sustainability Strategy - Materiality Analysis

Following the gap analysis, the company will get a general idea of what sustainability topics may be relevant to the industry to assist the materiality analysis. The material analysis is a methodology a company can use to identify ESG topics that impacts the business and the stakeholders. Top sustainability topics identified from the gap analysis are then listed for stakeholders to rank which topics are most important to them.

According to GRI, 'materiality' implies "those topics that have a direct or indirect impact on an organization's ability to create, preserve or erode economic, environment and social value for itself, its stakeholders and society at large". The GRI Standard is still the most adopted framework for undertaking materiality assessment, but some have started to complement it by considering Sustainability Accounting Standards Board (SASB) and Integrated Reporting (IR) guidelines as well.

Materiality analysis identifies which issues are the most important to be addressed by a company. These issues are directly relevant to the company's value change and are analyzed using two lenses:

1. What is the potential of each issue to positively or negatively impact the organization's growth, cost or trust; and
2. How important is each issue to its internal and external stakeholders.

This is conducted through various means including but not limited to, online surveys, one-on-one interviews and focus groups. If the company uses the GRI Standard, the topics are listed to be selected. The result of the analysis is a visual representation of which issues should be prioritized according to their importance of the organization's success and the stakeholders' expectations. The result of this analysis allows for companies to create their long-term sustainability strategy, targets and find the best methods to report their data found because of the process.

Fig. 2. Emirates NBD's materiality matrix 2020 [10].

The Materiality Matrix in Emirates NBD's 2020 Sustainability Report has two axis:

1. Importance to Stakeholders: identifies material topics that are of importance to the bank's external and internal stakeholders.
2. Importance to Business: material topics that are of significance to the bank's operations.

As presented in the Fig. 2, Exceptional Customer experience has a medium importance for stakeholders, instead has a very high importance to business. Climate change has a medium importance to stakeholders, but a low importance to business.

2.3 Step 3. Sustainability Strategy - Final Framework

Once the materiality analysis is completed, the material topics are then categorized using sustainability expertise and processes. This can be done through various methods depending on the company's requirements and needs. Some approaches include:

1. Logic Model - most utilized in the United Nations and private sector that identifies inputs, activities, outputs, outcomes and impact.
2. Theory of Change - an extensive illustration presenting how and why a desired change is expected to happen. Firstly, it involves identifying the desired long-term goals and then working back from these to identify all the conditions (outcomes) that must be in place for the goals to occur.

3. Creating Shared Value (CSV) - an integrated and interdependent framework that integrates ESG topics into a business strategy. CSV is the practice of creating economic value in a way that also creates value for society by addressing its needs and challenges.

The Fig. 3 contains the CVS model of Emirates NBD's Corporate Investment Framework 2020, which presents its key drivers, key focus and supporting areas, impact measurement and implementation mechanisms.

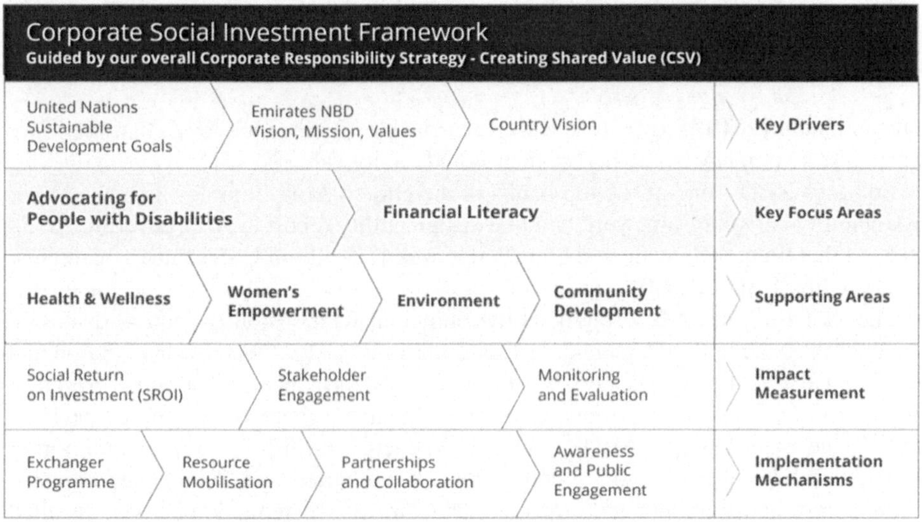

Fig. 3. Emirates NBD's Corporate Investment Framework 2020 [10].

Companies need to consider what their key drivers are, their"motivations", how they are measuring impact, and how they plan on implementing them. The strategy can also include key goals and targets (i.e. key performance indicators) for their material topics. International Standards like GRI and SASB and TFCD has suggested indicators in which organizations determine relevant targets against.

2.4 Step 4. Sustainability Report - on ESG targets and goals

Utilizing the gap analysis, companies can get an idea of how extensive their report can be in comparison to their peers, as well as identify the best methods to report relevant to their industry. The report is the main tool for a company to voluntarily disclose its non-financial performance and ESG impact - positive or negative. The material analysis ensures that the report is relevant to stakeholders.

The aim of the sustainability report is to provide transparency on the company's contribution on all ESG matters and is viewed as a means of accountability for stakeholders like investors, employees, market regulators, suppliers, civil society and customers. Sustainability reports define performance, goals and metrics, measure performance, evaluate performance and manage performance. Most reports disclose sections including but not limited: financial summaries, customer engagement, innovation and technology, human resources, supply chain, facilities management, and community engagement.

Sustainability reports highlight the links between ESG matters and the business by compiling information and updating it regularly (usually annually). The report can improve management of the organization by identifying ESG risks, discovering opportunities to save energy and avoiding compliance issues. Further, it strengthens the internal communications especially in large organizations where employees are often unaware of actions by the company they work for.

3 Impact of Implementing the Described Methodology at Emirates NBD Bank, UAE

A recent study (2019) proves that with a reasonable significance level, that the more sustainable a company is, the higher their brand equity value is [11].

Emirates NBD - one of the top banks in the United Arab Emirates, launched their sustainability banking framework and first sustainability report in 2017 covering 2015–2016. At that time, their evaluated brand value was 1.78 billion USD, value which grew to 4.13 billion USD in 2020.

The first Emirates NBD report's Environmental Responsibility section disclosed their energy consumption, Greenhouse Gas Emissions (GHG), Water Consumption and Waste Management practices. By identifying their performance and carbon footprint, it allowed for the Bank to determine easily implementable methods for energy savings. In 2018, the Bank reported AED 1.9 million in energy savings in 2017 by monitoring facilities used, installed LED lights at their main headquarters and 35 branches, adjusted the number of hours chillers are turned on (achieving a 20% reduction in use), installed atmospheric water generators, and introduced carbon-neutral events.

In their 5th report, in 2020, the company highlighted they had 46 people with disabilities, 5,216 h volunteered for the local community, a reduced by 13% Greenhouse Gas (GHG) emissions compared to 2019, and all this by improving their financial performance: 8% of net interest income increase (y-o-y), 4% of total income increase, 2% growth of total assets and a net profit of AED 6.965 million.

The bank's achievement in cost reduction is an example of how the company is mindful in its operations, thus providing evidence of its non-financial performance successes and opportunities for improvement.

The Bank entrusts Brand Finance to evaluate their brand value each year. Brand Finance is a platform that evaluates and values 5000 brands per year across all sectors and geographies through its Brand Strength Index (BSI), a proprietary methodology to calculate the brand value. The BSI is divided into three equal parts:

1. Finance - market share, market share growth, revenue, margin;
2. Security/Risk - visual identity, distribution, credit rating;
3. Brand Equity - functional performance, emotional connection, conduct, loyalty.

According to a report which collected data from both sources: Brand Finance and CSRHub, on the link between Brand Value and Sustainability, the most important drivers of correlation appear to be how well a company treats its employees and its environmental policies [12].

The research suggests that more stakeholders are increasing awareness of ESG and Sustainability topics. Therefore, for platforms like Brand Finance to maximize calculations of their BSI such as functional (non-financial performance) and identifying ESG risk, the significance of brand value for investors and consumers will continue to increase.

4 Conclusions

In the last decades, the awareness on the sustainability importance for national economies and companies has increased. Firms perform better as they integrate sustainability into their overall strategy and implement sustainable programs and operations.

The observed increase in interest in ESG & Sustainability related articles, as well as focus on ESG topics showcases suggests that consumers, market regulators, investors and relevant stakeholders are becoming more aware of sustainability performance (non-financial) of private sector companies. The increase of interest with government incentives rising such as the Green Deal suggests that sustainability awareness in the area will continue to grow rapidly with focus on human rights and supply chain on the circular economy. Brand Finance and CSR Hubs' results shows that brands may continue to see broad benefits for their brand strength by disclosing the company's ESG performance. Thus, companies have another reason to care more about ESG topics and their sustainability performance.

Sustainability disclosure helped Emirates NBD and other companies to significantly increase their brand value. This work describes the methodology and the most relevant steps to develop a strategy and start a sustainability reporting.

References

1. Mohammad, W.M.W., Wasiuzzaman, S.: Environmental, Social and Governance (ESG) disclosure, competitive advantage and performance of firms in Malaysia. Clean. Environ. Syst. **2**(100015), 1–11 (2021)
2. Baldassarre, B., Keskin, D., Diehl, J.C., Bocken, N., Calabretta, G.: Implementing sustainable design theory in business practice: a call to action. J. Clean. Prod. **273**(123113), 1–17 (2020)
3. Hardin, G.: The tragedy of the commons. Science **162**(3859), 1243–1248 (1968)
4. Carson, R.: Silent Spring. Crest Book (1962)
5. Science Direct homepage. www.sciencedirect.com. Accessed 26 June 2021
6. Zenya, A., Nystad, O.: Assessing corporate sustainability with the enterprise sustainability evaluation tool (E-SET). Sustainability **10**(12), 4661 (2018)
7. Governance & Accountability Institute (G&A) homepage. https://www.ga-institute.com/research-reports/flash-reports/2020-sp-500-flash-report.html. Accessed 26 June 2021
8. Robeco The Investment Engineers homepage. https://www.robeco.com/en/key-strengths/sustainable-investing/glossary/european-green-deal.html. Accessed 26 June 2021
9. Ökten, N.Z., Okan, E.Y., Arslan, Ü., Güngör, M.Ö.: The effect of brand value on economic growth: a multinational analysis. Eur. Res. Manag. Bus. Econ. **25**(1), 1–7 (2019)
10. Emirates NBD's homepage, Sustainability Report (2020). https://www.emiratesnbd.com/en/assets/File/Sustainability_Report_2020_1.pdf. Accessed 29 June 2021

11. Ajour El Zein, S., Consolacion-Segura, C., Huertas-Garcia, R.: The role of sustainability in brand equity value in the financial sector. Sustainability **12**(1) 254, 1–19 (2020)
12. Brand Finance homepage, Gidwani, B.: The link between brand value and sustainability, director notes (2013). https://brandfinance.com/wp-content/uploads/1/the_link_between_brand_value_and_sustainability.pdf. Accessed 29 June 2021

Effects of Using Combined Approach of Quality Circles and 7 Steps Method in Automotive Industry. A Case Study

Sebastian Candea, Cristina Veres(✉), Petruta Blaga, and Emil Nutiu

"George Emil Palade" University of Medicine, Pharmacy, Science, and Technology of Targu Mures, Gh.Marinescu str. 38, 540142 Targu Mureş, Romania
{cristina.veres,petruta.blaga,emil.nutiu}@umfst.ro

Abstract. Quality is strictly regulated in any industry, but an extra attention is paid to the automotive field. Thus, quality control is needed to keep costs down, to identify problems before a piece or product is delivered and to ensure they are safe for man's use.

This paperwork presents a case study from automotive industry, of applying a combined approach of both Quality Circles and the 7S Methodology to reduce the number of claims and improve quality, highlighting the importance of establishing the quality circle in production factories and the effects of the implementation. A systemic outlook and an appropriate involvement support the company's continuous improvement efforts.

The introduction presents general information on the quality circles and its advantages, and besides gives an insight on the seven steps tool. Going forward, the step-by-step combined approach of the mentioned tools are described, showing in parallel the evolution of the implementation in an automotive respectable company. The paper ends up with the formulation of conclusions regarding the deduced effects.

Keywords: 7 steps · Automotive industry · Quality circles · Quality

1 Introduction

1.1 Quality Circles

Any production company's efforts to improve quality involve human resources. Employees' potential is used in several ways to add value or streamline a process. It was noticed that one way to get people involved is to make them work in groups [1].

Quality circles, also called quality control circles, were introduced by the Japanese professor and organizational theorist Kaoru Ishikawa in 1962, when he organized a meeting of a working group with the aim to exchange ideas on specific company issues. Quality circles are organized voluntary, on a weekly or bi-monthly frequency, with 8–10 people to perform quality improvements at the workplace. Quality Control Circle is a management tool for quality improvement [2].

L. Moldovan and A. Gligor (Eds.): Inter-Eng 2021, LNNS 386, pp. 289–297, 2022.
https://doi.org/10.1007/978-3-030-93817-8_28

The initial objectives of quality circles were to make contributions to company development, respect human relations, develop a satisfactory working environment, use human capacities to the fullest [3, 4].

Quality circles bring a lot of advantages to a company: improves professional knowledge, promotes a communication climate and improves collaboration, opinions are listened to, operators get a better knowledge of the quality problems and ways of solving them.

1.2 Seven Steps Tool

Ongoing globalization and digitalization confront people with an increasingly complex environment that demands numerous problems to be solved in personal life as well as at the workplace [5, 6]. To face workplace challenges that arise, companies use different problem-solving tools. For the successful problem solution, it is appropriate to perform the planning of the particular steps that lead to its successful resolution [7].

Seven Steps tool is a systemic approach, which follow the succession:

- Step 1: Problem Clarification;
- Step 2: Problem analysis;
- Step 3: Problem description;
- Step 4: Root Cause Analysis;
- Step 5: Alternate solutions development;
- Step 6: Solution implementation;
- Step 7: Results measurement.

Each step will be described further in the paperwork.

2 Applying Quality Circles and 7 Steps Tool in Automotive Industry

In the automotive industry both the quality circles and the 7 steps tool are used as tools to improve processes. If both were proved to be beneficial for a company and for its employees, it is interesting to see the impact of the combination of the two tools.

Such a combined approach was applied in an automotive corporation, which choose to no disclose its name. The case study is based on real data. In the following subchapter the detailed process will be described. The 7 steps were adapted to the specific of the problem and industry.

2.1 Preparations for the Quality Circle

The quality circle started with defining the work team, which was formed of:

- the quality engineer, which was the coordinator/mediator;
- the team members: operators (volunteers), quality inspector;
- the support team: segment manager, process engineer, specialist engineers, regulators.

A shift out of the 3 ones was chosen for the analysis.

For the following activities, some guidelines were established:

- There would be 2–3 meetings/month for the quality circle;
- The segment team would be periodically informed on the quality circle activities status;
- Monthly meetings will be organized for presenting activities to the management.
- Feedback and guidance will be given during meetings.
- 7 quality circles started in the same time, each project would be scored and the best projects in the program would be nominated in the competition.

The 7 Step tool was using "One page A3" format. The 7 steps include:

1. Problem clarification;
2. Problem breakdown;
3. Target setting;

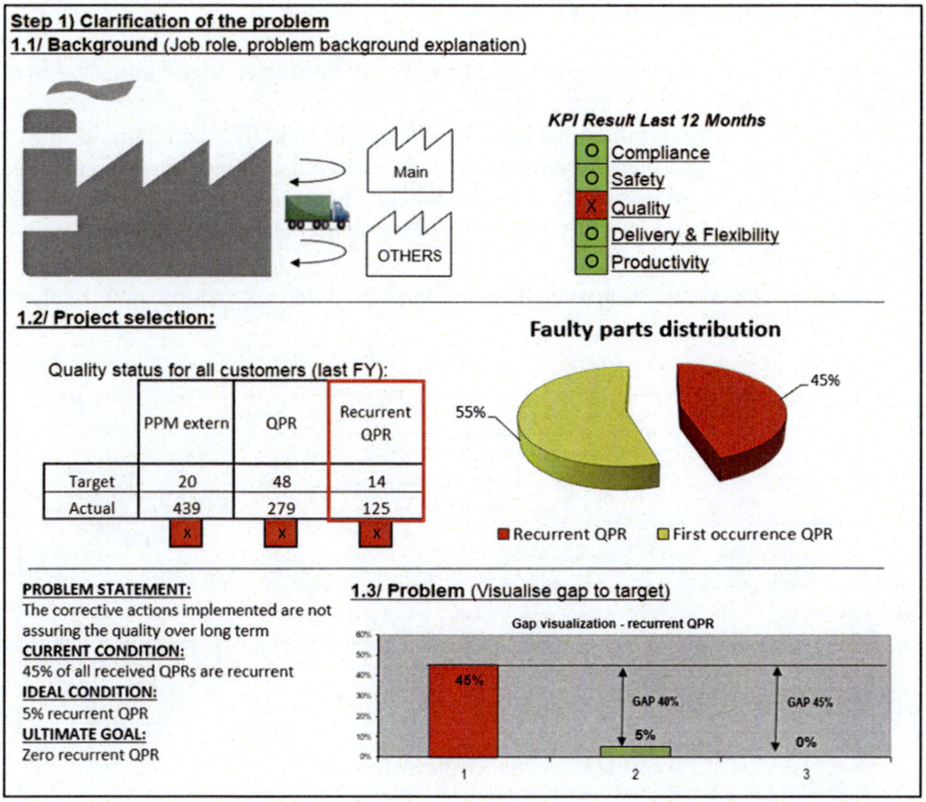

Fig. 1. Problem clarification.

4. Root cause analysis;
5. Countermeasure Development;
6. Results evaluation;
7. Standardizing the Successful Process.

2.2 Step 1. Problem Clarification

The first step consists in the clarification of the problem, in terms of the: Background,

Project Selection and Problem

In the Background part, the context of the job role and of the problem is explained. As the Recurrent QPR (Quality Problem Report) was far above the target, the problem statement highlighted that the corrective actions implemented are not assuring the quality over long term. 45% of all received QPRs were recurrent at the beginning of the project, so an ideal condition (5% recurrent QPR) and an ultimate goal (zero recurrent QPR) was established. A graphic was used to visualize the gap (see Fig. 1).

2.3 Step 2. Problem Breakdown

From this step on, each time the team meets to work on the project, they summarize first al the previous steps.

A breakdown of the problem happens in the second step. The company used 8D analysis tool. Each step of the 8 Disciplines from the received claims in the last 6 months (74 claims) was evaluated to identify the week points. The team found out that they have problems in the Root cause analysis, validation of corrective actions and standardization and containment measures.

The areas that need improvement are Supplier, Final Assembly and Leather Wrapping (see Fig. 2).

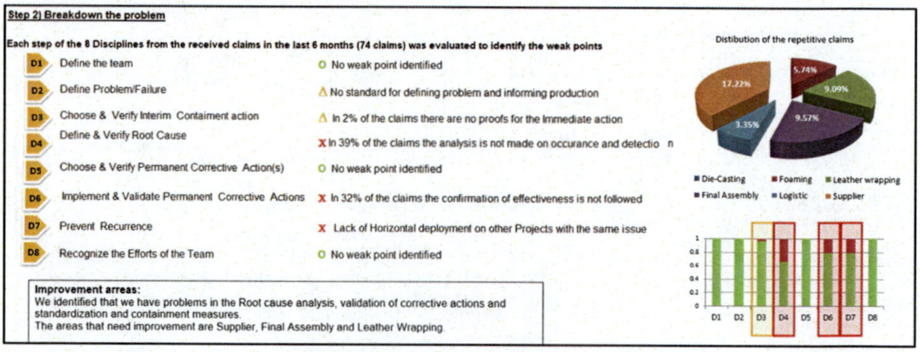

Fig. 2. Problem breakdown.

2.4 Step 3. Target Setting

When setting the target, the team would first make a recap of the steps 1 and 2.

This step defines the final objective of the project and the ramp down plan framed in unit of time (i.e. weekly/monthly), objective which should be SMART:

S - Specific (targets a specific area for development);
M - Measurable (quantified or at least shows an indicator of progress);
A - Assigned (achievable, who are responsible for their realization);
R - Realistic (as close to the truth as possible using the available resources);
T - Time frame (specifying the time when the objectives will be met).

Gradual targets were established and represented in a visual way (Fig. 3).

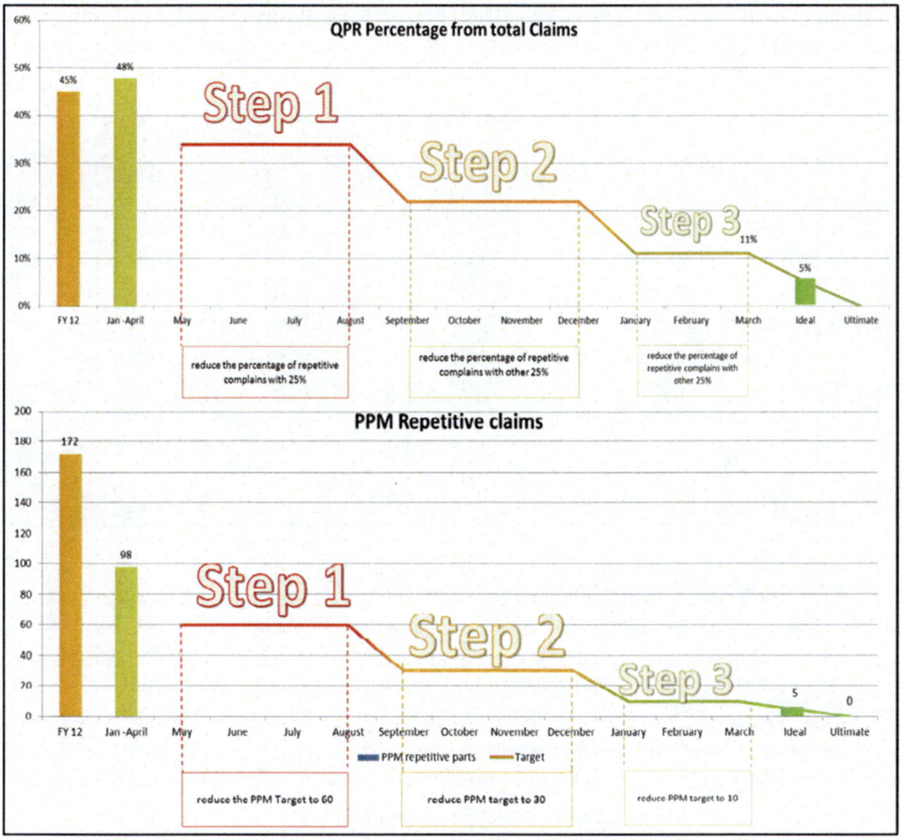

Fig. 3. Target setting.

The targets were related to 2 KPI-s:

- PPM (Parts per million) repetitive claims (step 1: 60, step 2: 30, step 3: 10, ideal: 5, ultimate goal: 0).
- QPR percentage from total claims (step 1: 35%, step 2: 22%, step 3: 11%, ideal: 5%, ultimate goal: 0%).

2.5 Step 4. Root Cause Analysis

A root cause analysis was done on the repetitive QPR (Fig. 4) using the "5 Why-s" method. The main purpose of the "5 Why-s" analysis technique is to determine the root cause of a defect or problem, by repeating the question "Why?". Each answer is the basis of the next question, one or more root causes are discovered as a result.

Weak point of the process were identified and written in form by the team. The weak points were then prioritized.

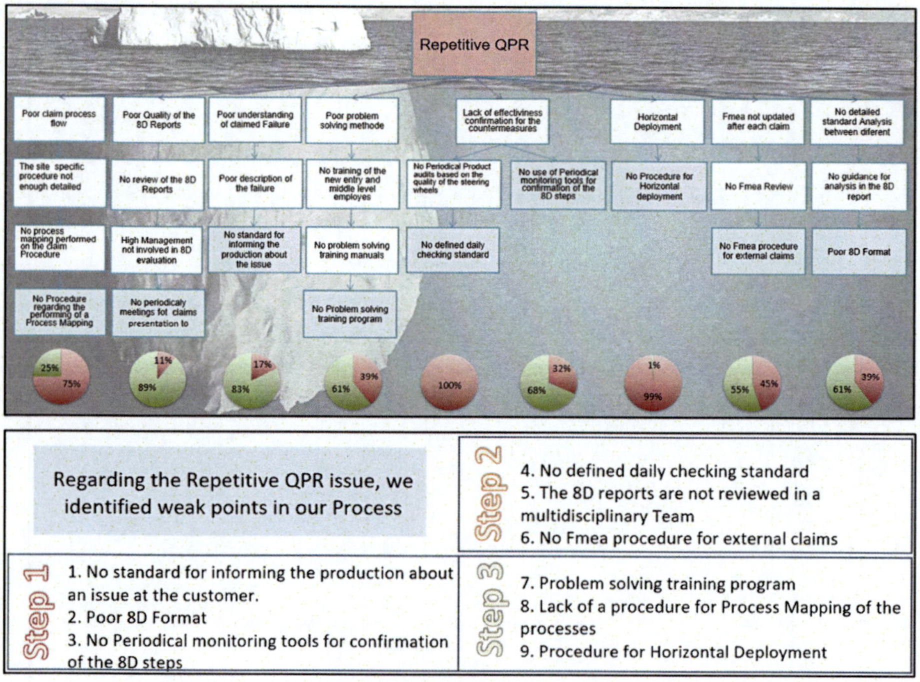

Fig. 4. Root cause analysis.

2.6 Step 5. Countermeasure Development

After identifying the causes that led to the problem, an implementation plan for improvement actions will be developed using the Gantt chart, prioritizing actions according to:

- Their effect;
- Implementation time;
- Implementation costs or costs avoided by implementation.

The action plan was then implemented for the following months (Fig. 5).

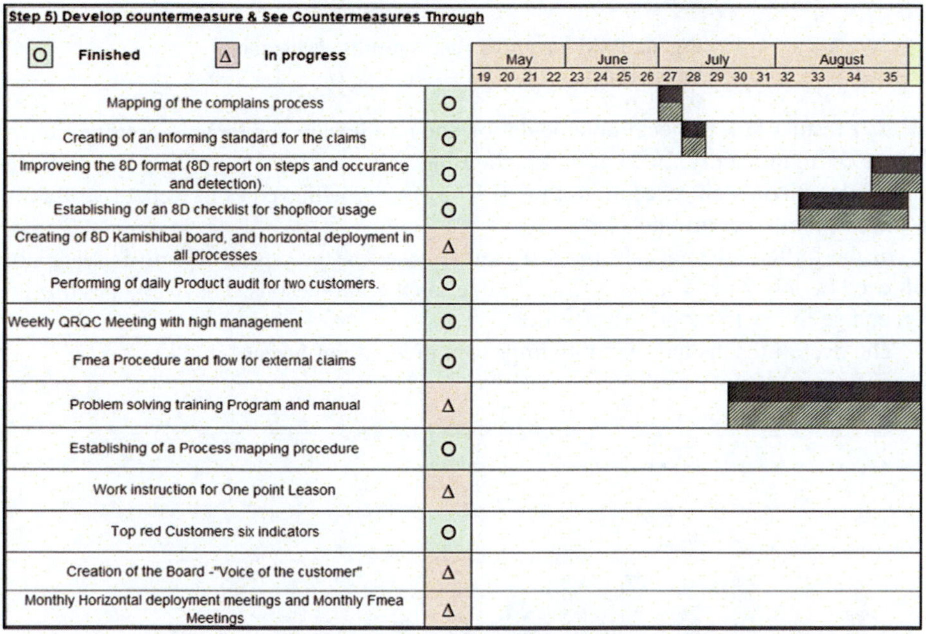

Fig. 5. Countermeasure development

2.7 Step 6. Results Evaluation

During and after the improvement actions implementation, a monitoring is performed with a certain frequency, monthly in the presented case study, with the aim to evaluate the effectiveness of the taken actions and the results/effects.

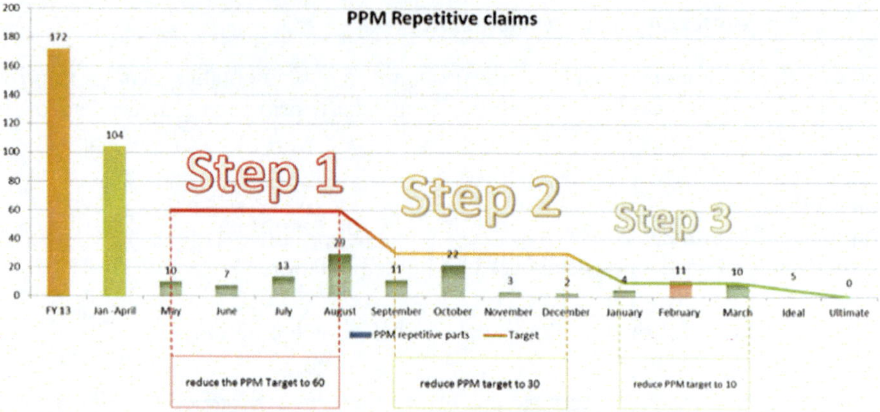

Fig. 6. PPM repetitive claims results evaluation

As a result of the project implementation and monitoring, the PPM Repetitive claims decreased significantly. In 11 months, the objectives for PPM repetitive claims were achieved, with only one exception (Fig. 6). The average achieved PPM Repetitive claims in the 11 month was around 11,09.

In the QPR Percentage from total claims case, the proposed objectives were not achieved in only 2 of 11 months (Fig. 7), due to external reasons. The average achieved percentage in the 11 month was 21,18%.

The overall results have notably improved in less than 1 year.

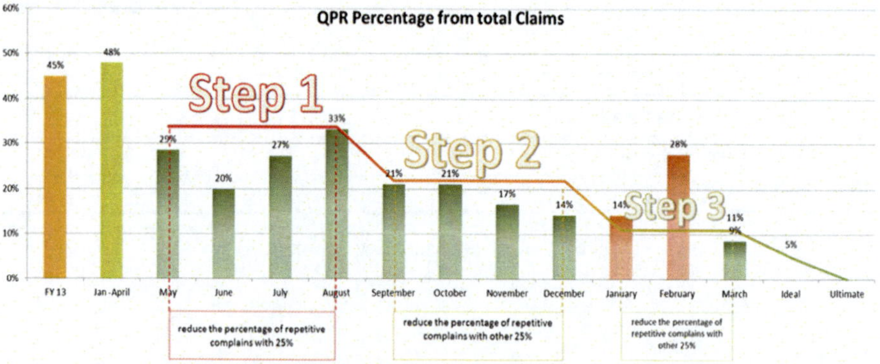

Fig. 7. QPR percentage from total claims results evaluation

2.8 Step 7. Standardizing the Successful Process

As the implemented solutions proved to have brought important improvements as a way to fix the existing issue, the change brought in the process was documented with the aim to replicate the change in other processes/departments/factories (Fig. 8).

Fig. 8. Standardize successful process

3 Conclusions

Companies, especially the ones in the automotive industry, are oriented on sharing the best quality and performance. Several tools are used in this regard. The presented case shows the effects of joining 2 instruments: the quality circles and the 7 Steps to improve results, describing also the taken steps to implement them. Two important quality indicators, PPM repetitive claims and QPR percentage from total claim, have improved significantly after implementing the described methodology. The implementation led to improved networking, raised quality awareness, better teams work, quality improvement through a systematic approach and development of the logical approach through problem solving.

Discipline, effectiveness monitoring, employees' involvement and commitment, consolidated by quality tools can notably improve company's KPI-s.

References

1. Moldovan, L.: Managementul calității. Editura Universității "Petru Maior", Targu Mures (2011)
2. Xiao, Z., et al.: Impact of a quality control circle on the incidence of catheter-associated urinary tract infection: an interrupted time series analysis. Am. J. Infect. Control **48**(10), 1184–1188 (2020)
3. Blaga, P., Jozsef, B.: Human resources, quality circles and innovation. Proc. Econ. Financ. **15**, 1458–1462 (2014). https://doi.org/10.1016/S2212-5671(14)00611-X
4. Richard, O.C., Johnson, N.B.: Strategic human resources management effectiveness and firm performance. Int. J. Hum. Resour. Manage. **12**(2), 299–310 (2001)
5. Eichmann, B., Goldhammer, F., Greiff, S., Pucite, L., Naumann, J.: The role of planning in complex problem solving. Comput. Educ. **128**, 1–12 (2019)
6. Fischer, A., Greiff, S., Funke, J.: The process of solving complex problems. J. Probl. Solving **4**(1), 19–42 (2012)
7. Jiří Dostál, J.: Theory of problem solving. Proc. Soc. Behav. Sci. **174**, 2798–2805 (2015)

The Challenges Brought by GDPR to the Use of Intelligent Systems

Andrea Kajcsa(✉) and Lucretia Dogaru

Department of Law and Public Administration, Faculty of Economics and Law, University of Medicine, Pharmacy, Sciences and Technology "G.E. Palade" of Târgu Mureş, 540142 Târgu Mureş, Romania
andrea.kajcsa@umfst.ro

Abstract. Intelligent systems and technologies have produced, without any doubt, many applications. These systems provide opportunities for economic and social development, energy sustainability, better health care, sustainable development and the spread of knowledge. However, as history shows us, all technological developments come with certain risks. One of these risks arises from the fact that many intelligent systems and technologies use large amounts of personal data. The area of personal data is covered by one of the most modern and comprehensive European legal instruments that regulates the right to data protection - The General Data Protection Regulation. GDPR is one of the most wide ranging pieces of legislation passed by the EU in recent years, and introduces many concepts that are yet to be fully discovered in practice, such as 'right to be forgotten', data portability, consent obligation. The GDPR does not mention intelligent systems in particular but many of its provisions are relevant for intelligent systems. We will analyze in our paper the clash between the traditional data protection framework that has as milestones concepts such as data minimisation, purpose limitation, special legal treatment of sensitive data, on the one hand, and, on the other hand, the revolutionary intelligent systems.

Keywords: Personal data · Intelligent systems · General Data Protection Regulation · Consent obligation

1 Introduction

A significant percentage of all human population interacts, on a daily basis, with intelligent systems, sharing with these information such as pictures, videos, emails, text messages, personal documents, browser history, financial data, location data, health records and more. However common these apps and smart devices may be, few of their users associate them with Artificial Intelligence or Intelligent Systems, concepts that perhaps seem more untangible to the average person.

The broadest definition of artificial intelligence (AI) characterises it as the attempt to build machines that "perform functions that require intelligence when performed by people" [1]. A more elaborate notion has been provided by the same the High Level

L. Moldovan and A. Gligor (Eds.): Inter-Eng 2021, LNNS 386, pp. 298–306, 2022.
https://doi.org/10.1007/978-3-030-93817-8_29

Expert Group on AI (AI HLEG), set up by the EU Commission, as follows: artificial intelligence (AI) systems are software (and possibly also hardware) systems designed by humans that, given a complex goal, act in the physical or digital dimension by perceiving their environment through data acquisition, interpreting the collected structured or unstructured data, reasoning on the knowledge, or processing the information, derived from this data and deciding the best action(s) to take to achieve the given goal. AI systems can either use symbolic rules or learn a numeric model, and they can also adapt their behaviour by analysing how the environment is affected by their previous actions [2].

AI can be defined, in a more understandable fashion, as a system that displays intelligent behaviour by analysing its environment and taking action—with some degree of autonomy—to achieve specific goals [3].

AI obviously simulates human decisional patterns in many aspects, including decision-making. It is therefore able to exhibit signs of rational thinking, it is capable of adapting to a detected change in circumstances and it is able to engage in autonomous actions [4].

AI enables automated decision-making even in areas that imply complex choices. These machine-made decisions could, at times, prove to be more precise and more impartial than human ones, since AI are not subject to psychology. It is nonetheless true that automated decisions can unintentionally discriminate and take blunt decisions that do not take into consideration the manny nuances real life has. Lets just imagine for one second what would we prefer as a judge: a machine that takes automated decisions based on alghoritms that we do not know and fully comprehend or a human persons that is susceptible to compassion and empathy?

Even when automated assessments of individuals are fair and accurate, they are not unproblematic: they may negatively affect the individuals concerned, who are subject to pervasive surveillance, persistent evaluation, insistent influence, and possible manipulation [5].

The General Data Protection Regulation (GDPR) does not mention intelligent systems in particular, and the definition given by European bodies to AI is later dated then the adoption of the GDPR. This could mean that, at the time of drafting the GDPR, EU did not give enough emphasis to the AI's undeniable relationship with data and Machine Learning, but rather focused on the intelligence and autonomy aspects of the AI systems [4].

2 AI and Personal Data

Intelligent systems (AI especially) have many abilities – they negociate, solve problems, recognize voices and faces, they can process language, they can even interact with persons in an intelligent manner, a personalized, even emotional way. Perhaps too often, users do not even realise how often they access AI services. We use AI when we translate texts online, when we use mobile apps to find the shortest route to our next destination, we use AI in our household and in the medical field, in traffic and marketing, we use it for pleasure and/or business. These abilities of AI systems are built upon data availability and data usage. How? Simply put, AI processes data in order to make predictions. Economy and consumerism are both based upon personal or personalized services and this can only mean significant amounts of personal data used as grounds for Machine Learning. As a consequence, personal data have become more valuable then ever before in human history.

Perhaps the most remarkable ability of AI is to predict, that is to foresee a certain outcome in a new case by means of jumping from certain known features of that case to an unknown feature of that case. This ability of prediction is based on models. These models may be designed by humans, but in the machine learning approach, machines discover the correlations that are the foundation of models, and then apply these models to make predictions in new cases.

When combining vast amounts of data with AI, automated predictions of a much greater variety become possible. For example, targeted advertising is based on linking certain characteristics (gender, age, social background) and behaviour of consumers (purchase history, web browsing) to their responses to ads. Or, the prediction of the likelihoods of recidivism by a particular offender may be based on records combining characteristics of past offenders (education, employment history, family status, criminal record, psychological tests, etc.) with data on their recidivism [2]. The examples are many, and we have all become somewhat familiar with these practices.

The amount of personal data used and storaged on the Internet, especially deriving from Social media, is huge. IDC analysts predict that in 2025, 175 zettabytes of data will be available in data storages such as clouds, smart phones, Internet of Things (IoT) devices, or cell towers [6]. The easiness with which we share personal data seems to show a minimum of interest in protecting it and raises the question whether we correctly asses (or asses at all!) the consequences these actions might have.

As a result of the need to learn by analysing vast amount of data, AI has become "hungry" for data, and this hunger has spurred data collection, in a self-reinforcing spiral [7]. Thus, the development of machine-learning AI systems necessarily implies and fosters the creation of vast data sets, i.e., *big data* [8]. Given this fact, it is imperative to fully understand how GDPR adresses the issue of personal data.

Regulation no. 2016/679 on the protection of natural persons with regard to the processing of personal data and on the free movement of such data (GDPR) represents one of the most wide ranging pieces of legislation passed by the EU in recent years, having been adopted in times when the internet and new technologies find themselves in a dual moment: on one side the development of new technologies and on the other hand the need of regulation [9].

AI can be compared to a framework that puts economy and the protection of human rights face to face. The law will have to prioritize! The regulations adopted in this field, at national, european or even international level, are of utmost interest for citizens, companies and public bodies since these are the three main stakeholders in the matter, each one pursuing its own agenda. Companies will align on the economic side, citizens on the human rights side and the public bodies will have to find some sort of balance in the legal provisions that they enact and carry out.

3 Legal Values and Legal Norms in the AI Field

From the very beginning, we can notice that legal regulations in the field of AI are scarce. Therefore, it is important to focus not only on existing regulations, but also on principles and ethical guidelines and rules.

On 8 April 2019, the High-Level Expert Group on AI presented Ethics Guidelines for Trustworthy Artificial Intelligence. This followed the publication of the guidelines' first draft in December 2018 on which more than 500 comments were received through an open consultation process. According to these Guidelines, trustworthy AI should be: lawful - respecting all applicable laws and regulations; ethical - respecting ethical principles and values; robust - from a technical perspective while taking into account its social environment.

According to the High-Level Expert Group, in order to implement and achieve trustworthy AI, seven requirements should be met:

- Human agency and oversight, including fundamental rights;
- Technical robustness and safety, including resilience to attack and security, fall back plan and general safety, accuracy, reliability and reproducibility;
- Privacy and data governance, including respect for privacy, quality and integrity of data, and access to data;
- Transparency, including traceability, explainability and communication;
- Diversity, non-discrimination and fairness, including the avoidance of unfair bias, accessibility and universal design, and stakeholder participation;
- Societal and environmental wellbeing, including sustainability and environmental friendliness, social impact, society and democracy;
- Accountability, including auditability, minimisation and reporting of negative impact, trade-offs and redress.

Some of the terms used in this document, that is not legally binding, fall well within the scope of GDPR and the protection of personal data, such as: respect for privacy, access to data, transparency, minimisation, reporting of negative impact.

A comparative analysis of documents on the ethics of AI has concluded on a global convergence on values such as transparency, non-maleficence, responsibility, and privacy, while dignity, solidarity and responsibility are less often mentioned [10]. However, substantial differences exists on how to balance competing requirements, i.e., on how to address cases in which some of the values just mentioned above are affected, but at the same time economic, administrative, political or military advantages are also obtained [5].

Moving towards a more legal approach, the most pronounced clash exists between AI and privacy and data protection. Protection of personal data and respect for private life are European fundamental rights. European bodies have always insisted on the need to strike a balance between enhancing security and safeguarding human rights, including data protection and privacy. This fundamental right has more then one legal basis, being provided for in Article 16 of the Treaty on the Functioning of the European Union and articles 7 and 8 of the EU Charter of Fundamental Rights.

Given the major impact it has on citizens' lives, AI falls under the scope of different legal regimes, including data protection law, consumer protection law, and competition law. It has been observed by the European Data Protection Supervisor (EDPS) in Opinion 8/18 on the legislative package 'A New Deal for Consumers', that there is synergy between the three regimes.

We can identify more then one points of interest concerning personal data and AI. First, there undoubtedly exists an interest in data protection, in ensuring a lawful and proportionate processing of personal data. This in itself is hardly compatible with the online environment, where every action, like, video watch, search is tracked, and the resulting data is used to extract further information about the preferences and dislikes individuals have. This information, that we as particular persons do not see as commodities and we do not attach any pecuniar value to them, are processed in ways that, too ofthen, are contrary to the individuals interests and well-being.

The need to keep the processing of personal data under control and to understand and possibly challenge the reasons behind determinations that affect individuals, raises concern from an algorithmic transparency. People will want to know how and why a certain algorithmic response has been given or a decision made, so as 'to understand and hold accountable the decision-making processes of AI [11].

AI systems have access to huge amounts of information about individuals and they can easily use this information to bring about a particular desired behaviour, for purposes that citizens may not share and most of the times not even be aware of. Individuals have an interest in being able to trust these systems, to trust that the controllers of those systems will not profit from the people's sharing of personal data. Most importantly, individuals need to know that they are not being subject to market-power abuses resulting from exclusive control over masses of data and technologies.

4 AI and Data Protection Principles

We have already established that AI uses big data in order to make predictions. This brings AI automatically in the scope of GDPR and the general principles of data protection. Article 5 of the GDPR sets out the principles governing the processing of personal data, as follows: lawfulness, fairness and transparency; purpose limitation; data minimisation; data accuracy; storage limitation; integrity and confidentiality.

Articles 13, 14 and 15 of GDPR provide that data controllers must present meaningful information related to data processing activities they carry out, including such activities in which the decision is made algorithmically. What constitutes meaningful information is however not defined. Who should determine what is meaningul? In an opinion, that we also share, if meaningfulness is to be determined by the data-subject, data subjects have no interest in any complex technical terms specific to the applied AI, and would prefer the simplest and clearest, most comon explanation; while other data subjects might prefer more detailed information in line with their level of AI knowledge [4]. If, on the other hand, meaningfulness is to be determined by data controlers and AI developers it could lead to understanding issues the same way, which generates more complications especially for people without technical background knowledge.

Recital 58 details the principle of transparency, stating that it requires that any information addressed to the public or to the data subject be concise, easily accessible and easy to understand, and that clear and plain language and, additionally, where appropriate, visualisation be used. Transparency is deeply intertwined with fairness, particularly informational fairness [5]. Informational fairness requires that data subjects are not deceived or misled regarding the processing of their personal data. Informational fairness raises specific issues in connection with AI and big data, because of the complexity of the processing involved in AI-applications, the uncertainty of its outcome, and the multiplicity of its purposes [5]. A specific aspect of transparency is put into question in relation to machine learning – characteristic to AI in particular. The problem of access to data arises in connection to the systems training or learning set.

Access to data may be needed to identify possible causes of unfairness resulting from inadequate or biased data or training algorithm. This is particularly important when the learned algorithmic model is opaque, so that possible flaws cannot be detected through its inspection [5].

The principle of purpose limitation is provided for in article Article 5(1)(b). According to this, personal data should be collected for specified, explicit and legitimate purposes and not further processed in a manner that is incompatible with those purposes; further processing for archiving purposes in the public interest, scientific or historical research purposes or statistical purposes shall, in accordance with article 89(1), not be considered to be incompatible with the initial purposes ('purpose limitation'). The notion of a purpose is mentioned in Article 6 in relation to the first legal basis, namely, consent, which should be given 'for one or more specific purposes', and for the last legal basis - 'the purposes of the legitimate interests pursued by the controller or by a third party'. Still, the need for legitimate purpose is implicit in the other legal bases, which consist in the necessity of the processing for performing a contract, complying with a legal obligation, protecting vital interests, performing a task in the public interest or exercising a legitimate authority.

AI and big data technologies and the purpose limitation requirement do not match easily, given that these technologies make it possible to reuse personal data for new purposes, purposes that are different from those for which the data were originally collected. For example, 'likes' that are meant to express and communicate one's opinion may be used to detect psychological attitudes, political or commercial preferences [5].

A key issue concerning AI is the possible repurposing of personal data. The rule is stated in article 5(1)(b) as a particularization of the principle of purpose limitation. According to this article, personal data shall be "not further processed in a manner that is incompatible" with the original purposes. Recitals 50 also refer to the interdiction of repursposing personal data, stating: "the processing of personal data for purposes other than those for which the personal data were initially collected should be allowed only where the processing is compatible with the purposes for which the personal data were initially collected".

Repurposing of personal data for new and different purposes then the original one, has become crucial in the era of AI and big data, since vast data are available and artificial intelligence are used to discover sometimes unexpected correlations between apparently unrelated data and identify possible causal links that are used to influence behavioral patterns of data subjects.

The repurposing of data is legitimate or not depending on whether the new purpose is 'compatible' or, on the contrary, 'not incompatible' with the purpose for which the data were originally collected. According to the Article 29 WP, the relevant criteria to be used for evaluating the legality of repurposing are: the distance between the new purpose and the original purpose, the alignment of the new purpose with the data subjects' expectations, the nature of the data and their impact on the data subjects' interests, and the safeguards adopted by the controller to ensure fair processing and prevent undue impacts [12].

We therefore know the criteria that determine compatibility, but we do not have a clear, definite answer concerning the possibility of reusing personal data in AI applications. Each individual case will have a different turnout, and there is no general, universal formula.

The principle of data minimisation, according to which personal data should be 'adequate, relevant and limited to what is necessary in relation to the purposes for which they are processed'. The very idea of big data and data analytics, which involves using AI and statistical methods to discover new unexpected correlations in vast datasets, infringe, at a first analysis, this principle. A more thorough analysis, however, shows that the idea of minimisation should be linked to the idea of proportionality. Minimisation does not exclude the inclusion of additional personal data in a processing, as long as the addition of such data provides a benefit. Even future processing may be justified, as long as adequate security measures are in place: pseudonymisation, in combination with other security measures, may contribute to limit risks and therefore ensure compatibility with minimisation. Furthermore, the processing of personal data for pure statistical purposes is subject to looser requirements. According to Recitals 162, statistical purposes mean any operation of collection and the processing of personal data necessary for statistical surveys or for the production of statistical results. Those statistical results may further be used for different purposes, including a scientific research purpose.

The right to be forgotten is quite a novelty in the field of AI, since computers are very good at remembering things. Article 17 of the GDPR defines the conditions for exercising the right to erasure. According to this legal provision, in case the data is being processed outside of the scope of initially indicated purposes and if there is no other legitimate basis available for the data controller to continue data processing, then data subjects erasure request must be fulfilled. In the context of AI systems, data subjects might desire to exercise this right if, for example, they do not consider themselves as part of a particular group decided on by an algorithm or might not want to disclose the entirety of their private choices to others.

However, both technically and legally, exercising this right is especially hard in todays data-driven AI systems [4]. The right to be forgotten has evolved in its legal and jurisprudential perspective in a way that data subjects may request search engines to hide information related to their past which is no longer public interest information, rather than erasure. It was interpreted as right not to be found or right not to be seen in different jurisdictions, such as Italy, because being fully forgotten is technically not possible and promising such a result could mislead the data subjects as well as the courts [13]. Applying privacy methods may make the algorithm learn slower than standard and may cause utility loss. This might also imply additional costs for developing AI. The

GDPR is quite well prepared in this sense, regulating in its article 25 the principle of Data Protection by Design and by Default.

5 Legal Basis for Processing Personal Data in Case of AI

Article 6(1) of GDPR sets out the conditions that must be satisfied for the processing of personal data to be lawful. It states that all processing of personal data requires a legal basis. The processing of personal data in the context of AI application raises some issues relating to the existence of a valid legal basis. The legal bases set forth in Article 6 GDPR are: consent of the data subject, or necessity for performing or entering into a contract, (for complying with a legal obligation, for protecting vital interests, for performing a task in the public interest or in the exercise of public authority, or for a legitimate interest.

A data subject's consent to the processing of personal data by an AI system can be envisaged in two possible scenarios: for the purpose of including such data in a training set, or for providing them to an algorithmic model meant to deliver individualised responses. The GDPR approaches consent more restrictively. In particular, it seeks to ensure that consent is specific to distinct purposes of processing. Consent has to be specific, granular and free. It is not easy for all these conditions to be met with regard to the AI-based processing of personal data. Thus, other legal basis might be needed.

The other legal basis provided for in article 6 of GDPR all involve establishing the necessity of the processing for a certain aim, such as: performing or entering (at the request of the data subject) into a contract, for complying with a legal obligation, in order to protect vital interests or for performing a task in the public interest or in the exercise of public authority. However, such legal bases do not apply to the AI-based processing that is subsequent to or independent of such aims. For example, the necessity of using personal data for performing or entering a particular contract does not cover the subsequent use of such data for purposes of business analytics or for developing predictive models.

Article 6(1)(f) of GDPR provides that the necessity of the processing is a general legal basis to the processing of personal data, stating that processing must be made: "(...) *for the purposes of the legitimate interests pursued by the controller or by a third party, except where such interests are overridden by the interests or fundamental rights and freedoms of the data subject which require protection of personal data* (...)". In those situations when the AI uses the data subjects' data for an algorithmic model, to make predictions concerning the data subject and its behaviour, in the light of the provisions of the GDPR, the priority should be placed on the data subject on his or her assessment. Thus, there should be consent and the data subject should be given the possibility to opt in. It is however important to admitt that there is no general guideline in the matter, since each and every application has its individuality and should therefore be individually assessed in terms of the degree in which it affects the individual and society in whole. In other words, the advantages and disadvantages should be weighed. If an application provides benefits that are outweighed by the disadvantages imposed on the data subjects, the conclusion should be that the application fails to have a basis according to article 6(1)(f) of GDPR [2].

6 Conclusion

The GDPR does not refer in an express manner to Intelligent System yet it generally provides meaningful indications for data protection relative to AI applications. Given the great development of AI application nowadays and since a number of AI-related data protections issues are not explicitly answered in the GDPR, this may lead to uncertainties and costs, and further regulations specific to AI may be needed in the near future. It is also true that the GDPR can be interpreted and applied in such a way that it does not hinder beneficial application of AI to personal data. For sure, controllers using AI-based processing should comply with the values of the GDPR and adopt a responsible and risk-oriented approach, not having to overlook economic profitability. In the same time, Data Protection Authorities should also provide guidance, given the lack of specific regulations in the field, the interpretability and contradictory interests of all stakeholders involved.

References

1. AI-HLEG, High-level expert group on artificial intelligence. a definition of AI: main capabilities and scientific disciplines (2019)
2. AI-HLEG, High-level expert group on artificial intelligence. Ethics guidelines for trustworthy AI (2019)
3. COM - Communication from the Commission to the European Parliament, the European Council, the Council, the European Economic and Social Committee and the Committee of the Regions - Coordinated Plan on Artificial Intelligence (2018)
4. Gultekin Varkonyi, G.: Operability of the GDPR's consent rule in intelligent systems: evaluating the transparency rule and the right to be forgotten. J. Ambient Intell. Smart Environ. 216–215 (2019). https://doi.org/10.3233/AISE190044
5. Sartor, G.: The impact of the General Data Protection Regulation (GDPR) on artificial intelligence. Study for the Panel for the Future of Science and Technology (2020)
6. Reinsel, D., Gantz, J., Rydning, J.: Data age 2025: the digitization of the world from edge to core, IDC (2018)
7. Cristianini, N.: Intelligence rethought: AIs know us, but don't think like us. New Scientist (2016)
8. Mayer-Schönberger, V., Padova, Y.: Regime change? Enabling big data through Europe's new data protection regulation. Columbia Sci. Technol. Law Rev. **17**(2), 315–335 (2016)
9. Şandru, D.M., Alexe, I. (eds.): Legislaţia Uniunii Europen privind protecţia datelor personale. Universitară Publishing House, Bucharest (2018)
10. Jobin, A., Ienca, M., Vayena, E.: Artificial intelligence: the global landscape of ethics guidelines. Nat. Mach. Intell. **1**, 389–399 (2019)
11. Floridi, L., et al.: Ai4people– an ethical framework for a good AI society: opportunities, risks, principles, and recommendations. Mind. Mach. **28**, 689–707 (2018)
12. WP29: Opinion 3/2013 on purpose limitation
13. Tiberi, G.: The right to be forgotten as the right to remove inconvenient journalism? An Italian perspective on the balancing between the right to be forgotten and the freedom of expression, e-conference on the right to be Forgotten in Europe and beyond. Blogdroiteuropeen 49–61 (2017). http://wp.me/p6OBGR-27k

European Union Strategies and Policies in the Current Context of Technologization

Lucretia Dogaru(✉) and Andrea Kajcsa

Department of Law and Public Administration, Faculty of Economics and Law, University of Medicine, Pharmacy, Sciences and Technology "G.E. Palade" of Târgu Mureş, 540142 Târgu Mureş, Romania

Abstract. Starting with the Fourth Industrial Revolution, which is characterized by the expansion of Internet, robotics, biotechnologies and advanced materials, by the emergence and rapid expansion of electric and smart cars and also by the promotion of green energy, a series of new economic, social and environmental challenges have become obvious. It has become undeniable that technology and its numerous new uses has become one of the most important distributors of power in the world. Clearly, the sustainable and qualitative development based on technological progress and the protection of natural resources and of the environment require not only appropriate policies in the field of industry and research but also in the field of environmental protection. In this context, and taking into consideration the mechanism that lies at the core of the UE, the strategies and policies that are adopted by the EU in this present context of technology will represent for Romania the main framework for its own strategies in this field. In this paper, we have tried to analyze the impact but also the consequences that the current technological progress generates on the economic, social and environmental field.

Keywords: Technological progress · Industrial revolution · Digitization · Robotization · Climate changes · UE strategies

1 Considerations on the Impact and Consequences of Technological Progress

The Fourth Industrial Revolution is characterized by the development of ecological energies, digitization and the expanding of robotics, by the development of biotechnology and advanced materials, by the appearance and use of electric cars, generates real economic, social and environmental challenges.

However, sustainable and quality development based on technological progress and designed to protect natural resources and the environment in general requires adequate and coherent industrial policies that need to be linked to education, research and innovation policies [1].

It is increasingly evident that the current industrial revolution is considerably based on the internet, the IoT, and green energies defined as energies that arise from renewable

L. Moldovan and A. Gligor (Eds.): Inter-Eng 2021, LNNS 386, pp. 307–315, 2022.
https://doi.org/10.1007/978-3-030-93817-8_30

sources (including wind, solar, hydroelectric and geothermal energy as well as biofuels and biomass).

If we take into account the internet, it is absolutely visible that it has had a major contribution to the cheapened marketing activities and that it has facilitated access to information and communication, for both companies and individuals. Besides, there are specialists [2] who have predicted, almost a decade ago, that by means of combining the Internet technology with green energies, the fourth industrial revolution will create a new economic paradigm and can be based on the following main pillars: the transition from fossil fuels to renewable energies; the conversion of buildings into "green" micro-power stations for the collection of renewable energies; the use of intermittent energy storage technologies; the use of internet technology to turn the power grid into an 'energy internet'; the switch to electric vehicles and fuel cells.

A relevant aspect is that this ambitious project of the future, where IoT, big data, AI, green energy and true sustainable development can intertwine harmoniously, has received support over time from the international community but also at a regional level, expressing not only the interest but also the concerns for the implementation of a new technological, integrated and interactive system.

It is important to note that the main directions of influence by the inevitable change generated by technological progress were addressed at the Davos Conference in 2016. In that context, it was also underlined that this new industrial revolution is a digital one, characterized by the fusion of new technologies that in essence diminish the boundaries between the physical, digital and biological worlds.

Reality has shown that new information technologies not only influence the biological world, but also make possible a number of products and services that facilitate personal and professional life through a transfer of information and interaction that create multiple opportunities. Of course, an interesting aspect is that revolutions in biotechnology and artificial intelligence influence even the human aspect, forcing the individual to reconsider his moral and ethical limits.

Speaking of artificial intelligence, we show that it starts from autonomous vehicles and drones to virtual assistants and software capable of investing on the stock market and contributing to the discovery of new drugs or algorithms capable of identifying both interests and social conduct. Thus, we find AI almost in every area of our lives, personal and professional.

In this context, a relatively recent and extremely curious concept is that of the Internet of Things, which involves using the Internet to connect different devices, services and automatic systems, thus forming a real network of objects [3]. Not only has it been expected, but it has also been true that it can enhance productivity, maintain the state of comfort and health of the population, but also reduce the need for energy with beneficial consequences for the environment.

We would like to emphasize again that the new information and communications technology, the Internet of Things and the synergy between technologies, machines and artificial intelligence, considerably diminishes the consumption of energy and natural resources and thus reduces the carbon footprint of production. All this, in a context where, the phenomenon of climate change has become equally obvious and both aggressive.

In the category of revolutionary technologies, 3D printing is a new, modern, completely revolutionary process, of genetic engineering, that rapidly delivers new products made of cells, or even other primary materials, that are completely unconventional. This technology is merely one of the many that offer particular perspectives to industrial development, although it has been proven to present certain major risks. However, 3D printing is a revolutionary technological process of wide application by which energy, materials and resources are saved at every stage of the digital manufacturing process.

In a constant and obvious manner, new techniques, technologies and processes are developed, which ensure greater adaptability and flexibility for current industrial production. We are witnessing an increasing use of advanced materials, much more reliable and durable, the robotization of many activities but also the production of modular cars (MQBs), a process that allows the production of all models on the same production line with beneficial consequences for the economy and the environment.

According to the Technology Tipping Points and Societal Impact Report from Davos in September 2015 [4], it is estimated that by 2025 over 90% of the world's population will connect and have access to the Internet, pharmacies will be served by robots in the USA; about 5% of consumer products will be printed in 3D technology by 2024 and the first car will also be made using this technology; about 90% of the population will use smartphones by 2023; over 50% of the traffic of homes and household appliances will be done via the Internet by 2023; 50% of cars will be electric by 2026 and about 10% of all cars in the US will be driverless cars; 3D-printed organ transplants will be performed by 2022; and the first city with more than 50,000 inhabitants and no traffic lights will be designed and operational [5]. Even SAP Community Networks admitted the major role of the Internet through both hyper connectivity and the Internet of Things. They are capable of determining a new cycle of global economic activity, based on sustainable solutions and with high potential in reducing dependence on fossil fuels. Technological progress, based on robotics and other innovations, allows the reduction of industrial waste and the redesign of production and consumption systems in order to make the use of resources more efficient, but with simultaneous job damage. However, both technical progress and the internet that offers many opportunities will benefit especially advanced societies.

At the level of 2016, a representative of Deloitte Consulting[1] recognized the significant impact of the fourth industrial revolution on the business environment and society, the digital transformation of manufacturing production favoring access to a number of exponential technologies (like robotics, artificial intelligence, nanotechnology, quantum computer). It is currently estimated that, from about 23 billion articles connected through the Internet of Things in 2020, it will reach almost 40 billion in 2025, a year in which even cutting-edge technologies will double their value to more than $100 billion. Besides, capital investment in robotics and artificial intelligence has also increased by more than 80% per year since 2015. Furthermore, it is appreciated that it is possible to accelerate the automotive innovation cycle through prototypes made in the virtual environment as well as with the help of sensors installed on cars. With the help of computers new business models can be developed, models designed to enable scientific analysis and forecasting. One of the obvious disadvantages is considered to be that technologies

[1] Gary Coleman, Managing Director, Global Industries, DTTL.

based on artificial intelligence do not have enough specialists and, furthermore, they can cause the restriction of the workforce [6], an aspect which has become reality.

According to recent data provided by General Motors, technological progress in the automotive industry is significant. Thus, it is evident that more and more environmentally friendly, more efficient, safer and smarter vehicles have been produced. The tendency of moving from autonomous vehicles (mechanically controlled and fuel-based) to interconnected, electronically controlled and multi-energy-powered vehicles with superior performance at consumption and distance travelled is seriously felt. Car electrification involves not only electric motors but also sophisticated cameras, radars and sensors that allow to achieve increased safety. For example, through the OnStar System, General Motors has so far responded to more than 1.5 billion requests from customers, and through 4G wireless connectivity the car is allowed to act as a Wi-Fi hotspot, which is why the company has distributed more than 75% of the total volume of actively connected vehicles by 2020. Another intelligent system is V2V (vehicle to vehicle), which involves a wireless communication service that allows traffic cars to share important information about speed and direction of travel, traffic flow, road and climatological conditions [7]. This intelligent system allows the detection of sudden braking, the danger of collision, and the stability of the car, the detection of vehicles that have left the roadway and the control of the braking and steering system. This aims to reduce the costs of road accidents considerably, which in developed countries reach up to 3% of Gross Domestic Product annually, but also to diminish the number of road accident victims.

Another smart system is that of vehicle-infrastructure communication system (V2I), which involves a wireless exchange of data between vehicles and road infrastructure. This type of connectivity will avoid road congestion, manage speed and thus ensure safety, of course with economic and environmental benefits [8]. Of course, for all this, the road infrastructure must move to digital messages designed so that the technology in automated cars can interpret the surrounding environment and respond quickly and accordingly.

However, despite all these positive aspects, there are economists [9–11] that argue that robots, artificial intelligence, 3D printer and other new technologies that revolutionize industrial production and ensure automation will significantly affect peoples' jobs. Moreover, at the World Economic Forum Annual Meeting held in Davos in 2016, a Report entitled The Future of Jobs was published, document that predicts that in 15 national economies covering over 60% of the global workforce, robots and other intelligent entities will eliminate over seven million jobs by the end of 2020, especially in field such as industry, energy and health.

It is estimated that the digitization and robotization process, which are the main features of the Fourth Industrial Revolution, will have the effect of significantly replacing people, but with the advantage of higher and more efficient productivity and better and faster communication [12, 13]. For example, US specialists have estimated that by 2020 about 60% of jobs in the USA and other developed countries will be affected as a result of the massive introduction of digitization. The massive process of robotization of industries such as consumer electronics, the automotive industry and the manufacturing sector has led to a considerable narrowing of the need for skilled workforce in such areas. Of course, the risk of high unemployment has also imposed a rapid adaptation to this paradigm.

However, reality shows that the economy of the digitized and robotic future will increasingly be based on data and the process of managing this data will of course generate new jobs in sectors such as 4IR technology, machine learning, robotics, nanotechnology and AI, 3D printing, genetics and biotechnology [14]. Within these sectors that will dominate the coming decades, employers and industries will turn to qualified data analysts who are able to interpret information and use it to increase the productivity and performance of companies, as well as managers of integrated management systems, with abilities in planning, budgeting and efficient management of companies' resources and operations.

Clearly, beyond data management, the economy of the future will also need specialists in technology, software development and support, which involves specialized engineers in design, development and testing of advanced manufacturing technologies, specialists in monitoring the electrical aspects of systems and specialists in monitoring and optimizing industrial processes, reducing bottlenecks, waste disposal, increased yield and maximizing efficiency. For example, according to a study conducted by Frames, between 2016 and 2019 the Romanian digital economy registered a growth rate almost five times faster than those of the main European economies, and in 2020, amid the economic crisis generated by the pandemic, the over 30,000 Romanian companies engaged in the technological business (CAEN codes 6201, 6202, 6203, 6209, 721, 7112, 4652, 4651) reported considerable business growths.

Another important issue generated by the current industrial revolution is that of the role and continuing importance of education in the industrial development of mankind and in technological progress in general. Of course that, in the current age of the Internet, it is imperative to combine traditional education with free online courses. For example, Massive Open Online Courses (MOOCs) benefited approximately 45 million young people in 2019 to become educated in many fields. Also, the MIT Sloan School of Management is dedicated to providing high education in the fields of technology, management and digital business strategy.

Of course, in order to improve the educational processes but also the educational results, it is necessary to reform and evaluate them constantly, in order to adapt to the new technological changes and opportunities. The fine lines of this new technological revolution require the fundamental revival of the educational system and a system of personalized learning, in order to meet the present and future changes [15, 16].

2 About European Strategies in the Technology Era with Impact on the Environment

The global economy, and in particular the European economy, are dependent on a flow of natural resources and materials, a dependency that can be a source of vulnerability for the environment, but also a source of insecurity for industry as well as economy.

The use of environmental natural resources, combined with the generation and treatment of waste cause considerable pressures on the environment both in the stages of extraction, production and use, and at the end of products life cycle. That is why, goals such as diminution of the amount of materials used in the economy, using resources rationally, reducing waste and turning it into new resources, have become key objectives of almost all national, regional and global environmental policies. Setting this as a starting point, the concept of circular economy was outlined and a lot of related policies were based on the issue of efficient use of natural resources, rational production and consumption as well as waste management [17].

At this time, modern technologies as well as the strategic incorporation of sustainability issues, allow industrial consumers to use natural resources more efficiently and sustainably throughout their life cycle. By optimizing and streamlining production processes and the flow of materials, materials, energy and waste can be significantly reduced.

In addition, it is a reality that a number of technological innovations as well as the development of alternative materials contribute significantly to improving competition for depleted and often non-renewable natural resources and to creating economic and environmental value [18]. An essential role is played by the development and implementation of intelligent alternative solutions, which help to optimize value processes, increase economic efficiency and also allow a responsible use of resources and materials.

The global environmental issues which humanity is facing today are considerably the result of overexploitation of natural resources and waste generation. It is becoming increasingly clear that the predominant model of economic development, that is based on the intensive use of natural resources in conjunction with pollution and waste generation, has become unsustainable.

Climate change and environmental degradation represent a real existential threat not only to Europe but to all of humanity [19]. That is why this complex phenomenon requires a comprehensive approach focused on concrete actions at international, regional and national level.

The European Union, proved in the recent period to be a major player in the strategic approach to climate change, has initiated, in order to counter this phenomenon, an important strategy, presented in a document entitled the European Green Deal [20], which aims the transition to climate neutrality [21] and thus the transformation of the European Union into a modern, competitive and efficient economy regarding the using of natural resources. Through this document, the climate emergency imposed the goal of carbon neutrality and at the same time credited the ecological transition as a continental strategy for sustainable growth [22].

In this context, we point out that, in order to ensure the achievement of the 2050 target, that of climate neutrality, in September 2020 the European Commission proposed the new Plan to increase the 2030 target to reduce greenhouse gas emissions by at least 55% compared to 1990. Subsequently, on 24 February 2021, the European Commission published the new EU Strategy on Adaptation to the Effects of Climate Change, which sets out a lot of objectives in this regard. In order to achieve these goals, the decarbonization of the energy system is necessary and essential, and to this ended the following steps and measures are required: the evaluation of the National Integrated Energy and Climate Change Plans; a Strategy for the smart integration of the energy sector in other sectors; revision of the TEN-E Regulation/Trans-European Energy Networks; initiative in the building sector – the so-called Renovation wave; wind energy strategy off shore.

Specifically, the European Green Deal provides an action plan in order to stimulate the efficient use of resources by moving towards a circular economy, restoring biodiversity and reducing pollution. Obviously, this document proposes a major change in the entire European society, by turning current environmental issues into real opportunities and transforming the European economic system into a sustainable and equitable one. We underline that radical changes are also needed in terms of energy efficiency to reduce global greenhouse gas emissions [23]. Of course, achieving these goals requires not only rules and standards but also funding and evaluation tools. In this respect, the required investments are indicated as well as the financing instruments available to ensure a fair and inclusive transition.

Obviously, a key role will be played by investments in green technologies, investments in innovation in the industrial sector, those aimed at decarbonising the energy sector and improving the energy efficiency of buildings, as well as those related to green transport. Through the Fair Transition Mechanism, the European Union will provide the financial support as well as the technical assistance for investments for the 2021–2027 period.

In order to ensure the transition to a green economy and also to fight the phenomenon of climate change, digital solutions are considered as they have proven to be extremely efficient. Such solutions can open up new opportunities for businesses, encourage the development of reliable technologies and contribute to a dynamic and viable economy. The aim of the European Union's Digital Strategy is to use this transformation to strengthen digital sovereignty with a clear focus on data, technology and infrastructure.

Thus, in March 2021, the European Council adopted the new Digital Europe Program, which will stimulate digital transformation by funding the implementation of cutting-edge technologies in a number of key areas, such as: artificial intelligence, supercomputing, cybersecurity, advanced digital skills and ensuring the widespread use of digital technologies at all levels of the economy and society. This Program, which has a budget of 7,5 billion euros, will run in the 2021–2027 period and aims to capitalize on digitalization for the benefit of Member States' companies and economies, to increase the autonomy of essential technologies and to strengthen competitiveness as well as the construction of latest generation, secure and high-performance digital services.

3 Conclusion

We can easily say that humanity is living in an age of globalization and advanced technology, that proves to be both beneficial as well as provocative.

Technological development is now present in the entire industrial landscape but also in various other sectors of society which are, of course, affected in one way or another. Certainly, new technologies are essential for economic development, as they generate a considerable impact on the economic activity of companies, in the context of intensive and global competition. However, advanced industrial technologies generate not only the emergence of complex and sophisticated products but also significant changes for markets and for the structure of industries as well as for the environment.

The new industrial revolution, through the significant transformations it generates in most fields, as well as through the felt impact, is a different one from the previous ones, just through the extended environment in which it is to take place. Advanced technologies generate a differential impact on important sectors such as production, transport, services and the environment. And despite the overwhelming benefits of technology, all these sectors will face major challenges in terms of their effective implementation and consequences.

Today, there is growing talk of the importance of low-carbon technologies as technologies that produce less carbon dioxide and are thus positive and desirable. Certainly, through these new technologies that are safe and reliable, the protection and sustainable development of the environment is intensified, and this trend contributes to the fight against climatic changes. On such a consideration, the NER300 Programme was set up, being one of the largest funding programmes worldwide for innovative demonstration projects in the field of "low carbon" technologies. This innovative CO2 funding programme, which brings together 2 billion €, combines low-carbon technologies and focuses on the large-scale demonstration of the capture and storage of green CO2 and innovative commercial-scale renewable energy technologies in the European Union.

From what is set out in the present work, it is clear that in the modern industrial landscape that exists now, interconnected, intelligent, uniform, predictable and proportionate strategies, policies and regulations are required, that are able to stimulate the resource efficiency of the whole European economy and industrial competitiveness, while respecting the environmental objectives.

Moreover, current environmental and industrial policies unfold in a coordinated manner, the sustainable industrial policy of the European Union being focused towards an increased efficiency of resources in industry, towards a progressive innovation and ecological technologies.

References

1. Prisecaru, P.: EU reindustrialization under the requirements of the fourth industrial revolution. Impact Socio-Econ. Technol. Transf. Natl. Eur. Int. Level **10**, 1–14 (2015)
2. Rifkin, J.: The Third Industrial Revolution: How Lateral Power is Transforming Energy, the Economy, and the World. Palgrave Macmillan (2013)
3. Boscoianu, E.-C., Popa, D.: Internet of things. AGIR Bull. **2**, 44–47 (2016)
4. Global Agenda Council on the Future of Software & Society, Survey Report, World Economic Forum, Geneva (2015)
5. Deep Shift-Technology Tipping Points and Societal Impacts, World Economic Forum, Geneva (2015)
6. Dumitrescu, A.L., Prisecaru, P.: The impact of the fourth industrial revolution on employment. J. Glob. Econ. **12**(1), 3–15 (2020)
7. At the International Consumer Electronics Show (CES), held in 2014 in Las Vegas, Ford Company unveiled the first functional prototype of V2V technology
8. Bento, L.C., Parafita, R., Racha, H., Nunes, U.: A study of the environmental impacts of intelligent automated vehicle control at intersections via V2V and V2I communications. J. Intell. Transp. Syst. Technol. Plann. Oper. **23**(1), 41–59 (2019)
9. Huges, J.: The End of Work. Wiley, Hoboken (2007)
10. McGaughey, E.: Will robots automate your job away? Full employment, basic income, and economic democracy. Centre for Business Research, University of Cambridge, Working Paper no. 496, pp. 1–34 (2018)

11. Ford, M.: The Rise of the Robots: Technology and the Threat of Mass Unemployment. Oneworld Publications (2017)
12. Wilkinson, A., Townsend, K.: The Future of Employment Relations. New Paradigms, New Developments. Palgrave McMillian (2014)
13. Frey, C.B., Osborne, M.: The future of employment: how susceptible are jobs to computerisation? Oxford Martin School, University of Oxford, Oxford, OX1 1PT (2013)
14. The Future of Jobs. Employment, Skills and Workforce Strategy for the Fourth Industrial Revolution, Report, World Economic Forum, Geneva (2016)
15. Penprase, B.E.: The Fourth Industrial Revolution and Higher Education, Higher Education in the Era of the Fourth Industrial Revolution. Palgrave Macmillan, Singapore (2018)
16. Hirschi, A.: The fourth industrial revolution: issues and implications for career research and practice. Career Dev. Q. **66**(3), 192–204 (2018)
17. Busu, M.: Adopting circular economy at the European union level and its impact on economic growth. Soc. Sci. **8**(5), 159 (2019)
18. Robinson, Sh., Stubberud, H.A.: Green innovation and environmental impact in Europe. J. Int. Bus. Res. **14**(1), 127–138 (2015)
19. Dogaru, L.: Environmental Law. Pro Universitaria Publishing House, Bucharest (2020)
20. European Commission (COM), European Green Deal - The New EU growth Strategy, (2019)
21. On 15 December 2020, the EU Council adopted the Regulation establishing the framework for achieving the goal of climate neutrality and amending Regulation (EU) 2018/1999 - Climate Law, which represents a European legislation on climate change, which enshrines the goal of achieving climate neutrality in 2050
22. Duţu, M.: Climate Law. Universul Juridic Publishing House, Bucharest (2021)
23. Napoli, G., Barbaro, S., Giuffrida, S., Trovato, M.R.: The European green deal: new challenges for the economic feasibility of energy retrofit at district scale. In: Bevilacqua, C., Calabrò, F., Della Spina, L. (eds.) NMP 2020. SIST, vol. 178, pp. 1248–1258. Springer, Cham (2021). https://doi.org/10.1007/978-3-030-48279-4_116

The Continuous Improvement Cycle Core Activities for the Sustainable Development of Healthcare Facilities

Flaviu Moldovan[1(✉)] and Petruta Blaga[2]

[1] IOSUD Doctoral School, "George Emil Palade" University of Medicine, Pharmacy, Science, and Technology of Targu Mures, Gh. Marinescu str. 38, 540142 Targu Mures, Romania
moldovan.flaviu95@yahoo.com

[2] "George Emil Palade" University of Medicine, Pharmacy, Science, and Technology of Targu Mures, Gh. Marinescu str. 38, 540142 Targu Mures, Romania

Abstract. Globally, health care systems and organizations seek to improve the performance of health systems by implementing patient-centered care models that can be supported by reference frameworks. The adoption of a sustainability assessment framework is the central part of a sustainable development strategy that directly influences the other competitive strategies adopted by a healthcare organization: the professional development strategy, the partnership and networking strategy, the community engagement strategy, the transfer strategy. The aim of this research is to identify the core activities related to a reference framework for sustainable development of healthcare facilities and to establish correspondences with other existing reference frameworks that are used in healthcare. The identified activities are employed to develop a new and innovative framework that will be structured on the 3 pillars of sustainable development: social, economic and environmental, incorporated in the 7 basic topics of social responsibility, which are mentioned in the ISO26000 - Social responsibility standard, adapted to the context of healthcare provision. The new San-Q (Sanitary Quality) quality reference framework is structured in 4 main phases, which correspond to the quality cycle Planning - Implementation - Evaluation - Review, each being divided into two basic activities. These are matched to the evaluation matrices set out in the updated framework for evaluating and improving the quality and safety in European hospitals. The paper describes the content of the 8 basic activities related to the reference framework for sustainable development of healthcare organizations so that they can be further used to build the indicators of the new San-Q reference framework.

Keywords: Healthcare facility · Quality improvement · Sustainable development · Reference framework

1 Introduction

Ensuring good or even excellent quality of healthcare care services requires healthcare systems to collect the health needs and expectations of the population, to ensure good

L. Moldovan and A. Gligor (Eds.): Inter-Eng 2021, LNNS 386, pp. 316–325, 2022.
https://doi.org/10.1007/978-3-030-93817-8_31

governance of the healthcare systems, to establish partnerships with other sectors in order to develop high-performance platforms in providing healthcare.

In Romania, the national legislation on quality assurance in healthcare has important references that have to be used as specifications in the development of reference frameworks of healthcare facilities [1]: Law no. 185/2017 on quality assurance in the healthcare system [2]; The standards, procedure and methodology for the evaluation and accreditation of hospitals, approved by the Order of the Minister of Health no. 446/2017 [3]; Government Decision no. 629/2015 on the composition, attributions, organization and functioning of the National Authority for Quality Management in Health [4]; Order no. 151 (ANMCS - National Authority for Quality Management in Health) dated 18.07.2017 on the approval of working tools used in the second cycle of hospital accreditation [5].

Hospital accreditation standards [3] include 3 references: 1 - Strategic and organizational management; 2 - Clinical management; 3 - Medical ethics and patient rights, which are supplemented by 64 checklists.

The European Commission launched a Youth Health Initiative in 2009, which aims to improve the health and well-being of young people [6]. Actions to promote health, reduce health inequalities and improve protection against health threats have required the development of a framework for interventions on major diseases.

The objectives of this research are to identify the core activities related to a reference framework for sustainable development in healthcare and to establish correspondences with other existing reference frameworks that apply to healthcare.

The research results will be used later in defining and developing an analysis and evaluation model, in the form of a reference framework, for the design, development and implementation, periodic evaluation and continuous improvement of quality management systems and sustainable development in healthcare, at micro system level, which will be also compatible with the applicable national and international standards in the field.

2 Material and Method

The scientific research in this paper has an exploratory character. It was used as research methodology the study of literature relevant to the research topic, in order to create an appropriate framework for conducting the empirical study, and the managerial approach of the researched field, including concepts, methods, techniques and tools specific to quality management in healthcare.

Currently, healthcare facilities face major challenges, as patients demand continuous improvement in the quality of healthcare, while health insurance companies demand lower costs. There are quality improvement programs inspired by the industrial environment, such as "lean manufacturing", which is an excellent tool to meet current health challenges [7]. Some hospitals tend to adopt lean quality improvement programs, which require a lean assessment from a much more critical perspective [8] and quantification of sustainability effects [9].

Healthcare systems administrators have a critical task to identify quality deficiencies in care and address system limitations. The frameworks that facilitate the supervisory function of administrators and the use of mechanisms to improve results remain underdeveloped. Despite a growing database on the effectiveness of certain mechanisms for improving the quality of care, the frameworks that facilitate the supervisory function of administrators and the use of mechanisms to improve results are also underdeveloped. Tello et al. [10] cataloged a wide range of quality care mechanisms, evidence of their effectiveness and mapped them into a two-dimensional framework: (i) governance sub-functions; and (ii) targets for the quality of care mechanisms. It has been identified 128 quality care mechanisms which show that they are more developed for some target areas, such as the healthcare workforce. Within the governance sub-functions, there are several mechanisms and evidence of effectiveness for setting priorities and standards, as well as organizing and monitoring actions.

In a study, Glasgow [11] identified 539 potential articles on quality improvement methodologies and found that the true impact of this approach is difficult to assess, given the lack of a rigorous evaluation or clearly sustained improvements provides little evidence to support widespread adoption. Future research is needed to gather evidence that can support a better understanding of health phenomena.

Globally, healthcare systems and organizations seek to improve the performance of the health systems by implementing patient-centered care models. Santana et al. [12] have used the Donabedian model in order to improve healthcare and have classified the areas of a proposed new model of person-centered care into 3 categories: structure, process, and outcome. In this way they provided the basis for the reference framework on person-centered care, which is made up of the following structural areas: the creation of a person-centered culture care during the provision of continuous healthcare; co-design of educational programs, as well as programs with patients for the promotion and prevention of health; ensure a supportive and accommodating environment; and the development and integration of structures to support health information technology and to measure and monitor the performance of patient-centered healthcare.

Buttigieg et al. [13] have developed an integrated, patient-oriented analytical frame of reference that aims to improve the quality of care in the event of accidents and emergencies. It can be part of Six Sigma and other quality initiatives. Kahn et al. [14] harmonized and organized the existing terms of data quality in a framework by defining three categories of: (1) Compliance (2) Completeness and (3) Plausibility in two contexts of data quality assessment: (1) Verification and (2) Validation.

In a case study from a hospital, Hovlid et al. [15] showed that organizational learning contributes to changing healthcare institutions. Improved understanding of the clinical system was a change in the mental patterns of employees that influenced how the organization changed its performance.

3 Results

3.1 Sustainable Development Strategy of the Healthcare Facility

In addition to environmental education, which is commonly claimed to be at the heart of sustainable development efforts, there is a need to develop appropriate frameworks in organizations that can support sustainability policies [16]. Implementation of a sustainability assessment framework in a healthcare facility is a strategic option, as it supports the sustainable development strategy, which ensures the competitiveness of the organization.

The adoption of a sustainability assessment framework is the central part of a sustainable development strategy that is positively linked to the other competitive strategies adopted by a healthcare organization. It has a central role in the competitive strategy of the organization determining the other competitive strategies: the professional development strategy, the partnership and networking strategy, the community engagement strategy, and the transfer strategy (see Fig. 1).

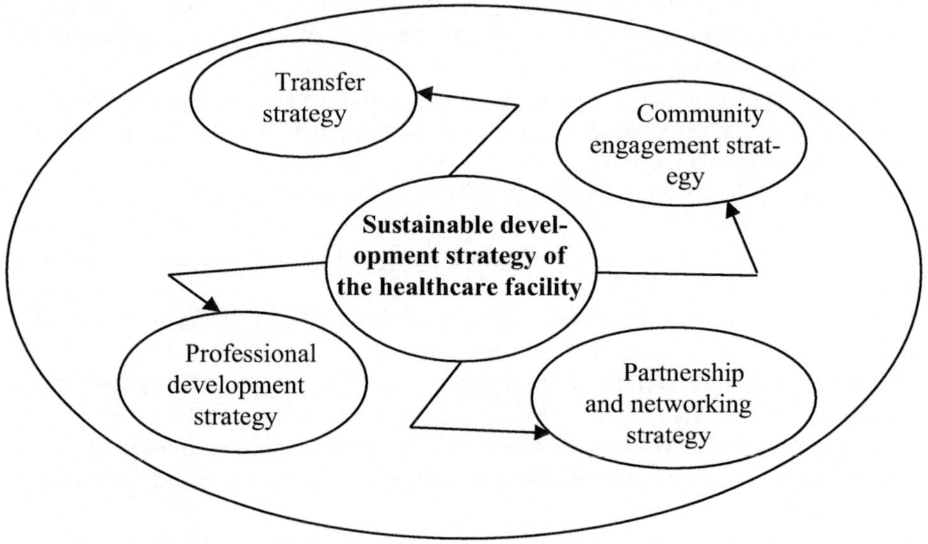

Fig. 1. The relationship between the sustainable development strategy and other competitive strategies of the healthcare facility.

Such strategies describe how the healthcare organization intends to gain advantages over competitors, for example by creating networks of healthcare facilities capable of providing a wide range of medical services.

3.2 Core Activities for the Sustainable Development of the Healthcare Facilities

Regarding the quality cycle references, in order to simplify the relationship with the quality frameworks already developed, the San-Q (Sanitary Quality) quality reference framework is structured based on the quality cycle developed by Deming: Plan Do Check Act (PDCA) – i.e., Planning, Implementation, Assessment and Review (PIAR), using health quality terminology and principles of health quality assurance [1].

Based on the global requirements, standards, procedure and methodology for the evaluation and accreditation of hospitals [3], as well as the ANMCS standards for outpatient health services [17] formulated at national level, but also the regional specificity, the quality reference framework San-Q will be developed, taking into account that the healthcare system can and should play a key role in promoting social cohesion and should also pursue financial sustainability and environmental responsibility.

Quality improvement can promote the association between health security and healthcare systems. By integrating the quality improvement approaches into healthcare systems, the global health security priorities are addressed, which has the effect of improving health outcomes at all levels [18].

The member bodies of the International Organization for Standardization consider that the impact of an organization on the environment has become a critical part of measuring its overall performance and its ability to continue to function effectively, as the economy is dependent on the health of the world's ecosystems. To this end, the ISO 26000 standard "Guidelines on social responsibility" is intended to help organizations to contribute to sustainable development [19]. They are encouraged to go beyond the law, recognizing that law enforcement is a fundamental duty of any organization and an essential part of their social responsibility. It is promoted a common understanding in the field of social responsibility and complementing it with other instruments and initiatives for social responsibility.

In applying ISO 26000, it is advisable for an organization to take into account the societal, environmental, legal, cultural, political and organizational diversity, as well as differences in economic conditions, while being in line with international rules of conduct.

The reason for adopting this conceptual model is based on the principles set out in ISO 26000, as regards the definition of sustainable development and social responsibility:

Sustainable development is "development that meets the needs of the present without compromising the ability of future generations to meet their own needs" [20];

Social responsibility is "responsibility of an organization for the impact of its decisions and activities on society and the environment, through transparent and ethical behavior that:

- contributes to sustainable development, including the health and well-being of society;
- takes into account the expectations of stakeholders;
- complies with applicable law and complies with international rules of conduct; and
- is integrated throughout the organization and practiced in its relationships" [19].

For this reason, the San-Q reference framework is structured on the 3 pillars of sustainable development: social, economic and environmental, incorporated in the 7 core topics of social responsibility that are mentioned in ISO26000 standard [19], adapted to the context of the provision of healthcare): 1 - Organizational governance, 2 - Human rights, 3 - Work practices, 4 - Environment, 5 - Good healthcare practices, 6 - Patient issues, 7 - Community involvement and development.

In this way, the San-Q quality reference framework will be structured in 4 main phases, each corresponding to the 4 phases of the quality cycle: Planning - Implementation - Evaluation - Review (PIER), in terms of healthcare provision: design of the medical services provision, provision of the medical services, the evaluation of the medical services, continuous improvement, as shown in Fig. 2.

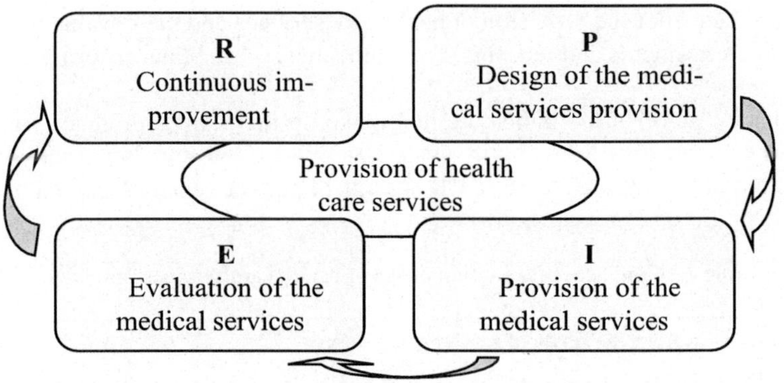

Fig. 2. The quality cycle in the provision of healthcare services.

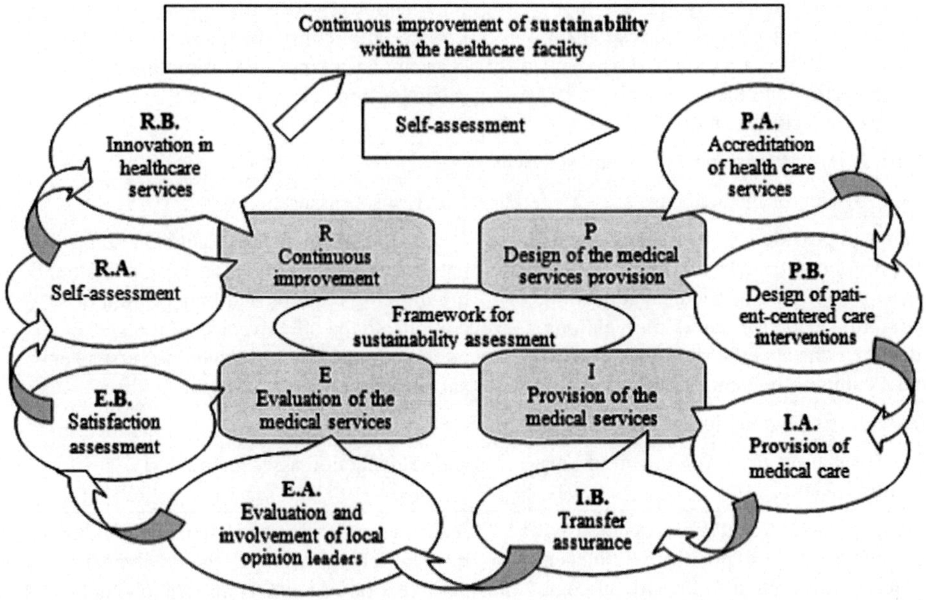

Fig. 3. The core activities of the continuous improvement cycle for the sustainable development of the healthcare facility.

In the Planning phase, the San-Q reference framework includes the design of the provision of healthcare services that covers all aspects related to the definition of the healthcare services offer by the healthcare facilities. The Implementation phase consists in providing healthcare services which in fact is the process of treating patients. In the Evaluation phase, the San-Q reference framework includes the evaluation of healthcare services that contains all aspects related to the evaluation of patient satisfaction, the effectiveness of treatment and the evaluation of medical staff satisfaction. The Review phase takes place after the provision of healthcare services and their evaluation, by conducting self-assessments and redesigning healthcare services, thus ensuring continuous improvement.

In order to establish the indicators that make up the San-Q reference framework, each of the four phases of the quality cycle was divided into two basic activities: P.A. Accreditation of health care services, P.B. Design of patient-centered care interventions,

Table 1. Core activities in each phase of health care services provision.

Phase 1: (P) – Design of the medical services provision	
P.A. Accreditation of health care services	P.B. Design of patient-centered care interventions
It aims to develop the project of accreditation of health care services by defining the organizational structures and responsibilities of medical staff, according to the real needs of the population and the community, as well as by creating a culture of quality	It aims to design patient-centered care interventions by selecting appropriate medical services after consulting patients, training clinical consultants, and promoting patient self-management
Phase 2: (I) – Provision of the medical services	
I.A. Provision of medical care	I.B. Transfer assurance
It aims to provide healthcare services with the support of computerized systems for clinical decisions, continuing medical education, promoting evidence-based interventions adapted to the specific situations of patients and a culture of patient safety	It aims to control critical transition points in patient care during and after hospitalization by highlighting features that may prevent or complicate the effectiveness of transfer interventions in order to avoid adverse events and adverse effects on patients
Phase 3: (E) - Evaluation of the medical services	
E.A. Evaluation and involvement of local opinion leaders	E.B. Satisfaction assessment
It aims to involve health professionals who have the potential to positively influence professional practice and health outcomes and who can provide support for quality improvement strategies, including through their community involvement	It aims to assess patients' satisfaction with the medical service received, the involvement and participation of partners and stakeholders, and to assess the satisfaction of medical staff in terms of working conditions, by ensuring appropriate monitoring mechanisms
Phase 4: (R) – Continuous improvement	
R.A. Self-assessment	R.B. Innovation in healthcare services
It aims to use qualitative and quantitative self-assessment tools, the audit and feedback, which are key tools for quality improvement and can be applied individually or as parts of combined interventions	It aims to use advanced methods to continuously improve the quality of care: Six Sigma, Lean, medical lists, incident reporting/root cause analysis that prevents the recurrence of adverse events and medical errors

I.A. Provision of medical care, I.B. Transfer assurance, E.A. Evaluation and involvement of local opinion leaders, E.B. Satisfaction assessment, R.A. Self-assessment, R.B. Innovation in healthcare services (see Fig. 3).

The core activities of each phase of the continuous improvement cycle for the sustainable development of the health unit and the innovation of the healthcare services represented in Fig. 3 are described in Table 1.

4 Conclusion

The San-Q reference framework is structured in 4 main phases, each of them corresponding to the 4 phases of the quality cycle, regarding the provision of healthcare services. Each phase and corresponding core activity is directly linked to the reference framework established by the Hospital Accreditation Standards.

The extended applicability of the new San-Q reference framework requires first of all a correspondence between the San-Q indicators and the indicators from the Hospital Accreditation Standards, published in the Standards, procedure and methodology for hospital evaluation and accreditation, provided in Order no. 446/2017 issued by the Ministry of Health - Annex 1 [3].

For reasons of compatibility and applicability, the new San-Q reference framework will also correspond to other reference frameworks for the evaluation of medical services. Thus the San-Q framework establishes a correspondence with the evaluation matrices established in the updated framework for the evaluation and improvement of quality and safety in hospitals in Europe DUQuE (Deepening our understanding of quality improvement in Europe) [21]:

Phase 1 - Design of the medical services provision within San-Q, has as correspondents within DUQuE: the evaluation matrix 1 - Accreditation of health care services; the evaluation matrix 8 - Patient-centered care interventions; the evaluation matrix 10 - Performance information. Phase 2 - Provision of the medical services within San-Q, has as correspondents within DUQuE: the evaluation matrix 3 - Continuing medical education; the evaluation matrix 4 - Patient safety culture; the evaluation matrix 5 - Computerized clinical decision support systems; the evaluation matrix 6 - Dissemination and implementation of guidelines. Phase 3 - Evaluation of medical services within San-Q, has as correspondents within DUQuE: the evaluation matrix 7 - Interventions to improve transfers. Phase 4 - Continuous improvement within San-Q, has as correspondents within DUQUE: the evaluation matrix 2 - Effectiveness of local management; the evaluation matrix 9 - Six Sigma and Lean; the evaluation matrix 11 - Audit and feedback; the evaluation matrix 12 - Reporting hospital incidents; the evaluation matrix 13 - Safety checklists; the evaluation matrix 14 - educational visits.

The content of the 8 core activities of the continuous improvement cycle for the sustainable development of healthcare facilities, described in this research, facilitates the indicators construction of the new San-Q reference framework for sustainable development.

References

1. Moldovan, F.: Framework specifications for evaluation of quality improvement and sustainable development in healthcare facilities. In: Multidisciplinary Digital Publishing Institute Proceedings, vo. 63, no. 1, p. 2 (2020)
2. Legislative Portal, Legea nr. 185/2017 privind asigurarea calitatii in sistemul de sanatate (Law no. 185/2017 on quality assurance in the health system). http://legislatie.just.ro/Public/DetaliiDocument/225750. Accessed 10 June 2021
3. Ministry of Health Homepage, Ordin Nr. 446/2017 privind aprobarea Standardelor, Procedurii si metodologiei de evaluare si acreditare a spitalelor (Order No. 446/2017 on the approval of the Standards, Procedure and methodology for the evaluation and accreditation of hospitals). https://anmcs.gov.ro/web/wp-content/uploads/2017/04/OrdinMS-446-2017-StandardeEd.II_.pdf. Accessed 10 June 2021
4. Legislative Portal, Hotararea Guvernului nr. 629/2015 privind componenta, atributiile, modul de organizare si functionare ale Autoritatii Nationale de Management al Calitatii in Sanatate (Government Decision no. 629/2015 on the composition, attributions, organization and functioning of the National Authority for Quality Management in Health). https://lege5.ro/Gratuit/g42donrzg4/hotararea-nr-629-2015-privind-componenta-atributiile-modul-de-organizare-si-functionare-ale-autoritatii-nationale-de-management-al-calitatii-in-sanatate. Accessed 10 June 2021
5. ANMCS Homepage, Ordin nr. 151/2017 privind aprobarea instrumentelor de lucru utilizate in cadrul celui de al II-lea ciclu de acreditare a spitalelor (Order no. 151/2017 on the approval of working tools used in the second cycle of hospital accreditation). https://anmcs.gov.ro/web/wp-content/uploads/2017/07/Ordin-instrumente-evaluare-ANMCS-2017.pdf. Accessed 10 June 2021
6. European Commission Homepage, 'Be Healthy, be yourself' – Commission launches youth health initiative. https://ec.europa.eu/commission/presscorner/detail/en/IP_09_1109. Accessed 10 June 2021
7. Van den Heuvel, J., Does, R., de Koning, H.: Lean six sigma in a hospital. Int. J. Six Sigma Competit. Adv. **2**(4), 377–388 (2006)
8. de Souza, L.B.: Trends and approaches in lean healthcare. Leadersh. Health Serv. **22**(2), 121–139 (2009)
9. Kruk, M.E., et al.: High-quality health systems in the sustainable development goals era: time for a revolution. Lancet Glob. Health **6**(11), 1–57 (2018)
10. Tello, J.E., Barbazza, E., Waddell, K.: Review of 128 quality of care mechanisms: a framework and mapping for health system stewards. Health Policy **124**(1), 12–24 (2020)
11. Glasgow, J.M.: Guiding inpatient quality improvement: a systematic review of lean and six sigma. Joint Commission J. Qual. Patient Saf. **36**(12), 533–540 (2010)
12. Santana, M.J., Manalili, K., Jolley, R.J., Zelinsky, S., Quan, H., Lu, M.: How to practice person-centred care: A conceptual framework. Health Expect. **21**(2), 429–440 (2018)
13. Buttigieg, S.C., Dey, P.K., Cassar, M.R.: Combined quality function deployment and logical framework analysis to improve quality of emergency care in Malta. Int J Health Care Qual Assur. **29**(2), 123–140 (2016)
14. Kahn, M.G., et al.: A harmonized data quality assessment terminology and framework for the secondary use of electronic health record data. EGEMS (Wash DC). **4**(1), 1244 (2016)
15. Hovlid, E., Bukve, O., Haug, K., Aslaksen, A.B., von Plessen, C.: Sustainability of healthcare improvement: what can we learn from learning theory? BMC Health Serv. Res. **12**, 235 (2012)
16. Moldovan, F.: New approaches and trends in health care. Proc. Manuf. **22**, 947–951 (2018)
17. ANMCS Homepage, Ordinul Presedintelui Autoritatii Nationale de Management al Calitatii în Sanatate nr. 353/09.10.2019 privind aprobarea Standardelor Autoritatii Nationale de Management al Calitatii în Sanatate pentru serviciile de sanatate acordate în regim ambulatoriu

(Order of the President of the National Authority for Quality Management in Health no. 353/09.10.2019 on the approval of the Standards of the National Authority for Health Quality Management for health services provided on an outpatient basis). https://anmcs.gov.ro/web/standarde-ambulatoriu/. Accessed 10 June 2021
18. Mensah Abrampah, N., et al.: Quality improvement and emerging global health priorities. Int. J. Qual. Health **30**(1), 5–9 (2018)
19. ISO Homepage, ISO26000 - Guidelines on social responsibility. https://www.iso.org/iso-26000-social-responsibility.html. Accessed 10 June 2021
20. Shah, M.M.: Sustainable development. In: Jorgensen, S.E., Fath, B.D. (eds.) Encyclopedia of Ecology, pp. 3443–3446. Elsevier Science (2008)
21. Groene, O., Kringos, D., Sunol, R.: on behalf of the DUQuE Project. Seven ways to improve quality and safety in hospitals. An evidence-based guide. DUQuE Collaboration (2014). http://www.duque.eu/uploads/DUQuE_Seven_Ways_To_Improve_Quality_And_Safety_2014.pdf. Accessed 10 June 2021

Organizational Governance Assessment of Healthcare Facilities for Sustainable Development

Flaviu Moldovan[1(✉)] and Petruta Blaga[2]

[1] IOSUD Doctoral School, "George Emil Palade" University of Medicine, Pharmacy, Science, and Technology of Targu Mures, Gh. Marinescu str. 38, 540142 Targu Mures, Romania
moldovan.flaviu95@yahoo.com

[2] "George Emil Palade" University of Medicine, Pharmacy, Science, and Technology of Targu Mures, Gh. Marinescu str. 38, 540142 Targu Mures, Romania

Abstract. The new San-Q (Sanitary Quality) framework for sustainable development evaluates the 3 pillars of sustainable development: social, economic and environmental, incorporated in the 7 basic topics of social responsibility adapted to the context of healthcare provision: Organizational governance, Human rights, Labor practices, Environment, Fair operating practices, Consumer issues, Community involvement and development. The objective of this study is to design performance indicators that assess the organizational governance aspects of the San-Q sustainable development framework and their validation in practice in a healthcare facility. The design of performance indicators was carried out following a qualitative study in the scientific literature for healthcare facilities that are representative as medical performance, having different levels of human capital and forms of public/private property, from which emerged the most relevant practices confirmed by metastudies published in the PubMed database. It was designed 8 indicators, which are evaluated on two combined scales that describe the fulfillment degrees of the performance indicators, respectively the significance of the indicators for the evaluated healthcare facility. Each scale has six evaluation steps, and the evaluation grid of each indicator describes the degrees of fulfillment, which allows an easy evaluation. The 8 performance indicators designed to assess responsibility for organizational governance, as a component of the new framework for sustainable development, are validated in practice in a hospital.

Keywords: Healthcare facility · Quality improvement · Sustainable development · Organizational governance · Reference framework

1 Introduction

The new San-Q (Sanitary Quality) sustainable development framework is structured in 4 main phases, each corresponding to the 4 phases of the quality cycle: Planning - Implementation - Evaluation - Review (PIER), in terms of healthcare provision: design of the medical services provision, provision of the medical services, the evaluation of the

L. Moldovan and A. Gligor (Eds.): Inter-Eng 2021, LNNS 386, pp. 326–347, 2022.
https://doi.org/10.1007/978-3-030-93817-8_32

medical services, continuous improvement, which evaluates the 3 pillars of sustainable development: social, economic and environmental, incorporated in the 7 basic topics of social responsibility adapted to the context of healthcare provision: organizational governance, human rights, labor practices, environment, fair operating practices, consumer issues, community involvement and development [1].

In order to establish the indicators that make up the San-Q reference framework, each of the four phases of the quality cycle was divided into two basic activities: P.A. Accreditation of health care services, P.B. Design of patient-centered care interventions, I.A. Provision of medical care, I.B. Transfer assurance, E.A. Evaluation and involvement of local opinion leaders, E.B. Satisfaction assessment, R.A. Self-assessment, R.B. Innovation in healthcare services [2].

The objective of this study is to design performance indicators that assess the organizational governance component of the San-Q sustainable development framework and their validation in practice by application in a hospital.

2 Material and Method

In order to design the performance indicators for the organizational governance component of the new reference framework for sustainable development, we conducted a qualitative study in the scientific literature, mainly from the PubMed database, for healthcare facilities considered representative as medical performance, which have different levels of human capital and forms of public/private ownership, from which we extracted the most relevant sustainable practices confirmed in various metastudies.

The collection of quality and sustainability approaches developed in the selected studies was performed through a comparative analysis to achieve a common reference framework.

We examined how healthcare facilities have implemented quality management systems and organizational governance activities for sustainable development, and we found different approaches that can help other healthcare facilities to find a common way to implement sustainability. The key aspects collected and used in our methodology are:

- Vision/mission/objectives related to quality and sustainability assessment;
- Previous experience of the healthcare facility;
- Institutional context and key issues on the transition from sustainability assessment to sustainable development; and
- Key success factors.

2.1 Organizational Governance Practices

The exploration of the scientific literature allowed us to identify the successful practices for organizational governance, which are designed and verified in practice.

In a meta-study [3] by Greenfield et al., which identified and analyzed 66 scientific articles examining the impact of accreditation, have classified the results into 10 categories. Of these, for two categories: promotion of change and professional development, consistent findings are presented regarding the impact of accreditation.

Inconsistent results were identified in five other categories: attitudes of the professions towards accreditation, organizational impact, financial impact, quality measures and program evaluation, while for the other three categories: patient satisfaction, public communication and inspectors' problems - there were no enough studies to draw a conclusion.

The study conducted by Flodgren et al. [4] could not draw firm conclusions on the effectiveness of the external inspection on compliance with the standards. The analysis performed by Brubakk et al. [5] found no evidence to support that hospital accreditation and certification leads to relevant changes in the quality of care, which is measured by quality metrics and standards.

The meta-study performed by Avia and Harivati [6], which includes 11 articles, indicates that accreditation has a positive impact on the quality of healthcare, the benefits of accreditation having the effects of improving teamwork and productivity, and also identifying real quality improvements related to leadership, commitment and support. The quality of information has also been improved through more efficient communication.

Hospital accreditation has a positive impact on efficiency, effectiveness, safety, and patient orientation [7]. The study conducted by Andres et al. [8] concludes that the hospital accreditation process can contribute to changes in staff perceptions of organizational culture, with different views of organizational culture between professional groups.

Accreditation is a critical and comprehensive review of the hospital, which includes areas that are often neglected. During the preparation and evaluation periods, it consumes some of the time that could be devoted to patient care, but creates organizational bases for future quality improvement initiatives [9]. There is considerable evidence that accreditation programs improve the clinical outcomes of a wide range of clinical specialties [10] and should be supported as tools to improve the quality of healthcare services.

Some studies compare the quality of care in accredited and non-accredited hospitals for different medical specialties and conclude that general hospital and subspecialty accreditation programs for: acute myocardial infarction [3, 10], traumatology [3], outpatient surgical care [4], infection control [4], improve the care process provided by healthcare services as it is improved the structure and organization of healthcare institutions.

A meta-analysis of 78 articles concludes that accreditation is increasingly used as a tool to improve the quality of healthcare in low- and middle-income countries, which have established national accreditation programs for hospitals, adapted to their national contexts [11].

There are a limited number of individual studies that have examined the effectiveness of public communication of performance information on patient and healthcare provider behavior or the quality of health care. Studies have confirmed that feedback on the physician's clinical performance results in improved performance [12]. Publication of these data may have differential effects on disadvantaged populations [13].

Several studies have shown that the use of feedback reports combined with other implementation strategies, such as education or the use of quality improvement plans, has led to improved outcomes in the healthcare process [14].

The effectiveness of quality patient-centered interventions was investigated in randomized controlled trials and controlled clinical trials. The effectiveness of training interventions has been demonstrated, the distribution of informative material that have positive effects on medical consultation processes regarding: increasing the detection of psychological suffering; increasing the proportion of medical consultations in which all the patient's health problems were analyzed; improving the patient's perception of disease-specific information [15].

Mixed effects on medical consultation processes were also identified, determined by medical staff regarding: patient-centered communication behavior; empathy skills; the use of various data collection skills, as well as co-decision-making through patient involvement [16].

There were positive effects of patient satisfaction on: the art of care; technical quality of care; assessing total satisfaction [17], developing measures to monitor safe and effective person- and family-centered transitions from hospital to home [18].

There has been a clear improvement in health in reducing emotional distress for patients with this diagnosis [19]. Patient-centered care has also reduced the length of stay in intensive care, but not mortality [20]. There were no effects on general health measures or mixed effects on physiological health measures [21].

Clinicians can select a patient-centered approach that best suits their patient's needs and ensure that it meets the three essential elements of patient-centered care: communication, partnership, and health promotion [22].

Computerized support systems for clinical decisions have been evaluated through numerous studies, often randomized controlled trials that provide strong arguments for their effectiveness [23, 24]. Studies have shown that computerized clinical decision support systems, both commercially developed and locally developed, lead to:

Substantial increases in identifying their events and adverse drug rates [25];

Consistent improvement of preventive care services [26];

Adequacy of treatment and therapy ordered by providers in terms of conducting clinical trials [27], reduction of costs and hospitalization costs [28, 29];

Improving morbidity outcomes, although studies have a limited ability to detect significant clinical differences in mortality [29].

A number of studies have evaluated the effectiveness of interventions aimed at improving hospital transfers. The transfer technology solutions used can lead to the prevention and reduction of adverse events and better satisfaction with the quality of the transfer [30].

The healthcare study reveals that supplementing verbal communication with a written environment leads to improved information retention [30]. The "White Papers" and the "Clinical Consensus Statements" [31] allow for effective verbal exchange, with minimal interruptions and the allocation of question time, focusing on sick patients and the actions needed to be taken. However, the contents must be continuously updated to ensure the communication of the latest clinical information [30].

2.2 Evaluation Indicators

Based on the study carried out in the specialized medical literature from the PubMed database, the indicators of the San-Q sustainable development reference framework were designed (see Table 1).

Table 1. San-Q sustainable development framework indicator matrix – regarding the responsibility for organizational governance.

The phases of the quality cycle in the healthcare facility	Performance indicators	
P Design of the medical services provision	P.A.1 Decision-making processes and structures	P.B.1 Design of quality assurance processes
I Provision of the medical services	I.A.1 Computerized support systems for clinical decisions	I.B.1 Evaluation mechanisms
E Evaluation of the medical services	E.A.1 Existence and recognition of local opinion leaders	E.B.1 Assignment of monitoring mechanisms
R Continuous improvement	R.A.1 Self-assessment tools	R.B.1 Changes to healthcare services

Each indicator has a description which is developed following the conclusions drawn from specialized medical studies. Then, by using the description of the indicator we built questions for its evaluation (Tables 2, 3, 4, 5, 6, 7, 8 and 9).

We have built the evaluation tool on a scale with 6 steps for values 0, …, 5 which correspond to the performance levels: not applicable, weak, satisfactorily, good, very good and excellent. For each fulfillment degree of the projected indicators there are developed descriptions of the contents, which allow an easy evaluation.

Table 2. Evaluation grid of the indicator P.A.1 - decision processes and structures.

Indicator	P.A.1 - Decision processes and structures
Description of the indicator	Definition of organizational structures, responsibilities among staff and schemes / references used for accreditation containing: scope and method of establishment, documentation and implementation in relation to accreditation requirements (standards) and assessment tools.
Evaluation questions	• Are the organizational structures involved in the accreditation of the healthcare facility defined? • Are the responsibilities of the persons involved in accreditation defined? • Are the normative requirements (accreditation / reference schemes) for accreditation identified? • Is the accreditation scope established? • Are the activities carried out to establish the quality management system according to the identified requirements (standards) and the evaluation tools applied? • Is the quality management system documented according to the identified requirements? • Is the quality management system implemented according to the elaborated documentation?

Value [G]	Fulfillment degree	Description
0	Not applicable	X
1	Weak	The organizational structures involved in the accreditation of the healthcare facility are defined. The responsibilities of the persons involved in the accreditation are defined.
2	Satisfactorily	The normative requirements, accreditation / reference schemes for accreditation are identified and the accreditation domain of the healthcare facility is established. The quality management system is documented.
3	Good	The healthcare facility is being accredited. The quality management system is implemented and evaluation and monitoring tools are applied.
4	Very good	The healthcare facility is accredited with low confidence. The quality management system is implemented and is certified.
5	Excellent	The healthcare facility is accredited with a high degree of confidence. The quality management system is implemented and certified.

Table 3. Evaluation grid of the indicator P.B.1 - design of quality assurance processes.

Indicator	P.B.1 - Design of quality assurance processes
Description of the indicator	Patient-centered care seeks to improve: – consultation processes; – the experiences of patients and caregivers; – knowledge of patients; – use of healthcare services; – health and well-being.
Evaluation questions	• Does the design of medical services consider patient-centered care? • Are there improvements in services as a result of the application of patient-centered care? • Is patient consultation improved through focused care? • Are patient experiences improved as a result of patient-centered care?

Value [G]	Fulfillment degree	Description
0	Not applicable	X
1	Weak	There are patient-centered care approaches but they are not input into the design of healthcare services.
2	Satisfactorily	The approach in the healthcare facility is patient-centered, by applying three basic conditions that create an atmosphere conducive to change and therapeutic growth: unconditional positive image, empathic understanding, and congruence. The design of medical services takes into account patient-centered care.
3	Good	The consultation is patient-centered, through communication, setting priorities and working in partnership to assess his circumstances. There are improvements in medical services as a result of the application of patient-centered care.
4	Very good	Patient-centered care creates a partnership between the healthcare staff and the patient, and by developing and applying the consultation specific to the general practice, effective doctor-patient relationships are obtained, respecting the principle of patient autonomy. Patient consultation is enhanced by the application of patient-centered care.
5	Excellent	Patient-centered care develops the skills of continuous care, which arise from the needs of the patient and provides continuous and coordinated care. Patient experiences are enhanced as a result of patient-centered care.

Table 4. Evaluation grid of the indicator I.A.1 - computerized support systems for clinical decisions.

Indicator	I.A.1 - Computerized support systems for clinical decisions
Description of the indicator	The various functions in the system have at their disposal computerized support systems for clinical decisions, related to: – diagnosis; – reminders for preventive care or disease management; – dosing and prescribing drugs.
Evaluation questions	• Do healthcare professionals have computer support systems available for clinical decisions? • Are computer support systems used to diagnose patients? • Do computer support systems provide reminders for preventive care or disease management? • Are computerized support systems used for dosing and prescribing medications?

Value [G]	Fulfillment degree	Description
0	Not applicable	X
1	Weak	Information systems are used to manage administrative and / or financial data.
2	Satisfactorily	Healthcare professionals have computerized support systems for clinical decisions that allow a retrospective analysis of clinical data.
3	Good	Computerized support systems provide reminders for preventive care and disease management. Computerized support systems are used to diagnose patients.
4	Very good	Clinical decision support systems are used to assist in the process of establishing the diagnosis and forming the conclusion. Based on the patient's condition data, which are provided by the doctor, the system offers the best solution.
5	Excellent	Computerized support systems are used for: • assisting the process of diagnosing patients; • monitoring the anamnesis, clinical manifestations; paraclinical investigations, treatment and rehabilitation; • planning the necessary resources for treating patients; • training of medical specialists; • prophylaxis.

Table 5. Evaluation grid of the indicator I.B.1 - evaluation mechanisms for transfers.

Indicator	I.B.1 - Evaluation mechanisms for transfers
Description of the indicator	Development and validation of self-assessment and peer quality assessment hospital transfers is important and can be incorporated into certification programs. Medical professional organizations can also serve as strong mediators of change, for example by increasing the visibility of transfers and by mobilizing research funds.
Evaluation questions	• Is the quality of hospital transfers assessed? • Are medical professional organizations involved in transfer changes? Examples are: increasing visibility, allocating research funding.

Value [G]	Fulfillment degree	Description
0	Not applicable	X
1	Weak	The healthcare facility self-assesses the quality of hospital transfers.
2	Satisfactorily	A peer review of hospital transfers is performed.
3	Good	Hospital transfers are incorporated into the unit's certification programs.
4	Very good	Medical professional organizations contribute to increasing the visibility of transfers.
5	Excellent	Medical professional organizations mobilize research funds for transfers.

Table 6. Evaluation grid of the indicator E.A.1 - existence and recognition of local opinion leaders.

Indicator	E.A.1 - Existence and recognition of local opinion leaders
Description of the indicator	There are local opinion leaders. Local opinion leaders are healthcare professionals named by their colleagues as "educationally influential".
Evaluation questions	• Does the institutional culture have as values the recognition of local opinion leaders? • Are there local opinion leaders? • Are they recognized in the community?

Value [G]	Fulfillment degree	Description
0	Not applicable	X
1	Weak	Institutional culture has as values the recognition of local opinion leaders.
2	Satisfactorily	Within the healthcare facility there are professionals who could become local opinion leaders.
3	Good	Within the healthcare facility there are professionals who are appreciated by the professional community as local opinion leaders.
4	Very good	Local opinion leaders are involved in the continuous improvement of medical services.
5	Excellent	Local opinion leaders are recognized in the community for their professionalism and contribution to improving medical services.

Table 7. Evaluation grid of the indicator E.B.1 - assignment of monitoring mechanisms.

Indicator	E.B.1 - Assignment of monitoring mechanisms
Description of the indicator	Designation of responsible persons and periodicity of the satisfaction assessment.
Evaluation questions	• Is there a specific procedure for measuring satisfaction? • Are persons / responsibilities designated to measure satisfaction? • Is the periodicity of evaluations established?

Value [G]	Fulfillment degree	Description
0	Not applicable	X
1	Weak	Measuring the satisfaction of patients and medical staff is a concern of the healthcare facility.
2	Satisfactorily	There is a specific procedure of the healthcare facility for measuring the satisfaction of patients and medical staff.
3	Good	Persons / responsibilities for measuring satisfaction are designated.
4	Very good	The periodicity of the evaluations is established and they are carried out according to the planning.
5	Excellent	Following the evaluations, conclusions and suggestions for improvement on monitoring mechanisms are formulated.

Table 8. Evaluation grid of the indicator R.A.1 - self-assessment tools.

Indicator	R.A.1 - Self-assessment tools
Description of the indicator	List of existing qualitative and quantitative self-assessment tools.
Evaluation questions	• Are the self-assessment tools used? • What are the quantitative tools for self-assessment? • What are the qualitative tools for self-assessment?

Value [G]	Fulfillment degree	Description
0	Not applicable	X
1	Weak	The healthcare facility has tools for quantitative self-assessment of economic, environmental and social performance.
2	Satisfactorily	The healthcare facility has tools for qualitative self-assessment of economic, environmental and social performance.
3	Good	Quantitative self-assessment tools of economic, environmental and social performance are used for periodic assessments.
4	Very good	Qualitative self-assessment tools of economic, environmental and social performance are used for periodic assessments.
5	Excellent	Quantitative and qualitative self-assessment tools are constantly being improved.

Table 9. Evaluation grid of the indicator R.B.1 - changes to healthcare services.

Indicator	R.B.1 - Changes to healthcare services
Description of the indicator	Changes are made based on suggestions from medical staff, patients, clients and stakeholders.
Evaluation questions	• What new medical services can be included in the health unit's portfolio? • What is the development potential of the medical services offered?

Value [G]	Fulfillment degree	Description
0	Not applicable	X
1	Weak	Medical staff, patients, clients and stakeholders make proposals to change medical services.
2	Satisfactorily	Proposals to change the medical services are analyzed by the management of the healthcare facility and proposals with potential for improvement are selected.
3	Good	New medical services are proposed that can be included in the portfolio of the healthcare facility.
4	Very good	There is potential for the development of the medical services offered in terms of how to issue a prescription and how to provide treatments.
5	Excellent	The use of digital media is encouraged in the various phases of an integrated vision of healthcare: disease prevention, early detection and diagnosis, treatment, healthcare and monitoring. Caregivers and health professionals are trained, and information is provided to patients.

The employment of the indicators presented in Tables 2, 3, 4, 5, 6, 7, 8 and 9 allow the evaluation of the organizational governance within the healthcare facility, following the four stages of the quality cycle (see Fig. 1): Planning - P.A.1 Decision-making processes and structures; P.B.1 Design of quality assurance processes; Implementation – I.A.1

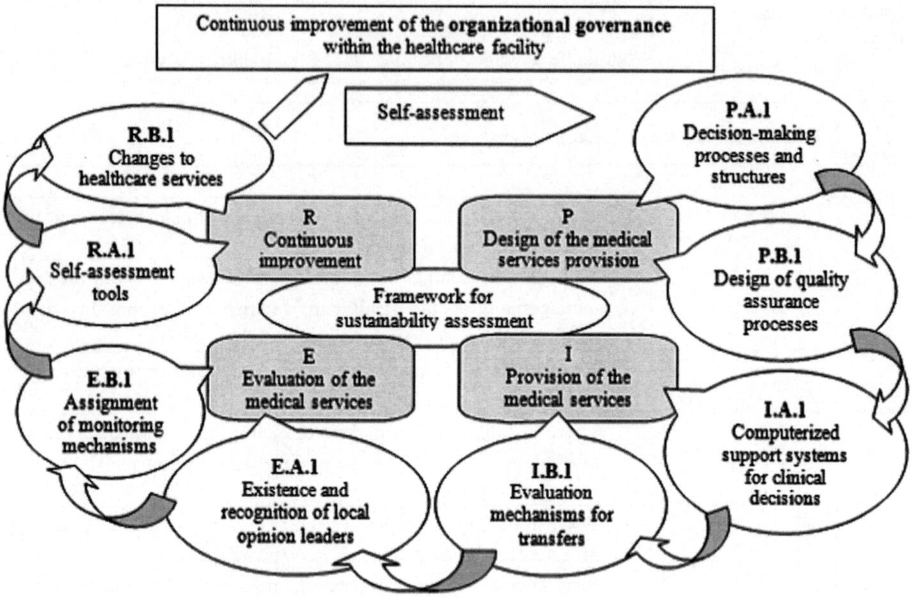

Fig. 1. The cycle of continuous improvement of organizational governance within the healthcare facility.

Computerized support systems for clinical decisions; I.B.1 Evaluation mechanisms for transfers; Evaluation - E.A.1 Existence and recognition of local opinion leaders; E.B.1 Assignment of monitoring mechanisms; Review - R.A.1 Self-assessment tools; R.B.1 Changes to healthcare services.

Indicators are also assessed by their significance for the healthcare facility, which may have the values 0,…, 5, corresponding to the significance categories: not applicable, insignificant, low significant, significant, very significant, high significant (see Table 10).

Table 10. Significance of indicators.

Value [S]	Significance category	Description
0	Not applicable	X
1	Insignificant	The subject is of little importance for the healthcare facility and there is a marginal tendency for evaluation
2	Low significant	Failure to comply with the requirement could adversely affect the activity of the healthcare facility
3	Significant	Failure to comply with the requirement could compromise the activity of the healthcare facility. It is essential to meet the requirement for the provision of healthcare
4	Very significant	Failure to meet the requirement could compromise the successful provision of healthcare Meeting the requirement is essential for the successful delivery of healthcare
5	High significant	Failure to comply with the requirement may even compromise the existence of the healthcare facility

3 Results

Validation in practice of the eight performance indicators designed to assess responsibility for organizational governance, as a component of the new framework for sustainable development San-Q, was achieved through a study at the Emergency County Clinical Hospital Targu Mures (ECCHTM) [32].

Regarding the assessments made following the evaluation of the indicators are:

Indicator - P.A.1 Decision-Making Processes and Structures – ECCHTM has defined the organizational structures involved in the accreditation of the hospital. On the organizational chart of the hospital there is a Service for the quality management of the medical services [33], being refined the responsibilities of the persons involved in accreditation. Targu Mures County Emergency Clinical Hospital was evaluated in 2018 based on the Order of the President of ANMCS no. 10/2018, being included in Category IV - Accredited with low confidence, according to the Order of the President of ANMCS no. 222 of 12.06.2019 [34]. The classification of the hospital in this category was due to the degree of compliance of 85.59% with the sanitary operation authorization.

ECCHTM has implemented and certified the quality management system in accordance with the requirements of the standard SR EN ISO 9001: 2015, for the domain [35]: *"Medical radiology and imaging laboratory, pathological anatomy service, and clinical trials with therapeutic benefit"*. The medical analysis laboratory is RENAR accredited according to SR EN 15189:2007 and is competent to perform medical analysis activities.

Indicator - P.B.1 Design of Quality Assurance Processes – The consultation is patient-centered and there are improvements in medical services as a result of the application of patient-centered care. The diagnostic and/or therapeutic protocols within the hospital are continuously reviewed supported on a system procedure regarding the implementation of work protocols formalized on medical activities in each section/compartment of the hospital, based on which decisions are made on the clinical management of a disease. A number of 177 diagnostic and therapeutic protocols are implemented.

Indicator - I.A.1 Computerized Support Systems for Clinical Decisions – Computer systems are used to manage administrative and financial data. The Information and Technology department installed fiber optic sections for back-up in important locations; expanded the local data network by more than 600 m, in redeveloped sections and in new locations, due to increasing computerization needs; the fleet of printers and multifunctionals has been supplemented; health card readers have been supplemented, installed and configured; new software licenses were purchased; over 400 computers were repaired. With this support, healthcare professionals have at their disposal computer systems for clinical decisions that allow a retrospective analysis of clinical data and allow preventive care and disease management.

Indicator - I.B.1 Evaluation Mechanisms for Transfers – In 2020, 63,220 cases were presented at Emergency department UPU-SMURD within ECCHTM, of which 11,289 cases were hospitalized and 10,207 cases were transferred. An evaluation of hospital transfers is performed.

Indicator - E.A.1 Existence and Recognition of Local Opinion Leaders – Recognition of local opinion leaders is achieved through administrative mechanisms. The head doctor of the department, the person in charge of quality management nominated within the department and the medical director, are appreciated as local opinion leaders with professional influences on the medical staff in the department. They are involved in the continuous improvement of medical services.

Indicator - E.B.1 Assignment of Monitoring Mechanisms – The measurement of patient satisfaction as well as medical staff is performed through a specific procedure of the healthcare facility based on the questionnaires accessible online, and it is published the results analysis of the satisfaction questionnaires.

Indicator - R.A.1 Self-assessment Tools – ECCHTM is a public institution financed entirely from its own revenues, which operates on the principle of financial autonomy. The own revenues come from the amounts collected for medical services, contract-based benefits and other sources, the amount of revenues for 2020 being 606,050,363 lei. The expenses recorded are personnel expenses (69.54% of the total expenses), expenses with goods and services.

Economic performance is assessed using financial indicators: equity 262,966,490 lei; the overall solvency ratio = 3.75 reflects the hospital's long-term ability to pay its debts at maturity; current liquidity = 2.30 reflect the hospital's ability to meet its short-term obligations; immediate liquidity = 1.41 shows the hospital's ability to pay current debts from current cash in accounts and short-term receivables.

Indicator - R.B.1 Changes to Healthcare Services – Implementing the rules and procedures of good medical practice and monitoring their compliance, implementing therapies tailored to the patient's objective needs, health parameters, pre-existing administrative and material conditions, monitoring their changes and real-time adaptation of therapy to the changed conditions is in charge of the ECCHTM Drug Commission.

At the request of the General Surgery Clinical Section II, the ECCHTM Antibiotic Therapy Commission has initiated the introduction of a new antimicrobial drug, Zolinef (Cefazolin), an antibiotic used to treat infections: skin and soft tissue infections, bone and joint infections, during or after surgery, in order to prevent possible infections.

The ECCHTM Pharmacovigilance Commission monitors, coordinates and methodologically guides the implementation of the principles of pharmacovigilance within the hospital, which ensure the achievement of the institution's objectives in a safe way for the patient, staff and institution in a traceable way. It aims to detect early side effects and drug interactions; monitoring of known adverse reactions; identification of risk factors and underlying adverse reactions; analysis and dissemination of information necessary for the correct prescription and regulation of drug circulation; rational and safe use of medicines. In 2020, there were no reports of adverse drug reactions from hospital sections/compartments.

The effectiveness of the quality management system in the field of blood transfusion, as well as the identification of nonconformities and potential risks in carrying out the transfusion activity in order to increase the quality of medical care and safety in the field of blood transfusions in the hospital is analyzed by the Hemovigilance Commission. The total number of transfused patients in 2020 is 2692.

The values of the indicators related to the responsibility regarding organizational governance are presented in the self-assessment tool in Table 11.

Table 11. Self-assessment tool regarding the responsibility for organizational governance.

No. crt.	Symbol and name of the indicator	Significance Si	Fulfillment degree Gi	Quality indicator $QI_i = S_i \times G_i$
1	P.A.1 Decision-making processes and structures	5	4	20
2	P.B.1 Design of quality assurance processes	3	3	9
3	I.A.1 Computerized support systems for clinical decisions	2	3	6
4	I.B.1 Evaluation mechanisms for transfers	3	2	6
5	E.A.1 Existence and recognition of local opinion leaders	3	5	15

(*continued*)

Table 11. (*continued*)

No. crt.	Symbol and name of the indicator	Significance Si	Fulfillment degree Gi	Quality indicator $QI_i = S_i \times G_i$
6	E.B.1 Assignment of monitoring mechanisms	2	3	6
7	R.A.1 Self-assessment tools	2	1	2
8	R.B.1 Changes to healthcare services	3	4	12

The fulfillment degree of the indicators related to the responsibility regarding organizational governance, on the scale from 0 to 5, is represented graphically in Fig. 2.

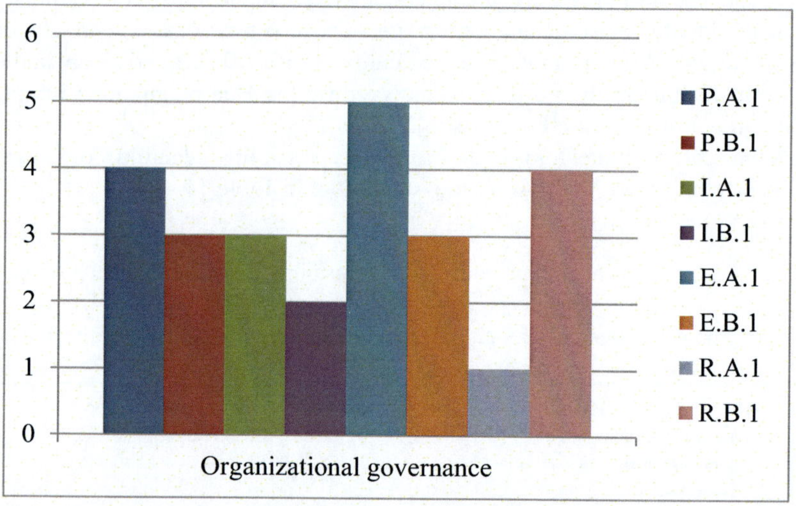

Fig. 2. The fulfillment degree of the indicators related to the responsibility regarding the organizational governance of the hospital.

The indicator RA1 Self-assessment tools, has the value for fulfillment degree 1, the lowest in this group, while the highest fulfillment degree 5 is recorded for the indicator EA1 Existence and recognition of local opinion leaders.

The results for the performance indicators regarding the organizational governance in terms of fulfillment degree and significance are represented in the assessment chart, which is shown in Fig. 3.

The Global Quality Indicator for organizational governance (GQI_{org_gov}) is obtained by summing the values for all the 8 evaluated quality indicators (Table 11):

$$GQI_{org\_gov} = \sum\nolimits_{i=1}^{8} QI_i = \sum\nolimits_{i=1}^{8} S_i \cdot G_i = 76 \tag{1}$$

The maximum value of the Global Quality Indicator for organizational governance (GQI-maxorg_gov) is obtained if the degree of fulfillment of each indicator is maximum and is evaluated with the value 5, situation in which it is calculated with the formula:

$$GQImax_{org\_gov} = 5 \cdot \sum\nolimits_{i=1}^{8} S_i = 5 \cdot 23 = 115 \quad (2)$$

The overall level of quality for organizational governance (OLQ_{org_gov}) is calculated as the ratio between the Global Quality Indicator for organizational governance (GQI_{org_gov}) and the maximum value of the Global Quality Indicator for organizational governance ($GQImax_{org\_gov}$), multiplied by 100 in order to be expressed as a percentage:

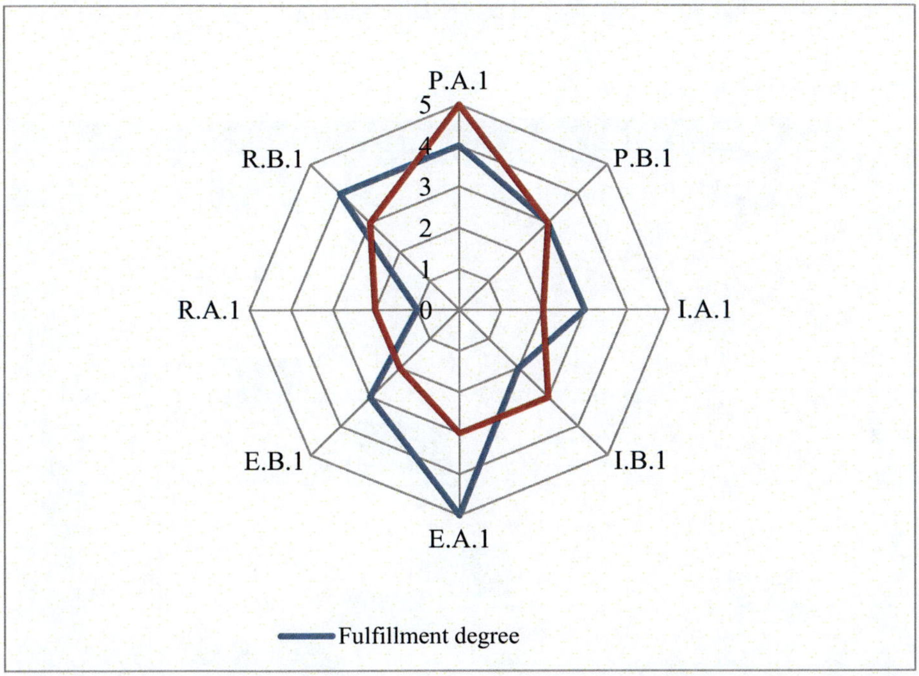

P.A.1 - Decision-making processes and structures	E.A.1 - Existence and recognition of local opinion leaders
P.B.1 - Design of quality assurance processes	E.B.1 - Assignment of monitoring mechanisms
I.A.1 - Computerized support systems for clinical decisions	R.A.1 - Self-assessment tools
I.B.1 - Evaluation mechanisms for transfers	R.B.1 - Changes to healthcare services

Fig. 3. Assessment chart related to the responsibility regarding the organizational governance of the hospital.

$$\mathrm{OLQ}_{\mathrm{org\_gov}} = \frac{\mathrm{GQI}_{\mathrm{org\_gov}}}{\mathrm{GQImax}_{\mathrm{org\_gov}}} \cdot 100 = \frac{76}{115} \cdot 100 = 66,08\% \qquad (3)$$

The overall level of quality of organizational governance provides an overview of the overall state of the organization in relation to the requirements of the San-Q reference framework on organizational governance.

Next, the results for the evaluated organizational governance are represented on the sustainability assessment diagram (see Fig. 4).

This is an Eisenhower matrix that provides an overview of the sustainability assessment regarding the organizational governance of the hospital, which helps to set priorities on a scale of 1 - high to 4 - low, as well as decision-making.

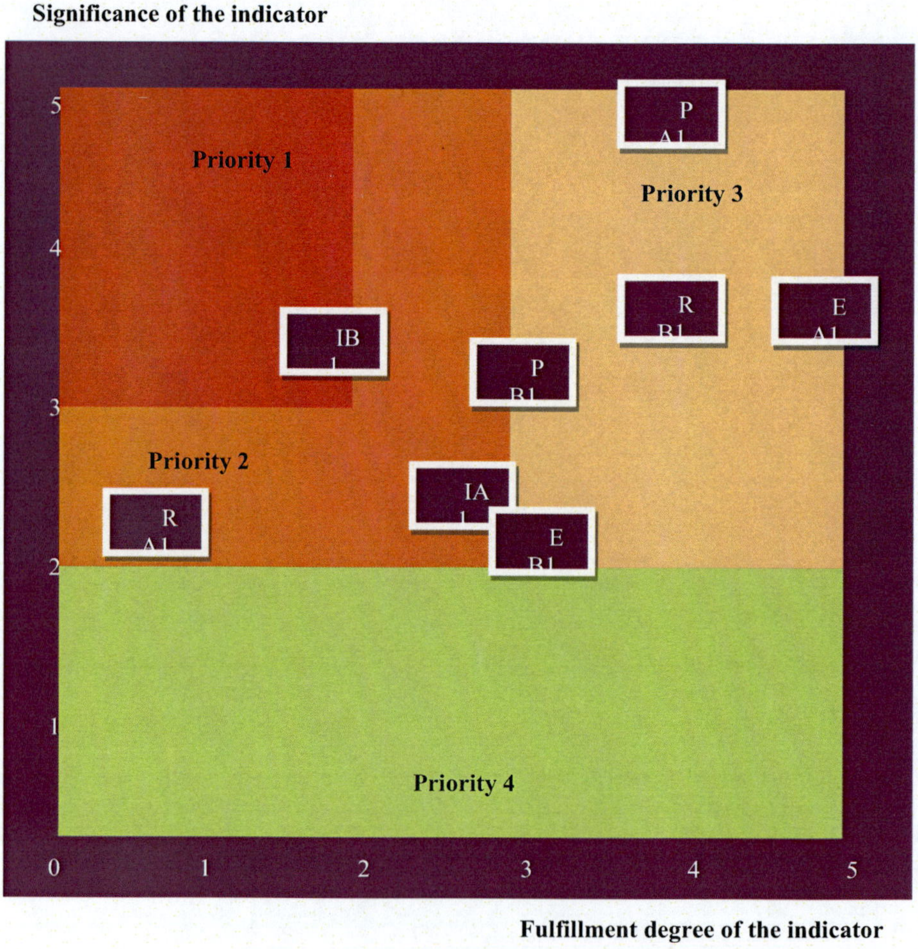

Fig. 4. Sustainability assessment diagram on organizational governance.

The analysis of the diagram shows that the highest priority should have been given to indicator I.B.1 - Transfer evaluation mechanisms.

4 Conclusion

This study allowed the validation in practice of the eight performance indicators designed to assess responsibility regarding organizational governance as a component of the new framework for sustainable development. Through the evaluation in practice, some corrections were made to the description of the indicators, so that they are as easy to apply as possible in the subsequent evaluations.

Regarding the result of the evaluation, through the improvement measures adopted, it is necessary for the hospital to make efforts for accreditation in a higher category. The hospital has a compliance plan that runs for more than 24 months, and which is attached to the operating license. It is necessary that day hospitalization to be separated from the area of continuous hospitalization.

Clinical decision support systems can be improved to better assist the patient's diagnostic process. Hospital transfer activities should be incorporated into certification programs and research funding should be allocated to improve them with the support of medical professional organizations. The assessment of staff satisfaction is not sufficiently publicized in terms of conclusions and suggestions for improvement. In the different stages of healthcare processes integration, it can be increased the degree of digital media use.

Regarding the experience of assessing the responsibility component on organizational governance that characterizes the sustainability of the hospital, this was an opportunity to take a different approach, at different levels of responsibility.

One of the positive results of the evaluation is the discovery of aspects related to the activities of the organization that have never been addressed before.

The evaluation showed several important aspects: the time resource is crucial as good planning and scheduling of activities is needed, and the key elements for the success of the evaluation process are a strong personal commitment and a sense of shared responsibility between management and staff.

References

1. Moldovan, F.: Framework specifications for evaluation of quality improvement and sustainable development in healthcare facilities. In: Multidisciplinary Digital Publishing Institute Proceedings, vol. 63, no. 1, p. 2 (2020)
2. Moldovan, F., Blaga, P.: The basic activities of the continuous improvement cycle for the sustainable development of health facilities. In: Lecture Notes in Networks and Systems, Currently Being Published (2021)
3. Greenfield, D., Braithwaite, J.: Health sector accreditation research: a systematic review. Int. J. Qual. Health **20**(3), 172–183 (2008)
4. Flodgren, G., Pomey, M.P., Taber, S.A., Eccles, M.P.: Effectiveness of external inspection of compliance with standards in improving healthcare organisation behaviour, healthcare professional behaviour or patient outcomes. Cochrane Database Syst. Rev. **11**, CD008992 (2011)

5. Brubakk, K., Vist, G.E., Bukholm, G., Barach, P., Tjomsland, O.: A systematic review of hospital accreditation: the challenges of measuring complex intervention effects. BMC Health Serv. Res. **15**, 280 (2015)
6. Avia, I., Hariyati, R.T.S.: Impact of hospital accreditation on quality of care: a literature review. Enferm Clin. **29**(S2), 315–320 (2019)
7. Araujo, C.A.S., Siqueira, M.M., Malik, A.M.: Hospital accreditation impact on healthcare quality dimensions: a systematic review. Int. J. Qual. Health Care **32**(8), 531–544 (2020)
8. Andres, E.B., Song, W., Schooling, C.M., Johnston, J.M.: The influence of hospital accreditation: a longitudinal assessment of organisational culture. BMC Health Serv. Res. **19**, 467 (2019)
9. Bogh, S.B., et al.: Hospital accreditation: staff experiences and perceptions. Int. J. Health Care Qual. Assur. **31**(5), 420–427 (2018)
10. Alkhenizan, A., Shaw, C.: Impact of accreditation on the quality of healthcare services: a systematic review of the literature. Ann. Saudi Med. **31**(4), 407–416 (2011)
11. Mansour, W., Boyd, A., Walshe, K.: The development of hospital accreditation in low- and middle-income countries: a literature review. Health Policy Plan. **35**(6), 684–700 (2020)
12. Brown, B., et al.: Clinical performance feedback intervention theory (CP-FIT): a new theory for designing, implementing, and evaluating feedback in health care based on a systematic review and meta-synthesis of qualitative research. Implement Sci. **14**(1), 40 (2019)
13. Metcalfe, D., et al.: Impact of public release of performance data on the behaviour of healthcare consumers and providers. Cochrane Database Syst. Rev. **9**(9), CD004538 (2018)
14. de Vos, M., Graafmans, W., Kooistra, M., Meijboom, B., van der Voort, P., Westert, G.: Using quality indicators to improve hospital care: a review of the literature. Int. J. Qual. Health Care **21**(2), 119–129 (2009)
15. Fossey, J., Masson, S., Stafford, J., Lawrence, V., Corbett, A., Ballard, C.: The disconnect between evidence and practice: a systematic review of person-centred interventions and training manuals for care home staff working with people with dementia. Int. J. Geriatr. Psychiatry **29**(8), 797–807 (2014)
16. Ding, X., Zhu, L., Zhang, R., Wang, L., Wang, T.T., Latour, J.M.: Effects of family-centred care interventions on preterm infants and parents in neonatal intensive care units: a systematic review and meta-analysis of randomised controlled trials. Aust. Crit. Care **32**(1), 63–75 (2019)
17. Conn, V.S., Ruppar, T.M., Enriquez, M., Cooper, P.S.: Patient-centered outcomes of medication adherence interventions: systematic review and meta-analysis. Value Health **19**(2), 277–285 (2016)
18. Backman, C., Chartrand, J., Dingwall, O., Shea, B.: Effectiveness of person- and family-centered care transition interventions: a systematic review protocol. Syst. Rev. **6**(1), 158 (2017)
19. Kim, S.K., Park, M.: Effectiveness of person-centered care on people with dementia: a systematic review and meta-analysis. Clin. Interv. Aging **12**, 381–397 (2017)
20. Goldfarb, M.J., Bibas, L., Bartlett, V., Jones, H., Khan, N.: Outcomes of patient- and family-centered care interventions in the ICU: a systematic review and meta-analysis. Crit. Care Med. **45**(10), 1751–1761 (2017)
21. Rubashkin, N., Warnock, R., Diamond-Smith, N.: A systematic review of person-centered care interventions to improve quality of facility-based delivery. Reprod. Health **15**(1), 169 (2018)
22. Constand, M.K., MacDermid, J.C., Dal Bello-Haas, V., Law, M.: Scoping review of patient-centered care approaches in healthcare. BMC Health Serv. Res. **14**, 271 (2014)
23. Bright, T.J., et al.: Effect of clinical decision-support systems: a systematic review. Ann. Intern. Med. **157**(1), 29–43 (2012)
24. Lobach, D., et al.: Enabling health care decisionmaking through clinical decision support and knowledge management. Evid. Rep. Technol. Assess (Full Rep.) **203**, 1–784 (2012)

25. Kaushal, R., Shojania, K.G., Bates, D.W.: Effects of computerized physician order entry and clinical decision support systems on medication safety: a systematic review. 2003. In: Database of Abstracts of Reviews of Effects (DARE): Quality-assessed Reviews. Centre for Reviews and Dissemination, York (1995). https://www.ncbi.nlm.nih.gov/books/NBK70091/
26. Chan, A.J., et al.: Order sets in health care: a systematic review of their effects. 2012. In: Database of Abstracts of Reviews of Effects (DARE): Quality-assessed Reviews. Centre for Reviews and Dissemination, York (1995). https://www.ncbi.nlm.nih.gov/books/NBK114399/
27. Main, C., Moxham, T., Wyatt, J.C., Kay, J., Anderson, R., Stein K.: Computerised decision support systems in order communication for diagnostic, screening or monitoring test ordering: systematic reviews of the effects and cost-effectiveness of systems. 2010. In: NIHR Health Technology Assessment programme: Executive Summaries. NIHR Journals Library, Southampton (2003). https://www.ncbi.nlm.nih.gov/books/NBK56829/
28. Sahota, N., et al.: CCDSS systematic review team. Computerized clinical decision support systems for acute care management: a decision-maker-researcher partnership systematic review of effects on process of care and patient outcomes. Implement Sci. **3**(6), 91 (2011)
29. Murphy, E.V.: Clinical decision support: effectiveness in improving quality processes and clinical outcomes and factors that may influence success. Yale J. Biol. Med. **87**(2), 187–197 (2017)
30. Arora, V.M., Manjarrez, E., Dressler, D.D., Basaviah, P., Halasyamani, L., Kripalani, S.: Hospitalist handoffs: a systematic review and task force recommendations. J. Hosp. Med. **4**(7), 433–440 (2009)
31. Roukis, T.S.: White papers, position papers, clinical consensus statements, and clinical practice guidelines: future directions for ACFAS. J. Foot Ankle Surg. **54**, 151–152 (2015)
32. Emergency County Clinical Hospital Targu Mures Homepage. https://www.spitalmures.ro/. Accessed 10 June 2021
33. Emergency County Clinical Hospital Targu Mures Homepage. General presentation / About the hospital. https://www.spitalmures.ro/despre-spital/prezentare-generala. Accessed 10 June 2021
34. Emergency County Clinical Hospital Targu Mures Homepage. Activity report of the Targu Mures County Emergency Clinical Hospital for 2020. https://www.spitalmures.ro/wp-content/uploads/2021/05/raport-de-activitate-2020_compressed.pdf. Accessed 10 June 2021
35. Emergency County Clinical Hospital Targu Mures Homepage. URS Certificate of registration. https://www.spitalmures.ro/wp-content/uploads/2021/03/Certificat-ISO-2015.pdf. Accessed 10 June 2021

An Innovative Project for Higher Education Leadership in Advancing Inclusive Innovation for Development

Liviu Moldovan(✉)

"George Emil Palade" University of Medicine, Pharmacy, Science, and Technology of Targu Mures, Gh. Marinescu str. 38, 540142 Targu Mures, Romania
liviu.moldovan@umfst.ro

Abstract. Inclusive innovation explicitly implies the reduction of the exclusion condition of a specific population. For sustainable operations research to be relevant in the context of emerging economies, it needs to incorporate social concerns and conditions of underserved populations, with an emphasis on inclusion and equity. Change has never been more critical in educational environments, such as universities, and the most significant challenges centered on the need for strategic leadership, flexibility, creativity and change-capability. In this context we have developed an Erasmus+ project for Capacity Building in Higher Education, which is a partnership between 4 Higher Education Institutions (HEIs) from Europe: Romania, Poland, United Kingdom, Italy, 1 Small Medium Enterprise from Bulgaria, 5 HEIs from Kenya, 2 HEIs from Tanzania and 3 HEIs from Uganda. The overall objective of the projects to enhance the management, governance, teaching, research and evaluation capacities of Partner Country (PC) HEIs in view of enabling them to implement the knowledge triangle and better integrate their research, education and innovation functions in support of innovation for sustainable development and inclusive growth. The activities are geared towards enhancing PC HEIs innovation capacities, with a focus on inclusive innovation, development and local challenges. The results of the project are described in terms of added value, the innovative character and the impact on target groups. This development creates a sustainable impact which is ensured by maintaining the activities of the innovation for development hubs.

Keywords: Inclusive innovation · Leadership · Higher education · Sustainable development · Inclusive growth

1 Introduction

Inclusive innovation explicitly implies the reduction of the exclusion condition of a specific population [1]. Consideration of inclusive innovation points to inequalities that may arise in the development and commercialization of innovations, and also acknowledges the inequalities that may occur as a result of value creation and capture [2] as well as fostering inclusive innovation ecosystems [3] supported by work-based learning models

L. Moldovan and A. Gligor (Eds.): Inter-Eng 2021, LNNS 386, pp. 348–357, 2022.
https://doi.org/10.1007/978-3-030-93817-8_33

[4] and industry 4.0 considerations [5]. For example, in higher education in Kenya there is a need to develop policies and strategies both at national and local level geared towards increasing women's participation in decision making and leadership [6].

For sustainable operations research to be relevant in the context of emerging economies, it needs to incorporate social concerns and conditions of underserved populations [7, 8], with an emphasis on inclusion and equity [9].Traditionally, universities have been governed on a collegiate basis [10], but change has never been more critical in educational environments, such as universities, and the most significant challenges centeredon the need for strategic leadership, flexibility, creativity and change-capability [11].

There is also a need for design methods on innovative products [12]. Hubs play critical roles in building and sustaining robust integrated networks [13]. The leader has tremendous control over the knowledge-processing environment and the role of leadership has broader influence than the resolution of knowledge gaps [14].

A traditional lack of engineering–global health collaborations, as well as limited faculty and inadequate science, technology, engineering, and mathematical research funding in low-income countries, has stifled progress. Current international university-level partnerships are working towards integrating engineering into global health research and strengthening science, technology, engineering, and mathematics innovation among universities in low-income countries, but more can be done [15]. Various innovative models for education can be used [16–18], for e-learning [19, 20] supported by innovative computer solutions [21] as well as for assessment [22–24], evaluation [25, 26] which can be incorporated in frameworks [27, 28].

In this context we have developed the African Higher Education Leadership in Advancing Inclusive Innovation for Development AHEAD project [29], in the framework of Erasmus+ programme, which is a partnership between 4 Higher Education Institutions (HEIs) from Europe: Romania (RO), Poland (PL), United Kingdom (UK), Italy (IT), 1 Small Medium Enterprise from Bulgaria (BG), 5 Higher Education Institutions (HEIs) from Kenya, 2 HEIs from Tanzania and 3 HEIs from Uganda. It is an Erasmus+ project for Capacity Building in Higher Education.

The overall objective of the projectis to enhance the management, governance, teaching, research and evaluation capacities of Partner Country (PC) HEIs in view of enabling them to implement the knowledge triangle and better integrate their research, education and innovation functions in support of innovation for sustainable development and inclusive growth.

2 Matherial and Method

2.1 Aims and Objectives

The project seeks to initiate a long-term partnership that mobilizes European Union (EU) expertise in support of building Kenyan (KE), Tanzanian (TZ) and Ugandan (UG) HEIs' capacities to lead and manage innovation that best fits their countries' inclusive and sustainable development needs.

The specific objectives of the research are:

- To enable PC HEIs to review and reform their institutional structures and agendas, build the vital human capital and strengthen the managerial expertise needed to lead and manage knowledge flows and innovation partnerships with a variety of stakeholders in the field of sustainable development, entrepreneurship and inclusive growth;
- To contribute to the PC HEIs' knowledge and teaching base in the field of innovation management and inclusive innovation, as well as to build PC HEIs' capacities for entrepreneurship and social entrepreneurship education;
- To create inclusive stakeholder-oriented university-led support structures and communication platforms with diverse partners in the innovation system in view of nurturing local entrepreneurship and inclusive development in the participating PCs.

AHEAD contributes to achieving positive change with regard to the identified needs of the PC HEIs. In particular, PC HEIs are:

- better acquainted and capable to assess and adapt existing good practices and workable knowledge triangle models;
- improve and mobilize their internal capacity to perform impactful research and to better align their teaching and research missions with a third mission of community engagement, knowledge transfer, innovation and start-up support in local economies, in particular by engaging in self-assessment, considering external assessment, designing internal institutional reforms and/or development plans, and creating a core of faculty and managerial staff committed to positive change;
- better align their research activities with pressing local socio-economic challenges and key national development goals related to sustainability and inclusive growth;
- take leadership and assume ownership of open innovation and knowledge exchange in both low-tech and high-tech sectors, in particular by following a multidisciplinary approach, involving a variety of stakeholders and building the necessary human capital and managerial skills;
- build human and technological capacities, support structures within the institution, expertise and the knowledge/teaching base necessary for promoting the upscaling of grassroots innovation, entrepreneurship at local level, and inclusive (pro-poor) innovation;
- become better integrated into their external environment and be better positioned to contribute to local economic growth.

2.2 Activities and Methodology

AHEAD activities are geared towards enhancing PC HEIs innovation capacities, with a focus on inclusive innovation, development and local challenges. The work programme follows the key stages of institutional transformation:

- Context analysis and self-assessment of existing PC HEI capacities;
- Development of knowledge base supported by training;
- Analyzing feasibility by understanding internal and external constraints to transformation;

– Developing constraints-sensitive actionable transformation designs;
– Implementing transformation on pilot basis and strengthening human resources capacity;
– Ensuring quality;
– Communicating and celebrating institutional transformation.

The framework encourages initiative and ownership on the part of PC HEIs, while EU partners provide technical leadership and support and promote knowledge and innovation management tools.

The main activities include:

- Analysis of national innovation systems and institutional innovation capacities;
- Developing learning materials in Innovation Management and Inclusive Innovation;
- Training of trainers and workshops;
- Improving the good-practice and evidence base for delivering (social) entrepreneurship education;
- Creation of Virtual Knowledge Gateways (VKGs) for structured cooperation between PC HEIs and key stakeholders and strengthening stakeholder collaboration and consultation;
- Establishment of Innovation for Development Hubs (IDHs) encouraging local and regional development and stakeholder co-creation in innovation.

2.3 The Consortium

The 4 HEIs from 4 Programme countries (RO, UK, IT, PL) and 1 SME (BG) are experienced in linking education institutions and enterprises. We have ensured variation in socio-economic context, innovation systems and models and intensity of university-business interactions. The UK provided many good practices for AHEAD due to the presence of wide-scale knowledge exchange, "strong innovator" economy and auspicious 'liberal' context and government leadership in support of (social) entrepreneurship. Partners from Poland, Romania and Bulgaria experience is applicable for PC contexts: while lagging behind best EU practices, these countries provide insights about the most effective ways to transfer and adapt best practices to less knowledge-intensive and innovating economies. The Italian experience is also relevant to the PCs due to context similarities related to stronger role of informal and family-based interactions in the economy and society.

The aspect that unites the partners from the 5 Kenyan, 3 Ugandan and 2 Tanzanian HEIs is the orientation of teaching and research to societal challenges, developmental issues and inclusiveness, as well as an expressed desire for reform or improvement. In other aspects there is difference and complementarity. With regard to HEI research specialization and staff profiles, a mix was sought between technological fields - which are powerful drivers of (inclusive) innovation - and non-technological fields such as agriculture, health, community development and inclusion.

Due to rapidly growing demand for higher education in the PCs, a large number of new HEIs were created very recently (usually by transforming former colleges). Such institutions have greater capacity-building needs in the areas of quality assurance, research and embedding into the external environment. Some of these new universities were invited in the proposal (Lira, Kibabii) in order to encourage not just EU-PC learning, but also peer learning within the PC region. Seeking strong project impact on institutional development, they chose to involve staff at top managerial positions in order to enable more effective and quick transformation. For established universities, high-level management was not recommended to be involved as staff due to already heavy administrative burdens.

Kenyatta and Mount Kenya University are the most advanced PC HEI in terms of innovation and research capacity, while Kenya itself provides a more advanced model of higher education ecosystem for innovation. Two private universities were included, too, under the assumptions that they operate in a slightly different environment (less government involvement) and warrant special attention. This would make product outputs more relevant to the needs of the wider PC HE sectors.

A gender balance among staff proved difficult to achieve, but efforts were made to include a high average share of women in the staff pool.

3 Results

3.1 Euroean Added Value

AHEAD improves and mobilizes PC HEIs' institutional capacities for Innovation for Development. Improving capacities is predicated upon exchange of experience, good knowledge triangle practices and innovation partnership models, which PC HEIs can use as inspiration when charting their own developmental plans. Participation of several EU partners ensures diversity of ideas, models, socio-economic contexts and lessons learned.

Mobilizing capacities and ultimately introducing new support structures and policies are facilitated by the core project teams in PC HEIs are backed up by a credible international (rather than national or local) team of EU practitioners and researchers.

The project provides EU cooperation in developing a knowledge base applicable to PCs: HEI leadership in grassroots and inclusive innovation are emergent issues. Good practices and models are not well documented yet even at EU level, which necessitates a diverse research team.

3.2 Innovative Character

AHEAD builds PC HEIs' capacities for innovation and local entrepreneurship support as vital parts of knowledge triangle practices. Key innovating elements are:

Context sensitivity: the capacity building (CB) approach which is aligned with the specific challenges and priorities of the innovation system context in PCs: importance of the citizen sector, the informal sector and grassroots knowledge, continuing significance of agriculture and low-tech sectors, focus on equality and poverty, etc. EU partners are tasked with building the knowledge and good practices base and providing continuous consultation, while PC HEIs are tasked with planning and implementing tailor-made innovation support structures and policies, thus ensuring sensitivity not just to national and local contexts but also to institutional development plans and existing coalitions and interests within HEIs.

Emphasis on institutional ownership: Apart from ensuring optimal relevance, the AHEAD CB approach relies on a work programme that creates a multi-stage process of internal and external-stakeholder consultations leading to the introduction of new or the reform of existing institutional policies and support structures for innovation and knowledge transfer. This is a key prerequisite for achieving the desired project impact: to both enable and encourage PC HEI leadership in innovation and entrepreneurship support attuned to local socio-economic challenges and inclusiveness.

Multi-stakeholder approach: AHEAD affirms an innovative understanding of HE research as an endeavor involving a variety of stakeholders in addition to enterprises and government: non-profits, communities, grassroots inventors and entrepreneurs.

Attention to grassroots innovation upscaling and inclusive innovation: these types of innovation are vital for PC economies yet they remain under-researched and under-promoted, especially within the HE sector. The attempt to change this is the main innovating element of AHEAD.

3.3 Impact

The impact is on:

Primary target groups:

Target groups within AHEAD PC HEIs: management, faculty, staff, postgraduate students, graduates (institutional level):

- directly or through stakeholder consultation are involved in: a) assessment of institutional innovation capacities and national innovation systems; b) planning, design and implementation of institutional support structures for innovation and knowledge triangle practices; c) enhanced communication with enterprises, research institutions, communities and non-profits; d) use of project OERs, good practices and Resource Pack; e) training of trainers;
- benefit from: a) enhanced knowledge base and improved human, technical and administrative capacities for innovation management, inclusive innovation and entrepreneurship support provision; b) peer support and guidance in knowledge triangle practices; c) improved quality of research and innovation collaboration with enterprises, research institutions and community/non-profit stakeholders; d) more opportunities for engaging in research or for receiving start-up support for entrepreneurial ideas.

Target groups outside AHEAD PC HEIs - through stakeholder consultation, dissemination activities and access to the virtual knwoledge gateways (VKGs) will be involved in the planning and implementation of support structures for innovation and knowledge triangle practices in AHEAD PC HEIs:

- Research institutions and academic communities of PC HEIs collaborating with AHEAD HEIs: will benefit from: a) enhanced knowledge and good practices in the fields of innovation, social entrepreneurship, knowledge triangle practices; b) enhanced opportunities for interinstitutional cooperation in research and innovation;

- Enterprises, entrepreneurs and relevant non-profit organizations in UG, KE and TZ (national, regional, local levels); grassroots innovators, community leaders in AHEAD regions (local level): will benefit from more opportunities for knowledge transfer or start-up support from HEIs.

Secondary TG sare involved through dissemination activities (events, website, online resources, VKGs):

- Academic communities of HEIs in the PCs not included as primary TG (national level);
- Public bodies and government agencies responsible for higher education, research, innovation and sustainable development (national, regional levels);
- Academic communities of HEIs in the PCs and Public bodies and government agencies will benefit from enhanced knowledge and good practices in the area of innovation, social entrepreneurship and the knowledge triangle;
- Donors (European level): will benefit from a) improved HEI capacities for international and interinstitutional collaboration; b) insight about national and institutional innovation contexts; c) increased salience of inclusive and grassroots innovation;
- Low-income social groups in UG, KE and TZ and potentially in EU (national, European levels);
- Local communities in AHEAD regions (local level).

Low-income social groupslocal communities benefit from more relevant and efficient HEI contribution to sustainable and inclusive development and pro-poor economic growth.

The participating HEIs and stakeholders are reached through the AHEAD results maintained post project in order to ensure sustainability – the virtual knowledge gateways; the Innovation for Development Hubs (research, awareness raising and consortium building activities), the OERs.

Primary target groups:

1. Management, faculty, staff, postgraduate students, graduates at AHEAD PC HEIs; as well as research institutions and academic communities in HEIs currently or potentially collaborating with AHEAD PC HEIs in the areas of research, innovation or education, are involved in:
   a) continuing structured communication with enterprises, research institutions and community/non-profit stakeholders through the virtual knowledge gateways, focused on identifying common challenges, setting research agendas, documenting and scaling grassroots innovations and developing interdisciplinary research portfolios able to promote inclusive growth and local entrepreneurship;
   b) planning and executing research and innovation projects supported by the Innovation for Development Hubs with direct impact on inclusive development and local entrepreneurship;
   c) using and enriching the OERs and good practices for promoting and entrepreneurship and social entrepreneurship education as part of a more general process of building and improving skills for innovation and research;

   d) continuing research and innovation cooperation involving the participating HEIs and their national and international network of partners;
   e) continuing start-up support for entrepreneurs and university spin-offs through the organization of start-up competitions, maintenance of mentor databases and other relevant IDHs activities.
2. By means of consultation or co-creation in innovation in Hubs, enterprises/industry and entrepreneurs in UG, KE and TZ, especially in the regions of AHEAD PC HEIs, are able to:
   a) provide input into HEIs' research agenda;
   b) provide researchers with knowledge related to markets;
   c) benefit from graduates' skilled in innovation and entrepreneurship.
3. Through the operation of the Innovation for Development Hubs, local innovators and community leaders in the regions of AHEAD PC HEIs are consulted on research priorities and projects and receive entrepreneurship support or support for scaling local innovations.

Secondary targer groups by publicizing successful and upcoming projects, building research and project consortiums and disseminating information at relevant events, the AHEAD network of Hubs will seek to attract new stakeholders, raise awareness and improve skills for inclusive innovation. It will seek to reach out to the academic communities of all HEIs in the PCs, public bodies and government agencies responsible for higher education, research, innovation and sustainable development, and donors providing development assistance.

The virtual knowledge gateways on their part will help involve a larger share of the secondary TGs, notably the local communities in the regions of AHEAD PC HEIs.

4 Conclusions

The development creates a sustainable impact:

- a core of staff skilled to maintain support structures for innovation and a multi-stakeholder approach to research;
- institutional and external stakeholder involvement and support, viability and testing of the new structures and approaches;
- basic material and knowledge-base capacity.

Sustainability of impact post project will be ensured by maintaining the activities of the Innovation for Development Hubs, including:

- providing physical working space for stakeholder collaboration and co-creation in research & innovation. Increasing stakeholder involvement in the Virtual Knowledge Gateways (VKGs) will be sought: target within 5 years post project - 90 stakeholders in TZ, 120 in UG, 200 in KE. Project website and VKGs will be maintained by the coordinator 5 years post project. Hubs will continue networking;

- providing start-up and entrepreneurship support (incl. for scaling of grassroots innovations);
- supporting consortium building and project development and implementation with a view to attracting funding. Fundraising will focus on: a) funding by business for applied research; b) public funding through national research support funds; c) funding for cutting edge innovation supported by donors, incl. EU Programmes, e.g. Horizon 2020;
- organizing 1 annual training in innovation management and inclusive innovation for new promising students, 1 annual start-up competition for students/alumni, and maintaining alumni mentor databases: for min. 3 years post project;
- implementing 1 annual review of institutional innovation capacity, with recommendations for improvement: for min. 5 years post project.

Staff and maintenance costs for Hub operations will be borne by partners - they consider investing in strong innovation & research structures and stakeholder management to be a comparative advantage rather than redundant cost. Responsibilities will be included in the portfolios of regular staff in departments involved in research.

Acknowledgements. This publication reflects the views only of the author, and the Commission cannot be held responsible for any use, which may be made of the information contained therein. My acknowledgements go to project partners who have participated in the AHEAD project development: Ludmil Manev, Evgenia Nikulina and others.

Funding. This research was funded by the European Commission, Erasmus+ programme, grant number 585919-EPP-1-2017-1-RO-EPPKA2-CBHE-JP.

Conflicts of Interest. The author declares no conflict of interest.

References

1. Bernardo, P.V., María Villalba-Morales, L., Acosta-Amaya, M., Villegas-Arboleda, C., Calderón-Sanín, E.: Towards the conceptual understanding of social innovation and inclusive innovation: a literature review. Innov. Dev. **12**, 1–22 (2020)
2. George, G., McGahan, A.M., Prabhu, J.: Innovation for inclusive growth: towards a theoretical framework and a research agenda. J. Manag. Stud. **49**(4), 661–683 (2012)
3. Rego, L., Gergen, C.: Fostering Inclusive Innovation Ecosystems. In: Breaking the Zero-Sum Game (Building Leadership Bridges), Emerald Publishing Limited, Bingley, pp. 43–57 (2017)
4. Moldovan, L.: Review of legislation framework in the field of work-based learning. Procedia Manuf. **32**, 302–308 (2019)
5. Moldovan, L.: State-of-the-art analysis on the knowledge and skills gaps on the topic of Industry 4.0 and the requirements for work-based learning. Procedia Manuf. **32**, 294–301 (2019)
6. Odhiambo, G.: Women and higher education leadership in Kenya: a critical analysis. J. High. Educ. Policy Manag. **33**(6), 667–678 (2011)
7. Moldovan, F.: New approaches and trends in health care. Procedia Manuf. **22**, 947–951 (2018)

8. Moldovan, F.: Framework specifications for evaluation of quality improvement and sustainable development in healthcare facilities. In: Proceedings, vol. 63, p. 2 (2020)
9. Kalkanci, B., Rahmani, M., Toktay, L.B.: The role of inclusive innovation in promoting social sustainability. Prod. Oper. Manag. **28**(12), 2960–2982 (2019)
10. Davies, J., Hides, M.T., Casey, S.: Leadership in higher education. Total Qual. Manag. **12**(7–8), 1025–1030 (2001)
11. Drew, G.: Issues and challenges in higher education leadership: engaging for change. Aust. Educ. Res. **37**, 57–76 (2010)
12. Moldovan, L.: QFD employment for a new product design in a mineral water company. Procedia Technol. **12**, 462–468 (2014)
13. Taylor, K.L., Kenny, N.A., Perrault, E., Mueller, R.A.: Building integrated networks to develop teaching and learning: the critical role of hubs. Int. J. Acad. Dev. **3**, 1–13 (2021)
14. Martin, J.S., Marion, R.: Higher education leadership roles in knowledge processing. Learn. Organ. **12**(2), 140–151 (2005)
15. Clifford, K.L., Zaman, M.H.: Engineering, global health, and inclusive innovation: focus on partnership, system strengthening, and local impact for SDGs. Glob. Health Action **9**, 1 (2006)
16. Moldovan, L.: Innovative tools and models for vocational education and training. In: Review of Management and Economic Engineering - first Management Conference: Twenty Years After - How Management Theory Works, no. 9, pp. 282–290 (2010)
17. Moldovan, L.: Design of a new learning environment for training in quality assurance. Procedia Technol. **12**, 483–488 (2014)
18. Moldovan, L.: Innovative models for vocational education and training in Romania. Procedia Soc. Behav. Sci. **46**, 5425–5429 (2016)
19. Moldovan, L.: Design and development of innovative tools and models for e-learning in central and western Romania. In: The 6th International Seminar Quality Management in Higher Education – QMHE2010, Tulcea, Romania Book II, pp. 543–546 (2010)
20. Moldovan, L.: Innovative method of peer assisted learning by technology and assessment of practical skills. Procedia Technol. **12**, 667–674 (2014)
21. Mikolajczyk, T., Moldovan, L., Chalupczak, A., Moldovan, F.: Computer aided learning process. Procedia Eng. **181**, 1028–1035 (2017)
22. Moldovan, L.: Sustainability assessment framework for VET organizations. Sustainability **7–6**, 7156–7174 (2015)
23. Moldovan, L., Moldovan, A.-M.: Green methodology for learning assessment. Procedia Technol. **22**, 1176–1183 (2016)
24. Moldovan, L.: The environmental pillar assessment in vocational education. Environ. Eng. Manag. J. **16**(3), 739–750 (2017)
25. Moldovan, L.: New evaluation model by means of mobile technology. Procedia Technol. **19**, 1094–1101 (2015)
26. Moldovan, L.: Training outcome evaluation model. Procedia Technol. **22**, 1184–1190 (2016)
27. Moldovan, L.: Framework development for European quality assurance in VET towards environmentally sustainable economy. Procedia Eng. **181**, 1064–1071 (2017)
28. Moldovan, L.: Practical implementation of a framework for European quality assurance in VET towards environmentally sustainable economy. Procedia Eng. **181**, 1072–1079 (2017)
29. Ahead project webpage. http://www.ahead-project.net/. Accessed 22 Aug 2021

An Innovative Trainers' Toolkit for Innovation Management in Low- and Middle-Income Countries

Liviu Moldovan(✉)

"George Emil Palade" University of Medicine, Pharmacy, Science, and Technology of Targu Mures, Gh. Marinescu str. 38, 540142 Targu Mures, Romania
liviu.moldovan@umfst.ro

Abstract. The character of the economies in low- and middle-income countries is traditional, production-oriented, prosperity being planned through extensive measures to increase agricultural production, housing construction, and the establishment of traditional factories. In the current context of economic globalization, it is necessary for these countries to focus on the real sources of prosperity and capital in the new era, which are not represented by material goods, but by human thinking, knowledge and innovation. This is a fundamental change in the economic model, in which the emphasis is on the development of intangible resources, inventions and know-how, and on their transformation into sources of innovation. The phrase "knowledge-based society" emphasizes this new orientation. The objective of this paper is to present the development of an innovative module "Innovation Management in Low- and Middle-Income Countries" which is composed of 8 lectures that describe specific aspects of innovation management in these economies. General objective of the module is to develop a supplementary training material in order to aid the delivery of innovation and entrepreneurship training and education, consisting in trainers' methodological toolkits accompanying the open educational resources.

Keywords: Inclusive innovation · Leadership · Higher education · Sustainable development · Inclusive growth

1 Introduction

We are living in the period of accelerated transition, marked by complex and profound transformations in all areas of activity. The scale of innovation is primarily reflected in the high pace of development of new products and technologies, but the changes are not just about tangible things. Within organizations, there are more and more innovation actions oriented towards business management methods, organization and configuration, which contribute to obtaining sustainable competitive advantages.

At the same time, innovation manifests itself in society in general, materializing in new strategies, concepts, ideas and organizations that address social needs - from the labor market and working conditions, to education, health and community development.

L. Moldovan and A. Gligor (Eds.): Inter-Eng 2021, LNNS 386, pp. 358–369, 2022.
https://doi.org/10.1007/978-3-030-93817-8_34

The current importance of the activities of introducing the new can be explained from the perspective of the transformations in the economy and society, determined by the increase of competition, by the technical progress and, especially, by the unimaginable development of information technologies.

An important and increasingly important role is played by the social economy in the market economy, functioning together with it. By putting economic efficiency at the service of social needs, the social economy creates a real interdependence between the economic and the social aspect, without subordinating it to each other. It is necessary to recognize and promote the social economy, a sector that is a real cornerstone for jobs but also for social cohesion and for building and consolidating a pillar of social rights.

The new context corresponds with fundamental changes in economic models that are an important support for the development of low- and middle-income countries. The character of the economies in these countries is traditional, production-oriented, prosperity being planned through extensive measures to increase agricultural production, housing construction, the establishment of traditional factories. In the current context of economic globalization, it is necessary for these countries to focus on the real sources of prosperity and capital in the new era, which are not represented by material goods, but by human thinking, knowledge and innovation. This is a fundamental change in the economic model, in which the emphasis is on the development of intangible resources, inventions and know-how, and on their transformation into sources of innovation. The phrase "knowledge-based society" emphasizes this new orientation.

From this perspective, innovation management is a new paradigm of approaching innovation, characterized by the application of specific models, rules and policies.

Social and inclusive innovation spheres address critical social, development and humanitarian needs through products designed to be pro-poor, equitable and just [1, 2]. Inclusive innovation lens accentuates the participation of marginalized actors and poverty reduction [3] supported by work based learning models [4], or even industry 4.0 considerations [5].

Instead of a grand theory of inclusive innovation that applies universally, there are several ways of enacting inclusive innovation [6]. As a policy recommendation, institutional weaknesses and failures that manifest in the public sector in Africa need to be urgently tackled so as to reduce corruption and its deleterious effects on firm innovation [7]. For sustainable operations research to be relevant in the context of emerging economies, it needs to incorporate social concerns and conditions of underserved populations [8, 9].

Persistent innovators enjoy large and sustained comparative advantages [10]. It is recommended that in transition economies policy stakeholders should ensure the creation of enabling environments structured around responsive micro and macro decision-making [11]. Firms with innovations based on research and development activities are more resilient to business cycle fluctuations than non-innovative firms [12, 13]. The link between knowledge and firm growth has been a core topic in economics of innovation for a long time and there is evidence on a curvilinear relation between knowledge bases and growth of firms [14].

In this context we have developed the African Higher Education Leadership in Advancing Inclusive Innovation for Development AHEAD project [15], in the framework of Erasmus+ programme, which is a partnership between 4 Higher Education Institutions (HEIs) from Europe: Romania (RO), Poland (PL), United Kingdom (UK), Italy (IT), 1 Small Medium Enterprise from Bulgaria (BG), 5 Higher Education Institutions (HEIs) from Kenya, 2 HEIs from Tanzania and 3 HEIs from Uganda. It is an Erasmus+ project for Capacity Building in Higher Education.

The objective of this paper is to present the development of an innovative module "Innovation Management in Low- and Middle-Income Countries" is composed of 8 lectures that describe specific aspects of innovation management in these economies. General objective of the module is to develop a supplementary training material in order to aid the delivery of innovation and entrepreneurship training and education, consisting in trainers' methodological toolkits accompanying the open educational resources OERs – activity performed by the EU research teams in the framework of the AHEAD project.

2 Material and Method

2.1 Methodological Guidelines

The "Innovation Management in Low- and Middle-Income Countries" course is designed to provide foundational knowledge on innovation management in Low- and Middle-Income Countries. It is a basic learning material in the topic, separated into sub-topics: 8 lectures. The course is intended for teaching staff, students, trainers, administrative staff, librarians and others who are interested in studying innovation tools to support the development of economies in countries with gaps.

The course is built on the rationale that long-term improvement in research and innovation capacity requires continuous investment into building specialized skills among staff, faculty and students. Skills for innovation management are universally relevant to all innovation and research activities of higher education institutions (HEIs). Universities in Partner Countries (PCs) are in excellent position to assume leading roles in innovation that addresses societal challenges and contributes to inclusive development.

Since enterprises in the three PCs - Kenya, Tanzania and Uganda - are less likely than European business to invest in innovation and research, PC universities receive large portions of their research budgets from public funds, which naturally are oriented towards the public good and national developmental priorities. Thus, PC HEIs are already expected and involved in contributing to national development and put efforts to address social challenges. Given that large groups of the population can be characterized as low-income consumers and economic growth in the three PCs contributes to rising inequality, the potential for the development of inclusive innovation is especially high in these countries. AHEAD project is developed with a view to improving HEI capacity to lead this process.

The course is a unitary whole composed of 8 lectures, which progressively approach the knowledge of innovation management, starting from the primary definitions of notions, continuing with the aspects of identifying learning methods and solutions, and concluding with the aspects of innovation exploitation.

The course is integrated in the AHEAD website in a form that is suitable for individual learning and is downloadable for the use in traditional courses. Each lecture is accompanied by a multiple-choice test and a list of resources for further reading. The materials can be further used by students and alumni, as well as researchers and faculty who wish to develop a course on the subject or to include innovation management as a sub-topic in other courses.

This toolkit is designed to aid the delivery (teaching/training) of the module "Innovation Management in Low- and Middle-Income Countries" based on the open educational resources (OERs) in a traditional learning environment: instructor-led or trainer-led course at an education or training institution.

Each lecture in the course addresses specific issues and topics. For convenience, these are indicated in the section on class structure. Various innovative models for training delivery can be used [16–18], by employment of e-learning [19, 20] which can be supported by innovative computer solutions [21].

2.2 Instruction and In-Class Participation

Instruction should respect the culture of instruction prevalent in the educational institution. However, teachers should note that this program is specifically designed to facilitate seminar-type training. In this context, lectures should be kept to a minimum, and classroom study should be based on discussions and the presentation of examples and case studies. It is desirable that the audience have read the necessary readings beforehand.

The purpose of the training activities is to encourage debate on the topics covered and to increase creative thinking and training in innovation management, with specific emphasis on the peculiarities of Low- and Middle-Income Countries. For students who have more advanced knowledge, bibliography for additional reading is suggested. It can be used for a deeper exploration of the topic.

It is desirable that students come to class prepared in order to have a participatory attitude and contribute to class discussions and learning activities proposed for group solving. It is recommended to involve in the debate guest professors recognized for their work in the field, as well as entrepreneurs who achieved outstanding results through the use of inventions and innovations in their enterprises.

Each training session should include a group task related to the topics covered, for instance each group presents a practical example, a case study or an opinion on the issue under discussion. The tasks assigned to each group should be able to be solved by studying the module and in addition should involve elements of creativity and/or debate, as well as the improvement of the students' oral presentation skills. It is also recommended that group tasks require the resolution of interdisciplinary issues, which involve contributions from different fields. Examples of learning activities that could be used as in-class group assignments are provided in the section "Class structure and learning activities".

2.3 Assessment and Grading

Various innovative models for assessment [22–24], evaluation [25, 26] which can be incorporated in frameworks [27, 28] can be used. Peer training must be comprehensive enough to equip peers with the knowledge and skills needed for their work [29].

Some assessment options, which are designed in the form of a list from which each teacher/trainer can choose the desired assessment elements based on their own assessment plan and taking into account the institutional context and the objectives of the course, are presented below.

Final exam: It is recommended that the final exam contains both multiple-choice and open-ended questions. The duration of such an examination can vary from 1 h to 1.5 h. For students of advanced level or in the case of practice-oriented instruction, an examination based on case studies is recommended. If this option is chosen, students can be asked to analyze a case study through open-ended and multiple-choice questions. The duration of the examination could vary from 1.5 h to 2.5 h, depending on the complexity of the case study and the number of aspects analyzed. Teachers/trainers can also opt to have an intermediate exam.

Individual/Group E-portfolios of Case Studies of Innovation Management: The selection of case studies can take place throughout the course, and the final e-portfolio should be due by the end of the course. Formal requirements regarding length are between 3,500 and 5,000 words. The electronic portfolio aims to introduce and familiarize the student with the process of exploration, research and analysis in the field of innovation management. One case is a synopsis of a real-world situation facing an enterprise that promotes innovation. The main result of the e-Portfolio should be the presentation, analysis and detailed evaluation of 2 case studies of innovation management, with a special focus on the economic impact achieved in each case and the challenges faced by this initiative. Papers that discuss high-quality case studies should present an interview with at least one company that manages innovation processes. In addition, multimedia artifacts should be presented, such as photograph and audio files from interviews. Teachers/trainers can develop the approach to implementing this assignment according to the needs of the students and their level of knowledge in the related fields.

Guidelines regarding the expected content of the case studies content:

- Areas of activity and major achievements of the enterprise in the field of innovation;
- Elements which distinguish the analyzed enterprise from other enterprises in the field that do not routinely apply innovation in their activities;
- Who are the innovators in the enterprise and what is their role in the organization;
- Main idea of the innovations and their estimated efficiency;
- The problems, challenges and opportunities that the company encounters in the development and application of innovations, related to, for example, financing, sustainability, management (operational or strategic), market access, marketing, personnel;
- The general evaluation of the economic impact and viability of the analyzed innovation solution, but not limited to financing, sustainability, management (operational or strategic), market access, marketing, personnel;

- Lessons learned and good practices that can be drawn from this case;
- Recommendations for business activities that can improve performance.

The assessment should comply with the rules of assessment and grading in the educational institution where this module is implemented. In order to stimulate creativity, it is recommended that project evaluation prioritize innovative and creative ideas: Individual in-class participation: 5%; Participation in in-class group assignments: 10%; Final exam: 20%; E-portfolios of case studies of innovation management: 15%; Managing the implementation of an innovation: 50%.

3 Results

The module is a unitary whole that integrates the 8 lectures by gradually presenting the concepts depending on the level of complexity and difficulty of understanding as well as application in practice (see Table 1). At the beginning, elements related to the structure and organization of innovation systems at different levels is presented. After that we move on to the presentation of the essential elements of innovation processes. The study continues by approaching the aspects of leadership and management in these processes, after which the innovation is discussed under all the defining aspects, with emphasis on the open innovation that has the potential to accelerate the transformation processes. This is facilitated by the widespread adoption of technologies in developing countries, supported by a good management that can generate good exploitation.

Table 1. Module structure and workload.

Lecture. 1 3 academic hours (2 h in-class and 1 h independent learning), 1 h online work load	National, regional, sectoral and technological innovation system *Annotation*: Lecture 1 analyzes and evaluates the role of the innovation system at national, regional, Innovation sectoral and technological level. It describes the innovation systems at these levels in terms of structure, elements and factors that determine their functioning
Lecture. 2 3 academic hours (2 h in-class and 1 h independent learning), 1 h online work load	Innovation processes and structures *Annotation*: Lecture 2 addresses the essential aspects of innovation processes by presenting the notion of innovation in a narrow and broad sense, which allows for a conceptual clarification while presenting the forms of research as constituent elements of the innovation process. Furthermore, the lecture introduces different models of the innovation process that can be adopted in low- and middle-income countries

(*continued*)

Table 1. (*continued*)

Lecture. 3 3 academic hours (2 h in-class and 1 h independent learning), 1 h online work load	Innovation leadership *Annotation*: Lecture 3 explores different approaches to leadership and discusses in more detail the qualification, behavioral and situational approaches. It explains differences between transactional and transformational leadership. Knowledge management is emphasized as one of the key functions of an effective leader. Such issues as management of learning organizations and leadership for innovation are also discussed
Lecture. 4 3 academic hours (2 h in-class and 2 h independent learning), 1 h online work load	Types of innovation *Annotation*: Lecture 4 analyses different types of innovation, including: product innovation, process innovation, marketing innovation, organizational innovation; radical innovations versus incremental innovations; sustaining innovations versus disruptive innovations; open innovation versus closed innovation. The difference between innovation and invention is also discussed
Lecture. 5 3 academic hours (2 h in-class and 1 h independent learning), 1 h online work load	Open innovation *Annotation*: Lecture 5 offers a perspective on open innovation as a recommended model for an organization that pursues accelerated progress and not only relies on its internal knowledge, sources and resources to innovate products, services, business models, processes, etc., but stimulates innovation through more external sources that integrate external partners such as customers, research institutes or suppliers in the innovation process
Lecture. 6 3 academic hours (2 h in-class and 1 h independent learning), 1 h online work load	Knowledge transfer *Annotation*: Lecture 6 emphasizes mutually beneficial collaborations between universities, companies and the public sector through the systematic process of knowledge exchange and learning from the experience of others that is generated by the transfer of practical or theoretical knowledge

(*continued*)

Table 1. (*continued*)

Lecture. 7 3 academic hours (2 h in-class and 1 h independent learning), 1 h online work load	Technological learning, technological catch-up, technological leapfrogging *Annotation*: Lecture 7 presents technology as a key tool for stimulating economic growth that plays a central role in achieving social welfare goals that can hardly be reached through the conventional economic measures. Through the rapid and widespread deployment of technologies in and between developing countries, they can rapidly reduce the gaps with economically advanced countries
Lecture. 8 3 academic hours (2 h in-class and 1 h independent learning), 1 h online work load	Innovations exploitation and management *Annotation*: Lecture 8 discusses practical aspects of exploiting innovation in order to transform innovative ideas into valuable economic and/or social results. This is achieved through the process, which includes three distinct steps: conception, exploration, and exploitation of innovation. The defining aspects of these steps are analyzed in detail. The lecture provides an overview of the STI priorities in the East Africa Community Countries, with a focus on Tanzania, Kenya and Uganda. In addition, the final section of the lecture provides general information on the management innovation that plays a crucial role in the framework of the development processes
Final project 6 academic hours(independent learning), 3 h online work load	Final exam / E-portfolio of case studies of innovation management / Essay: Managing the implementation of an innovation

The defining notions that integrate the 8 lectures can be summarized on the route: System → Model → Management → Innovation → Open innovation → Knowledge transfer → Technologies → Exploitation.

The module "Innovation Management in Low- and Middle-Income Countries" highlights the fact that innovation is not an uncontrollable force, linked exclusively to inspiration and creativity, it is not the prerogative of certain companies, elite, or developed countries, and it does not depend on magic formulas, accessible only to certain organizations or countries. In fact, it is about good management.

The module foresees 30 academic hours of workload, including 16 h of in-class learning and 14 h of independent learning. The hours of in-class and independent learning are equally distributed among the lectures (3 h in class and 1 h independent learning). The development of the final project requires about 6 h of preparatory work (independent learning).

4 Conclusion

As a result of engaging with the learning materials in this module, learners are expected to develop the following knowledge, skills and competences as presented in Table 2.

The implementation of this course will be possible, if the university (teacher/trainer) delivering this content manages to engage innovative companies and if these companies agree to involve students in the innovation process they carry out (or at least share information related to innovation management in their company with students).

The format of the assignment is recommended as an essay of 4,000–5,000 words. This project should be due by the end of the course. By implementing it, students will address the challenges of managing the development of an innovation in an enterprise. The aim of the activity is to make students reflection the challenges inherent in being innovative. The content of essays can include the following elements:

- Presentation of an enterprise, the field of activity and technological performance;
- Sectoral and technological innovation systems analysis;
- Designing innovation processes and related structures;
- Analysis of the type of leadership;
- Adopting the type of innovation and discussing its open versus closed character;
- Analysis of the knowledge transfer opportunity and of the technological options;
- Economic exploitation of innovation.

This list can be reconfigured by the teacher/trainer to take into account the needs of the students and their level of knowledge in the related fields.

Table 2. Knowledge, skills and competences.

Knowledge	Structure and elements of innovation system at national, regional, sectoral and technological level; Features, elements, phases and models of innovation process (supply model, demand model, coupled model, network model and open innovation model); Types of innovation: product innovation, process innovation, marketing innovation, organizational innovation; radical innovations vs. incremental innovations; sustaining innovations vs. disruptive innovations; open innovation vs. closed innovation; Benefits and tactics of open innovation; Qualification, behavioral and situational approaches to leadership, characteristics of transactional and transformational leadership, and leadership for innovation; Concept of knowledge transfer and common approaches and techniques used for knowledge transfer; Processes of technological learning, technological catch-up and technological leapfrogging; Innovation exploration and innovation exploitation processes; management innovation
Skills	Explain interdependence and interconnection among the elements constituting the innovation systems; Compare and contract different models of the innovation process; Explain the specifics of innovation (driving factors and constraints) in low- and middle-income countries; Argue for benefits of open innovation for an enterprise; Relate the processes of innovation leadership, knowledge management and management of learning organizations; Analyze the specifics of knowledge transfer in low- and middle-income countries; Identify and explain key factors supporting technological learning, catch-up and leapfrogging at country and enterprise level; Relate the priorities of the EASTECO Strategic Plan 2017/18 – 2021/22 to the developmental challenges faced by East African countries
Competences	Systemic understanding of innovation (incl. innovation systems at different levels, innovation process, innovation types, impact of science, technology and innovation policies on the uptake of innovation in low- and middle-income countries); Awareness of the value of innovation for sustainable and inclusive development in low- and middle-income countries

Acknowledgments. This publication reflects the views only of the author, and the Commission cannot be held responsible for any use, which may be made of the information contained therein. My acknowledgements go to project partners who have participated in the AHEAD project development: Ludmil Manev, Evgenia Nikulina and others.

Funding. This research was funded by the European Commission, Erasmus+ programme, grant number 585919-EPP-1-2017-1-RO-EPPKA2-CBHE-JP.

Conflicts of Interest. The author declares no conflict of interest.

References

1. Miklian, J., Hoelscher, K.: A new research approach for peace innovation. Innov. Dev. **8**(2), 189–207 (2018)
2. Papaioannou, T.: How inclusive can innovation and development be in the twenty-first century? Innov. Dev. **4**(2), 187–202 (2014)
3. Onsongo, E.K., Knorringa, P.: Comparing frugality and inclusion in innovation for development: logic, process and outcome. CFIA Work. Paper Ser. **7**, 1–35 (2020)
4. Moldovan, L.: Review of legislation framework in the field of work-based learning. Procedia Manuf. **32**, 302–308 (2019)
5. Moldovan, L.: State-of-the-art analysis on the knowledge and skills gaps on the topic of Industry 4.0 and the requirements for work-based learning. Procedia Manuf. **32**, 294–301 (2019)
6. Opola, F.O., Klerkx, L., Leeuwis, C., Kilelu, C.W.: The hybridity of inclusive innovation narratives between theory and practice: a framing analysis. Eur. J. Dev. Res. **33**, 626–648 (2021)
7. Bukari, C., Anaman, E.A.A.: Corruption and firm innovation: a grease or sand in the wheels of commerce? Evidence from lower-middle and upper-middle income economies. Eurasian Bus. Rev. **11**(2), 267–302 (2020). https://doi.org/10.1007/s40821-020-00164-8
8. Moldovan, F.: New approaches and trends in health care. Procedia Manuf. **22**, 947–951 (2018)
9. Moldovan, F.: Framework specifications for evaluation of quality improvement and sustainable development in healthcare facilities. In: Proceedings, vol. 63, p. 2 (2020)
10. Guarascio, D., Tamagni, F.: Persistence of innovation and patterns of firm growth. Res. Policy **48**(6), 1493–1512 (2019)
11. Meyer, N., de Jongh, J.: The Importance of entrepreneurship as a contributing factor to economic growth and development: the case of selected European countries. J. Econ. Behav. Stud. **10**(4), 287–299 (2019)
12. Spescha, A., Woerter, M.: Innovation and firm growth over the business cycle. Ind. Innov. **26**(3), 321–347 (2019)
13. Moldovan, L.: QFD employment for a new product design in a mineral water company. Procedia Technol. **12**, 462–468 (2014)
14. Grillitsch, M., Schubert, T., Srholec, M.: Knowledge base combinations and firm growth. Res. Policy **48**(1), 234–247 (2019)
15. Ahead project webpage. http://www.ahead-project.net/. Accessed 22 Aug 2021
16. Moldovan, L.: Innovative tools and models for vocational education and training. In: Review of Management and Economic Engineering - First Management Conference: Twenty Years After - How Management Theory Works, no. 9, pp. 282–290 (2010)

17. Moldovan, L.: Design of a new learning environment for training in quality assurance. Procedia Technol. **12**, 483–488 (2014)
18. Moldovan, L.: Innovative models for vocational education and training in Romania. Procedia Soc. Behav. Sci. **46**, 5425–5429 (2016)
19. Moldovan, L.: Design and development of innovative tools and models for e-learning in central and western Romania. In: The 6th International Seminar Quality Management in Higher Education – QMHE2010, Tulcea, Romania Book II, pp. 543–546 (2010)
20. Moldovan, L.: Innovative method of peer assisted learning by technology and assessment of practical skills. Procedia Technol. **12**, 667–674 (2014)
21. Mikolajczyk, T., Moldovan, L., Chalupczak, A., Moldovan, F.: Computer aided learning process. Procedia Eng. **181**, 1028–1035 (2017)
22. Moldovan, L.: Sustainability assessment framework for VET organizations. Sustainability **7–6**, 7156–7174 (2015)
23. Moldovan, L., Moldovan, A.-M.: Green methodology for learning assessment. Procedia Technol. **22**, 1176–1183 (2016)
24. Moldovan, L.: The environmental pillar assessment in vocational education. Environ. Eng. Manag. J. **16**(3), 739–750 (2017)
25. Moldovan, L.: New evaluation model by means of mobile technology. Procedia Technol. **19**, 1094–1101 (2015)
26. Moldovan, L.: Training outcome evaluation model. Procedia Technol. **22**, 1184–1190 (2016)
27. Moldovan, L.: Framework development for European quality assurance in VET towards environmentally sustainable economy. Procedia Eng. **181**, 1064–1071 (2017)
28. Moldovan, L.: Practical implementation of a framework for European quality assurance in VET towards environmentally sustainable economy. Procedia Eng. **181**, 1072–1079 (2017)
29. Tobias, C.R., Downes, A., Eddens, S., Ruiz, J.: Building Blocks for peer success: lessons learned from a train-the-trainer program. AIDS Patient Care STDS **26**(1), 53–59 (2012)

Neuromarketing Tools in Industry 4.0 Context: A Study on the Romanian Market

Alexandra Ioanid(✉) and Cezar Scarlat

University Politehnica of Bucharest, Bucharest, Romania
alexandra.ioanid@upb.ro

Abstract. Neuromarketing is a relevant tool in industry 4.0 context as it allows businesses to use advance technology in order to understand better the customer, and in this way to increase the effectiveness of their marketing campaigns. This research aims at analyzing how these tools and techniques are perceived by Romanian companies' representatives, if they are using any and with what results. A small number of Romanian companies rely on neuroscientific methods like electroencephalography (EEG), functional magnetic resonance imaging (fMRI) or eye-tracking tools (E-T) when designing a product, or when testing the potential success of an advertisement, however using neuromarketing increases the effectiveness of marketing campaigns, and indirectly the business profit.

Keywords: Neuromarketing · Consumer behavior · Industry 4.0 · Consumer Neuroscience

1 Introduction

Industry 4.0 changed the way companies produce, improve and deliver their goods taking advantage of all digital solutions available, leading to more flexibility and optimization of all business processes. The main advantage besides the improved efficiency based on better usage of resources is the implementation of a better marketing strategy that helps managers influence the consumer behavior [1–3].

Traditional methods to forecast the success of a marketing campaign are based on surveys/interviews where researchers base their results on the ability and willingness of respondents to describe their levels of preferences, emotions, attention and future buying behavior in relation to the marketing/advertisement campaign or product they have been exposed to [4, 5].

Neuromarketing is a research field at the intersection of marketing, decision-theory neuroscience, economics, and psychology [6, 7] that does not focus only on the declarative component of the traditional marketing studies, but combines neuroscience and physiological measurements in order to better understand the customer behavior and to be able to forecast the customer preferences in a decision-making process [6, 8].

Neuromarketing is seen as an innovation in marketing because it involves applying neuroscience notions in marketing research, to see which stimuli produce certain reactions in customers. In this way it "helps business to facilitate product development and marketing/advertisement by underlying more about the mind of customers" [7].

L. Moldovan and A. Gligor (Eds.): Inter-Eng 2021, LNNS 386, pp. 370–381, 2022.
https://doi.org/10.1007/978-3-030-93817-8_35

Current studies describe six classifications of neuromarketing intended results and research focus for each as presented in Table 1 [7, 9].

Table 1. Neuromarketing classification [adapted after 9].

Category	Research focus
Attention/arousal	Stimuli regarding attention and emotional arousal
Product/brand appraisal	Marketing-based judgements
Product/brand preference	Preferred vs. non-preferred brands
Purchase behavior	Factors influencing purchasing decisions
Memory	Factors contributing to recall/recognition of brands
Brand extension	Successful vs. non-successful brand extensions

It is worth mentioning that in 2011 the German government promoted a strategic plan called "Industry 4.0" designed to boost the local economy by using artificial intelligence, robots and digital technologies [1, 10], that included also understanding and implementing "consumer persuasion through sensory marketing" [1, 11].

In this context, the authors want to test how the neuromarketing concept is perceived by Romanian managers and entrepreneurs, if they understand what the neuromarketing techniques imply, if they applied neuromarketing techniques in marketing researches within their companies, and if yes with what benefits.

2 Neuromarketing Tools

The study of consumer behavior by psychophysiological measures began in 1920s with electro-dermal response, continued in 1960s with pupillary dilation, and later with heart-rate measures and eye-tracking [12, 13]. The next steps in developing this science were taken in 1999 by G. Zaltman from Harvard University, when he used the functional magnetic resonance imagining (fMRI) as a marketing tool. The "neuromarketing" term was defined in 2002 by A. Smidth as neuroscience techniques used to understand customer behavior in order to improve marketing strategies.

Advertising is considered according to [14] "the area that most benefits from neuromarketing because neuromarketing is capable to record and measure the impact of advertising on the neural mechanisms of consumer and decision".

Understanding customer behaviour, or as [15] states entering "human black box" goes beyond asking questions regarding beliefs, thoughts, feelings, decision-making process, and focus on the hidden psychological and biological processes.

In Table 2, the authors adapted after [14] the most used neuromarketing techniques.

The main four elements of consumer neuroscience are: attention, emotional processing, memory and reward processing [12]. In the last years, these elements have become the main indicators of advertisement effectiveness [16].

EEG is more accurate then fMRI in evaluating customers' responses to each moment of commercials because it has a higher temporal resolution [16], while fMRI offers

Table 2. Neuroscientific methods and tools (adapted after [14]).

Method	When it's used	Advantages	Disadvantages	Cost
Functional magnetic resonance imagining (fMRI)	Testing advertisements and brand performance, prices, quality, promotion, etc	Accurate Reliable for measuring cognitive processes	Expensive, inconvenient, non-scalable, raises ethical Issues, wide room needed	High
Electroencephalography (EEG)	Testing advertisements, logos, in-store environment, website design and usability, packaging design, identifying key moments of an video advertisement	Non-invasive, relatively inexpensive, statistical software available	Non-scalable, results can be influenced by the environment factors	Low
Magnetoencephalography (MEG)	Testing advertisements and brand performance, prices, new product, packaging design	Non-invasive, good temporal accuracy	Expensive, raises ethical issues	High
Eye-tracking (E-T)	Testing what raises the attention of the customer in terms of product placement or design of advertisements	Non-invasive, statistical software available	Non-scalable, results can be influenced by the environment factors	Moderate

much better special resolution then EEG and predicts with higher success the overall effectiveness of advertisement campaigns [17].

One of the hardest thing to do it to catch customers' attention, so companies/brands study how to be more competitive from this point of view. In order to measure the attention of the subject regarding a campaign or a product, best results were obtained when using eye-tracking tools or EEG [18, 19]. We are having nowadays more information from the environment that we are able to handle, so that deciding which information is relevant for us and worth giving our attention becomes critical [20].

The eye-tracking technology seems to be the most accesible from the cost point of view. This technique can be used to measure the eye movements of the research participants in response to certain marketing stimuli in either the physical or virtual environment. This implies using one camera in the direction of the user's eyes and a second camera facing the external environment to see what the subject is focusing on.

The first such study was conducted in 1999 at Stanford University in collaboration with the Polynter Institute. The research aimed to find out what are the reactions of visitors to a news website, more precisely what is the news that attracts their attention first, what is the order in which the news is read, etc. [21] The number of participants to this study - 67 - was small compared to traditional marketing research, but high compared to neuromarketing research using IMRf or EEG equipment. The methodology of such research assumes that after the number of targeted subjects visit a commercial space or a website, or simply look at an advertisement (image, not video), a heat map will be generated. The areas that are most often viewed will be represented in light colors (yellow or orange), while the areas that are not of interest will be represented in darker colors (blue, green, black). The result of such a study consists in identifying the "hot spots", which the businesses that invest in such research, can use to increase product visibility by placing in the respective area some products or messages that better appeal to the consumer.

3 Methodology

The research objective of this pilot study is to identify trends in the Romanian industry as far as use of neuromarketing technologies. The research is divided in two parts: the first is a questionnaire-based survey, and the second is a guided interview with selected respondents. The respondents filled in an online questionnaire and later only the ones that applied/used neuromarketing tools in their companies' marketing campaigns were invited to a second qualitative study that consisted of face-to-face or telephone/skype interview.

In this study a number of 331 respondents, aged between 24 and 67 were administrated an online questionnaire. This study was conducted between May and June 2021. As an introduction to the research, the respondents were given all details regarding the purpose of this study and also were assured of the confidentiality of the data shared with the researchers.

In July 2021 the interviews were conducted with 11 business owners or marketing managers and the final obtained data was interpreted. This paper is part on an on-going research and will be further developed.

3.1 Research Results and Interpretation

Based on the first quantitative study, demographic data was analized and the results are as follows: in the study participated 144 males and 182 females, 47% aged 24–35, 30% aged 36–45, 13% aged 46–55 and 10% older than 56. The participants are highly educated: 5% graduated highschool, 50% graduated faculty and 45% graduated a master or PhD.

The respondents represent companies from various sectors of the economy, the majority (about 90% being located in the capital city, Bucharest).

Respondents were asked first to evaluate their knowledge/understanding on the term "neuromarketing" and the methods/tools associated (fRMI, EEG, E-T, etc.) on a scale of 1 to 5 (5 being the best). The results are presented in Fig. 1 and show that only obout 21% of the respondents consider they know enough and understand how neuromarketing methods are used in marketing research.

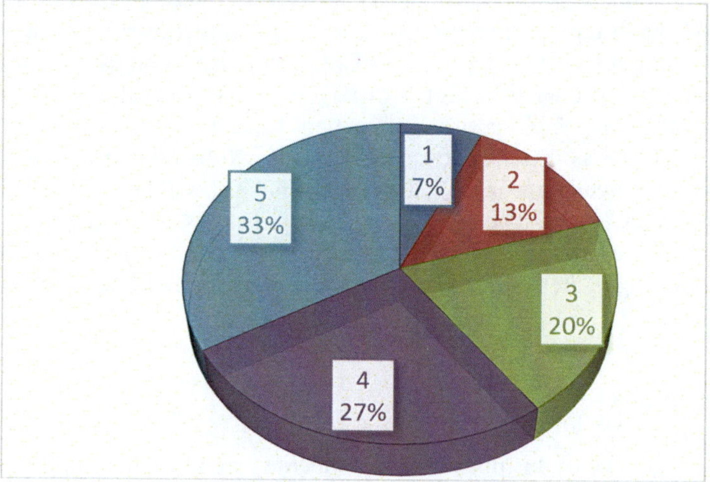

Fig. 1. Auto-evaluation on understanding neuromarketing and its' methods/tools.

The next question was applied to the respondents in order to find out if the neuromarketing methods/tools were used in the design of any campaign/product/packaging of the company they represent. Only 17 out of about 70 answered yes for this question and were asked for details on how these techniques were integrated in the marketing studies, if they relied their marketing decisions only on neuro-physiological measures or they combined these data with the traditional declarative component.

It was given more attention to these 17 questionnaires filled by respondents representing companies active on the Romanian market, either local companies or subsidiaries of global successful companies. 11 out of 17 respondents accepted to give more details on the marketing research of the company they represent, others did not have the acceptance to share this information.

The second component of the research was the interview held face-to-face in 3 cases and on the telephone/video call with 8 respondents in order to find out more about the purpose of integrating neuromarketing in their market research and what are the results.

The activity sectors of these companies are:

- Medical industry
- Hospitality industry
- Fashion sector

- Food industry
- Food supplements industry

The methods used were EEG and eye-tracking together with face coding for the following purposes:

- Select between two or more packaging options in terms of design, color, texture, etc.
- Understand what the customer needs and expects during the design of a new product
- Organize the space available in stores and place products of interest on the shelter according to points of most interest.
- Understand the reaction of customers to various advertisements in all forms (posters, video, online advertisements/campaigns)
- Understand what are the key elements that make the customer buy the product, make an appointment with a doctor (in the medical sector) or book a hotel room (in tourist industry) from the moment they visit the website or search for certain characteristics for a product/good or service.

Most significant results mentioned by companies from food and food supplements industry were obtained when working on the brand logo and product packaging. The effectiveness of changing/improving the logo or packaging according to EEG studies performed, resulted in sales growth on those particular products with up to 10–15% compared to previous version.

Also, the more research (including neuromarketing studies on carefully selected respondents) was performed before creating/launching a product in the fashion sector (clothing, watches, jewelry, etc.) the higher the sale volumes comparing to launching the products based on other criteria (fashion trends, declarative needs of customers, etc.). Again in this case EEG tool was used most, but in one case fRMI was mentioned. However, the results were similar, and EEG is a cheaper version.

In services case, for example in medical sector or in tourism, due to its rather intangible characteristic, companies focus on what makes the customer decide to make an appointment reservation and in the digital world this usually starts by visiting the website. There are certain aspects of the web page in terms of design, colors use, presentation of the service that makes the customer buy. Also, these service providers buy space in magazines and design and place the commercial in the best part of the page according again to exact measurements done with eye-tracking devices or with a combination of eye-tracking and EEG/fRMI. In these cases, it is not available a clear result of this measure, as it's never the only marketing related strategy of a company but it definitely improves the overall experience for the customer.

All 11 respondents say they test the reaction of potential customers to their advertismets video before they launch them, because TV commerial time is rather expensive and companies want to see what moments in the comercial get the most attention and if they actually create an emotion to the potential customer. Also, the actor selected to appear in the commercial/"the image of the company" is sometimes selected after deep research on the reaction of various options on the sample respondents.

Another finding worth mentioning is related to the trust managers or entrepreneurs have in agencies offering neuromarketing research services. There is a general concern regarding the ethical implications of neuromarketing, as undersanding how the customer thinks and makes decisions, gives companies more power that can be used to better meet customers' needs and desires, but at the same time that "might be used in a anegative way, to make it difficult for the customer to resit buying a certain product or service, or even worst create addictions". So, from this point of view, the authors understand the necesity of an ethical code to be followed by all the researchers applying neuromarketing methods and techniques. Respondents declared during interviews they are afraid of customer reaction on finding out they are using modern neuromarketing instruments, as these studies are still perceived as negative by the general public.

4 Conclusions, Limitation and Further Study

This paper addresses a relatively new topic in Romania, but which has a significant impact in the field of business. Even though, neuromarketing is a concept introduced in the scientific literature in 2002, being located on the border between marketing, psychology, neuroeconomics and neuroscience, it is a field of great interest for marketing specialists, because the research conducted reveals the mechanisms underlying the decision-making process and consumer behavior. Neuromarketing helps business understand better their customers, and have a clear image of what stimuli influence customers in certain ways.

The development of neurological and neuroimaging investigation techniques in recent decades and their application outside the medical field has opened new directions of research on consumer behavior and decision-making.

A major challenge for Romanian companies, especially in the context of the development of new social technologies, is to identify new ways to meet the needs and desires of the consumer. Even if it is used worldwide by many companies, in Romania not many companies want in invest in this innovative marketing tool mainly because it is much more expensive than traditional market studies, but also because until this moment it has no clear legal methodology available. The companies active in Romania that use neuro-science methods/tools in order to create marketing campaigns or to create/design products justify their decision by the fact that the cost of the campaign is high and so they want to make sure they invested correctly.

Companies with smaller marketing budget find this technique rather expensive and prefer to rely on the declarative component of the marketing studies.

On the one hand, the development of science in the field benefits industrial organizations that can better adapt their products and services to the needs and desires of the consumer, but on the other hand can use the knowledge gained and the results of neuromarketing research in a negative way. they are addictive or personalized according to consumption habits, consumer fears and the factors that can influence their decision-making.

Using the results of neuromarketing research together with the information held by social platforms, for example Facebook and associated products - WhatsApp, Instagram, Messenger can provide with high predictions about a user's preferences and decisions that the user can make.

According to Zaltman, 95% of the decisions a person makes on a daily basis are not conscious, but mechanical habits [22], so many of the products used daily also have an addictive component that we are often unaware of. The same issue is addressed by Kahneman, who in 2002 won the Nobel Prize in Economics for his research on decision-making under uncertainty [23].

Thus, it is recommended, both managers and entrepreneurs, to use new methods and techniques of neuromarketing, but in a responsible and ethical way, respecting the ethics guide promoted by the associations in the field.

The limitations of the study are related with the small sample size and with the fact that it refers only to companies active on the Romanian market, so the results can not be generalized. However, the authors will further develop this research as it is part of a wider study.

5 Perspectives

The neuromarketing tools will definitely be used in the future, under the unavoidable pressure of wide-scale practical applications (industry included) and technology development: computing power based on the next generation of supercomputers would develop, artificial intelligence devices will become kind of common place – cheaper as more usable by both businesses and their customers.

Computer science and information technology – exploring big data with appropriate algorithms – would use potential of internet to use mind genomics for digital marketing [24]. The new approach in targeted marketing offered by mind genomics allows to reach each prospect with a different personalized message. Supercomputers will allow significant progress in understanding the natural language, which is rapidly understanding individual's state and behaviour (as interest in certain product/s), and sentiment analysis [25].

In the coming years, progress in techniques for data flow control [26] as well as supercomputers development [27] are expected. Supercomputing used in the brain studies – study and also stimulation of the brain [28] would impact neuromarketing in ways that could only be subject to intuition today.

Big data is one important concept that goes together with neuromarketing, as researchers believe "the integrated cognitive construct facilitates the analysis of large data sets of human consumption and media use, which should show functions of human cognitive function". [29] The main idea is that it is not enough only to collect, store, or interrogate these type of data, but it's important to interpret and use the results obtained for best managerial decisions.

In Fig. 2, there is the graphical representation of the data used in neuromarketing studies by manual gathering during neurosciences studies and how the impact of the results would change if neuromarketing tools are used together with large amounts of available data (big data).

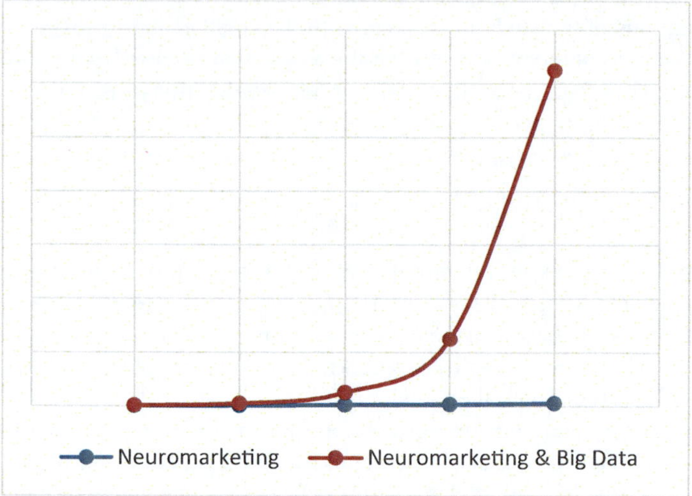

Fig. 2. Using neuromarketing tools alone versus using neuromarketing together with big data technologies.

The utility of interpreting large amounts of data gives a competitive advantage to companies willing to invest in artificial intelligence tools, but also in cloud services needed to store and at the same time protect personal data collected through neuroscience studies.

In Fig. 3, the authors emphasize on investing in big data software solutions and use them together with neuroscience tools in order to increase the overall accuracy of the research results. However, higher accuracy comes with more effort into selecting the needed data, as big amounts of data are usually raw data. Even if cloud solutions provide researches with secure ways of storing the data, the ethical concerns will probably increase with the volume stored. The implications for businesses, are presented as a conclusion of this research in Fig. 4.

It is a normal desire for every manager of entrepreneur to understand how the customer thinks, and nowadays there are available neuroscience tools that through collecting and interpreting data in neuromarketing studies, are able to predict with high accuracy what is the best package for a product or even how it should be design to meet the needs of customers, but the highest impact of neuromarketing on business refers to the efficiency of advertising, and so on growing profitability.

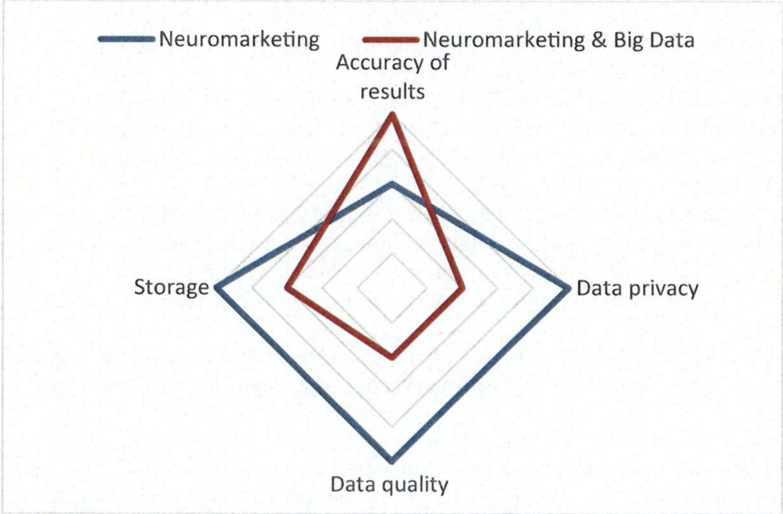

Fig. 3. Comparison between the benefits obtained by using neuromarketing tools alone versus using neuromarketing together with big data technologies.

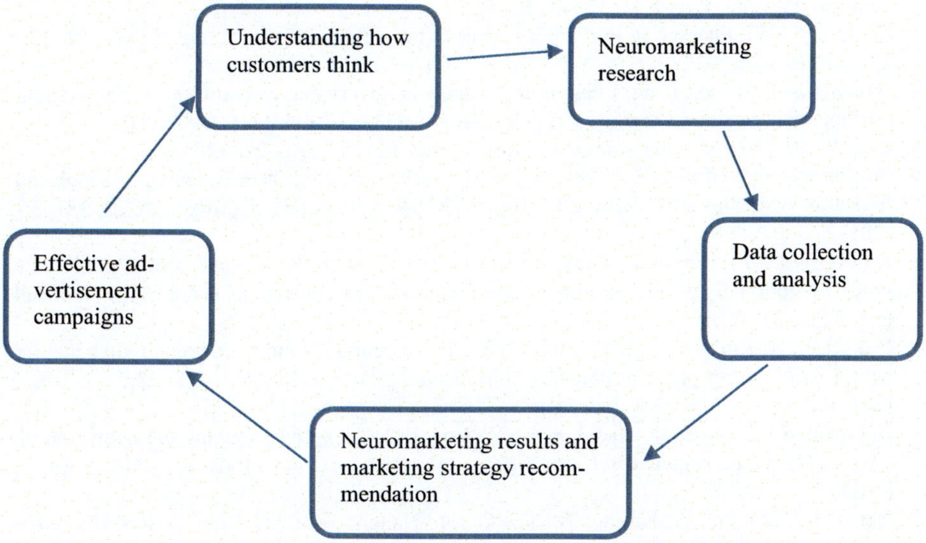

Fig. 4. The impact of neuromarketing research for business.

References

1. Jiménez-Marín, G., Elías Zambrano, R., Galiano-Coronil, A., Ravina-Ripoll, R.: Business and energy efficiency in the age of industry 4.0: the hulten, broweus and van dijk sensory marketing model applied to Spanish textile stores during the COVID-19 crisis. Energies **14**, 1966 (2021)

2. Ghosh, D., Sant, T.G., Kuiti, M.R., Swami, S., Shankar, R.: Strategic decisions, competition and cost-sharing contract under industry 4.0 and environmental considerations. Resour. Conserv. Recycl. **162**, 105057 (2020)
3. Kocsi, B., Maiko Matonya, M., Pusztai, L.P., Budai, I.: Real-time decision-support system for high-mix low-volume production scheduling in industry 4.0. Processes **8**, 912 (2020)
4. Fisher, R.J.: Social desirability bias and the validity of indirect questioning. J. Consum. Res. **20**(2) (1993)
5. Zaltman, G.: Consumer researchers: take a hike! J. Consum. Res. **26**, 423–428 (2000)
6. Nilashi, M., et al.: Decision to adopt neuromarketing techniques for sustainable product marketing: a fuzzy decision-making approach. Symmetry **12**, 305 (2020)
7. Morin, C.: Neuromarketing: the new science of consumer behavior. Society **48**, 131–135 (2011)
8. Yoon, C., Gutchess, A.H., Feinberg, F., Polk, T.A.: A functional magnetic resonance imaging study of neural dissociations between brand and person judgments. J. Consum. Res. **33**, 31–40 (2006)
9. Daugherty, T., Hoffman, E.: Neuromarketing: understanding the application of neuroscientific methods within marketing research. In: Ethics and Neuromarketing, pp. 5–30 (2006)
10. Reischauer, G.: Industry 4.0 as policy-driven discourse to institutionalize innovation systems in manufacturing. Technol. Forecast. Soc. Chang. **132**, 26–33 (2018)
11. Xu, L.D., Xu, E.L., Li, L.: Industry 4.0: state of the art and future trends. Int. J. Prod. Res. **56**(8), 2941–62 (2018)
12. Shaw, S.D., Bagozzi, R.P.: The neuropsychology of consumer behavior and marketing. Consum. Psychol. Rev. **1**(1), 22–40 (2018)
13. Ćosić, D.: Neuromarketing in market research. Interdiscip. Descr. Complex Syst. INDECS **14**(2), 139–147 (2016)
14. Alsharif, A.H., et al.: Neuroimaging techniques in advertising research: main applications, development, and brain regions and processes. Sustainability. **13**(11), 6488 (2021)
15. Lim, W.M.: Demystifying neuromarketing. J. Bus. Res. **91**, 205–220 (2018)
16. Pozharliev, R., Verbeke, W.J., Bagozzi, R.P.: Social consumer neuroscience: Neurophysiological measures of advertising effectiveness in a social context. J. Advert. **46**(3), 351–362 (2017)
17. Dmochowski, J.P., Sajda, P., Dias, J., Parra, L.C.: Correlated components of ongoing EEG point to emotionally laden attention–a possible marker of engagement? Front. Hum. Neurosci. **6**, 112 (2012)
18. Falk, E.B., Berkman, E.T., Lieberman, M.D.: From neural responses to population behavior: neural focus group predicts population-level media effects. Psychol. Sci. **23**(5), 439–445 (2012)
19. Boerman, S.C., Van Reijmersdal, E.A., Neijens, P.C.: Using eye tracking to understand the effects of brand placement disclosure types in television programs. J. Advert. **44**(3), 196–207 (2015)
20. Nobre, A.C., Nobre, K., Kastner, S.: The Oxford Handbook of Attention. Oxford University Press, Oxford (2014)
21. Pan, B., Hembrooke, H.A., Gay, G.K., Granka, L.A., Feusner, M.K., Newman, J.K.: The determinants of web page viewing behavior: an eye-tracking study. In: Proceedings of the 2004 Symposium on Eye Tracking Research & Applications, pp. 147–154 (2004)
22. Zaltman, G.: How Customers Think: Essential Insights into the Mind of the Market. Harvard Business Press, Boston (2003)
23. Kahneman, D., Slovic, S.P., Slovic, P., Tversky, A.: Judgment Under Uncertainty: Heuristics and Biases. Cambridge University Press, Cambridge (1982)
24. Salom, J.: Mind genomics with big data for digital marketing on the internet. In: Handbook of Research on Methodologies and Applications of Supercomputing, pp. 282–289 (2021)

25. Stanojevic, M., Alshehri, J., Obradovic, Z.: High performance computing for understanding natural language. Chapter In: Handbook of Research on Methodologies and Applications of Supercomputing, pp. 133–144 (2021)
26. Milutinović, V., et al.: The ultimate data flow for ultimate super computers-on-a-chip. In: Handbook of Research on Methodologies and Applications of Supercomputing, pp. 312–318 (2021)
27. Dalkilic, M.: What supercomputing will be like in the coming years. Chapter In: Handbook of Research on Methodologies and Applications of Supercomputing, pp. 309–311 (2021)
28. Dipietro, L., Elkin-Frankston, S., Ramos-Estebanez, C., Wagner, T.: Supercomputing in the Study and Stimulation of the Brain. Chapter In: Handbook of Research on Methodologies and Applications of Supercomputing, pp. 290–300 (2021)
29. Breiter, H.C., et al.: Redefining neuromarketing as an integrated science of influence. Front. Hum. Neurosci. **8**, 1073 (2015)

Particular Life Span of a Medical Rehabilitation Exoskeleton Device – The First Step in Implementing a Quality Management Norm in Real Life

Liviu Cristian Chiş(✉), Liviu Moldovan, and Monica Chiş

"George Emil Palade" University of Medicine, Pharmacy, Science, and Technology of Targu Mures, 540142 Targu Mures, Romania
liviu.chis@umfst.ro

Abstract. The medical exoskeletons are a recognized benefit in the rehabilitation domain. Still, there are no standardization norms for medical exoskeletons, except for Japanese Standard Association (JSA) that has provided norms for medical and military field. The F48 committee organized in 2018 by ASTM is responsible for the standards related to exoskeletons regulation. The main aim of our study was to determine the spam life or the fatigue life of a certain real life used medical exoskeleton. The exoskeletons are complex devices, so the hip joint piece that was affected by cracks in the user real life observation was the targeted appliance. The AnSysR15.0 software was used in order to virtual stimulate the forces applied to the hip joint device. Biomechanics showed us that the femur is the higher strength bone that support 30 times the weight of the individual, that's why the forces applied were of 700 N, 1000 N, 2000 N, 2250 N and 6500 N. Due to the complexity of the appliance and the biases induced by threaded surfaces – the data were imported from the manufacturer pdf files. The life spam represented by the fatigue life, the fatigue damage and the safety fatigue was observed. The material of the device resisted to increased forces applied on x and y axes. Intriguing, the final data showed us a decrease safety life when the tensions were applied laterally on the z ax. Standardization is needed for a good clinical practice. The manufacturer upgraded the hip device after the user's observations.

Keywords: Life Span · Exoskeleton · Standards

1 Introduction

The exoskeletons are devices that already showed their utility and versatility in the military, industrial and medical field [1, 2]. According to the definition, the exoskeletons are portable devices that generate forces on one or multiple joints in order to sustain the physical activity. The robots can be passive - storing and generating energy, active – by using an external energy source (e.g. battery) or according with the F48 ASTM's classification quasi-active or quasi-passive. The standardization committees for exoskeletons

L. Moldovan and A. Gligor (Eds.): Inter-Eng 2021, LNNS 386, pp. 382–391, 2022.
https://doi.org/10.1007/978-3-030-93817-8_36

aim to introduce them as personal protective equipment (PPE) as other robots are international standardized. This cause is currently under dispute. In order for a device to be recognized by international standardization in a certain group/class and promoted it should have the following properties: safety, quality and efficiency. Any robot device that uses energy (generating and storing it) corroborated with the human being can injure the user if not used properly. An unsafe device can induce musculoskeletal conditions such as muscle or ligaments tears, fractures. So, the safety of the exoskeleton is linked with the quality of its components such amplifying the efficacy if the first two standards are met [3–5].

In 2018 due to the lack of common and validated standards for exoskeletons and exosuits, ASTM organizes the F48 committee who has as primer objective the regulations of standards related to exoskeletons (exo standards).

In 2014, the International Organization for Standardization (ISO) has launched the 13482 certification that shows the requirements for primary robots, respectively the robotics industry (different main robotic components used in industry). This certification does not apply to the medical, military or public force robotics. Only the Japanese Standard Association (JSA) has provided norms for medical and military field [3, 6, 7]. The standardization of medical robots is still a bridge to pass.

The study will aim for the physical wear of the exoskeletons used in ambulatory setting.

The life span of an industrial robot can be between 5 to 15 years [8, 9], but generally the industry of the exoskeletons didn't offer a precise life span time frame. The span life or wear depends on different characteristics: the type of the material used in manufacturing, the storage of the robot, the way the exo was used. Apart from the moral wear of an exoskeleton or any other device that cannot be invalidated or/and improved, an individual user will wish for the highest life span/wear that the device can have [3, 4]. It does not exist generalized and individualized norms on different types of exoskeletons that can provides with the optimal number of cycles reported to the time unit and associated with the highest span life.

Any component of the exoskeleton can be affected by the storage and mode of employing; in time, the device being exposed to deterioration, cracking can harm the user. That's why the prediction of the spam life is very important. But, the life span estimation can be quite difficult due to the secreted data concerning the manufacturing of a complex device [3].

The prediction of an equipment can be limited to three engineering problems: 1) how to determine the span life of a device that wasn't turn on; 2) how to determine the time left of an already used device; 3) how to estimate the spam life or wear life of a device that was stored for a long period of timeframe between uses [3, 10, 11].

The main aim of our study was to determine in vitro the span life of an active exoskeleton used for medical rehabilitation. The secondary objectives targeted proposals/norms for enhanced the resistance and durability of the robot when used to its full potential.

2 Material and Methods

The exoskeletons are complex devices so to study them as a whole is quite challenging [12, 13]. An exoskeleton, a medical robotic is formed by different devices as in structures and as well as material. The exo that we are using most in daily living is the Phoenix MK1. It is a light, mobile, battery based, individual button control activated robot that aims the rehabilitation of lower members. It is an active exo used for posture, gain, adjustment of the balance, augmentation of the muscles and also is used in rehabilitation of the diabetic foot. It was initially dedicated to the spinal cord injury's patients starting from the dorsal root 7, so with maintenance of the functionality of the upper limbs. The spinal lesions situated at the upper spine as cervical spine the patient will experience quadriplegia/paresis so the inability to use the upper limbs and lower limbs. A spinal lesion situated at the dorsal or lumbar spine, clinically is associated with paraplegia/paresis – the inability of using the lower limbs with preserved of the upper limbs. The patients are using their arms to sustain themselves in the walker or crutches and to activate the on button in order to initiate the posture, the balance and the gain. During this maneuver different forces are activated including the overload forces that can produce damages to different parts of the medical robot. In the real-life conditions, the authors observed that the most affected part of the Phoenix MK1 was the hip joint device. The observations were noted by the manufacturer and the piece was modified.

The software AnSysR15.0 was used in order to perform a real-life simulation and notice the appearance of wear state of the Phoenix MK1 exoskeleton. The aim was to analyze only the device that represents the hip joint and not the entire robot due to its complexity. The piece was selected when the cracks appeared in real life – 468 sessions, 56160 min, and 234 days.

The hip joint device was 3D scanned in a pdf format and was imported in AnSysR15.0. A static condition analysis was performed, followed by a dynamic one.

In the static mode, the generated tensions when the strain occurred were monitored. The data collected helped us to make a projection of the future norms to be applied. The optimal conditions of using the exoskeleton were defined as the ones that didn't bring prejudices to the user.

The bias data were excluded by monitoring the device of the exoskeleton that was used by the same patient so the biometrics data were identical. An exception was performed. During the rehabilitation plan, the patient gain weight so the center of gravity changed.

The fatigue test was performed. This analysis is based on finding the number of cycles till crack appears.

The steps followed were to establish the stress life of the device. The stress life consists of the strain life (the crack initiation) corroborated with the fracture mechanics (the crack life) that will represent the total life of the piece. The strain life represents the number of the cycles realized till the appearance of the material's crack. The fracture mechanics is the so-called crack life.

The material's fatigue is expressed in cycles. There are materials with high cycle fatigue (HCF) $>10^5$ and with low cycle fatigue (LCF) $<10^5$ [14].

The weariness/the life spam of the device is represented by the number of cycles of it withstand. The number of cycles is equivalent with the alternative stress; an indirect relation exists between the two of them that defines the homogeneity or inhomogeneity of the device/piece. The alternative stress or the shear stress is defined as the stress on a material created by forces that are changing their direction or magnitude with respect to time. An Alternating Stress is associated with a cyclic force, torque, or bending being applied to a component. Its units are N/mm^2 or MPa [14].

In the pursuit of our objectives, the fatigue life, the fatigue damage and the fatigue safety factor were calculated.

The fatigue life is the life available for the given fatigue analysis. It can be applied to the whole device or strictly on certain parts, edges and surfaces. If the loading is of constant amplitude, the fatigue life will be represented by the number of cycles until the device will fail due to fatigue. If the loading is of non-constant amplitude, the fatigue life will be represented by the number of loading blocks until failure [14].

The fatigue damage or the design (available) life can be set manually and a value greater than 1 indicates failure before the design life is achieved.

The safety fatigue factor is a factor of safety with respect to a fatigue failure at a given design life. The maximum value is 15. A value lesser than 1 indicates failure before the design life is achieved.

The sensitivity of the fatigue shows how the result changes as a function of the loading at the critical point [14].

Further on, the equivalent stress represented by von Mises Stress was used. Von Mises Stress is a scalar indicator to determine the fatigue life of the device. It is applied to the flexible structures and the theory mentions that all the surfaces are undergoing the same overloading [14].

The properties of the material – the stretching, the deformities, the compression of the material were evaluated by the Young modulus, Poisson's ratio, the shear modulus and the bulk modulus.

In the virtual simulation the following forces were used: 700 N, 1000 N, 2000 N, 2250 N and 6500 N at a temperature of 22 °C on the x, y, z coordinates. The reason for choosing those parameters was imposed by the hip joint device. Biomechanics gives some important information about the anatomical region. The femur is the longest and the most powerful bone of the human body having a mean length of 48 cm with a diameter of 2.39 cm corresponding to a male with a body weight equivalent of 872 N. The femur can be uploaded up to 30 times with the body weight of the human being [14, 15].

In the graphic projection of the device, intentionally due to the biases the threaded parts of the piece weren't projected but their characteristics were imported from the pdf files. The threaded surfaces are the first ones to crack and the software generates non-real data if on a small area are a lot of threaded, coiled, sharp surfaces (in static mode) due to the fact that it generates a lot of tensions/forces.

3 Results

The characteristics of the device's material were: a density of .75e–006 kg/mm^3, thermal expansion coefficient of 1.2e–005 C^{-1}, specific heat of 4.34e+005 $mJkg^{-1}\,C^{-1}$, thermal conductivity of 6.05–002 $W\,mm^{-1}\,C^{-1}$ and a resistivity of 1.7e–004 Ωmm.

The properties of the material were expressed by: the Young modulus –2.23e+005, Poisson's ratio –0.272, the shear modulus –87657 and the bulk modulus – 1.6301e+005.

The compressive strength was 250 MPa with a tensile strength of 250 MPa at a force of 700 N with an increase to a constant of 1387 MPa. The maximum tensile strength was of 460 MPa at a 700 N with a constant of 1414 MPa maintained at 1000 N, 2000 N, 2250 N and 6500 N at a temperature of 22 °C. A drag coefficient of 920 and 0.213 was registered for ductility/elasticity at a cyclic force of 1000 MPa.

The data for the fatigue statistics are presented in Table 1.

Table 1. The fatigue life vectors of the hip joint device.

	700 N	1000 N	2000 N	2250 N	6500 N
The life (cycles) min	4333.1	3.8e+007	2.3557e+005	1.0213e+005	1.9162e+005
The fatigue damage (cycles) max	2.3078e +00.5	26.316	4245	9791.2	5218.7
The safety fatigue factor (cycles)	0.25102	1.2839	0.83998	0.74665	0.79951

The next figures - are representing the fatigue statistics for 700 N force that is more linked with the real life conditions of a patient using the exoskeleton (see Fig. 1, 2, 3, 4, 5, 6, 7, 8 and 9).

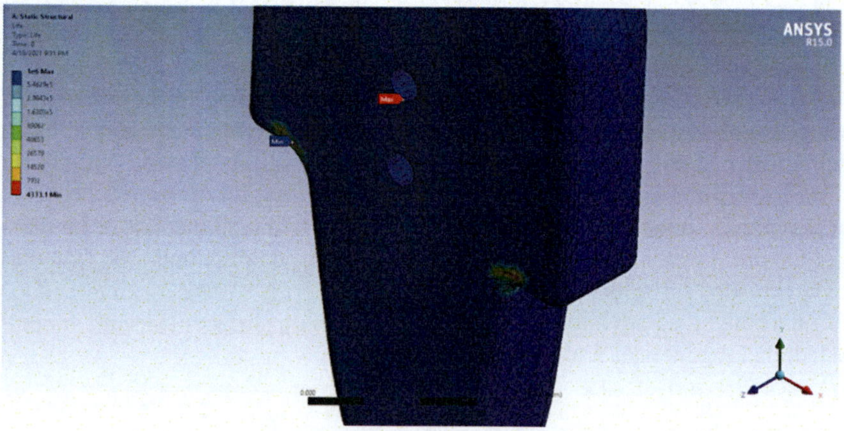

Fig. 1. The life of the device when applying a 700 N force on x ax.

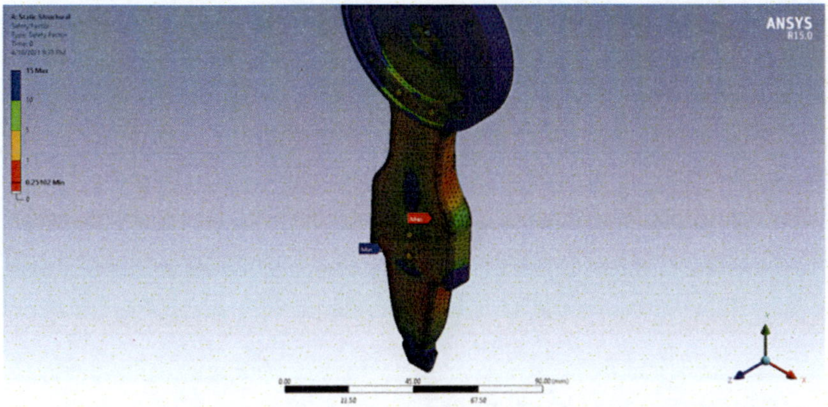

Fig. 2. The safety factor of the device when applying a 700 N force on x ax.

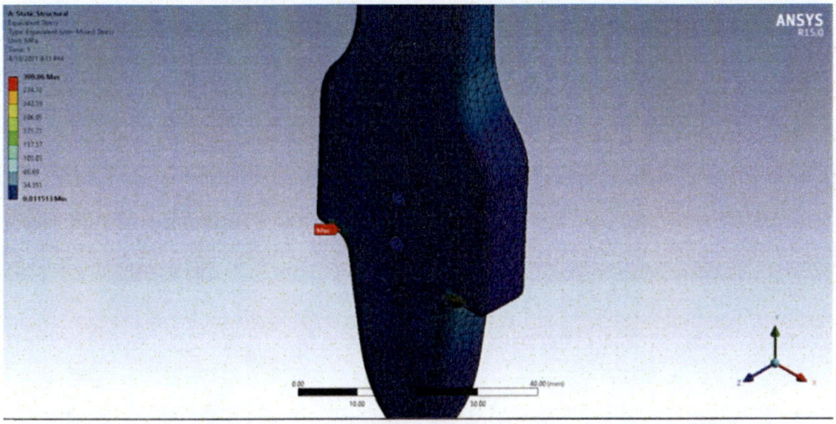

Fig. 3. The von Mises stress of the device when applying a 700 N force on x ax.

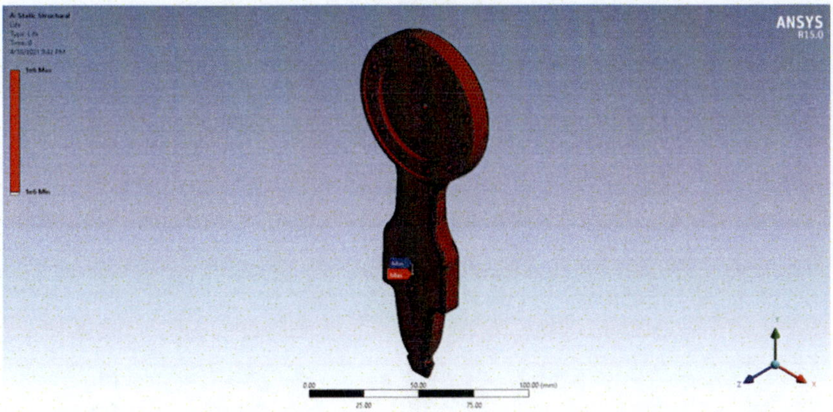

Fig. 4. The life of the device when applying a 700 N force on y ax.

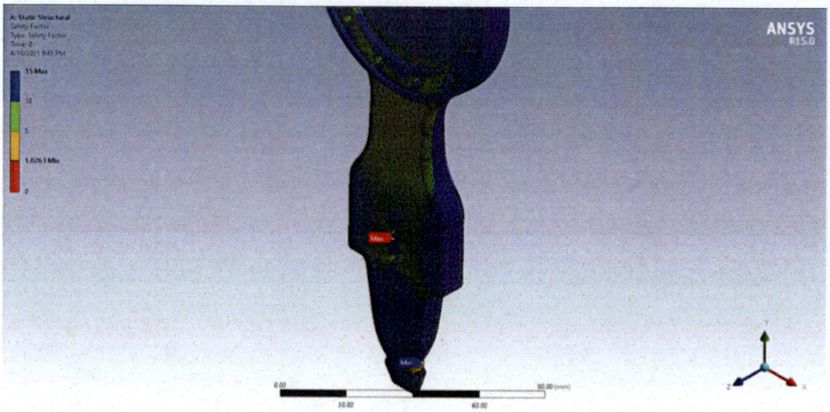

Fig. 5. The safety factor of the device when applying a 700 N force on y ax.

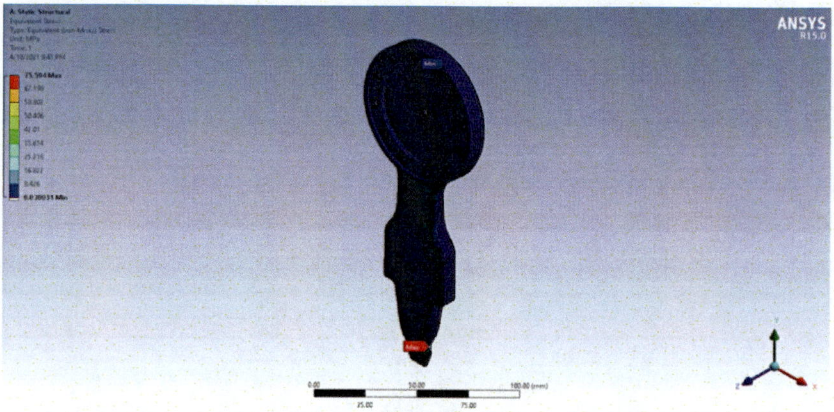

Fig. 6. The von Mises stress of the device when applying a 700 N force on y ax.

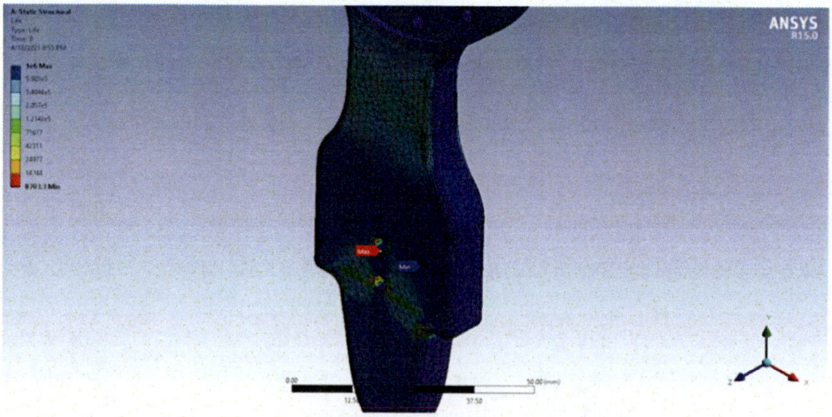

Fig. 7. The life of the device when applying a 700 N force on z ax.

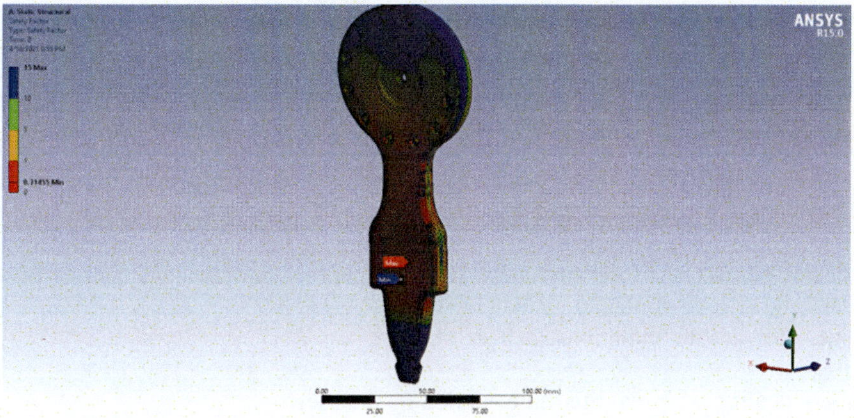

Fig. 8. The safety factor of the device when applying a 700 N force on z ax.

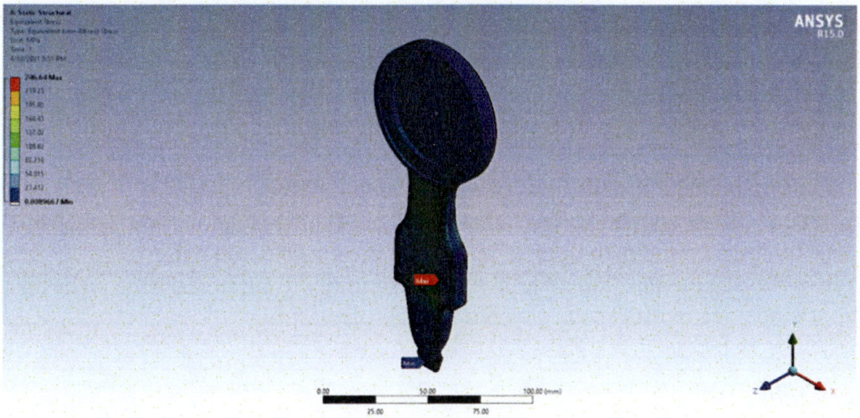

Fig. 9. The von Mises stress of the device when applying a 700 N force on z ax.

4 Discussions

The number of cycles obtained secondary to the virtual stimulation in AnSys R 15.0 showed a longevity of the fatigue life if the forces are applied on x and y axes. The surprise came when applying the forces on z ax. The device is compromised exhibiting the first cracks on applying a lateral force at 700 N such the yield stress and the tensile stress was exceeded. This can be interpreted in the context of changing the center of gravity, the pathological posture of the patient, the simulated data that couldn't predict the lateral forces strength.

Corroborating the data obtained with those already published in the journals and according with the S-N curve, the alloy is having a fatigue life at a 100.000 cycles with a mean force of 850 MPa. The stainless steel (e.g. industry) is having a fatigue life of 10000 cycles at 260 MPa decreasing the life spam at 1000 cycles at 400 MPa [16, 17].

In applied engineering, the cracks have a centripetal behavior, they develop from the surface to the depth. The majority of failures of the material depend on the surface microstructure. The upgrade of the surface and the underlying layer will increase the performance of the device.

The behavior of the device (the inhomogeneous differentiation of the metallic structure) make us believe that we are dealing with an over surface layer of titanium alloy such enveloping the previous stainless steel.

The fatigue study has identified the weak point of the hip joint device. This can be considered as one of the first steps in implementing new management of quality standards for a good clinical practice [18, 19].

The study has its limitations such as:

- lack of dealing with real life force applied by the user;
- lack of having access to the all characteristics of the device;
- lack of having access to the upgrades of the device due to confidentiality issues.

5 Conclusions

The standardizations of the medical exoskeleton are of major interests in order to accomplish a good clinical practice with minimum side effects.

The manufacturer improved the device according with the real-life observations made by the user about the life span of the hip joint device.

A force of 700 N applied on the z axes is still an Achilles tendon for the device. That may be explained by the pathological posture of the exoskeleton's users.

It is necessary the development of a software that will offer real time wear data of the hip joint device in order to prolonged its life span.

References

1. Chis, C.L., Copotoiu, M., Moldovan, L.: Different types of exoskeletons can improve the life of spinal cord injury's patients – a meta-analysis. Procedia Manuf. **46**, 844–849 (2020)
2. Ciobanu, I., et al.: The usability pilot study of a mechatronic system for gait rehabilitation. Procedia Manuf. **22**, 864–871 (2018)
3. Lowe, B.D., Billotte, W.G., Peterson, D.R.: ASTM F48 formation and standards for industrial exoskeletons and exosuits. IISE Trans. Occup. Ergon. Hum. Factors **7**, 10 (2019)
4. de Looze, M.P., Bosch, T., Krause, F., Stadler, K.S., O'Sullivan, L.W.: Exoskeletons for industrial application and their potential effects on physical work load. Ergonomics **59**(5), 671–681 (2016)
5. Lee, H., Kim, W., Han, J., Han, C.: The technical trend of the exoskeleton robot system for human power assistance. Int. J. Precis. Eng. Manuf. **13**(8), 1491–1497 (2012)
6. International Organization for Standardization: ISO 13482:2014: Robots and robotic devices -- Safety requirements for personal care robots. https://www.iso.org/standard/53820.html. Accessed 16 June 2021
7. Pesenti, M., Antonietti, A., Gandolla, M., Pedrocchi, A.: Towards a functional performance validation standard for industrial low-backexoskeletons: state of the art review. Sensors **21**, 808 (2021)

8. Moldovan, L., Gligor, A., Grif, H.S., Moldovan, F.: Dynamic numerical simulation of the 6-PGK parallel robot manipulator. Proc. Rom. Acad. Ser. A **20**(1) 67–75 (2019)
9. Moldovan, L., Grif, H.S., Gligor, A.: ANN based inverse dynamic model of the 6-PGK parallel robot manipulator. Int. J. Comput. Commun. Control **11**(1), 90–104 (2016)
10. Moldovan, L.: Trajectory errors of the 6-PGK parallel robot. In: Complexity in Artificial and Natural Systems, pp. 68–75 (2008)
11. Moldovan, L.: Geometrical method for description of the 6-PGK parallel robot's workspace. In: First International Conference on Complexity and Intelligence of the Artificial and Natural Complex Systems. Medical Applications of the Complex Systems. Biomedical Computing, pp. 45–51 (2008)
12. Mikolajczyk, T., Moldovan, L., Chalupczak, A., Moldovan, F.: Computer aided learning process. Procedia Eng. **181**, 1028–1035 (2017)
13. Mikolajczyk, T., Moldovan, F., Ciobanu, I., Chalupczak, A., Marin, A.G.: Brain research using computer test. Procedia Technol. **22**, 1113–1120 (2016)
14. Eltra – Elemental Analyzers Homepage. https://www.youtube.com/watch?v=LEHfQsu1I2Y. Accessed 16 June 2021
15. Centers for Disease Control and Prevention Homepage. https://www.cdc.gov/nchs/fastats/body-measurements.htm. Accessed 16 June 2021
16. Cao, X., Xu, L., Xu, X., Wang, Q.: Fatigue fracture characteristics of Ti6Al4V subjected to ultrasonic nanocrystal surface modification. Metals **8**(1), 77 (2018)
17. Jebieshia, T.R., Kim, J.M., Kang, J.W., Son, S.W., Kim, H.D: Microstructural and very high cycle fatigue (VHCF) behavior of Ti6Al4V—a comparative study. Materials **13**(8), 1948 (2020)
18. Mrozik, D., Mikolajczyk, T., Moldovan, L., Pimenov, D.Y.: Unconventional drive system of a 3D printed wheeled mobile robot. Procedia Manuf. **46**, 509–516 (2020)
19. Moldovan, F.: New approaches and trends in health care. Procedia Manuf. **22**, 947–951 (2018)

Study on the Short-Term Impact of the COVID-19 Pandemic on the Logistics Field

Ștefan Nagy-Bota(✉), Liviu Moldovan, Monica-Cristina Nagy-Bota, and Iulia E. Varga

"George Emil Palade" University of Medicine, Pharmacy, Sciences and Technology of Târgu Mureș, Gh. Marinescu Street 38, 540142 Târgu Mureș, Romania
nagy-bota.stefan.20@stud.umfst.ro

Abstract. For more than one year, the world we live in has changed and there is a greater degree of uncertainty regarding the future. From early 2020 onwards, the world has been going through an unprecedented wave of lockdowns, shutdowns, and preventive measures due to the COVID-19 pandemic. Many sectors were affected, but in this paper, our purpose is to analyze the implications that the pandemic had on logistics and supply chain activities. Therefore, we analyzed several bibliographic sources but also statistics published so far, to see which was the short-term impact of the pandemic. We also interviewed several companies in the logistics field from Transylvania region, Romania, to see which the consequences on their activity were, but also the main challenges they faced. According to our data, the most affected logistics service was the transport, where the delivery cost increased by up to 300%. Also, manufacturing/delivery time from Asian suppliers has increased from about 4 months to 8–10 months.

Keywords: Logistics · Supply Chain · COVID-19 · Pandemic Effects

1 Introduction

If we look back, we notice that people and societies tend to overreact to epidemics. This is on the one hand due to the media, which constantly transmits news about them, and on the other hand, to the survival instinct and fear, to the human tendency to think negatively and to exaggerate.

Although we have faced disease and pandemics throughout history, COVID-19 has been a real challenge because it was the first time in history that the entire world economy was put on hold, which added to the health challenges, various economic problems [1, 2].

From a health point of view, the effects around the world were not as severe as it had been predicted at the beginning. This can be explained by the radical measures applied by the governments. From an economic point of view, the consequences were severe, especially on developing countries [3].

L. Moldovan and A. Gligor (Eds.): Inter-Eng 2021, LNNS 386, pp. 392–401, 2022.
https://doi.org/10.1007/978-3-030-93817-8_37

According to the United Nations Conference report on Trade and Development (UNCTAD), in the first two quarters of 2020 the economy contracted more sharply than in 2008–2009, and in some cases record decreases were registered. For the entire year 2020, figures indicate a generalized global recession matching the Great Depression of the 1930s.

Particularly, enterprises are facing a variety of problems such as a decrease in demand, supply chain disruptions, cancelation of export orders, raw material shortage, and transportation disruptions, among others [4].

Lockdown has taken us into unfamiliar territory. In the midst of the global health crisis, putting lives before profits triggered a series of supply, demand and financial shocks. A full recovery means that the government will not only allocate as much money as needed until the private sector begins to reinvest, but also to handle the weaknesses that have already been exposed by the global financial crisis and then ignored for a decade [5].

The manufacturing operations worldwide were affected by the disruptions to manufacturing in China, causing many problems to the major industries around the world, from automotive industry to pharmaceuticals and health supplies or consumer goods [6].

1.1 Some Important Aspects of Global Pandemic Effects on Economy and Logistics Industry

In December 2019, we heard for the first time about coronavirus (COVID-19), an extremely contagious virus that affected the population of Wuhan, China and which soon spread all over the world. That was the reason why in March 2020, the World Health Organization declared the coronavirus outbreak a pandemic.

At the beginning of 2020, Europe was the second region in the world (after Asia) most affected by the COVID-19 pandemic. In order to reduce human contact and thus prevent the collapse of healthcare systems by stopping the spread of the virus, most European countries have implemented unprecedented restrictive measures [7].

The economic impacts of a virus can be categorized at a company level—e.g., using the division of businesses in the primary, secondary, and tertiary sectors [8]. In previous viruses epidemics, the tertiary sector, such as tourism and services, was the first sector affected, and the COVID-19 pandemic was no different.

Regarding the primary sector, in the current crisis, mining has been under lockdowns and shutdowns in different countries, and silver production was hit the most [9].

The secondary sector was also severely affected. This crisis has highlighted the vulnerabilities of global supply chains, which stem from a lack of diversification of suppliers or just-in-time strategies that led to declining safety stocks [10].

The study conducted by Statista on how the COVID-19 pandemic has affected the global retail market, shows the link between government restrictions and the blocking of certain economic sectors. The whole context we live in has changed, from going to school, work or leisure to shopping or traveling, many of them switched to online [11].

By mid-2020, specialists in both private and public spheres, expected a sharp contraction in the world economy, especially in the first half of the year, followed by a continuous recovery until 2021. Of course, the pace of recovery is dependent on the restrictions imposed by the authorities, but the global income is unlikely to reach the pre-pandemic level.

The onset and spread of Covid-19 brought about a dramatic fall in international trade flows. According to the Netherlands Bureau for Economic Policy Analysis, the world trade volume has dropped 17,7% in May compared to the same month in 2019. We've witnessed a similar situation during the Great Financial Crisis, when, in January 2009, was registered the largest single month drop of 19%.

Data provided by United Nations in their Trade and Development Report (see Table 1), show us that the volumes traded globally differ from one region to another, the most affected area being the euro one.

Table 1. Export and import volumes of goods between 2018 and 2020 (Percentage change over previous year).

Group/country	*Volume of exports*			*Volume of imports*		
	2018	*2019*	*2020a*	*2018*	*2019*	*2020a*
World	**3.1**	**-0.5**	**-8.8**	**3.8**	**-0.4**	**-8.5**
Developed Countries *of which:*	**2.6**	**0.0**	**-12.0**	**2.5**	**0.1**	**-10.5**
Japan	2.6	-1.6	-9.2	3.1	0.9	-4.4
United States	4.2	-0.5	-12.1	5.2	-0.3	-9.1
Euro area	1.9	-0.2	-13.3	2.2	0.0	-12.2
Developing Countries *of which:*	**3.8**	**-1.1**	**-4.7**	**5.7**	**-1.2**	**-5.6**
China	5.4	0.5	-4.4	6.9	-0.4	-2.0
Asia (excl. China)	3.7	-1.8	-4.4	6.9	-2.3	-6.6
Latin America	3.0	0.5	-8.7	4.8	-1.6	-12.0
Africa and Middle East	1.0	-3.9	-3.1	0.8	-0.2	-2.0

Source: UNCTAD secretariat calculations based on CPB *World Trade Monitor*, July 2020.

Here, both the volume of exports and of imports decreased by 13.3% and respectively 12.2% in 2020 compared to 2019. On the other hand, although China was the first country hit by the pandemic, it fell by only 4.4% in exports and 2% in imports.

Even though all industries have suffered from the COVID-19 pandemic and the global economy is declining, there are some sectors that have seen growth, such as e-commerce [12].

Due to restrictions and social distance recommendations, consumers made fewer visits to physical stores and turned more to e-commerce. As we can see in Fig. 1, in March 2020 compared to February 2020, more than half of respondents in Vietnam, India and China bought more goods online, compared to only 12% in Germany or 16% in France and Canada.

The impact of the COVID-19 pandemic includes also the disruptions experienced in transport links and supply chains across the globe. In order to stop the spread of the virus, the governments introduced severe restrictions that include closing down of ports and borders, constraints on air travel, as well as delays in customs processing (particularly relevant for the trade in perishable goods) [5].

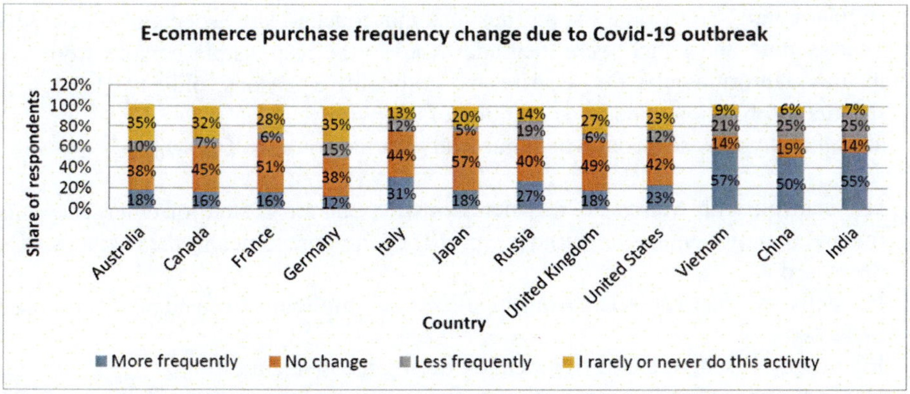

Fig. 1. E-commerce purchase frequency change due to Covid-19 in 2020. Source: statista.com [18]

According to the Institute of Shipping Economics and Logistics (ISL), air cargo volumes registered a contraction of 27.7% in April 2020 compared to the previous year, while global sea cargo fell 6.4% [19].

Large global supply chains, including the ones from the manufacturing sectors (i.e. automobile, electronics, pharmaceutical, etc.) and service sectors (i.e. airline, retail suppliers, transportations, etc.), have been substantially impacted because of their high dependence on China [13].

Several studies conducted in the manufacturing sector [14, 15], show that the pandemic will have the following three big consequences: a strong financial impact, supply disruptions would cascade down to other manufacturing sectors due to the supply-chain contagion effect, and third, the macroeconomic declines in aggregate demand, along with investment delays by companies, would undoubtedly generate demand disruptions [17].

As T.M. Choi said in his article "Risk analysis in logistics systems: A research agenda during and after the COVID-19 pandemic" [16], there are many categories of risks that we must consider in the field of logistics, especially now that COVID-19 caused so many problems. Among the challenges he mentioned are identifying a new business model, adapting to the new normality or enhancing transparency and supporting "end-to-end" visibility of the logistics systems.

2 Case Study - the Impact of COVID-19 on the Activity of Two Companies in the Logistics Sector from Transylvania Region, Romania

2.1 Methodology

Giving the fact that the purpose of this paper was to analyze the impact of the COVID-19 pandemic on the business environment and especially on the logistic sector, we thought it would be interesting to carry out a case study on several companies in Romania.

In this regard, we made a list of questions that touched key aspects of the activity they carried out, and which were then addressed to the responsible persons from five companies. Unfortunately, for security and competitive reasons, only two agreed to publish the information under the condition of anonymity.

The main hypotheses from which our study started were the following:

1. H1: starting with March 2020, when Romania entered a state of emergency, the main economic indicators (turnover, profit) of the companies in the logistics field decreased
2. H2: delivery times for orders from suppliers and, implicitly, to customers, have been increased
3. H3: supply costs increased significantly during the pandemic
4. H4: companies have increased their stocks to ensure the continuity of their activity
5. H5: customer consumption behavior has changed - fewer orders, rational consumption, and preference for certain products.

These hypotheses started from the analysis of the specialized literature, our desire being to see if the Romanian companies faced the same problems encountered in other countries and how they managed to get over the crisis period.

The interviews took place between July and August 2021, both by phone and by e-mail. Probably another factor that participated in the reluctance of the answers received was the overlap of the vacation period with that of the study.

2.2 Main Findings

Before presenting the summary of the discussions with the two companies, we must make the following clarifications:

- Company A is a company producing professional tools and hand tools, with a vast experience in the field, both in Romania and abroad;
- Company B is one of the main distributors on the market for the maintenance of industrial and agricultural equipment in Romania, operating in three central areas of Transylvania.

Considering that since March 2020, a state of emergency has been declared in Romania and many activities have been suspended, we wanted to see if this was reflected in the activity of the analyzed companies.

The data received from the first company show us that overall, the year 2020 was similar to 2019 in terms of turnover obtained in Romania, even if in April we were in full pandemic. On the other hand, the export registered a considerable decrease in turnover. On the profit side, in 2020 they had a higher profit because they decreased their expenses (fewer trips, events, promotions, etc.) and also received state aid.

The situation in the case of the second company is a bit surprising, in the sense that the total value of sales increased in 2020 compared to 2019 by over 70%, and the profit by over 80%. This can be explained by the increase in the number of individual customers, but also in the average value of a transaction (Figs. 2, 3).

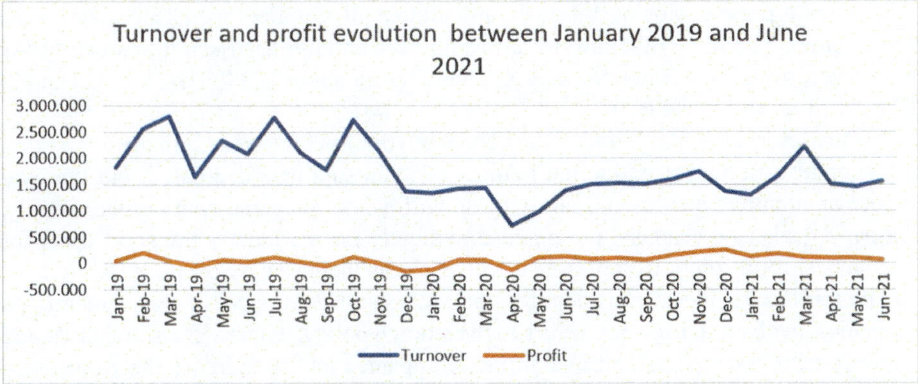

Fig. 2. Company A - turnover and profit evolution. Source: personal analysis

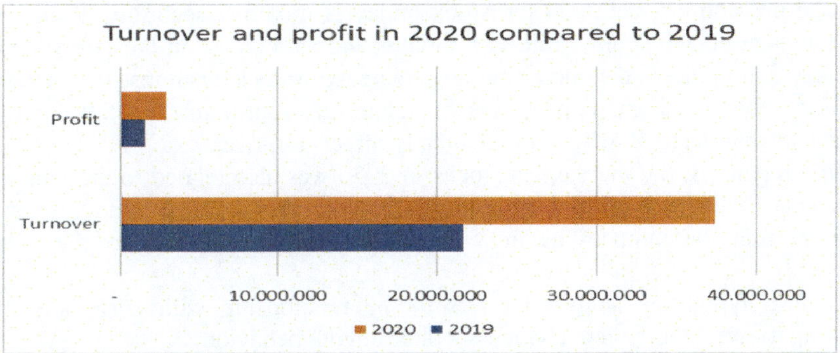

Fig. 3. Company B - turnover and profit comparation. Source: personal analysis

An important factor that led to these increases was the company's policy to sell its products online, this type of sales reaching 10% of total sales. Compared to 2019, online sales increased by 2 percentage points. On the other hand, we think the situation is similar with the construction market, which has seen fabulous growth because many people, being trapped in houses due to the lockdown, have chosen to spend their vacation money on rebuilding/modernizing their homes.

The second hypothesis tested concerned the delivery terms of orders from suppliers and, implicitly, to customers. In the case of the first company, the manufacturing/delivery time from Asia to Europe increased from about 4 months to 8–10 months. This interval also includes the duration of maritime transport, which is approximately 2–3 months. For the second company, the effects of the crisis were felt towards the end of 2020, when the terms of product delivery increased a lot, from 10–15 days to 30–60 days.

The third hypothesis refers to supply costs during the pandemic. The experience of the first analyzed company, which depended on Asian suppliers, results in a short period from 2019, but also from 2020, in which, due to the closure of factories in Asia (especially China), it could not be produced. The share of prices did not change significantly until

2021, when a price explosion followed and almost all suppliers increased their prices every quarter. The price of transport, especially sea and air, has risen dramatically. As an example, bringing a container from China will cost about 1,700 EUR in 2020 and in 2021 it will reach 15,000 EUR.

For the second company, due to the pandemic, things changed a bit in the interaction with suppliers, in the sense that the number of main suppliers decreased, but the total number of suppliers increased, mainly those with whom they had spot purchases only. Company B also experienced an increase in supply prices, mainly transport costs for containers/ships arriving from the east, whose prices increased by over 300%.

The fourth hypothesis considered in the analysis of the impact of the pandemic on companies involved in logistics, refers to the storage policy, where we assumed that due to government restrictions and quarantine, companies had to increase their inventory levels to could ensure continuity of deliveries to customers. This was confirmed in the case of both companies analyzed. Company A works with an automated management system, which introduces certain parameters such as: stock required, supply time, safety stock and promotion stock, and the stock policy is modified according to new data. For Asian suppliers, a supply time of 8 months and an increase in the safety stock by 2 weeks was introduced, the impact being an increase in stock by almost 10%. However, the major problem facing Company A is that it manages approximately 15,000 parts that are not available in stock, which automatically affects turnover.

For company B, the increase in stocks in 2020 was also related to the increase in storage costs, which were influenced only by the increase in wage costs, respectively energy and fuel, the company having enough storage space provided for the expected increase.

The last hypothesis analyzed refers to the relationship with customers, being interesting to see if there were changes in their buying behavior.

In the case of A company, as we can see in Fig. 4, the total number of invoices issued in 2020 compared to 2019 remained almost unchanged; however, there were decreases in terms of the average value of an invoice. From the data provided by the company for 2021, in the first 6 months of this year the total number of invoices from 2020 has already been reached, which means that the market is showing signs of recovery (Fig. 5).

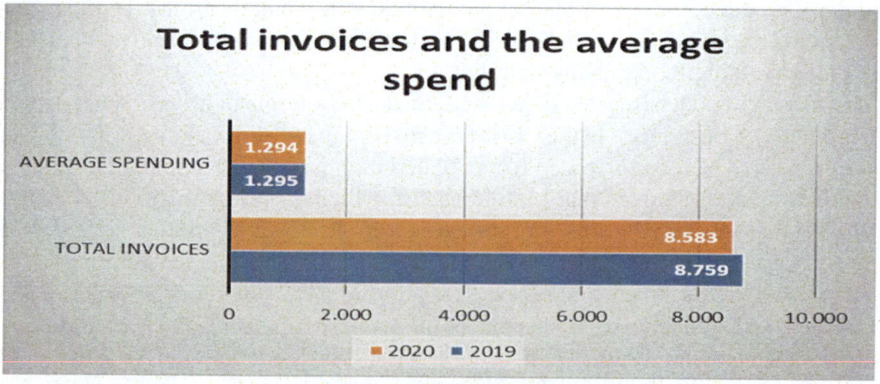

Fig. 4. Company A – total invoices and average spending. Source: personal analysis

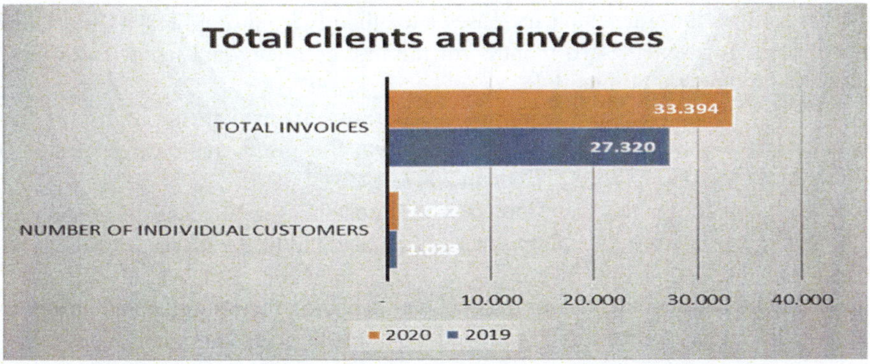

Fig. 5. Company B - clients and invoices. Source: personal analysis

In the case of company B, it is observed that the total number of individual customers increased in 2020 compared to 2019, by approximately 70 people. What is interesting is that the total number of invoices issued in 2020 is much higher than in 2019, which may mean that a customer bought more often during the pandemic than before. Also, the average value of an acquisition increased from almost 800 lei to over 1,100 lei.

One aspect that we considered important to mention is that in the case of both companies analyzed, the number of employees remained approximately the same, the differences being only two employees from one year to another. In the conditions of a global crisis, we can say that this is a success, the companies managing to organize their activity in such a way as to diminish the risks.

3 Conclusion

COVID-19 has certainly brought us in a situation for which we were not prepared, changing many aspects of daily life and putting many businesses in bankruptcy risk.

We tend to believe those who say that the COVID-19 pandemic will have an unprecedented impact on societies worldwide, as governments impose practices of social distancing and urge non-essential businesses to close, a major uncertainty about how these measures will change our lives and livelihoods appears.

In the current context, the importance of infrastructure and logistics is more evident than ever, ensuring the supply of food, medical supplies and other consumption goods.

The current crisis has also reminded us that logistics is a complex system, in which each company is a link in a chain, interconnected with each other globally, and when there is a disruption at a certain level, it turns into a domino that affects the entire supply chain.

Regarding our case study, we established several hypotheses and here are the conclusions that emerged from discussions with specialists:

Although many companies have been affected by the restrictions imposed by the pandemic, there are also areas that have seen increases in turnover and profit, as happened in the case of the parts distributor for the maintenance of industrial and agricultural equipment.

Delivery times for orders have increased a lot, mainly from suppliers in China, where from 4 months it reached 8–10 months for the first company and from 10–15 days it reached 30–60 days for company B.

It is also confirmed the hypothesis according to which the transport costs have changed dramatically due to the pandemic, the prices increasing from 1,700 EUR/container in 2020 to 15,000 EUR in 2021.

The analyzed companies were forced by circumstances to increase the level of their current stocks, but also of the safety ones, which also led to the increase of the related costs, such as the storage one.

There are no significant changes in consumer behavior, their number and the average value of an invoice being almost unchanged from one period to another.

References

1. Rickards, J.: The Road to Ruin: The Global Elites' Secret Plan for the Next Financial Crisis. Portfolio, New York, NY, USA (2016)
2. Handfield, R.B., Graham, G., Burns, L.: Corona virus, tariffs, trade wars and supply chain evolutionary design. Int. J. Oper. Prod. Manag. **40**(10), 1649–1660 (2020)
3. Radulescu, C.V., Ladaru, G.R., Burlacu, S., Constantin, F., Ioanas, C., Petre, I.L.: Impact of the COVID-19 pandemic on the Romanian labour market. Sustainability **13**, 271 (2021). https://doi.org/10.3390/su13010271
4. Shafi, M., Liu, J., Ren, W.: Impact of COVID-19 pandemic on micro, small, and medium-sized enterprises operating in Pakistan. Res. Globalization **2**, 100018 (2020). https://doi.org/10.1016/j.resglo.2020.100018
5. United Nations Conference on Trade and Development: Trade and Development Report 2020 - From global pandemic to prosperity for all: avoiding another lost decade. United Nations Publications, Geneva (2020). ISBN: 978-92-1-112992-2
6. Twinn, I., Qureshi, N., Conde, M.L., Guinea, C.G., Rojas, D.P.: The impact of COVID-19 on logistics, International Finance Corporation (IFC), June 2020
7. Fana, M., Tolan, S., Torrejón, S., Urzi Brancati, C., Fernández-Macías, E.: The COVID confinement measures and EU labour markets, EUR 30190 EN; Publications Office of the European Union: Luxembourg (2020). ISBN 978-92-79-18812-4. https://publications.jrc.ec.europa.eu/repository/bitstream/JRC120578/jrc120578_report_covid_confinement_measures_final_updated_good.pdf. Accessed 2 Aug 2021
8. Gong, B., Zhang, S., Yuan, L., Chen, K.Z.: A balance act: minimizing economic loss while controlling novel coronavirus pneumonia. J. Chin. Gov. **5**, 249–268 (2020)
9. Golubova, A.: COVID-19 Mining Shutdowns Hurt Silver Production the Most, Gold the Least—Report. Kitco News, 29 April 2020. https://www.kitco.com/news/2020-04-29/COVID-19-mining-shutdowns-hurt-silver-production-the-most-gold-the-least-report.html. Accessed 2 Aug 2021
10. Zhu, G., Chou, M.C., Tsai, C.W.: Lessons learned from the COVID-19 pandemic exposing the shortcomings of current supply chain operations: a long-term prescriptive offering. Sustainability **12**(5858) (2020). https://doi.org/10.3390/su12145858
11. Statista - Coronavirus: impact on the retail industry worldwide. https://www.statista.com/study/71767/coronavirus-impact-on-the-global-retail-industry/. Accessed 2 Aug 2021
12. Pantelimon, F.V., Georgescu, T.M., Posedaru, B.S.: The impact of mobile e-commerce on GDP: a comparative analysis between Romania and Germany and how covid-19 influences the e-commerce activity worldwide. Inform. Econ. **24**(2) (2020). https://doi.org/10.24818/issn14531305/24.2.2020.03

13. Belhadi, A., Kamble, S., Chiappetta Jabbour, C.J., Gunasekaran, A., Ndubisi, N.O., Venkatesh, M.: Manufacturing and service supply chain resilience to the COVID-19 outbreak: lessons learned from the automobile and airline industries. Technol. Forecast. Soc. Chang. **163**, 120447 (2021). https://doi.org/10.1016/j.techfore.2020.120447
14. NAM (National Association of Manufacturers): Manufacturers' Survey Reveals Current Industry Impact of COVID-19 (2020). https://www.nam.org/manufacturers-survey-reveals-current-industry-impact-of-covid-19-7411. Accessed 2 Aug 2021
15. Baldwin, R., Tomiura, E.: Thinking ahead about the trade impact of COVID-19. Economics in the Time of COVID-19, A CEPR Press VoxEU.org eBook, pp. 59–71 (2020). ISBN: 978-1-912179-28-2
16. Choi, T.M.: Risk analysis in logistics systems: a research agenda during and after the COVID-19 pandemic. Transp. Res. Part E **145**, 102190 (2021). https://doi.org/10.1016/j.tre.2020.102190
17. Hilmola, O.P., Lähdeaho, O., Henttu, V., Hilletofth, P.: Covid-19 pandemic: early implications for north European manufacturing and logistics. Sustain. MDPI Open Access J. **12**, 1–13 (2020). https://doi.org/10.3390/su12208315
18. Statista - Change in e-commerce usage to purchase products normally bought in-store due to coronavirus (COVID-19) worldwide as of March 15, 2020, by country. https://www.statista.com/statistics/1105597/coronavirus-e-commerce-usage-frequency-change-by-country-worldwide/. Accessed 2 Aug 2021
19. PwC's COVID-19 CFO Pulse Survey (2020). https://www.pwc.com/us/en/library/covid-19/pwc-covid-19-cfo-pulse-survey.html. Accessed 2 Aug 2021

An Interdisciplinary Approach to Ecological Education in the Language Class for Technical Students

Dana Rus(✉), Nicoleta Marcu, and Ymammuhammet Annagurbanov

UMFST "G.E. Palade", Targu Mures, Romania
dana.rus@umfst.ro

Abstract. The paper is the result of interdisciplinary study combining waste management and ecology studies on the one hand, and language studies on the other hand. This type of approach is relevant for a language teaching method known as Content and Language Integrated Learning (CLIL), which aims to introduce new information to students interested in particular fields of science by means of practicing language structures. The technical content of the paper is the result of research regarding packaging waste management, a category which is increasingly interesting to specialists, but also to operators involved in waste recovery and recycling. The paper includes theoretical presentations of the importance of ecological education and ecological awareness for students, an overview of the most important traits of CLIL, factual information regarding the status of packaging waste collection in the Mures county, Romania which is going to be used as language material. The paper also includes a suggestion of potential activities which can be designed starting from the content of packaging waste materials. The most significant benefits of such an approach are the increase in students' motivation and the awareness of ecological issues at a local, but also global level, as well as the practical opportunities of language practice.

Keywords: Specialized language · Ecological education · Waste management · Interdisciplinary approach

1 Introduction: Ecological Education and Language Learning

Certain basic needs are common to all people: food, water, shelter, education, and health. But the world has not yet reached the stage where basic needs are available to all people equally. Equity in the distribution of resources globally is another challenge. Certain needs of the population in developed countries endanger the needs of the population in countries with lower economic development. And, in this case, there is a strong need to educate people for a healthy life, for a healthy environment. Environmental education aims to improve the quality of life, can help people gain the knowledge, skills, motivations and values they need to efficiently manage the Earth's resources and take responsibility for maintaining the quality of the environment.

L. Moldovan and A. Gligor (Eds.): Inter-Eng 2021, LNNS 386, pp. 402–413, 2022.
https://doi.org/10.1007/978-3-030-93817-8_38

It is generally acknowledged, both at a European and at a global level, that ecological thinking and attitudes must be a priority of our educational systems. It is essential that we, educators, raise awareness regarding environmental hazards and investigate possibilities in which nature and the environmental balance be preserved in our immediate surroundings [1].

The present study aims to present an interdisciplinary method of achieving the purposes of environmental education in an academic environment. More specifically, we intend to present technical/engineering content with a strong ecological emphasis, used as language learning material. This approach is part of a method called Content and Language Integrated Learning (CLIL), which is an umbrella term including a wide range of teaching methodologies resulting to "dual focused education" where equal attention is given to both content material and the language of instruction [2]. Moreover, it has been shown that approaches which combine objectives belonging to several subject matters are likely to increase students' motivation [3, 4].

When using specialized material as language material, success is quantified by several factors: content (seen as progression of knowledge, skills and understanding related to elements of the curriculum); communication (the practical use of language in communicative contexts); cognition (development of reasoning skills linking concept formation, understanding and language) and culture (raising awareness with regard to different lifestyles and practices, accepting and celebrating difference, deepening awareness of otherness) [5].

We have selected content material focusing on a highly topical aspect of nowadays' world, namely waste management, which has a high impact on a local level, but is also a worldwide problem. In the case of teaching foreign languages, given the flexibility of the subject matter, it is relatively easy to incorporate ecological subjects, so as to meet several objectives: both linguistic ones – in terms of language acquisition and language skill development and content-based ones. In this particular case, the practical study consisted of research regarding the matter of packaging waste and its recycling suitability.

2 General Aspects Regarding the Generation of Packaging Waste

Today, the vast majority of products offered for sale are packaged. Production and consumption are increasing day by day, in parallel with industrial development. There are also increases in the materials used for packaging, such as glass, metal, plastic and paper. As the consumption of products increases, so does the amount of packaging waste. After consuming the products, the materials used for packaging remain and to many consumers they are seen as useless, and they are disposed of. Unfortunately, in this way, we create mountains of garbage in landfills or other places unsuitable for storage. About 20% of the waste we produce is packaging waste.

If the generation of waste is inevitable, then we must generate as little as possible, but also to reuse or recycle it. Society should be aware of this problem and used packaging should not be treated as waste. But when society does not act consciously on this issue, waste pollution will have a significant impact on the environment. If we are not aware of the value of packaging, we will destroy the world's natural, every day.

The European Union's policies in the field of waste management take into account, in particular, the energy recovery of waste, highlight the importance and, at the same time, the need for an integrated approach in waste management, which includes the construction of waste recovery facilities. Additionally, measures are taken to prevent/minimize the production of waste and its recycling, which comply with the following principles:

- prevention of waste production and its negative impact;
- waste recovery through recycling and reuse;
- safe final disposal of waste where there is no possibility of recovery.

3 Types of Packaging Waste

This subchapter briefly presents types of packaging waste, and for each type of packaging waste we mention the numbering system and abbreviations, according to Law no. 249 of 2015 regarding the management of packaging and packaging waste. The law regulates the management of packaging and packaging waste in order to prevent or reduce the impact on the environment.

Appendix 3 of Law 249/2015 regarding the management of packaging and packaging waste includes a numbering system for 7 groups of packaging waste. For each group, the numbering is summarized as follows:

- 1–19 – plastics
- 20–39 – paper and cardboard
- 40–49 – metal
- 50–59 – wood materials
- 60–69 – textiles
- 70–79 – glass
- 80–99 – composite [6–8]

Plastic Packaging. Plastic packaging waste is washed and melted and is often transformed into granules or flakes, so it is recovered. If a 2.5-L plastic bottle is recovered and used in production, it is possible to save 60 watts of electricity [9–11].

Metal Packaging. There are two types of metal packaging materials used in the packaging of food and beverages. These are tin and aluminum. Recycling of metal packaging by melting means the recovery of used aluminum, low energy consumption and raw materials. By producing waste aluminum from aluminum, greenhouse gas (GHG) emissions can be reduced by 95% and wastewater pollution can be reduced by 97% [10, 11].

Glass Packaging. Melting and recycling broken glass needs 32% less energy to be used than the original process. The glass powder resulting from the mechanical treatment of glass packaging mixed with sand, limestone and soda, at high temperature is molded and transformed into new products. Recycling a single bottle saves energy, which could light a 100 W light bulb for four hours [10–12].

Packaging of Paper and Cardboard/Fiberboard. Among the types of paper and cardboard, newspapers are an important source of household waste. Paper and board production organizations mix about 30% of paper waste with the original raw material and use it in production. If a significant part of household waste is collected selectively, then it is also possible to recycle it. As a result of recycling paper and cardboard packaging, the amount of waste is reduced, and energy and natural resources are saved. For example, recycling a ton of wastepaper saves 17 mature trees. To this end, public awareness should be raised on this issue and the necessary measures should be taken by ensuring cooperation between public institutions and organizations for the separate collection of waste [10, 12].

Composite Packaging. Composite packaging is packaging made of different materials that cannot be separated manually. The composite material represents a set of distinct materials, which has characteristics that the individual constituent materials lack. These materials were developed in the aerospace industry, out of the need to control and improve the properties of the materials, in accordance with the requirements imposed by the destination. When we say that packaging is a valuable material, we should not neglect this. If each type of waste is handled correctly, it will not turn into garbage.

4 Packaging and Packaging Waste Collected by Sanitation Operators and Other Authorized Economic Operators in Mures County, Romania

The present chapter contains detailed information collected for the purposes of the current study. The localization of the research at a community level is intended to narrow down the field of knowledge to aspects with which students are familiar. The purposes of such an approach are to motivate the students and raise awareness with regard to environmental topics in their close proximity. For linguistic purposes, a variety of techniques and activities will be suggested in following chapters starting from the content presented below.

The effective management of packaging waste is carried out by operators/collection agents and treatment operators (sorting stations, economic operators which carry out recovery, but also landfill operators, who take over unused packaging waste). According to the data presented by Environment Protection Agency in the Mureș County, Romania, in 2019 there were 22 economic operators who were authorized to collect packaging waste directly from generators (both from the population and from industry and trade). However, the monitoring of the collection of packaging waste by the economic operators who collect packaging waste is not done at a city or county level. Monitoring is done at the level of each operator, who sometimes collects waste from neighboring counties as well. This can lead to some problems with a potential impact on the environment.

In Table 1, one my notice a reduction in the amount of waste collected in the period 2015–2019 (with a small deviation in 2018, compared to the previous year), from 24,700 tons collected in 2015 to 10,689 tons collected in 2019. This is due to the fact that less packaging waste was generated, and, at the same time, less packaging entered the

Table 1. Quantities of packaging and packaging waste collected by sanitation operators and other economic agents in Mures county [7].

Mureş county/Year	Waste quantities (tons)				
	2015	2016	2017	2018	2019
Total	24,700	22,833	12,555	13,181	10,689
From individual households	22,069	20,450	11,461	12,052	10,378
From institutions/companies	2,631	2,383	1,094	1,129	311

collection area. Packaging waste is collected by both sanitation operators and economic operators authorized for the collection and recovery of this category of waste.

The situation of the quantities of waste collected in the last 3 years (2017–2019) only by the sanitation operators, from the population, but also from economic agents is presented in Table 2 and Figs. 1, 2, 3, 4 and 5.

Table 2. Quantity of packaging waste collected by sanitation operators in the period 2017–2018 [7].

Year	Waste code	Collected/recovered quantity (tons) from:		
		Population	Economic agents	Total
2017	15 01 01	614,719	161,536	**776,255**
	15 01 02	2367,310	609,135	**2976,445**
	15 01 03	0,000	0,000	**0,000**
	15 01 04	4,140	1,630	**5,770**
	15 01 07	97,390	21,380	**118,770**
Total-2017	–	3083,559	793,681	**3877,24**
2018	15 01 01	736,799	231,870	**968,669**
	15 01 02	3221,024	853,841	**4074,865**
	15 01 03	12,000	2,000	**14,000**
	15 01 04	105,908	29,243	**135,151**
	15 01 07	218,480	11,440	**229,920**
Total 2018	–	4294,211	1128,394	**5422,605**
2019	15 01 01	626,165	55,987	**682,152**
	15 01 02	1418,050	201,081	**1619,131**
	15 01 03	0,000	0,000	**0,000**
	15 01 04	177,086	14,914	**192,000**
	15 01 07	153,657	38,163	**191,820**

(continued)

Table 2. (*continued*)

Year	Waste code	Collected/recovered quantity (tons) from:		
		Population	Economic agents	Total
Total 2019	–	2374,958	310,145	**2685,103**
Total 2017–2019	–	9752,728	2232,22	**11984,948**

One may note that in 2017 and 2019 the wood packaging waste (code 15 01 03) was not collected by the sanitation operators, because the respective operators did not carry out recovery operations. During that period, the authorized economic operators also dealt with the collection of these types of waste in the recovery activity. It can also be observed that for the other 4 categories of waste, the quantities collected from the population were higher than those collected by the sanitation operators from the economic units, and the total quantity collected was 11984,948 tons for the entire period.

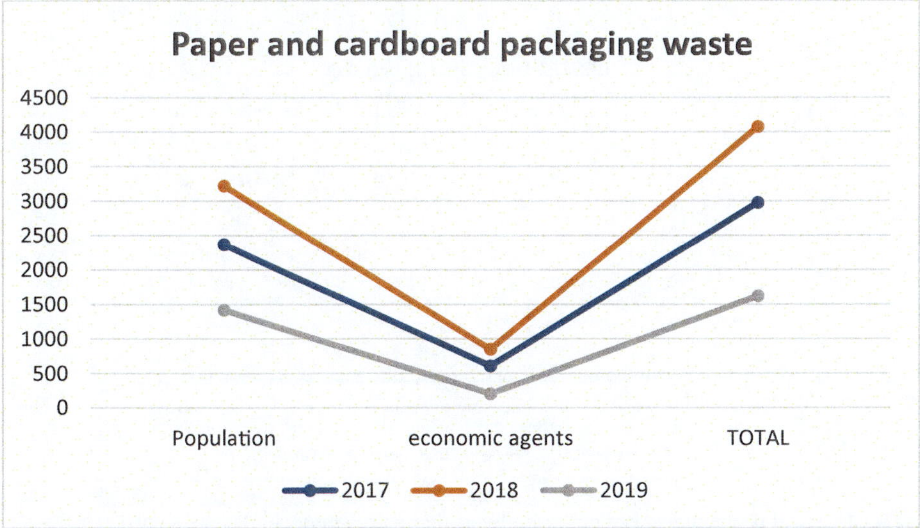

Fig. 1. Paper and cardboard packaging waste collected by sanitation operators in the period 2017–2019.

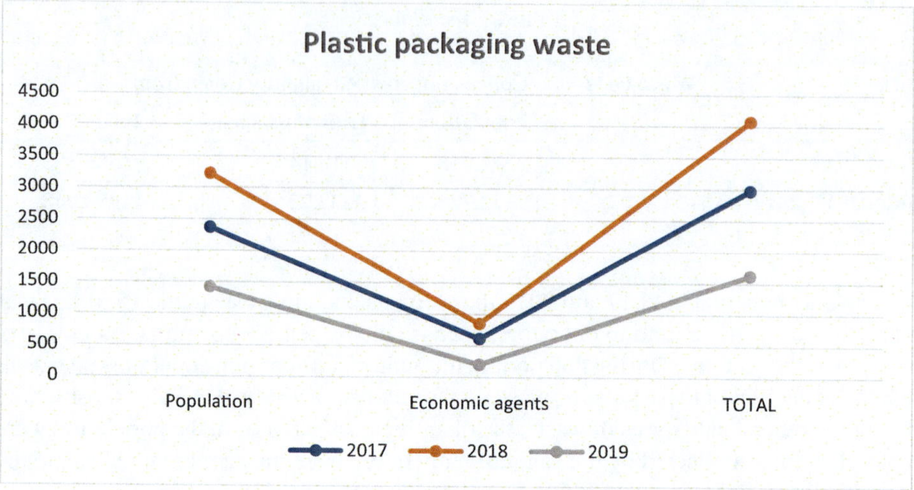

Fig. 2. Plastic packaging waste collected by sanitation operators in the period 2017–2019.

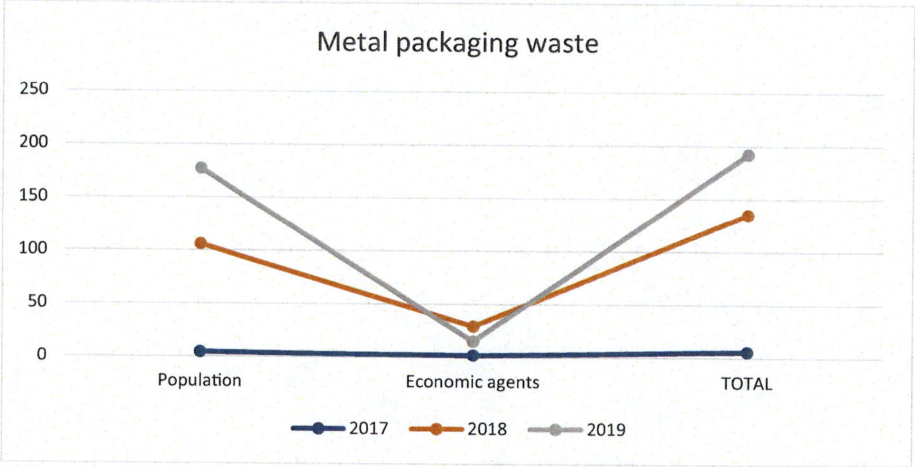

Fig. 3. Metal packaging waste collected by sanitation operators in the period 2017–2019.

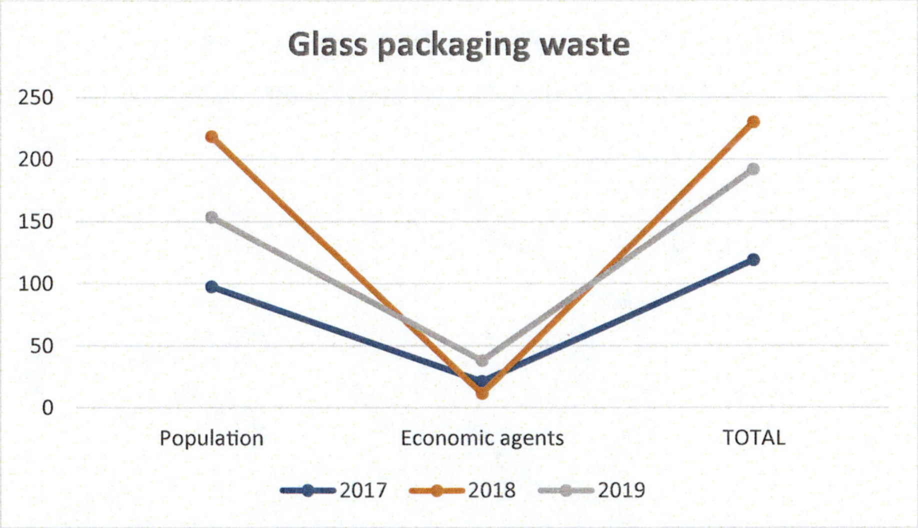

Fig. 4. Glass packaging waste collected by sanitation operators in the period 2017–2019.

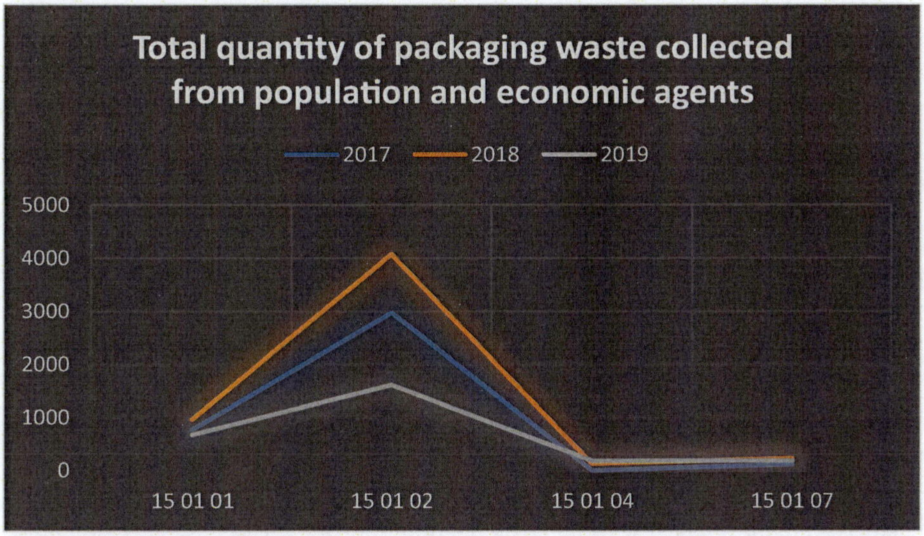

Fig. 5. Total quantity of packaging waste collected from population and economic agents.

Fig. 6. Quantities of recovered packaging waste and used via recycling, according to the type of material, over a period of 10 years, at a national level [13].

In Fig. 6 one may note a small difference between the amount of packaging waste recovered and recycled, by type of component material, in the period of the study; there is a total difference of 329,996 tons of waste that has not been recovered by recycling.

5 Language Learning Opportunities

Starting from the information above, language instructors can devise activities for language practice and more. There are undeniable advantages in using factual, topical information as language content. Among the most relevant ones, we mention:

- This method raises awareness with regard to serious worldwide topics and contributes to the sustainable management of these issues;
- Students are motivated intrinsically, due to the relevance of the selected themes in terms of their professional interests;
- CLIL-inspired methodologies create the sense of real-life, reducing a certain artificiality sometimes attached to language learning outside a specific epistemological area;
- The content used as language material can be used to practice virtually all language areas (vocabulary, grammar, pronunciation, language functions) in an experiential way, emulating reality;
- CLIL methods create realistic contexts for the practice of all language skills;
- CLIL activities favor group work involving collaboration and thus practicing soft skills;
- CLIL-based assessment allows project-based evaluation, with beneficial effects in developing students' professional abilities;

- CLIL facilitates the acquisition of authentic information on a variety of specialized fields, according to students' fields of interest.

In practical terms, a selection of the most suitable activities to be implemented in the language class starting from the information regarding packaging waste management may include the following. We mention in brackets which area of the language or language skills they are meant to practice:

- Based on the information in the paragraph describing packaging waste, identify the main types of such waste and their main characteristics (reading skills). Present them orally (speaking skills) or write a paragraph with the main findings (writing skills, summarizing skills).
- Look at chapter 3, "Types of packaging waste". Read the text and identify vocabulary which can be connected to environment protection. Write them under three separate headings: nouns, verbs, and adjectives (reading for detail, scanning, vocabulary). Follow up: make up sentences using a verb, a noun and an adjective in a logical way.
- In groups, discuss and compare the information provided in Table 2: Quantity of packaging waste collected by sanitation operators in the period 2017–2018. Report on the following data: the quantities of collected/recovered waste in each year of the interval 2017–2019, a comparison between the quantities collected from the population and the economic agents for the same period of time, most significant differences (scanning for information, information transfer, reading skill, summarizing skills, speaking skills, language for comparing figures).
- Look at Figs. 1–5. Describe the most significant findings and compare the trends of the quantities of recovered and recycled packaging waste (information transfer, speaking skills, language for describing graphs, language for describing trends).
- Project-based assessment: in groups, identify the most significant European Union policies regarding waste management and present them in class (research skills, specific vocabulary acquisition, presentation skills, collaborative skills).
- Project-based assessment: in groups, devise a questionnaire to be used in a survey regarding attitudes towards selective waste. Apply the questionnaire in your community. Present your findings in class (writing skills, specific vocabulary acquisition, presentation skills, collaborative skills).
- Group work: make a glossary of the most significant specialized vocabulary related to waste recycling. Give Romanian equivalents. Create a web page/make a diary entry to disseminate your work (research skills, vocabulary acquisition).
- Do some research to identify and understand the specific ways in which a local company of your choice is implementing waste collection and recovery. Ask the necessary questions to find detailed answers. Use the results in a presentation to the class. Include information, descriptions, figures, pictures, videos (research skills, specific vocabulary acquisition, presentation skills, collaborative skills).
- Write a letter to the city hall suggesting ways in which waste collection in your town can be improved (letter writing, argumentative language, descriptive language, specific vocabulary).

6 Conclusions

The interdisciplinary approaches to education are the trend of the future. More and more, educators all over the world are recognizing the advantages of this type of philosophy towards learning. It is an approach which emulates real life: in reality, we use the language as an instrument to communicate facts. On the other hand, ecological education is a trend which is going to be more and more present in all educational activities. While it is true that languages seem to be the perfect medium for the dissemination of its objectives, other subject matters can also be adjusted so as to incorporate topics and notions regarding environment protection. Education is, indeed, the means which is likely to result in the most visible and certain results. Cultivating the minds of young people in the spirit of respect, of inclusion, of protection for the world in which we live is a sustainable method to care for the wellbeing of the entire world.

In order to achieve this, it is necessary that educators realize their common goals and cooperate in order to create meaningful activities combining factual information and language notions. The task is complex insofar as it implies a revision of a lot of language teaching areas: from content to materials, to student and teacher's role, to teaching objectives and philosophies as to the essence of learning. However, this is a necessity in order to successfully meet educational purposes and to provide the necessary takeaway to the beneficiary of the educational process, i.e. the student [14]. Innovative approaches are required, together with a reconsideration of basic teaching goals: to equip the students with the best skills and practices to adjust to the world in a considerate, successful and respectful way.

References

1. Giolitto, P.: Environmental education in the European Union (1997). https://op.europa.eu/en/publication-detail/-/publication/1168d16b-e998-4ff9-a864-606268b15cb6. Accessed 20 May 2021
2. Kovács, J.: CLIL – early competence in two languages. In: Kovács, J., Benkő, É.T. (eds.) The World at Their Feet: Children's Early Competence in Two Languages through Education, pp. 15–97. Eötvös József Könyvkiadó, Budapest (2014)
3. Huang, K.-M.: Motivating lessons: a classroom-oriented investigation of the effects of content-based instruction on EFL young learners' motivated behaviours and classroom verbal interaction. System **39**, 186–201 (2011)
4. Arnó-Macià, E., Mancho-Barés, G.: The role of content and language in content and language integrated learning (CLIL) at university: challenges and implications for ESP". Engl. Specif. Purp. **37**, 63–73 (2015)
5. Coyle, D.: CLIL – a pedagogical approach. In: Van Deusen-Scholl, N., Hornberger, N. (eds.) Encyclopedia of Language and Education, 2nd ed., pp. 97–111 (2008)
6. http://www.mmediu.gov.ro/. Accessed 12 Feb 2021
7. http://www.anpm.ro/docments/24337/52297582/Raport+anual+privind+starea+mediului+-+MURES+2019.pdf/4e2bb36c-cc82-41f6-9c15-d50c02f960bd. Accessed 12 Feb 2021
8. https://ec.europa.eu/commission/presscorner/detail/ro/MEMO_07_7.Strategia Energetică a României 2016–2030, cu perspectiva anului 2050. Accessed 12 Feb 2021

9. https://www.eca.europa.eu/Lists/ECADocuments/RW20_04/RW_Plastic_waste_RO.pdf. Accessed 12 Feb 2021
10. http://www-old.anpm.ro/docfiles/view/%20144477-excel. Accessed 12 Feb 2021
11. http://www.anpm.ro/ro/deseuri-de-ambalaje/-/asset_publisher/slQIjt0nSvje/content/cantitati-de-ambalaje-introduse-pe-piata-nationala-si-obiective. Accessed 12 Feb 2021
12. https://eur-lex.europa.eu/legal-content/ro/LSU/?uri=CELEX:32018L0852. Accessed 12 Feb 2021
13. https://www.ambalaje-romania.com. Accessed 12 Feb 2021
14. Han, B.: What Else is New in Teaching Linguistics? On the Use of ELSE Methods and Tools. Acta Marisiensis. Philologia 2, University Press, Tg.-Mureş, ISSN: 2268-9537, pp. 96–101 (2020)

Smart Production Concept and Improvements in Automotive Industry

Cristina Veres(✉) and Roxana-Diana Dobrau

"George Emil Palade" University of Medicine, Pharmacy, Science, and Technology of Targu Mures, Gh.Marinescu str. 38, 540142 Targu Mureş, Romania
cristina.veres@umfst.ro

Abstract. Smart production is becoming more and more widespread in manufacturing industry, due to its several advantages, such as high adaptability, rapid design changes, increased production efficiency etc.

For manufacturing companies, the execution processes encountered in production are of the utmost importance and require careful control for quality assurance, timely order fulfillment and cost efficiency.

The article starts with an introduction to the Industry 4.0 and Smart Production concepts, then describes a smart solution implemented by an automotive manufacturer. The practical part of the work is represented by the improvement ideas brought in the assembly process to avoid the risk of mixing the products. The actions involved line optimization, as well as changes in the process flow, as the old process flow was allowing errors to be made by operators. The paper presents the assembly method on Spider type equipment, as well as the taken actions to streamline production and save costs.

Keywords: Automotive · Industry 4.0 · Manufacturing · Quality · Smart production

1 Introduction

The introduction of the Internet of Things and Services into the manufacturing environment is ushering in a fourth industrial revolution [1]. Industry 4.0 is a Program initiated by German Government and Industry as a new model of automatization of manufacturing technologies [2]. The concept describes digital systems in which all parts: operators, equipment, installations, logistics, raw materials and products are all interconnected. Industry 4.0 comes as an expected evolution of previous industrialization processes that stimulate the digitization of economic operators for efficiency and optimization of production, cost control and sustainable profit growth. Moving forward to new levels of industrialization, smart manufacturing represents systems integration and automation supported by Cyber-Physical Systems (CPS) to enable more autonomous, agile and sustainable production processes [3].

Internet of things (IoT) can provide real-time production events, which represent the change of material flows [4]. Due to the wide usage of digital devices and easy access

L. Moldovan and A. Gligor (Eds.): Inter-Eng 2021, LNNS 386, pp. 414–425, 2022.
https://doi.org/10.1007/978-3-030-93817-8_39

to the edge items in manufacturing industry, massive industrial data is generated and collected [5].

In the current socio-economic situation, smart products are essential for daily life [6]. Through Smart Production companies are able to gain competitive advantage with this new technology [7].

This study presents the concept of Smart Production inside an automotive factory, as well as its implementation on the global scale of the company.

Hirschmann Automotive is an Austrian company which manufactures components for the automobile industry. The company can be found in seven different countries in Europe, Asia and America.

The concept of digitalization started with the 4.0 Industrial Strategy, a project which is based on the technical vision of the products. This industrialization holds a series of interconnections for an organized production system, such as production processes, logistics, etc.

One of the most important reasons of Smart Production implementation is the concept of product traceability. Traceability is defined as "the ability to follow the movement of a food through specified stages of production, processing and distribution" [8, 9].

In this regards, Hirschmann Automotive has defined and implemented a traceability standard process. This concept brings a series of bonuses to the industry, such as sustainability, lower costs and high product quality.

The smart factory involves the transition from traditional automation to complex, flexible and interconnected systems. The new system adapts to the new standards, using its own statistics, based on the history of production processes. The industry integrates information for the whole system in order to have constant development in production engineering, digitalization and logistics material management.

2 Smart Production in Hirschmann Automotive

2.1 SAP Manufacturing Execution

For manufacturing companies it's important to have an overview of the quality in every process, taking into consideration the delivery time schedule and cost efficiency. In this regard, the planning and logistics management are controlled by a resource planning system.

SAP Manufacturing Execution, called shortly SAP ME, is a comprehensive manufacturing execution system (MES) solution for discrete manufacturing processes [10].

In Hirschmann Automotive the system SAP ME was implemented in 2019 at the prototype level and in this moment, it is being expanded to the series production level. The system was tailored according to company's requirements and needs to ensure a lean management and control of the production processes.

There are few challenges companies are facing at the production level from IT perspective:

- Design and control of manufacturing stages;
- Traceability;
- Non-compliance management/quality control;
- Real time data capture;
- Real time reporting;
- Production line control;
- Integration of an ERP system (Enterprise Planning System);
- Assistance.

Being properly and professionally addressed, these challenges turn into clear advantages for the company.

At another level, Smart Production brings lots of opportunities, but lots of challenges as well. A SWOT analysis of the Smart Production concept in Hirschmann Automotive company was developed to get a better overview (see Table 1).

Table 1. SWOT analysis of Smart Production in Hirschmann Automotive.

Strengths	*Opportunities*
• Customer orientation; • Solving competence a customer problems; • Market Knowledge; • Growing the niche market; • Global presence; • Technological leadership, • Special tools and machines; • Work equipment; • Cooperative culture of the company	• Risk management; • Expanding the client portfolio and products; • Lean management; • Smart factory; • Authentic brand of the employer; • Qualified and motivated employees; • Excellent management; • Business intelligence
Weaknesses	*Threats*
• Internal customer orientation; • Entrepreneurial thinking and actions; • Efficiency; • Target costs; • Capital fund management; • Transparency of costs in the indirect areas; • Product risk assessment and new processes; • Production and logistics processes stable and transparent; • Prototype management; • Knowledge transfer (internal); • Management of key positions	• Consistent fulfillment of all customer requirements; • Acquisition risk; • Transfer prices; • Information security

To shortly detail the presented SWOT analysis, Hirschmann evaluates the customer orientation as their strength, as they strongly focus on the customer satisfaction by ensuring quality of the needed products, but also data transparency. A smart factory can offer all these "features" due to the work equipment and the technical vision acquired.

2.2 Assembly Process in a Smart Factory

For a better understanding, more information will be provided on the assembly process.

Assembly is the process by which two or more parts, mechanisms of a system are combined.

For product groups in the "axial wiring" category, there is required an implementation of a Lean type of device to transport the wiring together with a housing by rotating them, so that the assembly step can be possible.

When starting the digitalization project, the old holder was transformed into a spider (see Fig. 1).

Fig. 1. Workstation before and after Smart Production implementation.

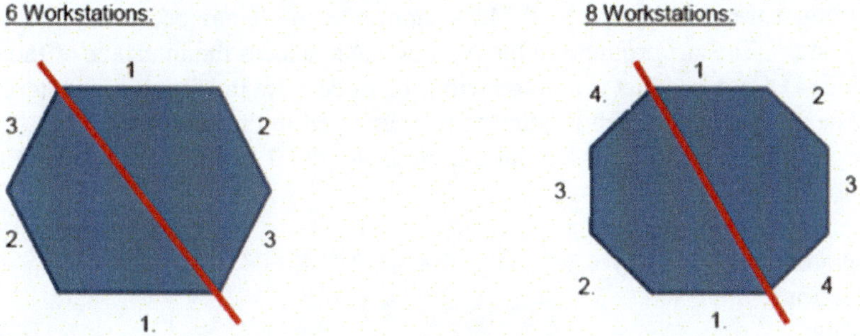

Fig. 2. Workstation models.

To be able to absorb the variety of products the equipment is divided into three or four workstations, as presented in Fig. 2.

With Spider type equipment two products can be manufactured at the same time, the quantity of steps can be six or eight. Used material must be scanned before production start to be approved as good parts and released by SAP ME system.

The Spider equipment consists of two parts, an upper and a lower part, as presented in Fig. 3.

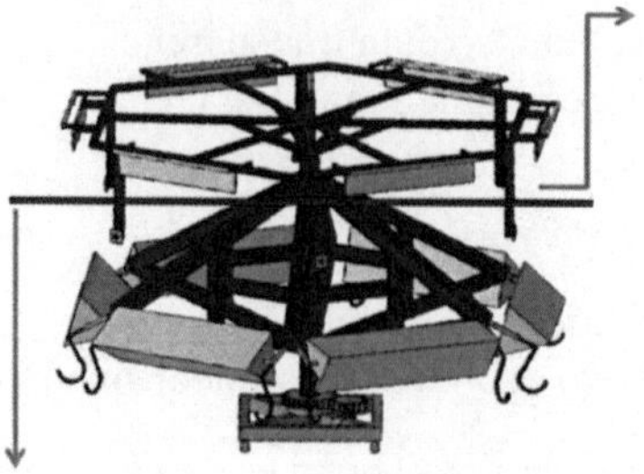

Fig. 3. Spider equipment.

The upper part is the fix part of the equipment and contains light; scanner and tablet; metal plate to display documents; signal lamp and start/stop button. The lower part represents the moving part, rotating in a circle, and it contains: LUB (Ergonomic work surface/adjustable tilt); input/output area; equipment engine and brake.

By dividing the workstations into 3/3 or 4/4 a double output is obtained, i.e. two products at the same time. By rotating the equipment, the operator does not have to move from one workstation to another. The equipment rotates at a time interval of 5–30 s, this processing time is adjusted by a responsible person according to the production process requirements.

3 Improving the Assembly Process

3.1 The Current Concept

In Hirschmann production the equipment is connected to the internet network.

For product traceability the SAP ME system was implemented, as an extension of the SAP system. To start production the worker has to access the interface of the digital system (Fig. 4). Each product is marked with a barcode from the first process step, called cutting. The worker scans each product to be registered in the system.

LU represents a product passed through process step. LU wiring must include most of the following process steps:

- Cutting and marking;
- Certification;
- Assembly;
- Injection;
- Dismantling;
- Testing.
- The products will be assembled on wooden assembly boards called LUB (lay-up board).

The assembly of LU products with housings is planned on these boards, the final product is called ZB. A ZB harness includes two or more LU harnesses, which consists of assembling LU products into enclosures on Spider-type equipment.

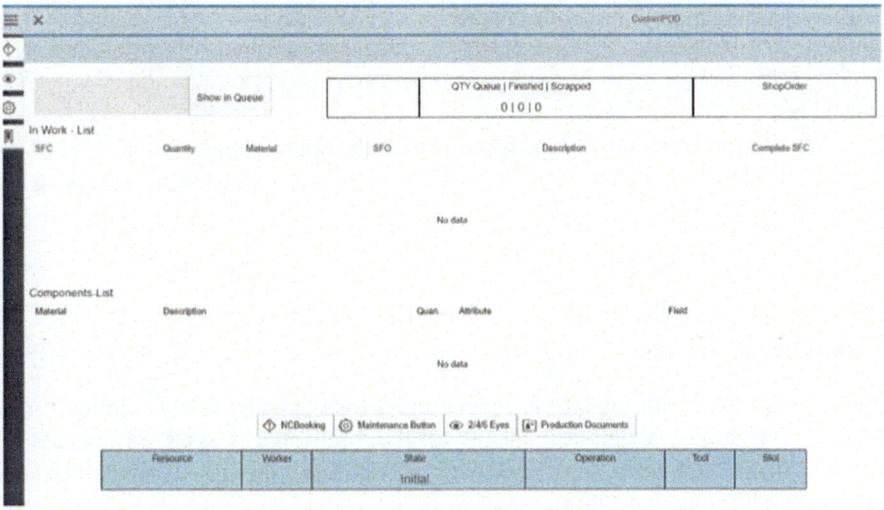

Fig. 4. SAP ME system interface.

ZBs include fewer process steps:

- Assembly;
- Testing and labeling;
- 100% verification.

But the current working way still faces some issues, as it has no measure to prevent mixing products and, because of this, some final products are affected (see Fig. 5).

LU – Product passed through process step

ZB – Final product of assembly

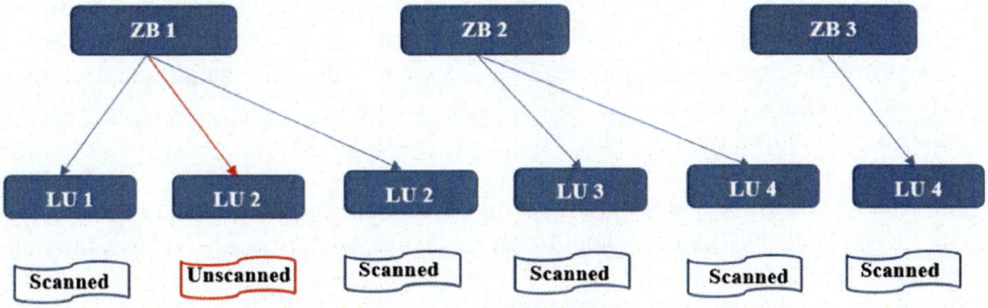

Fig. 5. SAP ME system interface

Some risks are emerging out of the process:

- Mixing ZB products that can only be detected at a special testing table, called TSK test table;
- Waste of time and material is done by reopening the housing and sorting LU products;
- Delivered products that don't comply with customer requirements are generating high costs and complaints.

To avoid the described risks some special actions were needed.

3.2 Improving Actions

A team was formed with the aim to reduce the error risks and find a solution for not mixing the products. 5 people responsible for the digitized processes participated to a brainstorming session, ideas were generated, gathered and discussed. Two of the ideas were selected for implementation.

Improvement action 1. The first improvement action involved procedure analysis. A new procedure in the scanning method was established, as a measure to prevent product mixing.

Steps of the new procedure implies:

- Using the "Complete SFC" button and the scan field to control each ZB assembled;
- The scan field will be disabled as soon as the second LU is scanned (Fig. 6);
- The "Complete SFC" button will be active only when both LU's have been scanned (Fig. 6);
- After each ZB assembled the worker must manually press the "Complete SFC".

It's important to mention that the new way of working increases the processing time, because the worker must manually enter the assembled product into the system. Still, the cost of an error is much higher.

If the worker tries to scan a third LU on the SAP ME interface, a suggestive message for warning "Please complete the ZB before starting the next assembly" will appear.

A new process flow was created (see Fig. 7) to standardize the change and operators were trained according to it.

Additional conditions:

- The scan field must be active until the worker scans the last product;
- The "Complete SFC" button should only be active when the ZB has assigned all components;
- The message must be implemented when the worker tries to scan an additional component.

Improvement action 2. The second improvement action consists in the implementation of one tablet at each workstation, leading to a better optimization and management of the process. In this case, the operator at the last workstation must no longer scan each

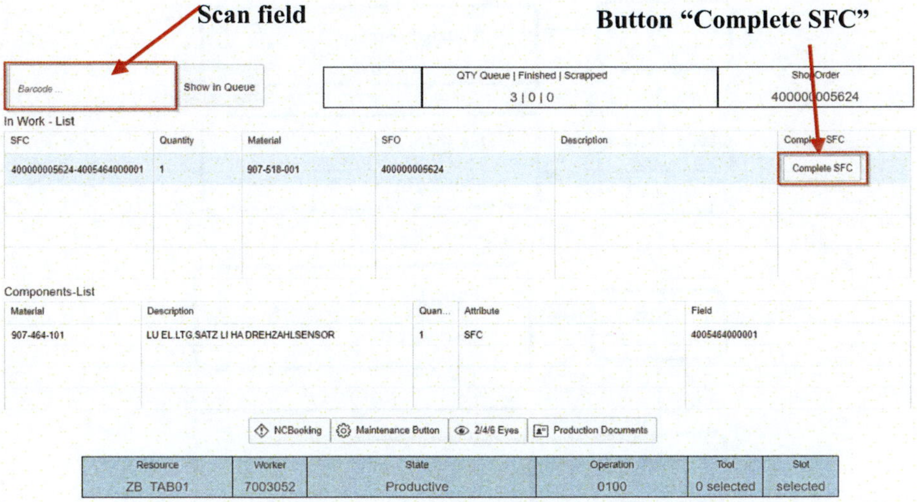

Fig. 6. Interface of SAP ME system.

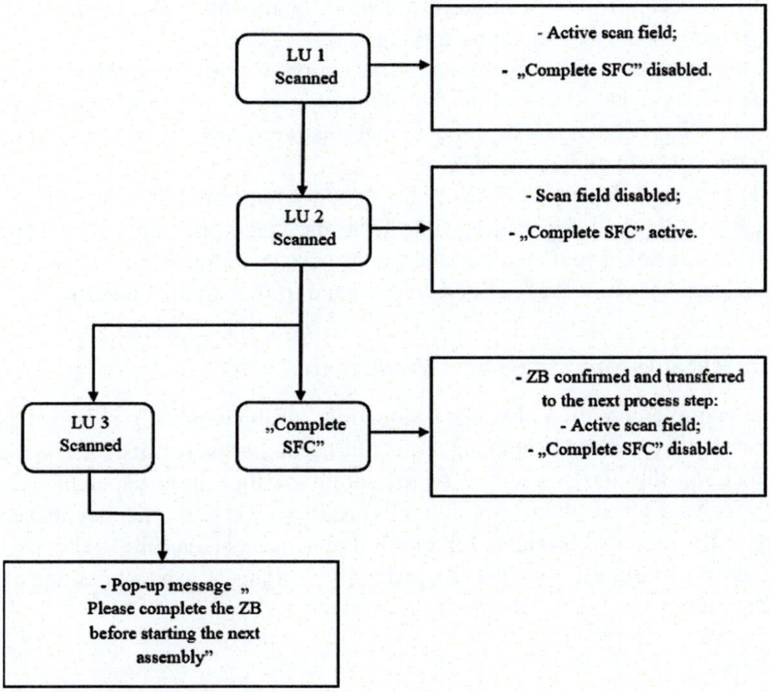

LU - Product passed through process step
ZB - Final product of assembly

Fig. 7. Process flow of the new way of scanning the products.

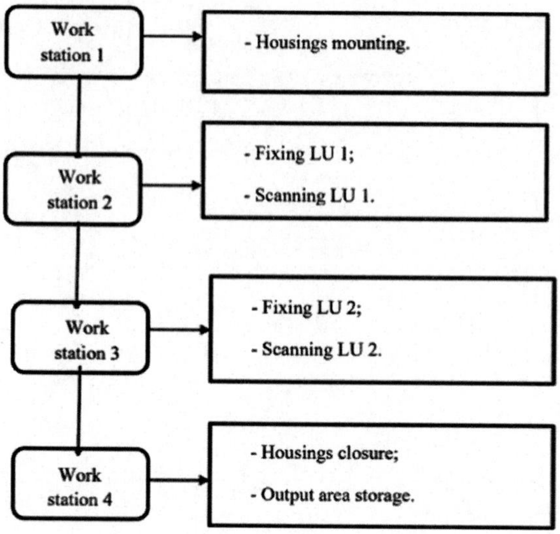

Fig. 8. Process flow of the new way of work.

LU wiring, because the operators responsible for mounting the wiring will scan their product themselves, thus avoiding mixing.

At the same time, by implementing this action, the annual number of products will increase, because the scanning time from the last workstation, which has been considered a bottleneck has been reduced, and the other operators will not have to wait until the last workstation has completed the assembly.

It is expected that in 2024–2025 the product requests will reach up to 1,000,000 products/year. Due to the waiting time from the last workstation which is around 5–6 s/product, during a year, an operator would have lost around 1,500 h/year.

A new process flow was defined (Fig. 8) and trained to the operators.

3.3 Effects of Introducing Smart Production

Due to the implementation of Smart Production starting with 2019 in Hirschmann Automotive company, several benefits in terms of traceability were noticed. The major aspect was changing the classic method of traceability, which used paper tags in past (see Fig. 9), with a digital system, which tracks production in real time, but also contains the history of the recorded products for a period of 15 years (see Fig. 10).

In terms of traceability, there is a specially designed report where one can identify information on:

- The operator;
- Production start date;
- The start date of a process step;
- The end date of a process step;
- Production end date.

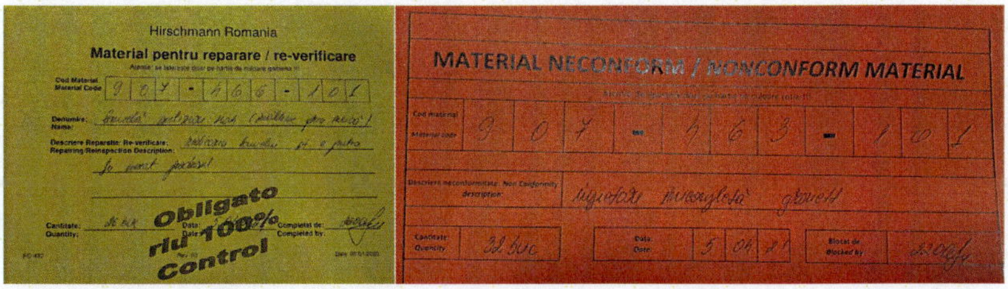

Fig. 9. Non-compliant material label/Non-compliant product label.

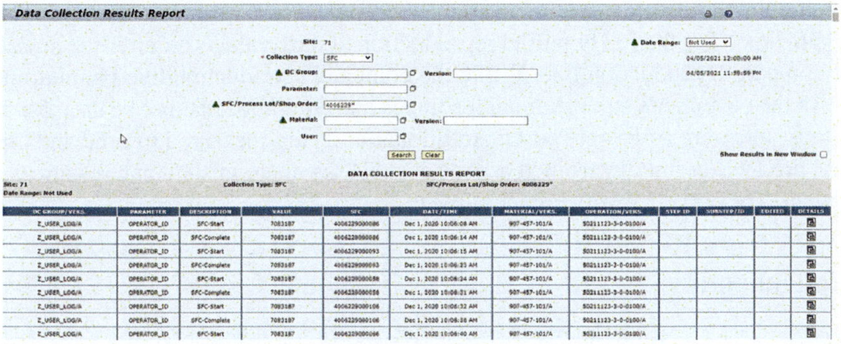

Fig. 10. Traceability report.

Smart Production contributed to:

a) Reduction of raw material costs;
b) Reducing the costs of identifying and sorting products;
c) Efficient use of available resources;
d) Reducing the risk of having customer complaints.

Out of the May 2021 production, 18% of products were mixed and 8.67% were unscanned in the previous step.

We applied the below formula (1)–(3) to calculate the cost benefit of the put in place improvements and reached a saving of tens of thousands of euro:

$$\text{Total expenditure} = \text{Housing expenditure} + \text{LU expenditure} \tag{1}$$

2 housings and 2 LU products are used for assembling the final product.

$$Housing\ expenditure = no.\,scrap\,products * 2\,pieces * Housing\ cost \tag{2}$$

$$LU\ expenditure = no.\,scrap\,products * 2\,pieces * LU\ cost \tag{3}$$

In the case of a product sorting requested by the customer following a complaint different sorting costs are involved.

The estimated costs vary between the minimum cost of approximately 10 euro/hour/person, in case of a visual sorting with own resources within company and can reach up to 100 euro/hour/person if contracting a third-party company, at the client.

Depending on the type of sorting, other expenses may be added. For example, for laboratory sorting, such as X-Ray, costs increase exponentially, involving additional risks, since this type of sorting is time consuming.

If defective products are distributed to a customer, he will initiate a complaint to the company. This complaint starts at a price of 10,000 euro and can increase depending on the severity of the defects. The total costs of the claims are borne by the company.

Regarding the efficient use of available resources, it is assumed that a customer complaint involving the full or partial reproduction of the products requires the cessation of the company's production line. While the company's production line I shut down to replenish the customer's stock, human resources and equipment cannot be used for their own stock, meaning additional costs. Additional costs are measured in time and profit, as these production shutdowns increase the costs of raw materials, human resources and equipment. These costs involve money losses.

4 Conclusions

In the study the Smart Production concept at Hirschmann Automotive was monitored and studied. The Smart Production comes with a series of benefits for the company but also for the staff employed, and it's also a customer requirement to meet the standard traceability. Using the SAP Manufacturing Execution system, Hirschmann team developed a spider equipment, and then 2 improvements were implemented to save costs and reduce waste.

Smart Production improvement contributed to a reduction of product identification and sorting costs, raw material costs, risk of customer complaints and an increase of efficient use of available resources. The projects is monthly bringing savings to the company and 2 more Smart Production project has been put in place.

References

1. Veza, I., Mladineo, M., Gjeldum, N.: Managing innovative production network of smart factories. IFAC Papers Online **48**(3), 555–560 (2015)
2. Majstorovic, V.D., Mitrovic, R.: Industry 4.0 programs worldwide. In: Monostori, L., Majstorovic, V.D., Jack Hu, S., Djurdjanovic, D. (eds.) AMP 2019. LNME, pp. 78–99. Springer, Cham (2019). https://doi.org/10.1007/978-3-030-18180-2_7
3. Machado, C.G., Kurdve, M., Winroth, M., Bennett, D.: Production management and smart manufacturing from a systems perspective. Adv. Manuf. Technol. XXXII **8**, 329–334 (2018)
4. Wang, C., Jiang, P., Lu, T.: Production events graphical deduction model enabled real-time production control system for smart job shop. Proc. Inst. Mech. Eng. C **232**(16), 2803–2820 (2018)
5. Zhang, Y., Cheng, Y., Wang, X.V., Zhong, R.Y., Zhang, Y., Tao, F.: Data-driven smart production line and its common factors. Int. J. Adv. Manuf. Technol. **103**(1–4), 1211–1223 (2019). https://doi.org/10.1007/s00170-019-03469-9

6. Bhuniya, S., Pareek, S., Sarkar, B., Sett, B.K.: A smart production process for the optimum energy consumption with maintenance policy under a supply chain management. Processes **9**(1), 19 (2021)
7. Gerekli, I., Celik, T.Z., Bozkurt, I.: Industry 4.0 and smart production. TEM J. Technol. Educ. Manage. Inform. **10**(2), 799–805 (2021)
8. FAO: Codex Alimentarius Principles for Traceability 2015. Available from: http://www.fao.org/ag/againfo/themes/en/meat/quality_trace.html
9. Henchion, M., De Backer, C.J.S., Hudders, L.: Ethical and sustainable aspects of meat production; consumer perceptions and system credibility. In: New Aspects of Meat Quality, pp. 649–666. Elsevier (2017). https://doi.org/10.1016/B978-0-08-100593-4.00027-8
10. www.systema.com. Accessed 25 July 2021

Integrating Environmental Issues in the Development of Communication Skills and the Importance of Environmental Sustainability Training

Nicoleta Aurelia Marcu(✉), Petru-Dragos Morar, and Dana Rus

UMFST "G.E. Palade", Targu Mures, Romania
nicoleta.marcu@umfst.ro

Abstract. The purpose of this study is to bring to the attention of the students, educators and specialists in the field of technical education the importance of the school subjects that contribute to raising awareness of the young generation about the quality of the environment. For this reason, the authors conducted a study about the way high-school and university students understand and approach some issues regarding the environment. The study was conducted through a questionnaire containing multiple choice questions. The questionnaire was made available to high-school and university students from different specializations.

This study also indicates how the specialist knowledge of the learners (in our case, the engineering students) can be used for learning specialized terminology by means of authentic learning materials. This teaching approach called CBI (Content Based Instruction) serves as a link between language and content and is considered to be essential in any ESP (English for Specific Purposes) course. By using authentic materials from the students' specialist area, CBI generates better learning outcomes as the students show more confidence when they need to learn languages on the basis of a content they are already familiarized with.

Keywords: Content Based Instruction · Waste selective collection · Environmental education

1 Introduction

The purpose of this study is to raise awareness regarding environmental issues among high-school and university students and to highlight the role of school subjects in promoting environmental friendly attitudes and behaviors. This study is also an interdisciplinary approach as it makes reference to a teaching method for learning languages, namely CBI (Content Based Instruction) by means of which the language instructor can use authentic materials, in our case, materials regarding environmental protection. The gain is twofold: the specialized terminology that needs to be acquired in the ESP (English for Specific Purposes) class is better acquired and internalized by the learners and, at the same time, the educational purpose of the authentic materials meant to inform and raise awareness regarding environmental issues among the students is also achieved.

L. Moldovan and A. Gligor (Eds.): Inter-Eng 2021, LNNS 386, pp. 426–439, 2022.
https://doi.org/10.1007/978-3-030-93817-8_40

Within little more than a decade there are likely to be around 8.5 billion people around [1] and this means more than doubling the resources that our planet can supply. Global demand for materials has increased ten-fold since the beginning of the 20th century and is set to double again by 2030, compared to 2010 [2]. The ecological, economic, cultural, spiritual needs of the global citizenry require not only awareness regarding these issues, but mainly action to address them in an effective way. As the 16-year old environmental activist, Greta Thunberg, put it, "the eyes of all future generations are upon you" [3]. Hence the responsibility of all of us to take a stand and act accordingly so as not to be remembered by the future generations for our legacy of inaction.

One way of addressing these issues, and probably the most effective way, is environmental education. This should start from an early age, within the family, and continue throughout our entire life [4]. Environmental education in the educational institutions should equip learners with the knowledge, skills, and motivation to handle the complex environmental challenges in the 21st Century. By teaching the students how to think and not what to think, educators develop their problem-solving skills that they need in order to make informed choices about the environment [5, 6]. These skills are high in demand on the 21st century work-force market which is faced with complex environmental issues such as climate change and energy.

2 Environmental Education and ESP

This study also indicates how ESP courses can take advantage of the specialist knowledge of the learners (in our case, technical/ engineering field) and can better mediate the process of learning specialized terminology by means of authentic learning materials.

Thus, devising questionnaires such as the one that is at the basis of our study could serve as a project-based method by means of which students can acquire terminology related to environmental protection. The authenticity of the study and the research students need to do in order to design, administer and interpret the questionnaire guarantee that the specialized terminology will be acquired appropriately and will be correctly transferred to similar contexts.

Consequently, this method of teaching languages by using authentic materials from the students' specialist area generates better learning outcomes as the students show more confidence when they need to learn languages on the basis of a content they are already familiarized with in their maternal language. The application of this teaching approach called Content-Based Instruction (CBI) to wider and varied language contexts has become increasingly popular [7]. This approach engages students in activities that develop their learning and communicative strategies, as well as their critical thinking skills [8].

CBI serves as a link between language and content and it is considered to be essential in any ESP (English for Specific Purposes) course. ESP learners expect to use the language meaningfully and to complete tasks that are cognitively demanding. According to Y.G. Lo, a successful ESP syllabus should include, among others, meaningful tasks [9]. Therefore, a course for adult learners should go beyond the traditional focus on notional or functional aspects of language and use language as a means of acquiring information, rather than as an end in itself [10]. What is more, many students tend to search for information regarding their specialist area in English; the ESP teacher can take advantage of this and design meaningful tasks that meet not only students' language needs but also their needs for specific information.

In the case of our study, drawing up a questionnaire about environmental issues as a project-based activity for the ESP course can also render the CBI method as a formative one due to the fact that, apart from linguistic competency, such an activity will raise students' awareness regarding environmental protection.

Learning about environmental problems, doing research and making projects on these issues involve the development of essential skills for the 21st century such as questioning, investigating, identifying problems, analyzing, interpreting, reaching conclusions, and solving problems [11]. At the same time, if the students need to make a group project such as the questionnaire that we used in our study, they can develop leadership and team working skills; this can be an effective way of developing cooperative learning with their colleagues, of promoting constructive discussions and debates, and of focusing on action strategies with real-world applications.

It turns out that an ESP course is more than a language course. According to Hutchinson and Waters, content focus is an essential component of the ESP materials design [12]. As we have seen so far, combining content and language in ESP can be an efficient approach to develop language proficiency as well as to build up the skills required in the students' future professions.

3 Anthropic Activities and Their Dangerous Impact on the Environment

Human activities have always been characterized by consumption of raw materials and energy (in various forms). Thus, at a turning point in the evolution of the relationship between man and nature, the environmental science and environmental protection emerged. Following the countless and, at the same time, recurrent damages manifested upon nature, man has understood that the survival of all species, the quality of life and the success of all human activities depend on the integrity of the biological processes of the terrestrial ecosystem. What man is currently doing on Earth is in complete contradiction with this essential condition. In short, we are in a crisis that threatens the possibility of inhabiting the ecosphere and the survival of the human species. In 1971, the reputed biologist Barry Commoner warned us about this environmental crisis: "Because the global ecosystem is a connected whole, in which nothing can be gained or lost and which is not subject to over-all improvement, anything extracted from it by human effort must be replaced. Payment of this price cannot be avoided; it can only be delayed. The present environmental crisis is a warning that we have delayed nearly too long" [13].

Economic development, and implicitly economic security, depend on natural resources. They allow the world states to provide for their consumption needs in full or in a significant proportion. Therefore, many specialists who are really concerned about the security 'phenomenon' appreciate that when discussing economic security one must take into account several components, such as: food security, trade security, financial security, security of goods and services, energy security, resource security [14].

It is known that any activity results in a product or service. But it also results in one or even more wastes, which (sometimes not all of them!) can be considered by-products, depending on the technology, the energy and the materials that are used. The increasing raise in the volume of waste is a problem that is attracting more and more the attention of specialists, and not only. Preventing the increase in waste volume is part of the Integrated Product Policy (IPP) [15]. That is why measures are being considered to stimulate recycling, recovery and, only as a last resort, waste disposal by incineration. Various documents that complement the environmental policy have as main objective the reduction of the amount of waste. Thus, it is predicted to reduce the amount of waste by 50% by 2050 (compared to 2000) [16].

Recycling and waste prevention – a strategy launched in 2003 by the European Commission ("Towards a Thematic Strategy on Waste Prevention and Recycling") – is the first separate approach to the aspects of recycling and waste prevention. This study analyzes ways to promote product recycling (where appropriate) and analyzes the most suitable options in terms of cost-effectiveness. We should not forget that when choosing the method of exploitation and recovery of waste, the characteristics, the physical and chemical composition of the waste must be taken into account [17].

Waste management must be carried out without endangering human health and the environment. Precisely for this reason, the European Union's policy on waste management is based on three principles:

- prevention of waste production;
- recovery of waste (recycling, reuse);
- improving final disposal techniques and monitoring final disposal.

The devastating aggressions that are currently taking place on the natural environment (and which are making human life more and more difficult, endangering its existence), have led to the need to take firm measures both at national and global level in order to mitigate the negative effects determined by anthropic activities such as deforestation and pollution, and to establish man-biosphere relations that are up to the level of the scientific and technical developments of this era.

4 Materials and Methods

In order to carry out our study, which mainly aimed at identifying the level of information and knowledge about environmental issues, a questionnaire with 12 questions (multiple answer questions with one choice) was sent to a group of young people (high-school and university students). It was considered necessary to request some data about the respondents (level of education: high school/university; the respondent's area of residency) in order to confirm the importance of including among the technical disciplines, of school subjects whose main object of study is environmental protection.

Apart from elements like education level or area of residency, the development of such programs that include environmental education disciplines should take into consideration various other factors, for example students' learning styles as mentioned in the paper "Project-based Learning in Technical Higher Education" [18].

The processing and the analysis of the data collected by means of the questionnaire involved studying the answers obtained from the respondents. In order to analyze and interpret the given answers, we first processed data and then created a database meant to support the analysis of the study with graphical and statistical methods. The questionnaire can be accessed at the address: https://docs.google.com/forms/d/e/1FAIpQLSc8NWm_fqhZ3U4zeQbT_bk7OB_szGJ0Ehv8pujSNdzJfbov_w/formResponse. The following 12 questions were included in the questionnaire:

1. Did you study, within the didactic activities, disciplines regarding the protection of the environment, environmental legislation, ecology, waste, etc.?

   • Yes/• No

2. Do you think it is necessary to raise present and future generations' awareness regarding environmental problems?

   • Yes/• No

3. In what ways do you think the information could be better transmitted and assimilated? (express your option in the order of the importance you assign to the respective means).

• Subjects in the educational process/ debates and conferences/ mass-media/ camps or educational workshops
• Camps or educational workshops/ debates and conferences/ subjects in the educational process/ mass-media
• Debates and conferences/ mass-media/ camps or educational workshops/ subjects in the educational process

4. Do you consider that selective waste collection takes place in the institution where you study?

   • Yes/• I don't know/• No

5. Do you consider that you have enough information in order to be responsible consumers of natural resources?

   • Yes/• No

6. Do you consider that you are a person with a sense of responsibility regarding natural resources consumption?

   • Yes/• No

7. Are you familiar with the concept "selective waste collection"?

   • Yes/• I do not have enough information/• No

8. Are you an active participant in your daily activities to the process of selective waste collection?

   - Yes, I am an active participant
   - I desire to be, but I do not have the necessary means
   - No, I have a busy schedule
   - No, I am not an active participant

9. Do you prefer using recycled or disposable products?

   - I prefer recycled products
   - I prefer disposable products

10. How do you think waste reduction can be achieved?

    - By using recycled products
    - By using disposable products

11. Is environmental protection a trend or a necessity nowadays?

    - A trend
    - A necessity
    - Both

12. Assess the level of public awareness regarding issues of environmental protection and conservation.

Very low						Very high
1	2	3	4	5	6	7

5 Results and Discussion

The answers given by the respondents to the questions of the questionnaire, as well as their interpretation, will be presented as follows.

5.1 Data About Respondents

Out of the 121 respondents, 40.4% are attending or have attended a high school with a technical profile, and 59.6% are attending or have attended a high school with a theoretical profile. Out of the respondents who have a university degree (or are still pursuing), 57.0% chose engineering sciences and 43.0% other fields (Fig. 1).

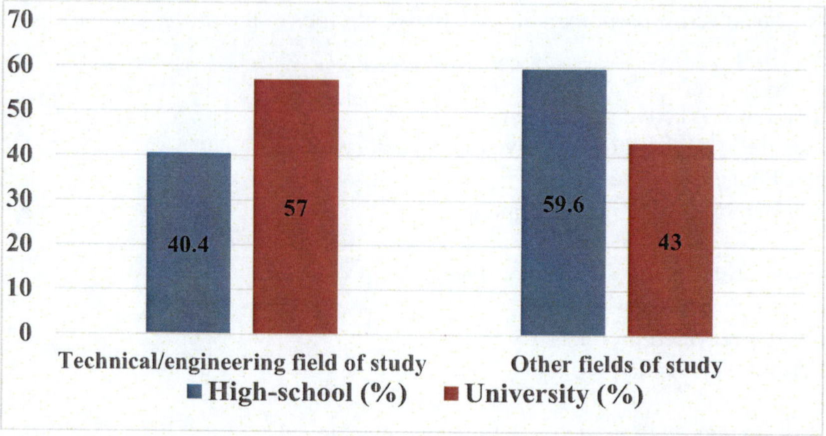

Fig. 1. Share of technical education (high-school, university) among respondents.

Regarding the question about area of residency, the authors found that 51.2% of respondents come from urban areas and 48.8% from rural areas (Fig. 2).

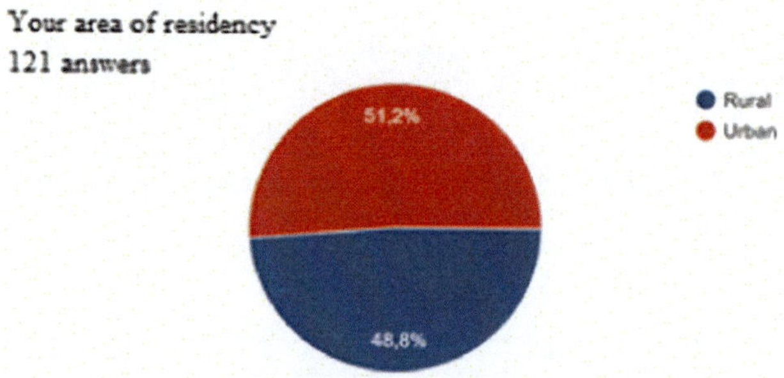

Fig. 2. Respondents' area of residency.

5.2 The Environment and the Respondents

The other questions (12) are addressed to the respondents in order to analyze their perception of environmental problems (selective waste collection, consumption of natural resources, waste recycling, etc.), following the information the respondents received formally during their studies or the information they gathered because of their need to acquire knowledge. Thus, the first 4 questions refer to the respondents' contact with the information about the environment during their studies.

To question number 1, "Did you study, within the didactic activities, disciplines regarding the protection of the environment, the environmental legislation, ecology, waste, etc.?" (Fig. 3), 33.9% of the respondents did not study in the educational process the disciplines mentioned in the question.

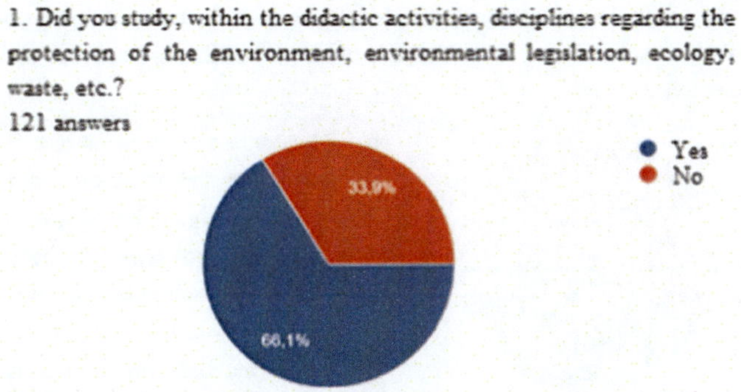

Fig. 3. Share of respondents that have studied or not environmental protection subjects.

Even if more than half of the respondents (66.1%) studied in school the subjects mentioned in the question (Fig. 3), all 121 respondents consider it necessary to continue to raise awareness and educate current and future generations on environmental issues (Fig. 4).

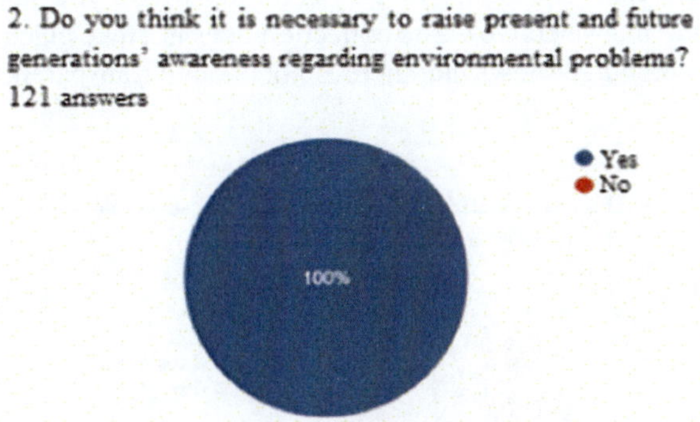

Fig. 4. The necessity of educating present and future generations on environmental protection.

To the question, "In what ways do you think the information could be better transmitted and assimilated?" 45.5% chose the second option, followed closely by the first option (42,1%) (Fig. 5). This indicates that extracurricular activities are welcomed by the respondents for environmental education, and, at the same time, the integration in the curricula of school subjects meant to raise awareness regarding environmental issues is also perceived as being very effective.

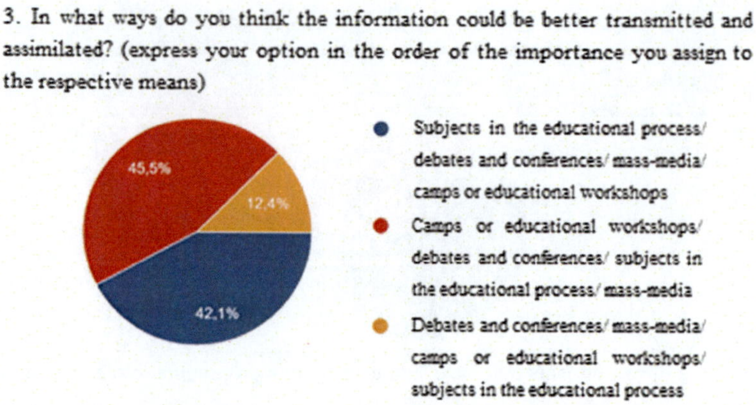

Fig. 5. Ways of sharing information on environmental protection.

By Law no. 132 of June 30, 2010, on the selective collection of waste, public institutions and educational institutions must make a selective collection of waste [19]. In this regard, the following question was formulated and asked: "Do you consider that selective waste collection takes place in the institution where you study?" 24% of the respondents consider that selective waste collection does not take place in the institution they attend. This percentage shows that there is not enough transparency of information in this area (Fig. 6).

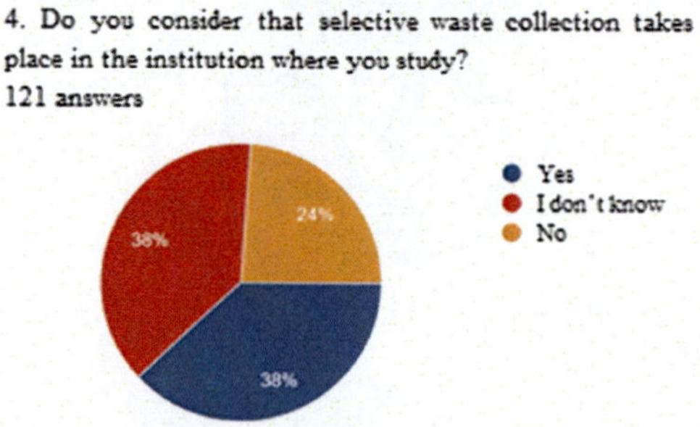

Fig. 6. Waste collection within the respondents' educational institution.

Analyzing the answers to the following two questions (Fig. 7 and Fig. 8), the authors found that the interviewed respondents consider that, by informing the population about reducing the consumption of natural resources, they can become more responsible in this regard.

Questions 7 and 8, respectively Figs. 9 and 10 bring to the attention of the respondents the problem of selective waste collection. Most respondents (81.8%) (Fig. 9) are familiar with the term 'selective collection', but, for various reasons (lack of containers, busy work schedule, etc.), only 21.5% identify themselves as active participants in the waste collection process (Fig. 10).

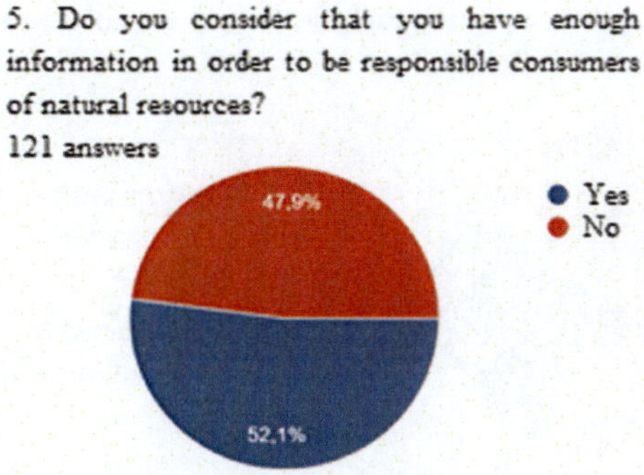

Fig. 7. Respondents' level of information that can qualify them as responsible users of natural resources.

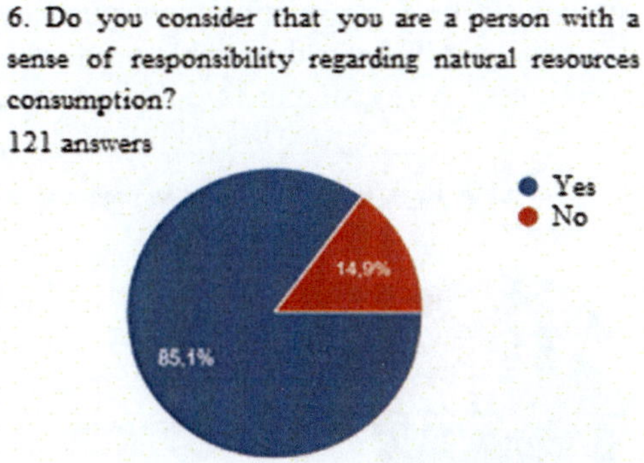

Fig. 8. Respondents' sense of responsibility regarding natural resources consumption.

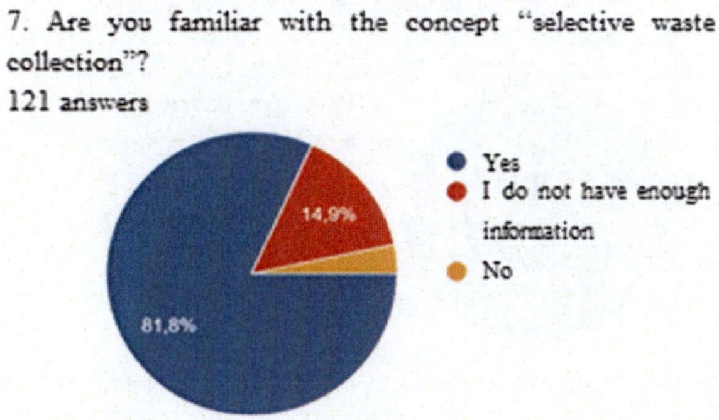

Fig. 9. Respondents' knowledge of the concept "selective waste collection".

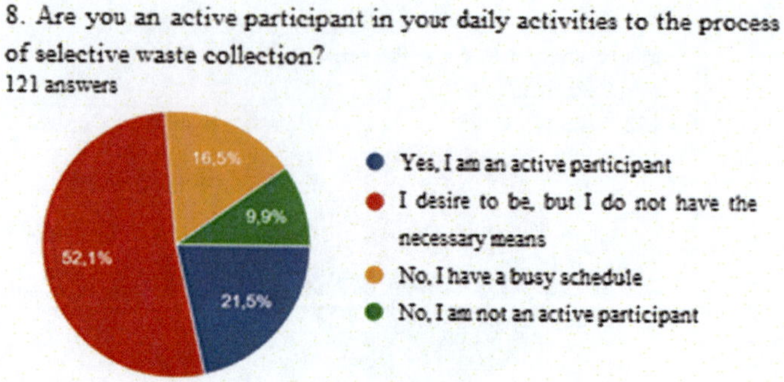

Fig. 10. Respondents' daily involvement in the process of selective waste collection.

A very important issue related to the consumption of natural resources and, implicitly, to the generation of waste (resulting from the exploitation and processing of resources) is the recovery of waste through recycling. Questions 9 and 10 (Figs. 11 and 12) reflect the above. For some reasons (lack of information, convenience, lack of trust in the manufacturer, etc.), some people still prefer disposable products; this involves higher consumption of natural resources, energy, so more waste is generated (Figs. 11 and 12).

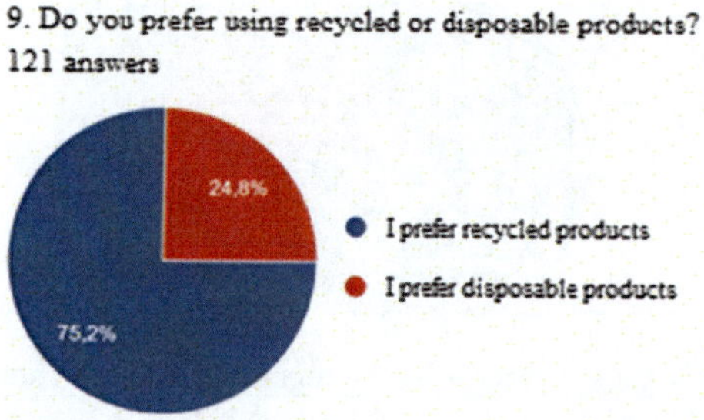

Fig. 11. Respondents' preference for recycled or disposable products.

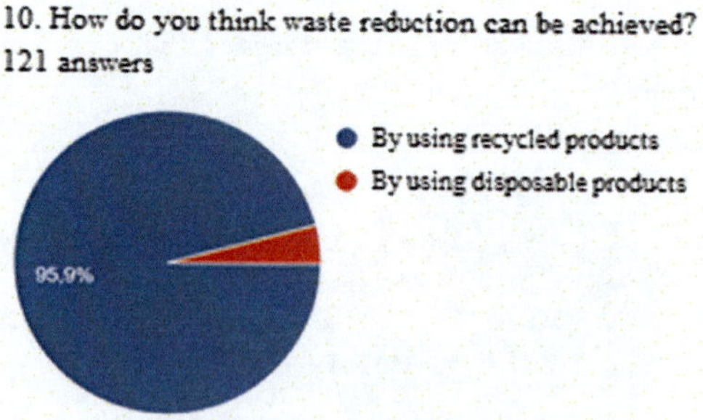

Fig. 12. Ways of reducing waste: recycled or disposable products?

At the end of the questionnaire (the last two questions: 11 and 12, respectively Fig. 13 and 14) the respondents could express their opinion, as a conclusion regarding the environmental issues. A percentage of 66.1% of them consider that it is necessary to protect the environment; 3.3% of respondents consider that environmental protection is a trend, and the others (30.6%) consider that environmental protection is both a trend and a necessity (Fig. 13). This indicates the need of more information regarding environmental protection because the critical issues it entails do not qualify it as a mere trend.

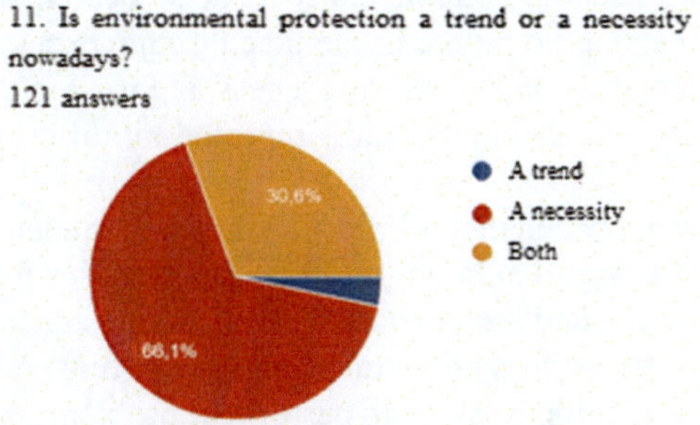

Fig. 13. Environmental protection: a trend or a necessity?

To the question regarding public awareness of these critical issues (Fig. 14), most respondents gave an average score (on the evaluation scale); this indicates that there is room for more information and better training meant to raise our awareness as global citizens and consumers in matters affecting the health of the environment.

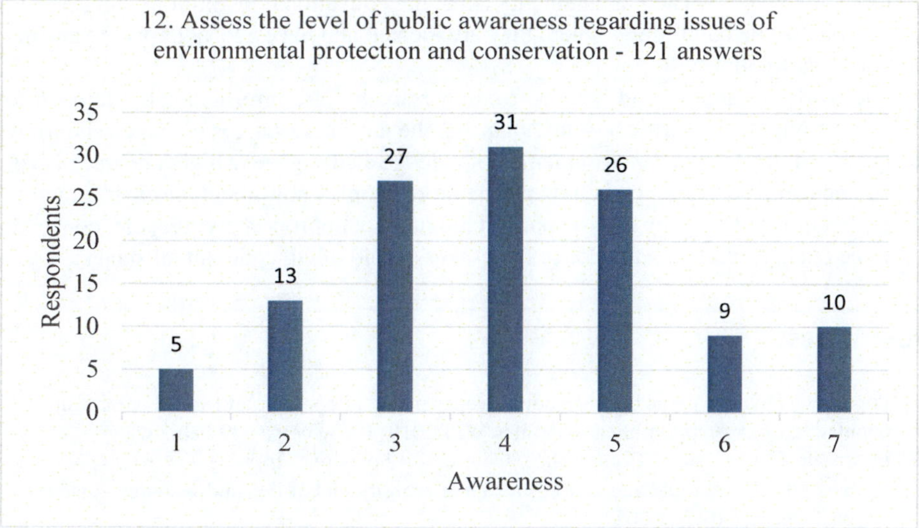

Fig. 14. Public awareness regarding environmental problems.

6 Conclusions

Following the study, we have reached the following conclusions: there is a need for the consistent introduction in the educational process of some disciplines that deal with environmental issues, supplemented with the organization of extracurricular activities, which should complete and help young people to deepen the theoretical notions acquired during their courses; high-school and university students need to be informed from the first years of study also about the problems we face through excessive waste generation and poor waste management; consumers' awareness and trust in the products obtained by recycling waste should be promoted and supported by different means (social and economic).

Environmental education benefits all parties involved: students, educators, educational institutions, and the world at large. It offers opportunities for students to become engaged in real world issues and see beyond the walls of the educational institutions. If the students see a match between their studies and the existing environmental issues of our planet, then they can acquire the skills to perform as problem solvers and as powerful promoters of ecologically friendly solutions.

By means of environmental education the institutions would benefit from a greener and healthier learning environment. The students 'motivation is higher because of the authentic learning materials which inspire them better than the traditional pedagogy. The educators' confidence is also increased by matching the learning content with concrete results, by designing more engaging and interactive learning activities that motivate students to be not only passive recipients of knowledge but also accountable users of knowledge. Another benefit for educational institutions could be the actual cost reduction of electricity, water, waste management. Thus, the educational institutions can serve as an example for the community where they are located and have a local impact regarding environmental policies.

The world at large would benefit from citizens that are informed and equipped to make decisions that are critical to safeguarding the natural resources on which economy and quality of life depend. Environmental knowledge also engages a higher degree of pro-environment and conservation behavior. The more people know about the environmental issues, the more likely they are to take a stand and act appropriately: recycle, conserve water, be energy efficient, etc. Hence the essential role of education at all stages.

References

1. Population Division of the United Nations Department of Economic and Social Affairs. https://population.un.org/wpp/Publications/Files/WPP2019_10. Accessed 09 Feb 2021
2. European Commission. Competence Centre on Foresight. https://knowledge4policy.ec.europa.eu/foresight/topic/aggravating-resource-scarcity/global-demand-resources-materials_en. Accessed 09 Feb 2021
3. Firstpost. 'Eyes of all future generations are upon you': Climate activist Greta Thunberg slams world leaders for failing to tackle greenhouse gas emissions. Agence France-Presse (2019). https://www.firstpost.com/world/you-have-stolen-my-dreams-my-childhood-at-un-teen-activist-greta-thunberg-slams-world-leaders-for-failing-to-act-on-climate-change-7390021.html. Accessed 09 Feb 2021

4. Boca, G.D., Saraçlı, S.: Environmental education and student's perception, for sustainability. Sustainability, **11** (2020)
5. Leal Filho, W. (ed.): Implementing sustainability in the curriculum of universities. WSS, Springer, Cham (2018). https://doi.org/10.1007/978-3-319-70281-0
6. Lozano, R., Merrill, M.Y., Sammalisto, K., Ceulemans, K., Lozano, F.J.: Connecting competences and pedagogical approaches for sustainable development in higher education: a literature review and framework proposal. Sustainability **9**, 1889 (2017)
7. Grabe, W., Stoller, F.L.: A six-T's approach to content-based instruction. In: Brinton, D.M., Snow, M.A. (eds.) Content-Based Classroom: Perspectives on Integrating Language and Content, pp. 78–94. Longman, New York (1997)
8. Morioka, A., Hayashi, A.T., Ushida, E.: Developing web-based multilevel materials for Japanese content based instruction. Jpn. Lang. Lit. **42**(2), 361–387 (2008)
9. Lo, Y.G.: ESP versus EGP: a case study of an ESP program for vocational high school students of tourism. Taiwan Int. ESP J. **3**(2), 71–100 (2012)
10. Richards, J.C., Rodgers, T.S.: Approaches and Methods in Language Teaching. Cambridge University Press, New York (2001)
11. https://www.plt.org/about-us/why-environmental-education-is-important. Accessed 05 Feb 2021
12. Hutchinson, T., Waters, A.: English for Specific Purposes: A Learning Centered Approach. Cambridge University Press, Cambridge (1987)
13. Commoner, B.: The Closing Circle: Nature, Man, and Technology. Random House, New York (1971)
14. Stănişteanu, A.I.: Energy Security and Energy Corridors in the Balkans. Bulletin of "Carol I" National Defense University, pp. 260–267 (2015)
15. Kristensen, P.: Integrated Product Policy 2020 – A Nordic Discussion Paper Regarding a Coherent European Product Policy (2020). https://norden.diva-portal.org/smash/get/diva2:1384958/FULLTEXT01.pdf. Accessed 09 Feb 2021
16. Ministry of Energy – Nov. 2016 – Romania's Energy Strategy 2016–2030, with the perspective of 2050. https://www.anre.ro/ro/eficienta-energetica/rapoarte/rapoarte-de-monitorizare-a-implementarii-planului-national-de-actiune-in-domeniul-eficientei-energetice-pnaee. Accessed 09 Mar 2021
17. Apostol, T., Mărculescu, C.: Managementul deşeurilor solide. AGIR Printing Press, Bucharest (2006)
18. Peterlicean, A., Morar, F.: Project-based learning in technical higher education. Appl. Mech. Mater. **371**, 739–743 (2013)
19. Law No. 132 of June 30, 2010 on the Selective Collection of Waste in Public Institutions. https://www.ecotic.ro

On the Use of Technology in Education. A Case Study on the Application of Interdisciplinarity in Technical Education: The ECORE Tool

Bianca Han(✉)

"George Emil Palade" University of Medicine, Pharmacy, Science, and Technology of Targu Mures, Gh. Marinescu str. 38, 540142 Targu Mures, Romania
bianca.han@umfst.ro

Abstract. After almost 2 years of Covid-19 pandemic-imposed online teaching, educators might wonder whether technology in education is indeed the future. Before the pandemic struck the world, technology has been part of the teaching process for a while already, by means of online manuals, gadgets, use of the internet, etc. but never before has technology become such an integrated part to the teaching process as in this period, in which courses were taught almost exclusively online. The article offers arguments to support the need of adapting and adopting technology to the new classroom (online or on-site) and to the digital-native student; it also describes some examples of good practices proving the advantages of digital education while underlining the idea that the students need to be in the centre of educators' preoccupations, who ought to permanently adjust to the novelty-imposed technological changes to teaching, in view of Industry 4.0 requirements.

The present paper presents the way in which technology has interfered with education, in the case of an educational project that focuses upon designing digital tools to support innovation in the classroom. The practical part of the paper presents a model of implementing digital technology into the classroom in order to introduce technical content. The methodology of the research contains a questionnaire testing and analysing students' attitudes with regard to this new teaching medium.

Keywords: Innovation · Teaching · Adaptability · Virtual teaching tools

1 Introduction: Theoretical Background on the Use of Technology in Education

Recent studies reveal that digital skills are needed on the topic of Industry 4.0 [1] even if new educational models [2, 3] or digital assessment methods have been developed [4, 5].

The purpose of the present paper is to suggest a model of implementing digital technology as means of delivering technical education in higher education. Nowadays overviews on education perceive it as an essentially interdisciplinary approach making

L. Moldovan and A. Gligor (Eds.): Inter-Eng 2021, LNNS 386, pp. 440–452, 2022.
https://doi.org/10.1007/978-3-030-93817-8_41

use of a wide variety of technological tools in order to achieve its scope. This is in accordance with contemporary views on the concept of Industry 4.0, for which education plays a major role. Some of the connections between the principles of Industry 4.0 and education refer to notions of interdisciplinarity, interconnectivity, collaborative work and the resort to technology and means of instruction. Considering the current developments in all educational areas, technological education included, it is necessary to adjust and innovate methods, means and contents in order to meet both student needs and the requirements on the labour market.

In view of the above, the present paper presents the way in which technology has interfered with education, in the case of an educational project that focuses upon designing digital tools to support innovation in the classroom. The practical part of the paper presents a model of implementing digital technology into the classroom in order to introduce technical content. The methodology of the research contains a questionnaire testing and analysing students' attitudes with regard to this new teaching medium.

2 Good Practices in the Implementation of Digital Learning Environments

We are living a time in which we need to acknowledge that on the one hand, the students need to be placed in the centre, as the direct beneficiary of the learning process, and on the other hand that education cannot longer be performed outside a technology-marked environment. This would be a good moment to salute projects that acknowledge this reality and take initiatives that develop and support innovation in the classroom. A good example seems to be, for instance, the ELSE Erasmus+ Project: Eco/Logical Learning and Simulation Environments in Higher Education, which is a project developed between 2018–2021, strongly supporting the idea that education is in a continuous need of technology-backed tools, that aim at creating and maintaining a challenging and rewarding atmosphere for all the actants involved. Therefore, the project initiated by the UNITUS University of Viterbo, Italy and conducted together with other 10 partners (9 European universities and a software company) proposed an eco-system composed of three tools: EVOLI, EDASH and ECORE.

EVOLI, developed by the POLIMI University, Milan, Italy, is a tool that exploits video tagging, and it may be perceived as a preliminary stage to the flipped classroom teaching method. The teacher uses certain YouTube videos, as knowledge clips, to provide students with content supporting the teaching aim, and the students are invited to watch the suggested videos, react to it, even leave comments, before the class is actually taking place. By doing that, the teacher is able to gather the feedback before class, and offer further required explanations, focusing especially upon the problems or issues identified, signalled by the students, as it is described in the ELSE Project – A handbook for Teacher.

The next tool developed by the University of Cyprus within the ELSE project is the EDASH tool is a platform, of the Moodle type, that aims at helping teachers and students monitor the latter's assessments, in an objective way, by offering info on the level of progress and performance, thus getting the pulse of the class, and initiating the necessary interventions and possible actions to be taken in remedy.

Last, but not least, we approach the ECOREtool, developed by the Entropy Knowledge Network company, Rome, Italy, which is a type of simulation tool, used to create serious (video) games that convey content, promote reflection in students, and stimulate learning by doing, the so-called experiential learning. This tool will be further debated in the research part of our paper.

3 Practical Case Study: The Serious Games as Digital Learning Tool

The above-mentioned examples, inscribed into the general trend towards technologising educational acts, led to the formulation of the hypothesis according to which presenting technical content by resorting to technology is a motivational factor in the successful acquisition of interdisciplinary knowledge in higher education. By interdisciplinary knowledge we imply technical knowledge related to the automotive industry field and linguistic knowledge, in the form of specialised terminology.

In order to test this hypothesis, we used a practical application of 'serious games', which is an ELSE-branded ECORE tool to create a video game in which students studying technical content for specialised technical translations were exposed to technical content and terminology from the automotive engineering field.

The idea of 'serious games' appears to have gained ground in education, and not only, since it brings together the necessities of the learning act, i.e. to inform and educate, and the needs of the current digital native learner, who expects to be both instructed and entertained, in the process. Referring to the latest definitions of the term 'serious games', we read that this type of game is a situation, an instance in which the players (in our case, the learners) are invited to become active in a teacher pre-established virtual framework, and play the game, implying that they need to chose from a set of actions and steps, leading to a goal, while taking decisions that affect the outcome of the game [6].

It is of utmost importance for the objectives and the rules of the game to be clearly set from the beginning, for the tool to prove its efficiency in the educational act. This idea only makes us more aware of the rather contradictory association of the two terms of the collocation we are analysing 'serious' and 'games', since they seem to belong to two different connotative, pragma-semantic areas. Therefore, analysing the picture in Fig. 1, might bring more light onto the idea of 'serious games' being highly connected

Fig. 1. Presentation of the relation of the 'serious games' to other fields [8].

to the learning process as much as it is linked to entertainment, thus, supporting one of the basic requirements of the current learner. To support that even more, we also appeal to the ideas underlining the fact that, according to recent studies, training and learning with the aid of serious games proved to be more engaging and effective, contributing to the improvement of cognitive skills, knowledge, and retention of information taught, in comparison to traditional or conventional methods of instruction [7].

Therefore, the ELSE-branded ECORE tool aims at re-creating real life situations, in that it offers context and environment which resemble natural settings, in which the player (the student) faces certain problems that need to be solved, step by step. The problems that the student needs to solve may vary, according to the objectives of the lessons, type of discipline, pedagogical aims of the teachers, etc. The tool develops an editor, which allows the teacher to play while building the serious game into a number of frames which are meant to lead the student through the game. At each step, the student is invited to perform certain tasks (read certain materials, watch certain videos, consult certain links, solve certain situations, exercises, etc.) and then choose from 3 pre-designed answers, out of which only one is correct. The beauty of the game lies in the fact that the students can seriously play as many times as they need to, in order to achieve good results, and thus, learn from/while doing so.

To exemplify, the ECORE tool was used to create the serious game *Know my car*, which aims to teach students certain specialised terms, in the field of automobiles. The group of students was represented by applied modern languages students enrolled to specialised terminology courses, where they study to become specialised translators, or interpreters, in different fields.

The teacher needs to have a clear view upon the goal of the game, in order to be able to design the frames accordingly. Thus, the storyboard that fills the framework prepared by the editor of the tool is an important part of the game, and it is created in such a manner that the teacher is carefully guided into its production. Needless to point out that the digital skills of the creator of the game are highly exploited, proving once again the teacher's open-mindedness in using technology in the classroom. Therefore, the game editor opens with the frame in which the teacher introduces the title, the description,

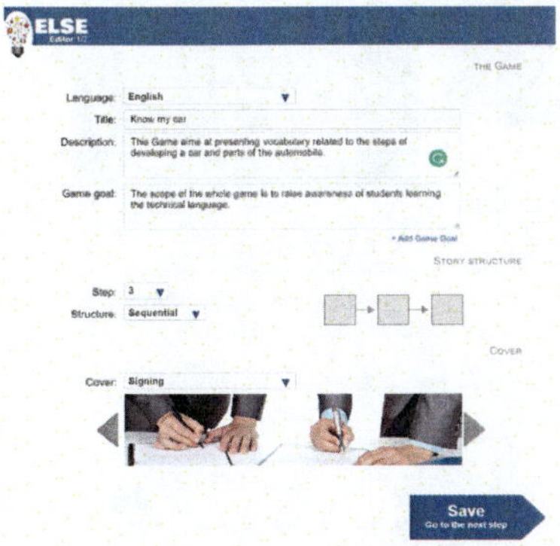

Fig. 2. Sample from the ECORE game Know my car, presenting game description and objectives, as set by the teacher.

and the goal of the game, and also chooses other details referring to the way the game is going to develop: the number of steps, the structure (sequential or parallel), or the cover (for aesthetic and pragmatic purposes) (Fig. 2).

In the serious game we created, the character, George Smith, is looking for a job as a mechanic at an auto repair shop. The student playing the game needs to choose the best answer out of the three offered, to provide the character the best continuation to the next step. The answers are scored according to their suitability to fit the needs of the character: score 2 being the least likely and score 6 being the most likely to fit (Fig. 3).

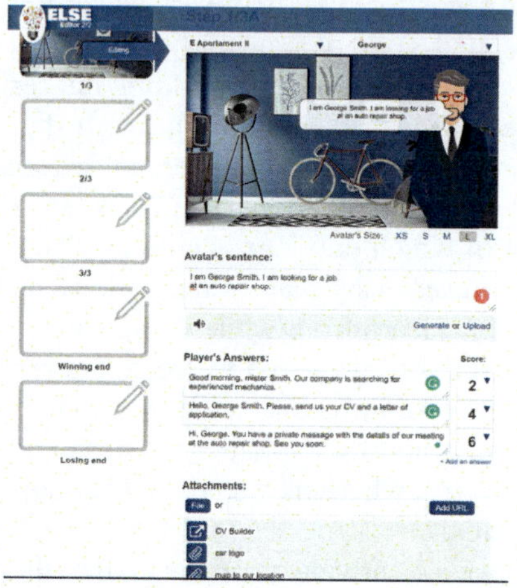

Fig. 3. Sample from the ECORE game Know my car, presenting one of the initial phases of the game.

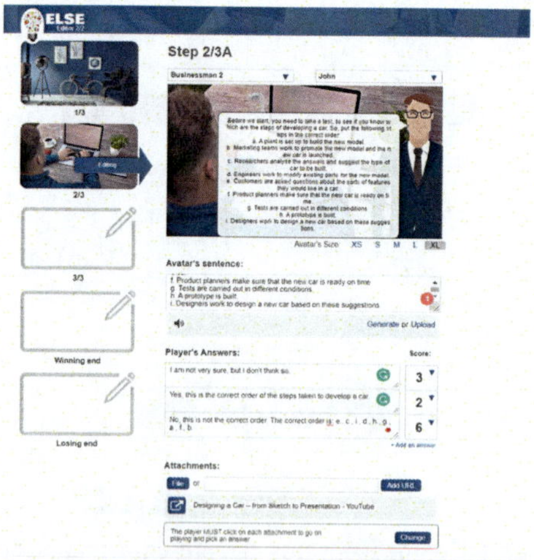

Fig. 4. Sample from the ECORE game Know my car, presenting the core exercises of the game.

The next step takes the character to a test that needs to be solved to prove his skills. The test refers to the ability of the character to arrange the steps of developing a car in the right order [9]. In order to solve the task, the students must watch a Youtube video on designing a car from sketch to presentation. By doing that, they will be exposed to the specific technical language, which should help them achieve the task successfully (Fig. 4).

The next step is another test that the character needs to pass, in order to get the job, and that entails matching parts of the car with their definitions. To help students become aware of the specialised technical vocabulary, they are invited to consult a drawing on the topic, presenting the parts of the car [9] (Fig. 5).

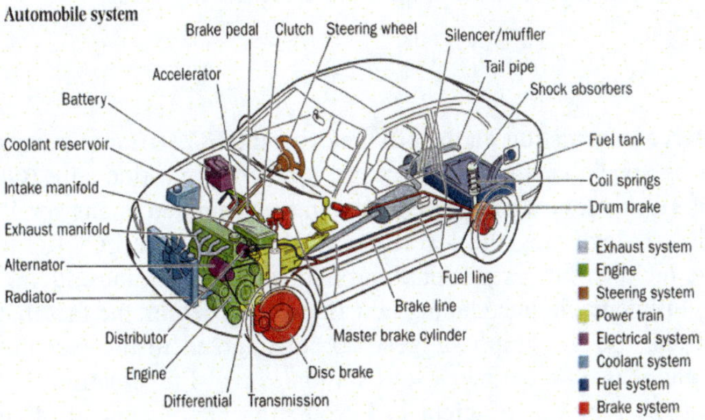

Fig. 5. Automobile parts [9].

The game ends after the students pass all the steps and solve all the tasks. As mentioned previously, the difficulty of the task may vary, according to the teacher's pedagogical objectives and the purpose of the lesson. Yet, regardless of its difficulty, the students will always seriously play serious games, becoming aware of the fact that they are actually learning by doing something entertaining, that fits their expectations. More than that, the ECORE game offers feedback under the form of percentages of correct answers for each frame, and of the total score, so that the students can see which frame was problematic and play again, for as many times as they need to, until they get the desired score, which will definitely entail retaining the necessary knowledge on the learning topic (Fig. 6).

The ECORE tools provides students with the possibility to play the serious game on their own, in their free time, to acquire the information in a safe and stress-free environment, after studying the material suggested as support (videos, textbooks, dictionaries) or in groups, or in the classroom, with the teacher. It is important to underline the fact that, even if the ECORE tool implies a certain amount of individuality on the part of the students, since they can play alone, the teacher can and ought to monitor students' performance and progress, and provide more explanation, if needed, in class. That is to say, the tool is to be perceived as a complementary alternative to the direct teaching, belonging to the flipped classroom model of teaching methods.

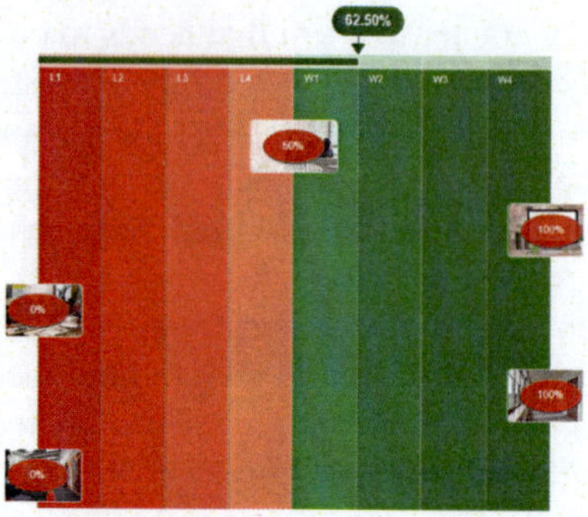

Fig. 6. Sample of the score in the ECORE game.

4 Results

In order to receive information on the efficiency of the ECORE tool upon the linguistic students learning specialised terminology, we applied a questionnaire (Google forms) to a number of 19 students. The group of students was invited to answer 10 questions:

For question 1, regarding the level of appreciation of the ECORE tool, the suggested answers included: its user-friendliness, its efficiency in introducing new technical content, its efficiency in introducing new linguistic content, the fact that it simulates real-life situations. As the graph suggests, 52.6% agreed to the fact that the tool is user-friendly, and it is efficient because it simulates real-life situations. Also, students considered that the tool was efficient in introducing new technical (47.4%) and linguistic (42.1%) content, which rightfully serves their purposes in acquiring specialised vocabulary (Fig. 7).

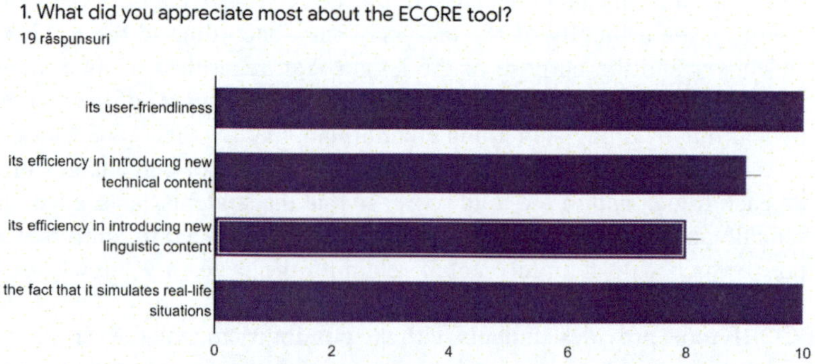

Fig. 7. Statistics of answers to Question 1 related to ECORE tool.

In the case of question 2, on the areas in which students felt they made progress after using the ECORE-based activity, 63.2% admitted that they improved their digital skills, 47.7% of the responses showed improvement in linguistic knowledge, 42.1% showed improvement in the technical knowledge and only 5.3% considered that not much progress was made (Fig. 8).

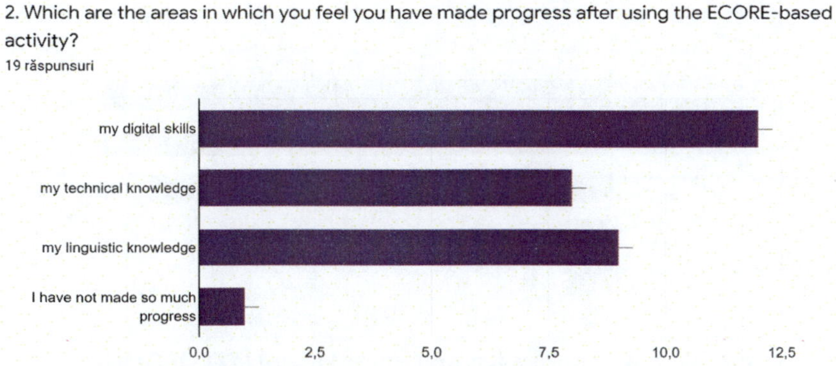

Fig. 8. Statistics of answers to Question 2 related to ECORE tool.

Question 3, asking about the familiarity with this type of learning technologies before using ECORE showed that the majority was between somewhat familiar (47.7%) and very familiar (31.6%), leaving the minority in a less familiar zone, which proves that the technology-based learning has been gaining ground in education (Fig. 9).

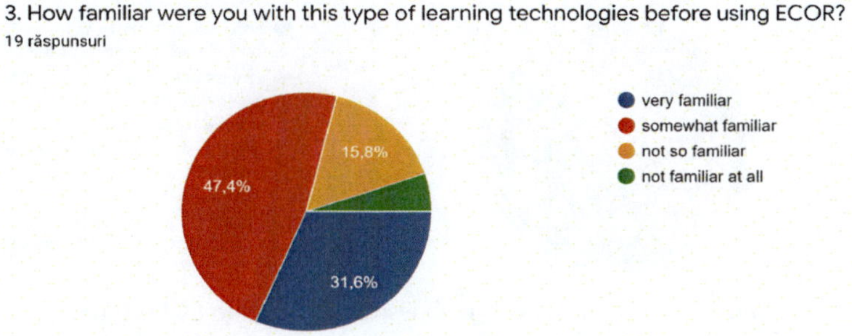

Fig. 9. Statistics of answers to Question 3 related to ECORE tool.

Question 4 aimed at finding the main advantages of ECORE compared to traditional learning methods. It showed that more than half of the answers (57.9%) appreciated that the ECORE tool ensures a natural way of acquiring the technical information and also that it makes use of modern, interactive and intuitive technology, while a small percentage (5.3%) considered that there were no significant advantages to the tool. Also, the result on the development of soft and hard skills is just as significant, since it rounds up to 52.6% of the responses, proving that the tool was well perceived by the users (Fig. 10).

Question 5 aimed to discover the level of suitability of the tool for introducing technical vocabulary from other engineering branches and the answers were quite balanced amongst the suggested answers; thus, 31.6% of the respondents considered it extremely suitable, because it incorporated modern technology, 36.8% believed that the tool was somewhat suitable, but it also had some disadvantages and 21.1% considered that the tool was suitable in general, depending on the technical content, and that not all technical contents could be adapted to this method. This raises awareness on the fact that extended

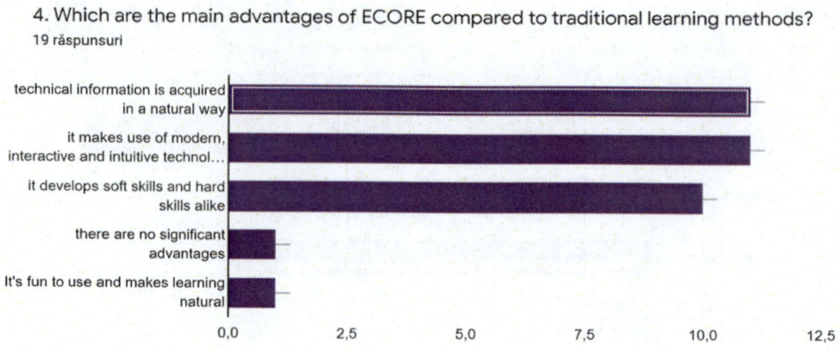

Fig. 10. Statistics of answers to Question 4 related to ECORE tool.

care needs to be taken when choosing the right content material to fill the game, in order for it to be effective (Fig. 11).

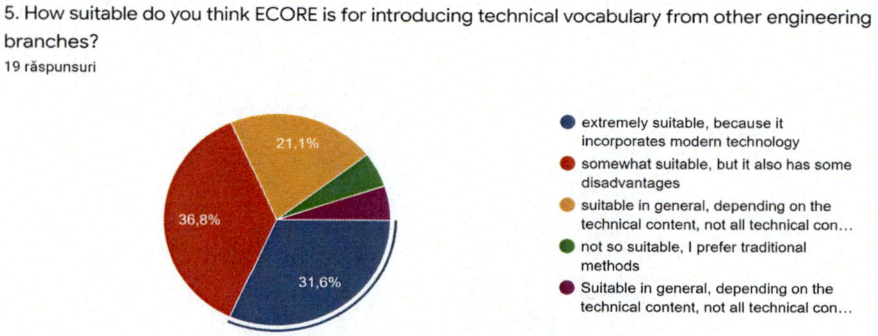

Fig. 11. Statistics of answers to Question 5 related to ECORE tool.

Questions 6 to 10 invited the students to rate certain statements of appreciation of the application of the tool, by using a scale from 1 to 5, where 1 means to a very little extent and 5 means to a great extent. Thus, question 6 checked whether it was easier to identify, label and learn automotive terminology while playing, because learning motivation was increased. Following the charts, it was interesting to observe the balanced answers, as only the rate-point 4 gathered 36.8% of the responses, while rate-point 2,3 and 5 equally showed 21.1%, implying that the opinions of the students are fairly spread on the rate-scale, leaving out completely only the rate-point 1. These results seem rather expected since, even if the students in the sample group have different linguistic backgrounds in terms of specialised terminology, varying from very little to some knowledge, the game provided them with enough content material to support their search in an entertaining, thus motivating way (Fig. 12).

Question 7 aimed at a more emotional aspect, checking the increase in content knowledge regarding car parts and operating procedures in an entertaining and interactive way. As the chart shows, the largest majority of answers were the 4-point rated ones,

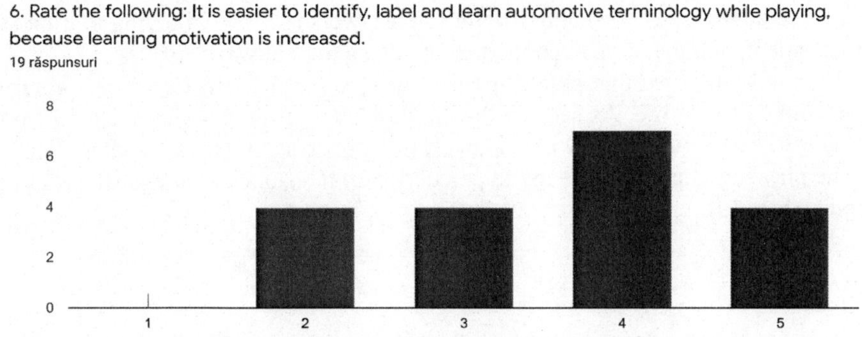

Fig. 12. Statistics of answers related to Question 6.

representing 52.6% of the total answers, implying that the students enjoyed the activity that helped them find new specialised vocabulary (Fig. 13).

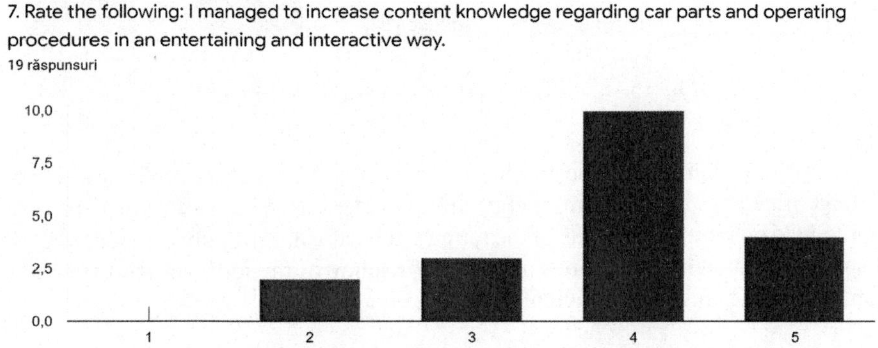

Fig. 13. Statistics of answers related to Question 7.

Question 8 verified whether the language skills were practised in a natural way, and the results favoured again the 4-point rate answers (47.4%), followed closely by the

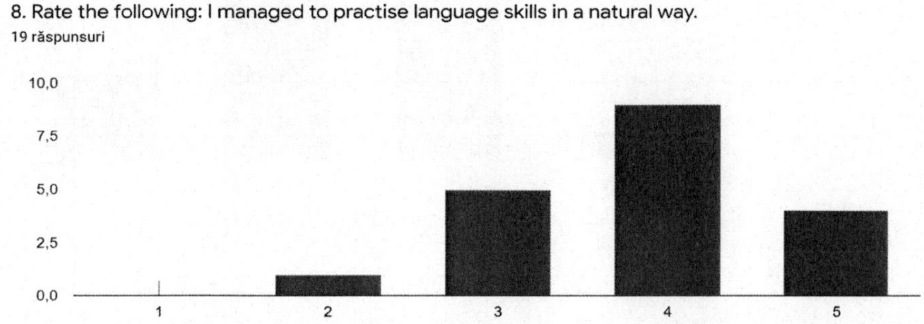

Fig. 14. Statistics of answers related to Question 8.

3-point rate answers (26.2%) and 5-point rate answers (21.2%) implying that most of the respondents felt they practiced the language in a natural way (Fig. 14).

Question 9 observed that practical tasks, such as the ECORE game, are preferred, as they simulate real life. As seen in the chart, 84.2% of the answers were of 4-point rate (57.9%) and 5-point rate (26.3%), underlining once again the idea that the real-life attribute plays an important role in the perception of the students in regards to learning activities (Fig. 15).

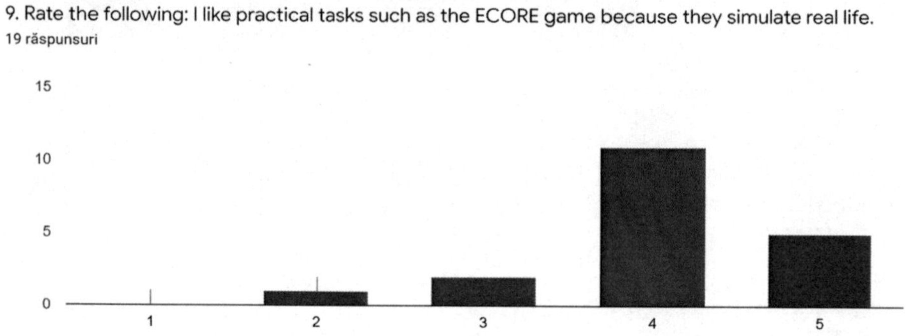

Fig. 15. Statistics of answers related to Question 9.

The last question of the questionnaire looked for the students' opinion regarding their belief that engineering terminology should make use of interactive methods using technology. Apparently, the limits of the rating scale were not strongly covered, implying that students had their reservations as to the efficiency of the tool, yet with a tendency towards the tool being more efficient than not (Fig. 16).

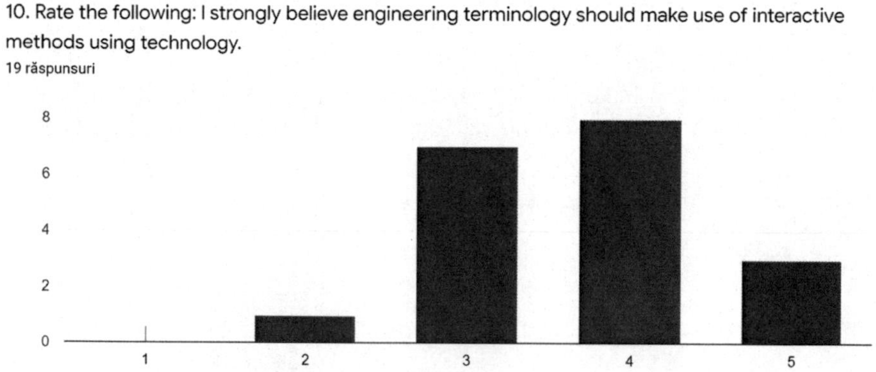

Fig. 16. Statistics of answers related to Question 10.

5 Discussion

The results of the survey confirmed the hypothesis that the resort to technology stimulates learning motivation and acquisition success. Both components of the hypothesis were confirmed: the technical part and the linguistic one, thus making proof of the efficiency of interdisciplinary approaches in education.

It is obvious that the current and future trend in education ought to ensure easy access to teaching and learning resources for the teachers, who need to adapt their teaching methods and techniques to the expectations and requirements of the young generation of digital-native students. Education can no longer continue to entail only frontal teaching, when the teacher lectures in front of the classroom and the students (rather passively) sit in their desks taking notes; teaching today needs to have the teacher (literally and figuratively speaking) step down from the teacher's desk and join the students, meet their expectations, adapt to their style and comply to their preoccupations.

Therefore, we can observe how, after almost two years of Covid-19 pandemic-imposed online teaching, educators have reached the conclusion that technology in education is indeed, the future. Before the pandemic afflicted the world, technology has been part of the teaching process, for a while already, by means of online manuals, gadgets, use of internet, etc., but never before has technology become such an integrated part to the teaching process as in this period, in which courses were taught almost exclusively online. Keeping this in mind, we ask ourselves: Does technology bring about more added value to teaching, or not? Does it imply re-formatting all that the educators knew and applied in the classroom? Does it mean that old-school schooling is obsolete? After applying the ECORE tool, we may answer that indeed, more added value appears to be imprinted to teaching, once technology is applied in the classroom, and that educators do need to adjust their teaching styles, their soft and hard skills to meet the requirements of the new age of education. Lastly, old-school schooling is not necessarily obsolete, but it ought to be complemented and adjusted to the novelty imposed to teaching, in view of Industry 4.0 requirements.

6 Conclusion

The tools are out there, the possibilities of applying them to teaching exist, all there is left seems to be the educators' ability and availability to embark on this new adventure. Teachers need to reconsider their role in the educational process, from providers of information to facilitators [10]. It is no longer enough for educators to design lessons following (or shadowing, really) the official teaching syllabus, making sure that students are being delivered all the right information, but it is of paramount importance for the educators to acknowledge and value their students' personality, needs and interests. The curriculum needs to be adapted, the teaching methods and tools must be adjusted having in mind the receivers, the beneficiaries of the act of education.

It goes without saying, this is a tough process, which entails, more than anything else, readiness and availability, willingness to adapt. It implies de-coding the educational needs of the current student, who was born digital-native, who can easily operate gadgets almost all the time, who has quick and direct access to information. It implies re-coding all this data into 'tailor-made' methods, which stand chances to entertain and instruct the new student. Last, but not least, it means that the teachers ought to be ready and willing to learn and train themselves on the new technologies, in order to keep up and perform. This brings us back to the idea of adaptability, which should not be considered lightly, since it involves employing and developing technology skills and competencies, together with the harsh awareness that the 'what' and the 'how' in teaching today might be obsolete tomorrow. It only leaves us with the realisation that in order to survive, we need to adapt to always adapting.

Acknowledgements. This paper ensues from the implementation of the Erasmus+ project Eco/Logical Learning and Simulation Environments in Higher Education (ELSE), 2018-1-IT02-KA203-048006 – Programme KA2: Cooperation for Innovation and the Exchange of Good Practices, Strategic Partnerships for Higher Education, project duration: September 2018 – March 2021; the G.E. Palade University UMPhST of Târgu Mureş, Romania was one of the 11 partners.

References

1. Moldovan, L.: State-of-the-art analysis on the knowledge and skills gaps on the topic of Industry 4.0 and the requirements for work-based learning. Procedia Manuf. **32**, 294–301 (2019). https://doi.org/10.1016/j.promfg.2019.02.217
2. Moldovan, L.: Design of a new learning environment for training in quality assurance. Procedia Technol. **12**, 483–488 (2014)
3. Moldovan, L.: Innovative models for vocational education and training in Romania. Procedia Soc. Behav. Sci. **46**, 5425–5429 (2016)
4. Moldovan, L., Moldovan, A.-M.: Green methodology for learning assessment. Procedia Technol. **22**, 1176–1183 (2016)
5. Moldovan, L.: New evaluation model by means of mobile technology. Procedia Technol. **19**, 1094–1101 (2015)
6. Sauve, L., Renaud, L., Kaufman, D.: Games, simulations, and simulation games for learning: definitions and distinctions. In: Kaufman, D., Sauve, L. (eds.) Educational Gameplay and Simulation Environments, Information Science Reference, p. 3. Hershey, NY (2010)
7. van der Linden, A., van Joolingen, W.: Supporting conceptual change in physics with a serious game. In: Cai, Y., van Joolingen, W., Walker, Z. (eds.) Simulations and Serious Games for Education, p. 18. Springer, Singapore (2019)
8. ONSERIOUSGAMES Homepage. http://www.onseriousgames.com/differences-between-e-learning-gamification-and-serious-games/. Accessed 28 September 2021
9. Brieger, N., Pohl, A.: Technical English. Vocabulary and Grammar. Summerton Publishing, Oxford (2008)
10. Rus, D.: English for specific purposes in the context of the shifting educational paradigm triggered by industry 4.0. In: Moldovan, L., Gligor, A. (eds.) Proceedings 63, 64. MDPI, Basel (2020)

Power Systems, Energy Efficiency and Renewable Technologies

Investigations on the Hygrothermal Properties of Aerogel Insulation Blankets

Adrian Bucur, Ligia Mihaela Moga(✉), and Daniela Lucia Manea

Technical University of Cluj-Napoca, 28 Memorandumului, 400114 Cluj-Napoca, Romania
ligia.moga@ccm.utcluj.ro

Abstract. The European Union's long-term strategy is that of a decarbonized economy that will ensure a better and longer life for our planet. In the building industry, important participants in this strategy are the construction materials. Thereby, the paper presents investigations done on the aerogel blanket, a nano insulation material which may become a better alternative to the usual thermal insulations, due to its superior thermal insulating properties at reduced thickness. In this research, hygrothermal determinations are done on aerogel samples exposed to different humidity conditions in order to observe their influence on the material thermal conductivity. The results are compared to the declared values for two of the most common thermal insulations: expanded polystyrene and mineral wool. The measurements reveal that the aerogel has a better hygrothermal behavior than these thermal insulations, which is beneficial against mold development, for waterproofing and also for maintaining the thermal properties of the material even in humid conditions.

Keywords: Water absorption · Immersion · Thermal conductivity · NZEB · Nano insulation material

1 Introduction

The nearly Zero Energy Buildings (nZEB) is a concept of thermal performance that is compulsory for all new buildings, as it is established by the European Union's revised Directives 844/2018 [1] and 2002/2018 [2]. The nZEB performance criteria follow the European Union's Energy Strategy which implies that greenhouse gas emissions must decrease by 2050 by 80 to 95%, when compared to pre-industrial levels. At the same time, there is a large building stock across the European Union, which is 50–100 years old. Most of these buildings are energy-intensive, needing deep retrofitting and lacking a proper level of thermal comfort. By deep retrofitting, all the building stock will have to be eventually converted to nZEB.

Towards improving the thermal performance of a building, the first step consists in the thermal optimization of its envelope elements, which in most situations is obtained by using a thermal insulation system which has to cover completely these elements. An increased level of thermal performance is obtained by using common thermal insulations (e.g. expanded polystyrene, mineral wool, etc.) with large thicknesses, which may represent a problem in several situations.

L. Moldovan and A. Gligor (Eds.): Inter-Eng 2021, LNNS 386, pp. 455–465, 2022.
https://doi.org/10.1007/978-3-030-93817-8_42

In this matter, aerogel may be an alternative solution since it has about 2.5 lower thermal conductivity compared to the usual thermal insulations. Therefore, for the same thermal performance, aerogel has a reduced thickness compared to common solutions. Among other characteristics, aerogel has also a specific hygrothermal behavior, which is analyzed in this study. There are made different determinations such as the short term and long-term water absorption capacity of the material through partial and total immersion, and the water absorption capacity in order to reach saturation. At the same time, using a laboratory heat flow meter, there is determined the thermal conductivity of the aerogel samples in dry state and in a water saturated state. The results are compared to the declared values for expanded polystyrene and mineral wool.

2 Material

Aerogel is a nano insulation material with the structure composed of spherical particles connected by thin links, its nanometric scale model being similar to the macroscopic one. It is produced by extracting the liquid component from a common gel and filling the resulted pores with air, the solid structure of the material representing only 0.13%–5% of its total volume. This makes the aerogel the lightest solid material used in practice with a density of only 3 kg/m^3 [3]. A detailed description regarding the aerogel synthesis can be found in several works [4, 5].

The bulk aerogel is characterized by a low mass, soundproofing qualities – a sound velocity through its structure of only 100 m/s (3.5 lower than through the air), transparency in the visible light domain with a reduced refraction index (glare reduction capacity) and non-flammable qualities. At the same time, the material is well known for its thermal insulation properties. The declared value of the aerogel thermal conductivity is around 0.015 W/(m·K), about 2.5 lower than that of the common thermal insulations [3]. Therefore, this material may represent a viable alternative to the traditional thermal insulations, especially when it is required an increased thermal performance of the building envelope elements altogether with a reduced thickness of their finishing layers.

The material is produced with different shapes and compositions, the most common ones being the monolithic and granular aerogel. For thermal insulation solutions, the monolithic aerogel is used as blankets, which are reinforced with fibers because the bulk material is very brittle. The granular aerogel is used in highly thermal efficient rendering products with a thermal conductivity of about 0.030 W/(m·K). Using this type of renderings may be beneficial especially for heritage buildings, where a conventional thermal insulation cannot be used at the exterior side, in order to preserve its façade. At the same time, both monolithic and granular aerogels are used as filling material for glazing units (i.e. windows). Compared to common glazing units, the aerogel-based ones have an increased thermal performance and also a reduced glare, providing an uniform distribution of sunlight, which may increase the interior comfort [3]. At the same time, as Cha et al. [6] suggest in their study, aerogel-based films can be applied on the existing windows too, in order to increase their thermal performance.

The material produces large quantities of dust, which may be inconvenient in several situations. This type of dust is not considered dangerous, being a non-carcinogenic material. However, protection gear should be used because direct exposure may affect the respiratory system for shorter or longer periods of time [3].

Fig. 1. Aerogel blanket.

One of the most common monolithic aerogel insulation material is the silica aerogel blanket, which is analyzed further in this study (Fig. 1). But the production of silica nanoparticles requires a large quantity of energy, which implies high costs for the final product. Therefore, research efforts are needed toward determinations of alternative production technologies or less demanding base materials [3]. An example in this matter is the research study of Horvat et al. [7], in which are developed polysaccharide aerogels through alcohol gelation method, measuring at the same time its thermal performance. The best result was obtained for low-methoxyl pectin aerogel, having a thermal conductivity of about 0.021 W/(m·K) [7]. A special type of aerogels are the structurally graded ones with the pores' diameter changing. The change of pores diameter through the material thickness is reversely correlated to its density and it affects the effective thermal conductivity. Regarding this subject, Noroozi et al. [8] evaluates through simulation the thermal performance of these aerogels with the variation of their density: from higher to lower values (HtL) and from lower to higher values (LtH). Results indicate that the HtL type aerogel have the best thermal insulation behavior.

The paper presents the hygrothermal investigations done on the fiber reinforced silica aerogel blanket (i.e. Spaceloft product – from Aspen Aerogels), a solution widely used in practice.

3 Methods

The research focus on the material exposed to humidity levels above 95%, a situation in which the sample is in direct contact with the water in the liquid phase. Following this idea, in some specific exterior conditions, there are measured the short term and long-term water absorption capacity of the material through partial and total immersion (Fig. 2). At the same time, other samples are kept in total immersion in water until they reach saturation phase. On the respective samples are determined: the water absorption capacity so as to reach saturation and the thermal conductivity after being dried to a constant mass. Besides these measurements, the research aims to evaluate the influence of the humidity on the aerogel thermal performance in the specific given conditions.

The short-term water absorption W_p is determined through partial immersion following the EN ISO 1609:2013 requirements [9]. The sample is placed into a vessel, having its inferior surface immersed in water, without a temperature gradient. Its mass is determined in the initial phase and then after 24 h, in this manner being simulated the water absorption of the sample being exposed to rain drops during the construction works. As a side note, in the current case, the determinations were done also in the first three hours of the time interval thus to observe the accelerated water absorption trend in this stage of exposure.

The short-term water absorption W_p is computed with the following formula:

$$W_p = \frac{m_{24} - m_0}{A_p} \left[\frac{kg}{m^2}\right] \quad (1)$$

where:

- m_{24} – mass of the sample after 24 h [kg];
- m_0 – mass of the sample mass in initial state [kg];
- A_p – surface of the sample [m^2].

The long-term water absorption is determined through partial or total immersion following the EN ISO 12087:2013 requirements [10]. The sample is placed into a vessel, being partially or totally immersed in water, without a temperature gradient. Its mass is determined in the initial phase and then at regular periods of time, the final measurements being made after 28 days, according to the standard. However, in our case, the measurement period for the total immersion testing was extended to 60 days. The partial immersion testing lasted 28 days, as it is recommended by the standard. In this manner, through partial immersion is simulated the water absorption of the sample being in long term water exposure. Through total immersion, the absorption is independent to the exterior conditions. For both determinations, intermediate measurements at regular time intervals were taken too therefore to observe the water absorption evolution.

The long-term water absorption through partial immersion W_{lp} is determined with the following formula:

$$W_{lp} = \frac{m_{28} - m_0}{A_p} \left[\frac{kg}{m^2}\right] \quad (2)$$

where:

- m_{28} – mass of the sample after 28 days [kg];
- m_0 – mass of the sample mass in initial state [kg];
- A_p – surface of the sample [m^2].

The long-term water absorption W_{lt} through total immersion is determined with the following formula:

$$W_{lt} = \frac{m_{28} - m_0}{V} \times \frac{100}{\rho_w} [\%] \quad (3)$$

where:

- m_{28} – mass of the sample after 28 days [kg];

- m_0 – mass of the sample mass in initial state [kg];
- V – initial volume of the sample [m^3];
- ρ_w – water density; $\rho_w = 1000$ kg/m^3.

The water absorption coefficient W_w is determined through partial immersion following the EN ISO 15148:2004 requirements [11]. The sample is placed into a vessel, having its inferior surface immersed in water, without a temperature gradient. Its mass is determined in the initial phase and then at regular time periods, but not less than 24 h, according to the standard. However, in our case, the measurements have been done at regular daily intervals for 28 days (672 h), in order to obtain a stable value.

The water absorption coefficient W_w is determined with the following formula:

$$W_w = \frac{m_t - m_i}{A \times \sqrt{t}} \left[\frac{kg}{m^2 \times h^{0.5}} \right] \quad (4)$$

where:

- m_t – mass of the sample at the moment t [kg];
- m_i – mass of the sample mass in initial state [kg];
- A – surface of the sample [m^2];
- t – time interval [h].

The exterior conditions for all tests were: T = 23 ± 2 °C and RH = 25–30%. One should note that the exterior relative humidity was lower than that imposed by the above norms: RH = 50 ± 5% and therefore the results obtained may be slightly different than the standard based ones.

Fig. 2. Aerogel in partial immersion – close view (left), total immersion – view from above (right).

The thermal conductivity was measured with the means of the Lasercomp 200 Heat Flow Meter, following the EN ISO 12667:2002 [12] prescriptions. In the heat flow meter, the sample is placed between two temperature-controlled plates. In this way, a temperature difference across the sample (ΔT) is established. In this case, the temperatures were set to 0 °C and 20 °C. The specimen thickness (d) is measured by the optical encoders of the device. The sample surfaces are covered by two thin film heat flux transducers which measure the heat flux (q) from steady state transfer (Fig. 3). Then, the thermal conductivity is determined with the following formula:

$$\lambda = \frac{q \times d}{\Delta T}\left[\frac{W}{m \times K}\right] \tag{5}$$

where:

- q – heat flux [W];
- d – sample thickness [m];
- ΔT – temperature difference between the plates [K].

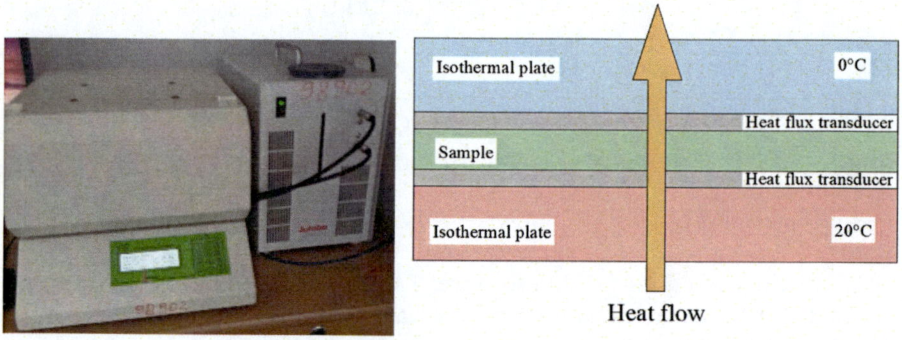

Fig. 3. Lasercomp Fox 200 Heat Flow Meter.

The determinations were made on a single layer (1 cm) and a triple layer pack (3 cm) of aerogel. Another measurement is done on a sample which was previously totally immersed in water for 80 days. All these samples were dried to constant mass in the laboratory oven before testing. A last determination was done on a sample which was previously exposed to a relative humidity of 80% for 24 h and which was not dried before testing. The samples had a surface area of 200 × 200 mm.

This study determines the thermal behavior of the aerogel only, but there are several studies which examine the thermal performance of the building elements insulated with this material. For example, Lakatos evaluates in his work [13], the thermal behavior (heat fluxes, surface temperatures, thermal conductivity, and retardation time) of a wall thermally insulated with aerogel, placed either at its exterior or its interior surface.

4 Results and Discussion

When the aerogel blanket is in dry state, its surface is hydrophobic, as it can be seen in the first figure below. However, after it was partially or totally immersed in water for several days, it absorbed water in its volume. Also, in its initial state, the color of the product is dark blue. When it is immersed in water, the color of the material becomes paler and the fiber network in the blanket degrades slowly (Fig. 4).

Fig. 4. Aerogel blanket surface: in dry state (left), after several days of water immersion (center), after 80 days of water immersion (right).

Regarding the short-term water absorption by partial immersion (W_p), the aerogel has a value of 0.09 kg/m^2 after 24 h of partial immersion in water. (Fig. 5) As a comparison, the mineral wool has $W_p \leq 1$ kg/m^2, a much greater value and therefore the aerogel has a much lower water absorption capacity, which is beneficial for any insulation material.

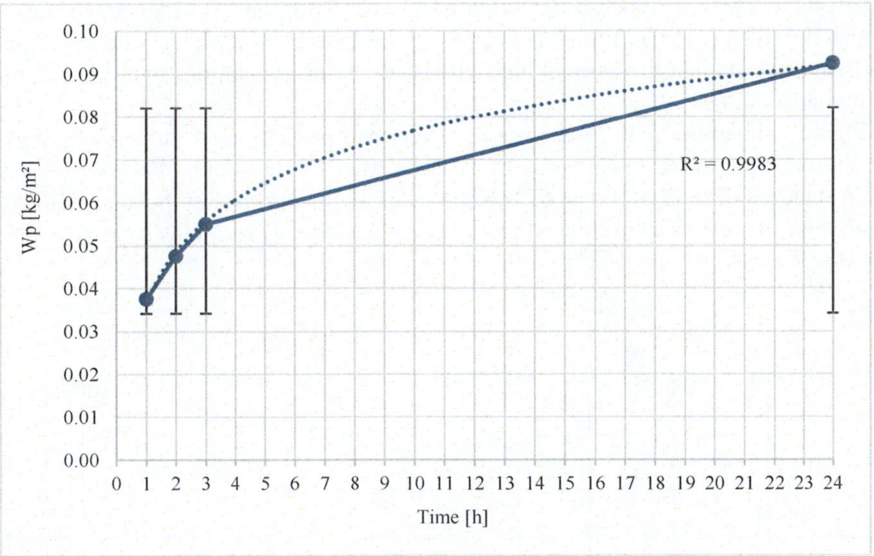

Fig. 5. Short term water absorption by partial immersion W_p.

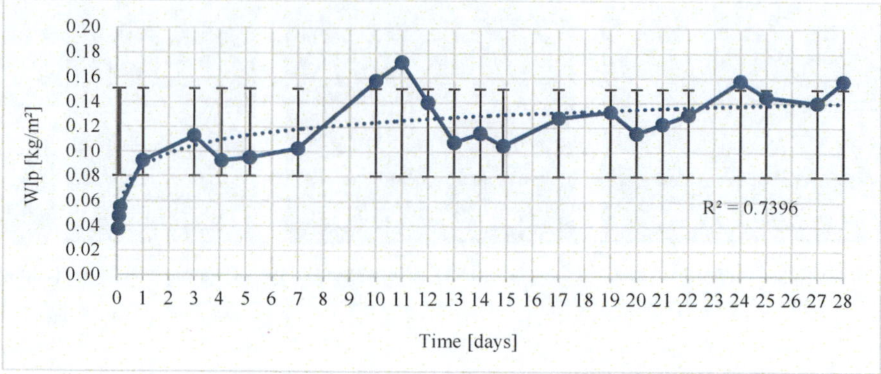

Fig. 6. Long term water absorption by partial immersion W_{lp}.

As shown in Fig. 6, the long-term water absorption by partial immersion (W_{lp}) of aerogel has a measured mean value of 0.12 kg/m^2. After 28 days, W_{lp} is equal to 0.16 kg/m^2 with a standard deviation of $\sigma = 0.036$ kg/m^2. At the same time, the expanded polystyrene has $W_{lp} \leq 0.5$ kg/m^2 and mineral wool $W_{lp} \leq 3$ kg/m^2. Again, it is demonstrated that the aerogel has better waterproofing qualities than the common thermal insulations.

Figure 7 illustrates that the long-term water absorption by total immersion (W_{lt}) of aerogel has a measured mean value of 12–14% after 28 days and 14% after 60 days, with a standard deviation of $\sigma = 6.28\%$. The measurements were stopped after the day 60, because the results have shown that the probe has already reach the saturation phase after 30–35 days.

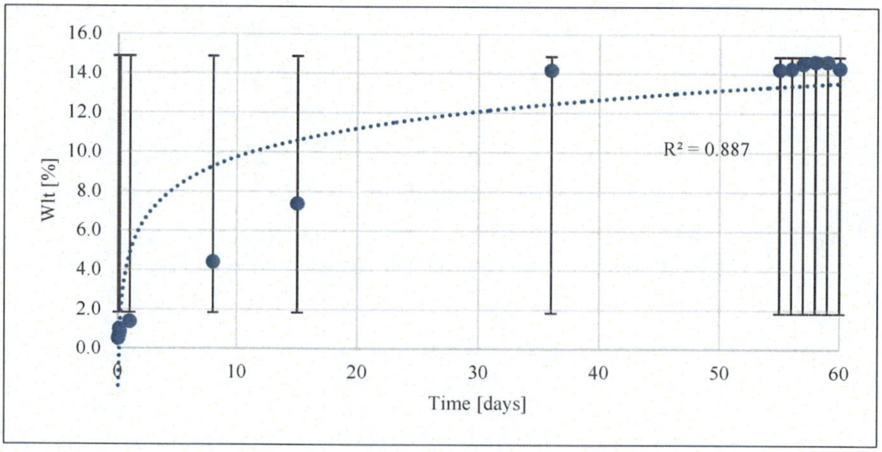

Fig. 7. Long term water absorption by total immersion W_{lt}.

Also, the measured water absorption coefficient (W_w) for aerogel has a stable average value after 28 days around 0.005 kg/m$^2\sqrt{}$h (see Fig. 8).

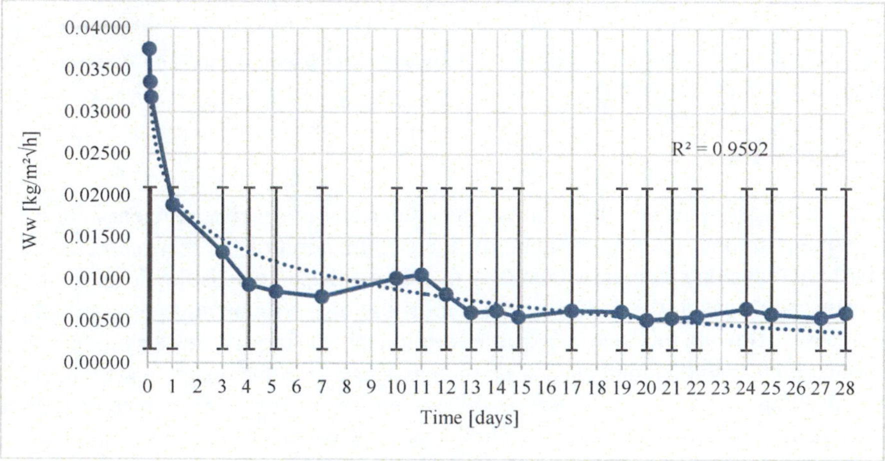

Fig. 8. Water absorption coefficient W_W.

As shown in Table 1, the studied situations for thermal conductivity revealed that multiple layers of aerogel in direct contact have a slightly decreased thermal performance compared to the single layer of this material. Also, an interesting fact is that after it was kept in total immersion for 80 days, after drying, the resulted thermal conductivity of the respective sample was slightly lower compared to the sample which was not previously exposed to potential damaging factors. Therefore, the direct contact with water in liquid form does not damage the internal structure of the material in such a way that it affects its thermal conductivity. Also, one should note that for the product used (i.e. Spaceloft blanket – from Aspen Aerogels), the declared thermal conductivity is 0.015 W/(m·K), a slightly lower value than all the measured ones, but of the same magnitude order.

Table 1. Measured thermal conductivity.

Material	Thickness [cm]	λ [W/(m·K)]	Standard error (SEM)
Aerogel – single layer	1	0.01766	3.99E−05
Aerogel – triple layers in direct contact	3	0.01815	4.21E−05
Aerogel – after 80 days of immersion	1	0.01648	1.67E−05
Aerogel – at RH 80%	1	0.01892	2.25E−05

As it was previously presented, the first three measurements were done only after the samples were dried to a constant mass at a specific temperature and relative humidity. The last measurement was done on a sample which was exposed previously at a relative humidity of 80%. The result in this case shows that when it is in wet state or is it exposed to a high relative humidity, the aerogel thermal conductivity tends to increase. This is in accordance with previous studies such as Hoseini, Bahrami [14] and Lakatos [15]. This is due to the fact that the air contained by the dry material is gradually replaced by water which creates new bonds between the molecules of its solid structure, which otherwise are dispersed. This creates a continuity for the thermal transfer path, increasing in this way the thermal conductivity.

5 Conclusion

The current regulations and the future targets regarding the energy efficiency for buildings require, among other improvements, the optimization of the thermal behavior of the building envelope elements. One way to achieve these goals is the development of alternative solutions to the common thermal insulations, with better hygrothermal performances. Following this idea, a possible solution may be the aerogel, a nano insulation material which is studied and used at an increasing scale in the recent years. One of its main advantages is given by a very good thermal insulating performance at reduced thickness, due to its very low thermal conductivity. The material has other qualities too (soundproofing, fireproofing, lightness), but at the same time must be improved in order to become more affordable.

In this study, there are presented the results of the laboratory hygrothermal determinations done on the silica aerogel blanket. The measurements reveal that the aerogel has a lower short- and long-term water absorption capacity, both through partial and total immersion than the most used thermal insulations: expanded polystyrene and mineral wool. This is an important outcome that shows that aerogel protects better against humidity and water infiltrations the construction element on which is installed. It represents a feature that reduce the risk of mold development and is also important for the building elements which need waterproofing (flat roofs, grade slabs, walls in contact with earth etc.).

Regarding the aerogel thermal performance, we observe that, in dry state, the material maintains a lower thermal conductivity both in a single layer as well as in multiple layers being in direct contact. A similar outcome is obtained even after the material was subjected to a damaging factor such as full contact with water in liquid form for a large period of time (i.e. 60 days). This shows the material conservation qualities that ensure a large lifespan. However, this is obtained only if the material is dried after it was exposed to high levels of humidity. If it is in wet state, its thermal conductivity tends to increase, as the results have shown. Therefore, when the aerogel is put in place as a thermal insulation for buildings, it is recommended to protect it against humidity, because it may be difficult to dry it to a satisfactory level.

To conclude, the aerogel nano-insulation material could represent a feasible solution for obtaining the decarbonized building stock required through the Renovation Wave but also in respect to the 2050 energy and climate target. Obtaining reduced energy consumption levels in some cases it is very difficult from both design and technological approach to be obtained using conventional thermal insulation materials. Therefore, the aerogel-based materials will represent a convenient alternative for "The Make-or-Break Decade" that has started.

References

1. European Parliament and Council of the European Union, "Directive 2018/844/EU amending Directive 2010/31/EU on the energy performance of buildings" (2018)
2. European Parliament and Council of the European Union, "Directive 2018/2002/EU amending Directive 2012/27/EU on the energy efficiency" (2018)
3. Moga, L., Bucur, A.: Nano insulation materials for application in nZEB. Proc. Manuf. **22**, 309–316 (2018). https://doi.org/10.1016/j.promfg.2018.03.047
4. Kuchta, L., Fajnor, V.Š: About the synthesis and thermal stability of SiO_2-aerogel. J. Therm. Anal. **46**, 515–520 (1996). https://doi.org/10.1007/BF02135029
5. Baetens, R., Jelle, B.P., Gustavsen, A.: Aerogel insulation for building applications: a state-of-the-art review. Energy Build. **43**(4), 761–769 (2011). https://doi.org/10.1016/j.enbuild.2010.12.012
6. Cha, J., Kim, S., Park, K.-W., Lee, D.R., Jo, J.-H., Kim, S.: Improvement of window thermal performance using aerogel insulation film for building energy saving. J. Therm. Anal. Calorim. **116**(1), 219–224 (2013). https://doi.org/10.1007/s10973-013-3521-5
7. Horvat, G., et al.: Thermal properties of polysaccharide aerogels. J. Therm. Anal. Calorim. **127**, 363–370 (2017). https://doi.org/10.1007/s10973-016-5814-y
8. Noroozi, M., Panahi-Sarmad, M., Bahramian, A.R., Sharif, A.: Theoretical investigation of heat transfer in structurally graded silica aerogels with pores diameter changing. J. Therm. Anal. Calorim. **135**(3), 1713–1721 (2018). https://doi.org/10.1007/s10973-018-7315-7
9. EN 1609:2013 – Thermal insulating products for building applications. Determination of short term water absorption by partial immersion
10. EN 12087:2013 – Thermal insulating products for building applications. Determination of long term water absorption by immersion
11. EN ISO 15148:2004 – Hygrothermal performance of building materials and products. Determination of water absorption coefficient by partial immersion
12. EN 12667:2002 – Thermal performance of building materials and products. Determination of thermal resistance by means of guarded hot plate and heat flow meter methods. Products of high and medium thermal resistance
13. Lakatos, Á.: Effect of the placement of aerogel insulation in the heat transfer properties. J. Therm. Anal. Calorim. **133**(1), 321–327 (2017). https://doi.org/10.1007/s10973-017-6745-y
14. Hoseini, A., Bahrami, M.: Effects of humidity on thermal performance of aerogel insulation blankets. J. Build. Eng. **13**, 107–115 (2017). https://doi.org/10.1016/j.jobe.2017.07.001
15. Lakatos, Á.: Investigation of the moisture induced degradation of the thermal properties of aerogel blankets: measurements, calculations, simulations. Energy Build. **139**, 506–516 (2017). https://doi.org/10.1016/j.enbuild.2017.01.054

Improving the Thermal Insulation of Cement-Based Composites Using Tea Waste Aggregates

Othmane Horma(✉), Mouatassim Charai, and Ahmed Mezrhab

Laboratory of Mechanics and Energy, Mohammed First University, 6000 Oujda, Morocco
horma.othmane@ump.ac.ma

Abstract. In recent decades, energy sustainability in buildings has become a global concern due to the fast-rising energy demand and its disastrous effects on the environment. One of the promising solutions is recycling waste materials and by-products (vegetable fibers, etc.) through innovative low-environmental impact building materials to promote passive green buildings. This work evaluates the thermal insulating potential of tea waste within a cement matrix using the Hot Disk Method. The experimental results indicate the positive influence of adding 10% in weight of tea waste to a cement matrix. The thermal conductivity and diffusivity dropped from 0.67 to 0.27 W/mK and 0.36 to 0.23 mm^2/s, respectively. In addition, a significant decrease of 25% in density was observed, highlighting the lightweight aspect of the developed mixture. Finally, annual thermal simulations were carried out through Energy-Plus software, while introducing a thin layer of TW-based composite in the external building envelope to illustrate the energy savings. The results indicated, as expected, the thermal insulating ability of this proposed bio-based material, as the annual heating loads were reduced by 14.5%.

Keywords: Bio-sourced materials · Tea waste · Cement · Passive heating · Thermal characteristics · Thermal insulation · Energy-plus simulation

Nomenclature and Abbreviations

λ	Thermal conductivity
A	Thermal diffusivity
c_p	Specific heat capacity
ρ	Density
wt.%	Binder's weight percentage
TW	Tea Waste
TPS	Transient Plane Source
RTCM	Moroccan Construction Thermal Regulation
TMY	Typical Meteorological Year

L. Moldovan and A. Gligor (Eds.): Inter-Eng 2021, LNNS 386, pp. 466–476, 2022.
https://doi.org/10.1007/978-3-030-93817-8_43

1 Introduction

Recently, following the climate crisis growing concern and the International Panel on Climate Change reports [1–3], people are urging governments to move towards renewable energy and energy efficiency. Thus, scientists are requested to look for ways to attain energy sustainability, which affect positively country's economic stability along with its environmental preservation.

Building and construction industries play a major role in the current energy and climate issues, consuming third of world global energy, which leads to an even bigger share of the global carbon footprint [4]. Conventional insulation materials and techniques are not affordable for most countries. Therefore, innovative, low-carbon and local construction materials with improved thermal insulating performance appear to be a promising solution to this predicament. Moreover, studies on adding agricultural, industrial and municipal wastes to either clay bricks or cement-based bricks have been extensively recorded [5–10].

Tea waste results from the infusion of tea as a soft drink and is considered as biomass. At the moment it is mainly used in animal food production and as a fertilizer [11, 12]. From the literature, studies on Tea Waste (TW) in the field of construction materials seem very limited, despite tea being the most consumed beverage worldwide, except for water, reaching five million tons per year [13]. Moreover, the tea market was valued at nearly 50 billion US dollars in 2017 expected to exceed 73 billion in 2024 [14], with China being the number one producer with over 2.5 million metric tons in 2018 [15]. This rapidly growing commodity will undoubtedly produce enormous amounts of unused waste. Therefore, taking advantage of TW should be economically and environmentally beneficial.

Ozturk et al. [16] studied the influence of up to 12.5%wt TW on fired clay brick's thermal, mechanical and physical properties and observed an improvement in the thermal performance with increasing porosity while decreasing compressive strength. Demir et al. [17] worked on the addition of processed waste tea (PWT) into burnt and unburnt clay bricks; the findings were an increase in water uptake and porosity for both clay bricks with increasing waste content (0%, 2.5% and 5%).

While the few existing papers cited before on TW are focused on clay matrix, this study identifies the impact of TW content on the thermal characteristics of cement-based composite. Along with promoting the recycling of tea waste aggregate, especially for high tea consuming countries like Morocco [18]. The thermal behavior was assessed using the Hot Disk Method, while the more TW additives were added, the lower the density of the composites. From the tested properties, the application of TW composite in the building industry was briefly discussed.

This work also aims to expand the limited data on TW in the construction field. Finally, annual dynamic thermal simulations were carried out for a virtual building in eastern Morocco using Oujda City's weather file. Calculations were performed using Energy Plus software, one of the most reliable and accurate energy analysis programs to establish the effect of incorporating TW additives into the cement-based building components.

2 Materials and Methods

2.1 Sample Preparation

Chinese green gunpowder tea, commonly commercialized in Morocco, was infused in boiling water for about five minutes, as of usual in tea-making process, leaving wet tea waste (TW) behind, after consumption. First, the TW was filtered and stored, then sun-dried for about 8 h to remove the adsorbed water. Finally, TWs were mixed in Portland cement matrix using 0.5 of water to cement ratio. An increment of 2 wt.% waste was chosen to make different specimens ranging from 0 to 10% of TW additives. Samples were prepared in a cylindrical shape of 5 cm diameter and 2 cm height. Figure 1 shows the components involved in the preparation process and the final product.

Fig. 1. Tea waste composites preparation

Finding how much waste is produced from boxes of gunpowder tea can help estimate the expected amount of waste generated by the raw substance. Thus, during the sample preparation process, TW was weighted after sun drying. By comparing the dry TW mass and the mass of gunpowder tea, noted beforehand. We found that about 70% of the original gunpowder tea mass is conserved as waste.

2.2 Experimental Method

The heat conduction process has two main states, transient, where the temperature is dependent on time, while temperature variations in the steady-state are negligible. Although steady-state measurement techniques are usually based on simple calculations, yielding accurate results can be tedious due to the time-consuming character. Transient heat measurement methods are more complex. However, they are distinguished by a fast measurement time. The hot disk method (HDM) was used in this present study. It is a transient plane source (TPS) technique that has gained popularity, as a rapid and accurate thermal properties measurement device. In addition to determining multiple properties at the same time, the tests are quick, the software is user-friendly and the sensor flexibility allows measurement on liquids, powder along with solids. Further theoretical explanations on the operating principles of HDM are given by Yi He [19].

The HDM revolves around and is named after the disk-shaped sensor. It is composed of two nickel spirals, wrapped in a thin insulation material layer, in our case it was Kapton.

The main feature of the sensor is its ability to act simultaneously as a heat source and a temperature sensing element for the monitoring the time dependent response of samples.

Measurement Protocol
In this work a TPS 2200 (Hot disk, Sweden) apparatus was used. it yields thermal information according to ISO 22007–2 [20]. The method setup is shown in Fig. 2, and the measurement process consists of the following steps:

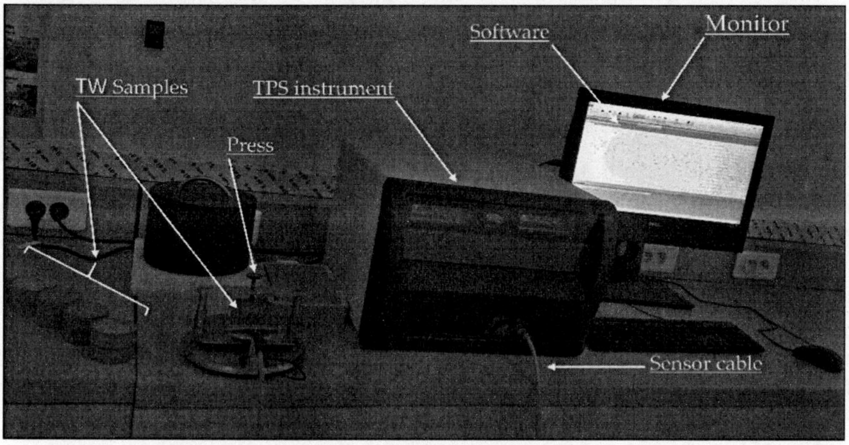

Fig. 2. The hot disk method setup

- Placing the probe between two identical samples, the faces facing the sensor must be plane.
- Pressing the samples to reduce the sensor's air contact.
- Inputting on the software, the sensor's radius (6.4 mm) and resistance (11.8 Ω), room temperature (16–20 C°), measurement time (80 s) and heating power (80–100 mW).
- The software (Hot Disk Thermal Analyzer) calculates the thermal properties, based on the parameters mentioned above and recorded temperature variations by the sensor.
- A time condition is required, depending on the sample, there should be an interval of 7 to 15 min between two consecutive measurements, allowing thermal stabilization to be reached.

3 Results and Discussion

3.1 Thermal Properties

Table 1 presents the average of five measurements for each specimen. The apparent density was derived from samples masses and dimensions.

The finding results (Table 1) show the positive influence of TW additives on the thermal properties of the cement-based composite. A remarkable decrease in thermal

Table 1. Thermophysical characteristics of the produced specimens

TW content (wt.%)	Conductivity [W/mK]	Diffusivity [mm^2/s]	Vol. specific heat [MJ/m^3K]	Density [kg/m^3]
0	0.671 ± 0.001	0.361 ± 0.002	1.858 ± 0.01	1775.6 ± 2
2	0.531 ± 0.001	0.316 ± 0.004	1.684 ± 0.03	1652.3 ± 3
4	0.426 ± 0.003	0.297 ± 0.003	1.438 ± 0.02	1565.9 ± 3
6	0.363 ± 0.007	0.268 ± 0.011	1.350 ± 0.04	1479.6 ± 1
8	0.295 ± 0.005	0.239 ± 0.009	1.242 ± 0.06	1370.7 ± 1
10	0.272 ± 0.003	0.225 ± 0.003	1.207 ± 0.01	1331.7 ± 2

conductivity of 59.5% from 0.671 to 0.272 W/mK is observed with 10% in weight of TW incorporation. It represents an improvement of more than 50% compared to TW integrated into fired clay bricks [16].

Figure 3 shows the smart lightweight aspect of the TW composite, as more additives lead to less thermal conduction and density. The addition of 10 wt.% TW reduces nearly 60% of the conductivity and 25% of density. This decrease indicates that compared to the ordinary Portland cement (0%), the composite is significantly lighter and its thermal insulating ability has strongly improved. The green strip in the Fig. 3 represents the smart lightweight zone where the specimens with TW content higher than 4wt.% are situated, representing their enhanced thermal and physical properties over the control specimen (0 wt.% TW).

The Specific heat capacity c_p was calculated using Eq. 1, where λ and α are, respectively, the thermal conductivity and diffusivity, and ρ the density.

$$c_p = \frac{\lambda}{\alpha \times \rho} \tag{1}$$

The variation of the thermal properties for each TW content, shown in Fig. 4, represents the main finding of this work as they describe the thermal behavior of the material. The specific heat, derived from Eq. (1), declines rapidly between 0 and 4 wt.% of TW addition, it stabilizes around 910 J/kg K for upward TW additions. Both thermal conductivity and diffusivity decrease continuously to reach ultimately 59,5% and 37% reduction, respectively. The specific heat decreases by roughly 13%, which could indicate a lower aptitude in thermal inertia for the material. However, the great improvement of thermal conductivity and diffusivity strongly suggests an enhancement of the overall thermal performances for the composite with 10% TW addition by weight.

Even though no in-depth mechanical strength analysis was conducted, the compressive strength of the specimen with 0% and 10wt.% of TW was examined. As it is the case in most plant-based additives, the strength decreased with the introduction of TW into the binder. However, the minimum requirements of >0.5 MPa and <0.3 W/m.K for compressive strength and thermal conductivity [21], respectively, were met. Concluding that cement incorporating 10wt.% of TW can be effectively used in the construction industry as a thermally insulating, non-structural component.

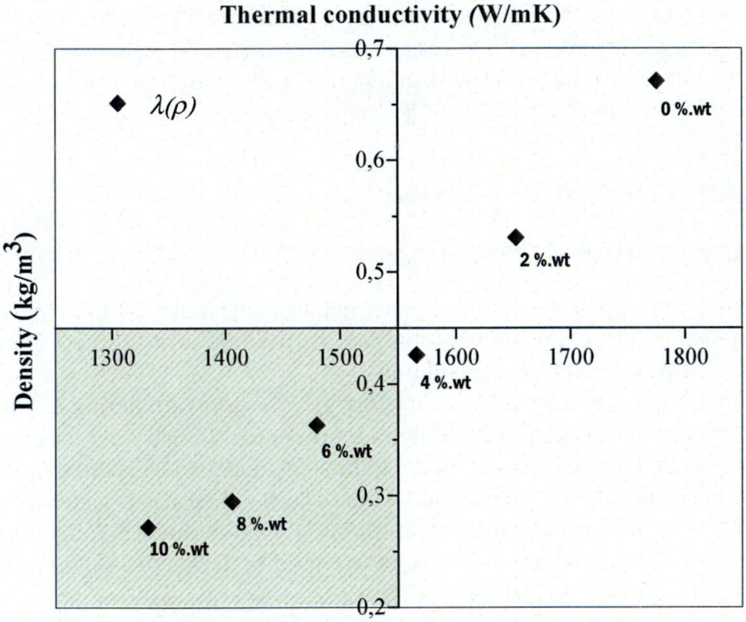

Fig. 3. Density versus thermal conductivity

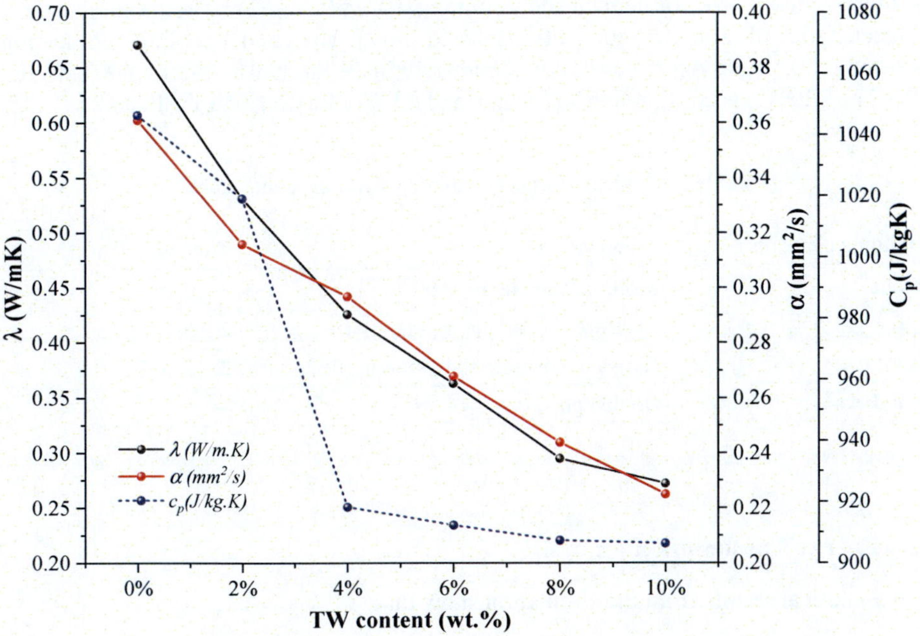

Fig. 4. Thermal properties evolution versus TW content

The above-mentioned results indicate the potential of TW in the development of low-cost, energy-efficient, and non-load-bearing construction materials. Nevertheless, further researches are needed to assess and optimize the impact of TW additives on the physical, and mechanical properties of the cement matrix.

4 Dynamic Thermal Simulation

4.1 Simulation Process and Parameters

Although the core of this paper was the thermal characterization of the TW composite seen previously, this section aims to showcase the usefulness of this new material as a thermal insulating component in a building.

The simulation was conducted using Energy Plus software during a TMY under Eastern Moroccan climate. The building is a residential ground floor space of 70 m^2 facing east, located in Oujda city, with the following geographical coordinates: Latitude 34.8°N, Longitude −1.9°E, Elevation 470 m. The building has a window-wall ratio of 23%. We assume that there is no air infiltration with no internal gains (unoccupied buildings). It should be noted that the software used to design the building, and as an interface to Energy Plus was VI-suite [22].

Two buildings are tested to show the impact of TW composites on the heating and cooling demands. The construction components for the conventional building (Ref.) are given in Table2. In addition, Fig. 5 shows a visual representation of the second building (TWC). Both buildings are identical, except for the exchange of 2 cm external fraction of mortar into a layer of cement incorporating 10 wt.% of TW. The thermophysical characteristics of all the components are listed in [7]. The annual cooling and heating demand for the buildings to constantly remain within the thermal comfort zone (between 20 °C and 26 °C) according to Moroccan standard RTCM [23], was evaluated.

Table 2. Typical Moroccan building envelope components

Building elements	Components	Thickness (cm)
Walls	Mortar, concrete, mortar	2, 20, 2
Floor	Reinforced concrete, mortar, tile	20, 5, 2
Roof	Gypsum, concrete slab, mortar, tile	2, 16, 5, 2
Windows	Double glazing	–

4.2 Energy Consumption Analysis

The numerical results from the simulation show that:

- TW has a significant influence when it comes to heating loads, as displayed in (a), Fig. 6. The annual heating loads reduction from Ref to TWC is 668 kWh, representing

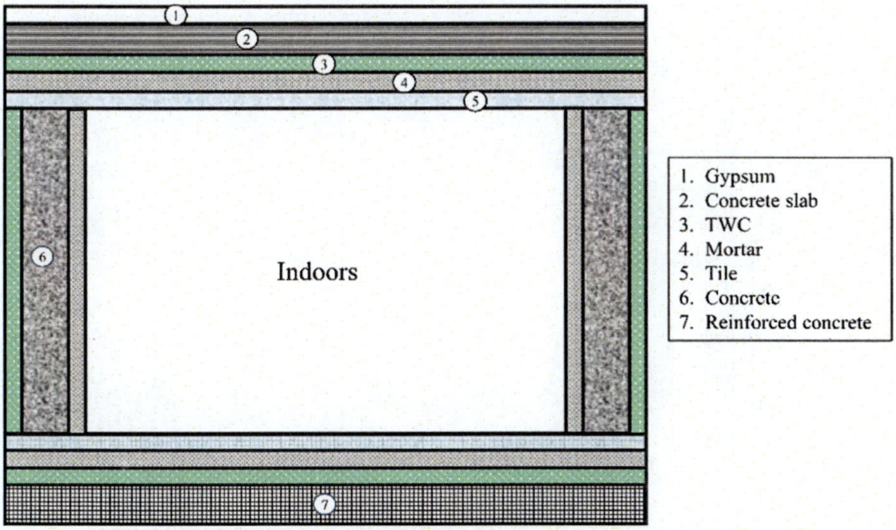

Fig. 5. Visual representation of TWC and its construction components

almost 15%. This is the most relevant result from this simulation, proving that TW composite indeed improves the thermal insulation of the building envelope.

- The cooling loads are shown in (b), Fig. 6. Unfortunately, from Ref to TWC, a slight increase in the annual cooling energy demand of 4.5% has been noticed from 3672 to 3836 (kWh). This is result was expected mainly due to the assumptions made earlier as without air infiltration the insulation can have the opposing effect. Moreover, the relatively lower density and specific heat in the TWC layer can affect negatively its ability to store heat. This can lower the thermal lag inside the building, causing it to heat up faster during the day.
- The total passive energy saving adds up to 6% per year, which is remarkable for such a minimalistic change represented by an insulating thin layer of tea waste-cement.

According to these results, this material has a great thermal insulation potential and seems to be better suited for cold climate.

In the light of these findings, the same simulation was carried out using London climate, to confirm that the TW composite is more effective in colder regions. Without going into details, the answer was affirmative, as the total energy savings were around 8%, From Ref to TWC, instead of 6% in comparison with the earlier simulation.

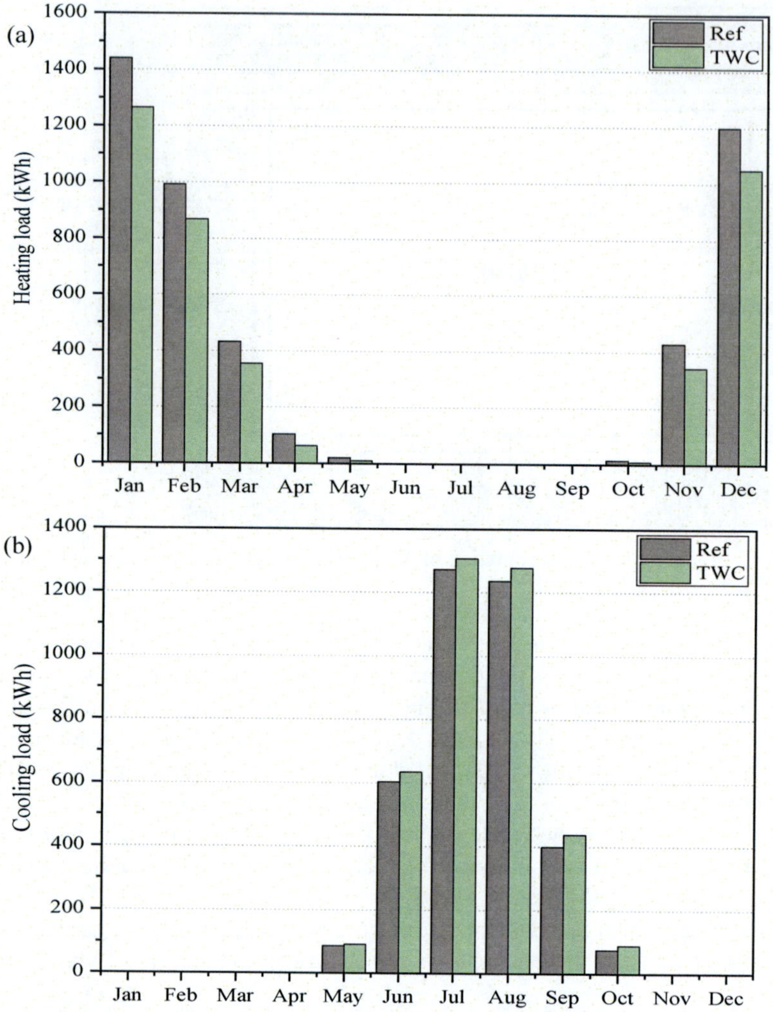

Fig. 6. Annual energy demand: (a) Heating loads; (b) Cooling loads

5 Conclusion

In this work, the thermal properties of Portland cement incorporating TW aggregates were investigated. The study aims to develop an innovative, low-cost lightweight construction material with improved thermal characteristics and showcase its potential application as a thermal insulating building component through an annual dynamic thermal simulation. The experimental results showed that increasing TW content leads to decreased density along with an increase in the thermal resistance of the composite. The last specimen containing 10% of TW content resulted in the reduction of density, thermal conductivity, and diffusivity by 25%, 60%, and 37%, respectively. Moreover, its mechanical strength was examined. Even though the compressive strength decreased from the control sample, it was still within the acceptable range for insulating applications. The thermal simulation results were on par with the experimental ones. The comparative analysis of the thermal performance shows that tea waste composite (TWC) implementation in the building envelope can be highly effective for passive heating, reducing 14.5% of the heating loads from the reference case.

These findings allow the conclusion that tea waste aggregates can indeed contribute to developing non-structural, eco-friendly, bio-composite materials to enhance the energy efficiency in buildings.

References

1. Masson-Delmotte, W.V., et al.: IPCC, 2018: Global warming of 1.5°C (2018)
2. Shukla, J.M.P.R., et al.: Climate change and land: an IPCC special report on climate change, desertification, land degradation, sustainable land management, food security, and greenhouse gas fluxes in terrestrial ecosystems (2019)
3. Moira, F., Huang, C.: A look at how people around the world view climate change, Pew research center, Fact tank. Pew Res. Cent. (2019). https://pewrsr.ch/2UpGcq7
4. IEA, Buildings a source of enormous untapped efficiency potential, (2019), https://www.iea.org/topics/buildings.
5. Zhang, Z., Wong, Y.C., Arulrajah, A., Horpibulsuk, S.: A review of studies on bricks using alternative materials and approaches. Constr. Build. Mater. **188**, 1101–1118 (2018). https://doi.org/10.1016/j.conbuildmat.2018.08.152
6. Chen, F., et al.: Fly ash based lightweight wall materials incorporating expanded perlite/SiO2 aerogel composite. Towar. Low Therm. Conduct. **249**, 118728 (2020). https://doi.org/10.1016/j.conbuildmat.2020.118728
7. Horma, O., Charai, M., Mezrhab, A., Karkri, M.: Thermal characterization of cement-plaster-expanded polystyrene composites. In: 2020 5th International Conference on Renewable Energies for Developing Countries (REDEC), pp. 1–6 (2020). https://doi.org/10.1109/REDEC49234.2020.9163883
8. Charai, M., Sghiouri, H., Mezrhab, A., Karkri, M.: Thermal insulation potential of non-industrial hemp (Moroccan cannabis sativa L.) fibers for green plaster-based building materials. J. Clean. Prod. **292**, 126064 (2021). https://doi.org/10.1016/j.jclepro.2021.126064
9. Liuzzi, S., Rubino, C., Martellotta, F., Stefanizzi, P., Casavola, C., Pappalettera, G.: Characterization of biomass-based materials for building applications: the case of straw and olive tree waste. Ind. Crops Prod. **147**, 112229 (2020). https://doi.org/10.1016/j.indcrop.2020.112229
10. Charai, M., Sghiouri, H., Mezrhab, A., Karkri, M., Elhammouti, K., Nasri, H.: Thermal performance and characterization of a sawdust-clay composite material. Procedia Manuf. **46**, 690–697 (2020). https://doi.org/10.1016/j.promfg.2020.03.098
11. Wang, H., Xu, C.: Utilization of tea grounds as feedstuff for ruminant. J. Anim. Sci. Biotechnol. **4**(1), 54 (2013). https://doi.org/10.1186/2049-1891-4-54
12. Chowdhury, A., Sarkar, S., Chowdhury, A., Bardhan, S., Mandal, P., Chowdhury, M.: Tea waste management: a case study from West Bengal. India. Indian J. Sci. Technol. **9**(42), 1–6 (2016)
13. Celik, M.Y., Celik, F.: Turkish tea: the only tea washed with snow that snow falls on. Int. J. Basic Clin. Stud. **3**(2), 1–15 (2014)
14. Bedford, E.: Global tea market size 2017–2024 (2020). https://www.statista.com/statistics/326384/global-tea-beverage-market-size/. Accessed 24 Jan 2020

15. Shahbandeh, M.: Tea production worldwide from 2006 to 2018, by leading country (2019). https://www.statista.com/statistics/264188/production-of-tea-by-main-producing-countries-since-2006/. Accessed 24 Jan 2020
16. Ozturk, S., Sutcu, M., Erdogmus, E., Gencel, O.: Influence of tea waste concentration in the physical, mechanical and thermal properties of brick clay mixtures. Constr. Build. Mater. **217**, 592–599 (2019). https://doi.org/10.1016/J.CONBUILDMAT.2019.05.114
17. Demir, I.: An investigation on the production of construction brick with processed waste tea. Build. Environ. **41**(9), 1274–1278 (2006). https://doi.org/10.1016/J.BUILDENV.2005.05.004
18. Research, D.S.: Global per capita tea consumption as of 2016, by country (2020). https://www.statista.com/statistics/507950/global-per-capita-tea-consumption-by-country/. Accessed 24 Jan 2020
19. He, Y.: Rapid thermal conductivity measurement with a hot disk sensor: Part 1. Theor. Considerations. Thermochim. Acta **436**(1–2), 122–129 (2005). https://doi.org/10.1016/J.TCA.2005.06.026
20. I. 22007–2: Plastics—Determination of thermal conductivity and thermal diffusivity—Part 2: Transient plane heat source (hot disc) method (2015). https://www.iso.org/obp/ui/#iso:std:iso:22007:-2:ed-2:v1:en
21. Rilem, L.: Functional classification of lightweight concrete. Mater. Struct. **11**, 281–283 (1978)
22. Southall, R., Biljecki, F.: The VI-Suite: a set of environmental analysis tools with geospatial data applications. Open Geospatial Data, Softw. Stand. **2**(1), 23 (2017). https://doi.org/10.1186/s40965-017-0036-1
23. ADEREE: Règlement Thermique de Construction au Maroc (RTCM) (2014)

Thermal Effects on Dynamic Control of FGM Kirchhoff Plate with Bonded Piezoelectric Materials

Loukmane El Khaldi(✉), Khadija Messaoudi, and Mustapha Sanbi

Advanced Science and Technology Team, National School of Applied Sciences, Abdelmalek Essaadi University, Sebta Avenue, 93002 Tétouan, Morocco

Abstract. To develop a mathematical modeling of the smart structure, the study consists in using, in first time, the finite element method and using linear thermopiezoelastic theory and Kirchhoff assumptions applied to a smart plate divided into a finite number of elements. For this purpose, we define the plate represented by the FG material, bonded on both surfaces with thermopiezoelectric patches used as sensors/actuators. The effects of temperature on the dynamics and the active control are studied and discussed. The equations of motion are obtained via Hamilton principle. The investigation is performed on natural frequencies of Functionally Graded Material (FGM) plates under temperature conditions. The outcomes show that the vibration characteristics are greatly influenced with the thermal parameters concerned in the study.

Keywords: Thermopiezoelectric plate · Kirchhoff plate theory · Finite elements · Functionally graded material

1 Introduction

The use of continuous sandwich structures has been developed and used for 50 years because of its high bending stiffness, low density and good insulating qualities. The sandwich structures represent a particular form of layered structure composed of two thin bodies of high rigidity and strength and a core of flexible material and relatively light. Due to the wide application of Functionally graded materials (FGM), several studies have been conducted to analyze the behavior of FGM structures.

Sanbi et al. in [1] have studied the thermoelastic and piezoelectric coupling effects on dynamics and active control of piezolaminated intelligent beam that is modeled by the finite element method.

In [2], Tounsi et al. presented a refined trigonometric shear deformation theory for thermoelastic bending of functionally graded sandwich plates while S.H. Shen in [3] performed a numerical examinations of the thermal post-buckling of laminated piezoelectric plate.

Zenkour and Aljadani in their work [4] investigated the thermo-electrical buckling of a functionally graded piezoelectric nanoscale plate via a refined hyperbolic higher-order shear deformation theory.

L. Moldovan and A. Gligor (Eds.): Inter-Eng 2021, LNNS 386, pp. 477–486, 2022.
https://doi.org/10.1007/978-3-030-93817-8_44

Torabi et al. in [5] studied buckling analysis of a functionally graded conical shell integrated with piezoelectric layers that is subjected to combined action of thermal and electrical loads is presented.

Mantari and Granados presented a thermoelastic bending analysis of functionally graded sandwich plates by using a new quasi-3D hybrid type higher order shear deformation theory (HSDT) [6].

Zenkour and Sobhy in [7] used the sinusoidal shear deformation plate theory to study the thermal buckling of functionally graded material (FGM) sandwich plates. El Harti et al. presented a dynamic study of sandwich functionally graded beam with piezoelectric layers that are used as sensors and actuators [8].

Yaghoobi et al. investigated in their work [9] the thermal buckling analysis of hybrid laminated plates made of two-layered functionally graded materials (FGMs) that are integrated with surface-bonded piezoelectric actuators referred to as (P/FGM)s, while Nourmohammadi and Behjat in [10] studied the static response of the functionally graded piezoelectric (FGP) plates under mechanical, electrical and thermal with piezoelectric layers (sandwich FGPM) based on the first order shear deformation plate theory.

This work deals with the thermoelastic effects on the mechanical properties of the Thermopiezoelectric plate FG using Kirchhoff plate theory with finite element techniques.

2 Mathematical Modeling

Let consider a FGM plate with the integrated sensors and actuators as shown in Fig. 1. Properties for the two piezoelectric materials used in the simulation can be found in the literature as in [11] and in Table 1.

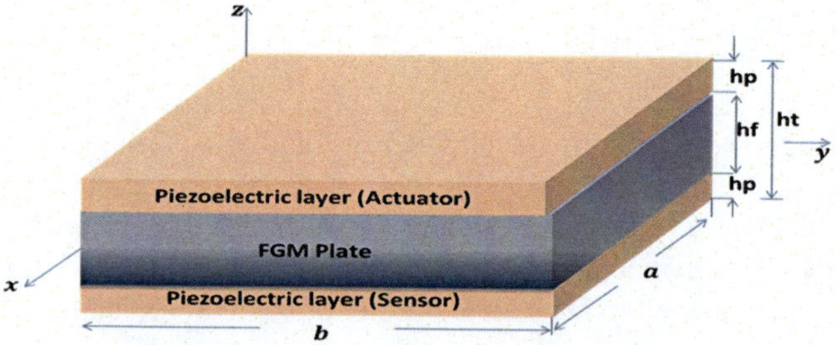

Fig. 1. Geometry and coordinate system of a rectangular FGM plate with upper and lower piezoelectric layers

The top layer is the piezoelectric actuator, while the bottom layer is the piezoelectric sensing layer. The intermediate layer is a FGM plate made up of a mixture of ceramic and metal. The piezoelectric FGM plate has the length a, width b and the total thickness $ht = hf + 2hp$, where hf and hp are the thicknesses of the core and piezoelectric layers, respectively. The rectangular Cartesian coordinate system (x, y, z) has the plane $z = 0$, coinciding with the mid-surface of the plates.

Table 1. Properties of the FGM plate, sensor, and actuator

Properties (units)	Labels (units)	FGM plate	Sensor PVDF	Actuator PZT-5H32
Length	*Lp*, s, a (m)	0.6	0.25	0.25
Width	lp, s, a (m)	0.4	0.25	0.25
Thickness	tp, s, a (m)	0.001	0.2	0.2
Density	ρ (kg/m^3)	$\rho_m = 2780$ $\rho_c = 3800$	$\rho_s = 1780$	$\rho_a = 7700$
Young modulus	*E* (GPa)	$E_m = 70$ $E_c = 380$	$E_s = 5.04$	$E_c = 62$
Piezoelectric stress constant	$g31 \times 10^{-3}$ (Vm/N)	–	0.15	−9.11
Piezoelectric strain constant	$d31 \times 10^{-12}$ (m/V)	–	4.34	−274
Pyroelectric constant	$p \times 10^{-5}$ (Cm^{-2} K^{-1})		−1.25	–
Dielectric constant	K × 8.85410^{-12} (F/m)		12	3300
Thermal expansion	$\alpha \times 10^{-6}$ (K^{-1})		140	3

The properties of the FGM fluctuate consistently because of the progressive change in the volume portion of the constituent materials (ceramic and metal) shown in Fig. 2, by and large toward thickness. The force law work is normally used to depict these varieties of material properties. The volume fraction of the FGM is assumed to vary with the thickness of the plate according to a simple power law function:

$$E(z) = (E_c - E_m)V_c + E_m, \tag{1}$$

$$\rho(z) = (\rho_c - \rho_m)V_c + \rho_m \tag{2}$$

$$V_c(z) = \left(\frac{2z + hf}{2hf}\right)^k = 1 - V_m(z) \tag{3}$$

where:

- *hf* is the thickness, and *z* is the thickness coordinate ($-hf/2 \leq z \leq hf/2$) and *n* is the power-law exponent $0 \leq n \leq \infty$;
- *E*(*z*), ρ(*z*), and *n* are respectively Young's modulus, mass density, and the power law index;
- the subscripts *m* and *c* represent the metallic and ceramic constituents; *Vc*(*z*) is the volume fraction of the ceramic, and Vm(z), is the volume fraction of the metallic.

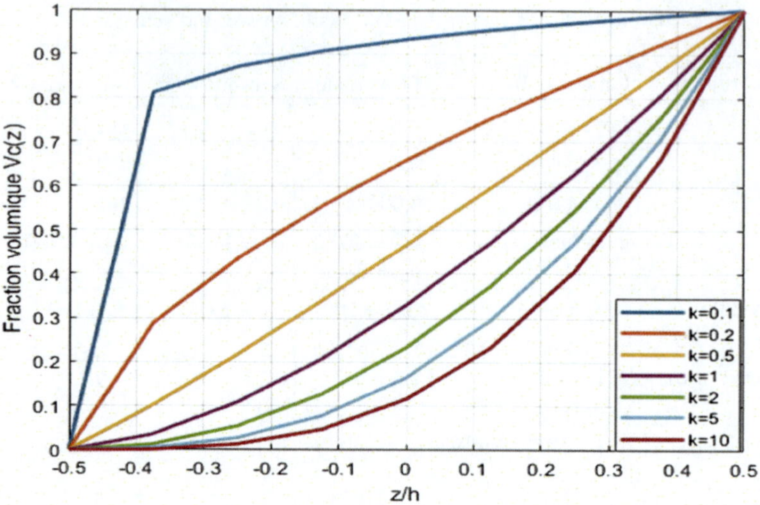

Fig. 2. Variation of the volume fraction function versus the nondimensional thickness for different power law indices

According to this distribution, the bottom surface ($z = -hf/2$) of the functionally graded plate is pure metal, and the top surface ($z = hf/2$) is pure ceramic. Thus, different volume fractions of metal can be obtained for different values of k. Therefore, the material properties of the plate, the Young's modulus and the mass density vary continuously in the direction of the thickness.

3 Temperature Dependent Properties (TD)

Functionally gradient materials are mainly used in high temperature environments, and the constituent materials possess temperature-dependent properties. The material properties can be expressed as follows [12]:

$$\mathrm{P(T)} = \mathrm{P_0}\left(\mathrm{P_{-1}T^{-1}} + 1 + \mathrm{P_1T} + \mathrm{P_2T^2} + \mathrm{P_3T^3}\right) \tag{4}$$

in which T = T0 + T and T0 = 300 K (room temperature), T(z) is the temperature rise through the thickness direction, and P0, P-1, P1, P2, and P3 are temperature-dependent coefficients that are unique to the constituent materials.

The mechanical and thermal properties of FGMs are determined from the volume fraction of the material constituents. We assume that the material properties, such as the modulus of elasticity (E), coefficient of thermal expansion (α), and Poisson's ratio(υ), and mass density $\rho(z)$ can be determined by:

$$\mathrm{E(z, T)} = \mathrm{Em(T)} + [\mathrm{Ec(T)} - \mathrm{Em(T)}]\left(\frac{2z + hf}{2hf}\right)^n \tag{5}$$

$$\rho(\mathrm{z, T}) = \rho\mathrm{m(T)} + [\rho\mathrm{c(T)} - \rho\mathrm{m(T)}]\left(\frac{2z + hf}{2hf}\right)^n \tag{6}$$

$$\alpha(z, T) = \alpha m(T) + [\alpha c(T) - \alpha m(T)]\left(\frac{2z + hf}{2hf}\right)^n \tag{7}$$

$$\nu(z, T) = \nu m(T) + [\nu c(T) - \nu m(T)]\left(\frac{2z + hf}{2hf}\right)^n \tag{8}$$

where m and c represent the metallic and ceramic constitutes and n is the index valve of power law.

4 Constitutive Relations

The displacement field of the Kirchhoff plate theory is given as.

$$\begin{aligned} u &= u0 + z\theta \\ v &= v0 + z\theta y \\ w &= w0 \end{aligned} \tag{9}$$

where:

- (u_0, v_0, w_0) denote the displacements along the (x, y, z) coordinate directions of a point on the midplane of the plate.
- $\theta_x = \frac{\partial w_0}{\partial y}$ and $\theta_y = -\frac{\partial w_0}{\partial x}$ are the rotations of normal transverse about the y and x axis, correspondingly, shown in Fig. 1.

The linear thermopiezoelectric constitutive equations coupling the piezoelectric and thermal effects, are given by [13, 14]:

$$\sigma_{ij} = C_{ijkl} - e_{kij}E_k - C_{ijkl}\alpha_{kl}\Delta T \tag{10}$$

$$D_i = e_{ikl}\mathcal{E}_{kl} + k_{ik}E_k - p_i\Delta T \tag{11}$$

where: σ_{ij}, $\mathcal{E}_{kl}$, D_i and E_k denote the stress, strain, electric displacement and electric field components, respectively; α_{kl} and ΔT are thermal expansion coefficient and temperature change C_{ijkl}, e_{kij}, p_i and k_{ik} are elastic, piezoelectric, pyroelectric and dielectric constant, respectively. the elastic stiffness of the FGM layer given by:

$$\mathcal{Q}_{11} = \mathcal{Q}_{22} = \frac{E}{\left(1 - \nu^2\right)}, \mathcal{Q}_{12} = \mathcal{Q}_{21} = \frac{\nu E}{\left(1 - \nu^2\right)}, \mathcal{Q}_{66} = \frac{E}{2(1 + \nu)}$$

5 Thermal Investigation

The behavior of FGM plate in thermal condition is considered for this study. The one-dimensional temperature distribution through the thickness direction is assumed. In this case three thermal conditions are considered: uniform, linear and nonlinear temperature distributions.

5.1 Temperature Distribution Uniformly (UTD)

In uniform temperature field, the temperature rise through the thickness is given as

$$T = T_0 + \Delta T \tag{12}$$

where:

- $\Delta T = T_c - T_m$, represents the temperature gradient.
- T_m and T_c are temperature at metal surface and at ceramic surface respectively.

5.2 Temperature Distribution Linearly (LTD)

The linear variation of temperature distribution along the thickness can be viewed as follows:

$$T = T_m + \Delta T\left(\frac{Z}{h} + \frac{1}{2}\right) \tag{13}$$

5.3 Temperature Distribution Nonlinearly (NTD)

The nonlinear variation of temperature distribution along the thickness can be viewed as follows:

$$T = T_m + \Delta T\left(\frac{Z}{h} + \frac{1}{2}\right)^{\gamma} \tag{14}$$

where γ is the exponent of temperature ($0 < \gamma < \infty$).

Note that the value of γ equals with the unit, represents a linear temperature change through the thickness, while the value of γ excluding the unit, represents a non-linear temperature change through the thickness.

6 Finite Element Analysis

Finite element modelling of FGM plate is carried out by a rectangular element with four nodes. The rectangular component has one node at each corner and seven degrees of freedom for every node. The in-plane displacements are u, v and w is the transverse displacement, θx, θy, $\frac{\partial w}{\partial x}$ and $\frac{\partial w}{\partial y}$ represent the rotations and slopes about x and y axes, respectively:

$$\begin{gathered} u = \sum\nolimits_{i=1}^{4} N_i u_i,\ v = \sum\nolimits_{i=1}^{4} N_i v_i,\ w = \sum\nolimits_{i=1}^{4} N_i w_i \\ \theta_x = \sum\nolimits_{i=1}^{4} N_i \theta_x^i,\ \theta_y = \sum\nolimits_{i=1}^{4} N_i \theta_y^i \\ \frac{\partial w}{\partial x} = \sum\nolimits_{i=1}^{4} N_i \frac{\partial w_i}{\partial x},\ \frac{\partial w}{\partial y} = \sum\nolimits_{i=1}^{4} N_i \frac{\partial w_i}{\partial y} \end{gathered} \tag{15}$$

Ni, $i = 1, 2, 3, 4$ represents the element nodal shape functions.

7 Governing Equations of Motion

The equation of motion for a FGM plate element in thermo mechanical environment is established by applying Hamilton's principle.

$$\delta \int_{t1}^{t2} \left(U^{(e)} - T^{(e)} + W^{(e)} \right) dt = 0 \tag{16}$$

The potential energy and the kinetic energy for plate element can be written in terms of displacement vector as

$$U^{(e)} = U_e^{(e)} - U_{Th}^{(e)} = \frac{1}{2}\left[\left\{q^{(e)}\right\}^T \left[K^{(e)}\right]\left\{q^{(e)}\right\}\right] - \frac{1}{2}\left[\left\{q^{(e)}\right\}^T \left[K_{Th}^{(e)}\right]\left\{q^{(e)}\right\}\right] = \ldots$$
$$\frac{1}{2}\left[\left\{q^{(e)}\right\}^T \left[K_{eff}^{(e)}\right]\left\{q^{(e)}\right\}\right] \tag{17}$$

where, $\left[K_{eff}^{(e)}\right] = \left[K^{e)}\right] - \left[K_{Th}^{(e)}\right]$, $\left[K_{eff}^{(e)}\right]$ is the effective stiffness matrix of the plate element and {q(e)} is the displacement vector Plate element.

The kinetic energy at the element level is defined as:

$$T^{(e)} = \frac{1}{2}\left[\left\{\dot{q}^{(e)}\right\}^T \left[M^{(e)}\right]\left\{\dot{q}^{(e)}\right\}\right] \tag{18}$$

where $\left[M^{(e)}\right]$ is the mass matrix of the plate element.

The governing equation of motion of the axially loaded FGM plate element in matrix form can written as:

$$\left[M^{(e)}\right]\left\{\ddot{q}^{(e)}\right\} + \left[K_{ef}^{(e)}\right]\left\{q^{(e)}\right\} - P(t)\left[K_g^{(e)}\right]\left\{q^{(e)}\right\} = 0$$

where P(t) is the time dependent dynamic load.

For the free vibration, the finite element technique gives the governing equations of motion which is solved as an eigenvalue/eigenvector issue can be written as

$$\left[\left[K_{eff}^{(e)}\right] - \omega^2\left[M^{(e)}\right]\right]\left\{q^{(e)}\right\} = 0 \tag{19}$$

here ω is natural frequency.

8 Discussion

The eigenfrequencies of the thermopiezoelectric plate fields FG are performed for a numerical analysis. The power law indices n = 1 and 5 with uniform, linear and non-linear temperature conditions are illustrated in Fig. 3, 4 and 5, illustrating the variation of the fundamental frequency parameter with respect to temperature. Observations from these plots revealed that increasing the temperature difference, reduced the natural frequency parameters of FGM plates.

This is due to the increase in temperature which reduces the resistance of the properties of the FGM plate, so it reduces the frequency parameters.

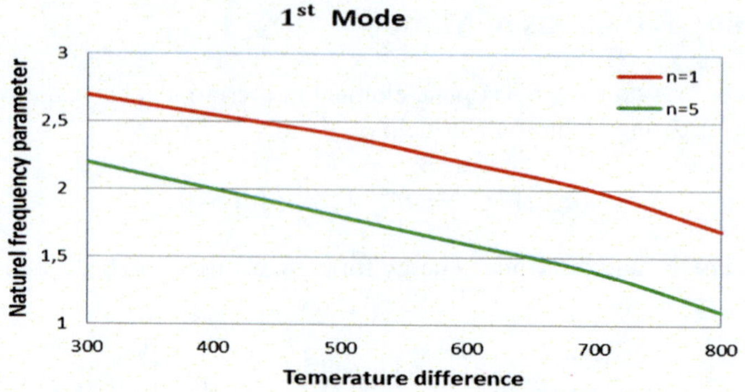

Fig. 3. Variation first mode frequency parameter with temperature rise. UTD

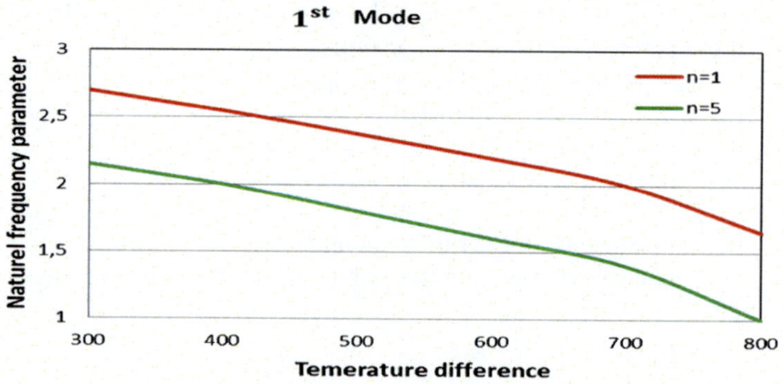

Fig. 4. Variation first mode frequency parameter with temperature rise. LTD

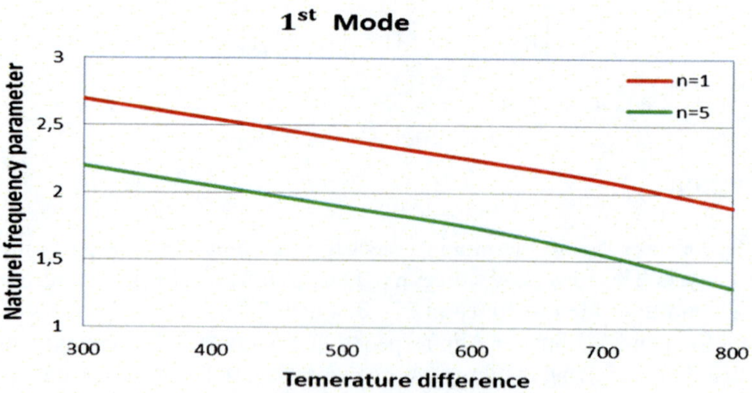

Fig. 5. Variation first mode frequency parameter with temperature rise. NTD

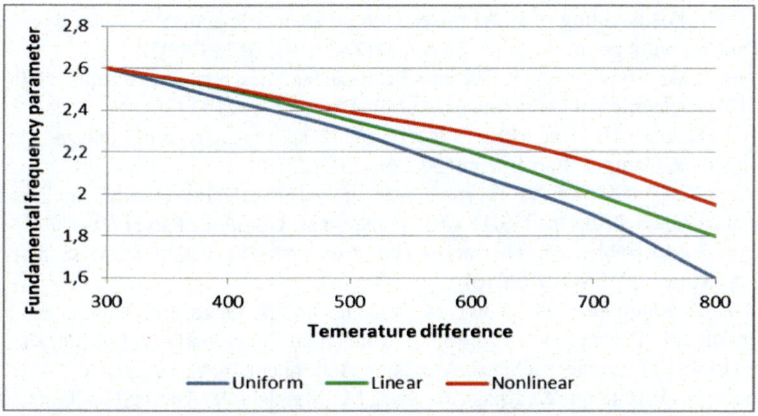

Fig. 6. Fundamental frequency parameter verses temperature change with UTD, LTD and NTD. (n = 1);

Figure 6 describes the increase of temperature in thermal environment of FGM plate with various thermal fields like uniform, linear and nonlinear temperature conditions.

The effect of fundamental frequency parameter is more prominent in uniform temperature field compared with linear and nonlinear fields. the effect of the fundamental frequency parameter is larger in the uniform temperature field than in the linear and nonlinear fields.

9 Conclusion

In this work, A parametric study of the vibration and instability of FGM thermopiezoelectric plates with a power-law property distribution along the thickness in thermal environments has been performed based on Kirchhoff plate theory using the finite element method in conjunction with Hamilton's principle. The natural first frequencies of the FGM plate in uniform temperature, linear and nonlinear environments are reduced by increasing the temperature difference. With the increase of the power index value, the natural frequencies are reduced. The natural frequencies of FGM plates decrease with increasing temperature. The FGM plates in thermal environment with higher values of power law index are more responsive to change of the temperature rise than those with lower values of index.

References

1. Sanbi, M., Saadani, R., Sbai, K., Rahmoune, M.: Thermoelastic and pyroelectric couplings effects on dynamics and active control of smart piezolaminated beam modeled by finite element method. Smart Mater. Res. (2014)
2. Tounsi, A., Houari, M.S.A., Benyoucef, S., Adda Bedia, E.A.: A refined trigonometric shear deformation theory for thermoelastic bending of functionally graded sandwich plates. Aerosp. Sci. Technol. **24**, 209–220 (2013)

3. Shen, S.H.: Postbuckling of FGM plates with piezoelectric actuators under thermo-electro-mechanical loadings. Int. J. Solids Struct. **42**(23), 6101–6121 (2005)
4. Zenkour, A.M., Aljadani, M.H.: Thermo-electrical buckling response of actuated functionally graded piezoelectric nanoscale plates. Results Phys. **13**, 102192 (2019)
5. Torabi, J., Kiani, Y., Eslami, M.R.: Linear thermal buckling analysis of truncated hybrid FGM conical shells. Compos. Part B: Eng. **50**, 265–272 (2013)
6. Mantari, J.L., Granados, E.V.: Thermoelastic analysis of advanced sandwich plates based on a new quasi-3D hybrid type HSDT with 5 unknowns. Compos.: Part B **69**, 317–334 (2015)
7. Zenkour, A.M.: Sobhy, M: Thermal buckling of various types of FGM sandwich plates. Compos. Struct. **93**, 93–102 (2010)
8. El Harti, K., Rahmoune, M., Sanbi, M., Saadani, R., Bentaleb, M., Rahmoune, M.: Finite element model of vibration control for an exponential functionally graded timoshenko beam with distributed PiezoelectricSensor/Actuator. Actionneurs **8**(1), 19 (2019)
9. Yaghoobi, H., Fereidoon, A., Khaksari Nouri, M., Mareishi, S.: Thermal buckling analysis of piezoelectric functionally graded plates with temperature-dependent, properties. Mech. Adv. Mater. Struct. **22**(10), 864–875 (2015)
10. Nourmohammadi, H., Behjat, B.: Optimum response of functionally graded piezoelectric plates in thermal environments. Mater. Sci.-Poland **35**(3), 606–617 (2017)
11. Sanbi, M., Saadani, R., Sbai, K., Rahmoune, M.: Thermal effects on vibration and control of piezocomposite kirchhoff plate modeled by finite elements method. Smart Mater. Res. (2015)
12. Touloukian, Y.S.: Thermophysical Properties of High Temperature Solid Materials. Macmillian, New York (1967)
13. Bajoria, K.M., Jadhav, P.A.: Buckling of simply supported piezoelectric FGM plates subjected to electro-mechanical loading using higher-order shear deformation theory. Adv. Mater. Res. **622–623**, 200–205 (2013)
14. Sanbi, M., El Marhoune, M., Essoufi, E.H., Faik, L.A., Rahmoune, M.: Centrifugal and thermal influence on the dispersion of surface waves propagating on a thermopiezoelectric half-space. Int. J. Mech. Mechatron. Eng. IJMME-IJENS **16**(2), 70–78 (2016)

Simulation of a Solar Driven Air Conditioning System for Mitigating the Cooling Demand of Buildings Located in Semi-arid Climates: A Case Study

Sara El Hassani(✉), Mouatassim Charai, Mohammed Amine Moussaoui, and Ahmed Mezrhab

Mechanics and Energy Laboratory, Mohammed First University, BV Mohammed VI, BP 717, 60000 Oujda, Morocco

Abstract. Nowadays, research on thermal comfort in buildings has become progressively more interested in renewable energy resources, mainly solar energy. In this paper, the modeling of a LiBr/H_2O absorption solar cooling system using TRNSYS software was discussed. The motive behind this is to evaluate the energy performance of single-effect LiBr/H_2O absorption chillers systems coupled with vacuum solar collectors in meeting the cooling needs of a real case study building located in semi-arid climate conditions. In order to better understand and predict the factors affecting the solar absorption system, the numerical model was developed taking into account three specific areas of work handling: the identification of the cooling demand, the design model of the adopted absorption cooling system and the dimensional optimization of the components. The simulation of the absorption chiller performance reveals that an area of 50 m^2 of evacuated collectors with a 2-m^3 tank of hot water storage permits to achieve an optimum solar fraction of over 86%. Hence, improving the reliability of the system allows to meet the air conditioning requirements of buildings located in semi-arid climates.

Keywords: Solar cooling technology · Single-effect absorption chillers · Simulation · TRNSYS

1 Introduction

Over the last decade, the rise of the standard of living, the increased demand of comfort and the high outdoor summer temperatures have led to a strong development of the air-conditioning systems. Besides, the energy consumption in the buildings has increased with the development of the world economy, accounting for 30% of the total energy used [1, 2].

Today, the energy management of buildings is a major challenge, in order to optimize energy consumption while improving their comfort and quality of use and to meet national commitments to reduce CO_2 emissions and greenhouse gases. Especially since air conditioning in the industrial, commercial and domestic sectors represents between

L. Moldovan and A. Gligor (Eds.): Inter-Eng 2021, LNNS 386, pp. 487–496, 2022.
https://doi.org/10.1007/978-3-030-93817-8_45

40 to 50% of the of global energy consumption [3], in which the residential sectors alone are responsible for more than 20% of CO_2 emissions in the European Union [4].

In this context, the contribution of renewable energies is gradually becoming indispensable to achieve the reduction targets set by the various authorities to improve the energy efficiency of buildings. For that, the solar-powered cooling systems appear to be an attractive alternative to conventional electric compressor units, as they use a natural refrigerants that are not harmful to the ozone layer so virtually no environmental impact, besides the main advantage of solar cooling lies in the simultaneous demand for cold and sunshine. When the heat necessary for the operation of the refrigeration machine is provided by the sun, the cold provided is free (i.e. no cost, no pollution). Simulations of solar cooling systems especially LiBr/ H_2O absorption systems have been studied, but a general model for all conditions has not yet been developed. Burckhart et al. [5] described the simulation of a solar cooling system using TRNSYS software with real data, consisting of a 250 m^2 field of evacuated tube solar thermal collectors and a lithium bromide/water absorption chiller with a capacity of 95 kW, to cover the thermal loads of a 4000 m^2 building. Mateus and Oliveira [6] carried out a comprehensive energy and economic analysis of the use of solar cooling for various buildings and climatic conditions. Based on their study, they concluded that the usage of evacuated tube collectors lead to a lower collector area of approximately 15% and 50% compared to flat plate collectors. Buonomano et al. [7] investigated the applicability of a solar-powered absorption cooling system based on a novel type of flat-plate evacuated tube collector (ETC) embedded in a double-effect LiBr-H_2O absorption cooler. The simulation results show that the systems coupled with a flat plate ETC achieve a higher solar fraction of 77% versus 66.3% for PTC collectors. Wang et al. [8] analyzed the efficiency of a double effect absorption solar cooler combined with a parabolic trough collector (PTC) to cover the cooling load. They found a high COP near to 1.195 and a solar energy utilization efficiency of 61.98%. In a recent study, Al-Falahi et al. [9] evaluated the energy effectiveness of a solar absorption cooling system combined with ETC using weather Baghdad data (Iraq). The seasonal collector efficiencies and solar fractions were determined to be 54% and 58% respectively. The COP was found 0.44. Elahi et al. [10] conducted a computational investigation to examine the effect of using solar plasma to promote improved absorption cycles using solar power. In terms of thermo-economic analyses, the absorption energy efficiency can exceed 90%.

From the literature, a variety of solar collector types and configurations has been employed for this application. The most commonly used type is the evacuated tube collector (ETC). Besides, it was remarked that LiBr-H_2O took part in most of the research activities in this field.

In this regard, the objective of this work is to master the utilization of solar energy in the field of air conditioning of buildings in order to evaluate the dimensions of the installation for the specific climate of Oujda city (eastern Morocco), which is characterized by a semi-arid climate. To this end, this paper presents simulations of the different factors that can affect the solar cooling system to meet the cooling demands of a building located in hot and dry semi-arid climates.

2 Brief System Description

The solar cooling system, in this investigation, is based on a single effect absorption-cooling machine coupled with an evacuated tube collector. The model was created in TRNSYS to satisfy the cooling needs of a building with a total ground area of 1022 m^2 located in the Oujda city of eastern Morocco. The real view as well as the sketch model of the case study building are presented in Fig. 1. For more details about the case study building, please refer to [11].

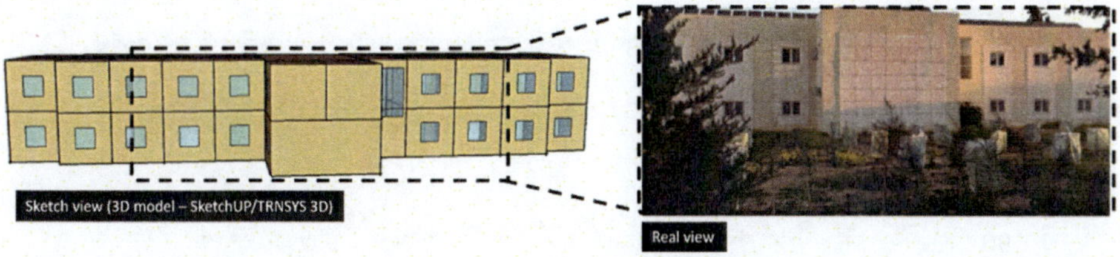

Fig. 1. Case study building

The solar absorption system under study consists of two main loops as shown in Fig. 2, the first loop is the solar loop in which in an evacuated tube solar collector, the heat transfer fluid (HTF) is heated and the heat energy collected is passed to a sensible heat storage tank. The second loop is a cooling loop for the absorption chiller. It contains the absorption chiller and the space to be cooled. It is pertinent to mention that through the second loop an auxiliary boiler has been added and controlled to supply hot fluid to the absorption unit at the required levels (i.e $T_{g,i}$ = 108.89 °C) in the presence of insufficient solar energy.

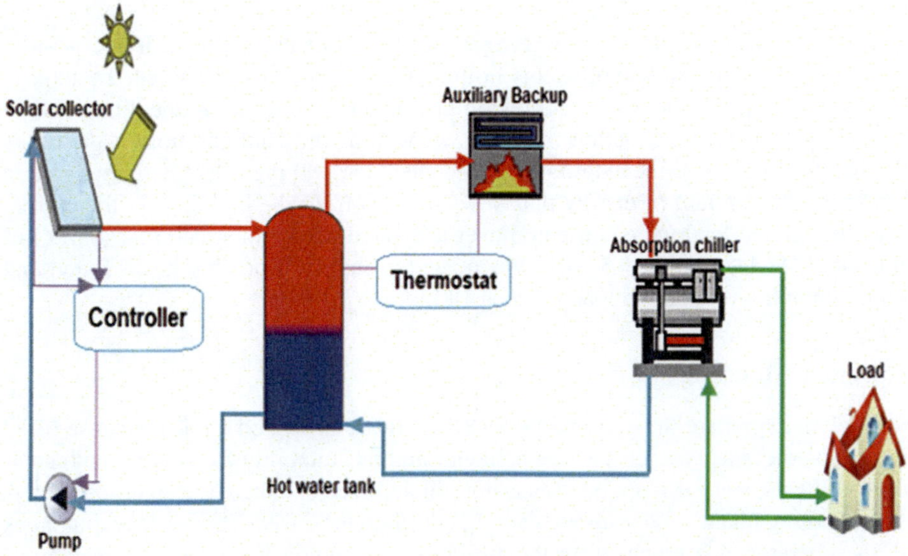

Fig. 2. Schematic of the configuration adopted for solar absorption cooling system

In this context, the required generator input temperature is 108.89 °C. This value is the minimum value for the operation of the absorption cycle. Therefore, it is very important to adjust the generator inlet temperature in order to control the primary energy consumption and to guarantee the safety of the system. In addition, due to the exchange between the temperature of the working fluid coming from the chiller and the temperature of the fluid in the storage tank, the generator outlet temperature ($T_{g,o}$) seems to decrease (see Fig. 3).

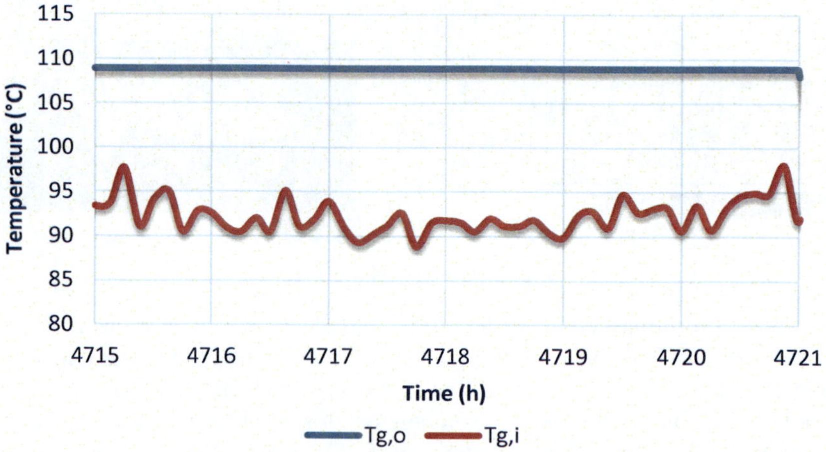

Fig. 3. Variation of $T_{g,i}$, $T_{g,o}$ on a typical summer day 16 July

2.1 Thermal Loads Calculation

In order to adequately establish the refrigeration needs of the building, it is crucial to take into account the load components in the particular area to be cooled. To achieve this, it is essential to consider the following parameters: the location, the dimension of the building and the construction materials, as well as the configuration of the doors and windows, mainly of the external building envelope. All these factors play a role in determining the amount of energy that flows in and out of the building. In the current study, the thermal load were computed over a Typical Meteorological Year (TMY) of Oujda city (34° 41′ 12″ N, 1° 54′ 41″ W) located in eastern Morocco. It was found that the maximum hourly cooling load reaches 18 kW.

2.2 System Modeling

To model and simulate the system, TRNSYS software is a good choice because of its ability to predict the dynamic system behavior and its suitability for dealing with solar cooling systems, since this application is not standardized and there are no usable experimental cooling plants with accurate data. In addition, the TRNSYS interface dialogues with the operator as a graphical programming tool. Moreover, no prior knowledge of the programming language is required to run a simulation. For all these advantages, TRNSYS has attracted the interest of researchers in recent years.

Besides, in the newest release of TRNSYS (version 18.1) [12], we can find the type 56 (multizone building) which gives the possibility to call a 3D model created by sketchup (Fig. 1). So from a 3d Google sketchup model we can create an executable TRNSYS simulation which can be run directly from Google Sketchup, hence joining the strength of the TRNSYS simulation environment with the ability and simplicity of using Google sketchup.

The system is made up of several items where the evacuated tube solar collectors employed were Sunstar Olymp 2470 [13], which were represented by "Type 71" in TRNSYS. The absorption system was a Yazaki WFC SC 50 with the specified default capacity and the COP of the chiller was 1494 kW and 0.53 respectively expressed as "Type 107". The typical characteristics and description of the main components, which were used as input parameters in the simulator developed in TRNSYS, are specified in detail in Table 1. Finally, the climate database of Oujda city used was obtained from a high precision MHP station located at Mohamed Premier University [14] built as a TMY2 format (i.e. Typical Meteorological Year) and this data was read using "Type 109". As the purpose of this study is to analyze the effect of the main design keys of the system configurations on the overall performance, the following simplifying assumptions are made in the TRNSYS modeling and simulation [14]:

- Power losses in valves and/or connected tubes are not taken into account.
- The effects of boiling and/or freezing of the working fluid have not been assumed.

3 Results and Discussion

The study of the case envisaged in this work represents a system of refrigeration by solar absorption responding to a request for air-conditioning under the dry and hot climate of Oujda city (Eastern Morocco). Based on the simulation performed with TRNbuild (Type 56), we tend to identify the cooling demand in the studied building. The obtained results indicate that this demand begins from April to September, with critical periods for the months of June, July and August in which the maximum load of 18 kW occurs.

Using the TRNSYS model, a number of simulations considering the weather data for Oujda are performed to identify the key parameters that provide insight into the ideal size of the most important components of the solar absorption refrigeration system and to evaluate those critical variables on the system performance. Such factors include the effect of solar collector area, thermal tank volume on useful energy gain as well as auxiliary energy consumption are studied.

An auxiliary heater with natural gas as a back-up fuel, which has a maximum capacity of 10^6 kJ/hr was carried out. the effect of solar collector area on the useful collector energy and auxiliary energy required by the system is shown in Fig. 4 and Fig. 5. The domain of analysis for the collector area variable ranges from 5 m^2 to 50 m^2, with increments of 5 m^2. It is observed that an increase in collector area results in an increase in useful collector energy as expected, which translates into a decrease in the auxiliary heating

Table 1. Typical operating parameters of the absorption chiller used in the TRNSYS modeling

Parameter	Description
Hot water absorption chiller (Type 107)	
Chilled water set point temperature	6.67 °C
Cooling water inlet temperature	29 °C
Minimum driving temperature of the absorption chiller generator	108.89 °C
Hot water flow rate between storage tank and chiller	48 $m^3\ h^{-1}$
Chilled water flow rate between chiller and building load	45 $m^3\ h^{-1}$
Cooling water flow rate for chiller's heat rejection	148 $m^3\ h^{-1}$
Specific heat of hot water, cooling water, and chilled water	4.18 $kJ\ kg^{-1}\ K^{-1}$
Evacuated tube collector (Type 71)	
Optical or zero loss efficiency	0.802
Efficiency slope	5.4 $kJ\ h^{-1}\ m^{-2}\ K^{-1}$
Efficiency curvature	0.00576 $kJh^{-2}\ m^{-2}\ K^{-2}$
Mass flow rate through the solar collector	64 $kg\ h^{-1}\ m^{-2}$
Incidence angle modifier longitudinal	IAM_L.at 50° = 1
Incidence angle modifier transversal	IAM_T.at 50° = 0.93
Thermal storage (Type 4a)	
Tank type	Stratified
Tank loss coefficient	0.83 $W\ m^{-2}\ K^{-1}$
No. of nodes	10
Water density	1000 $kg\ m^{-3}$

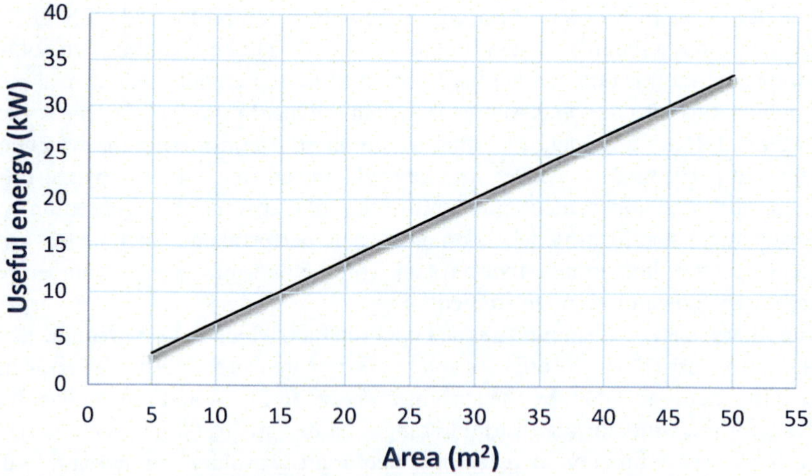

Fig. 4. Effect of the collector area on the gain of useful energy

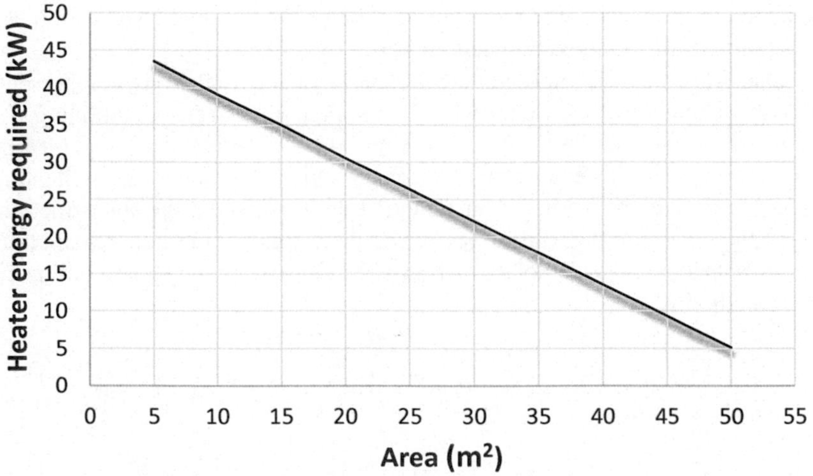

Fig. 5. Effect of the collector area on the auxiliary heat of the system

energy required by the system. Therefore, the optimum value needs to be decided by following an economic analysis.

Another parameter that affects the performance of the system is the size of the storage tank volume on the auxiliary energy consumption of the backup boiler. The result in Fig. 6 shows that by increasing the storage volume, the auxiliary energy is maximized. Besides, the minimum value of the obtained auxiliary heat energy was for a tank volume of 2 m^3 as can be seen from the results of this simulation.

In addition, Fig. 6 clearly shows that the oversized solar tank leads to an increase in auxiliary energy. Therefore, it is not surprising that the optimal capacity of the storage

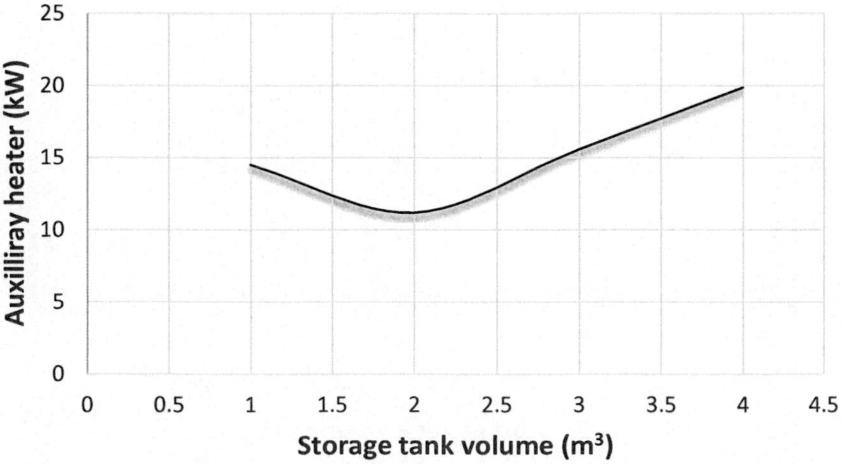

Fig. 6. Effect of tank volume on the consumption of auxiliary energy of the system

tank is in the lower values of the recommended range. The finding results are in a good agreement with those obtained by Djelloul et al. [15].

The solar fraction is an important coefficient in solar installations; it represents the proportion of solar energy in the total energy consumption of an air-conditioning system. In our case, the solar fraction is equal to the ratio of the thermal energy gained by the collector to the sum of this energy and the energy supplied by the auxiliary heating.

In order to enrich our work, we have plotted the variation of the solar fraction as a function of the surface of the solar collector and the volume of the storage tank in Fig. 7 and Fig. 8, respectively. This solar fraction is estimated for the same cooling power during the studied year.

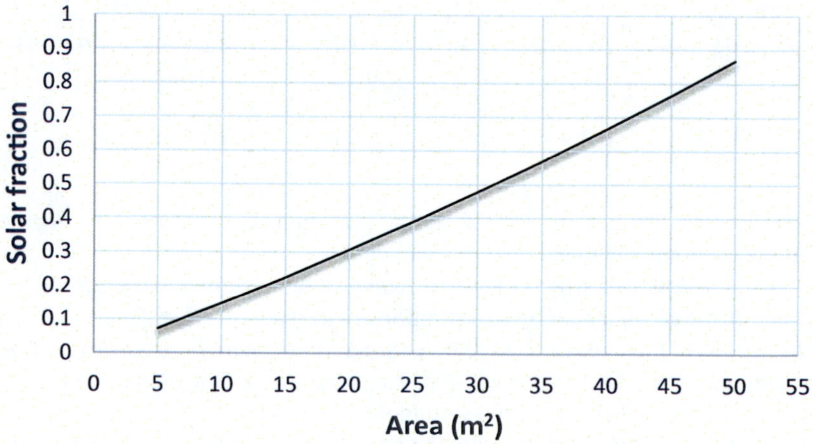

Fig. 7. Effect of total area on the solar fraction.

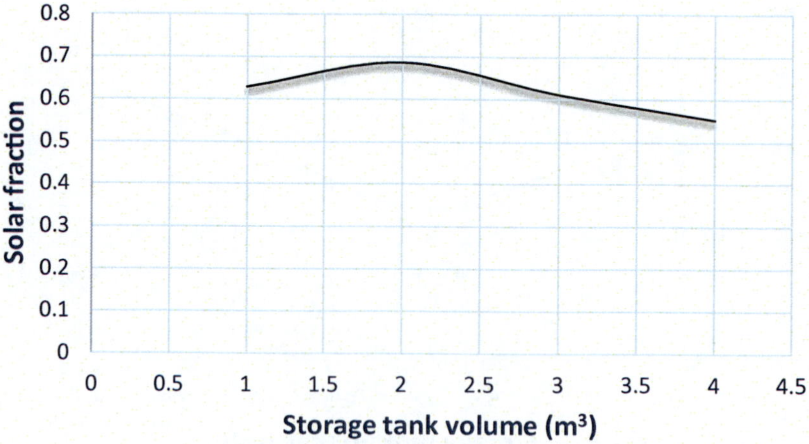

Fig. 8. Effect of tank volume on the solar fraction

Figure 7 shows the solar fraction for a collector with a surface area ranging from 5 to 50 m^2. As was predicted, the larger the collector area, the greater the amount of heat collected and the less strain on the boiler. This observation is clearly shown in the Fig. 7 where the solar fraction gets to increase when using a higher collector area.

While Fig. 8 shows that by increasing the storage tank capacity, the solar fraction tends to decrease. Therefore, we can conclude that the oversized solar tank has a tendency to decrease the solar fraction due to the increase in thermal losses. However, a thermo-economic analysis it is required for the optimization process of the important components such as the storage tank capacity and the solar collector surface.

4 Conclusion

This paper proposes a detailed analysis of the performance of a solar driven absorption cooling system as an alternative technology for the air conditioning of buildings under semi-arid climates of Morocco. Different parameters influencing the solar cooling performance of the proposed system were discussed.

Firstly, the Sketchup software was used to model the building under study, taking into account the different factors that can affect the air conditioning demand. Secondly, the developed 3D model was called by the TRNSYS TRNbuild, which permits to provide realistic simulations of the building's air conditioning needs.

The simulation of a solar absorption chiller with the thermal COP is 0.53 coupled with evacuated tube collector (ETC) was carried out on a real case building with a peak cooling load of 18 kW. The results generally indicate that the greater the area of the evacuated tube collector, the greater the solar fraction obtained. While, the larger the capacity of the storage tank, at a certain level, the solar fraction becomes less. The results show that an evacuated tube collector with a surface area of 50 m^2 and a hot water tank of 2 m^3 can successfully meet the air-conditioning demand of the building studied (total floor area of 1020 m^2.) However, a techno-economic analysis must be done to decide the final optimum configuration that can be employed in any circumstances.

Finally, due to the global pollution problem from burning fossil fuels, switching to solar energy as an energy source for absorption chillers is an alternative that should not be overlooked.

References

1. Aebischer, B., Catenazzi, G., Jakob, M.: Impact of climate change on thermal comfort, heating and cooling energy demand in Europe. In Proceedings ECEEE, pp. 859–870, June 2007
2. Grignon-Massé, L.: Développement d'une méthodologie d'analyse coût-bénéfice en vue d'évaluer le potentiel de réduction des impacts environnementaux liés au confort d'été: cas des climatiseurs individuels fixes en France métropolitaine. Doctoral dissertation, École Nationale Supérieure des Mines de Paris (2010)
3. Song, Z., Wang, R., Zhai, X.: Solar-driven high temperature radiant cooling. Chin. Sci. Bull. **54**(6), 978–985 (2009)
4. Henning, H.M., Wiemken, E.: Solar cooling. In: Proceedings of ISES World Congress 2007 (Vol. I–Vol. V), pp. 60–67. Springer, Heidelberg (2008)

5. Burckhart, H.J., Audinet, F., Gabassi, M.L., Martel, C.: Application of a novel, vacuum-insulated solar collector for heating and cooling. Energy Proc. **48**, 790–795 (2014)
6. Mateus, T., Oliveira, A.C.: Energy and economic analysis of an integrated solar absorption cooling and heating system in different building types and climates. Appl. Energy **86**(6), 949–957 (2009)
7. Buonomano, A., et al.: Experimental analysis and dynamic simulation of a novel high-temperature solar cooling system. Energy Convers. Manag. **109**, 19–39 (2016)
8. Wang, J., Yan, R., Wang, Z., Zhang, X., Shi, G.: Thermal performance analysis of an absorption cooling system based on parabolic trough solar collectors. Energies **11**(10), 2679-1-2679–17 (2018)
9. Al-Falahi, A., Alobaid, F., Epple, B.: A new design of an integrated solar absorption cooling system driven by an evacuated tube collector: a case study for Baghdad, Iraq. Appl. Sci. **10**(10), 3622:1-3622:23 (2020)
10. Elahi, R., Nagassou, D., Mohsenian, S., Trelles, J.P.: Enhanced solar radiation absorption by carbon dioxide in thermodynamic nonequilibrium: a computational study. Sol. Energy **195**, 369–381 (2020)
11. Sghiouri, H., Charai, M., Mezrhab, A.: Hemp plaster and passive cooling techniques for retrofit: a case study. In: Eco-efficient Materials for Reducing Cooling Needs in Buildings and Construction, pp. 139–167. Woodhead Publishing (2021)
12. TRNSYS 18 A TRaNsient System Simulation program. User's Manual Solar Energy Laboratory, University of Wisconsin-Madison
13. Solar collector fact sheet olymp sunstar 2470. SPF Testing, Institut für Solartechnik SPF, Hochschule für Technik Rapperswil HSR, CH-8640, Rapperswil, Switzerland (2019). 14.03.2016/SCFv3.0en. www.solarenergy.ch
14. El Hassani, S., Lahoussine, H.A., Amaoui, H., Moussaoui, M.A., Mezrhab, A.: TRNSYS simulation of a solar cooling system under Oujda climate. In: Masrour, T., El Hassani, I., Cherrafi, A. (eds.) A2IA 2020. LNNS, vol. 144, pp. 91–98. Springer, Cham (2020). https://doi.org/10.1007/978-3-030-53970-2_8
15. Djelloul, A., Draoui, B., Moummi, N.: Simulation of a solar driven air conditioning system for a house in dry and hot climate of Algeria. Present at the Courrier du Savoir, vol. 1, no. 15, pp. 31–39 (2013)

Analysis of the Power Demand in Romania During the COVID-19 Pandemic

Lucian-Ioan Dulău(✉) and Dorin Bică

"George Emil Palade" University of Medicine, Pharmacy, Science and Technology of Târgu-Mureş, 540139 Târgu-Mureş, Romania
lucian.dulau@umfst.ro

Abstract. The COVID-19 pandemic has caused changes in people habits which were amplified by the lockdown and restrictions taken by many countries in order to prevent the spread of COVID-19. These changes had a direct effect on the power system, namely on the operation of facilities and on the power demand. So, in this paper is presented the variation of the power demand in Romania. The power demand is obtained from the data provided by the transmission system operator in Romania (Transelectrica). The variation is presented for the year prior to the COVID-19 pandemic (2019), respectively in the years of the COVID-19 pandemic (2020 and 2021), in comparison with other studies that presented the power demand especially for 2019 and 2020 in different countries. The power demand is presented mainly for the months of March, April and May, considering the fact that Romania was in lockdown during these months of 2020. A comparison of the power demand during these months is also presented.

Keywords: COVID-19 pandemic · Lockdown · Power demand · Power variation · Power demand comparison

1 Introduction

The coronavirus (COVID-19) that appeared at the end of 2019 has a significant and deep impact on different aspects of the life for the people around the world [1].

Many measures were and are still taken by the world countries governments, such as quarantine, social distancing and lockdown in order to reduce the degree of infection of the population and slow the spread of the virus. The COVID-19 pandemic has affected many industries, such as agriculture, manufacturing, finance, education, healthcare, sports, tourism, and food [1]. Therefore, many companies had to limit or reorganize their production, or in several cases close their production for an amount of time due to the lack of raw materials. Also, the fear of infection at workplace or school caused employees and students to work from home, especially at the beginning of the COVID-19 pandemic. As a consequence, the power demand was also affected [1–3].

The researches performed analyzed the impact of the COVID-19 on the power demand as it follows. In [1] the analysis was performed for the most important countries

L. Moldovan and A. Gligor (Eds.): Inter-Eng 2021, LNNS 386, pp. 497–507, 2022.
https://doi.org/10.1007/978-3-030-93817-8_46

around the world considering the lockdown and after the lockdown. In [2] the analysis was performed for Spain, while in [3] for China. In [4], it was analyzed the power demand, power generation and power price for Germany.

In [5] it was analyzed the electric and thermal power demand for a university building, also considering the ambient factors (such as temperature), while in [6] was analyzed a residential building during the months of March, April and May before and during the pandemic.

In [7], it was analyzed the power demand, power balance and power market price in Poland during the COVID-19 pandemic. In [8], it was analyzed the effects of the pandemic in Hungary and Ukraine considering the changes power demand patterns, amount of renewables in power balance and if the national power systems of the respective countries can handle future crises.

In [9], it was analyzed the impact of the pandemic on the power demand, power supply and power interexchange for the U.S. In [10], it was analyzed the power demand, respectively the CO2 emissions for several countries, considering the available data and projections until 2024. In [11], it was analyzed the impact of the pandemic on power demand and load forecasting for major cities in Jordan.

In this paper, compared with the other researches, we will analyze the power demand for Romania, before and during the pandemic for 2019, 2020 and 2021, especially for the months of March, April and May.

The remainder of the paper is structured as follows. In Sect. 2 will be presented the methodology, in Sect. 3 will be presented the analyses, while in Sect. 4 will be presented the conclusions.

2 Methodology of the Analysis

Romania was in lockdown between March 16, 2020 and May 14, 2020. During this period many people worked from home. After May 14, 2020 Romania is in an alert stage, many people still working from home. During this stage, considering different months, for example students were not allowed to go to school (from May to June 2020) or allowed to go to school (in October 2020). In Fig. 1 is presented how the power demand and work changed before and during the pandemic.

So, before the pandemic people worked in offices and in factories, students went to schools. During the pandemic, especially in lockdown, people and students worked from home (e-work and e-school). The power demand is interrelated, so a decrease in power demand in offices lead to an increase in power demand at home.

The power demand can be calculated as:

$$P_D = P_{Dresidential} + P_{Dnon\text{-}residential}[MW] \tag{1}$$

where P_D is the total power demand in MW, $P_{Dresidential}$ is the power demand for the residential loads in MW, while $P_{Dnon\text{-}residential}$ is the power demand for the non-residential loads such as industrial loads in MW.

Considering the fact that the power demand from offices and schools is different compared with the one from homes, people used their home equipment instead of those at their work, factories, commerce and other activities had to reduce their working hours

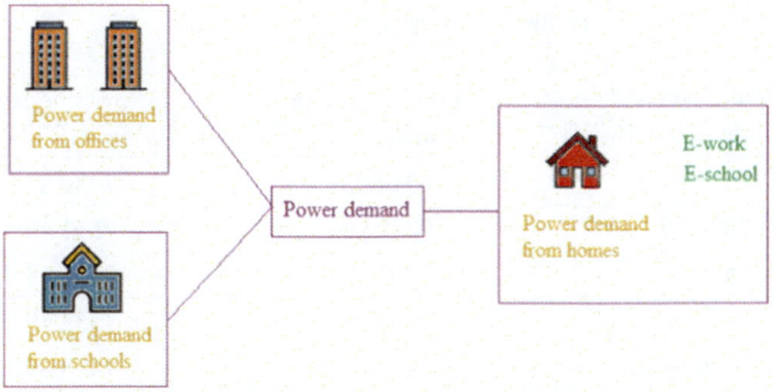

Fig. 1. Power demand and work before and during pandemic

or work at a limited capacity due to the restrictions and COVID-19 cases, the power demand analyses will be performed based on following.

The analyses will be performed considering the statistical data from the Transmission System Operator in Romania (Transelectrica) [12], and we will analyze the power demand before and during the pandemic.

So, the years 2019, 2020 and 2021 will be considered. The focus will be mainly for the months of March, April and May, months in which in 2020 Romania was mostly in lockdown. Therefore, we can observe the changes in this period and make comparisons.

3 Analysis of Power Demand

The power demand analyses before and after COVID-19 for Romania are presented further, based on the statistical data from [12]. The raw data from [12] is processed, being presented a multitude of information (power demand, power supplied by different types of power plants, power import and power export, each for an hour with an interval of 10 min), so a single data for the power demand is obtained for the months in which Romania was in lockdown. The data comprises only the total power demand in Romania and is not divided into categories or by type of loads (consumers). In Table 1 is presented the average power demand for the month of March.

Table 1. Power demand for March

Day	Power demand March 2019 [MW]	Power demand March 2020 [MW]	Power demand March 2021 [MW]
1	7525.59	6314.23	7488.8
2	7024.59	7249.96	7531.49
3	6541.2	7262.79	7501.98
4	7364.66	7300.82	7535.79
5	7486.89	7411.35	7478.49
6	7454.22	7226.89	7073.54
7	7518.12	6623.14	6544.02
8	7221.94	6187.99	7375.83
9	6643.45	7173.23	7596.09
10	6150.22	7319.14	8123.17
11	6962.36	7275.36	7961.95
12	7391.18	7033.29	7897.62
13	7292.43	6960.12	7263.58
14	7342.2	6484.15	6482.14
15	7321.23	6123.62	7544.28
16	6976.62	7015.66	7792.91
17	6147.65	7077.28	7803.21
18	6949.84	6980.38	7802.8
19	7155.58	6885.59	7762.35
20	7249.54	6697.35	7118.61
21	7252.28	6085.1	6597.49
22	7035.5	5897.37	7747.93
23	6648.48	7002.5	7804.14
24	5972.48	7096.04	7887.76
25	6728.18	6935.78	7685.26
26	7270.59	6834.27	7483.33
27	7136.99	6507.94	6817.02
28	7157.7	5926.72	6224.77
29	7101.62	5367.19	7183.94
30	6543.81	6156.91	7293.72
31	5898.98	6639.66	7187.04
Total	216466.12	209051.82	229591.05

From Table 1 it can be observed the fact that the power demand for March was in the second half of the month in 2020, during the lockdown period, lower compared with the demand in 2019. Also, it can be observed that the power demand in 2021 was higher compared with 2019 in most cases, so it recovered at is usual.

In Fig. 2 is presented a comparison of the total power demand for March by year (2019, 2020 and 2021).

In March the power demand in 2020 was 4% lower compared with 2019, while the power demand in 2021 was 6% higher compared with 2019 and 10% higher compared with 2020.

In Table 2 is presented the average power demand for the month of April.

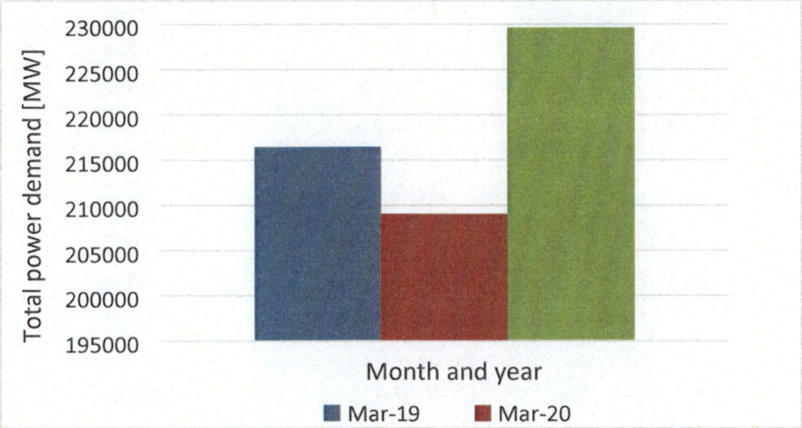

Fig. 2. Total power demand comparison (March 2019, March 2020 and March 2021)

Table 2. Power demand for April

Day	Power demand April 2019 [MW]	Power demand April 2020 [MW]	Power demand April 2021 [MW]
1	6763.06	6556.85	7036.01
2	7090.05	6452.11	6997.65
3	7060.4	6292.42	6629.77
4	7074.11	5936.81	6149.04
5	7053.39	5486.92	6951.39
6	6643.94	6096.63	7196.54
7	6053.11	6128.04	7549.69
8	6911.06	6052.85	7527.09

(continued)

Table 2. (*continued*)

Day	Power demand April 2019 [MW]	Power demand April 2020 [MW]	Power demand April 2021 [MW]
9	7016.93	5994.21	7388
10	6927.71	5878.1	6713.06
11	7151.86	5604.32	6077.32
12	7047.07	5111.7	6860.33
13	6772.08	5760.89	7081.19
14	6194.84	6018.57	7483.09
15	6990.28	6073.6	7479.43
16	7166	5856.44	7315.99
17	7235.59	5300.76	6778.32
18	7290.23	5306.32	6327.4
19	7119.15	4539.95	7234.46
20	6595.33	4605.08	7369.17
21	5879.92	5334.28	7260.22
22	6857.75	5758.4	7140.15
23	7133.1	5567.41	7004.79
24	7142.56	5515.72	6579.74
25	6905.95	5421.93	5876.15
26	6098.38	4992.27	6908.06
27	5879.79	5597.49	7195.04
28	4928.54	5775.04	7334.29
29	4935.23	5744.27	7040.66
30	5608.24	5690.36	6194.99
Total	199525.65	170449.74	208679.03

From Table 2 it can be observed the fact that the power demand for April in 2020 was also lower compared with the demand in 2019. Also, it can be observed that the power demand for April in 2021 was higher compared with 2019.

In Fig. 3 is presented a comparison of the total power demand for April by year (2019, 2020 and 2021).

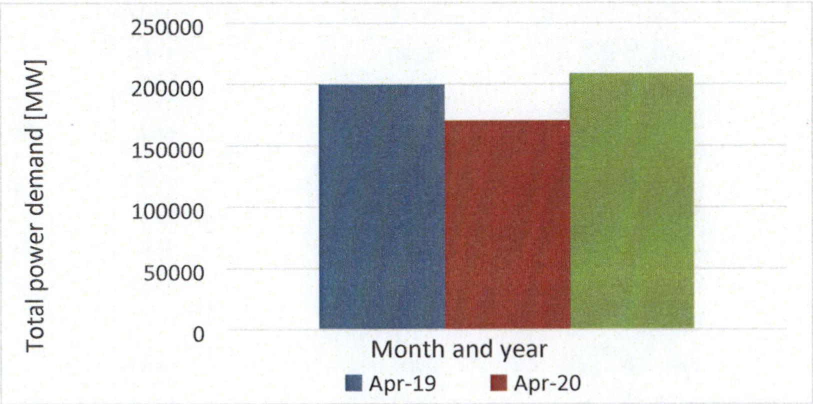

Fig. 3. Total power demand comparison (April 2019, April 2020 and April 2021)

In April the power demand in 2020 was 15% lower compared with 2019, while the power demand in 2021 was 4% higher compared with 2019 and 22% higher compared with 2020.

In Table 3 is presented the average power demand for the month of May.

From Table 3 it can be observed the fact that the power demand for May in 2020 was also lower compared with the demand in 2019, especially in the first half of the month. Also, it can be observed that the power demand in 2021 was higher compared with 2019.

In Fig. 4 is presented a comparison of the total power demand for May by year (2019, 2020 and 2021).

In May the power demand in 2020 was 12% lower compared with 2019, respectively 2% lower compared with 2021. The causes for the lower power demand during May 2021 were that an important holiday was at the start of the month, namely Easter and several important consumers (car manufacturers) had to reduce production due to the semiconductor shortage.

Table 3. Power demand for May

Day	Power demand May 2019 [MW]	Power demand May 2020 [MW]	Power demand May 2021 [MW]
1	5922.54	5178.76	5977.48
2	6348.48	5100.65	4896.49
3	6274.1	4972.7	4991.1
4	6068.25	5790.9	5968.27
5	5640.27	5870.02	6655.47
6	6565.74	5975.16	6644.25
7	6938.12	6004.21	6596.7
8	6837.75	5850.02	6231.3
9	6838.71	5403.42	5584.87
10	6928.42	4931.97	6412.21
11	6290.24	5719.78	6606.36
12	5583.16	5907.02	6643.71
13	6430.33	5852.05	6716.63
14	6829.47	5977.43	6614.08
15	6798.38	5836.79	6146.11
16	6824.97	5389.45	5618.91
17	6705.75	5082.3	6492.02
18	6142.47	5803.89	6744.95
19	5543.65	5948.46	6683.51
20	6413.85	6131.33	6929.38
21	6427.45	5920.4	6505.34
22	6469.4	5906.02	6140.86
23	6536.7	5356.52	5534.68
24	6511.23	5080.63	6390.47
25	5993.23	5894.31	6563.16
26	5375.78	6059.09	6580.07
27	6297.14	6138.97	6622.59
28	6532.44	5955.65	6611.64
29	6590.6	6016.48	6069.31
30	6609.71	5473.64	5475.68
31	6543.47	5000.75	6254.94
Total	197811.8	175528.77	193902.54

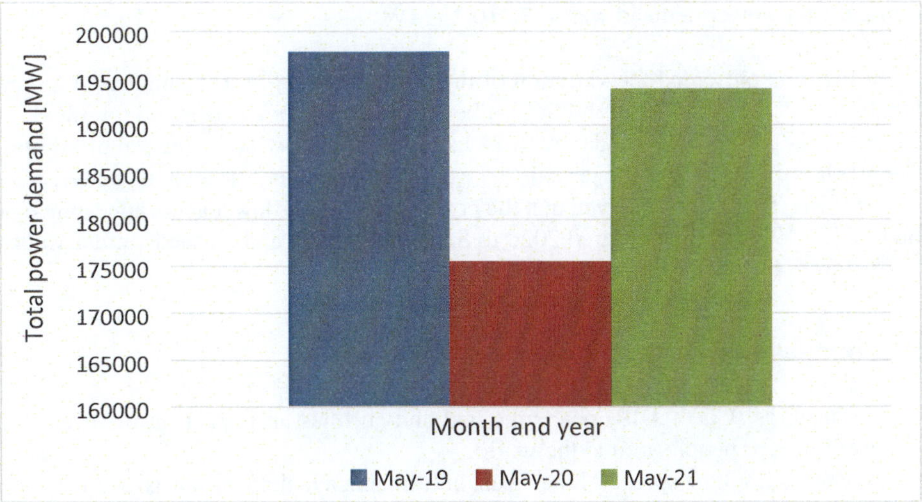

Fig. 4. Total power demand comparison (May 2019, May 2020 and May 2021)

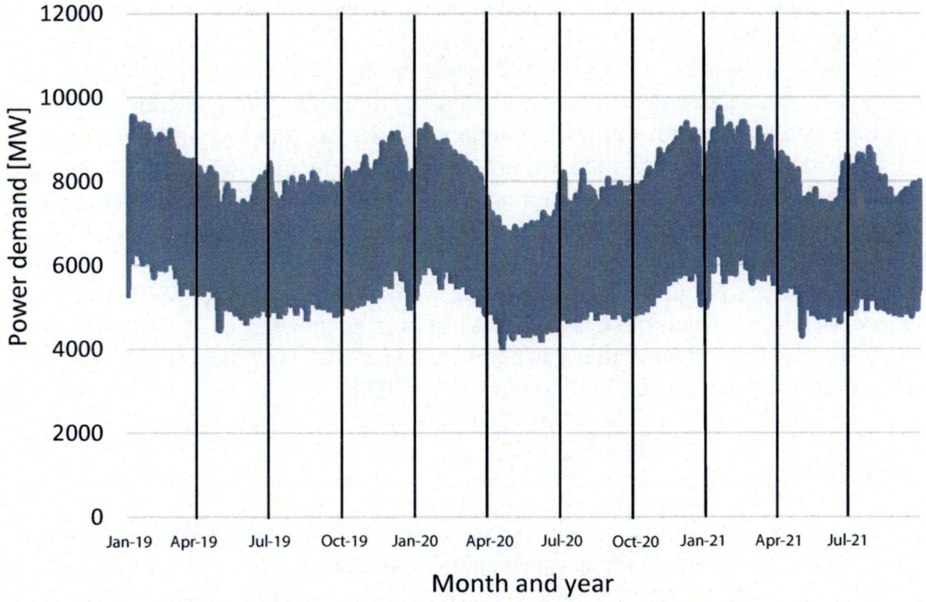

Fig. 5. Power demand in Romania between January 2019 and September 2021 [12]

If we compare the corresponding power demand during the lockdown period (March 16 and May 14) the following results are obtained:

- in 2019 the power demand was 398,247.07 MW;
- in 2020 the power demand was 354,089.57 MW;

- in 2021 the power demand was 413,410.23 MW.

In Fig. 5 is presented the variation of the power demand in Romania based on the data from [12], between 2019 and September 2021. So, the power demand for 365 days of the year is presented, each day with 24 h, and each hour with an interval of 10 min. The graph is divided into time periods, so we can compare the variation of power.

From Fig. 5 it can be observed that the power demand was lower at the appearance of the COVID-19 virus, from March 2020 to September 2020, and reached similar values of 2019 from October 2020.

4 Conclusions

The coronavirus (COVID-19) pandemic had and still has an impact on the everyday aspects of life for people around the world.

In order to limit the spread of the virus, in Romania a lockdown was issued between March 16 2020 and May 14 2020. Many people had to work from home and also attend school from home during this period, but also in the following months in the alert s. Also, due to number of COVID-19 cases and restrictions taken by the government many businesses closed or worked at a limited capacity, such as factories or hotels. This had an influence on the power demand in Romania.

The power demand was determined based on the data for 2019, 2020 and 2021 provided by the transmission system operator in Romania, Transelectrica, compared with other studies that focused on different countries and mainly on the demand for 2019 and 2020. The data provided was processed for the months in which Romania was in lockdown, therefore a daily and a monthly power demand was obtained. So, during the lockdown period, after analyzing the data it can be stated that the power demand was on average 8.33% lower if we compare 2019 and 2020.

If we compare the power demand between 2019 and September 2021, the power demand was lower when the COVID-19 virus was acknowledged in March 2019 in Romania and several months after that compared with 2019, and recovered to similar or higher values compared with 2019 from October 2020.

References

1. Jiang, P., Fan, Y.V., Klemes, J.J.: Impacts of COVID-19 on energy demand and consumption: challenges, lessons and emerging opportunities. Appl. Energy **285**, 116441 (2021)
2. Santiago, I., Moreno-Munoz, A., Quintero-Jiménez, P., Garcia-Torres, F., Gonzalez-Redondo, M.J.: Electricity demand during pandemic times: the case of the COVID-19 in Spain. Energy Policy **148**(A), 111964 (2021)
3. Huang, L., Liao, Q., Qiu, R., Liang, Y., Long, Y.: Prediction-based analysis on power consumption gap under long-term emergency: a case in China under COVID-19. Appl. Energy **283**, 116339 (2021)
4. Halbrügge, S., Schott, P., Weibelzahl, M., Buhl, H.U., Fridgen, G., Schöpf, M.: How did the German and other European electricity systems react to the COVID-19 pandemic? Appl. Energy **285**, 116370 (2021)

5. Mokhtari, R., Jahangir, M.H.: The effect of occupant distribution on energy consumption and COVID-19 infection in buildings: a case study of university building. Build. Environ. **190**, 107561 (2021)
6. Krarti, M., Aldubyan, M.: Review analysis of COVID-19 impact on electricity demand for residential buildings. Renew. Sustain. Energy Rev. **143**, 110888 (2021)
7. Czosnyka, M., Wnukowska, B., Karbowa, K.: Electrical energy consumption and the energy market in Poland during the COVID-19 pandemic. In: 2020 Progress in Applied Electrical Engineering (PAEE), pp. 1–5 (2020)
8. Morva, G., Diahovchenko, I.: Effects of COVID-19 on the electricity sectors of Ukraine and Hungary: challenges of energy demand and renewables integration. In: 2020 IEEE 3rd International Conference and Workshop in Óbuda on Electrical and Power Engineering (CANDO-EPE), pp. 41–46 (2020)
9. Agdas, D., Barooah, P.: Impact of the COVID-19 pandemic on the U.S. electricity demand and supply: an early view from data. IEEE Access **8**, 151523–151534 (2020)
10. Bertram, C., Luderer, G., Creutzig, F., Bauer, N., Ueckerdt, F., Malik, A., Edenhofer, O.: COVID-19-induced low power demand and market forces starkly reduce CO_2 emissions. Nat. Clim. Chang. **11**, 193–196 (2021)
11. Alasali, F., Nusair, K., Alhmoud, L., Zarour, E.: Impact of the COVID-19 pandemic on electricity demand and load forecasting. Sustainability **13**(3), 1435 (2021)
12. Transelectrica Homepage. https://www.transelectrica.ro. Accessed 01 Oct 2021

Energy Losses Estimation in the Electric Distribution Networks Using Clustering-Based Selection of the Representative Feeders

Ecaterina Chelaru, Livia Noroc, Gheorghe Grigoras(✉), and Bogdan-Constantin Neagu

Faculty of Electrical Engineering, Iasi, Romania
ggrigor@tuiasi.ro

Abstract. The energy loss estimation is essential to evaluate the technical and economic performance corresponding to electric distribution networks (EDNs). The development of some energy loss estimation methodologies represents one of the main objectives included in the strategies implemented by the Distribution Network Operators (DNOs) for an optimal operation and planning of the EDNs. The paper presents an improved methodology for the energy losses estimation in the EDNs using a clustering-based selection of the representative feeders (RFs), considering both the energy and technical features. The annual transmitted energy, total length, cross-sections of each branch, and the capacity of the distribution transformers represented the input data in the clustering process. The RFs are identified for each feature class assigned to the energy patterns obtained from the clustering process. Testing the proposed methodology has been done considering the 20 kV EDNs from a rural area belonging to a Romanian DNO. The results emphasized the accuracy of the proposed method to determine the energy losses in the EDNs. The percentage error between the proposed method and the Backward-Forward Sweep-based method, considered the reference, was 3.5%, compared with 12.99% associated with the average element method.

Keywords: Energy losses · Representative feeders · Clustering · Electric distribution networks

1 Introduction

Reducing the power/energy losses represents an essential measure integrated into the energy efficiency strategies of the Distribution Network Operators (DNOs). So far, the electricity and gas consumptions account for 43% of final energy consumption in the European Union (EU), and the losses in the network elements have average values of 1.5% and 0.3%, respectively. These losses have a level that varies from one country to another, and through their reduction, the DNOs can obtain a performing distribution process. Therefore, the potential for energy efficiency involves minimizing losses at all distribution levels (high - HV, medium - MV, and low voltage - LV) [1].

L. Moldovan and A. Gligor (Eds.): Inter-Eng 2021, LNNS 386, pp. 508–521, 2022.
https://doi.org/10.1007/978-3-030-93817-8_47

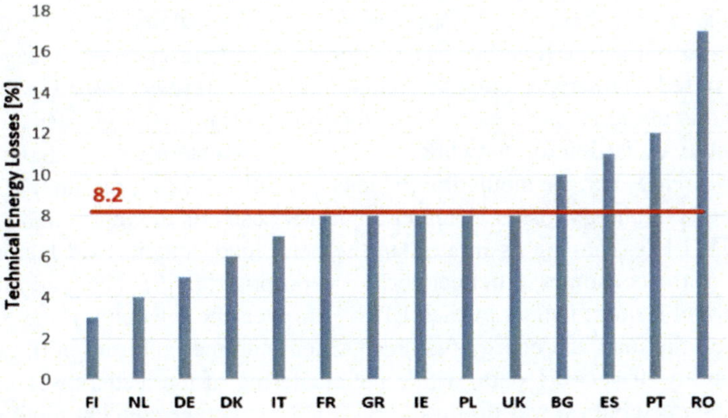

Fig. 1. The technical energy losses in various countries from the UE, [1]

A detailed classification of losses in the EDNs must be done to determine the critical areas and the reasons for the occurrence. Also, the most significant factors with a great impact on the losses' level in the EDNs have been identified, including:

- loading (considering, also, maximum demand);
- highest number of MV/LV transformers;
- lengths of the overhead lines from the rural areas and cables from the urban areas;
- power quality;
- integration degree of the renewable energy sources.

Figure 1 shows that the level of losses varies significantly, from 3% (Finland) to 17% (Romania), suggesting that the potentials for improving energy efficiency are different and depend on local conditions (aging of the networks; voltage level; or the number of transformers).

The main potential for reducing the energy losses is at the distribution level, where most of the losses occur. Implementing the Smart Metering System (SMS) in the MV/LV electric distribution substations (EDSs) and consumers represents the first step in identifying the areas with issues and supervising the energy efficiency measures. They have mainly aimed at the evaluation with increased accuracy of the energy losses. High accuracy and reduced computational time are challenges of the DNOs due to the high number of distribution feeders (DFs) and a reduced implementation degree of the smart meters in some supplied areas (mainly in the rural areas). Thus, there are approaches proposed in the literature to estimate the technical energy losses in the EDNs considering scarce data. A deep-learning loss model used in the LV EDNs considering the data provided by the smart meters installed at the some consumers has been proposed in [2]. Others methods used the regression models [3, 4], or approximated model of load profiles [5, 6]. The power losses have been calculated recursively using a method based on a nodal distribution loss analysis in [7], knowing the loads and distributed generation sources.

On the other hand, the estimation of power/energy losses depends on several parameters/variables that follow both the design criteria and the operating conditions of the

EDNs, such as rated voltage, length, sections, the capacity and number of the transformers from the EDSs, the type (overhead, cable), or the supplied consumers' categories. The approaches based on the representative feeders/networks were developed in the last few years using this initial information. The methodology proposed in [8] leads to building the representative distribution networks for three large-scale networks and six feeder-type networks. However, the methodology does not include the representative feeders in the calculations of energy losses. Another approach based on the representative feeders from the MV EDNs, having as input data the peak load, length, load factor, and the distribution of the consumer in the network, was proposed in [9]. The load information is provided by the smart meters in each EDS. However, the selection of the RF is done only based on the load and energy features. A heuristic method used in the LV EDNs having a scarcity of network data, where the clustering of the feeders was based only on the load characteristics, was proposed in [10]. The average element method applied in the EDNs, depending on the distribution installation category, by Romanian DNOs is presented in [11]. The average element is chosen considering an EDS having the closest annual transmitted energy by the average value. For this EDS, the average elements associated with the lines and MV/LV transformers are identified. The method can lead to high errors in estimating the energy losses due to large deviations, which can be at the annual transmitted energies through the EDSs and the reduced possibilities to choose the average elements associated with the feeders supplied from the average EDS.

To remove the weaknesses of the presented approaches, the paper aims to identify the MV RFs belonging to a Romanian DNO based on the clustering considering both the energy and technical features. The proposed methodology can be applied for all distribution installation categories according to the voltage level (20, 10, and 6 kV) and the supply areas (urban and rural). The annual transmitted energy length, cross-sections of each branch, and the capacity of distribution transformers represent the input data. The RF is chosen for each feature class assigned to the energy patterns obtained from the clustering process.

The structure of the paper is the following: Sect. 2 treats the main aspects of clustering analysis, Sect. 3 details the steps of the proposed methodology to estimate the energy losses in the EDNs, Sect. 4 presents the results obtained for the 20 kV EDNs from a rural area supplied by a Romanian DNO, and Sect. 5 highlights the conclusions and the future work.

2 Clustering Analysis

Clustering analysis is a data mining technique used to identify how different elements can be grouped due to their similarities. When they belong to the same cluster, the membership degree is maximum, and when they do not belong, the degree is minimal.

This kind of analysis is usually used in the exploratory stage of knowledge discovery when the Decision-Maker (DM) wants to understand the analyzed processes. It is usually not the only data mining method used but rather done in the early stages to assist the DM in the performed analysis. Figure 2 presents the main steps of the clustering analysis [12, 13].

There are several different types of clustering analysis. The two most commonly used are K-means clustering and hierarchical clustering [12]. The hierarchical clustering gives the best results when are several up to a few dozen input data, and K-means clustering when the initial number is high (over hundred).

K-means clustering treats the data as vectors having positions and distances from each other. The most used distance is Euclidean distance. The algorithm divides the elements into K clusters as close as possible to each other. At the same time, these elements are as far away from the elements in other clusters as possible [14]. Each cluster is characterized through its average value of the features associated with the assigned elements (see Fig. 3).

The validation of the clusters is based on the various indexes, but one of the most used is the Silhouette Global Index [16].

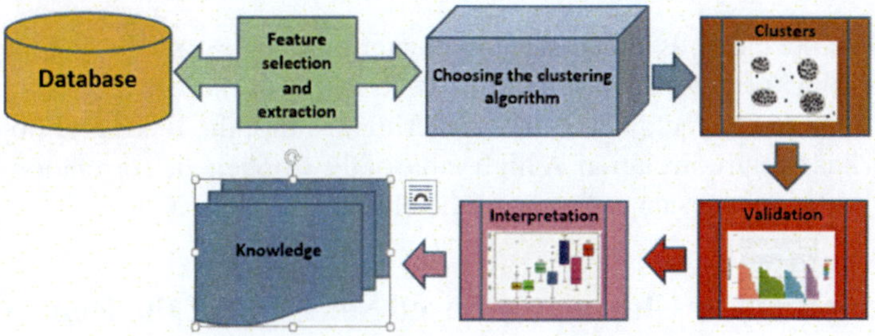

Fig. 2. The main steps of the clustering analysis

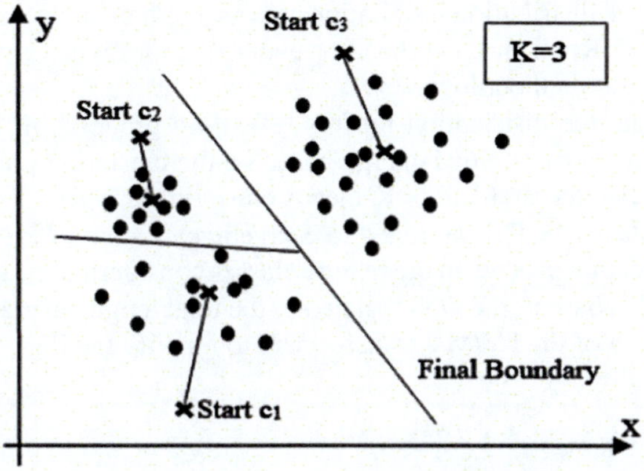

Fig. 3. The evolution of the clustering process using K-means algorithm [15]

Hierarchical clustering represents a simple way to investigate at the same time data clusters on a wide range of scales and distances based on a cluster tree with different levels, namely dendrogram (see Fig. 4). Unlike the K-means group, the dendrogram does not offer a single solution. Thus, the tree represents a multi-level hierarchy, where the clusters from a level are joined at the next higher level. The clustering process starts with each case or variable in a separate cluster and then combines them until only one remain. This reasoning allows the DM to choose which level of grouping is most appropriate.

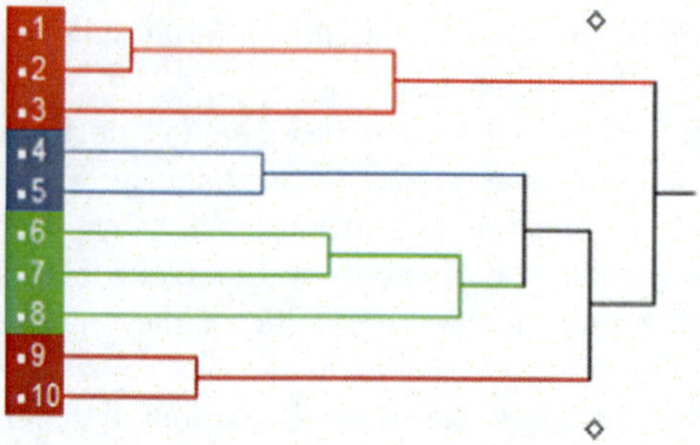

Fig. 4. The dendrogram associated with the hierarchical clustering

There are more methods included in this clustering category: single linkage, complete linkage, centroid, average, or Ward. The difference between these consists of the distance calculated between clusters [17]. If the clustering is valid, the linking of objects in the dendrogram has a high correlation. A high value of the Cophenetic correlation coefficient (CC), close to 1, indicates an accuracy of the clustering solution [18].

3 Methodology for Clustering-Based Selection of the Representative Feeders

The methodology consists of aggregating the values calculated using the representative feeders (RFs) for each distribution installation category: urban or rural, the voltage levels (6, 10, and 20 kV, in Romania), overhead or underground. Figure 5 presents the main stages of the proposed methodology.

In the first stage, the distribution feeders (DFs) are grouped in "energy patterns" according to the annual transmitted energy based on the clustering process. Depending on the number of DFs, hierarchical or K-means clustering is used.

In the second stage, the DFs assigned to each energy pattern, P_i, $i = 1, \ldots, N_P$, are subject to a classification process in the feature classes considering the technical features resulted from the database of the DNO related to the total length, average cross-section, and the total capacity of the LV/MV transformers to identify the RFs.

$$L_T^{(k)} = \sum_{b=1}^{B^{(k)}} L_b^{(k)}, \quad s_{av}^{(k)} = \frac{\sum_{j=1}^{B_s^{(k)}} s_j^{(k)} \cdot l_j^{(k)}}{L_T^{(k)}}, \quad S_T^{(k)} = \sum_{t=1}^{N_t^{(k)}} S_{r,t}^{(k)}, \quad k = 1, \ldots, N_f \tag{1}$$

where: ${L_T}^{(k)}$ – the total length of the feeder k; ${s_{av}}^{(k)}$ – the average cross-section of the feeder k; ${S_T}^{(k)}$ – the total capacity of the LV/MV transformers supplied by the feeder k; $B^{(k)}$ – the number of the branches; ${L_b}^{(k)}$ – the length of the branch b; ${B_s}^{(k)}$ – the total number of the branches with different cross-sections; ${s_j}^{(k)}$ – the cross-section of the branch j; ${l_j}^{(k)}$ – the length of branch j with the section ${s_j}^{(k)}$; ${N_t}^{(k)}$ – the number of the transformers from the MV/LV EDSs supplied by the feeder k; ${S_{r,t}}^{(k)}$ – the rated power of the transformer t.

The features identified for each feeder are recorded in a matrix $[F]$ to be easy processed in a new clustering process:

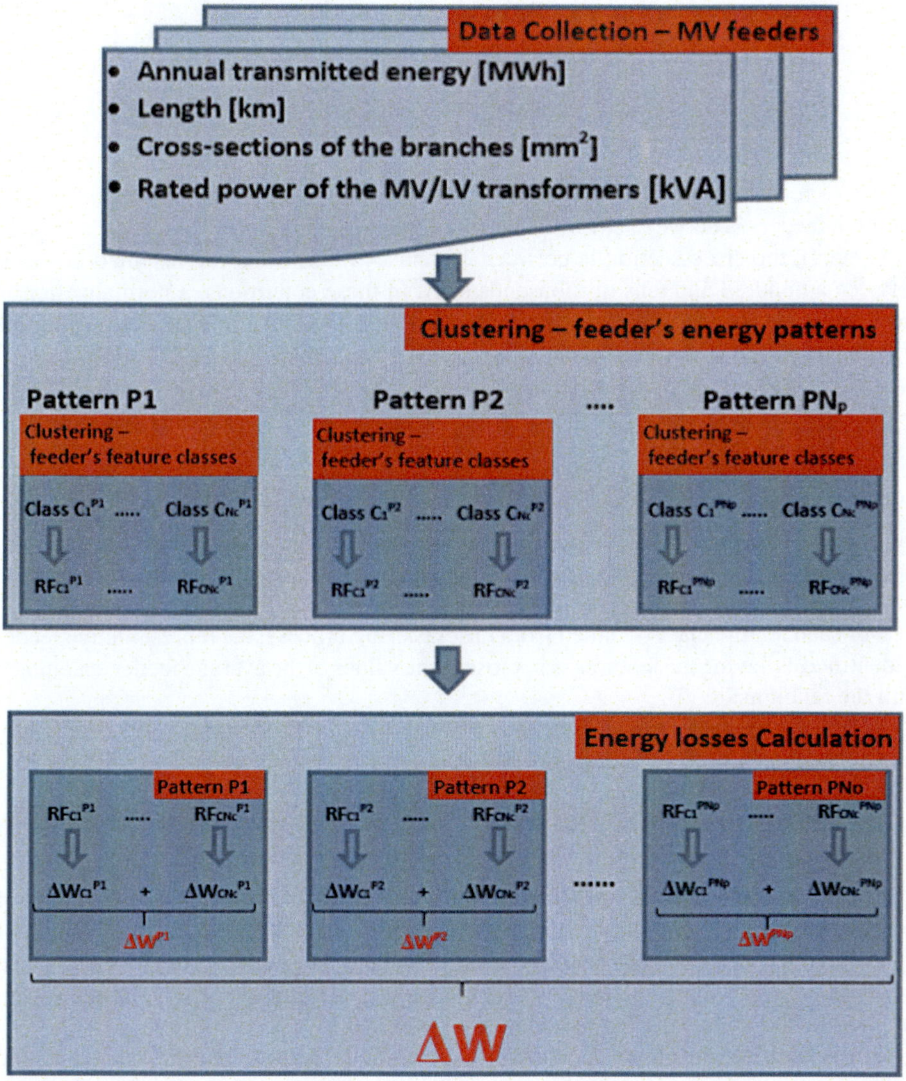

Fig. 5. The methodology for energy losses calculation with clustering-based selection of the RFs

$$F = \begin{bmatrix} f_{L_T}^{(1)} & f_{s_{av}}^{(1)} & f_{S_T}^{(1)} \\ \vdots & \vdots & \vdots \\ f_{L_T}^{(k)} & f_{s_{av}}^{(k)} & f_{S_T}^{(k)} \\ \vdots & \vdots & \vdots \\ f_{L_T}^{(N_f)} & f_{s_{av}}^{(N_f)} & f_{S_T}^{(N_f)} \end{bmatrix} \quad (2)$$

These features have different metrics. For example, the total length of the MV distribution feeders is between a few to tens of kilometers (sometimes it exceeds a hundred kilometers), the cross-section is between 16 and 240 mm^2, and the installed power is between a hundred and tens of thousands kVA. In these conditions, a normalization of the features is necessary to have the same weight in the clustering process, using the relation [16]:

$$f_w^{(k)*} = \frac{f_w^{(k)} - f_w^{\min}}{f_w^{\max} - f_w^{\min}}; \quad k = 1, \ldots, N_f, \quad w \in \{L_T, s_{av}, S_T\} \quad (3)$$

where: ${f_w}^{(k)}$ – the value of the feature $w \in \{L_T, s_{av}, S_T\}$ associated with the feeder k, recorded in matrix $[F]$; ${f_w}^{min}$ – the minimum value of the feature $w \in \{L_T, s_{av}, S_T\}$ determined for all feeders, recorded in matrix $[F]$; ${f_w}^{max}$ – the maximum value the feature $w \in \{L_T, s_{av}, S_T\}$ determined for all feeders, recorded in matrix $[F]$.

For each feature class obtained from the clustering process, the representative feeder is identified as having the features very close to the values of the average feeder, calculated with the relations:

$$L_T^{(c)} = \frac{\sum_{l=1}^{N_l^{(c)}} L_{T,l}^{(c)}}{N_l^{(c)}}, \quad s_{av}^{(c)} = \frac{\sum_{l=1}^{N_l^{(c)}} s_{av,l}^{(c)}}{N_l^{(c)}}, \quad S_T^{(c)} = \frac{\sum_{l=1}^{N_l^{(c)}} S_{T,l}^{(c)}}{N_l^{(c)}}, \quad c = 1, \ldots, N_c \quad (4)$$

where: N_c – the number of the feature classes resulted from the clustering process of the feeders; ${N_l}^{(c)}$ – the number of the feeders from the class c; ${L_T}^{(c)}$ – the total length of the average feeder associated with the class c; ${s_{av}}^{(c)}$ – the section-cross of the average feeder associated with the class c; ${S_T}^{(c)}$ – the total capacity of the MV/LV transformers supplied by the average feeder associate with the class c.

This classification aimed to identify as accurately as possible the representative feeders in terms of primary features and annual transmitted energy.

In the final stage, the energy losses for each representative feeder associated with each class are determined based on the loss factor method [19] (where the calculation of power losses at the maximum load is done with the iterative Backward-Forward Sweep-based power flow algorithm [20, 21]) and aggregated at the energy patterns of each distribution installation category.

$$\Delta W = \sum_{P_i=1}^{N_P} \sum_{c=1}^{N_c P_i} N_l^{(c)} \cdot \Delta W_{c,RF}^{P_i} \quad (5)$$

where: $\Delta W^{Pi}{}_{c,RF}$ – the energy losses calculated for the RF from feature class c associated with the pattern P_i; $N^{Pi}{}_c$ – the number of the features classes included in the pattern P_i; N_P – the number of the energy patterns corresponding to the DFs from the analysed distribution installation category.

4 Case Study

Testing the proposed methodology has been done considering the 20 kV rural EDNs from an area belonging to a Romanian DNO. The EDN contains 53 feeders supplied from 17 HV/MV EDSs. Figure 6 present the distribution of the feeders on the EDSs. The analysis of the distribution emphasizes an average number by three DFs, which leave from an EDS. There is a single EDS with over five DFs, six EDSs have four DFs, 5 EDSs supply three DFs, and 5 EDSs contain over two DFs.

In the first stage, the annual transmitted energy of each DF has been uploaded from the database of the DNO to be prepared for the clustering process. The JMP software [22] was used in the clustering-based data processing. All values were over 0, such that the clustering process has been initialized for all DFs. Due to the reduced number of DFs, the Ward method has been chosen from the hierarchical clustering category. Figure 7 offers information on the clustering process through the dendrogram and the representation of the obtained feeder's energy patterns.

Table 1 presents the statistical indicators associated with each pattern regarding the mean, standard deviation, and quantiles ($Q1$ – 25%, $Q2$ – 50%, and $Q3$ – 75%). They summarize the information contained in each energy pattern, giving the possibility of the DM to make an overall assessment of all DFs. By combining several indicators, relevant and synthetic information about each pattern can be obtained.

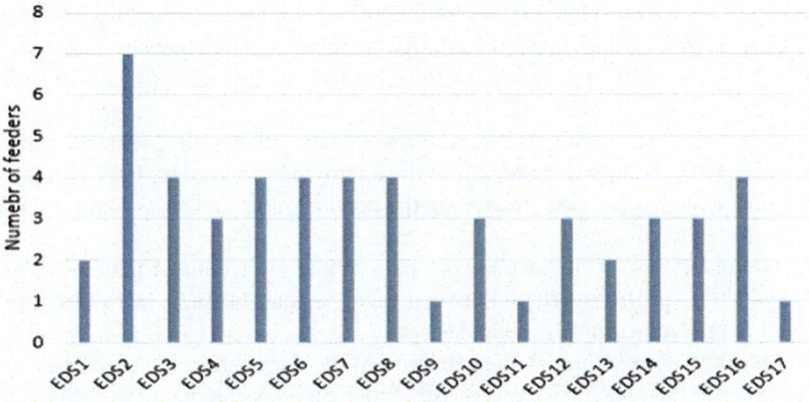

Fig. 6. The distribution of the feeders on the EDSs

The analysis of the results highlighted the quality of the clustering process characterized through the representative energy patterns with a weight between 19% (pattern P1) and 47% (pattern P2).

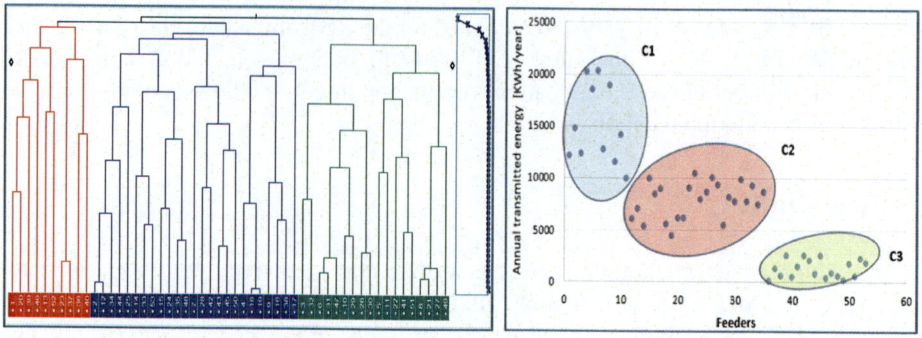

a. b.

Fig. 7. The results of the Ward clustering method a – The dendrogram; b – Representation of the feeder' energy patterns

Table 1. The statistical indicators of the feeder' energy patterns, [MWh]

Pattern	No. of feeders	Mean	Standard deviation	Quantiles		
				Q1	Q2	Q3
P1	10	15893	3543	12256	14409	18885
P2	25	7860	1718	6126	8082	9218
P3	18	1210	880	477	988	1970

In the next stage, the DFs from each energy pattern have been subject to a classification process in the feature classes considering the technical features resulted from the database of the DNO related to the total length, average cross-section, and the total capacity of the LV/MV transformers to identify the representative feeders. The Ward hierarchical method was applied for each pattern having as input data the above technical features.

Table 2 presents the statistical indicators associated with the three features of the DFs corresponding to each class integrated into the energy patterns. Also, Figs. 8 and 9 offer details on the clustering process presenting the dendrograms and the 3-D representation of the DFs from each energy pattern.

The obtained classes from each energy pattern had the technical features very well defined, and there are no overlaps between them, as can be seen from the graphical representations presented in Figs. 8 and 9.

The next stage corresponds to the selection of the RF from each class. They should have the technical features as close as possible to the means calculated in Table 2 for the average feeder (AF). Table 3 presents the feeders chosen to represent each class.

The energy losses on the line and MV/LV transformers for each RF have been calculated using the loss factor method, where the power losses at the maximum load have been calculated with the iterative Backward-Forward Sweep-based power flow

Table 2. The statistical indicators of the feeder' feature classes

Statistical indicators	Feeder's features	Mean	Standard deviation	Quantiles		
				Q1	Q2	Q3
Class C_1^{P1}	L_T [km]	60.6	44.8	21.8	65.8	90.3
	s_{av} [mm^2]	60.9	15.0	52	53	72.8
	S_T [kVA]	10893.4	1310.1	9731.5	11329	11961.7
Class C_2^{P1}	L_T [km]	38.3	9.3	31.5	43.3	43.9
	s_{av} [mm^2]	43.7	5.5	40.3	41	47.8
	S_T [kVA]	5627.7	783.9	5136.8	5259	6210.8
Class C_1^{P2}	L_T [km]	74.4	49.0	51.9	62.7	93.1
	s_{av} [mm^2]	49.5	11.1	40.5	47	54.8
	S_T [kVA]	8542.1	2917.1	6430.8	9183	10162.2
Class C_2^{P2}	L_T [km]	22.7	14.4	14.4	24.0	27.3
	s_{av} [mm^2]	51.9	10.7	41	54	60
	S_T [kVA]	3479.7	1221.4	2546	3717.5	4559
Class C_1^{P3}	L_T [km]	15.2	8.4	10.6	14.1	16.5
	s_{av} [mm^2]	93.8	13.7	84	92	106
	S_T [kVA]	5426.8	3281.2	3356	6291.5	9713
Class C_2^{P3}	L_T [km]	33.4	44.7	7.6	12.9	41.5
	s_{av} [mm^2]	54.3	10.8	47.5	51.5	62.5
	S_T [kVA]	3415	2893.8	1495	2030	4858

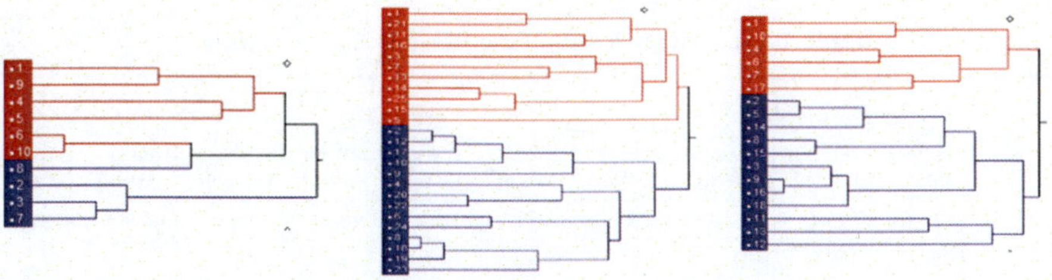

a. b. c.

Fig. 8. The dendrograms of the clustering process applied to each energy pattern; a. Pattern P1; b. Pattern P2; c. Pattern P3

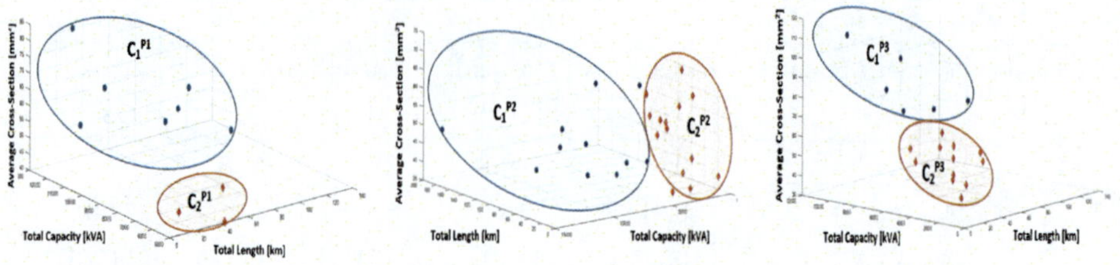

a. b. c.

Fig. 9. The 3-D representation of the feeder' features classes; a. Pattern P1; b. Pattern P2; c. Pattern P3

Table 3. The technical features of the RFs

Class	RF	Technical features of RF			Technical features of AF		
		L_T [km]	s_{av} [mm^2]	S_T [kVA]	L_T [km]	s_{av} [mm^2]	S_T [kVA]
$C_1{}^{P1}$	#52	75.8	52	10897	60.6	60.9	10893.4
$C_2{}^{P1}$	#20	43.3	40	5096	38.3	43.7	5627.7
$C_1{}^{P2}$	#15	79.7	47	9183	74.4	49.5	8542.1
$C_2{}^{P2}$	#35	25.6	36	3535	22.7	51.9	3479.7
$C_1{}^{P3}$	#49	16.52	106	5983	15.2	93.8	5426.8
$C_2{}^{P3}$	#29	66.9	50	3448	33.4	54.3	3415

Table 4. The energy losses of the RFs and feature classes

Class	No. DFs	Representative feeder					Feature Class				
		ΔW_{line} [MWh]	$\Delta W_{Tr,L}$ [MWh]	$\Delta W_{Tr,NL}$ [MWh]	ΔW_{Tr} [MWh]	ΔW_{Total} [MWh]	ΔW_{line} [MWh]	$\Delta W_{Tr,L}$ [MWh]	$\Delta W_{Tr,NL}$ [MWh]	ΔW_{Tr} [MWh]	ΔW_{Total} [MWh]
$C_1{}^{P1}$	7	320.6	95.4	333.1	428.5	749.1	2243.9	667.8	2331.9	2999.7	5243.6
$C_2{}^{P1}$	3	263.3	150.6	177.8	328.4	591.7	789.9	451.8	533.4	985.2	1775.1
$C_1{}^{P2}$	11	220.5	29.9	295.9	325.8	546.3	2425.1	328.4	3255.3	3583.7	6008.8
$C_2{}^{P2}$	14	108.2	131.1	125.1	256.3	364.5	1515.4	1836.0	1751.5	3587.5	5102.9
$C_1{}^{P3}$	6	12.8	6.4	159.1	165.4	178.3	76.9	38.3	954.3	992.6	1069.6
$C_2{}^{P3}$	12	2.6	1.0	135.5	136.5	139.1	31.1	12.3	1625.6	1637.9	1669.0

algorithm. Table 4 gives the energy losses for each RF and aggregate at the level of feature classes, where $\Delta W_{Tr,L}$ and $\Delta W_{Tr,NL}$ are the load and no-load losses in the transformers.

Figure 10 shows a comparison between the energy losses at the level of the analyzed rural area calculated with the proposed methodology considering the clustering-based selection of the representative feeders (CSRF), the average feeder (AF)-based method [AF], and the Backward-Forward Sweep (BFS)-based method for each DF. The BFS method has been considered a reference in the comparison due to the best results obtained in the power flow calculations in the MV EDNs.

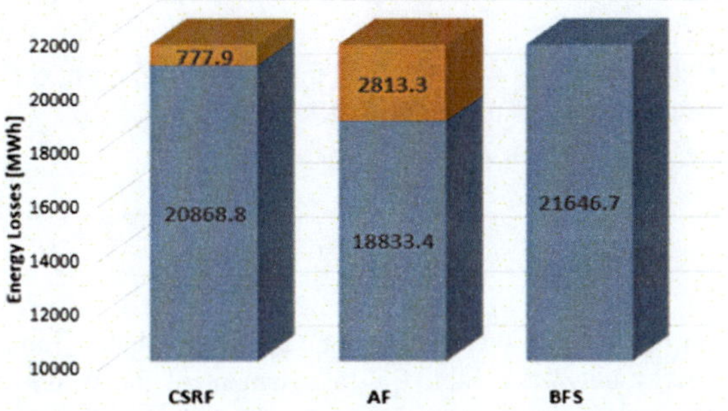

Fig. 10. Comparison between methods for the calculation of energy losses

The blue colour highlights the calculated values of the energy losses, and the brown colour the estimation error expressed in MWh compared with the BFS method. There is a difference by 2035.4 MWh between the CSRF and AF methods, representing approximately 9.5% compared with the BFS method. The percentage error between the proposed method (CSRF) and the BFS method is 3.5%, compared with 12.99% associated with the AF method.

These results emphasized the accuracy of the proposed method to determine the energy losses in the EDNs compared with the AF method.

5 Conclusions

The paper aims to identify the MV RFs belonging to a Romanian DNO based on the clustering considering both the energy and the technical features to estimate the energy losses in the EDNs.

The proposed methodology can be applied for all distribution installation categories, according to the voltage level (20, 10, and 6 kV) and the supply areas (urban and rural). In the first stage, the DFs were grouped in "energy patterns" according to the annual transmitted energy based on the Ward hierarchical clustering method. In the second stage, the DFs assigned to each energy pattern are subject to a classification process in the feature classes considering the technical features resulted from the database of the DNO related to the total length, average cross-section, and the total capacity of the LV/MV transformers to identify the RFs. In the final stage, the energy losses for each RF associated with each class are determined based on the loss factor method and aggregated at the energy patterns of each distribution installation category.

Testing the proposed methodology has been done considering the 20 kV rural EDNs from an area belonging to a Romanian DNO.

The comparison with the average feeder-based method [11], considering the Backward-Forward Sweep-based method as reference [20], highlights the performance of the proposed methodology. The percentage error between the proposed method and the BFS method is 3.5%, compared with 12.99% associated with the AF method.

The future work will consider as input data the presence of the distributed generation sources, which have a significant impact on the energy losses in the EDNs.

Acknowledgment. This work was supported by a publications grant of the "Gheorghe Asachi" Technical University of Iasi, project number GI /P19 /2021.

References

1. Tractebel Engineering, Report on identifying energy efficiency improvements and saving potential in energy networks, including assessment of the value of demand response. In support of the implementation of article 15 of the Energy Efficiency Directive (2012/27/EU). https://ec.europa.eu/energy/sites/ener/files/docments/GRIDEE_4NT_364174_000_01_TOTALDOC%20-%2018-1-2016.pdf
2. Velasco, J.A., Amaris, H., Alonso, M.: Deep Learning loss model for large-scale low voltage smart grids. Int. J. Electr. Power Energy Syst. **121**, 106054 (2020)
3. Queiroz, L., Cavellucci, C., Lyra, C.: Evaluation of technical losses estimation in LV power distribution systems. In: 20th International Conference on Electricity Distribution, Prague, 8–11 June 2009 (2009)
4. Grigoraş, G., Cârţină, G.: The fuzzy correlation approach in operation of electrical distribution systems. Int. J. Comput. Math. Electr. Electron. Eng. **32**(3), 1044–1066 (2013)
5. Iqteit, N., Arsoy, A.B., Çakır, B.: A simple method to estimate power losses in distribution networks. In: 10th International Conference on Electrical and Electronics Engineering (ELECO), Bursa, Turkey, 30 November–2 December 2017 (2017)
6. Jayawardhana, H., Hemapala, K., Bandara, A., De Silva, P.S.N.: Defining of normalized load profile curves for domestic customer groups to estimate feeder power loss. In: 3rd International Conference on Information Technology Research (ICITR), Moratuwa, Sri Lanka, 5–7 December 2018
7. Hongmei, L., Hantao, C., Hunjie, L.: Distribution network power loss analysis considering uncertainties in distributed generations. Sustainability **11**(1311) (2019). https://doi.org/10.3390/su11051311
8. Mateo, C., et al.: European representative electricity distribution networks. Int. J. Electr. Power Energy Syst. **99**, 273–280 (2018)
9. Ibrahim, K.A., Au, M.T., Gan, C.K., Tang, J.H.: System wide MV distribution network technical losses estimation based on reference feeder and energy flow model. Int. J. Electr. Power Energy Syst. **93**, 440–450 (2017)
10. Dashtaki, A.K., Haghifam, M.R.: A new loss estimation method in limited data electric distribution networks. IEEE Trans. Power Deliv. **28**(4), 2194–2200 (2013). Article number 6571263

11. Romanian Energy Regulatory Agency, Technical energy norm regarding the determination of own technological consumption in electricity networks of public interest - NTE 013/16/00 (in Romanian). www.anre.ro
12. Jain, A.K.: Data clustering: 50 years beyond K-means. Pattern Recogn. Lett. **31**(8), 651–6661 (2010)
13. Cartina, G., Grigoras, G., Bobric, E.C.: Clustering techniques for energy losses evaluation in distribution networks. In: IEEE Bucharest PowerTech, pp. 1–5 (2009)
14. Sinaga, K., Yang, M.S.: Unsupervised K-means clustering algorithm. IEEE Access **8**, 80716–80727 (2020)
15. Grigoras, G., Cartina, G., Rotaru, F.: Using K-means clustering method in determination of the energy losses levels from electric distribution systems. In: Proceedings of the 12th WSEAS International Conference on Mathematical Methods and Computational Techniques in Electrical Engineering, pp. 52–56 (2010)
16. Li, T., Ma, Y., Endoh, T.: Normalization-based validity index of adaptive K-means clustering for multi-solution application. IEEE Access **8**, 9403–9419 (2020)
17. Carbonera, J.L., Abel, M.: A subspace hierarchical clustering algorithm for categorical data. In: IEEE 31st International Conference on Tools with Artificial Intelligence, Portland, OR, USA, pp. 4–6 (2019)
18. Grigoras, G., Cartina, G., Istrate, M., Rotaru, F.: The efficiency of the clustering techniques in the energy losses evaluation from distribution networks. Int. J. Math. Models Methods Appl. Sci. **5**(1), 133–141 (2011)
19. Queiroz, L.M.O., Roselli, M.A., Cavellucci, C., Lyra, C.: Energy losses estimation in power distribution systems. IEEE Trans. Power Syst. **27**(4), 1879–1887 (2012)
20. Díaz, G., Gómez-Aleixandre, J., Coto, J.: Direct backward/forward sweep algorithm for solving load power flows in AC droop-regulated microgrids. IEEE Trans. Smart Grid **7**(5), 2208–2217 (2016)
21. Eremia, M.: Electric Power Systems. Electric Networks, vol. 1. Romanian Academy Publishing House, Bucharest (2006)
22. JMP software. https://www.jmp.com/en_us/software/data-analysis-software.html

Analysis of the Charging Price and Travel Time of Battery Electric Vehicles in Romania

Lucian-Ioan Dulău(✉) and Dorin Bică

"George Emil Palade" University of Medicine, Pharmacy, Science and Technology of Târgu-Mureş, 540139 Târgu-Mureş, Romania
lucian.dulau@umfst.ro

Abstract. In this paper is presented the analysis of the charging power price for battery electric vehicles in Romania. The electric vehicles can charge from regular or dedicated charging stations, so the charging duration is different. Also, the distance that has to be covered by these vehicles can be different. The charging is currently in most cases supported by economic operators like malls and hypermarkets, but in the future the users will have to pay the charging. There are considered different electric vehicles, from various manufacturers, the cost of a kWh and various distances covered in Romania. The results will be compared with the fueling cost of a petrol vehicle.

Keywords: Plug-in electric vehicles · Power cost · Charging duration · Charging cost

1 Introduction

The number of electric vehicles (EVs) is growing in Romania, especially due to the measures taken by the government to stimulate the acquisition of this type of vehicles. So, people can receive a voucher of 45000 lei (approximately 9150 euro), therefore the cost of buying an electric vehicle can be lowered and is attractive to many people.

Currently the charging of electric vehicles in Romania is supported by economic operators, such as shopping malls, hypermarkets or operators that installed charging stations and promote this type of vehicles. There are also charging stations were users have to pay the charging. In the nearby future, the charging of electric vehicles will not be free and everyone will have to pay for it. So, in this paper we will study the charging cost of different electric vehicles, considering the cost of a kWh, charging duration and the travel distance in Romania. The results will be compared with the cost of a fueling a petrol vehicle.

The researches regarding the charging of electric vehicles focused on the following. The smart charging services considering the users activity, travel schedules and charging choices were analyzed in [1]. An EV charging cost model was developed in [2] considering electric power rates, EV mobility and battery, charger specifications, so the user cost was minimized. The charging behavior of EV in Australia was studied in [3]. In [4]

L. Moldovan and A. Gligor (Eds.): Inter-Eng 2021, LNNS 386, pp. 522–532, 2022.
https://doi.org/10.1007/978-3-030-93817-8_48

was analyzed the EV charging network in Europe, the number of charging stations and type.

The optimization of EV charging and discharging was studied in [5] considering the power cost and user behavior and response, while in [6] was developed a strategy for the reduction of charging cost. The impact that the EV charging has on the cost competitiveness of generation and storage technologies in the power system was studied in [7]. In [8] the smart charging of EVs was analyzed by means of a fuzzy logic controller considering the power price so the impact on the distribution utility was minimum and the user benefits were maximized. The coordinated charging of EVs in order to reduce the operation cost considering time of use and power cost was studied in [9, 10]. The planning of EVs charging stations was studied in [11] considering traffic destination heat data and usage data of chargers.

2 Methodology of the Analyses

There are three types of electric vehicles: hybrid electric vehicles (HEV), plug-in hybrid electric vehicles (PHEV) and battery-powered electric vehicles (BEV). In this paper will be considered BEV due to the fact that they use only the electric power stored in a battery to power an electric motor and rotate the wheels. When discharged, the batteries are recharged using the power grid, either from a power outlet or a dedicated charging unit. The number of BEV has increased in Romania, from 1050 in 2018, 2372 in 2019 to 4483 in 2020 [12].

The map of the charging stations in Romania is presented in Fig. 1.

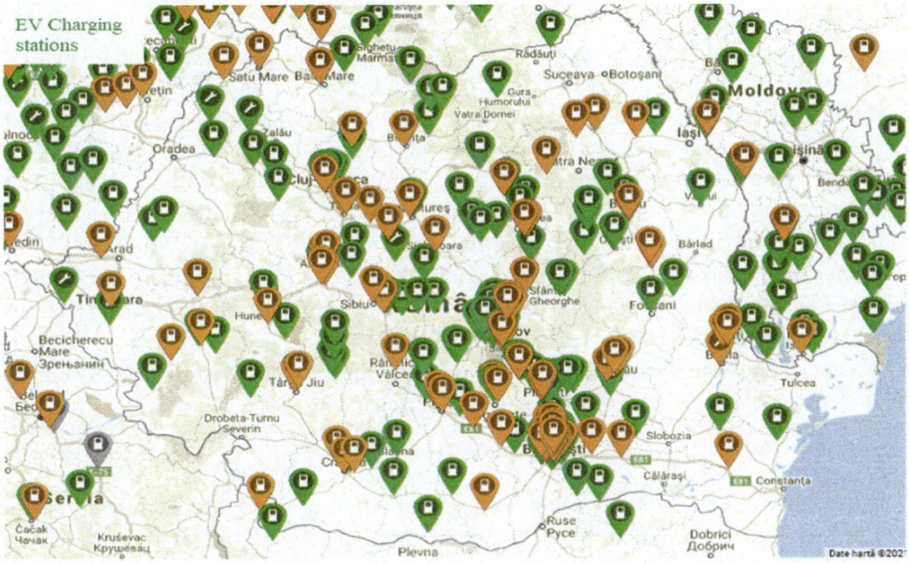

Fig. 1. Charging stations map in Romania [13].

The following data for the BEV are presented further in Table 1.

Table 1. Main electric vehicles characteristics.

Electric vehicle	Battery capacity [kWh]	Regular charging duration at home (2.3 kW outlet) [h]	Regular charging duration at station (22kW outlet) [h]	Fast charging duration at station (50 kW DC outlet) [h]	Distance that can be covered without recharging in mixt conditions (urban and extra-urban) [km]
Tesla Model 3	55	33	5.5	0.8	340
BMW i3	33.2	14	3	0.25	170
Hyundai Kona	42	20.25	6.5	0.8	251
Nissan Leaf	40	19	7	0.5	245
Volkswagen e-Golf	35.8	16.5	5.25	0.7	203
Renault Zoe	41	19	2	0.75	235

The charging cost can be calculated as:

$$C_{EV} = P_{EV} \cdot P_{cost}[lei] \qquad (1)$$

where C_{EV} is the charging cost of the EV in lei, P_{EV} is the power required by the EV in kWh, while P_{cost} is the power cost in lei/kWh. The power cost paid by the residential consumer in Romania is 0.6382 lei/kWh. However, the prices that are requested by the charging stations owners are higher. They start from 1 lei/kWh to 1.45 lei/kWh for regular charging at 22 kW outlet, respectively start from 1.35 lei/kWh to 1.95 lei/kWh for fast charging at 50 kW outlet. In this paper will be considered the power cost of 0.6382 lei/kWh (grid power cost), respectively an average of 1.26 lei/kWh for charging at a private regular station, respectively 1.79 lei/kWh for charging at a private fast station.

Considering the fact that an EV can be charged at a regular or fast charging station, the charging time varies.

We can also obtain the cost for a km by dividing the charging cost to the BEV range (the distance that can be covered without recharging):

$$C_{BEVkm} = C_{EV} \,/\, BEV\ range[lei/km] \qquad (2)$$

In the following case studies will be considered different travel distances, from under 100 km, between 100 km and 300 km, respectively between 300 km and 600 km. So, in the last case, at least two charging periods are required. The total charging cost is then compared with the cost of fueling a petrol vehicle. This cost is, considering a fuel consumption of 8 l/100 km in a mixt environment and petrol cost of 5.55 lei, 0.444 lei/km.

3 Case Studies

The travel route in Romania and travel distances that will be covered are presented in Table 2.

Table 2. Travel route and distances.

Travel route	Distance [km]
Târgu Mureş - Sighişoara	50
Târgu Mureş - Cluj-Napoca	110
Târgu Mureş - Braşov	168
Târgu Mureş - Oradea	256
Târgu Mureş - Bucharest	341
Târgu Mureş - Constanţa	569

The analyses are performed with the MATLAB software for the BEV from Table 2. From Table 2 we obtain the charging cost, respectively the cost for a km for each BEV, which is presented in Table 3.

Table 3. Charging cost from grid and station for BEV.

Electric vehicle	Charging cost from grid [lei]	Charging cost from grid for a km [lei/km]	Charging cost from private regular charging station [lei]	Charging cost from private regular charging station for a km [lei/km]	Charging cost from private fast charging station [lei]	Charging cost from private fast charging station for a km [lei/km]
Tesla Model 3	35.101	0.1032	69.3	0.2038	98.45	0.2895
BMW i3	21.1882	0.1246	41.832	0.2460	59.428	0.3495
Hyundai Kona	26.8044	0.1068	52.92	0.2108	75.18	0.2995
Nissan Leaf	25.528	0.1042	50.4	0.2057	71.6	0.2922
Volkswagen e-Golf	22.8476	0.1125	45.108	0.2222	64.082	0.3156
Renault Zoe	26.1662	0.1113	51.66	0.2198	73.39	0.3122

The results of the analyses are presented further, in Table 4, Table 5, Table 6, Fig. 2, Fig. 3 and Fig. 4 the fuel cost, respectively in Table 7 and Fig. 5 the travel time.

In the simulations are considered the cases when the battery capacity is not enough in order to cover the entire travel distance, so it needs to be charged during the travel again to 100% either at a regular charging station (RCS) or at a fast charging station (FCS). The recharging is done as the same type of station.

Table 4. Petrol fuel vehicle cost compared with fuel cost considering power grid cost.

Vehicle	Târgu Mureş - Sighişoara fuel cost [lei]	Târgu Mureş - Cluj-Napoca fuel cost [lei]	Târgu Mureş - Braşov fuel cost [lei]	Târgu Mureş - Oradea fuel cost [lei]	Târgu Mureş - Bucharest fuel cost [lei]	Târgu Mureş - Constanţa fuel cost [lei]
Petrol fueled vehicle	22.2	48.84	74.592	113.664	151.404	252.636
Tesla Model 3	5.1619	11.3562	17.344	26.429	35.2042	58.7426
BMW i3	6.2318	13.71	20.939	31.907	42.5011	70.9183
Hyundai Kona	5.3395	11.7469	17.9408	27.3384	36.4155	60.7638
Nissan Leaf	5.2098	11.4616	17.5049	26.6742	35.5308	59.2875
Volkswagen e-Golf	5.6275	12.3805	18.9083	28.8127	38.3794	64.0407
Renault Zoe	5.5673	12.248	18.706	28.5045	37.9688	63.3556

Considering the results, the following can be observed. For the routes which are on distance lower than 170 km, Târgu Mureş - Sighişoara, Târgu Mureş - Cluj-Napoca, respectively Târgu Mureş - Braşov, the BEV don't require an additional charging. However the BMW i3 is very close to reach the destination for Târgu Mureş - Braşov route.

For the routes which are on distance above 170 km, Târgu Mureş - Oradea, Târgu Mureş - Bucharest, respectively Târgu Mureş - Constanţa, the BEV require one or two additional charging of battery.

So, for the Târgu Mureş - Oradea route, the Tesla Model 3 is the only one that doesn't require an additional charging, the other BEV require one additional charging. For the Târgu Mureş - Bucharest route, the BMW i3 require two additional charging, while the other BEV require one additional charging. For the Târgu Mureş - Constanţa route, the Tesla Model 3 require one additional charging, the BMW i3 require three additional charging, while the other BEV require two additional charging.

The fuel cost is lower for the BEV compared with the petrol vehicle in all cases.

If we compare the travel time, then if the route is under 170 km the travel duration is the same as with a petrol vehicle. If the route is above 170 km, then the BEV need to be charged, so the travel duration increases and is higher compared with a petrol vehicle.

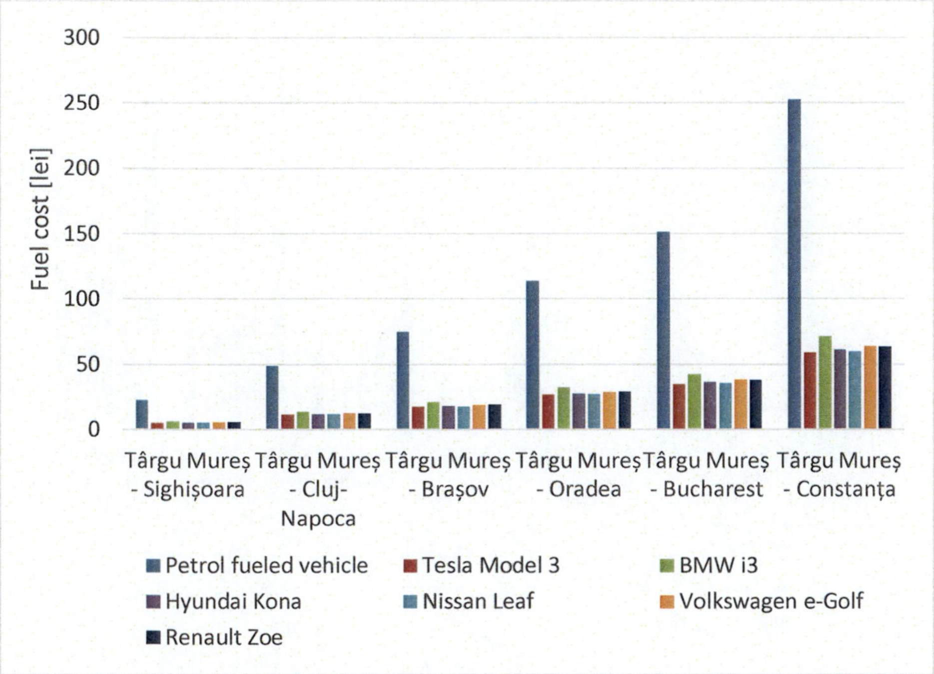

Fig. 2. Petrol fuel vehicle cost compared with fuel cost considering power grid cost for the considered routes.

Table 5. Petrol fuel vehicle cost compared with fuel cost considering private regular charging station.

Vehicle	Târgu Mureş – Sighişoara fuel cost [lei]	Târgu Mureş – Cluj-Napoca fuel cost [lei]	Târgu Mureş – Braşov fuel cost [lei]	Târgu Mureş – Oradea fuel cost [lei]	Târgu Mureş - Bucharest fuel cost [lei]	Târgu Mureş - Constanţa fuel cost [lei]
Petrol fueled vehicle	22.2	48.84	74.592	113.664	151.404	252.636
Tesla Model 3	10.1911	22.4205	34.2423	52.1788	69.5038	115.9755
BMW i3	12.3035	27.0677	41.3398	62.994	83.91	140.0141
Hyundai Kona	10.5418	23.192	35.4205	53.9741	71.8952	119.966
Nissan Leaf	10.2857	22.6285	34.56	52.6628	70.1485	117.0514
Volkswagen e-Golf	11.1103	24.4427	37.3307	56.8849	75.7725	126.4357
Renault Zoe	10.9914	24.1812	36.9314	56.2764	74.9619	125.0831

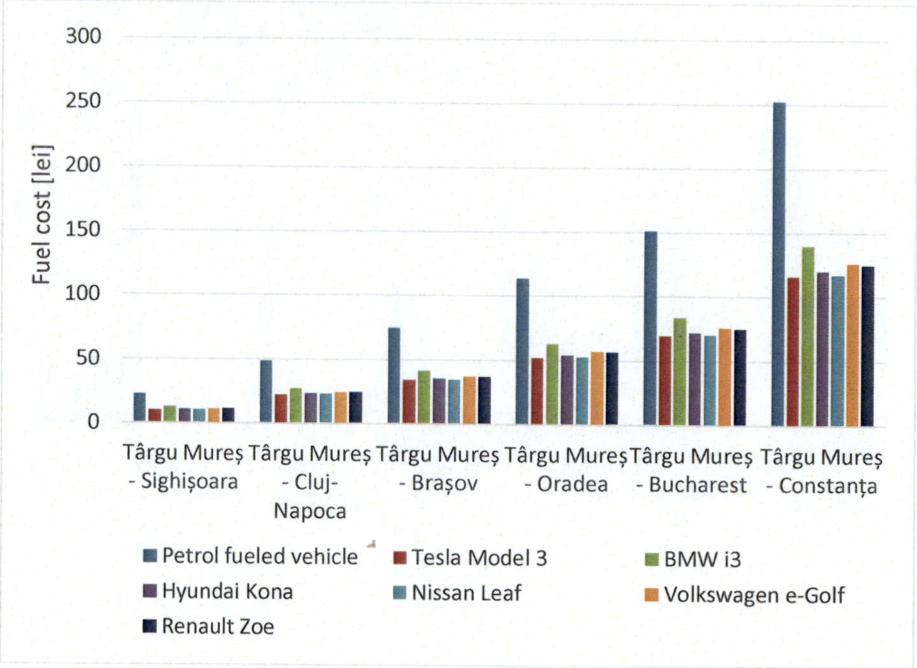

Fig. 3. Petrol fuel vehicle cost compared with fuel cost considering private regular charging station cost for the considered routes.

Table 6. Petrol fuel vehicle cost compared with fuel cost considering private fast charging station.

Vehicle	Târgu Mureş – Sighişoara fuel cost [lei]	Târgu Mureş – Cluj-Napoca fuel cost [lei]	Târgu Mureş – Braşov fuel cost [lei]	Târgu Mureş – Oradea fuel cost [lei]	Târgu Mureş - Bucharest fuel cost [lei]	Târgu Mureş - Constanţa fuel cost [lei]
Petrol fueled vehicle	22.2	48.84	74.592	113.664	151.404	252.636
Tesla Model 3	14.4779	31.8514	48.6458	74.127	98.7395	164.7589
BMW i3	17.4788	38.4534	58.7288	89.4915	119.2055	198.909
Hyundai Kona	14.976	32.9474	50.3196	76.6776	102.1369	170.4279
Nissan Leaf	14.6122	32.1469	49.0971	74.8146	99.6555	166.2873
Volkswagen e-Golf	15.7837	34.7242	53.0333	80.8127	107.6451	179.619
Renault Zoe	15.6148	34.3527	52.466	79.9482	106.4935	177.6974

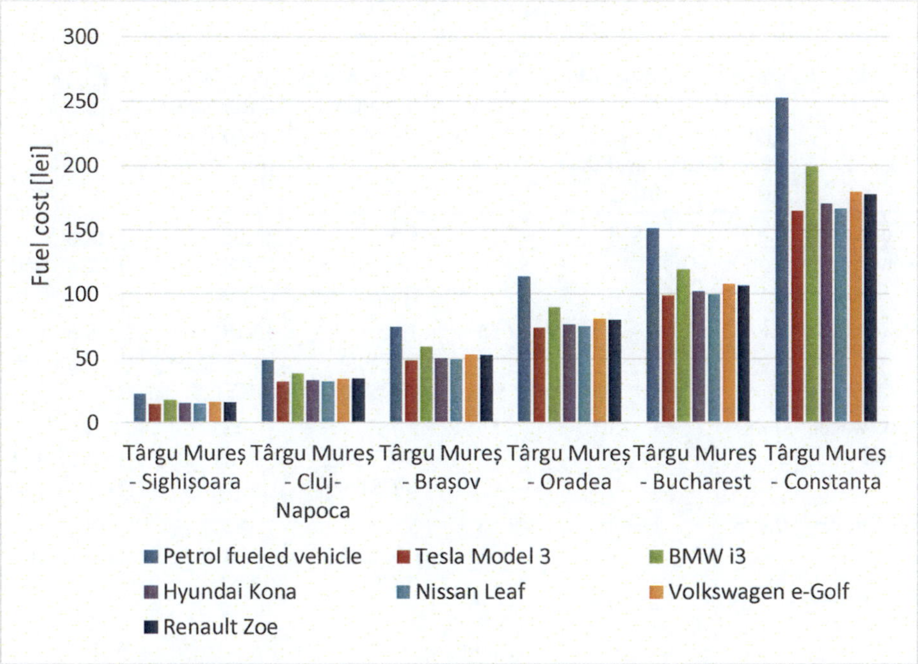

Fig. 4. Petrol fuel vehicle cost compared with fuel cost considering private fast charging station cost for the considered routes.

Table 7. Travel time results.

Vehicle	Târgu Mureş - Sighişoara travel time [h]	Târgu Mureş - Cluj-Napoca travel time [h]	Târgu Mureş – Braşov travel time [h]	Târgu Mureş - Oradea travel time [h]	Târgu Mureş – Bucharest travel time [h]	Târgu Mureş – Constanța travel time [h]
Petrol fueled vehicle	0.883	1.833	2.55	3.5	5.183	7.066
Tesla Model 3	0.883	1.833	2.55	3.5	10.683 (RCS)/5.983 (FCS)	12.566 (RCS)/7.866 (FCS)
BMW i3	0.883	1.833	2.55	6.5 (RCS)/3.75 (FCS)	11.183 (RCS)/5.683 (FCS)	16.066 (RCS)/7.816 (FCS)

(*continued*)

Table 7. (*continued*)

Vehicle	Târgu Mureş - Sighişoara travel time [h]	Târgu Mureş - Cluj-Napoca travel time [h]	Târgu Mureş – Braşov travel time [h]	Târgu Mureş - Oradea travel time [h]	Târgu Mureş – Bucharest travel time [h]	Târgu Mureş – Constanţa travel time [h]
Hyundai Kona	0.883	1.833	2.55	9.5 (RCS)/4.3 (FCS)	11.683 (RCS)/5.983 (FCS)	20.066 (RCS)/8.666 (FCS)
Nissan Leaf	0.883	1.833	2.55	10.5 (RCS)/4 (FCS)	12.183 (RCS)/5.683 (FCS)	21.066 (RCS)/8.066 (FCS)
Volkswagen e-Golf	0.883	1.833	2.55	8.75 (RCS)/4.2 (FCS)	10.433 (RCS)/5.883 (FCS)	17.566 (RCS)/8.466 (FCS)
Renault Zoe	0.883	1.833	2.55	5.5 (RCS)/4.25 (FCS)	7.183 (RCS)/5.933 (FCS)	11.066 (RCS)/8.566 (FCS)

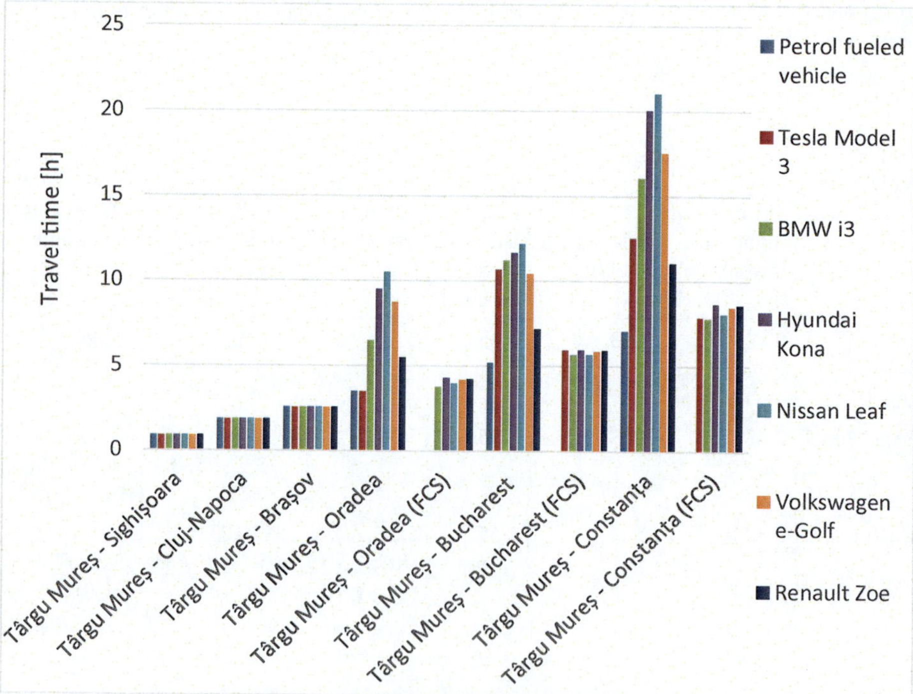

Fig. 5. Travel time for the considered routes.

The greater the distance, the greater the travel duration with a BEV, especially if the BEV is charged at a regular charging station.

4 Conclusions

The number of BEV has increased in Romania and across the globe, almost being double in Romania if we compare 2020 with 2019. This naturally lead to an increase of the number of charging stations, either regular or fast charging stations or the home outlet. The charging time is different considering the charging scenario.

In this paper we compared the fuel cost of several BEVs with the petrol fuel vehicle. Therefore, we considered different routes (from Târgu Mureş to Sighişoara, Cluj-Napoca, Braşov, Oradea, Bucharest and Constanţa) and travel distances (from 50 km to 569 km) in Romania. Also, we considered different charging options. Considering the analyses performed the fuel cost was lower for any of the BEV compared with the petrol fuel vehicle.

The travel time was the same for the BEV and petrol fuel vehicle if the travel distance was under 170 km. After this distance the travel time with the BEV increased compared with the petrol vehicle due to the charging that was required by BEV.

The battery electric vehicle with the highest cost, respectively the highest travel time was the vehicle with the largest battery and the highest range, Tesla Model 3. The Model 3 also required the fewest charges of battery due to the high battery capacity.

If we compare the average fuel cost for a km of the BEV with the petrol vehicle, the fuel cost for a km of the BEV is currently approximately 75% lower compared with the petrol vehicle if it is considered the grid cost. If it is considered the fuel cost of the owner of the charging station, then the fuel cost for a km of the BEV charging at a standard station is approximately 35% lower compared with the petrol vehicle, respectively approximately 20% lower compared with the petrol vehicle if the charging is done at a fast station.

As a future study, we will analyze routes in Romania and abroad that have a distance higher than 600 km and will consider the possibility for the user to wait at the charging stations due to the fact that there are no free outlets.

References

1. Dain, N., Sivakumar, A., Polak, J.W.: Electric vehicle charging choices: modelling and implications for smart charging services. Transp. Res. Part C: Emerg. Technol. **81**, 36–56 (2017)
2. Zhang, W., Dreise, C., Shao, R., Chang, L.: An improved minimum-cost charging schedule for large-scale penetration of electric vehicles. In: 2018 IEEE Applied Power Electronics Conference and Exposition (APEC), pp. 3411–3417 (2018)
3. Jabeen, F., Olaru, D., Smith, B., Braunl, T., Speidel, S.: Electric vehicle battery charging behaviour: findings from a driver survey. In: Australasian Transport Research Forum 2013 Proceedings, pp. 1–15 (2013)
4. Falchetta, G., Noussan, M.: Electric vehicle charging network in Europe: an accessibility and deployment trends analysis. Transp. Res. Part D: Transp. Environ. **94**, 102813 (2021)

5. Qu, Z., et al.: Optimization model of EV charging and discharging price considering vehicle owner response and power grid cost. J. Electr. Eng. Technol. **14**, 2251–2261 (2019)
6. Li, X., et al.: price incentive-based charging navigation strategy for electric vehicles. IEEE Trans. Ind. Appl. **56**(5), 5762–5774 (2020)
7. Taljegard, M., Walter, V., Göransson, L., Odenberger, M., Johnsson, F.: Impact of electric vehicles on the cost-competitiveness of generation and storage technologies in the electricity system. Environ. Res. Lett. **14**, 124087 (2019)
8. Nour, M., Said, S.M., Ali, A., Farkas, C.: Smart charging of electric vehicles according to electricity price. In: 2019 International Conference on Innovative Trends in Computer Engineering (ITCE), pp. 432–437 (2019)
9. Rawat, T., Niazi, K.R.: Coordinated charging of electric vehicles for reducing operation cost under TOU electricity prices. In: 2018 20th National Power Systems Conference (NPSC), pp. 1–5 (2018)
10. Xu, X., Niu, D., Li, Y., Sun, L.: Optimal pricing strategy of electric vehicle charging station for promoting green behavior based on time and space dimensions. J. Adv. Transp. **2020**, 8890233 (2020)
11. Danping, Z., Juan, L., Yuchun, C., Zhongjian, C., Liping, L.: Research on electric vehicle charging stations planning based on traffic destination heat data and charger usage data. In: 2020 IEEE 5th International Conference on Intelligent Transportation Engineering (ICITE), pp. 407–410 (2020)
12. https://insse.ro/cms/. Accessed 10 May 2021
13. https://www.plugshare.com/. Accessed 10 May 2021

Load Modeling Approaches in Smart Grids: An Overview

Bogdan-Constantin Neagu(✉), Gavrilas Mihai, Ovidiu Ivanov, and Gheorghe Grigoras

"Gheorghe Asachi" Technical University, Iasi, Romania
bogdan.neagu@tuiasi.ro

Abstract. The load modeling (LM) process is considered difficult due to the latest technologies developments and the newest classes of loads in the power system field; the electricity distribution network (EDN) is changing them from passive to active. The penetration of the small-scale renewable energy sources (SSRES) and the large utilization of power electronic devices supplementary to the participation of active end-users in EDN perform traditional techniques of load modeling uncertainty. It is assumed that more accurate modeling of the load will successfully solve the new challenges that inevitably arise in the operation and control of traditional EDN and modern Smart Grids (SGs). A global trend of the SGs researchers is to gain accurate load models. This paper outlined a survey of methods and strategies for load modeling from traditional to modern techniques for future SGs. This paper cover some of the usual methods found in the literature. Both advantages and limitations of each presented method are pointed out. More than that, the review is completed with the latest techniques in load modeling for key problems that must be solved in the SGs field, as demand response, prosumer behavior, transactive load sharing, energy storage systems (ESS), or electric vehicles (EVs) integration.

Keywords: Smart grids · Load modeling · Load profiling · Demand response · Load aggregation · Load shaping · Static and dynamic model

1 Introduction

Since the use of electricity plays a fundamental role in modern lifestyles, electrical load modeling (LM) must reflect the lives of people at work and leisure. The volume of information available and necessary for the operation, management, planning and security of the traditional power networks has increased with each decade, with each stage of technological development, making it necessary to introduce the calculation technique and gradually the intelligent solutions. In the last two decades, the technological avalanche as well as the exponential development of technologies in all energy sectors has allowed the collection and storage of information, opening new possibilities for monitoring, diagnosis, control, command and management of equipment and systems. In this context, the LM process become essential in the power systems studies. As a definition, the means of determining a load curve, profile, shape, or pattern of a consumer is known as the LM

L. Moldovan and A. Gligor (Eds.): Inter-Eng 2021, LNNS 386, pp. 533–561, 2022.
https://doi.org/10.1007/978-3-030-93817-8_49

concept [1]. An important point that needs to be highlight is that the load model does not represent reality, but is a projection of it in a simplified space, whose dimensions were chosen depending on the problem to be solved [2]. The model cannot be good or bad as such but only adapted or not to the problem requirements.

Most experts agree that the development and upgrading of traditional EDN are based on long-term development studies carried out in different pathways. The problem of demand evolution and forecast is fundamental in updating the networks planning of each stage in accordance with the technical and economic realities [3]. From other point of view [4], the operation and planning of SGs rely profoundly on modeling accurately its constituting components (generators, power lines, power transformers, or consumer loads). Probably only LM is the most difficult due to its variety and stochastic behavior [5]. It must be noted that the modeling of generation, transmission, and distribution process was long mostly concentrated on precisely representing, whereas the consumption modeling did not develop at the same dynamism [6].

The LM process can be performed in numerous ways. Not long ago, modeling meant systems of differential equations, transfer functions, and so on. The introduction of new technologies in the power system fields was perceived as a powerful means of computation and of pushing the limits of load model complexity. In the power engineering field, LM is a scientific endeavor that seeks to estimate the power demand based on the analysis, processing, and interpretation of several types of data, intending to obtain minimal deviations between the estimated and the realized power demand. In this context, as seen in the literature [7–12], it is widely recognized that load forecast and LM plays a major role in the optimization of distribution network structure and operation, being also a critical component in future SGs development and planning.

From the traditional perspective, in 1992 and 1995 the IEEE society published a comprehensive statement regarding the LM ways for practical studies in power systems [13, 14], which covered basic theories, LM strategies, and directions. Another relevant paper [15–17] was released in 2002, 2006, and 2010 using new methodologies, and it crosses numerous perspectives on LM process. Mainly in the last decade, once appearing smart grid technologies such as the concepts of SSRES, EVs, or demand response (DR), it's required to modernize existing LM techniques and refine the accuracy and versatility of load patterns [2, 18]. Consequently, accurate LM has reached one of the main provocations in modern SGs simulations [19].

More recently, the Working Group of CIGRE society acted the conclusions of a review on LM practices by worldwide EDNs [20]. The aforementioned review was stimulated by an intensified interest in LM due to the emerging of new classes of loads based on power electronics, and the increment of renewable energy sources (RES). Three years later the same society published a far-reaching article regarding LM [21], which comprises load patterns, modelling techniques and procedures, directions, and guidelines for complex load aggregation.

The development of SGs also grants formidable challenges. The load in the modern EDN must be regularly balanced to store power surplus when low demand spells and deliver it during the demand increases [22]. An up-to-date report published by the European Union - Smart Grids Task Force [23] adds an explicit direction to build a smart meter (SM) roadmap when these intelligent devices should be capable to satisfy the necessities of the future energy markets, through a modular and resilient structure for the metering architecture. In these circumstances, the SM outcomes contain useful data regarding the shape of the end-user's load [24]. The aforementioned data is remarkably relevant to classify the end-users into several categories that describe how active power is used from different user categories over time. The organization of the paper are briefly presented in Fig. 1.

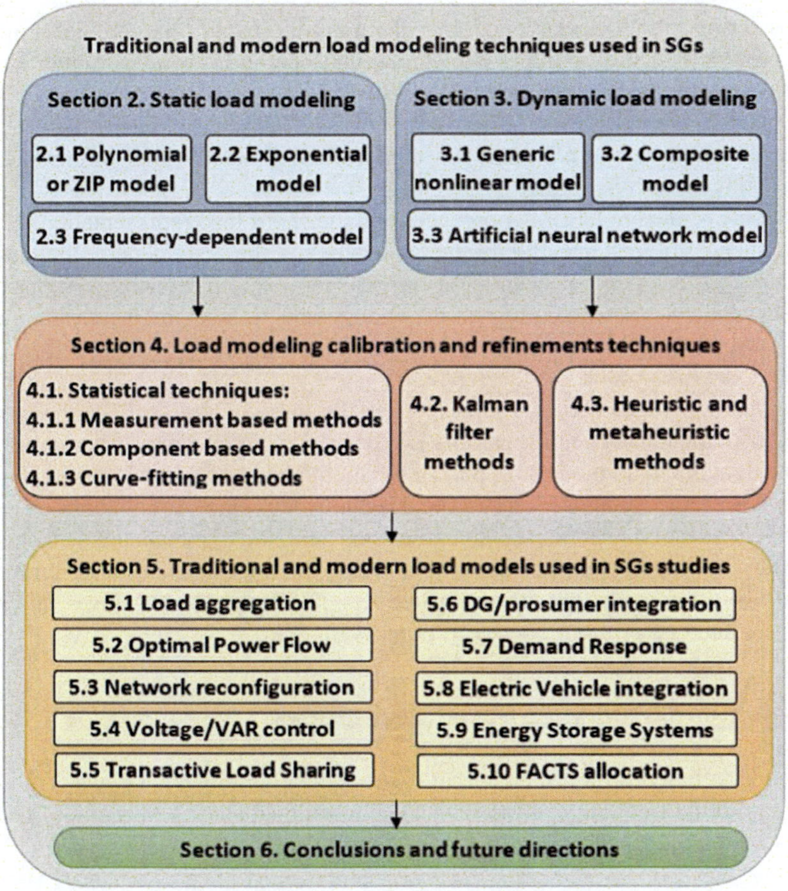

Fig. 1. The proposed organization of the paper.

The LM action can be listed as static and dynamic processes [5]. The static process reproduces the load with time-constant functions for voltage or frequency. In this LM process, the load value at each time is given as a voltage-dependent or frequency-dependent function at that corresponding time. From the opposing point of view, the dynamic LM process uses the present and preceding voltage or frequency states to reproduce the load. Regularly applied static LM process uses the exponential and the polynomial functions [14]. The first one describes the load as an exponential function, while the last applies a second-degree polynomial function, both of them being reported to the voltage. The LM frequency-dependent can be assessed to the above-specified structures, by increasing with an exponential frequency function, whose exponent indicates the load regularity sensitivity [25]. It must be mention that static LM is usually sufficient to reproduce the response only for a short time (1–5 s) after an event, but for more extended interval events the dynamic LM may be used. Dynamic LM expresses the active and reactive loads as voltage and time-based functions [26]. All these LM techniques have their peculiarities in terms of accuracy.

Considering the aforementioned consideration, the paper focuses on a number of critical problems regarding the SGs planning and operation, being different from the published ones by the following treated aspects:

- The reordering of the traditional and modern LM generally used in the traditional EDN and modern SGs studies.
- The update of the model LM approaches and the smart load representation.
- The influences of load uncertainty, energy incentives, the customers' willingness, and instantaneous reserves on accurate LM process for transactive energy systems (TES).
- The suggestions regarding future research directions to meet the increasing interests from power systems specialists.

This paper is organized as follows. After the introduction section, the next two presents fundamental concepts regarding LM process, pointed the most commonly used static and dynamic load models. In Sect. 4 the methodology for obtain the load models accuracy is presented, considering statistical techniques as curve-fitting or measurement- and component-based methods, Kalman filter and heuristic or metaheuristic techniques, followed by a comprehensive analysis of applying the traditional and modern methodology in SGs load modelling are given in Sects. 5. Finally, Sect. 6 gives this review conclusions and observations about LM process for SGs.

2 Static Load Modeling-Based Techniques

This section is based on a principle that relates active and reactive power both to the voltage and to the frequency of the EDN. The traditional static LM methods becomes modern in SGs studies.

2.1 Polynomial Load Modeling (PLM)

The PLM method is also known as ZIP model (Fig. 2) with constant values of impedance, current, and power [27–35]. The traditional ZIP model is expressed by a second-order polynomial of operating voltage magnitude (V0) in electric network bus [28].

The active and reactive power of the traditional ZIP load can be put into mathematical model as:

$$\begin{cases} P^{ZIP} = P_0 + I_a\left(\frac{V_0}{V_n}\right) + R_z\left(\frac{V_0}{V_n}\right)^2 \\ Q^{ZIP} = Q_0 + I_r\left(\frac{V_0}{V_n}\right) + X_z\left(\frac{V_0}{V_n}\right)^2 \end{cases} \quad (1)$$

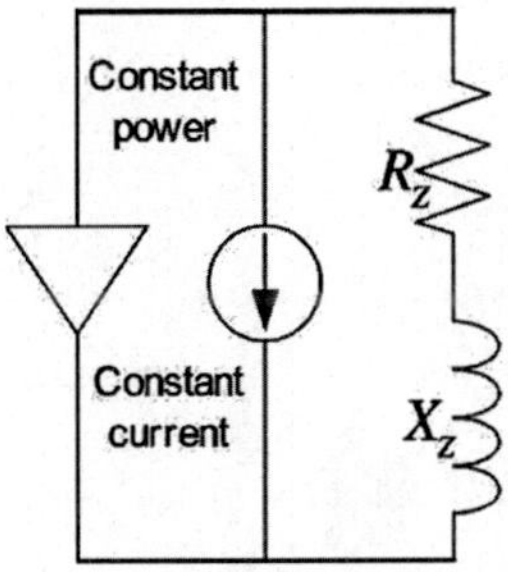

Fig. 2. The traditional ZIP (impedance-current-power) model (adapted after [29]).

$$\begin{cases} \underline{Z}_{const} = R_z + jX_z \\ \underline{S}_{const} = P_0 + jQ_0 \\ \underline{I}_{const} = I_a + jI_r \end{cases} \tag{2}$$

$$\underline{Z}_{const}^{(p.u.)} + \underline{I}_{const}^{(p.u.)} + \underline{S}_{const}^{(p.u.)} = 1 \tag{3}$$

In (1), (2) and (3), P^{ZIP} and Q^{ZIP} are the active and reactive load of the ZIP model, P_0 and Q_0 are the active and reactive constant load ($\underline{S}_{const}$.), I_a and I_r are the active and reactive component of bus currents ($\underline{I}_{const}$.); R_z and X_z are the two components of the constant impedance ($\underline{Z}_{const}$.), V_n is the nominal voltage of the analysed EDN. The modern BIG load model (Fig. 3) was recently proposed in [30] as a different option to the ZIP model that apprehends in addition to voltage magnitude, the angle sensitivities.

In contrast to ZIP, the BIG model was shown to be linear within the current injection and equivalent circuit formulations as [31]:

$$\begin{cases} I_R^{BIG} = \theta_R + G\left(\frac{V_{0,R}}{V_{n,R}}\right) - B\left(\frac{V_{0,I}}{V_{n,I}}\right) \\ I_I^{BIG} = \theta_I + G\left(\frac{V_{0,I}}{V_{n,I}}\right) + B\left(\frac{V_{0,R}}{V_{n,R}}\right) \end{cases} \tag{4}$$

In Fig. 3, and Eq. (4), the θ_R and θ_I are parameters of the constant current, G and B, represent the conductance and susceptance sensitivity elements for the two real (R) and imaginary (I) circuit. Important to highlight that the BIG model is linear in an I/V formulation, while the ZIP model is nonlinear in both I/V and traditional load model formulations. The BIG model is able of obtaining angle value compatible with the constant voltage magnitude devices, during which the ZIP models are not [32, 34, 35].

2.2 Exponential Load Modeling (ELM)

The ELM is also known as PQ model [5, 19, 31, 36, 37]. Paper [2] highlight that approximatively 68–70% of the EDN utilities use static LM for loads representation, and over 80% of the EDN operators or particular utilities use constant power in different load flow studies. The formulation for the exponential LM process is:

$$\begin{cases} P^{PQ} = P_0\left(\frac{V_0}{V_n}\right)^{\beta_P} \\ Q^{PQ} = Q_0\left(\frac{V_0}{V_n}\right)^{\beta_Q} \end{cases} \tag{5}$$

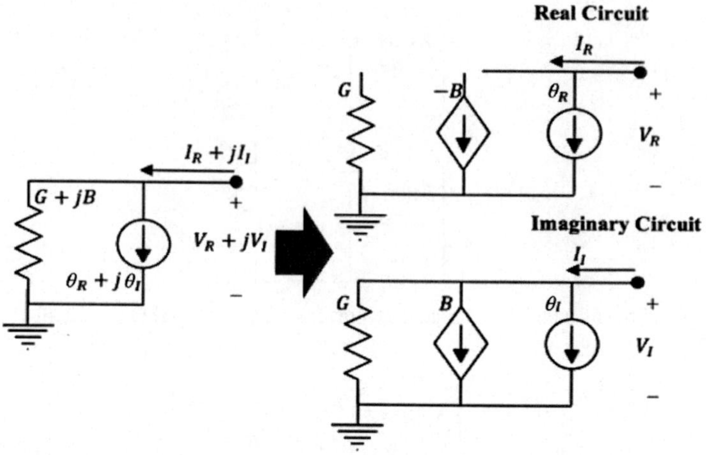

Fig. 3. The modern BIG (susceptance-current-conductance) model (adapted after [30]).

where: P^{PQ} and Q^{PQ} – are the active and reactive load of the PQ model; β_P and β_Q – are the voltage exponents for PQ model.

Regarding (5), the following aspects can be underlined [36] if: $\beta_P = \beta_Q = 2$ the model become as a constant impedance load characteristic; $\beta_P = \beta_Q = 1$ as constant current characteristic; and $\beta_P = \beta_Q = 0$ as constant power characteristic.

Some authors [19, 38–40] explains that in specific situations, which as the presence of capacitors for power factor (PF) corrections, the standard method of active and reactive power modelling as exponentially conditioned on voltage may lead to unrealistic values of β_Q exponents.

2.3 Frequency-Dependent (FD) Load Modelling

This active and reactive FD load model is obtained by multiplying the exponential model (P^{PQ} and Q^{PQ}) with a coefficient that depends on the frequency from the considered load [41]. The general mathematical formulation is [2]:

$$\begin{cases} P^{FD} = P^{PQ}\left[1 + \alpha_{fs}(f - f_0)\right] \\ Q^{FD} = Q^{PQ}\left[1 + \alpha_{fs}(f - f_0)\right] \end{cases} \tag{6}$$

where f is the frequency of the bus voltage, $f0$ is the rated frequency, and αfs is the frequency sensitivity coefficient.

From another viewpoint, it can be concluded that adding the frequency effective value to the ZIP model there is no physical significance because the component associated with the constant impedance becomes a frequency function [42].

3 Dynamic Load Modeling-Based Techniques

A dynamic traditional (or modern) LM is a time-dependent process, it states the dependence of operating voltage (or frequency) values, active and reactive power at any time interval, as a function of voltage and the frequency-time history, including automatically the instantaneous time. The dynamic LM is especially suitable in voltage and angular stability network analyses [2, 19, 43, 44].

3.1 Generic Nonlinear Dynamic (GND) Load Modelling

The most popular GND load model as indicated in literature [18–21, 31, 40–44] is the Induction Motor (IM), taking into account the over 70% of the total energy of the EDN utilities are consumed by electric machines. The IM load model is pictured in Fig. 4 and symbolizes the active and reactive power as a function of the previous and I-V including frequency of the power network bus [45–47]. Indeed, the exponential recovery LM can additionally be used to reproduce the GND-LM [45].

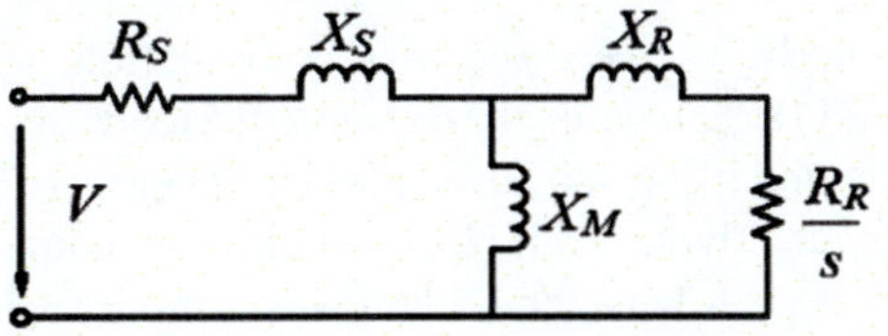

Fig. 4. The IM model (adapted after [46]).

In Fig. 4, al parameters are in p.u. (per units): R_S and R_R is the resistance of motor stator/rotor; X_S, X_R and X_M is the reactance of motor stator and rotor leakage, and motor magnetizing respectively; V is the input measured voltage value; s is defined as the motor slip which is reported to the IM rotation speed [48]. The IM load model can be described by the following nonlinear first order differential equations [2, 44]:

$$\begin{cases} T_P \frac{dX_P}{dt} = -X_P + P_0\left(\frac{V_0}{V_n}\right)^{\psi_{Pss}} - P_0\left(\frac{V_0}{V_n}\right)^{\psi_{Ptt}}; & P^{IM} = X_P + P_0\left(\frac{V_0}{V_n}\right)^{\psi_{Ptt}} \\ T_Q \frac{dX_Q}{dt} = -X_Q + Q_0\left(\frac{V_0}{V_n}\right)^{\psi_{Qss}} - Q_0\left(\frac{V_0}{V_n}\right)^{\psi_{Qtt}}; & Q^{IM} = X_Q + Q_0\left(\frac{V_0}{V_n}\right)^{\psi_{Qtt}} \end{cases} \tag{7}$$

where the parameters from (7) are defined as follows [2, 44]:

- P^{IM} and Q^{IM} are the active and reactive load of the IM model;
- X_P and X_Q are the variables based on the dynamics of active and reactive load;
- T_P and T_Q are the time constants of the exponential recovery response;
- Ψ_{Pss} and Ψ_{Qss} are the exponents coefficients for steady-state ("ss") load response;
- Ψ_{Ptt} and Ψ_{Qtt} are exponents coefficients related to the transient ("*tt*") load response.

3.2 Composite Load Modelling (CLM)

Another LM process is represented by a combination of the above two presented techniques, i.e., the static ZIP model and the dynamic IM model which become as Composite Load Model (CLM). Some papers of literature have noted that CLM provides a more precise load response (LR) than the singly used statics or dynamic LM [2, 44, 49]. A

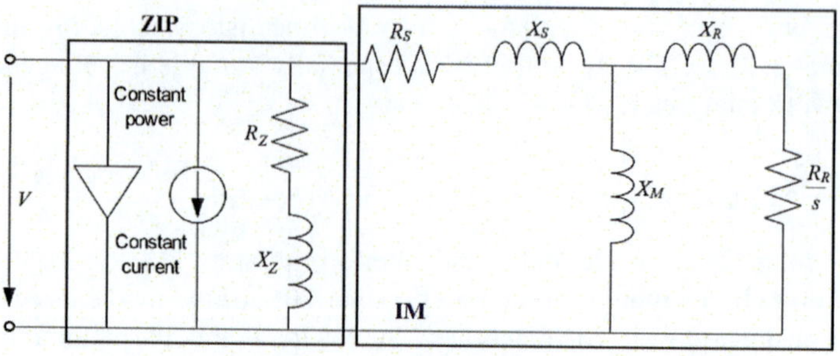

Fig. 5. The CLM model.

standard CLM is a combination of ZIP (Fig. 2) and IM (Fig. 4) models, as is shown in Fig. 5.

The mathematical model is expressed as a combination between static ZIP load represented by relations (1) to (3), while the dynamic IM load uses the system of Eqs. (7). In this case, is essential to underline that all standard load center of a modern SGs includes simultaneously statically and dynamically loads in different combinations [2, 32, 44–46, 50–52]. This is the real reason why the CLM gives a more realistic estimation of the loads, helping in the transitions from traditional to modern LM techniques.

3.3 Artificial Neural Network Load Modeling (ANN)

Taking into account the first two dynamic load models (IM and CLM) are traditionally used in the power system simulation, the LM based Artificial Neural Networks (ANN) is considered as a modern technique of load modelling, because was used until now strictly for short-term load forecasting (STLF) studies [9, 12, 53–56]. The authors of

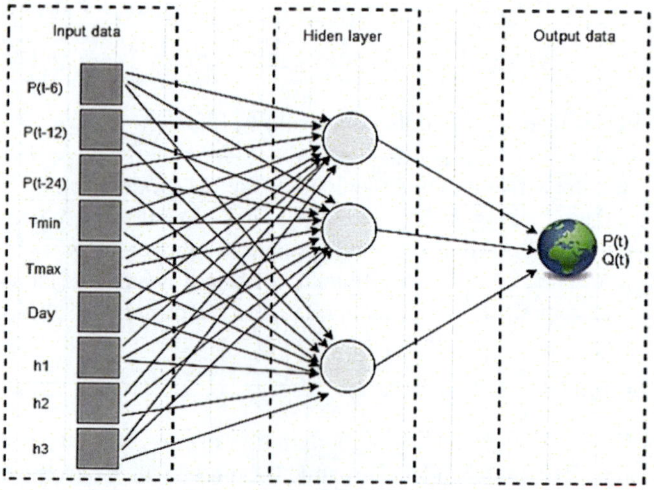

Fig. 6. The ANN architecture for LM process, adapted from [54].

this paper propose a ANN architecture in [54] with only three neurons on the hidden layer for daily active load modelling as can be seen in Fig. 6.

The set of data for modeling the energy consumption (active and reactive power) considered as input data, the loads from consumption history, the maximum and minimum temperature, the type of day (working or rest day) and the hours at which the LM process is desired. The modeled loads represent the output data. On the other hand, following [55] an undesirable ANN architecture, such as wrong numbers of perceptron's or learning iterations in the hidden layer, can lead to overfitting of the LM problem. This drawback is a result of an interpretation that agrees too strictly or specifically to a more specific set of load data, with a real reason for apply a validation process [57].

4 The Fine-Tuned of Load Model Coefficients

Once a satisfying LM technique is proposed, the mathematical expressions need to be calibrated to obtain the best solutions. For that, each model coefficients or exponents must be fine-tuned, as is presented in Fig. 7. With other words, in the following the most used techniques was critically presented, taking into account the literature overview. More than that, the selected fine-tuning parameters approaches was selected considering both traditional or modern based methods which have been used successfully in the power system simulations.

4.1 Statistical Techniques (ST)

Statistical load modeling techniques have been widely utilized in the literature [6, 14, 58–60]. In facts, the active and reactive loads from traditional EDNs (or even modern SGs) are changing in time, and the past recorded values are values of a random function of time [8, 61]. In order to obtain the statistical series of the random variable P (the

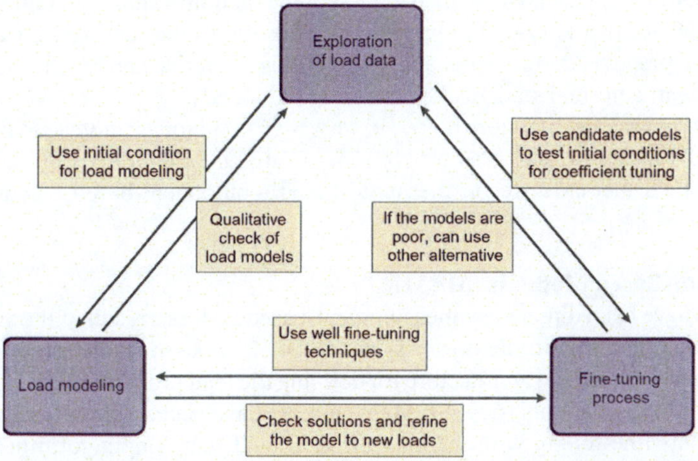

Fig. 7. The schematic diagram regarding the process of fine-tuning LM coefficient.

active load from a bus) in a given moment, the entire range of values $[P_{max}, P_{min}]$ is divided into w intervals, with a desired precision.

In each interval, the number w_k of the values of P_0 which satisfy the condition:

$$P_{w-1} < P < P_w \tag{8}$$

where $P_{w\text{-}1}$ and P_w are the values in the extremities of interval w.

Figure 8 shows the statistical series of the random variable P and the normal statistical distribution [62].

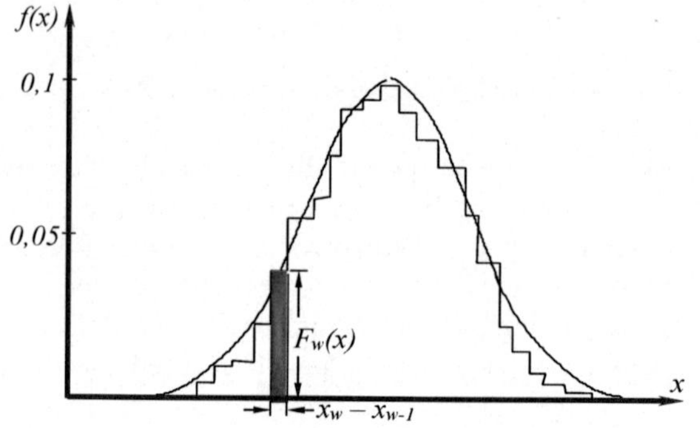

Fig. 8. The statistical variation of the random variable P_0.

The statistical frequency of the P_0 is given, for each interval, as:

$$F_w(P) = \frac{w_k}{w}, \quad w = \overline{1, W} \tag{9}$$

If a high number of measurements is used, it can be considered that the statistical series coincides with the statistical distribution of the random value. The same is true for the reactive power modelling – the variable Q. This allows the reasonable assumption, satisfactory in practice, that for rural and urban EDNs the active and reactive loads have, in each moment, a normal statistical distribution [8, 44, 63].

In the power systems literature [2–5, 19, 44, 61–65], a lot of accurate load fine-tuning statistical (LFTS) approaches was proposed by using historical data values. Thus, the most used LFTS methods are on measured-based, component-based or curve-fitting model formulation.

Measurement-Based Methods (MBM)

Most studies have been directed on measurement-based LM, as a result of the availability of smart meters mounted in the power systems. For LM based on measurements, two conditions must be fixed: an architecture for defining the load properties plus a procedure to assess the LM coefficients [66–68]. Many authors have stated various techniques for LM coefficients fine-tuning with the help of traditional searching algorithms [69–73]. These methods have the advantage of direct measurement of the loads. For example, a more recent papers [67] and [71] update the existent techniques designed for assessing

LM coefficients from the three-phase symmetrical event (fault) data recorded by the Phasor Measurement Unit (PMU) by LM of unsymmetrical faults in high voltage SGs. Other authors consider that the real measurement can provide a more accurate loads in low voltage (LV) EDN [68, 69]. The LM measurement-based modelling approach for parameter tuning is synthetically indicated in Fig. 9, [74].

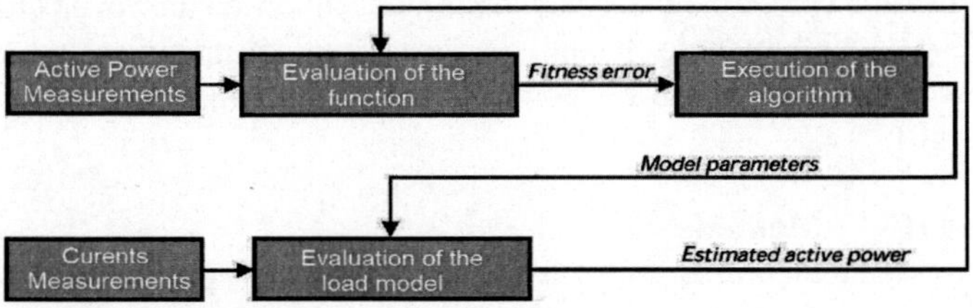

Fig. 9. The LM measurement-based modelling approach for parameter fine-tuning.

The LM coefficient are tuned using a Gauss–Newton method in [75], Genetic Algorithm (GA) in [70], Artificial Neural Networks (see Sect. 3.3) method in [72] and Simulated Annealing in [73].

Component-Based Methods (CBM)

The component-based method involves the tuning of the LM coefficients from data on its constituent elements [76]. Traditionally, consumers can be classified into four types such as residential, commercial, industrial, and agricultural. Each of them represented in terms of individual elements of load consumption. The characteristics of the individual load component have been rigorously considered and representative data have been generated for load structure and components for each consumer. This kind of method is used to create a composite load model. The CBM approach for parameter tuning is shown in Fig. 10, [77].

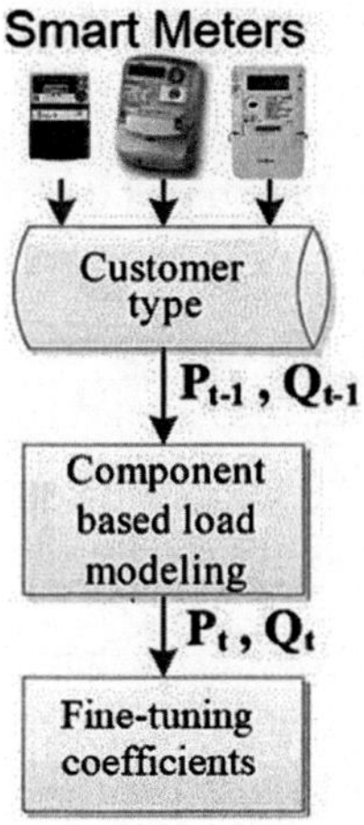

Fig. 10. The CBM approach for LM parameter fine-tuning.

In opposite with the measurement-based approach, this technique develops different mathematical load models. Thus, in [78] the LM process was tuning for a paper mill connected to the national system. However, it has some obvious deficiencies: the search speed is slower, the search area is random, and may have an unconventional result in a complex experiment. Other studies consider the component-based methods used for tuning the composite load models [2, 17, 18], based on GA in [26], an improved Particle Swarm Optimization (PSO) in [49], ANN in [56], or a modified improved clone selection algorithm (CSA), which includes adaptive-adjust Gaussian mutation operators and the directional evolution mechanism for power quality disturbance modelling in [79] and Petri Nets in [80].

Curve-Fitting (C-V) Methods
Generally, parameter identification of load in LM is a C-V problem, that applied the real measurement and tries to compare the real pattern with the simulated. C-V is a mathematical method for creating a plan that outlines the discrete input data set in the constant model. Various ways of curve fitting have been realized which describe the discrete input data set. The method of linear regression and arbitrage smoothing process has been explained in [81–83] used as a C-V simulation tool. Paper [84] apply the Curve-Fitting Method to obtain the optimal size of FACTS loads (DSTATCOM) at each level, expressed in the form of a simplistic quadrature equation.

4.2 Kalman Filter (KF) Technique

For optimize the ZIP model a KF technique is used [2, 19, 44]. The KF based on a parameter identification methodology track the unknown parameters of the exponential dynamic LM as in [19], and offers an advantage of fast parameter determination by removing the traditional iterative calculation, as is written in [85]. The results from [86] show that, the extended KF algorithm converges in low time and distinguishes LM parameters correctly with immeasurable security. It gives an efficient approach to the state estimation and LM process in the SGs. From the classical perspective, the KF technique was applied by authors of [87] for a harmonic modeling approach of linear and nonlinear loads. The aforementioned approach uses directly available samples of the load voltage and current to compute, and estimate the variation of LM parameters [88]. The synthetic ZIP parameter tuning based on KF approach is shown in Fig. 11.

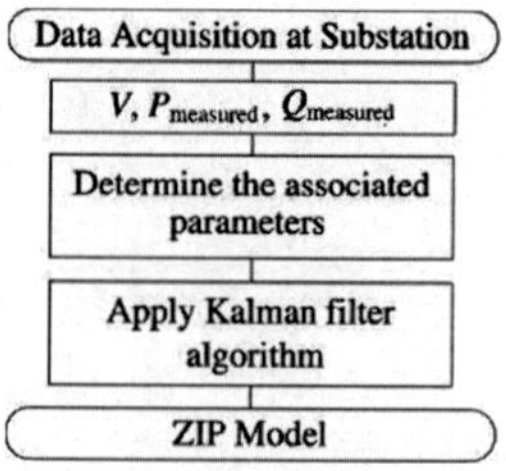

Fig. 11. The LM approach for parameter fine-tuning based on Kalman filter, adapted from [86].

4.3 Heuristic and Metaheuristic Techniques (MHT)

Many of the existing aforementioned conventional techniques face with difficulties such as: avoiding trapping into local optima during the search of the global optimum; conveniently handling constraints; risk of divergence, the exponential growth of the size of the search space for high-dimensional problems, or numerical computation of derivatives used by the search mechanism [89, 90]. To overcome such problems, heuristic and metaheuristic techniques were proposed in [91–96]. Unlike traditional methods, metaheuristics have a simple and compact theoretical support, and are often designed and tuned using an empirical approach. Recent research was directed towards a special category of heuristic search methods, based on nature-inspired mechanisms.

Unimodal optimization problems can be efficiently approached by conventional optimization methods, such as exact methods. For multimodal optimization problems, when the search space is non-convex and contains a large number of local optima, conventional optimization methods often fail to find a good solution in a reasonable amount of time. In response to these challenges, researchers have proposed alternative approaches designed to find a compromise between the accuracy with which the global optimum is determined and the numerical efficiency of the computational model.

These methods, called heuristics, are largely based on knowledge and experience of human experts and aim at exploring the search space in a more convenient way. Several of the most popular heuristics are Best First Search (BFS), Beam Search (DS) and the A* Search Algorithm [97]. Other heuristics are the so-called direct search methods, which start from an initial guess of the optimal solution and improve it by moving to a new position, where the objective function has a better value, chosen based on specific search strategies [98]. All heuristic search techniques, including those mentioned above, are characterized by a low numerical efficiency of the search process, especially for high dimensional problems. As a result, new methods have been proposed and developed based on stochastic computing techniques, known as metaheuristic methods. The most attractive feature of any metaheuristic method is that its use does not require detailed knowledge on the optimization problem and can therefore be used to define a general framework for solving optimization problems [95].

In the LM context, since their introduction in the mid-1980s until now, metaheuristics have been continuously developed to address and solve an increasingly number of complex optimization problems, previously considered difficult or impossible to solve. Some of these metaheuristics, such as Simulated Annealing (SA) or Tabu Search (TS) maintain a unique approximation of the required solution, which is improved by applying correction terms based on criteria aiming at moving the system to a lower energy state [99] or assessing the attractiveness of the new point in the search space based on an aspiration criterion [100]. Other metaheuristics maintain a population of solutions, called particles or agents, with parallel evolution guided by measures assessing their degree of optimality. The general flowchart of the metaheuristic approach used in different studies by the authors is briefly presented in Fig. 12.

The performance of different optimization techniques for parameter fine-tuning of LM uses in the last time the traditional and modern (meta) heuristic algorithms discussed in more detail in the last section. The defined LM structure can be used to characterize active and reactive power considering a set of nonlinear voltage functions [102–104].

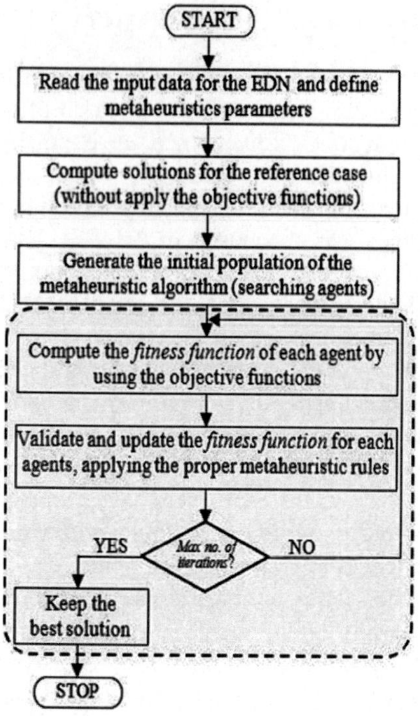

Fig. 12. The general flowchart of the metaheuristic methods, adapted from [101].

The LM overview from this paper consider the metaheuristics as modern approaches applied in SG operation and control.

5 The Traditional and Modern Load Models for Solving Key Problems in the Smart Grids Control and Operation

5.1 Traditional vs. Modern Load Models

Traditionally, the EDN has intentionally been designed to maintain unidirectional power flow with radial configuration, aimed at economic power delivery to the end-user. Nevertheless, intensifying load demands over wide geographic distances have occurred in critical technological issues in EDNs, predominately-higher losses, lower voltage control, compromised power quality, safety concerns, and expensive planning options [105]. In the discussed paper context, the transition from classical EDN to modern SGs is illustrated in Fig. 13. It can be seen in the presented figure, that load modelling for accurate SGs simulation must be a comprehensive process, using the single or composite models, and indubitably the fine-tuning of the obtained results.

Obviously, in the traditional EDN for the four class (residential, commercial, industrial, or aggregated) of consumers only two load type was used (dynamic or static). At the same time, because of the increased demands and the complicated dynamic response due to the integration of renewable energy sources (RES) in the SGs, a lot of load types

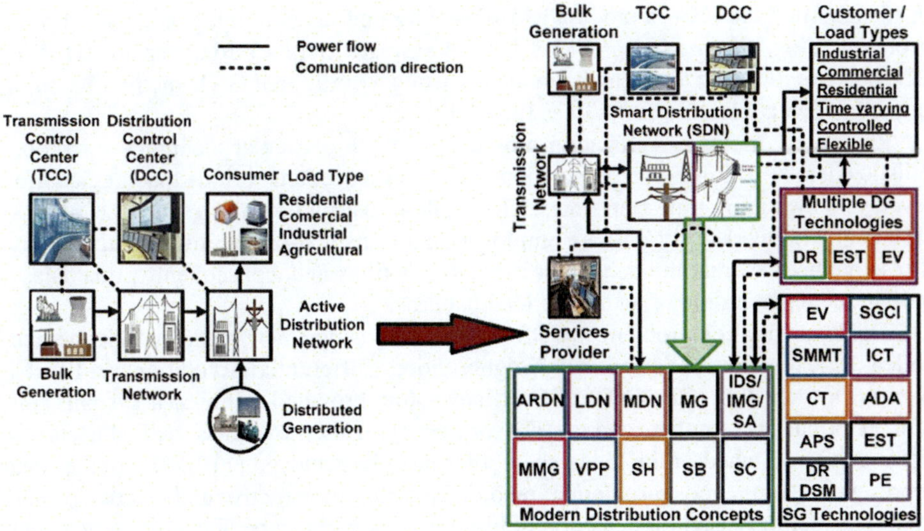

Fig. 13. The transition from traditional power systems to smart grids, adapted from [105, 106].

are featured (i.e. electric vehicle, energy storage systems, flywheels, prosumers, photovoltaic panels, wind turbines, batteries, etc.). We can define the two load models as traditional and modern, as is considered in Fig. 14.

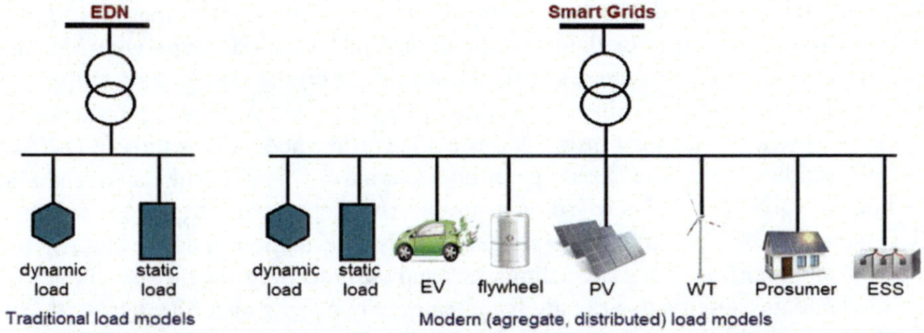

Fig. 14. The traditional vs. modern load models used in SGs simulations.

5.2 The Application of Load Models in SGs Operation and Control

Load Profiling and Aggregation (LPA)

The load models are specific, depending on the consumer's class (residential, commercial, industrial, or aggregated), the type of day (working, non-working or holidays), the season (cold or hot season), respectively the tariff used (monomial or differentiated tariff) [54]. A common name for such load models is called load type profiles (TLP) [3, 8,

9, 12, 107–110]. In [107] a modern LM is used for end-user grouping and consumption profiling using information provided from the Smart Meters. Moreover, the TLP for SGs power substation level can be modelled using a traditional GND method based on limited MBM calibration techniques [8, 108].

Regarding the load aggregation problem in [81] a particular technique for probabilistic characterization of aggregate load taking into account various load aggregation levels and averaging time step times. The Weibull distribution and generalized regression neural networks (GRNN) are employed in a statistical-probabilistic methodology. These methods extract the behaviour of load variations and, consequently, this dataset help in the regeneration of aggregate load patterns.

The load variation problem was solved by [109] using a new metaheuristic method, called as Cross-Entropy Variable Neighborhood Differential Evolutionary Particle Swarm Optimization (CE-VNDEPSO). Many values provided by the SMs systems contain abnormal data (outlier) and the TLP or the aggregation load loss the LM accuracy. For a good fine-tuned of the loads, the outlier must be removed [110]. The outlier data represent one of the most problem of recorded consumption (active and reactive power) data. The methods used for detection of load curve outlier can be divided into two main categories, based on data mining and state estimation.

Optimal Power Flow (OPF)

For the SGs model, it is assumed that all currents, powers, and voltages have sinusoidal waveform with constant amplitude [4, 14]. This operation is named steady-state SGs operation and leads to the remarkable fact that the states can be modeled with composite or aggregated variables. Should be pointed that the correlation between PQ and larger complex models (BIG models), which achieve a remarkable aspect of load impedance, hence sensitivity, is used to explain that the choice of load model type with Machine Learning Algorithm (MLA) can result in various SGs working steady-state at the OPF solution [31]. The complex values for voltages, currents, and powers are variations of the analogous steady-state quantities [59]. The power flow model of the three-phase SGs assumes so-called symmetric based operation and control [77, 87]. The day-ahead load modeling for solve the OPF is based on a simple AutoRegressive eXogenous variables (ARX) method. The ARX model has lagging hourly information of active and reactive loads from the same hour in the preceding week and the preceding day the same hour load since available measurements are only from the previous day's metering. The OPF was performed with the backward-forward method. In another research [88], the OPF analysis was carried out using multiple linear regression (MLR) method, to consider the filtered load components. The LM tuning use KF technologies [86]. Without going details of this assumption, it allows the modeling of the SGs with more accurate load models [111]. In this context, a new OPF was proposed in [112] by phase load balancing algorithm in the LV EDN with hybrid structures of the load consumption (switchable and non-switchable consumers). Different load models were proposed in the literature for OPF problem, solved with the metaheuristic modern algorithms as Adaptive Multiple Teams Perturbation-Guiding Jaya (AMTPG-Jaya) in [113], Coyote Optimization algorithm (COA) in [114], Spotted Hyena Optimization (SHO) in [115] or Kriging Assisted by Genetic Algorithm (KAGA) and by Particle Swarm Optimization (KAPSO) in [116]. The characteristics of the active and reactive load modeling process are summarized in the next section. The latest metaheuristics outperforms other algorithms, especially in terms of the computation time [109].

Optimal Network Reconfiguration (ONR)

The responsibility of operating efficiently an EDN has been traditionally aided by the application of algorithms for choosing the optimal configuration. In the classical approaches, the optimal configuration was limited by considering one or a few load models of the grid, especially by adopting load information associated with a high-loading state [28]. The distribution network operator (DNO) establishes a precise EDN configuration as the recommendation one and adopt this configuration as the conventional one, trying to avoid configuration modifications to improve the maintenance supervisors make their operations starting from a well-known EDN structure. Maintaining the above-mentioned reference configuration is also helpful for the DNO to present information to the regulatory authorities concerning the continuity of supply for safety applications based on the effects anticipated on that EDN. Today, the expansion of distribution automation and the dissipation of digital two-way SMs, protection, and control devices are allowing the association of enhanced functions for large distributed control, driving to the SG model. [117]. In this context, SGs analysis and optimization was extended to consider time-dependent loads in a wide energy-based assessment [118]. Furthermore, the growth of EDN electronic automation is performing for the DNO the possibility of switching from one to another configuration more easily than in the past. In the perspective of the SGs, the ONR could be run more times during a day, leading to the possibility to operate the distribution system closer to optimal conditions in different periods. A ZIP model has been expanded in [28] for voltage-dependent ONR problem using a Second-Order Conic Programming (SOCP) algorithm. An original approach that uses emergent theories of multi-criteria decision-making and anticipatory knowledge to operate the SGs in an efficient way is proposed in [119]. This strategy mixes the representation of a pseudo-ONR number recognized in the intra-day periods with DR resources anticipatory scheduling to drive load changes during operation.

Voltage – VAR Control (VVO)

The DNOs take into account the implementation of smart solutions to improve both the voltage level in the subordinate EDNs and the power factor, with the aim to maintain the balance between power generation and consumption while meeting the quality of supply standards and regulations. The bus voltage profile are key parameters in the EDN operation, both for traditional operating conditions or in modern microgrids with RES. In this context, the LM process was applied for voltage control considering different traditional or modern optimization techniques as indicated in [25, 27, 40, 43, 71, 76, 112]. Following the example of the LM in VVO process, in the last decade a more accent was pointed on the concept of Conservation Voltage Reduction (CVR), which signifies the load decrease resulting from a reduction of the operating voltage [27]. To efficiently evaluate CVR, the loads at each of the end-users uses the ZIP model-based on CLM techniques, the reason for that their voltage responsive function becomes accurate. A comparative study of several metaheuristic algorithms adapted for solving the VVO problem with the objective of power losses minimization in MV EDN was proposed by the authors in [120]. During the analysis, the well-known GA and PSO are tested against two newer metaheuristics that have seen previous uses in power engineering applications, the Bat Optimization algorithm BOA [121] and Whale Optimization algorithm (WOA) [122], and another recent but much less used method, the Sperm Whale algorithm (SWA), [120]. Traditionally, VVC is carried out by implementing the reactive power control using a synchronous generator, capacitor banks, and the latest FACTS (Flexible AC transmission systems) devices [123–125]. Currently, the voltage is controlled in the MV EDN bus by operating the switch of the branches of an overloaded power transformer of high voltage and medium voltage trough the on-load tap changers (OLTC). When power is supplied from modern loads as distributed generators, this method does not ensure proper voltage values in network nodes because voltage significantly rises, [120, 126, 127].

Transactive Load Sharing (TLS)
The traditional unidirectional of the passive EDN are rapidly developing into bidirectional, interactive, and multi-coordinated SGs that cover distributed power generation along with advanced information communications and electronic power technologies [2, 19, 105]. Due to this, following [128], the modern Transactive Energy concept is defined as not only loads of the SGs but also to provide resolutions to control the rate of load production in both network and end-users' demand. It is true that the loads are also modelled as in the traditional approaches, considering the new type of loads (see Fig. 14). In 2019, an extensive study was completed that evaluated the influence of the SSRES modeling (as prosumers) on the SGs power losses, [129]. At the same time, paper [130] presents a new trading platform based on smart peer-to-peer (P2P) contracts for prosumers energy surplus trading in a microgrids. The loads are modeled for give the possibility to perform trading based on peers' locations, instantaneous active power demand, maximum daily energy demand, and the principle of first come first served implemented in an anonymous blockchain trading ledger. These commercial types of loads are modelled with the traditional static models, and the problem is solved using a prosumer electricity surplus trading (PEST) algorithm. Other vision uses trust-region Newton method (TRN), with traditional LM techniques, [131].

Distributed Generation Integration (DGI)
The proliferation of small-scale distributed generation sources in medium voltage (MV) and low voltage (LV) distribution networks changed the operation conditions and management requirements of these networks, that must now integrate new technologies and management procedures. DNO need to convert the existing supply infrastructure into smart grids with enhanced monitoring and communication capabilities. For several decades, a number of external factors have led to some radical changes in the power industry model, which have led to a partial shift from centralized generation to distributed generation (DG). As mentioned in [132], the main factor of influence was the impact on the environment and its associated legislation and, by ricochet, the focus on RES such as wind, solar, geothermal, biomass or fuel cells. Another factor that led to DG solutions was the increase in electricity consumption, and the effect of this increase, the high costs of expanding and developing existing infrastructures. Last but not least, the broad development of new energy generation and storage technologies have been the critical factor for the implementation of the DG concept. The LM process used

in the DGI problem uses both traditional and modern techniques as in the literature [22, 51, 81, 104, 105]. For example, the identification of composite load model (CLM) with distributed photovoltaic generation and energy storage was proposed in [51], and a Fuzzy-Self Organizing Maps (FSOM) approach for detection.

Demand Response (DR)
The literature offers a broad spectrum of research carried out in this field. Regarding the methodologies and hypotheses used for DR, multi-objective optimization is the preferred choice. Papers [132, 133] use a multi-objective approach, carried out applying load management (LM) strategies and evolutionary methods. Here, the problem combines DR and real-time price estimation. Two objectives are considered: the minimization of the daily cost of active power and the minimization of customer dissatisfaction. Furthermore, it is expected to obtain similar and more flattened consumer demand profiles with sporadic peaks and a more secure and reliable network operation and control. In paper [134], a multi-objective theoretical optimization methodology for the energy payments is proposed, based on the interpretation of the correlation between price flexibility and probabilistic consumption, which considers stochastic load models. The investigation aims to improve the ability to accommodate wind production uncertainty. Moreover, the correlation between the quantity of DR programs and their efficiency is realized by actively engaging end-users in power networks.

Electric Vehicle Integration (EVI)
Modeling of EVs charging management, charging LM, the charging influence on the SGs, orderly charging approach, and planning of loading types of equipment have been centers of residential and different investigations regarding the regulation of EVs charging load pattern, and EVs charging types should be modelled accurately as in [74, 135]. From the previous researches, it is possible to conclude that it is essential the EVs charging type as the key to the regulation of EVs charging load patterns. Considering that drivers' driving times and distances are constant, it is common to study the modeling of the EVs charging modes with fuel vehicle data [136, 137]. In [142–144], the load considers the voltage-dependence through an ELM, which is used for EVs allocation. Particularly in [138], voltage-dependent load models are used to examine the effect of EV integration into SGs. For SGs studies, the use of a constant power LM to represent EV is not reasonable and not grant accurate results. Preferably of the previous, the ELM and the ZIP models satisfactorily represent the voltage depend on the characteristics of EV loads. For this reason, in [139] an EV based LM for SGs studies is stated as an aggregate of constant and ELM operation with a negative index value. The authors of [140] and [141] propose the use of PLM to interpret EVS integration into the SGs. The above use EV ZIP load models in the OPF to confirm that the power losses, peak demand, and voltage profiles are better correlated with those obtained in the case of using a constant loads model. Notwithstanding [141], states that EV ZIP load model is not suitable as a GND model, due to the different power type observed for some EV's trademarks analyzed under the same voltage fluctuations.

Energy Storage Systems (ESS)
In the literature, the ESS optimal placement in EDNs was studied with different objective

functions for solve particular problems [22]. The studies from the literature [51, 101] and [131] show that the ELM techniques can show the SGs optimization as the energy losses minimization and the voltage profile improvement, both of them corroborated with the reduction of the energy supplied during the peak periods by the main network, which suggests the benefits of ESS optimal placement. In addition, the results of [101] reveal the capability of ESS to supply the prosumers during the peak load that lead to energy savings, because the energy is not supplied from the system and the losses are reduced. In the future works, the energy saving benefits through optimal ESS placement will be analyses.

FACTS Allocation
The static and dynamic load models are classified according to the effect of voltage on the load. The load variation depends only on the instantaneous voltage input and related to the preceding voltage input. The static load model is a polynomial-based model composed of constant impedance characteristics, constant current characteristics, and constant power characteristics [145]. This static model is also known as ZIP load model and is often expanded to static load model with frequency characteristics using a proportion coefficient. An investigation of the Newton Raphson power flow incorporating ZIP load model with Static var compensator (SVC), Thyristor Controlled Series Compensator (TCSC), and Hybrid power flow controller (HPFC) presented. HPFC is a newly introduced member of flexible AC transmission system (FACTS), based on ZIP load modeling on the transmission system [146].

5.3 Summary of LM Process in the SGs Literature

Taking into account that the study presents an overview of the LM approaches this subsection highlight the key aspects of the paper. The above presented approaches were applied for solving the most occurred key problem in SGs control and operation are presented by emphasizing the load type, the load class (residential - Res, commercial - Com, industrial - Ind, or aggregated - Agg), the used model (PLM, ELM, CLM, etc.), the parameters tuning technique of the considered models, from the perspective of the traditional or modern approach. Thus, Table 1 summarizes the outcomes of the proposed review regarding the traditional and modern LM approaches developed in the extensive literature for SGs operation and control. The head of the table uses some abbreviations as "T" for the traditional approach and "M" for the modern approach. For a combination of load classes, the terms "All" is used for common consumers type (residential + commercial + industrial). Besides, in the paper end, the used abbreviation is explained.

6 Conclusions. Challenges and Future Directions

This paper interprets the overview of LM and parameter tuning methods. Different traditional or modern approaches for modelling EDN and SGs are also examined. LM process is challenging due to a considerable number of various load elements, the absence of accurate load structure knowledge, and the stochastic, time-varying or weather-dependent load functions. Currently, considering the proposed review must

Table 1. Summary of the LM approaches proposed for SGs analyses.

Problem solved	Load			LP parameters tuning technique	Proposed algorithm	Approach (T/M)	Reference number
	Type	Class	Model				
LPA	Dynamic	Res	CLM	C-V/MBM	Step-by-step	T	[5, 32]
	Static	Agg	ELM	MBM	Heuristic	T	[8, 108]
	Dynamic	Res	CLM	CBM	ANN-MLP	M	[9, 12, 53–56, 73, 107]
	Static	Res	FD	CBM	ANN	T	[72]
	Dynamic	Agg	ANN	C-V	GRNN	M	[81]
	Dynamic	Ind/Res	CLM	KF	MLR	M	[88]
	Dynamic	Agg	CLM	MHT	CE-VNDEPSO	M	[109]
OPF	Static	Agg	BIG	CBM	MLA	T	[31]
	Dynamic	Res	ANN	MBM/CBM	ARX	M	[77]
	Dynamic	Agg	GND	MHT	DSA	M	[111]
	Dynamic	Agg	GND	C-V/MBM	Heuristic	T	[112]
	Dynamic	Agg	PLM	CBM	AMTPG-Jaya	M	[113]
	Dynamic	Agg	PLM/CLM	CBM	COA, SHO	M	[114, 115]
	Dynamic	Agg	GND	MHT	KAGA,APSO	M	[116]
ONR	Static	All	PLM	CBM	SOCP	T	[28]
	Dynamic	Agg	CLM	CBM	PSO	M	[117, 118]
	Dynamic	Agg	CLM	MHT	Heuristic	M	[119]
VVO	Static	All	PLM	CBM	Heuristic	T/M	[27, 123]
	Dynamic	Res	GND	MHT	BA, WHA, SWA	M	[120–122]
	Dynamic	Agg	PLM/GND	CBM	LDC	M	[124, 125, 127]
TLS	Dynamic	Com	CLM	CBM	PEST	T	[129, 130]
	Static	All	PLM/FD	MBM	TRN	M	[131]
DGI	Dynamic	Res	CLM	CBM	GRNN	T	[51]
	Dynamic	Agg	ANN	C-V		M	[81]
DR	Dynamic	Res	CLM	CBM	NSGA-II,	M	[132, 133]
	Dynamic	Agg	GND	MBM	PSO	T	[134]
EVI	Dynamic	Res, Com	CLM	MBM	Heuristic	T, M	[74, 137]
	Dynamic	Res	CLM	CBM	Heuristic	M	[135, 136]
	Dynamic	Res	PLM/ELM	MBM	SA, ICA,	M	[139–144]
ESS	Dynamic	Agg	CLM	MBM	PSO	M	[22, 101, 131]
FACTS	Static	Agg	PLM	MBM	Heuristic	M	[145]
	Static	Agg	PLM	CBM	N-R	M	[146]

mention that the ZIP + IM composite model is one of the most widely used patterns in the power application. The ZIP + IM is less complex. However, it is inefficient to obtain the entire SGs features. Moreover, ZIP + IM cannot reproduce the latest modern loads behaviors. Although LM has been widely studied, more research being imperative to update existing LM and understand the components of modern loads (see Fig. 14) with developing technologies. For parameter tuning, the MBM is accepted as modern devices for recording information from PMUs and SM. In other words, it is still inviting to recognize a considerable number of remote LM exponents or coefficients.

Future research is required to generate advanced methods to show real-time LM using online data. Later, by promoting modern load models, they should be desegregated into SGs analysis performances. The growing insertion of DGs and the implementation of demand-response models new provocations to LM process. DGI and power electronic measurements may have complex SGs, taken into account in the model development. Customer behavior-driven and DR-enabled LM need to be built to facilitate DR studies.

The future analysis directions are recommended in terms of accurate modelling and classifying the SGs technologies. For LM composition improvement, more refined models that consider versatility and complexity are required. Load consumption is time-varying due to individual behaviors and climate conditions. In this context, several load models may be discovered in various stages. The CLM methods using measurement data at a specified time may not be able to achieve time-varying load behaviors and lack generalizability.

To diminish the number of parameters and classify them in the significant ones, a sensibility study has been proposed by the literature. Extreme sensitivities could lead to the SGs collapse of the LM with small variations in working circumstances. Thus, further research on sensitivity review is required. The deployment of SMs devices grants a great opportunity to outline mixed strategies that integrate the MBM and CBM.

The future direction must consider the LM process more accurate because of the SSRES integration. Also, the literature does not consider loads for distorting state, or harmonic and unbalance states. In this context, must consider modern based approaches for power quality, load sharing and emergent power factors. The load models must use the policy and recommendations of the energy regulators. The modified and improved models of the load must be combined for use accurate load models in SGs simulation. The artificial intelligence techniques can be applied bot for load modelling process and calibration parameters. From objective reasons, the paper considers the most studied problem in SGs. The traditional based new technologies load models will be considered in simulation of the SGs as: transient studies, slow voltage variations, harmonics analysis, grid resiliency, load sharing, etc.

References

1. Shafiei, M., Nourbakhsh, G., Ledwich, G., Femando, T., Iu, H., Arefi, A.: Dynamic load modelling using measured data in distribution networks. In: 3rd International Conference on Power Generation Systems and Renewable Energy Technologies (PGSRET), Malaysia, pp. 1–4 (2017)
2. Arif, A., Wang, Z., Wang, J., Mather, B., Bashualdo, H., Zhao, D.: Load modeling—a review. IEEE Trans. Smart Grid **9**(6), 5986–5999 (2018)
3. Cho, H., Goude, Y., Brossat, X., Yao, Q.: Modeling and forecasting daily electricity load curves: a hybrid approach. J. Am. Stat. Assoc. **108**, 7–21 (2013)
4. Bacher, R.: Power system models, objectives and constraints in optimal power flow calculations. In: Frauendorfer, K., Glavitsch, H., Bacher, R. (eds.) Optimization in Planning and

Operation of Electric Power Systems, pp. 217–263. Physica, Heidelberg (1993). https://doi.org/10.1007/978-3-662-12646-2_8
5. Perez Tellez, A.: Modelling aggregate loads in power systems, Dissertation thesis, KTH Royal Institute of Technology, Sweden (2017)
6. Vaahedi, E., Fl-Kady, M.A., Libaque-Esaine, J.A., Carvalho, V.F.: Load models for large-scale stability studies from end-user consumption. IEEE Trans. Power Syst. **2**(4), 864–870 (1987)
7. Liao, H.: Review on distribution network optimization under uncertainty. Energies **12**, 3369 (2019)
8. Neagu, B.C., Georgescu, G.: Posibilities of load curves modelling in electric energy distribution networks. Bull. Polyt. Instit. Iasi **60**, 89–102 (2014)
9. Hernández, L., et al.: Experimental analysis of the input variables' relevance to forecast next day's aggregated electric demand using neural networks. Energies **6**, 2927–2948 (2013)
10. Smida, M.B., Miled, K., Khalgui, M., Frey, G., Li, Z.: Modeling and verification of a reliable multi-agent solution promoting the autonomy and self-sufficiency of microgrids in an isolated location. IEEE Access **7**, 55090–55107 (2019)
11. Ugwudike, D.U., Madueme, T.C.A.: Theoretical approach to forecasting of electric power of distribution transformers. IOSR J. Electr. Electron. Eng. **12**(6), 20–31 (2017)
12. Gavrilas, M., Ivanov, O., Gavrilas, G.: Electricity load forecasting based on a mixed statistical-neural-computational intelligence approach. In: 12th Symposium on Neural Network Applications in Electrical Engineering (NEUREL), Serbia, pp. 61–66 (2014)
13. Load representation for dynamic performance analysis (of power systems). IEEE Trans. Power Syst. **8**(2), 472–482 (1993)
14. Bibliography on load models for power flow and dynamic performance simulation. IEEE Trans. Power Syst. **10**(1), 523–538 (1995)
15. Advanced Load Modeling, Electrical Power Research Institute (EPRI), Technical report 1007318 (2002)
16. Measurement-Based Load Modeling, Electrical Power Research Institute (EPRI), Technical report 1014402 (2006)
17. End-use load composition estimation using smart meter data, Electrical Power Research Institute (EPRI), Technical report 1020060 (2010)
18. Huang, Q., et al.: A generic modeling and development approach for WECC composite load model. Electr. Power Syst. Res. **172**, 1–10 (2019)
19. Rouhani, A., Abur, A.: Real-time dynamic parameter estimation for an exponential dynamic load model. IEEE Trans. Smart Grid **7**(3), 1530–1536 (2016)
20. Yamashita, K., Villanueva, S.M., Milanovic, J.V.: Initial results of international survey on industrial practice on power system load modelling conducted by CIGRE WG C4. 605. In: Proceedings of the CIGRE Symposium, vol. 4 (2011)
21. Yamashita, K., et al.: Modelling and aggregation of loads in flexible power networks–scope and status of the work of CIGRE WG C4. 605. IFAC Proc. **45**(21), 405–410 (2012)
22. Diahovchenko, I., Kolcun, M., Čonka, Z., Savkiv, V., Mykhailyshyn, R.: Progress and challenges in smart grids: distributed generation, smart metering, energy storage and smart loads. Iran J. Sci. Technol. Trans. Electr. Eng. 1–15 (2020)
23. European Smart Grids Task Force -Expert Group 3, Demand Side Flexibility -Perceived barriers and proposed recommendations, Final Report (2019)
24. Chicco, G., Labate, D., Notaristefano, A., Piglione, F.: Unveil the shape: data analytics for extracting knowledge from smart meters. Energia Elettrica Suppl. J. **96**, 1–16 (2020)
25. Saeidi, M., Rahmani, S., Pirayesh, A.: Modeling microgrids with voltage and frequency dependent loads. In: 7th International Conference on Power and Energy Systems (ICPES), Toronto, pp. 101–105 (2017)

26. Xie, J., Ma, Z., Ma, S., Wang, Z.: Data-driven based method for power system time-varying composite load modeling. IEEE Trans. Smart Grid **14**(8), 1–8 (2019)
27. Schneider, K.P., Jason, C.F., Chassin, D.: Evaluating conservation voltage reduction: an application of GridLAB-D: an open source software package. In: IEEE Power and Energy Society General Meeting 1–6 (2011)
28. Rasheed, S., Abhyankar, A.R.: Investigating the effect of load modeling on network reconfiguration of a distribution system. In: North American Power Symposium (NAPS), Wichita, KS, USA, pp. 1–6 (2019)
29. Korunović, L.M., et al.: Overview of existing load models and their applications. In: Proceedings of the CIGRE SC C4 Colloquium in Japan, Hakodate, Japan, pp. 121–126 (2012)
30. Jereminov, M., Pandey, A., Song, H.A., Hooi, B., Faloutsos, C., Pileggi, L.: Linear load model for robust power system analysis. In: IEEE PES Innovative Smart Grid Technologies Conference Europe (ISGT-Europe), pp. 1–6 (2017)
31. Jereminov, M., Hooi, B., Pandey, A., Song, H., Faloutsos, C., Pileggi, L.: Impact of load models on power flow optimization. In: 2019 IEEE Power & Energy Society General Meeting (PESGM), Atlanta, GA, USA, pp. 1–5 (2019)
32. Pandey, A., Jereminov, M., Li, X., Hug, G., Pileggi, L.: Aggregated load and generation equivalent circuit models with semi-empirical data fitting. In: IEEE Green Energy and Systems Conference (IGSEC), pp. 1–6 (2016)
33. Nashrullah, E., Halim, A.: Polynomial load model development for analysing residential electric energy use behaviour. In: MATEC Web of Conferences, vol. 218, p. 01007 (2018)
34. Wen, J.Y., Jiang, L., Wu, Q.H., Cheng, S.J.: Power system load modeling by learning based on system measurements. IEEE Trans. Power Deliv. **18**(2), 364–371 (2003)
35. Fungyai, K., Sangmeg, N., Pichetjamroen, A., Dechanupaprittha, S., Somakettarin, N.: Determination of ZIP load model parameters based on synchrophasor data by genetic algorithm. In: 8th International Electrical Engineering Congress (iEECON), pp. 1–4 (2020)
36. Gutierrez-Lagos, L., Ochoa, L.F.: OPF-based CVR operation in PV-Rich MV–LV distribution networks. IEEE Trans. Power Syst. **34**(4), 2778–2789 (2019)
37. Hossan, M.S., Mesbah Maruf, H.M., Chowdhury, B.: Comparison of the ZIP load model and the exponential load model for CVR factor evaluation. In: 2017 IEEE Power & Energy Society General Meeting, Chicago, pp. 1–5 (2017)
38. Milanović, J.V.: On unreliability of exponential load models. Electr. Power Syst. Res. **49**, 1–9 (1999)
39. Korunovic, L.M., Milanovic, J.V., Djokic, S.Z., Yamashita, K., Martinez-Villanueva, S., Sterpu, S.: Recommended parameter values and ranges of most frequently used static load models. IEEE Trans. Power Syst. **33**, 5923–5934 (2018)
40. Georgiev, G., Zicmane, I., Antonovs, E., Sergey, K.: Elimination of nodes with voltage dependent load characteristics in electrical network models. In: 10th International Conference on Environment and Electrical Engineering, pp. 1–4 (2011)
41. Wang, Q., Zhao, B., Tang, Y., Liu, L.: Modeling of load frequency characteristics in the load model for power system digital simulation. In: 2017 IEEE 7th Annual International Conference on CYBER Technology in Automation, Control, and Intelligent Systems (CYBER), Honolulu, pp. 538–542 (2017)
42. Ge, Y., Flueck, A.J., Kim, D.K., Ahn, J.B., Lee, J.D., Kwon, D.Y.: An event-oriented method for online load modeling based on synchrophasor data. IEEE Trans. Smart Grid **6**(4), 2060–2068 (2015)
43. Overbye, T.J.: Effects of load modelling on analysis of power system voltage stability. Int. J. Electr. Power Energy Syst. **16**(5), 329–338 (1994)
44. Mohammed, M., Abdulkarim, A., Abubakar, A.S., Kunya, A.B., Jibril, Y.: Load modeling techniques in distribution networks: a review. J. Appl. Mater. Technol. **1**(2), 63–70 (2020)

45. Regulski, P., Vilchis-Rodriguez, D.S., Djurovic, S., Terzija, V.: Estimation of composite load model parameters using an improved particle swarm optimization method. IEEE Trans. Power Deliv. **30**(2), 553–560 (2015)
46. Barzegkar-Ntovom, G.A., Ceylan, O., Papadopoulos, T.A.: Optimization techniques for parameter estimation of dynamic load models. In: 52nd International Universities Power Engineering Conference (UPEC), pp. 1–6 (2017)
47. Li, H., Chen, Q., Fu, C., Yu, Z., Shi, D., Wang, Z.: Bayesian estimation on load model coefficients of ZIP and induction motor model. Energies **12**, 547 (2019)
48. Jin, M.A., Dong, H., Renmu, H.E., Zhaoyang, D.O.N.G., Hill, D.J.: Research on identifiability of equivalent motor in composite load model. In: 2007 IEEE Lausanne Power Tech, pp. 1015–1020 (2007)
49. Jahromi, M.E.N., Ameli, M.T.: Measurement-based modelling of composite load using genetic algorithm. Electr. Power Syst. Res. **158**, 82–91 (2018)
50. Wang, C., Wang, Z., Wang, J., Zhao, D.: SVM-based parameter identification for composite ZIP and electronic load modeling. IEEE Trans. Power Syst. **34**(1), 182–193 (2019)
51. Zhang, X., Hill, D.J., Lu, C.: Identification of composite demand side model with distributed photovoltaic generation and energy storage. IEEE Trans. Sustain. Energy **11**(1), 326–336 (2019)
52. Wang, C., Wang, Z., Wang, J., Zhao, D.: Robust time-varying parameter identification for composite load modeling. IEEE Trans. Smart Grid **10**(1), 967–979 (2019)
53. Hernandez, L., Baladrón, C., Aguiar, J.M., Carro, B., Sanchez-Esguevillas, A.J., Lloret, J.: Short-term load forecasting for microgrids based on artificial neural networks. Energies **6**, 1385–1408 (2013)
54. Neagu, B.C., Ivanov, O., Grigoras, G., Gavrilas, M.: Short term load forecast approach based on artificial neural networks. Bull. Polyt. Instit. Iasi **69**(2), 9–21 (2019)
55. Kontis, E.O., Papadopoulos, T.A., Syed, M.H., Guillo-Sansano, E., Burt, G.M., Papagiannis, G.K.: Artificial-intelligence method for the derivation of generic aggregated dynamic equivalent models. IEEE Trans. Power Syst **34**(4), 2947–2956 (2019)
56. Park, R.-J., Song, K.-B., Lee, K.: The load model composition method in power systems using artificial neural network. J. Electr. Eng. Technol. **15**(2), 519–526 (2019). https://doi.org/10.1007/s42835-019-00335-2
57. Sheela, K.G., Deepa, S.N.: Review on methods to fix number of hidden neurons in neural networks. Math. Probl. Eng. **2013**, 425740 (2013)
58. Chong, C., Malhami, R.: Statistical synthesis of physically based load models with applications to cold load pickup. IEEE Trans. Power Apparatus Syst. PAS-**103**(7), 1621–1628 (1984)
59. Price, W., Wirgau, K., Murdoch, A., Nozari, F.: Load Modelling for Power Flow and Transient Stability Computer Studies, vol. 2: Load-Modeling Reference Manual. EPRI, Technical report (1987)
60. El-Férik, S., Malhamé, R.P.: Identification of alternating renewal electric load models from energy measurements. IEEE Trans. Autom. Control **39**(6), 1184–1196 (1994)
61. Tsagarakis, G., Collin, A.J., Kiprakis, A.E.: A statistical survey of the UK residential sector electrical loads. Int J. Emerg. Elect. Power Syst. **14**(5), 509–523 (2013)
62. Kharin, Y., Zhuk, E.: Robustness in statistical pattern recognition under "contaminations" of training samples. In: 12th IAPR International Conference on Pattern Recognition, vol. 3-Conference C: Signal Processing, vol. 2, pp. 504–506 (1994)
63. Shandilya, S., Thakur, T., Nagar, A.K.: Handbook of Research on Emerging Technologies for Electrical Power Planning, Analysis and Optimization; Engineering Science Reference (an Imprint of IGI Global), pp. 1–410. IGI Global, Hershey (2016)
64. Chabouni, N., Belarbi, Y., Benhassine, W.: Electricity load dynamics, temperature and seasonality Nexus in Algeria. Energy 117513 (2020)

65. Feinberg, E.A., Hajagos, J.T., Genethliou, D.: Load pocket modeling. In: Proceedings of the 2nd IASTED International Conference: Power and Energy Systems, Las Vegas, NV, USA, pp. 50–54 (2012)
66. Shang, X., Li, Z., Ji, T., Wu, P.Z., Wu, Q.: Online area load modeling in power systems using enhanced reinforcement learning. Energies **10**, 1852 (2017)
67. Han, S.: Three-phase-measurement-based load modeling technique using unbalanced fault data. IEEE Access **8**, 83655–83662 (2020)
68. Marchgraber, J., Xypolytou, E., Lupandina, I., Gawlik, W., Stifter, M.: Measurement-based determination of static load models in a low voltage grid. In: PES Innovative Smart Grid Technologies Conference Europe (ISGT-Europe), pp. 1–6 (2016)
69. Rodríguez-García, L., Pérez-Londoño, S., Mora-Flórez, J.: Measurement-based exponential recovery load model: development and validation. Dyna **82**(192), 131–140 (2015)
70. Ju, P., Handschin, E., Karlsson, D.: Nonlinear dynamic load modeling: model and parameter estimation. IEEE Trans. Power Syst. **11**(4), 1689–1697 (1996)
71. Liu, Q.S., Chen, Y.P., Duan, D.F.: The load modeling and parameter identification for voltage stability analysis. In: International Conference of Power System Technology, pp. 2030–2033 (2002)
72. Hiyama, T., Tokieda, M., Hubbi, W.: Artificial neural network based dynamic load modeling. IEEE Trans. Power Syst. **12**(4), 1576–1583 (1997)
73. Knyazkin, V., Cañizares, C., Söder, L.: On the parameter estimation and modeling of aggregate power system loads. IEEE Trans. Power Syst. **19**(2), 1023–1031 (2004)
74. Gil-Aguirre, J., Perez-Londono, S., Mora-Flórez, J.: A measurement-based load modelling methodology for electric vehicle fast-charging stations. Electr. Power Syst. Res. **176**, 105934 (2019)
75. Hiskens, I.A.: Nonlinear dynamic model evaluation from disturbance measurements. IEEE Trans. Power Syst. **16**(4), 702–710 (2001)
76. Lim, J.Y., Ji, P.S., Ozdemir, A., Singh, C.: Component-based load modeling including capacitor banks. In: 2001 Power Engineering Society Summer Meeting, Vancouver, BC, Canada, vol. 2, pp. 1199–1204 (2001)
77. Degefa, M., Millar, R., Koivisto, M., Humayun, M., Lehtonen, M.: Load flow analysis framework for active distribution networks based on smart meter reading system. Engineering **5**(10A), 1–8 (2013)
78. Wong, K., Haque, M.E., Davies, M.: Component-based dynamic load modeling of a paper mill. In: 22nd Australasian Universities Power Engineering Conference (AUPEC), pp. 1–6 (2012)
79. Yuan, R.F., Ai, Q., He, X.: Research on dynamic load modelling based on power quality monitoring system. IET Gener. Transm. Distrib. **7**(1), 46–51 (2013)
80. Kounev, S.: Performance modeling and evaluation of distributed component-based systems using queueing petri nets. IEEE Trans. Softw. Eng. **32**(7), 486–502 (2006)
81. Afzaal, M.U., Sajjad, I.A., Khan, M.F.N., Haroon, S.S., Amin, S., Bo, R.: Inter-temporal characterization of aggregate residential demand based on Weibull distribution and generalized regression neural networks for scenario generations. J. Intell. Fuzzy Syst. 1–13 (2020)
82. Cleveland, W.S., Devlin, S.J.: Locally weighted regression: an approach to regression analysis by local fitting. J. Am. Stat. Assoc. **83**, 596–610 (1988)
83. Desmet, L., Gijbels, I.: Curve fitting under jump and peak irregularities using local linear regression. Commun. Stat.-Theory Methods **40**, 4001–4020 (2011)
84. Yuvaraj, T., Ravi, K., Devabalaji, K.: DSTATCOM allocation in distribution networks considering load variations using bat algorithm. Ain Shams Eng. J. (2015)

85. Bian, S., Wang, Z., Ji, N., Chu, X., Du, Z., Yin, Q.: Input interaction and output fusion of adaptive interacting multiple load modelling. Int. J. Electr. Power Energy Syst. **117**, 105674 (2020)
86. Weijin, Z., Qian, A.: Application of extended Kalman filter in parameter identification of dynamic load model. Electr. Power Autom. Equipment **27**(2), 47–50 (2007)
87. Soliman, S.A., Alammari, R.A.: Harmonic modeling of linear and nonlinear loads based on Kalman filtering algorithm. Electr. Power Syst. Res. **72**(2), 147–155 (2004)
88. Takeda, H., Tamura, Y., Sato, S.: Using the ensemble Kalman filter for electricity load forecasting and analysis. Energy **104**, 184–198 (2016)
89. Fletcher, R.: Practical Methods of Optimization. Wiley, New York (2000)
90. Ravindran, R., Ragsdell, K.M., Reklaitis, G.V.: Engineering Optimization: Methods and Applications. Wiley, New York (2006)
91. Dorigo, M., Birattari, M., Stutzle, T.: Ant colony optimization - artificial ants as a computational intelligence technique. IEEE Comput. Intell. Mag. **6**, 28–39 (2006)
92. Farmer, J.D., Packard, N., Perelson, A.: The immune system, adaptation and machine learning. Physica D **22**, 187–204 (1986)
93. Goldberg, D.: Genetic Algorithms in Search, Optimization and Machine Learning, Reading. Addison-Wesley, Boston (1989)
94. Kennedy, J., Eberhart, R.: Particle swarm optimization. In: Proceedings of IEEE International Conference on Neural Networks, vol. 1, pp. 1942–1948 (1995)
95. Lee, K.Y., El-Sharkawi, M.A.: Modern Heuristic Optimization Techniques with Applications to Power Systems. IEEE Press Series on Power Engineering: Wiley (2008)
96. Nakrani, S., Tovey, S.: On honey-bees and dynamic server allocation in Internet hosting centers. Adapt. Behav. **12**(3–4), 223–240 (2004)
97. Zhou, R., Hansen, E.A.: Breadth-first heuristic search. Artif. Intell. **170**(4–5), 385–408 (2006)
98. Lewis, R.M., Torczon, V., Trosset, M.W.: Direct search methods: then and now. J. Comput. App. Math. **124**(1–2), 191–207 (2000)
99. Kirkpatrick, S., Gelatt, C.D., Vecchi, M.P.: Optimization by simulated annealing. Science **220**(4598), 671–680 (1983)
100. Glover, F.: Tabu search: a tutorial. Interfaces **20**(4), 79–94 (1990)
101. Neagu, B.C., Gavrilaş, M., Pentiuc, R.D., Hopulele, E.: Optimal placement of energy storage systems in microgrids using a PSO based approach. In: IEEE PES Innovative Smart Grid Technologies Europe (ISGT-Europe), Bucharest, Romania, pp. 1–5 (2019)
102. Metallinos, K.S., Papadopoulos, T.A., Charalambous, C.A.: Derivation and evaluation of generic measurement-based dynamic load models. Electr. Power Syst. Res. **140**, 193–200 (2016)
103. Karlsson, D., Hill, D.J.: Modelling and identification of nonlinear dynamic loads in power systems. IEEE Trans. Power Syst. **9**(1), 157–166 (1994)
104. Vahabzadeh, A., Kasaeian, A., Monsef, H., Aslani, A.: A fuzzy-SOM method for fraud detection in power distribution networks with high penetration of roof-top grid-connected PV. Energies **13**, 1287 (2020)
105. Kazmi, S.A.A., Shahzad, M.K., Khan, A.Z., Shin, D.R.: Smart distribution networks: a review of modern distribution concepts from a planning perspective. Energies **10**, 501 (2017)
106. International Energy Agency. Technology Road Map Smart Grids, April 2011. https://webstore.iea.org/download/direct/642. Accessed 24 June 2020
107. Grigoras, G., Ivanov, O., Gavrilas, M.: Customer classification and load profiling using data from Smart Meters. In: 12th Symposium on Neural Network Applications in Electrical Engineering (NEUREL), pp. 73–78 (2014)

108. Neagu, B.C., Georgescu, G., Gusa, M.D.: Load curves characteristics of consumers supplied from electricity repartition and distribution public systems. Bull. Polyt. Instit. Iasi **52**(1), 141–157 (2011)
109. Dabhi, D., Pandya, K.: Metaheuristic optimization algorithm for day-ahead energy resource management (ERM) in microgrid environment of power system. Recent Adv. Commun. Infrastruct. 115–125 (2020)
110. Neagu, B.C., Grigoraş, G., Scarlatache, F.: Outliers discovery from Smart Meters data using a statistical based data mining approach. In: 10th International Symposium on Advanced Topics in Electrical Engineering (ATEE), pp. 555–558 (2017)
111. Bouchekara, H.R.E.H., Abido, M.A.: Optimal power flow using differential search algorithm. Electr. Power Compon. Syst. **42**(15), 1683–1699 (2014)
112. Grigoraş, G., Neagu, B.-C., Gavrilaş, M., Triştiu, I., Bulac, C.: Optimal phase load balancing in low voltage distribution networks using a smart meter data-based algorithm. Mathematics **8**, 549 (2020)
113. Warid, W.: Optimal power flow using the AMTPG-Jaya algorithm. Appl. Soft Comput. 106252 (2020)
114. Li, Z., Cao, Y., Dai, L.V., Yang, X., Nguyen, T.T.: Optimal power flow for transmission power networks using a novel metaheuristic algorithm. Energies **12**, 4310 (2019)
115. Balasubbareddy, M., Dwivedi, D., Sathish, D.: Optimal power flow solution using spotted hyena optimization algorithm. Pramana Res. J. **9**(3), 147–158 (2019)
116. Deng, Z., Rotaru, M.D., Sykulski, J.K.: Kriging assisted surrogate evolutionary computation to solve optimal power flow problems. IEEE Trans. Power Syst. **35**(2), 831–839 (2019)
117. Moradi, H., Esfahanian, M., Abtahi, A., Zilouchian, A.: Modeling a hybrid microgrid using probabilistic reconfiguration under system uncertainties. Energies **10**, 1430 (2017)
118. Flaih, F.M.F., Lin, X., Abd, M.K., Dawoud, S.M., Li, Z., Adio, O.S.: A new method for distribution network reconfiguration analysis under different load demands. Energies **10**, 455 (2017)
119. Coroamă, I., Chicco, G., Gavrilaş, M., Russo, A.: Distribution system optimization with intra-day network reconfiguration and demand reduction procurement. Electr. Power Syst. Res. **98**, 29–38 (2013)
120. Ivanov, O., Neagu, B.-C., Grigoras, G., Gavrilas, M.: Optimal capacitor bank allocation in electricity distribution networks using metaheuristic algorithms. Energies **12**, 4239 (2019)
121. Neagu, B.C., Ivanov, O., Gavrilaş, M.: A comprehensive solution for optimal capacitor allocation problem in real distribution networks. In: Proceedings of the Conference on Electromechanical and Power System (SIELMEN), Iaşi, Romania, pp. 565–570 (2017)
122. Neagu, B.C., Ivanov, O., Gavrilas, M.: Voltage profile improvement in distribution networks using the whale optimization algorithm. In: International Conference on Electronics, Computers and Artificial Intelligence (ECAI 2017), Targoviste, Romania (2017)
123. Stanelyte, D., Radziukynas, V.: Review of voltage and reactive power control algorithms in electrical distribution networks. Energies **13**, 58 (2020)
124. Matei, G.G., Neagu, B.C., Gavrilas, M.: Optimal voltage control based on a modified line drop compensation method in distribution systems. In: Proceedings of the 2018 IEEE International Conference on Environment and Electrical Engineering and 2018 IEEE Industrial and Commercial Power Systems Europe, Palermo, Italy (2018)
125. Neagu, B.C., Gavrilas, M., Grigoras, G., Ivanov, O.: Voltage control in microgrids in the presence of small-scale renewable energy source. In: 2019 International Conference on Electromechanical and Energy Systems (SIELMEN), pp. 1–4 (2019)
126. Brenna, M., et al.: Automatic distributed voltage control algorithm in smart grids applications. IEEE Trans. Smart Grid **4**, 877–885 (2012)

127. Toma, R., Gavrilas, M.: Wind farm optimal grid integration based on voltage stability assessment. In: 2019 11th International Symposium on Advanced Topics in Electrical Engineering (ATEE), Bucharest, Romania, pp. 1–6 (2019)
128. Ozmen, O., et al.: Power grid simulation testbed for transactive energy management systems. Sustainability **12**, 4402 (2020)
129. Neagu, B.C., Grigoraş, G., Ivanov, O.: An efficient peer-to-peer based blockchain approach for prosumers energy trading in microgrids. In: 8th International Conference on Modern Power Systems (MPS), pp. 1–4 (2019)
130. Neagu, B.-C., Ivanov, O., Grigoras, G., Gavrilas, M.: A new vision on the prosumers energy surplus trading considering smart peer-to-peer contracts. Mathematics **8**, 235 (2020)
131. González-Romera, E., et al.: Secondary control for storage power converters in isolated nanogrids to allow peer-to-peer power sharing. Electronics **9**, 140 (2020)
132. Cortes-Arcos, T., Bernal-Agustín, J.L., Dufo-López, R., Lujano-Rojas, J.M., Contreras, J.: Multi-objective demand response to real-time prices (RTP) using a task scheduling methodology. Energy **138**, 19–31 (2017)
133. Veras, J.M., et al.: A multi-objective demand response optimization model for scheduling loads in a home energy management system. Sensors **18**, 3207 (2018)
134. Yang, S., Zeng, D., Ding, H., Yao, J., Wang, K., Li, Y.: Multi-objective demand response model considering the probabilistic characteristic of price elastic load. Energies **9**, 80 (2016)
135. Kriukov, A., Gavrilas, M.: Smart energy management in distribution networks with increasing number of electric vehicles. In: 2014 International Conference and Exposition on Electrical and Power Engineering, pp. 1039–1044 (2014)
136. Hershkovitz, B., Cohen, M., Sadot, E., Straschnov, Y.: Estimation and management of loads in electric vehicle networks. U.S. Patent Application No. 14/238,709 (2015)
137. Hasan, K.N., Muttaqi, K.M., Borboa, P., Scira, J., Zhang, Z., Leishman, M.: Measurement-based electric vehicle load profile and its impact on power system operation. In: 2019 9th International Conference on Power and Energy Systems, Perth, Australia, pp. 1–6 (2019)
138. Haidar, A., Muttaqi, K.: Impact assessment of electric vehicle demand through load modeling. In: Australasian Universities Power Engineering Conference (2013)
139. Dharmakeerthi, C., Mithulananthan, M., Saha, T.: Modeling and planning of EV fast charging station in power grid. In: IEEE PES General Meeting (2012)
140. Haidar, A., Muttaqi, K.: Behavioral characterization of electric vehicle charging loads in a distribution power grid through modeling of battery chargers. IEEE Trans. Ind. Appl. **52**(1), 483–492 (2016)
141. Sortomme, E., Negash, A., Venkata, S., Kirschen, D.: Voltage dependent load models of charging electric vehicles. In: IEEE PES General Meeting, pp. 1–5 (2013)
142. Rahmani-Andebili, M.: Distributed generation placement modeling feeder's failure rate and customer's load type. IEEE Trans. Ind. Electron. **63**(3), 1598–1606 (2016)
143. Rahmani-Andebili, M.: Reliability and economic-driven switchable capacitor placement in distribution network. IET Gener. Transm. Distrib. **9**(13), 1572–1579 (2015)
144. Rahmani-Andebili, M.: Simultaneous placement of DG and capacitor in distribution network. Electr. Power Syst. Res. **131**, 1–10 (2016)
145. Sairam, S.: Analysis of ZIP load modeling in power transmission system. Int. J. Control Autom **11**, 11–24 (2018)
146. Ram, S.S., Daram, S.B., Venkataramu, P.S., Nagaraj, M.S.: Comparative study of SVC, TCSC and HPFC using ZIP load modeling. Int. J. Electron. Electr. Comput. Syst. **7**(4), 595–611 (2018)

Analytical Modelling Approach of Photovoltaic Curves: Analysis and Comparison

Mohamed Louzazni[1](✉) and Brahim Belmahdi[2]

[1] Science Engineer Laboratory for Energy, National School of Applied Sciences, Chouaib Doukkali University, El Jadida, Morocco
[2] ETEE, Faculty of Science, Abdelmalek Essaadi University, B.P. 1717, Tetuan, Morocco

Abstract. This paper aims to present a set of approximate analytical solutions for solar cells and the I-V curve. This study is based on the Shockley equation to approaching the parameters and a power equation to estimate the characteristics curve. The proposed technique needs only data from solar cell characteristics, i.e., open-circuit voltage, short circuit current and maximum power point current and voltage. The accuracy of the presented theoretical I-V and P-V curves is verified by experimental data for solar cell RTC France Company, Photovoltaic module Photowatt-PWP 201, Polycrystalline panel STP6-120/36 and Monocrystalline panel STM6-40/36. In the end, the approximated curves are close to the experimental data and the power model can be used in other technologies.

Keywords: Solar cell · Shockley equation · Maximum power point · Approximated curve · Photovoltaic module

1 Introduction

Recently, several models have been presented to describe the behaviour of the current-voltage equation of solar cells. Furthermore, the equation characteristics of solar cells under illumination are complex and nonlinear. Besides, different models and technical methods were developed to modelling the nonlinear behaviour of current-voltage [1–3]. The implicit equation of the current-voltage characteristic of the solar cell, with the inclusion of parasitic phenomena like space charge limited current, series and shunt resistances [4] is presented in Fig. 1.

The mathematical equation of the solar cell single diode model is chosen and modelled by [4, 5].

$$I = I_{ph} - I_s\left[\exp(\frac{V + I \cdot R_s}{\alpha V_{th}}) - 1\right] - \frac{V + I \cdot R_s}{R_p} \tag{1}$$

where I_{ph} present the photo-current and collects the contributions of the two quasi-neutral zones and the depletion zone, I_d is the diode current, R_p and R_s the series and shunt resistance. The quality factor of the diode is represented by α, while $V_{th} = KT/q$ present the thermal voltage, K is the Boltzmann constant, T is the temperature and q are the electron charge.

L. Moldovan and A. Gligor (Eds.): Inter-Eng 2021, LNNS 386, pp. 562–572, 2022.
https://doi.org/10.1007/978-3-030-93817-8_50

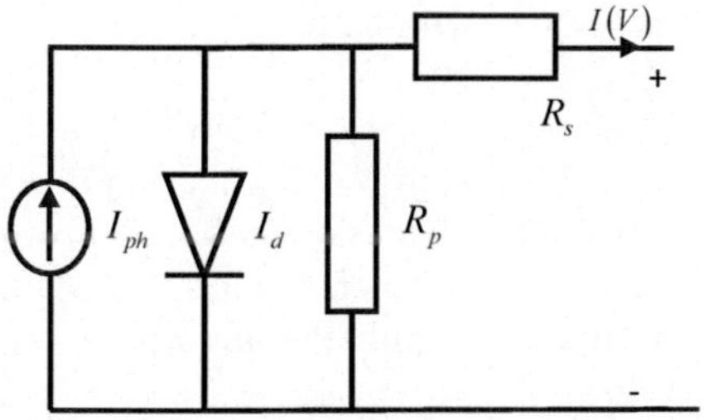

Fig. 1. The standard model of a single junction solar cell.

The presented models vary from simple assumptions to advanced models incorporating many physical variables and complexity. Moreover, Eq. (1) is widely used to explain the behaviour and to optimize the parameters of the solar cell. The relation between solar cell current and voltage has both implicit and nonlinear equations [6–9], depends on several factors such as module temperature, solar radiation and its distribution, cable losses, dust accumulation, shading and soiling [10, 11]. Moreover, many mathematical models have been developed in the literature to describe the electric behaviour of solar cells with a different level of complexity. They can be classified into implicit and explicit models [12, 13]. A different physical model is compared on solar cell power output prediction [14] and an available model of solar cell is presented [15]. These models differ mainly by the number of diodes, the presence or absence of a parallel resistor. Furthermore, the single diode model is the most commonly used. The five and seven parameter model evaluates the photocurrent, the saturation current, the series and shunt resistors and the quality factor of the diode.

The explicit form of Eq. (1) has been presented in [16, 17] to transform and simplify the calculation. Furthermore, some semi-empirical approach is also used on solar cells [18]. Recently, an explicit model was proposed in [19–22], which is validated with a wide variety of solar cells. An explicit expression of Eq. (1) has been presented in [23] based on analytical formulation and Lambert W-function. The Taylor's series expansion [4] padé approximants [20] has been used to approximate the implicit equation and to extract the solar cell parameters. A simple analytical solar cell *I-V* curve model has been presented and study [6] determined by four parameters and validated in dark and in illumination conditions. Moreover, a neural-analytical method [24] was proposed to determine the silicon and plastic solar cells and module parameters. The presented methods and models have the drawback and the modelling of the solar cell need more precision.

In this paper, we present an analytical modelling approach to solar cell parameters. While the *I-V* curve is approximated by two equations based on open circuit voltage, short circuit current and shape parameters. The predicted curves of current-voltage and power-voltage have been compared with experimental data for each technology. The shape parameters vary from solar cell to photovoltaic module and photovoltaic panel, also with the technologies used. In the end, the individual absolute error is calculated. The compared results show the performance of the used model and the predicted curve and are close to the experimental.

2 Analytical Parameters Approach

The characteristic current-voltage of solar cell is implicit and nonlinear. Usually, the numerical iterative methods are needed to solve the implicit equation. Therefore, an explicit mathematical approximation expression of Eq. (1) is convenient and necessary in simulation and to approach the parameter. The Eq. (1) can be translated into simple explicit model by series of algebraic manipulations with assuming $I_{ph} \gg I_s$, it can be argued the following equations:

$$I = I_{ph} - \frac{R_s}{R_p} I - \frac{V}{R_p} - I_s \left[\exp\left(\frac{V + R_s I}{\alpha V_t} \right) - 1 \right] \tag{2}$$

$$I = I_{ph} - \frac{R_s}{R_p} I - \frac{V}{R_p} - I_s \exp\left(\frac{V + R_s I}{\alpha V_t} \right) \tag{3}$$

$$I_{pv} = \frac{I_{ph}}{1 + \frac{R_s}{R_p}} - \frac{V_{pv}/R_p}{1 + \frac{R_s}{R_p}} - \frac{I_s}{1 + \frac{R_s}{R_p}} \exp\left(\frac{V_{pv} + R_s I_{pv}}{\alpha V_t} \right). \tag{4}$$

At the short circuit point: $I_{pv} = I_{sc}$ and $V_{pv} = 0$. I_{sc} Eq. (1) can thus be represented as follows:

$$I_{sc} = \frac{I_{ph}}{1 + \frac{R_s}{R_p}} - \frac{I_s}{1 + \frac{R_s}{R_p}} \exp\left(\frac{R_s I_{sc}}{\alpha V_t} \right) \tag{5}$$

$$I_{sc} = \frac{I_{ph}}{1 + \frac{R_s}{R_p}} \left[1 - \left(\frac{I_s}{I_{ph}} \right) \exp\left(\frac{R_s I_{sc}}{\alpha V_t} \right) \right]. \tag{6}$$

Neglecting I_s / I_{ph}:

$$I_{sc} \approx I_{ph} \left(1 + \frac{R_s}{R_p} \right)^{-1} \tag{7}$$

At the open circuit point: $V_{pv} = V_{oc}$ and $I_{pv} = 0$ Substituting the values into Eq. (1) yields:

$$0 = I_{ph} - \frac{V_{oc}}{R_p} - I_s \exp\left(\frac{V_{OC}}{\alpha V_t} \right) \tag{8}$$

$$I_{ph} = \frac{V_{oc}}{R_p} + I_s \left[\exp\left(\frac{V_{OC}}{\alpha V_t} \right) - 1 \right]. \tag{9}$$

The simplify I_{ph}:

$$I_{ph} \approx I_{sc} \left(1 + \frac{R_s}{R_p} \right). \tag{10}$$

Substituting Eq. (6) into Eq. (9) leads to an equation for I_s:

$$I_s \approx \frac{I_{sc}\left(1+\frac{R_s}{R_p}\right)-\frac{V_{oc}}{R_p}}{\exp\left(\frac{V_{OC}}{\alpha V_t}\right)-\exp\left(\frac{R_s I_{sc}}{\alpha V_t}\right)}. \tag{11}$$

Since, $\exp\left(\frac{V_{OC}}{\alpha V_t}\right) \gg \exp\left(\frac{R_s I_{sc}}{\alpha V_t}\right)$, and $1 \gg \frac{R_s}{R_p}$ Eq. (11) is simplified to:

$$I_s \approx \left(I_{sc}-\frac{V_{oc}}{R_p}\right)\exp\left(-\frac{V_{OC}}{\alpha V_t}\right). \tag{12}$$

We consider the differential equation of Eq. (1), then:

$$\frac{dV}{dI} = -R_s - \frac{\alpha V_{th} R_p}{\alpha V_{th} + I_s R_p \cdot \exp\left(\frac{V+I\cdot R_s}{\alpha V_{th}}\right)}. \tag{13}$$

Substitute the values at the short circuit $(0, I_{sc})$ and open circuit $(V_{oc}, 0)$ points into Eq. (13) (5), thus:

At the short circuit point:

$$\left.\frac{dV}{dI}\right|_{\cdot\cdot V=0} = -R_s - \frac{\alpha V_{th} R_p}{\alpha V_{th} + I_s R_p \cdot \exp\left(\frac{I_{sc}\cdot R_s}{\alpha V_{th}}\right)} \tag{14}$$

At the open circuit point:

$$\left.\frac{dV}{dI}\right|_{I=0} = -R_s - \frac{\alpha V_{th} R_p}{\alpha V_{th} + I_s R_p \cdot \exp\left(\frac{V_{oc}}{\alpha V_{th}}\right)}. \tag{15}$$

Because of the single peak characteristics of a photovoltaic module output at STC conditions, the value of the output power differential equation at the maximum power point is equal to zero. Therefore,

$$\left.\frac{dP}{dV}\right|_{MPP} = I_m + V_m \cdot \left.\frac{dV}{dI}\right|_{MPP} = 0 \tag{16}$$

that is

$$\left.\frac{dV}{dI}\right|_{MPP} = -\frac{V_m}{I_m} \tag{17}$$

Combine Eqs. (13) and (17) then,

$$\frac{V_m}{I_m} = R_s + \frac{\alpha V_{th} R_p}{\alpha V_{th} + I_s R_p \cdot \exp\left(\frac{V_m + I_m \cdot R_s}{\alpha V_{th}}\right)}. \tag{18}$$

3 Current-Voltage Curve Estimation

A simple expression for modeling and representing the behavior of solar cell has been proposed. Where an explicit relationship between the current and voltage is given by the following mathematical equation [6].

$$I = I_{sc}\left[1 - \left(\frac{V}{V_{oc}}\right)^{n}\right]^{\frac{1}{m}} \tag{19}$$

$$V = V_{oc}\left[1 - \left(\frac{I}{I_{sc}}\right)^{m}\right]^{\frac{1}{n}} \tag{20}$$

where: m and n represent the shape parameter, I_{sc} and V_{oc} are the short circuit current and open circuit voltage depends on the cell temperature and can be written as [25]

$$I_{sc} = I_{sc}(STC) + \alpha(T_c - 25) \tag{21}$$

$$V_{oc} = V_{oc}(STC) + \beta(T_c - 25) \tag{22}$$

where: $I_{sc}(STC)$ and $V_{oc}(STC)$ are the open-circuit voltage at STC, α and β are the current and voltage temperature coefficient at STC and T_c the cell temperature.

At the maximum power point, we can write:

$$I_{mpp} = I_{sc}\left(\frac{n}{n+m}\right)^{\frac{1}{m}} \tag{23}$$

$$V_{mpp} = V_{oc}\left(\frac{m}{n+m}\right)^{\frac{1}{n}} \tag{24}$$

and the fill factor is given by:

$$FF = \frac{I_{mpp} \cdot \mathrm{V}_{mpp}}{I_{sc} \cdot V_{oc}} = \left(\frac{m}{n+m}\right)^{\frac{1}{n}} \cdot \left(\frac{n}{n+m}\right)^{\frac{1}{m}}. \tag{25}$$

The value of n and m are determined from empirical data to plot the characteristics of current voltage. The Eq. (19) is tested in Fig. 2 with different variations of shapes parameters n and m.

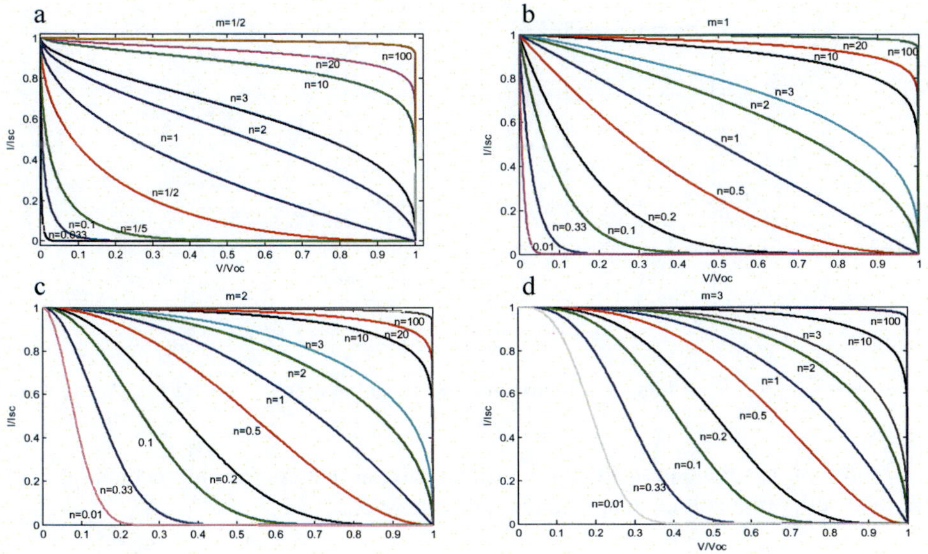

Fig. 2. Effect of variation of shape parameters m and n of Eq. (20), for (a) m = 1/2, (b) m = 1, (c) m = 2, (d) m = 3.

4 Discussion and Validation

To determine the good and the validation of the Eq. (19), we tested and compared it with experimental and theoretical data. The Fig. 3 (a), show the characteristics current-voltage of solar cell using Eq. (19) with different values of shape parameters. Furthermore, the characteristic curve of a-Si, a-SiGe and polymer cells with bias-dependent photocurrent for different varieties is plotted in Fig. 3 (b) [13]. Moreover, a comparison from two the two-plotted curve shows the numerical calculation is agreed with the measured data.

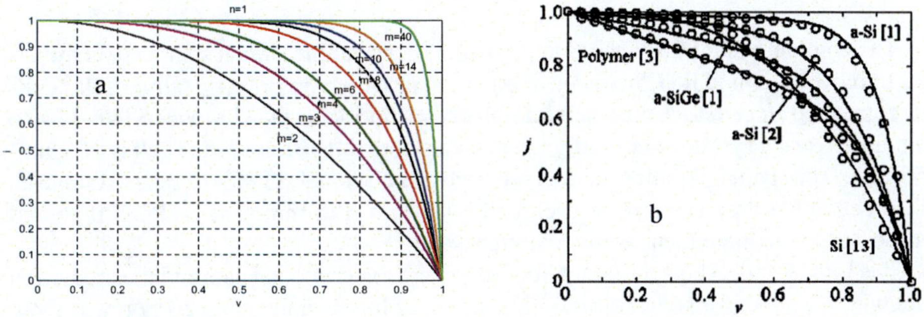

Fig. 3. Compared results of the curves for different values of m and, (a) Results by Eq. (19, (b) Results proposed [13].

Figure 4 shows the current-voltage curve approximated with n = 12 and m = 0.9, of a-Si [13].

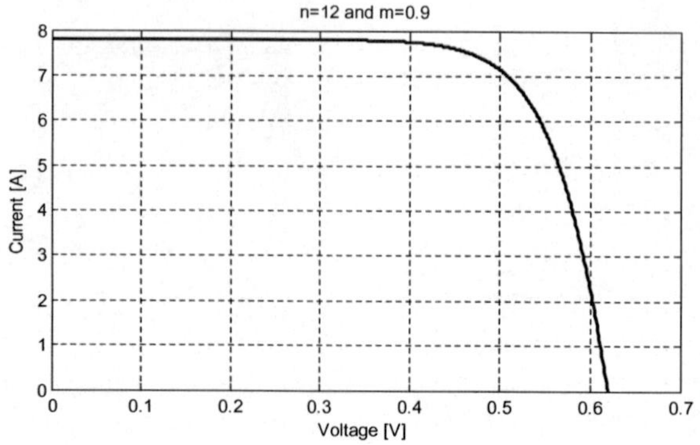

Fig. 4. I-V curve for n = 12 and m = 0.9.

Table 1 shows the datasheet and the calculated parameters for four technologies of the solar cell, photovoltaic module and photovoltaic panel.

Table 1. Electrical characteristics of the used PV modules at STC.

Technologies	I_{sc} (A)	V_{oc} (V)	n	m	Fill factor	
					Datasheet	Estimated
Solar cell RTC France Company [29]	0.7603	0.5728	11	0.9	0.7135	0.72462
Photovoltaic module Photowatt-PWP 201 [29]	1.03	16.778	11	0.5	0.668	0.68801
Polycrystalline: STP6–120/36 [30]	7.48	19.21	10	0.7	–	–
Monocrystalline: STM6–40/36 [30]	1.663	21.02	10	1.3	–	–
GaSb solar cells 1991 [31]	7.8	0.62	11	9	–	–

The commercial RTC France Company mono-crystalline silicon solar cell, commercial photovoltaic module (Photowatt-PWP 201) in which 36 polycrystalline solar cells are connected in series. A commercial polycrystalline solar panel model STP6-120/36 contains 36 polycrystalline cells aligned in series, the dimension of each cell is 156 mm-156 mm. This panel. A solar monocrystalline panel model STM6-40/36 is composed of 36 cells in series. The size of each cell is 38 mm–128 mm. The STP6-120/36 and STM6-40/36 are manufactured by Schutten Solar.

The Figs. 5 and 6 show the compared curve with experimental data and the prediction data with Eq. (19), for solar cell RTC France a, d Photovoltaic module Photowat-PWP 201. From the two figures, we can show the predicted curves are close to experimental data.

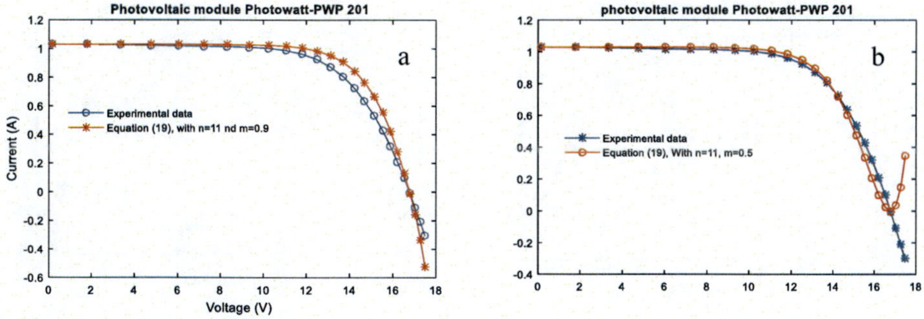

Fig. 5. Comparison of current-voltage: (a) RTC France Company solar cell, (b) photovoltaic module Photowatt-PWP 201.

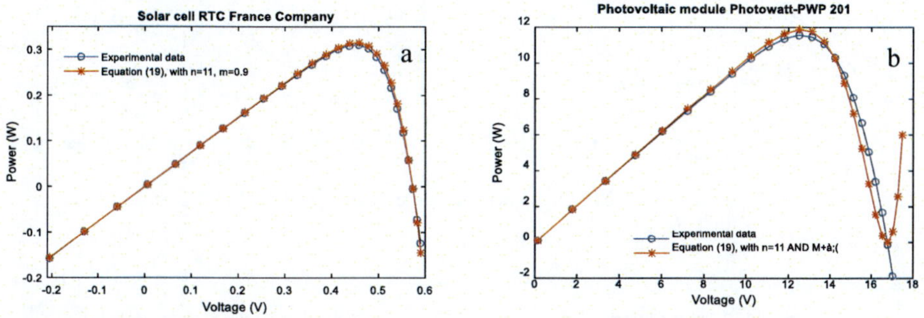

Fig. 6. Comparison of P-V curve, (a) RTC France Company solar cell, (b) photovoltaic module Photowatt-PWP 201.

Figure 7 show the compared current-voltage and power-voltage for; (a) commercial polycrystalline solar panel model STP6-120/36, (b) solar monocrystalline panel model STM6-40/36. The results from Fig. 6 show the validity of the Eq. (19) in the prediction of photovoltaic panel curves.

The Individual Absolute Error (Eq. 26) is evaluated and calculated and plotted in Fig. 8.

$$IAE = I_{cal} - I \exp \tag{26}$$

with I_{cal} and I_{exp} presents the calculated and experimental current.

The used equation is tested for GaSb solar cells 1991 [26] curve calculated by Fraas et al. Figure 9 shows the compared curve using Eq. (19) with m = 11 and n = 9. The curve plotted in Fig. 9 appears the curve for negative values of current-voltage, while Fraas et al. ignored it.

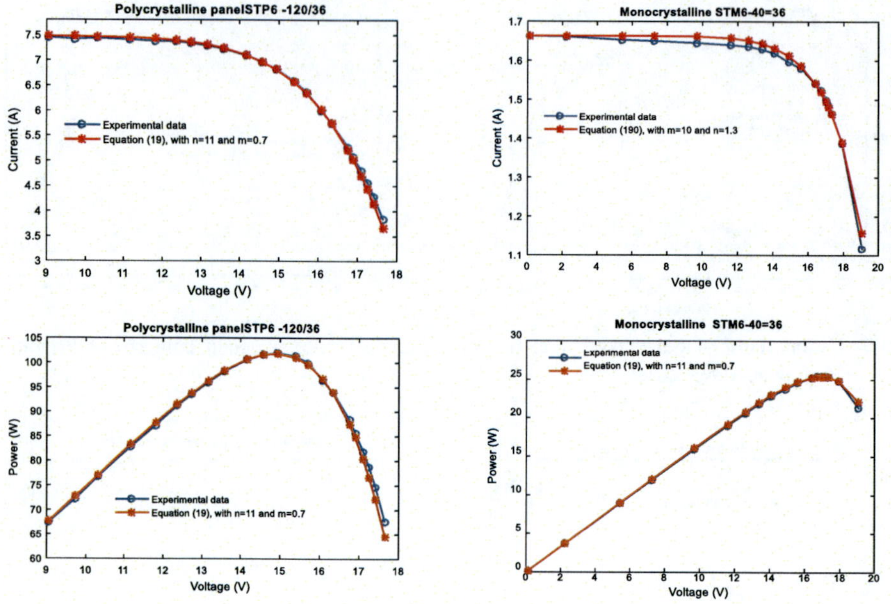

Fig. 7. Power-Voltage curve.

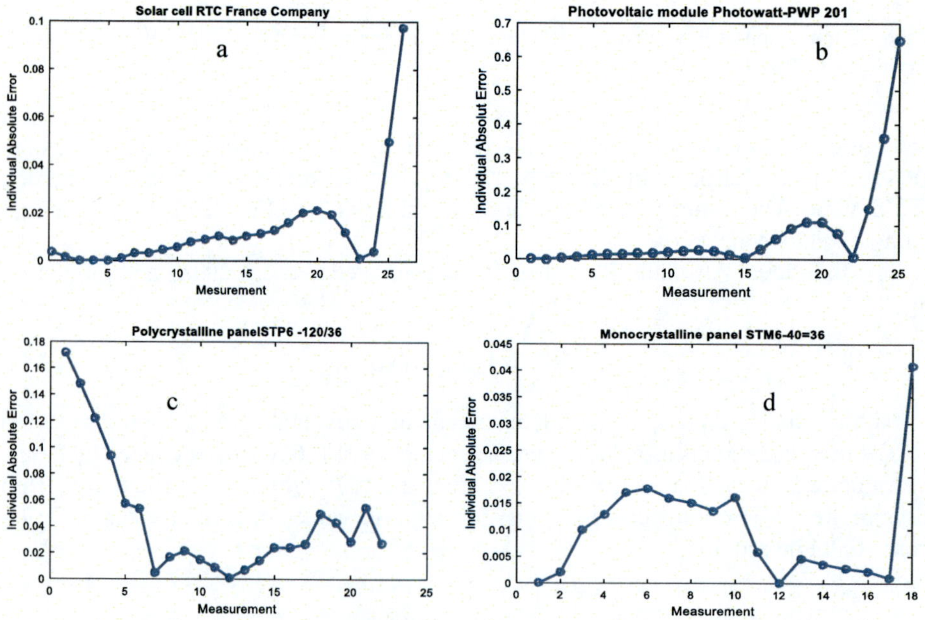

Fig. 8. Individual Absolute Error, (a) RTC France Company solar cell, (b) photovoltaic module Photowatt-PWP 201, (c) STP6-120/36, (d) STM6-40/36.

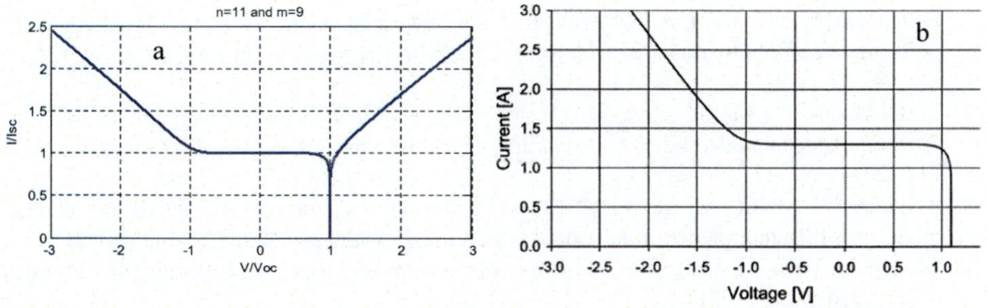

Fig. 9. I-V curve of GaSb solar cell modelled by, (a) Eq. (19), (b) curves given by Fraas et al. [31].

5 Conclusion

In this paper, we study an analytical mathematical implicit model for predicting the current-voltage and power voltage curves for different technologies. Furthermore, the equivalent electrical model of solar cells has been presented to approaching the parameters. Moreover, the analytical model is based on open-circuit voltage, short circuit current and the shape parameters capable of representing solar cells also in the negative voltage ranges. On the other way, and to add more validity of the used model is tested for solar cell, photovoltaic module and photovoltaic panel of different technologies. In the end, the calculated results are close to the experimental data. So, the used model can be present the behavior a variety of solar cell technologies at any given temperature and irradiance.

References

1. Ghani, F., Rosengarten, G., Duke, M., Carson, J.K.: The numerical calculation of single-diode solar-cell modelling parameters. Renew. Energy **72**, 105–112 (2014)
2. Toledo, F.J., Blanes, J.M.: Geometric properties of the single-diode photovoltaic model and a new very simple method for parameters extraction. Renew. Energy **72**, 125–133 (2014)
3. Pallarès, J., Cabré, R., Marsal, L.F., Schropp, R.E.I.: A compact equivalent circuit for the dark current-voltage characteristics of nonideal solar cells. J. Appl. Phys. **100**(8), 84513 (2006)
4. Xian Lun, S., et al.: An explicit approximate I-V characteristic model of a solar cell based on padé approximants. Sol. Energy **92**, 147–159 (2013)
5. Karatepe, E., Boztepe, M., Çolak, M.: Development of a suitable model for characterizing photovoltaic arrays with shaded solar cells. Sol. Energy **81**(8), 977–992 (2007)
6. Fathabadi, H.: Novel neural-analytical method for determining silicon/plastic solar cells and modules characteristics. Energy Convers. Manag. **76**, 253–259 (2013)
7. Mares, O., Paulescu, M., Badescu, V.: A simple but accurate procedure for solving the five-parameter model. Energy Convers. Manag. **105**(Suppl. C), 139–148 (2015)
8. Pelap, F.B., Dongo, P.D., Kapim, A.D.: Optimization of the characteristics of the PV cells using nonlinear electronic components. Sustain. Energy Technol. Assess. **16**(Suppl. C), 84–92 (2016)
9. Maghami, M.R., Hizam, H., Gomes, C., Radzi, M.A., Rezadad, M.I., Hajighorbani, S.: Power loss due to soiling on solar panel: a review. Renew. Sustain. Energy Rev. **59**, 1307–1316 (2016)

10. Fouad, M.M., Shihata, L.A., Morgan, E.I.: An integrated review of factors influencing the performance of photovoltaic panels. Renew. Sustain. Energy Rev. **80**(Suppl. C), 1499–1511 (2017)
11. Massi Pavan, A., Mellit, A., Lughi, V.: Explicit empirical model for general photovoltaic devices: experimental validation at maximum power point. Sol. Energy **101**(Suppl. C), 105–116 (2014)
12. Boutana, N., Mellit, A., Lughi, V., Massi Pavan, A.: Assessment of implicit and explicit models for different photovoltaic modules technologies. Energy **122**, 128–143 (2017)
13. Dolara, A., Leva, S., Manzolini, G.: Comparison of different physical models for PV power output prediction. Sol. Energy **119**, 83–99 (2015)
14. Hasan, M.A., Parida, S.K.: An overview of solar photovoltaic panel modeling based on analytical and experimental viewpoint. Renew. Sustain. Energy Rev. **60**(Suppl. C), 75–83 (2016)
15. Karmalkar, S., Haneefa, S.: A physically based explicit J-V model of a solar cell for simple design calculations. IEEE Electron. Device Lett. **29**(5), 449–451 (2008)
16. Jain, A., Kapoor, A.: Exact analytical solutions of the parameters of real solar cells using Lambert W-function. Sol. Energy Mater. Sol. Cells **81**(2), 269–277 (2004)
17. De Soto, W., Klein, S.A., Beckman, W.A.: Improvement and validation of a model for photovoltaic array performance. Sol. Energy **80**(1), 78–88 (2006)
18. Das, A.K.: An explicit J-V model of a solar cell for simple fill factor calculation. Sol. Energy **85**(9), 1906–1909 (2011)
19. Das, A.K., Karmalkar, S.: Analytical derivation of the closed-form power law J-V model of an illuminated solar cell from the physics based implicit model. IEEE Trans. Electron. Devices **58**(4), 1176–1181 (2011)
20. Saetre, T.O., Midtgård, O.-M., Yordanov, G.H.: A new analytical solar cell I-V curve model. Renew. Energy **36**(8), 2171–2176 (2011)
21. Das, A.K.: Analytical derivation of equivalent functional form of explicit J-V model of an illuminated solar cell from physics based implicit model. Sol. Energy **103**, 411–416 (2014)
22. Cubas, J., Pindado, S., de Manuel, C.: Explicit expressions for solar panel equivalent circuit parameters based on analytical formulation and the lambert W-function. Energies **7**(7), 4098–4115 (2014)
23. Lun, S., Du, C., Guo, T., Wang, S., Sang, J., Li, J.: A new explicit I-V model of a solar cell based on Taylor's series expansion. Sol. Energy **94**, 221–232 (2013)
24. Luque, A., Hegedus, S.: Handbook of Photovoltaic Science and Engineering. Wiley, Chichester (2011)
25. Easwarakhanthan, T., Bottin, J., Bouhouch, I., Boutrit, C.: Nonlinear minimization algorithm for determining the solar cell parameters with microcomputers. Int. J. Sol. Energy **4**(1), 1–12 (1986)
26. Fraas, L.M., Avery, J.E., Gruenbaum, P.E., Sundaram, V.S., Emery, K., Matson, R.: Fundamental characterization studies of GaSb solar cells. In: The Conference Record of the Twenty-Second IEEE Photovoltaic Specialists Conference, vol. 1, pp. 80–84 (1991)

Pitch Angle Control of the PMSG Wind Turbine

Sabra Ahyaten[1,2](✉), Jalal El Bahaoui[1], Narjisse Amahjour[1], and Francisco Ortegón Gallego[2]

[1] Abdelmalek Essaadi University, 93 000 Tetouan, Morocco
ah.sabraahya@alum.uca.es
[2] Cádiz University, 11 510 Puerto Real, Cádiz, Spain

Abstract. The applications of wind generation progress are faster than any other renewable energy. During this rapid development of the wind power, the majority of the wind turbines has been operated with diverse generators. One way to generate an efficient power is to control the pitch angle of the wind turbine. This paper aims at studying the pith angle and turbine management control strategy with Permanent Magnet Synchronous Generator-based different conditions. This research has investigated a model of the PMSG Wind Turbine Pitch Angle Control technique. Three different controllers namely: no controller, PI and PID control techniques are compared. On the other hand, the PID parameters have been adjusted. A PI/ PID pitch angle controller is studied in this paper to ensure the stability of mechanical torque and response of wind power. The model is implemented in MATLAB/SIMULINK software to investigate the outlined concept and control strategy. The simulation with PID controller shows good result when comparing with the others cases of study, in terms of the advanced stability of the parameters.

Keywords: Wind turbine · Permanent Magnet Synchronous Generator · Control strategy

1 Introduction

Among the subjects that were investigated one finds identifying atmospheric pollution areas, evaluating wind energy and improving the installation of wind energy structures [1]. The wind power and solar energy are now being the focus of investigation as safe, economical and hygienic renewable energies. The properties of the airflow and the conditions of the atmospheric boundary layer (ABL), affect the performance of a wind turbine. To precisely calculate the amount of wind energy, the exact height distribution of wind speed for the proposed wind turbine installation site is essential [2]. The wind turbine has been the quickest increase due to the progress in greenhouse gas concentrations in the atmosphere [3], which has negative impact to our globe. Morocco is a big country with high natural resources. Among the several renewable sources, solar and wind are two of the significant sources for electric power production. It is possible to improve Morocco's electricity more by effectively utilizing wind potential.

L. Moldovan and A. Gligor (Eds.): Inter-Eng 2021, LNNS 386, pp. 573–583, 2022.
https://doi.org/10.1007/978-3-030-93817-8_51

In Wind Energy System (WES), the kinetic energy from wind is harvested using wind turbine blades and transformed into electrical energy. Among the most commonly used topologies for wind turbines is the Doubly Fed Induction Generator (DFIG) or Permanent Magnet Synchronous Generator (PMSG). Further, wind energy systems using PMSG have high power factors because there is no magnetizing current. In order, the use of PMSG has a high effect on the design of wind power applications. However, the PMSG system is not easy to control since as wind turbine is a non-linear in nature, therefore, whatever the type of wind turbine is considered, the control technique has a great influence on the overall performance of WES [3]. Fixed gain linear controllers such as Proportional-Integral (PI), Linear Quadratic Regulator (LQR), etc. The proportional-integral (PI) and proportional-integral-derivative (PID) controller is usually implemented firstly to control the Permanent Magnet Synchronous Machine (PMSM) systems in engineering applications due to its easy, clear operating and efficacy [4].

The researchers investigated the three types of classifications of control PID for machine systems [5]. The first category of PID gain tuning methods based on intelligent control schemes like fuzzy control neural-net or genetic methods. The second category of research is on how to pick PID gain using alternative control schemes such as dynamics controls or inverse optimal controls. The third category focuses on a direct stability study based on Lyapunov stability for PID gain selection.

According to [6], the proposed advanced pitch angle control strategy with neural network for variable speed wind turbine comparing to fuzzy logic and proportional integral control techniques show better control performances. Fuzzy-PID control based an improved pitch control algorithm in [7] progress the performance static and dynamic of wind turbine, also avoid the wind turbine speeding and occurred off-grid.

This paper provides a comparison of three cases studies for controlling the pitch angle such as: without controller, PI and PID controller to ensure the stability of mechanical torque and the performance of wind power quality.

2 Mathematical Model

2.1 Wind Turbine Modeling

The $C_p(\lambda, \beta)$ curve, which is normally provided by the manufacturers, describes the aerodynamics of the wind turbine. The power coefficient C_p, relates to the greatest mechanical power extracted from wind at its maximum value and, which is a function of the tip-speed ratio λ and pitch angle β. The mechanical power (P_m) and mechanical torque (T_m) extracted from the wind by the wind turbine can be represented in terms of for a given Cp [8].

$$P_m = \frac{1}{2} C_p(\lambda, \beta) \rho A V_w^3 \tag{1}$$

$$C_p(\lambda, \beta) = C_1 \left(\frac{C_2}{\lambda_i} - C_3 \beta - C_4 \beta^7 - C_5 \right) e^{\left(-\frac{C_6}{\lambda_i} \right)} \tag{2}$$

$$\frac{1}{\lambda_i} = \frac{1}{\lambda + 0.08\beta} - \frac{0.035}{1 + \beta^3} \quad (3)$$

$$\lambda = \frac{w_r}{V_\omega} R \quad (4)$$

$$T_m = \frac{P_m}{w_r} \quad (5)$$

where:

P_m: nominal mechanical output power of the turbine (W);
$\rho = 1.225$: air density (kg/m^3);
A: Turbine swept area (m^2);
V_w: Wind speed (m/s);
$C_p(\lambda, \beta)$: Power coefficient depends to pitch angle (deg) and tip speed ratio;
$C_1 = 0.5176$; $C_2 = 116$; $C_3 = 0.4$; $C_4 = 5$; $C_5 = 21$; $C_6 = 0.0068$;
λ_i: Tip speed ratio of the rotor blade tip speed of wind speed;
β: Pitch angle to display wind turbine power characteristics (deg);
W_r: Rotational velocity (p.u);
R: Radius of the rotation of each blade;
T_m: Mechanical torque (N·m) (Table 1).

Table 1. Wind turbine and generator parameter's.

Parameters	Value
Air density (kgm^2)	1.225
Cut-in wind speed (m/s)	4
Cut-out wind speed (m/s)	25
Base wind speed (m/s)	12
Generator rated power (W)	9.10^3

Figure 1 depicts the variation of the power coefficient Cp as the tip speed ratio (TSR) are illustrated below.

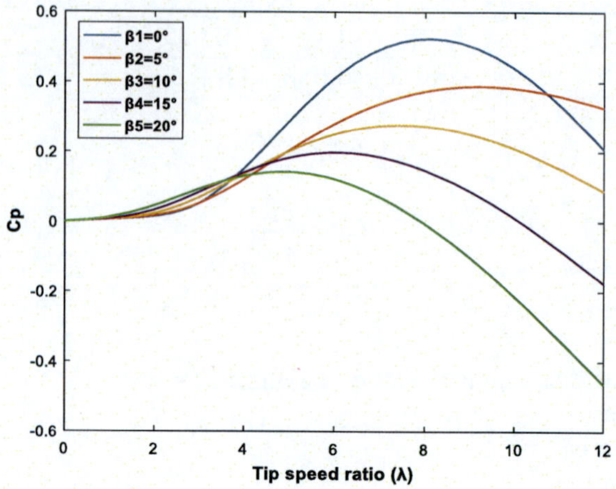

Fig. 1. C_p versus – λ curve.

2.2 Permanent Magnet Synchronous Generator

The advantages of a permanent magnetic synchronous generator (PMSG) are numerous and significant. Because of its small size, high power density, excellent dependability, and resilience, the PMSG is becoming increasingly popular. The gearless PMSG generator becomes more suited than others. since our study is centered on offshore wind power installation. This model is depicted in the d-q synchronous reference frame.

PMSG's voltage equations are as follows in Eqs. 6 to 8 [9]:

$$\frac{d}{dt}i_d = \frac{1}{L_d}V_d - \frac{R}{L_d}i_d + \frac{L_q}{L_d}p\omega_m i_q \tag{6}$$

$$\frac{d}{dt}i_q = \frac{1}{L_q}V_q - \frac{R}{L_q} - \frac{L_d}{L_q}p\omega_m i_d - \lambda p\frac{\omega_m}{L_q} \tag{7}$$

$$T_e = 1.5p\left[\lambda i_q + \left(L_d - L_q\right)i_d i_q\right] \tag{8}$$

where:

i_d and i_q are q and d axis currents;
L_d and L_q are d and q axis inductances;
V_d and V_q are d and q axis voltages;
R is the resistance of the stator windings;
p is number of pole pairs;
ω_m is the angular velocity of the rotor;
λ is the amplitude of the flux induced by PMSG of the rotor and the stator phase;
T_e is the electromagnetic torque.

3 Control of Pitch Angle

Wind turbine (WT) control objective dictate the moment of operation of each WT control system. These control objectives should be clearly defined to avoid any misunderstanding when studying WT control systems. The operational regions of wind turbines are dependent on control objectives [10]. Pitch angle controller is the most ways for improving many parameters. It is important to choose some variables, such as wind speed, generator speed and generator power. Among of this parameter in this study the mechanical torque and response of the power. Therefore, a model of the PMSG Wind Turbine's Pitch Angle Control technique studied with three operational controllers (Fig. 2) can be defined depending directly on wind speed. In part I, without controller, the wind speed is low so there's no electrical power production. Therefore, the losses in turbine operation are greater than the generated power. In this operational part the turbine should be stopped or set in inactive mode.

In part II, the wind speed increases with PI control: it increases to a power generation between the cut-in 4 m/s and the rated wind speed 12 m/s. this takes place when wind turbine is in partial-load mode. In part III, the wind exceeds rated speed with PID control as the wind turbine starts in full-load mode at 25m/s speed cut-off. To ensure operating within the safety constraints of generator speed and WT mechanical loads, power generation should be limited to the WT rated power [11].

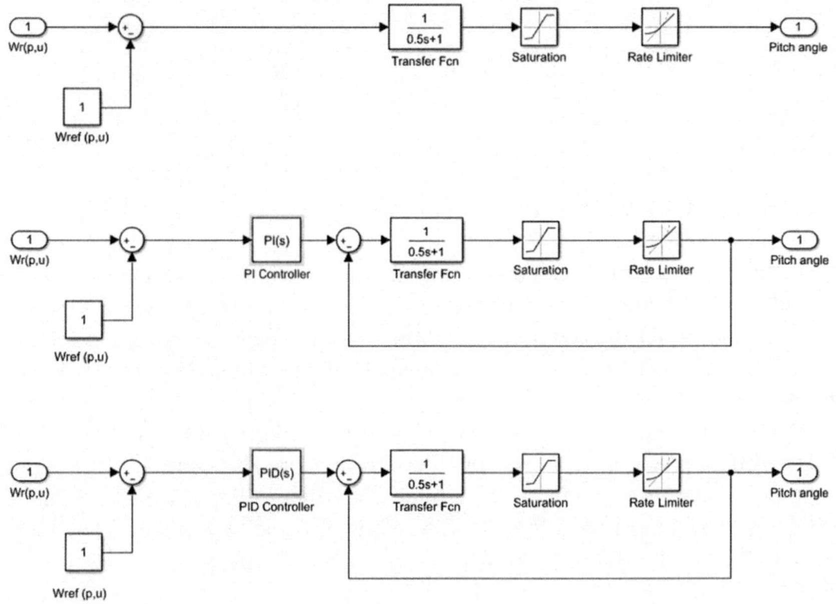

Fig. 2. Strategy of Pitch Control without controller (upper), PI controller (middle) and PID controller (lower).

Table 2. PI controller parameters.

Parameters	Value
k_p	41.8804097308638
k_i	12.7153520745356

Table 3. PID controller parameters.

Parameters	Value
k_p	79.3404493603136
k_i	37.0844551855851
k_d	−15.357125378536

Table 4. Time domain specification for unit step input without, PI and PID controller.

Time domain	Without	PI	PID
Rise Time (s)	10.135	4.852	2.143
Slew Rate (/ks)	2.088	190.728	387.213
Peak Overshoot (%)	5.851	9.259	21.642
Steady Overshoot (%)	0.531	14.198	27.612
State Error	0.75	0.25	0

4 Results and Discussion

First, it simulated the wind turbine system without controller to get the output. As the Fig. 2 pitch control without controller (upper). The unit step response of wind turbine pitch without controller is show in Fig. 3, which are the peak overshoot is 5% and the steady overshoot is very low, the time domain specification observed from the response in Table 4.

Second, it simulated the wind turbine system with PI controller is shown in Fig. 2 (middle) and the control parameters for PI controllers are shown in Table 2 and Table 4. The unit step response of wind turbine pitch controller in Fig. 4 we found that the PI controller has a short rise time 4.8%, peak overshoot 9% and high steady overshoot 14% compared to the wind turbine model without any controllers.

Third, the Simulink model of wind turbine pitch control system with PID controller is shown in Fig. 2 (lower). The control parameters for PID controller are shown in Table 3 and Table 4. The unit step response of wind turbine pitch control system is shown in Fig. 5. The time domain specification is observed from the response graphs lists in Table 4.

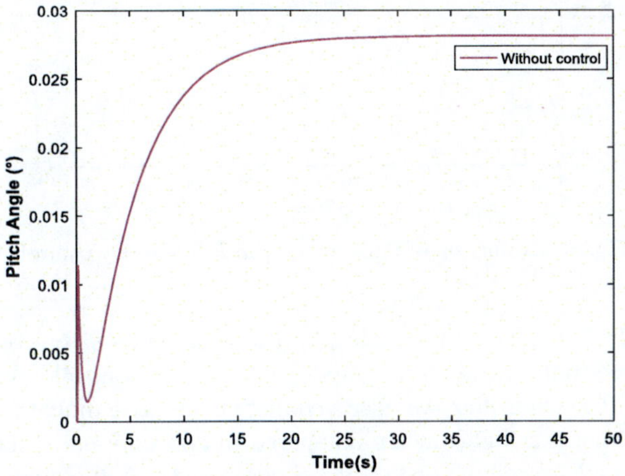

Fig. 3. Evaluation of Pitch angle without controller.

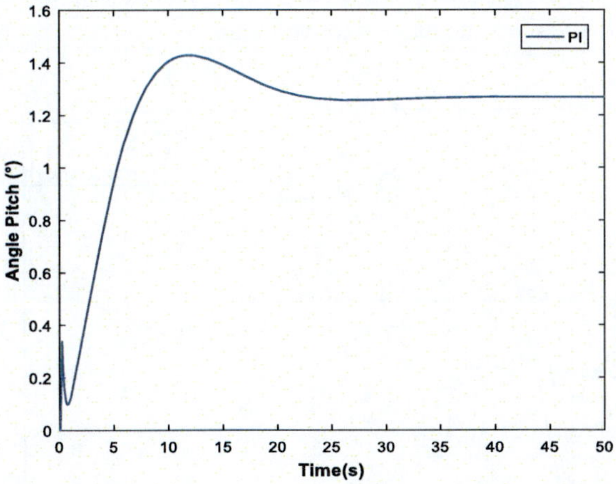

Fig. 4. Evaluation of Pitch angle with PI strategy control.

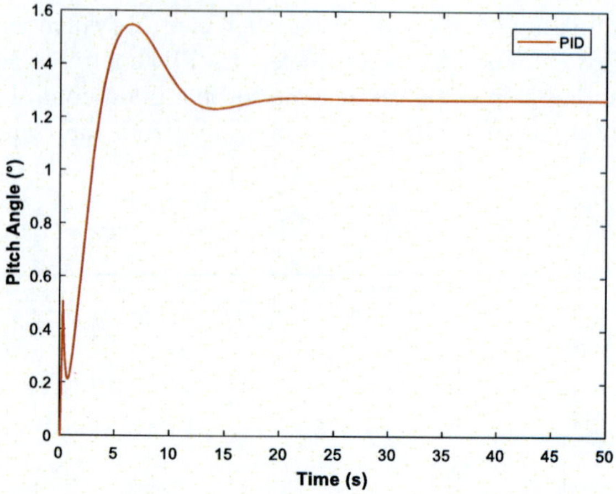

Fig. 5. Evaluation of Pitch angle with PID strategy control.

The PID controller reduced rise time to 2.1 s, peak and steady overshoot are very higher as compared to the wind turbine model without any controller and PI controller.

In the case of PID controller, we observed that the less rise time, the slew rate, peak and steady overshoot are higher when compared to conventional PI controller, and a very high peak, steady overshoot and slew rate compared to without any controller. The Fig. 6 illustrate the unit step responses of wind turbine without, with PI and PID pitch control system.

The following curves shows in Fig. 7 the mechanical torque without controller, PI and PID controller for the base wind speed of 12 m/s. it sees that the green line without controller the mechanical torque as in inactive mode.

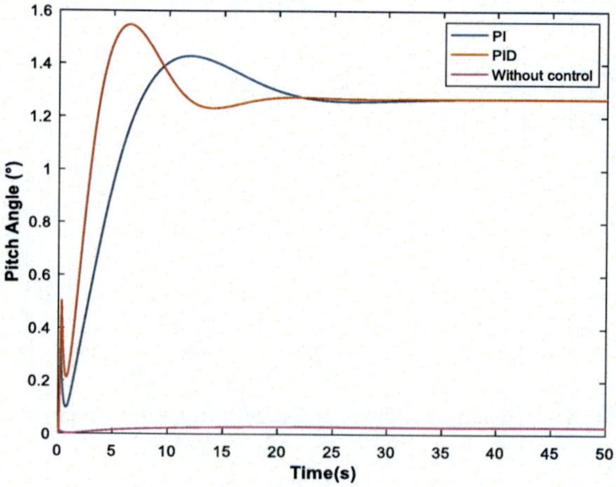

Fig. 6. The response of WT without, with PI and PID controller.

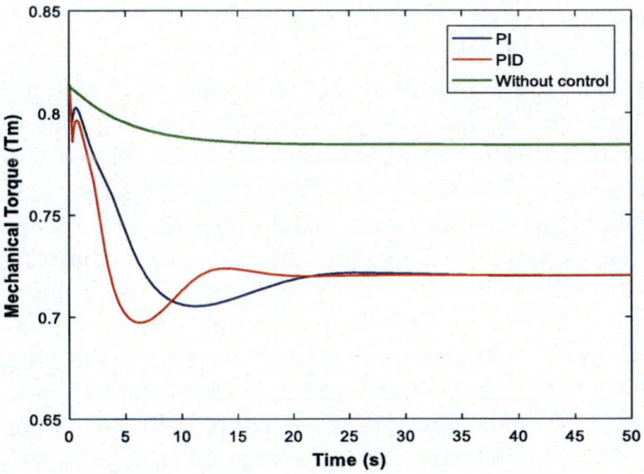

Fig. 7. The response of Mechanical torque for the different case of control strategy.

While the blue line represents the PI controller was started fluctuates until it comes to stable state after 12 s. The red line represents the PID controller, after some fluctuations, the mechanical torque become stable after 5 s.

As expected, the starting PID torque is lower than the running PI mechanical torque, it means the response of PID is quicker compared to the PI controller with time, that ensure the good performance of pitch control proposed in this system and other parameters.

Figure 8 depicts the output of AC power under three different situations. When compared to both the yellow scope without controller and the green scope with PI controller, the PID controller used for starting the machine initially provides electricity at 5 s. The power then rises over the rated value, and after 12 s, it reaches steady-state.

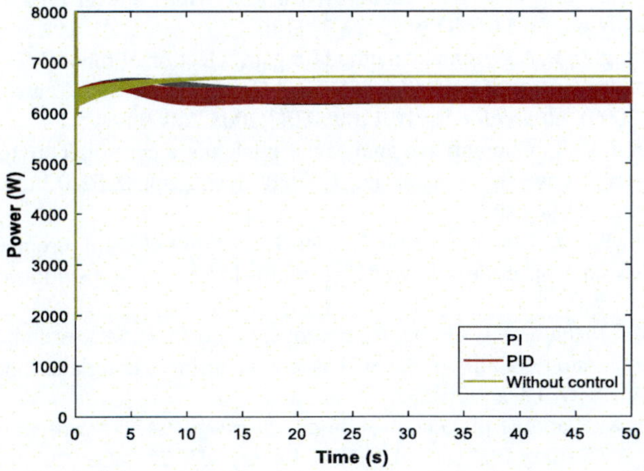

Fig. 8. The response of Power (W) for the different case of control strategy.

5 Conclusion

This paper has established a complete model of Pitch Angle Control strategy of the PMSG Wind Turbine. Three case studies have been presented: without controller, PI and PID control, respectively. Besides, PID parameters have been adjusted for controlling the pitch angle. A comparison between three different controllers has been carried out to evaluate the stability of mechanical torque and the response of wind power.

The PI controller recorded results in blue line shows the stability of pitch angle start from 20 s with state error 0.25. On the other hand, the PID controller recorded results in the red line shows a better stability from 15 s with state error 0. Additionally, the PID controller has shown a good result in terms of the rise time in comparison with PI controller with time of 2.7 s, and without controller with 7.9 s. It is important to keep in mind that sometimes the little discrepancies count, especially when it comes to stability.

Generally, both PI and PID controllers produced satisfactory results when compared to the case when control is not implied. The tuning parameters in MATLAB-Simulink that have been adjusted present a successfully performance for the analysed system.

References

1. Narjisse, A., Abdellatif, K.: Assessment of RANS turbulence closure models for predicting air flow in neutral ABL over hilly terrain. Int. Rev. Appl. Sci. Eng. IRASE (2021). https://doi.org/10.1556/1848.2021.00264
2. Narjisse, A., Khamlichi, A.: Computational fluid dynamics simulation to predict the airflow and turbulence in a wind farm in open complex terrain. In: Proceedings of the MDPI, vol. 63, p. 33 (2020)
3. Ahyaten, S., El Bahaoui, J.: Modeling of wind turbines based on DFIG generator. In: Proceedings, vol. 63, p. 16 (2020). https://doi.org/10.3390/proceedings2020063016
4. Jung, J.W., Leu, V.Q., Do, T.D., Kim, E.K., Choi, H.H.: Adaptive PID speed control design for permanent magnet synchronous motor drives. IEEE Trans. Power Electron. **30**, 900–908 (2015). https://doi.org/10.1109/TPEL.2014.2311462
5. Chang, P.H., Jung, J.H.: A systematic method for gain selection of robust PID control for non-linear plants of second-order controller canonical form. IEEE Trans. Control Syst. Technol. **17**, 473–483 (2009). https://doi.org/10.1109/TCST.2008.2000989
6. Tiwari, R., Babu, R.N.: Comparative analysis of pitch angle controller strategies for PMSG based wind energy conversion system. Int. J. Intell. Syst. Appl. **9**, 62–73 (2017). https://doi.org/10.5815/ijisa.2017.05.08
7. Guo, R., Du, J., Wu, J., Liu, Y.: The pitch control algorithm of wind turbine based on fuzzy control and PID control. Energy Power Eng. **05**, 6–10 (2013). https://doi.org/10.4236/epe.2013.53b002
8. Rashid, T.H.M.S., Routh, A.K., Rana, R., Ferdous, I., Sayed, R.: A novel approach to maximize performance and reliability of PMSG based wind turbine: Bangladesh perspective. Am. J. Eng. Res. (AJER) **7**, 17–26 (2018)
9. Pan, L., Wang, X.: Variable pitch control on direct-driven PMSG for offshore wind turbine using Repetitive-TS fuzzy PID control. Renew. Energy **159**, 221–237 (2020). https://doi.org/10.1016/j.renene.2020.05.093

10. Novaes Menezes, E.J., Araújo, A.M., Bouchonneau da Silva, N.S.: A review on wind turbine control and its associated methods. J. Clean. Prod. **174**, 945–953 (2018). https://doi.org/10.1016/j.jclepro.2017.10.297
11. Camblong, H., Vechiu, I., Etxeberria, A., Martínez, M.I.: Wind turbine mechanical stresses reduction and contribution to frequency regulation. Control Eng. Pract. **30**, 140–149 (2014). https://doi.org/10.1016/j.conengprac.2014.03.007

Study and Simulation of a Broken Induction Motor Rotor Bar Caused Motor Vibration

Ágoston Katalin(✉)

University of Medicine, Farmacy, Science and Tehnologie, Târgu Mureş, Romania
katalin.agoston@umfst.ro

Abstract. The paper presents the importance of motor vibration study and simulation; different faults which can appear during induction motor running and how broken rotor bar influences the whole system vibration. The environmental and internal, electrical and mechanical noises influence the normal functioning of the motor, increase the power consumption, deform the current spectrum, more heat is generated and vibrations are growing. Broken rotor bar changes the magnetic field distribution in the air gap between rotor and stator, unbalanced force appears which cause unwanted vibration and malfunctioning. Electrical equation is written for one rotor loop and the influence of the broken rotor bar is showed. The induction motor vibration can be studied through a simple system model with two degrees of freedom. The system model changes accordingly to the broken rotor bar and this change is studied through the transfer function written for rotor.

Keywords: Broken rotor bar · System vibration · Transfer function · System simulation

1 Introduction

Vibrations and noises are present in all areas of our lives. The vibrations are dynamic phenomena whose intensity varies with time. They may be generated and transmitted by different equipments and devices like motors, machine-tools, turbines, cars.

Some of the vibrations are desirable; others may be disturbing or even destructive. For this reason it is important to study and understand the causes of vibrations and to develop different types of methods to measure and analyze them.

To analyze the vibrations they will be measured in more points of the system: to compare the measured values with the maximum admissible values the vibration is measured at the output of the system, this is the case of noise and perturbation detection. The characteristics of the vibration can be measured at the input of the system to set these at the established values; this is the case of desired vibrations with well-defined parameters. To test how the system reacts to the different external forces and noises, vibrations must be measured both at the input and the output of the system. In each case different type of sensors and analysis methods will be used [2, 3].

Vibrations can be characterized with amplitude and frequency. The amplitude shows how strongly the vibration is, and the frequency shows the oscillation rate of the vibration.

L. Moldovan and A. Gligor (Eds.): Inter-Eng 2021, LNNS 386, pp. 584–590, 2022.
https://doi.org/10.1007/978-3-030-93817-8_52

The amplitude is related to the speed of the movement and to the force which causes the vibration. The peak of the amplitude shows the highest speed and the rms value of the amplitude shows the vibration energy [4].

Measuring and analyzing the vibrations at different places the cause of the vibration can be discovered. Vibration monitoring helps to find any problems that might develop and so damages and quality losses can be prevented [1, 5].

One of the important issues is monitoring the vibration of induction motors. Induction motors are part of different equipments and have great influence on the system reliability. Induction motors run in noisy environment and produce also noise and vibrations. The interaction of vibration and repeating forces cause damages, usages and/or electrical breakdowns.

In vibration monitoring it is important to know the most vulnerable places and the type of the forces which cause the vibration to select and use the suitable sensor. It is also important to know the basic principles of the appearance of vibrations to make the right measurement.

For accurate monitoring the frequency domain of the vibration, respectively of the vibration transducer must be known. The correlation of these two results from the frequency response of the accelerometer. The vibration transducer has to take over perfectly the vibration of the system, therefore must be attached firmly to the vibrating pieces.

The generated signal can be analyzed through different methods; the most used is the spectral analysis method. The spectrum of the vibration shows the frequencies at which the vibration occurs. The information from a spectrum depends on the maximum frequency and resolution of the measuring system. The measured signal is displayed as a velocity spectrum to find out the component of the vibration signal [6, 7].

2 System Vibration

Vibrations analysis can be important in motor fault detection. Induction motor is coupled with other equipment and/or is fixed to a base which can also vibrate, therefore its vibration can be approximated or modelled through systems with several degrees of freedom, for example the system consisting of motor and generator. The simplest model to study the motor vibration is a model with two degrees of freedom. For each system component a second degree equation is written, which describes the motion of the system component and depends on the characteristics (mass, damping and stiffness coefficient, displacement) of the component with which it is coupled.

The vibration sources are multiple, but they can be grouped into two categories, external vibrations due to environment and internal vibrations due to defects or faults. During operation there is a background vibration which is overlapped by the vibration generated by faults or malfunction. In vibration measurement and analysis it is important to separate these two components.

External vibrations transmitted through environment are caused by other equipment, motors nearby or at a greater distance.

Internal vibrations can be caused by bad fit, worn bearing, loose straps or screws, etc. Each fault generates its own vibration.

In vibration analysis it is not enough to obtain the frequency spectrum of the vibration which indicates the components of the vibration, the intensity of vibration must be determined too. As a conclusion the degree and type of fault must be determined.

In vibration measurement and analysis is very important the proper selection and correct mounting of the transducer in the right place.

2.1 Vibration Sources

Vibration measurement has an important place in electrical motor fault detection. To perform correct measurements the possible causes of the vibration must be studied. Electrical motor is a complex system in which there are electrical and mechanical parts that interact during operation. So electrical and mechanical forces interact that can change over time. Without going into details causes or sources of vibration can be: improper motor fixing, weak base, imbalance of the field between stator and rotor, thermal unbalance, by rigid coupling the driven machine unbalance can be transmitted to the motor, the type and size of the coupling, missing or broken rotor bar, non symmetrical air-gap between stator and rotor, eccentric rotor, used or improper fixed bearing, etc. [8].

2.2 Vibration Data Acquisition

The vibration can be measured through displacement, speed or acceleration of the studied system. Each unit highlights different characteristics of the vibration. Usually displacement is measured at low frequency vibrations and acceleration at high frequency vibrations. There are equations to make conversion from one unit to another.

3 Broken Rotor Bar Effect on the System Vibration

Electrical motors works in a variety of unwanted conditions which can cause the induction motor to go into a failure. If the failure of induction motor is not detected at its early stage of the failure it may result in the total loss of the machine and in addition downtime of the whole equipment.

The induction motor failures can be categorized according to the main components of the motor: stator related faults, rotor related faults, bearing related faults and other faults. Rotor related faults are broken rotor bars, cracked rotor end rings, which are 10% of all failures [10, 11].

The cause of rotor related failures can be thermal, electrical, mechanical or combination of these stresses. Thermal stress is usually a rise in temperature above the allowed value or non-uniform heat distribution. The electrical stresses can be transient, corona, or over-voltage and Foucault current. Mechanical stresses can be due to coil movement or rotor striking the stator.

When a rotor bar is broken no current flows through the bar, no magnetic flux is generated around the bar, which generates an asymmetry in the magnetic field around the rotor. A three-phase symmetrical stator winding supply with symmetrical voltage having f_1 frequency, produces a rotating magnetic field with the same frequency. If there is a perfect symmetry, no backward rotating field results. If any asymmetry appears in the supply or in windings; a backward rotating field will result.

Applying this rotating magnetic field to the rotor, the frequency of the induced electro-magnetic force and current will be at slip frequency sf_1. The rotor currents produces also a three-phase magnetic field which rotates at slip frequency $f_2 = sf_1$ with respect to the rotating rotor. At symmetrical cage only a forward rotating field exists. If a rotor asymmetry appears (broken bar) a backward rotating field will be result which frequency is f_2 with respect to the forward rotating rotor. This backward rotating field induces an EMF and current in the stator, which appear in the form of a lower sideband frequency in the stator current spectrum. Due to the cyclic variation of the current and oscillation of the rotor an upper sideband frequency appears. So broken rotor bar induces in the stator windings current with frequencies:

$$f_s = (1 \pm 2s)f_1 \text{ [Hz]} \tag{1}$$

This can be followed on the stator current spectrum [10].

Motor Current Signature Analysis (MCSA) is based on the fact that stator current contains harmonics related to the rotating flux caused by broken rotor bar. These sideband frequencies can be detected relatively easy at high power motors, but in case of low power induction motors the amplitude of sideband harmonics are small and due to the low signal-noise relation, the harmonics become difficult to detect. The sideband frequencies are also very close to the fundamental frequency [12].

Due to the change in magnetic field distribution caused by the broken rotor bar, an eccentricity occurs in the rotating magnetic field, which leads to the appearance of uncompensated forces. These forces act on the movement of the rotor, on the bearings and cause an unwanted vibration of the induction motor. Continuous vibration monitoring and comparing the amplitude or power spectrum with the healthy motor spectrum the broken rotor bar fault can be detected.

The stator and rotor of an electrical motor is coupled through the magnetic field from the air gap. Analyzing the simple two degree of freedom motor model it can be observed that the magnetic field change means the change of k_R spring coefficient (Fig. 1b). For the free body diagrams the following equations can be written [13]:

$$\begin{aligned} M_S\frac{d^2x_2}{dt^2} + x_2(k_S + k_R) - k_R x_1 = F_2(t) \\ M_R\frac{d^2x_1}{dt^2} + x_1 k_R - k_R x_2 = F_1(t) \end{aligned} \quad \text{[Hz]} \tag{2}$$

where $F_1(t)$ represent the uncompensated force due to magnetic field change caused by the broken bar.

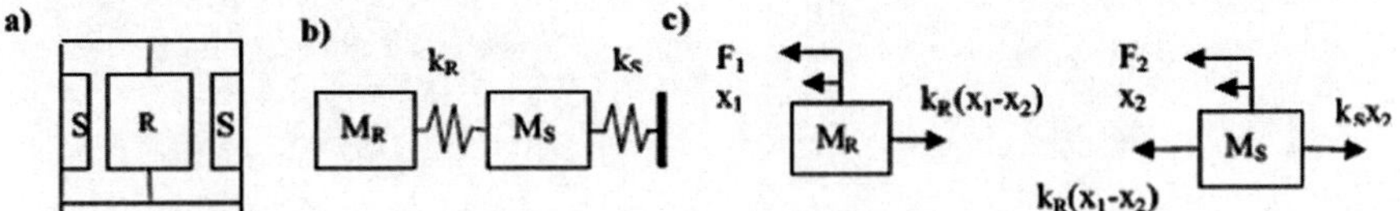

Fig. 1. Motor structure (a); equivalent system (b); free body diagrams (c).

The $F_2(t)$ representing the influence of the stator frame is neglected in this case.

Using the Laplace transformation of the system Eqs. (2) and replacing $X_2(s)$, the transfer function for the rotor subsystem can be obtained:

$$H_R(s) = \frac{X_1(s)}{F_1(s)} = \frac{M_S s^2 + k_S + k_R}{M_R M_S s^4 + (M_R k_S + M_R k_R + M_S k_R)s^2 + k_R k_S} \tag{3}$$

In this equation the k_R coefficient is influenced by the magnetic field change.

The equivalent circuit for two consecutive bars is presented on Fig. 2; R_b is the bar resistance, R_e is the end ring resistance, $d\Phi/dt$ represents the induced voltage.

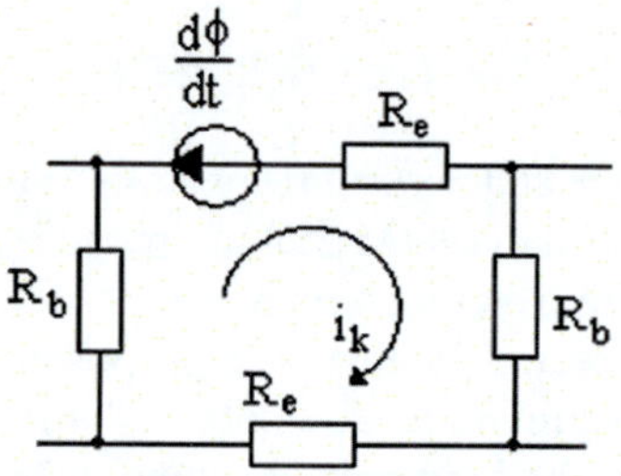

Fig. 2. Equivalent circuit for two consecutive bars.

For this equivalent circuit the following equation can be written:

$$R_{tot} \cdot i_k + \frac{d\Phi}{dt} = 0 \tag{4}$$

where $R_{tot} = 2R_e + 2R_b$. If a bar is broken R_{tot} is changing.

Figure 3a and 3b shows the current variation in the equivalent circuit for different R_{tot} values, the simulation was made in Simulink.

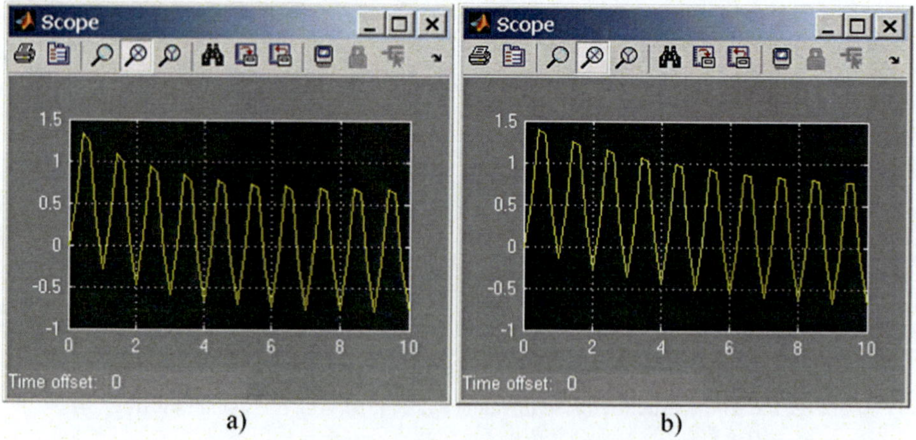

Fig. 3. Current variation for different R_{tot} values, $R_{tot}a > R_{tot}b$.

The current variation influences the flux and magnetic field variation through a linear function, $\Phi = Li$, H = f(i). The magnetic field has the same variation, only it is out of phase behind the current. The current change produces a magnetic field change.

Making the k_R coefficient dependent from the magnetic field through a linear function, $k_R = a \cdot H + b$ (where *a*, *b* are coefficients depending on the rotor characteristics), the influence of the R_{tot} change can be highlighted. The amplitude-frequency and phase-frequency characteristic variations for the rotor subsystem in two situations, for k_R variation, is presented in Fig. 4.

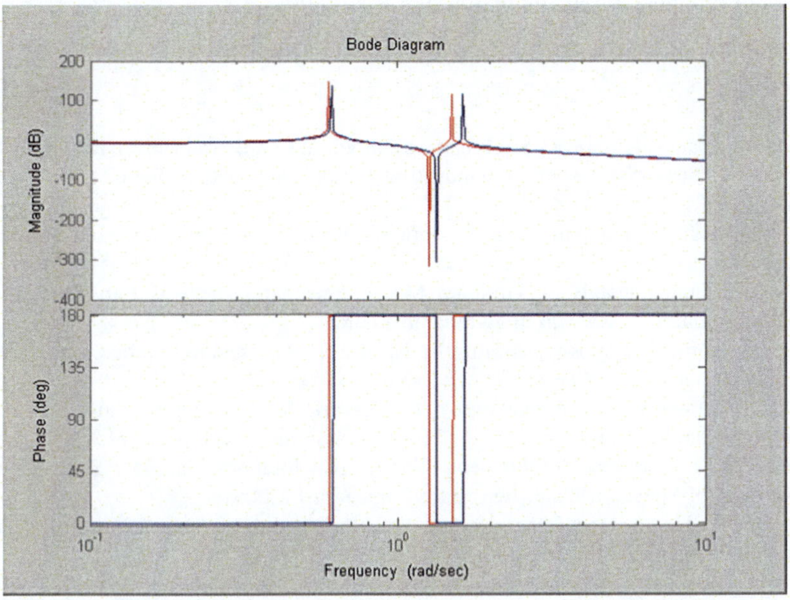

Fig. 4. Amplitude-frequency and phase-frequency characteristics for k_R variation.

From the amplitude-frequency characteristic can be observed that a small change in rotor total resistance produces a resonance frequency shifting of the rotor subsystem and through this of the whole motor system.

4 Conclusion

In this paper was highlighted that rotor related fault, broken rotor bar, has a big influence on induction motor working. Broken rotor bar changes the magnetic field distribution around the rotor and uncompensated forces appear. Due to these forces unwanted vibration of the induction motor appears.

Broken rotor bar causes the change of the whole rotor resistance and inductivity which influences the current through the rotor. To show this influence an equivalent circuit for two consecutive bars was presented. The influence of R_{tot} change on the rotor current was simulated in Simulink, based on differential equations written for the equivalent circuit.

A transfer function for the rotor subsystem was deduced, using a simple motor model. The coupling between stator and rotor is expressed by the k_R coefficient. This coefficient is influenced by the magnetic field in the air gap. It presumed that a linear dependency exists between k_R coefficient and the magnetic field, and the total resistence of the rotor, respectively. The amplitude-frequency and phase-frequency characteristic variations for the rotor subsystem in two situations, for k_R variation shows an influence of rotor total resistance change. Through the simulation it was highlighted that broken rotor bar has an important influence on the induction motor vibration.

References

1. Bower, A., Franck, J., Kim, K.-S.: Dynamics and Vibrations, Brown University, Spring (2012)
2. Agoston, K.: Senzori in automatizari industrial, Editura Universitatii "Petru Maior" Tg. Mures (2004)
3. Agoston, K.: Introduction to Shock & Vibration, Bruel & Kjaer. www.bk.com. Accessed 06 June 2021
4. Brown, P.: Fundamentals of Vibration Measurement and Analysis Explained, Lifetime Reliability Solutions. www.lifetime-reliability.com
5. Venkatasubramanian, V., Rengaswamy, R., Kavuri, S.-N.: A review of process fault detection and diagnosis. Comput. Chem. Eng. **27**, 293–313 (2003)
6. Venkatasubramanian, V., Rengaswamy, R., Kavuri, S-N.: Beginer's Guide to Machine Vibration, Commtest Instruments Ltd. (2006). www.commtest.com. Accessed 07 June 2021
7. Furman, B.J.: Vibration Measurement Experimental Methods, San Jose State University, Department of Mechanical and Aerospace Engineering (2005)
8. Finley, W.R., Hodowanec, M.M., Holter, W.G.: An Analytical Approach to Solving Motor Vibration Problems, IEEE, Paper no. PCIC-99–20
9. Reimche, W., Sudmersen, U., Pietsch, O., Scheer, C., Bach, F-W.: Basics of Vibration Monitoring for Fault Detection and Process Control, PANNDT, Rio-Brasil (2003)
10. Sonje, D.M., Munje, R.K.: Rotor cage fault detection in induction motors by motor current signature analysis. ICCIA, Int. J. Comput. Appl. (2011)
11. Bachir, S., Tnani, S., Champenois, G., Trigeasson, J.-C.: Induction Motor Modeling of Brocken Rotor Bars and Fault Detection by Parameter Estimation (2014). www.researchgate.net/publication/228459830
12. Cusido, J., Ortega, J.A., Garcia, J.R., Romenal, L.: Fault detection techniques for induction motors. IEEE Compat. Power Electron. **2005**, 85–90 (2005). https://doi.org/10.1109/CPE.2005.1547550
13. Agoston, K.: Fault detection of the electrical motors based on vibration analysis. Procedia Technol. **19**, 547–553 (2015)
14. Buryak, S.Y.: Mathematical Modeling of AC Electric Point Motor, Наука та прогрес транспорту. Вісник Дніпропетровського національногоуніверситету залізничного транспорту, № 2(50) (2014). ISSN 2307-3489 (Print), ISSN 2307-6666 (Online)

Designing and Development of an Intelligent Energy Supply and Powertrain Systems for Automotive Sector to Reduce Pollution and Health Risk

Doru-Laurean Băldean[1](✉), Lavinia Andrei[2], and Viorel Chindea[1]

[1] Automotive Engineering and Transportation Department, Faculty of Automotive Engineering, Mechatronics and Mechanics, Technical University of Cluj-Napoca, Muncii 103-105, Cluj-Napoca, Cluj, Romania
doru.baldean@auto.utcluj.ro

[2] Public Health and Management, Faculty of Medicine, University of Medicine and Pharmacology, Victor Babeş, Cluj-Napoca, Cluj, Romania

Abstract. The scientific paper shows the research and development results of a practical application and experiment with power and energy system implemented on road vehicle platform through digital control. Its overall configuration, inherent structure, mathematical modelling and practical application, digital programs, software controls and electronic control units are properly set-up to deliver solid results. The research paper main objective is to show the design of energy supply system and the electronic control developed for the improvement of efficiency and pollution reduction, which allow both researchers and common users to obtain relevant data regarding level of fuel consumption and emissions. The practical application of the paper has been implemented through integrated features like remote measurement, drive with automated systems activated on board, and autonomy definition, using the real model of the road motor vehicle as adapted tool in the experimental research and for validation of the scientific hypothesis. The later ones are stating that power and energy systems with digital management and control are proper solutions for increasing efficiency and pollution reduction, in order to diminish health risks and environmental impact. The physical road vehicle model has multiple mechatronic systems for powertrain management and the electronic kit for energy supply system with an embedded software package. The research describes also its primary objectives, real operation performance, powertrain design, energy supply system, automation challenges and environmental vulnerabilities which were studied and tackled so far by this paper. Energy supply system design, powertrain testing, emission determination and experimental development phases cover the aspects concerning various operating regimes (automatic and manual), and interdisciplinary engineering innovative solutions for achieving the main objective of the present research.

Keywords: Automation · Automotive · Energy supply · Pollution reduction · Powertrain

L. Moldovan and A. Gligor (Eds.): Inter-Eng 2021, LNNS 386, pp. 591–602, 2022.
https://doi.org/10.1007/978-3-030-93817-8_53

1 Introduction

Pollution and health risks are evolving together nowadays which is why a part of research must provide data and results to create a solid baseline with solutions and optimal procedures to meet the challenges and to provide safety measures in case of health hazards [1]. Metropolitan polluted areas will cause more and more health risks if there will not be put in place specific measures and methodologies for monitoring, diagnosing and controlling emissions [2, 3]. Injection system and energy supply to the engine influences the performances, as power output and torque on the crankshaft, and the emission characteristics in relation with fuel composition and spray quality [4, 5]. Air quality in dense road traffic areas is influenced by soot emissions, particulate matter, carbon compounds and other toxic chemicals that are exhausted through tail pipes of the vehicles and ventilated in the atmosphere from the brakes and tires [6, 7]. Some of the particulate matter and carbon deposits are temporarily retained by Diesel Particulate Filter [7, 8]. Both Diesel and petrol supply systems have spots suitable for optimization with electronic interfaces and control [9, 10].

Automation and electronic control are successfully applied nowadays in many areas regarding the fuel supply and emission after-treatment, as well as in the road traffic surroundings [11, 12]. If there is intended to put into practice an automated driving system based on artificial intelligence [13], then it must contain a feature for injection and fuel supply to the engine, as well as a program for lubrication and operational control [14, 15]. Electronic control, diagnosis and investigation of operating parameters during the air intake, compression, injection, expansion and exhaust processes facilitates rapid reading of values and characteristics in order to improve and optimize loading conditions [16, 17]. Important data and info regarding technical and loading conditions during engine operation may be displayed on dashboard panel to allow users and drivers to adapt their demands and commands for each driving situation [18]. Alternative solutions in composite materials may be as well implemented, as innovative measures, into the power-train for exhaust gas after-treatment and pollution control [19]. Electronic Diesel Control (EDC) system has a digital processing power based upon mechanical signals and gas flow characteristics. In this case, the fuel supply is monitored and calculated for each individual cycle of operation, based on the data collected from sensors [20]. Management system receives real time data from engine speed sensor, camshaft sensor, temperature sensor, air intake mass flow sensor, accelerator sensor and lambda sensors [21]. Automated system for engine and machinery management uses fuel level data to remotely asses efficiency and work productivity [22].

The present work shows the testing cycle results and data points of vehicle powertrain performance in a practical experiment. Having the power and energy systems placed on a car with digital control makes it more plausible to design and fully implement an intelligent energy supply interface for reducing the pollution and thus the health risk factors. The most important objective of the paper is to present a structure design for the energy supply system and the related connections in electronic architecture of the car developed for efficient control and for pollution reduction. This target once reached is the baseline for improved measures against health risk factors related to car operation generated pollution. The method of the research gives a simplified schematic of the car.

2 Materials and Methods

Research materials applied in the present study consist in the electronic application which is used for designing the virtual model of the car with an intelligent energy and powertrain systems, rendered in Unity 5. The CAD apps are also considered for schematic representations of the architecture. Method of the research is mainly based in applied experiment, firstly the design is made in virtual reality program and secondly there is applied in the real size car to outline the real data points and emission limits of most important gas in the exhaust pipeline. Three cycles are made.

Technical specification and mean values of important data are given also in Table 1.

Table 1. Technical specifications and average values regarding the testing of the vehicle.

Average/Test	Cycle 1	Cycle 2	Cycle 3
Engine displacement	1.6 [dm^3]	1.6 [dm^3]	1.6 [dm^3]
Fuel type	Diesel	Diesel	Diesel
Engine speed	1578 rpm	1402 rpm	1430 rpm
Fuel avg. consumption	2 l/h	2.35 l/h	3 l/h
Vehicle speed	21 km/h	30.8 km/h	44.97 km/h

For the control of powertrain operation there is used artificial intelligence as a strategy based on neural networks. Virtual Reality application, Unity 5, offers the possibility to implement a program with artificial intelligence features to control the energy supply and vehicle power-train. In operation during driving maneuvers and obstacle avoidance is virtually applied the artificial intelligence with the simplified scheme of neural networks for complex links as defined in Fig. 1.

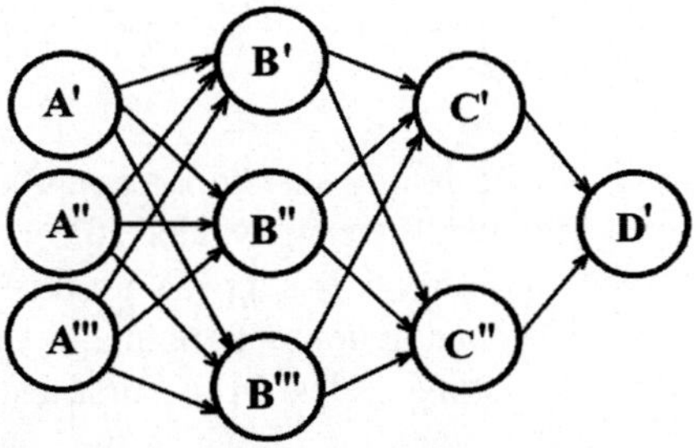

Fig. 1. Simplified scheme for control method with Artificial Intel. support of neural networks [13].

Simplified structure of the vehicle model used for designing and development of an intelligent energy supply and powertrain control is shown in Fig. 2. It has specific components for pollution control, fuel supply to the engine (5) and health risks mitigation through digital control of operation and powertrain outputs monitoring. Electrical energy supply and storage unit (1) gives the electricity needed by the electronic control unit (ECU) (2). The later one controls the engine (5) operation. It receives the electronic signals from the sensor's hub (6). The sensors monitor the engine crankshaft, the camshaft, engine temperature, accelerator position and air intake mass flow, sending all the information to the ECU. The later one processes the information received and gives a response through the actuators (4) that are connected to the engine digital controller (2). The actuators are strictly and precisely controlling the fuel/energy supply system (3) of the engine. The common rail injection system is basically controlled through the opening time interval of the injectors and the high-pressure value generated by the high-pressure pump in the system. The clutch (7) is mechanically connecting the engine (5) and the gear box or transmission (8). The kinematic and dynamic parameters are thus transferred toward the vehicle wheels (10). The smart/intelligent interface (9) connects and monitors all digital systems of the car in real time allowing data transfer.

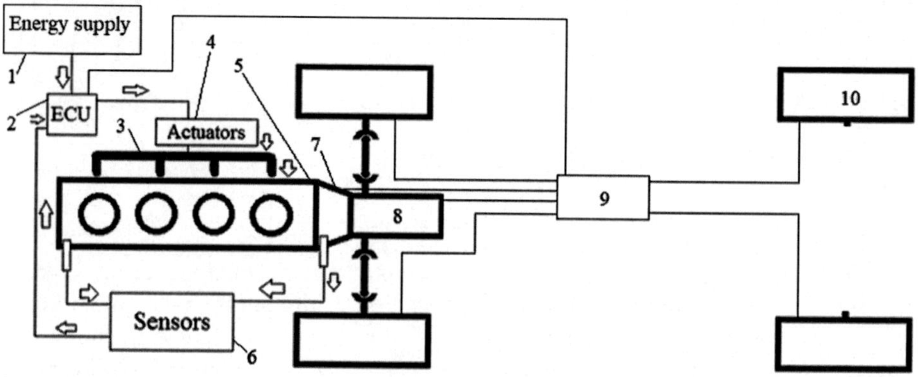

Fig. 2. Simplified structure of the powertrain with the basic systems and connections for control.

3 Investigation and Results

Applied results gained through the designing, development and experiment in real conditions have been represented graphically versus vehicle speed. Important findings of the applied engineering research on the energy supply with intelligent system attached to the car's powertrain show fuel consumption data points in each gear ratio engaged in transmission for three distinct testing cycles. In the first testing cycle (Cycle 1) is developed the model of cold engine operation, in which the vehicle is driven in urban environment at cold temperatures, below 20 °C. Thus, the optimal safe operation is provided by the first three gear ratios in the transmission of the car. The majority of the data points collected are placed around 12 l/100km regarding fuel consumption in this scenario. The average fuel consumption in the second gear ratio is around 15 l/100km. In the third gear ratio the average fuel consumption is placed around 3 l/100km. The overall fuel supply average for the Cycle 1 is almost 10 l/100km, which is corresponding to 2 l/h.

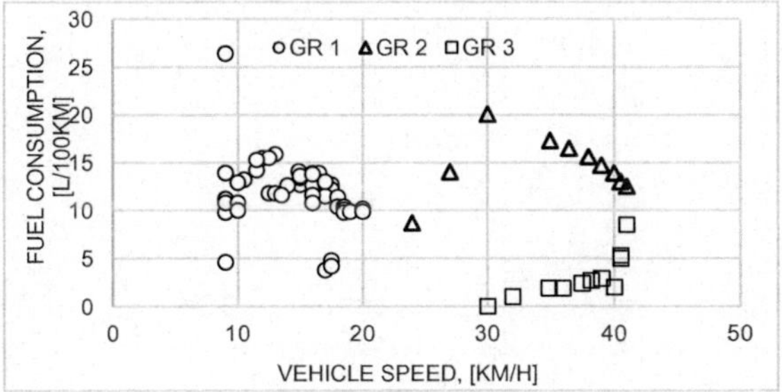

(a)

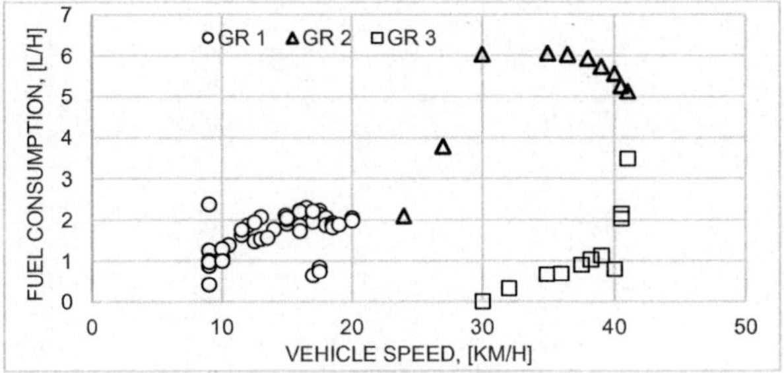

(b)

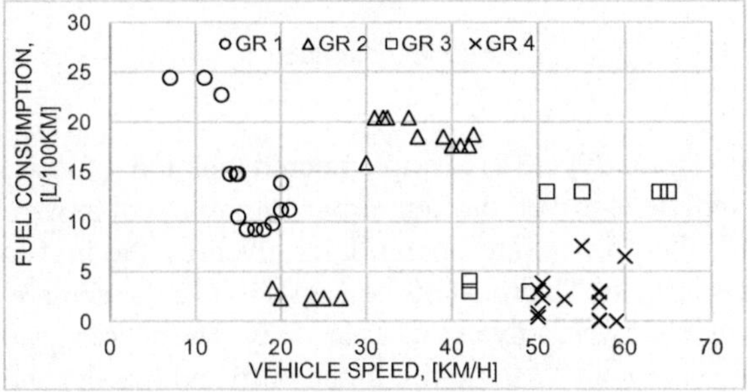

(c)

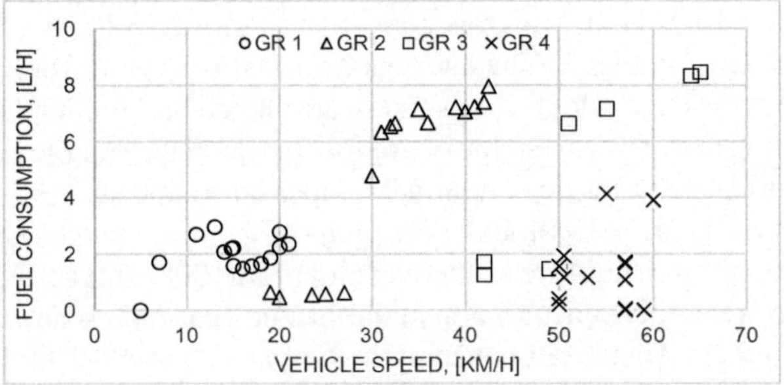

(d)

Fig. 3. Results of energy/fuel supply to the engine during the research: (a) fuel consumption in l/100 km during Cycle 1; (b) fuel consumption in l/h during Cycle 1; (c) fuel consumption in l/100 km during Cycle 2; (d) fuel consumption in l/h during Cycle 2; (e) fuel consumption in l/100 km during Cycle 3; (f) fuel consumption in l/h during Cycle 3 in extra-urban traffic and warm engine conditions.

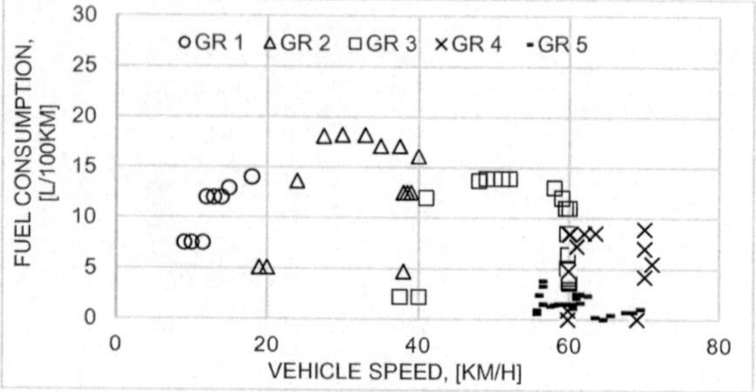

(e)

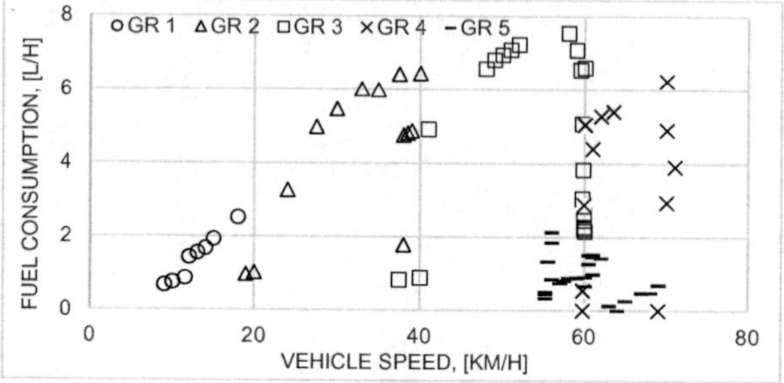

(f)

Fig. 3. continued

In the second test cycle (Cycle 2) is implemented the model for warm engine operation, in which the vehicle is driven through urban environment at regime temperatures, around 90 °C. Thus, the proper safe operation is given by the first four gear ratios in the mechanical transmission. The most of the data points collected are scattered around 10 l/100km regarding energy supply in this case study. The average fuel consumption in the second gear ratio is around 17 l/100km. In the third and fourth gear ratio the average fuel consumption is placed around 7 l/100km. The overall fuel supply average for the Cycle 2 is almost 7.1 l/100km, which is corresponding to 2.35 l/h at an average speed of 30.8 km/h, as shown in Fig. 3. The later one shows the graphics for both units.

In the third test cycle (Cycle 3) is designed, developed and implemented the testing model for mixed traffic and warm engine operation, in which the vehicle is driven through urban and extra-urban road-traffic at nominal temperatures, meaning 90 °C. The proper operation in this case is given by all five gear ratios of the mechanical transmission. The most of the data points collected are scattered below 15 l/100km regarding fuel supply to the engine. The average fuel consumption in the second gear ratio is around 16 l/100km. In third gear ratio the average fuel consumption is placed around 13.5 l/100km and in fourth gear ratio the average fuel consumption is placed around 8 l/100km. The overall fuel supply average for the Cycle 3 is almost 6.5 l/100km, which is corresponding to 3 l/h at an average speed of 44.97 km/h.

Evaluation and study of strength, weakness, opportunity, and threat aspects shows few of the interesting results, findings and vulnerabilities discovered, as are centralized and shown in Table 2.

Table 2. Panel of strengths, weaknesses, opportunities, and threats (SWOT) evaluation.

1 Strengths	2 Weaknesses	3 Opportunities	4 Threats
C1 – safe speeds	Cold operation	High amplitude values	High pollution
C2 – regime temperatures	Limited gear and speed	Good acceleration	Limited performance in operation
C3 – all gear ratios	Speed risk factor	Higher speeds[a]	Speed risks

[a] All the speeds and gear ratio are precisely controlled to be in the legal limits.

CO_2 emissions are presented in Fig. 4 along with the engine speed variation versus speed.

The main probability equation used for the AI which has been considered to be applied in the case study is expressed based on a training group as follows:

$$P_I(x|y) = [P_I(y|x) \cdot P_I(x)]/P_I(y) = \left[A_I(x) \cdot \Pi_{i-1}^{N} \cdot P_I(y_i|x)\right]/P_I(y), \quad (1)$$

4 Discussions and Conclusions

Some contributions and achievements in the process of visualizing, designing and development of an intelligent energy supply system and power train control program made for cars and automotive sector to reduce pollution and mitigate health risks consists in the vehicle virtual modeling and practical testing on the real mobile platform. Technical specifications of the vehicle type VW Golf 6 chosen for testing and real time data measurement are supporting the idea of digitalizing the control for energy supply and pollution control, to innovate the study of power train intelligent control. Thus, original part of the research comprises the case study with three cycles of testing on a modern vehicle (VW Golf 1.6 l engine displacement and Diesel fuel supply) adapted to cold and warming up operation in urban traffic, mixt and extra-urban operation with different gear ratios engaged.

Designing and developing intelligent energy supply systems for automotive sector to reduce pollution and mitigate health risks with instruments and complex devices is based and corelates quite strongly with other studies in the recent years [1–6]. Finding those aspects, elements and relations between power train components, such as Diesel Particle Filters and common rail injection system, allow complete planning of state-of-the-art vehicles to operate, controlled by EDC, in complex thermo-dynamic regimes. The studies already made are including the evaluation of the vulnerabilities corelated with the air quality and pollution control system [6]. Using digital instruments and artificial intelligence, eventually with virtual reality, in designing and development of an intelligent energy supply and powertrain system for automotive sector to reduce pollution and health risks is the complex process that brings to the present study a close relation with the other research papers [7–13]. With the Unity 5 program there are to be

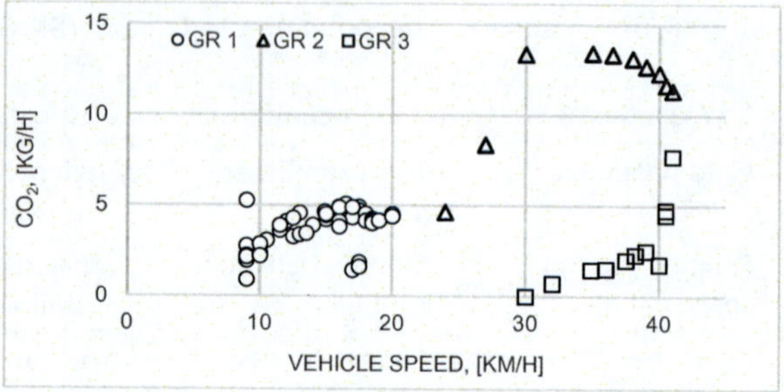

(a)

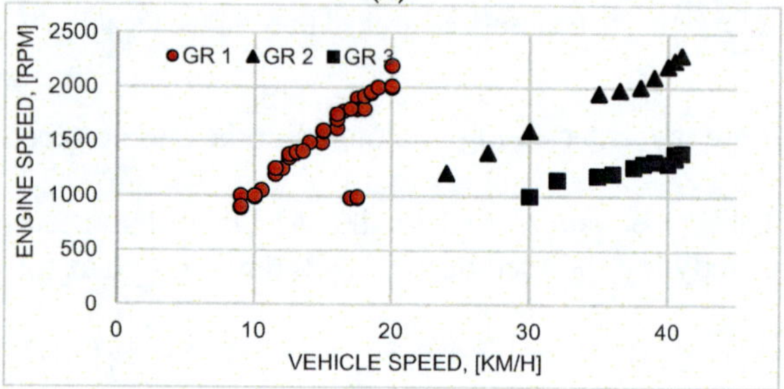

(b)

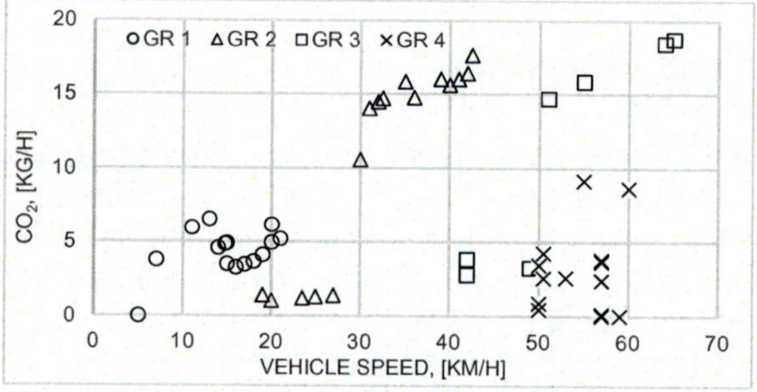

(c)

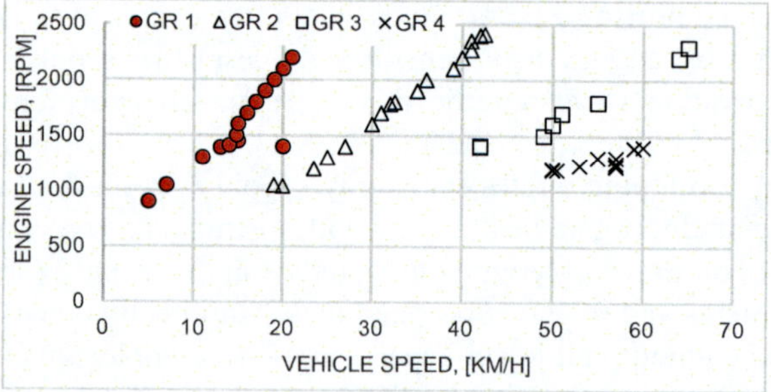

(d)

Fig. 4. CO_2 emissions and engine speed versus vehicle speed: (a) CO_2 vs. vehicle speed in Cycle 1; (b) rpm vs km/h in Cycle 1; (c) CO_2 in Cycle 2; (d) rpm in Cycle 2; (e) CO_2 in Cycle 3; (f) rpm in C 3.

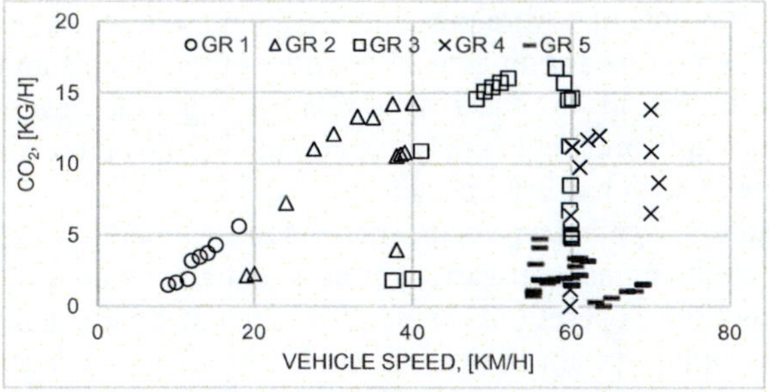

(e)

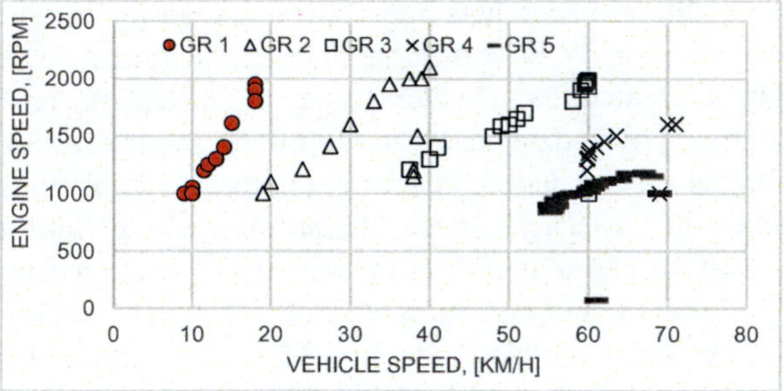

(f)

Fig. 4. continued

considered three important testing cycles for using AI in a real car: as an urban vehicle driven in cold short trips, as an urban traffic vehicle used for long hours in regime temperatures, and as a mixt extra-urban vehicle with in-city and inter-city longer trips. In this context the present study has been considered all the three instances.

Moreover the research staff has designed advanced method of data acquisition for improving the research program by recording multiple parameters (mechanical and digital, part of them taken from the electronic control unit of the engine, other recorded directly). The main part of the paper is developed with the support of the specialized software apps, such as CAD and ESITronic. The second part is developed in collaboration with specialists in public health management and air quality standards, due to the influence of the road traffic emissions on the environmental conditions as well as ambient hygiene.

Development, tests, and experiments for controlling the amount of fuel with electronic programs in real driving conditions of tested vehicle has been successfully implemented and gave us the insight of the aspects that must be considered (such as temperatures, loads, air intake) in future studies, both in mathematical modeling and in applied testing. The practical experiments in road traffic conditions on the public roads raise some challenges because there are multiple other factors that must be calculated and considered as important influence. The existing engine control systems are standard configured, but they allow further optimization possibilities for different kinds of applications, with the dynamic control of the powertrain. Using AI may allow full dynamic programming and adjustment of the fuel supply to any conditions and performance.

Most significant points that were presented in the present research are: CAD phase and testing of digital control for an actual car with common rail fuel injection system in actual road conditions. Main objective of this investigation was the designing and development of an intelligent fuel supply program and a powertrain system that may reduce the pollution, which has been presented.

Entire presentation, applied protocol, mathematical models and practical experiments, electronic applications, software interfaces and engine control units were configured to record and store operational values. The scientific paper has shown the CAD schematics for fuel supply system and the electronic control. It was developed to improve the efficiency of fuel use and to reduce pollutant emissions, especially CO2. The proposed protocol allows other researchers and vehicle users to gain relevant results and actual values of fuel supply and CO2 emissions. The application of the research protocol was implemented by integrated features like remote measurement, drive with automated systems (cruise control) activated on board, and autonomy range definition, using the real size road and the actual car as specific instrument in the practical engineering research and for the validation of initial hypothesis. These ones are showing that fuel supply system and exhaust after treatment with electronic management are practical ways for improving energy efficiency and pollution control, to manage health risks and environmental impact. The actual vehicle model has different mechatronic units for engine management and the electronic control for fuel injection system with an embedded software program. The research paper shows alongside its objectives, the practical output of the engine, powertrain CAD scheme, fuel supply system, automation challenges and environmental vulnerabilities which were outlined and detailed so far by this study. Fuel supply system CAD scheme, powertrain operation, CO2 emissions levels and experiment steps cover the topics regarding multiple working regimes (automatic and manual operation), and interdisciplinary solutions for reaching the proposed objective of the present research.

References

1. Andrei, L., et al.: Researching diesel car CO2 emissions in cold and warm-up transient test cycle. In: Grebenisan, G. (ed.) The IOP Conference Series: Materials Science and Engineering, Proceedings of the Annual Session of Scientific Papers "IMT ORADEA 2021", Oradea, Felix SPA, Romania, 27–28 May 2021, pp 1–6, vol. 1169. IOP Publishing, Philadelphia (2021). Accessed 16 Aug 2021
2. Andrei, L., Băldean, D., Borzan, A.I.: Applied measurements and instrumentation for improving diagnostic devices and systems in metropolitan polluted environments with nitric and carbon oxides. In: Vlad, S., Roman, N.M. (eds.) 6th International Conference on Advancements of Medicine and Health Care through Technology; 17–20 October 2018, Cluj-Napoca, Romania. IP, vol. 71, pp. 45–49. Springer, Singapore (2019). https://doi.org/10.1007/978-981-13-6207-1_8
3. Andrei, L., et al.: Analytic considerations regarding the link between air pollution with benzene and nasopharyngeal cancer. S I, 34(18), eISSN 2359-828X. http://stiintasiinginerie.ro/34-18
4. Barabas, I., et al.: Automated test bench for study of the fuel injection process. Solid State Phenom. **166**, 39–44 (2010). https://doi.org/10.4028/www.scientific.net/SSP.166-167.39

5. Barabas, I., et al.: Performance and emission characteristics of an CI engine fueled with diesel–biodiesel–bioethanol blends. Fuel **89**(12), 3827–3832 (2010). https://doi.org/10.1016/j.fuel.2010.07.011
6. Bec, P., et al.: Investigation of failures and vulnerabilities in road traffic air quality management system during 2020 pandemics. In: IOP Conference Series: Materials Science and Engineering, vol. 898, no. 1, p. 012009 (2020)
7. Borza, E.V., Băldean, D.-L., Jurchiş, B.-M.: Research of diesel particle filter soot impact and regeneration time trend-line. In: Dumitru, I., Covaciu, D., Racila, L., Rosca, A. (eds.) SMAT 2019, pp. 203–208. Springer, Cham (2020). https://doi.org/10.1007/978-3-030-32564-0_24
8. Borza, E.-V., Băldean, D.-L., Borzan, A.-I.: Research concerning fuel economy coefficient and carbon foot print in various conditions for a city compact size vehicle with digital control for a green solution and method at Technical University from Cluj-Napoca. In: Burnete, N., Varga, B.O. (eds.) AMMA2018 2018. PAE, pp. 181–189. Springer, Cham (2019). https://doi.org/10.1007/978-3-319-94409-8_22
9. Borzan, A.I., et al.: Experimental design for diesel supply control in order to improve fuel efficiency. In: IOP Conference Series: Materials Science and Engineering, vol. 568, no. 1, p. 012038
10. Borzan, A.I., et al.: The development of a new interface for intelligent control of energy supply in dynamic environment with process digitization. Procedia Manuf. **46**, 914–921
11. Chereches, I.-A., Băldean, D.-L., Borzan, A.-I.: Research of intelligent control of injection systems for subaru competition car. In: Dumitru, I., Covaciu, D., Racila, L., Rosca, A. (eds.) SMAT 2019, pp. 82–87. Springer, Cham (2020). https://doi.org/10.1007/978-3-030-32564-0_10
12. Chereches, A.I., et al.: Applied research on some data regarding air quality in Mărăşti area from Cluj-Napoca city. Sci. Eng. **34**(60), eISSN 2359-828X. http://stiintasiinginerie.ro/34-60
13. Covaciu, F., Băldean, D.-L.: Contribution to research the applied engineering protocol to implement a fuzzy regulator for autonomous driving of an automotive model implemented in virtual reality. In: Dumitru, I., Covaciu, D., Racila, L., Rosca, A. (eds.) SMAT 2019, pp. 468–476. Springer, Cham (2020). https://doi.org/10.1007/978-3-030-32564-0_54
14. Crisan, M.A., et al.: The study of some definitory characteristics for specific phenomena of lubricating process in tractor compression ignition engine. In: International Symposium of ISB-INMA TEH. Agricultural and Mechanical Engineering, Bucharest, Romania, 29–31 October 2015, pp. 599–606. ref. 6
15. Crisan, M.A., et al.: Researches concerning modeling of lubrication system in 3ld510 Lombardini engine. Sci. Eng. **30**(50), eISSN 2359-828X. http://stiintasiinginerie.ro/30-50
16. Ferenti, I., et al.: Experimental research in mechanic's field of control for further innovation in the hydraulic and electronic operated injection system. In: E3S Web of Conferences, vol. 85, p. 08006 (2019)
17. Iosif, F., Baldean, D.L.: Research of some operating parameters and the emissions level variation in a spark ignited engine through on-board investigation methods in different loading conditions. Central Eur. J. Eng. **4**(2), 192–208 (2014). https://doi.org/10.2478/s13531-013-0167-9
18. Jovrea, S., et al.: Researching on-board display of essential information concerning technical conditions in operation and fuel-economy of a motor-vehicle in operation. S I, **31**(67), eISSN 2359-828X. http://stiintasiinginerie.ro/31-67
19. Jurchiş, B.-M., Băldean, D.-L.: Research of alternative composite solutions in gas after-treatment system for diesel cars. In: Dumitru, I., Covaciu, D., Racila, L., Rosca, A. (eds.) SMAT 2019, pp. 160–164. Springer, Cham (2020). https://doi.org/10.1007/978-3-030-32564-0_19

20. Marincas, C., et al.: Contributions to the experimental research of Electronic Diesel Control (EDC) module operation in relation with supply of the N47 engine from BMW 320d (E90) automobile. Sci. Eng. **31**(82), eISSN 2359-828X (2017). http://stiintasiinginerie.ro/31-82
21. Moldovan, A., et al.: Experimental research of the management system from the Peugeot 4007 Sport Utility Vehicle. Sci. Eng. **31**(67), eISSN 2359-828X (2017). http://stiintasiinginerie.ro/31-71
22. Popescu, D., et al.: Development of an automated system for fuel tank level checking and machinery location management to optimize remote accessibility and mobile tracking. In: Proceedings, vol. 63, no. 1, p. 17 (2020). https://doi.org/10.3390/proceedings2020063017

Enhancing the Energy Efficiency of Photovoltaic Cells Through Water Cooling

Marius Brănoaea(✉), Andrei Burlacu, Marina Verdeş, Marius Costel Balan, and Robert Ştefan Vizitiu

Faculty of Civil Engineering and Building Services, "Gheorghe Asachi" Technical University of Iasi-Romania, Blvd. Mangeron, No. 1, 700050 Iasi, Romania
marius.branoaea@student.tuiasi.ro

Abstract. Hybrid systems not only generate a considerable amount of electricity but also provide heat. The thermal energy produced by photovoltaic thermal systems (PVTs) can support the needs of the building and can limit the gas consumption in the heating system.

The research aims to design and study a hybrid photovoltaic system, through cooling a photovoltaic panel using water as a coolant, thus increasing the efficiency of the photovoltaic cells. At the same time hot water is obtained, that can be used directly or as a preheated primary source for the production of hot drinking water or heating agent, thus obtaining a successive cost reduction by reducing fuel consumption.

By cooling a photovoltaic panel with water as a cooling agent, the efficiency of the photovoltaic cells is increasing from 15.74 in the case of the uncooled panel to 17.1 in the case of the water-cooled panel at flow rate v = 10 l/min, obtaining at the same time hot water with temperatures between 19.93 and 54.86 which can either be used directly or can be used as a preheated primary source for the production of hot water or heating agent, thus obtaining a further reduction in costs by reducing fuel consumption. The power production of the photovoltaic cells through cooling is increased by 4.2%, 7.46%, and 7.92%, for the 1 l/min, 6 l/min, and 10 l/min flow rates for the cooling agent.

Keywords: Photovoltaic cooling · Photovoltaic cell efficiency · Energy efficiency · Cooling methods · Computational fluid dynamics

1 Introduction

Energy consumption in the current world has been increasing over the past decades with the only exception being the year 2020 which was lower due to the outbreak of the SARS-CoV-2 viral pandemic. In terms of energy demand, the buildings sector is a net consumer of energy, consuming over 40% of the global energy, and is a major producer of net greenhouse gasses, producing over 30% of the global CO_2 emissions. The major factors for these high percentages are the chaotic urbanization coupled with deforestation as well as global climate change and the surge of indoor residence time and the subsequent increasing demand for indoor thermal environments. These factors

L. Moldovan and A. Gligor (Eds.): Inter-Eng 2021, LNNS 386, pp. 603–615, 2022.
https://doi.org/10.1007/978-3-030-93817-8_54

were accentuated by the pandemic situation which imposed strict circulation restrictions and led to a significant increase in the indoor residence time [1].

The energy requirements for space heating, air-conditioning, and household sanitary water preparation in a building will continue to rise with the developing economies around the world. At the present a significant portion of the global energy production is based on fossil fuels; this is the case due to the fact it is the cheapest form of energy production. Unfortunately, thing has a negative impact on the environment due to the high emissions of CO_2 and other greenhouse gases.

The growth of the gross domestic product (GDP) of a country is dependent on the energy consumption and the adoption of renewable energies, even though it has increased significantly in recent years, it is still slow and has a significantly larger cost per unit in comparison to traditional sources, a viable and impactful solution for the future is increasing the energy efficiency and recovering the waste energy [2–4].

The desire to reduce the energy consumption with the aim to reduce the drawbacks and the effects towards climate change has been highlighted on a regional scale through the implementation by the European Parliament a legislation in the year 2012 that imposed the reduction of CO_2 emissions by 20% by the year 2020. After further studies and analyses that highlighted the fact that these quotas would not be met, the target year was moved to 2030 but the values of the required reduction were set to a reduction of 27% in terms of the energy consumption and a reduction of 40% in terms of the emissions of greenhouse gasses. Apart from the reduction in energy consumption and emissions the aim was to reshape the European Union (EU's) economy in order to provide jobs and growth to the European citizens through the transition process [5].

The adopted legislation carries a very important impact on the EU economy and promotes actions that have the main effect of increasing the energy performance of the buildings sector. This implies that new types of heat exchangers are necessary, with increased performance in comparison to previous ones and harnessing the waste energy from the processes in industrial buildings or from the buildings in the public, commercial or residential sector [6–14].

Sustainable and economic development of the energy sector in recent years was centered around green energy. Renewable energy sources include wind power, hydropower, geothermal power, solar photovoltaic and thermal power, liquid or solid biomass, biogas, tidal and wave power as well as inorganic or organic waste. Almost all of these sources of renewable energy, except geothermal and tidal power, are related directly or indirectly to solar insolation, even biomass is sunlight dependent. In comparison to other types of green energy sources, solar energy distinguishes itself through the potential of being one of the largest suppliers of green energy, being more accessible, and has the potential to be implemented in every location in the world. Furthermore, the conversion from solar radiation to electricity is realized directly through the use of photovoltaic (PV) modules, which are more reliable and more accessible [15].

2 Photovoltaic Panels

Photovoltaic (PV) systems consist of PV cells that convert solar radiation into electrical power (EP) directly. However, the absorbed solar radiation that is not converted into electricity increases PV cell temperature, leading to a reduction of PV conversion efficiency, highlighted in Fig. 1 and Fig. 2. Thereby, PV cooling is necessary in order to keep the electrical efficiency at a satisfactory level and it can be conducted for example by means of liquid or air heat extraction [16].

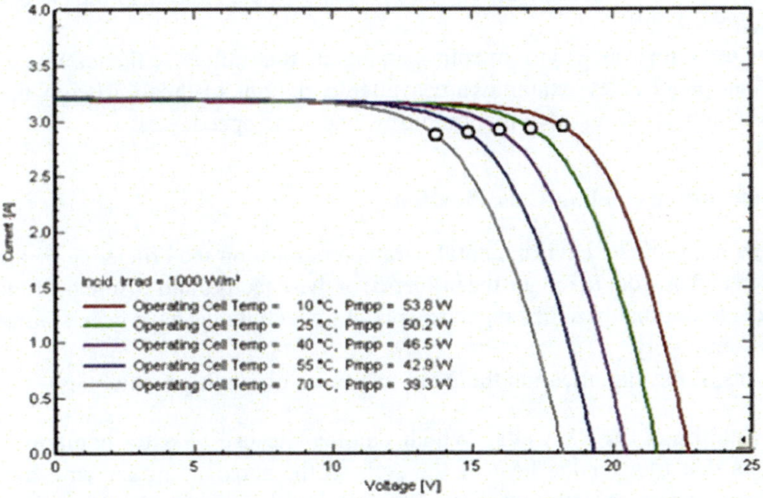

Fig. 1. Photovoltaic Cell current intensity and voltage variation based on cell temperature.

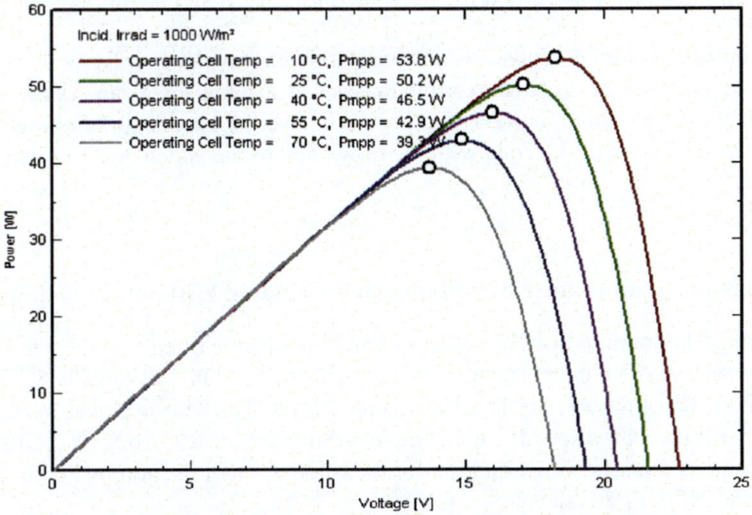

Fig. 2. Photovoltaic Cell power output and voltage variation based on cell temperature

The current efficiency of most commercial solar cells is typically about 17–18% [17]. Complex solar cells have become more accessible and can achieve higher maximum efficiency, the theoretical maximum efficiency achievable in a single p-n junction cell is about 29.4% [18].

The operating temperature of the PV cells in a photovoltaic panel has been studied by several researchers. Under normal conditions, the operating temperature of a PV cell is higher than the ambient temperature. In their analysis researchers determined that the temperature of the PV cells is dependent on the solar irradiance and the environmental temperature [19, 20].

There are numerous environmental parameters that influence the parameters of the photovoltaic panel. A few examples are: available sunlight, installation heigh, the outside and the cell temperatures, shading, humidity, and wind speed.

2.1 Photovoltaic Cooling Classification

Due to the high impact environmental parameters have on the efficiency of the photovoltaic cell and in order to reduce the influence of these factors and ensure the functioning of the cells at optimal parameters, various cooling techniques have been studied by the research community.

In terms of the classification there are two types of cooling techniques:

Passive Cooling of the PV Cells. Which employ the use of extra components, heat exchangers that transfer the heat of the cells to the surrounding air with heat pipes, heat sinks, or phase change materials (PCMs) placed on the backside of the panel in order to increase the natural convection. Heat pipes have a wide range of applications from the IT to the buildings sector [21–23]. These types of cooling techniques have the advantage of being easy and cost-effective, the only costs being the initial ones [24, 25].

Active Cooling of the PV Cells. Which cool the photovoltaic cells with a cooling agent, such as water or air, that require the use of a pump or a fan. These types of systems are more effective than passive ones but have the advantage of being easier to control the parameters of the cells with the drawback of the added cost of operating the pump/fan [26–28].

2.2 Operating Temperature of a Photovoltaic Module Without a Cooling System

The operating temperature of PV cells mounted on a frame tilted normally to the solar noon sun was measured under open-circuit conditions, with no load attached, following conditions of the nominal terrestrial environment (NTE), with a global solar flux of 800 W/m^2, average wind speed: 1 m/s, ambient temperature: 293.16 K (20 °C) and with these recordings the following equation for the calculation of the nominal operating cell temperature (NOCT) was developed [19]:

$$NOCT = (T_c - T_a)_{NTE} + 20\ ^\circ\text{C} \tag{1}$$

T_c = temperature of the PV cell (°C).
T_a = ambient temperature (°C).

The energy transferred by convection has the following expression:

$$Q_c = h \times A \times (T_c - T_a) \tag{2}$$

A well-known expression for the PV cell operating temperature in terms of the NOCT is [29]:

$$T_c = T_a + \left(\frac{G_T}{G_{NOCT}}\right)\left(\frac{U_{L,NOCT}}{U_L}\right)(T_{NOCT} - T_{a,NOCT}) \tag{3}$$

G_T = solar radiation flux on module plane at the cell temperature (W/m^2)
G_{NOCT} = solar radiation flux at the NOCT ($\mathrm{W/m^2}$)
U_L = overall thermal loss coefficient at the cell temperature ($\mathrm{W/m^2K}$)
$U_{L,NOCT}$ = overall thermal loss coefficient corresponding to the NOCT ($\mathrm{W/m^2K}$)
T_{NOCT} = nominal operating cell temperature (°C)
$T_{a,NOCT}$ = ambient temperature corresponding to the NOCT (°C)

The previous equation can be simplified by removing the thermal loss coefficients resulting:

$$T_c = T_a + \left(\frac{G_T}{G_{NOCT}}\right)(T_{NOCT} - T_{a,NOCT}) \tag{4}$$

2.3 Operating Temperature of a Photovoltaic Module with a Cooling System

The inclusion of a cooling system is beneficial owing to the variable characteristics of the photovoltaic cell under higher operating temperatures, resulting in various performance inefficiencies of the photovoltaic system.

In this section, heat transfer between the cooling fluid, seawater in this study, and the PV cells is evaluated. An integrated cooling system provides an important technique to keep PV modules at lower operating temperatures so that the recovered thermal energy can be used to enhance the PV module efficiency. Also, the electrical energy output is improved to increase module effectiveness. The coolant is seawater from the PV module inlet, driven through a pump, with the temperature at 20 °C. This forced convection is added through the side of the PV module to extract heat. The heat transferred from solar radiation within the PV module is the same as which without the coolant [29].

The energy transferred by convection and the energy transferred by radiation are defined as [19]

$$Q_c = h \times A \times (T_c - T_a) \tag{5}$$

$$Q_r = \varepsilon \times \sigma \times A \times \left(T_c^4 - T_a^4\right) \tag{6}$$

ε = PVT module surface emissivity = 0.9
σ = Stefan-Boltzmann Constant = $5.67 \times 10^{-8}(\mathrm{W/m^2})$
A = module surface area (m^2)

The heat transfer of the coolant convection (h_c) is given as a function of Nusselt number (Nu):

$$h_c = \frac{Nu \times k}{D} \tag{7}$$

k = water thermal conductivity = 0.58(W/m K)
D = water pipe diameter (m)

Electric energy (E_c) is calculated by solar radiation energy (Φ) and PV electric efficiency (η_{ref}) and given by the producer with the following formula:

$$E_c = \Phi \times \eta_{ref} \tag{8}$$

The actual temperature of the PV cell at any instant can be expressed by simply applying the energy balance equation:

$$T_c = \frac{1}{\rho V C_p}(\alpha\Phi - Q_c - Q_r - E_c) \tag{9}$$

ρ = photovoltaic polycrystalline silicon density ($\mathrm{kg/m^3}$)
V = module volume ($\mathrm{m^3}$)
C_p = module average specific heat (J/kg K)
α = absorptivity = 0.9

According to the literature, a 1 °C increase in cell temperature of polycrystalline silicon, monocrystalline, and amorphous silicon PV unit will decrease electrical efficiency by nearly 0.45%, 0.45%, and 0.25%, respectively [30, 31].

When the module is heated, the bandgap of the PV cell will decrease resulting in a significant reduction in the open-circuit voltage. The open-circuit voltage is one of the most influenced electrical parameters which affect the electric efficiency. Thus, by adjusting the operating temperature of the PV module with the cooling system, the electric efficiency of the module can be improved. In other words, the decreased thermal energy loss results in an enhanced effectiveness in the PV system.

The pronounced effect that PV operating temperature has upon PV electrical efficiency (η_c) is well documented [19]

$$\eta_c = \eta_{ref} - \left[\beta_{ref} \times \eta_{ref} \times \left(T_c - T_{ref}\right)\right] \tag{10}$$

β_{ref} = solar radiation coefficient ≈ 0.004 K
T_{ref} = reference temperature (°C)

The values of T_{ref} and β_{ref} are normally provided by the module's manufacturer.

The circulating cooling fluid results in a lower PV module operating temperature and a higher production rate of electrical power. The basic correlation for the output power is the following:

$$P = G_T \times l_{pv} \times \eta_c \tag{11}$$

l_{pv} = transmittance of the PV cell

An effective strategy for improving the electrical efficiency of a PV unit is through the attaching of a heat recovery system to the PV unit because through the absorption of excessive heat it decreases the surface temperature of the cell. This hybrid system is named a photovoltaic thermal system (PVTs). The PVT as a cogeneration system converts photons into both thermal and electrical energy. Concurrent generation of low-grade energy (thermal) and high-grade energy (electrical) in PVTs can be utilization for houses, air conditions, and other domestic applications [32].

Coolant fluid type plays a fundamental role in the PVTs performance. Air, pure water, and refrigerants are the most common coolant fluids which are utilized in PVTs [33].

3 Research Methodology

The research addresses multiple numerical simulations performed using CFD numerical simulations performed in the Autodesk CFD program as a means of studying the thermal distribution of both systems during the operation of the photovoltaic panels, in order to demonstrate the viability of the cooling system in improving energy performance and increasing the electricity production of the photovoltaic cells.

For this study, two models were designed in Autodesk Inventor software, a classic photovoltaic panel model and a model with a photovoltaic panel cooled by a perforated metal plate through which water circulates as a coolant.

The photovoltaic panel consists of 60 photovoltaic cells of 156 × 156 mm with a thickness of 5 mm made of monocrystalline silicon. The efficiency of the photovoltaic cells is 17.20% at the normal fractionation temperature of, T_{NOCT} = 46.9 °C. The decrease in the efficiency of the photovoltaic cells is 0.467%/°C.

The photovoltaic cells are cooled by means of a metal plate inside which are holes through which the coolant (cold water) flows.

Water enters the metal plate through a 20 mm diameter pipe, extracting heat from the photovoltaic cells.

In order to make a more conclusive comparison, we analyzed the cases in which the agent flow is 6 l/min, a lower flow rate of 1 l/min, and a higher flow rate of 10 l/min.

In the first phase, it was necessary to model each individual part using the Autodesk Inventor Professional 2021 software (Fig. 3).

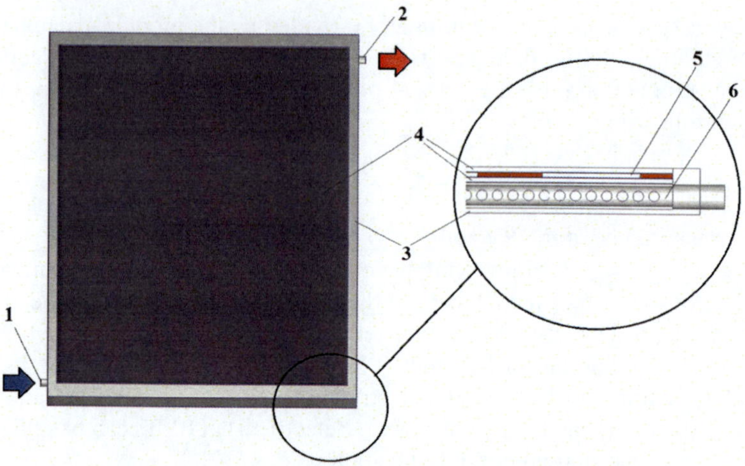

Fig. 3. Composition of the cooled photovoltaic panel

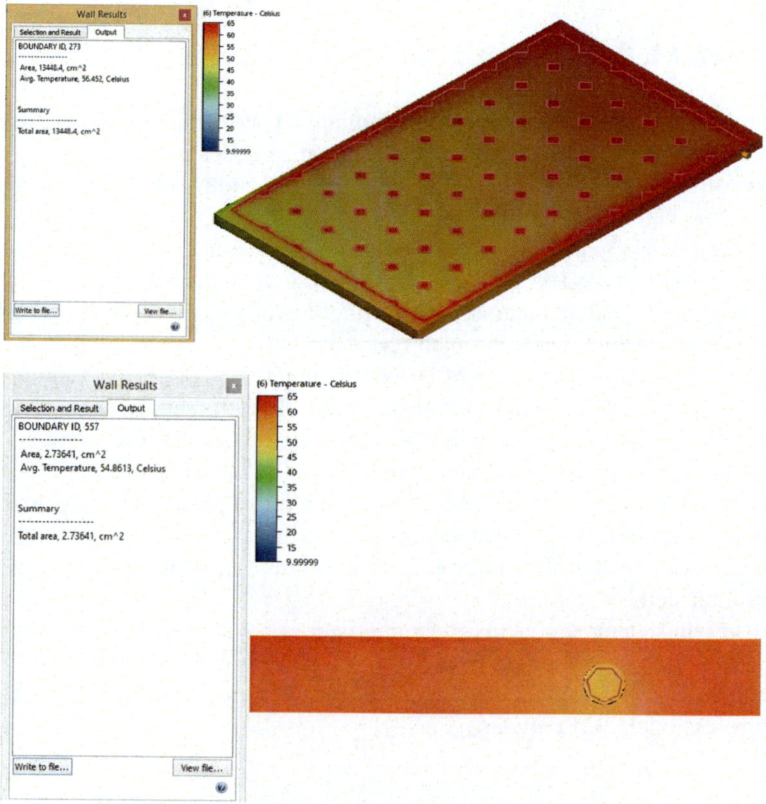

Fig. 4. The temperature of the cells of the water-cooled panel at the flow rate of v = 1 l/min

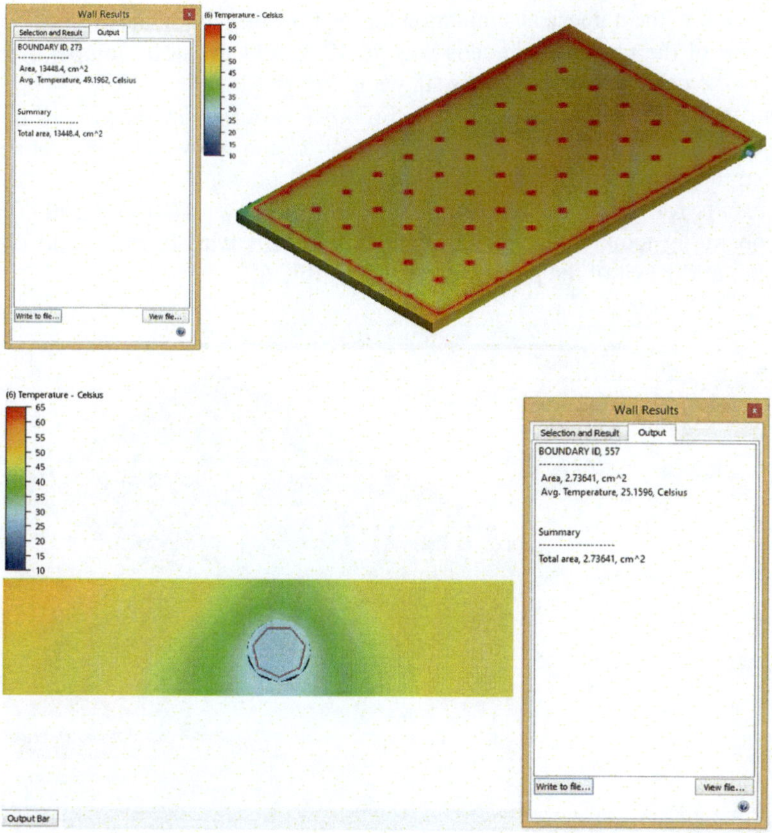

Fig. 5. The temperature of the cells of the water-cooled panel at the flow rate of v = 6 l/min

The panel, shown in Fig. 2, consists of the following elements:

1. Coolant inlet, water, D = 20 mm
2. Coolant outlet, water, D = 20 mm
3. Steel frame, L = 1677 mm, l = 990 mm, h = 45 mm
4. Protective layer, L = 1560 mm, l = 936 mm, h = 5 mm
5. Photovoltaic cells, L = 156 mm, l = 156 mm, h = 5 mm
6. Heat recovery metal plate, L = 1560 mm, l = 936 mm, h = 25 mm

The second stage consists in assigning the simulation parameters and boundary conditions.

Because photovoltaic panels are under the influence of solar radiation, they can heat up significantly, the temperature of 65 °C was used for the temperature of solar cells as an initial condition during operation.

Subsequently, the water inlet flow in the heat recovery plate was imposed for each of the studied cases, 6 l/min, 1 l/min, and 10 l/min and the water temperature was considered 10 °C.

After running the numerical simulation for the flow of 1 l/min, the average operating temperatures of the photovoltaic cells was 56.45 °C and the water temperature at the outlet from the panel was 54.86 °C, as shown in Fig. 4.

After running the numerical simulation for the flow of 6 l/min, the average operating temperatures of the photovoltaic cells was 49.19 °C and the water temperature at the outlet from the panel was 25.16 °C, Fig. 5.

The results for the numerical simulation with the flow of 10 l/min, highlight the average operating temperature of the photovoltaic cells was 48.15 °C and the water temperature at the exit of the panel was 19.93 °C, Fig. 6.

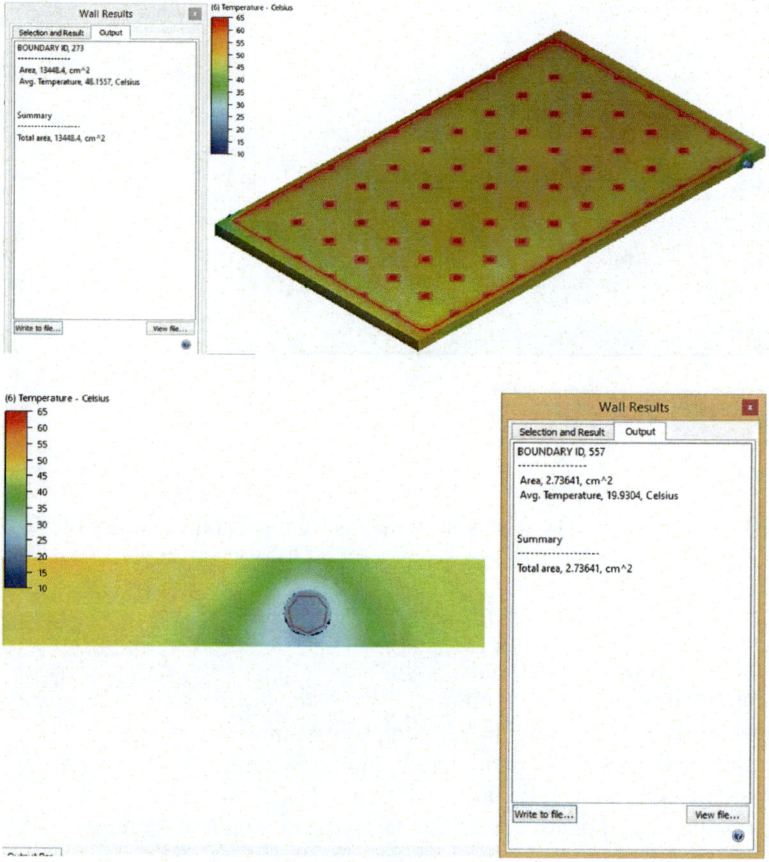

Fig. 6. The temperature of the cells of the water-cooled panel at the flow rate of v = 10 l/min

Regarding the efficiency of the photovoltaic panel, in the case of operation without a cooling system, the efficiency of the cells is 15.75, and in the case of the water-cooled panel at a flow rate v = 10 l/min, the efficiency of the cells is 17.1. as can be seen in Fig. 7.

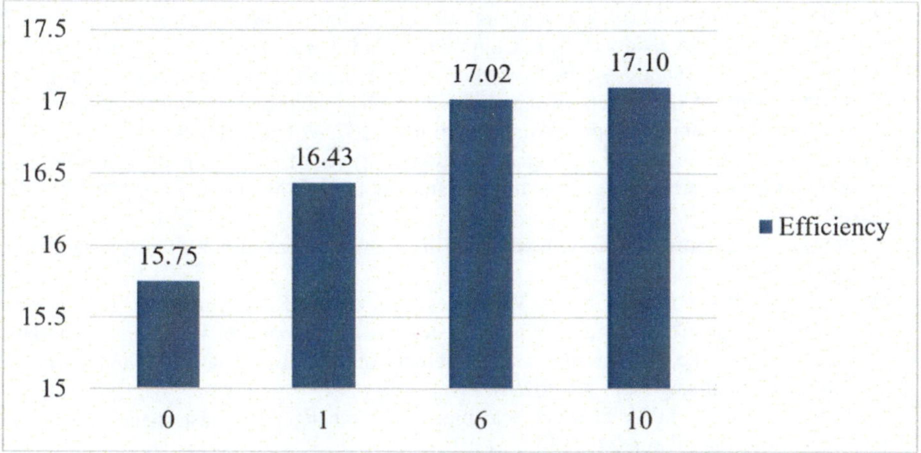

Fig. 7. The efficiency of the photovoltaic panel cells

Through cooling the photovoltaic panel, the efficiency rose by 4.2% using water with the flowrate of 1 l/min, 7.46 for the flowrate of 6 l/min, and 7.92% for the water flow rate of 10 l/min respectively.

4 Conclusions

Hybrid systems not only generate a considerable amount of electricity but also provide heat. The thermal energy produced by PVTs can support the needs of the building and can limit the gas consumption for the heating system or for producing domestic hot water.

This study analyses the performance and efficiency of an original design hybrid photovoltaic system by cooling a photovoltaic panel using water as a cooling agent.

By cooling a photovoltaic panel with water as a cooling agent, the efficiency of the photovoltaic cells is increasing from 15.74 in the case of the uncooled panel to 17.1 in the case of the water-cooled panel at flow rate $v = 10$ l/min, obtaining at the same time hot water with temperatures between 19.93 and 54.86 which can either be used directly or can be used as a preheated primary source for the production of hot water or heating agent, thus obtaining a further reduction in costs by reducing fuel consumption. The power production of the photovoltaic cells through cooling is increased by 4.2%, 7.46%, and 7.92%, for the 1 l/min, 6 l/min, and 10 l/min flow rates of the cooling agent.

Acknowledgments. This work was supported by a grant of the Romanian Ministry of Education and Research, CCCDI-UEFISCDI, project number PN-III-P2-2.1-PED-2019-3112, within PNCDI III.

References

1. Abdullah, M.F., et al.: Research and development efforts on texturization to reduce the optical losses at front surface of silicon solar cell. Renew. Sustain. Energy Rev. **66**, 380–398 (2016)

2. Akalpler, E., Shingil, M.E.: Statistical reasoning the link between energy demand, CO2 emissions and growth: evidence from China. Procedia Comput. Sci. **120**, 182–188 (2017)
3. AL-Musawi, A.I., Taheri, A., Farzanehnia, A., Sardarabadi, M., Passandideh-Fard, M.: Numerical study of the effects of nanofluids and phase-change materials in photovoltaic thermal (PVT) systems. J. Therm. Anal. Calorimetry **137**, 623–636 (2019)
4. Al-Waeli, A.H., Chaichan, M.T., Sopian, K., Kazem, H.A.: Modeling and experimental validation of a PVT system using nanofluid coolant and nano-PCM. Sol. Energy **177**, 178–191 (2019)
5. Bambrook, S.M., Sproul, A.B.: Maximising the energy output of a PVT air system. Sol. Energy **86**(6), 1857–1871 (2012)
6. Brănoaea, M., Burlacu, A., Ciocan, V., Verdeş, M., Vizitiu, R.Ş.: Numerical investigation of a novel heat pipe radiant floor heating system with integrated phase change materials. In: Proceedings of the 14th International Conference on Interdisciplinarity in Engineering—INTER-ENG 2020, no. 63, p. 15 (2020)
7. Burlacu, A., et al.: Energy efficient heat pipe heat exchanger for waste heat recovery in buildings. Procedia Manuf. **22**, 714–721 (2018)
8. Burlacu, A., Lăzărescu, C.D., Ciocan, V., Verdeş, M., Balan, M.C., Şerbănoiu, A.A.: CFD heat transfer analysis for heat pipes integration into buildings with Glazed Façades. Procedia Engineering **181**, 658–665 (2017)
9. Cambeiro, F.P., Armesto, J., Bastos, G., López, J.I., Barbeito, F.P.: Economic appraisal of energy efficiency renovations in tertiary buildings. Sustain. Cities Soc. **47**, 101503 (2019)
10. Cao, X., Dai, X., Liu, J.: Building energy-consumption status worldwide and the state-of-the-art technologies for zero-energy buildings during the past decade. Energy Build. **128**, 198–213 (2016)
11. Dubey, S., Solanki, S.C., Tiwari, A.: Energy and exergy analysis of PV/T air collectors connected in series. Energy Build. **41**(8), 863–870 (2009)
12. Fesharaki, V.J., Dehghani, M., Fesharaki, J.J.: The effect of temperature on photovoltaic cell efficiency. In: Proceedings of the 1st International Conference on Emerging Trends in Energy Conservation – ETEC, Tehran (2011)
13. Fotis, P., Karkalakos, S., Asteriou, D.: The relationship between energy demand and real GDP growth rate: the role of price asymmetries and spatial externalities within 34 countries across the globe. Energy Econ. **66**, 69–84 (2017)
14. Gong, G., Zeng, W., Wang, L., Wu, C.: A new heat recovery technique for air-conditioning/heat-pump system. Appl. Therm. Eng. **28**(17–18), 2360–2370 (2008)
15. Han, C., Zou, L.: Study on the heat transfer characteristics of a moderate-temperature heat pipe heat exchanger. Int. J. Heat Mass Transf. **91**, 302–310 (2015)
16. Hudişteanu, S.V., et al.: Effect of cooling on power generated by photovoltaic panels. IOP Conf. Ser.: Mater. Sci. Eng. **1141**, 012008 (2021)
17. IEA (2020), Global Energy Review 2020. (n.d.). (IEA, Paris). https://www.iea.org/reports/global-energy-review-2020
18. Jouhara, H., Chauhan, A., Nannou, T., Almahmoud, S., Delpech, B., Wrobel, L.C.: Heat pipe based systems - Advances and applications. Energy **128**, 729–754 (2017)
19. Jouhara, H., Khordehgah, N., Almahmoud, S., Delpech, B., Chauhan, A., Tassou, S.A.: Waste heat recovery technologies and applications. Therm. Sci. Eng. Progr. **6**, 268–289 (2018)
20. Kalogirou, S.A., Tripanagnostopoulos, Y.: Hybrid PV/T solar systems for domestic hot water and electricity production. Energy Convers. Manag. **47**(18–19), 3368–3382 (2006)
21. Kopnina, H.: Energy policy in the European union: renewable energy and the risks of subversion. In: Governance and Security Issues of the European Union, pp. 167–184. M.C. Asser Press, The Hague (2016)
22. Reay, D.A., McGlen, R.J., Kew, P.A.: Heat pipes, 6th Edn. Theory, Design and Applications. Elsevier Ltd. (2014). https://doi.org/10.1016/C2011-0-08979-2

23. Richter, A., Hermle, M., Glunz, W.: Reassessment of the limiting efficiency for crystalline silicon solar cells. IEEE J. Photovolt. **3**, 1184–1191 (2013)
24. Saghafifar, M., Omar, A., Mohammadi, K., Alashkar, A., Gadalla, M.: A review of unconventional bottoming cycles for waste heat recovery: part I – Analysis, design, and optimization. Energy Convers. Manag. **198**, 110905 (2019)
25. Sahota, L., Tiwari, G.N.: Review on series connected photovoltaic thermal (PVT) systems: analytical and experimental studies. Sol. Energy **150**, 96–127 (2017)
26. Salari, A., Taheri, A., Farzanehnia, A., Passandideh-fard, M., Sardarabadi, M.: An updated review of the performance of nanofluid-based photovoltaic thermal systems from energy, exergy, economic, and environmental (4E) approaches. J. Clean. Prod. 124318 (2020)
27. Skoplaki, E., Palyvos, J.A.: On the temperature dependence of photovoltaic module electrical performance: a review of efficiency/power correlations. Sol. Energy **83**(5), 614–624 (2009)
28. Skoplaki, E., Palyvos, J.A.: Operating temperature of photovoltaic modules: a survey of pertinent correlations. Renew. Energy **34**(1), 23–29 (2009)
29. Srimuang, W., Amatachaya, P.: A review of the applications of heat pipe heat exchangers for heat recovery. Renew. Sustain. Energy Rev. **16**(6), 4303–4315 (2012)
30. Tarabsheh, A.A., et al.: Performance of photovoltaic cells in photovoltaic thermal (PVT) modules. IET Renew. Power Gener. **10**(7), 1017–1023 (2016)
31. Vasiliev, L.L.: Heat pipes in modern heat exchangers. Appl. Therm. Eng. **25**, 1–19 (2005)
32. Vizitiu, R.Ş., et al.: Experimental investigation on the optimum filling ratio of heat pipes used for heat recovery systems. In: Proceedings of International Conference Building Services and Energy Efficiency, Sciendo, pp. 154–161 (2020)
33. Wu, S., Xiong, C.: Passive cooling technology for photovoltaic panels for domestic houses. Int. J. Low-Carbon Technol. **9**, 118–126 (2014)

Multicriterial Assessment of Power Losses in Electricity Distribution Grid Considering the Profile Consumers Analysis

Adrian Gligor[1], Cristian-Dragoș Dumitru[1(✉)], and Ilie Vlasa[2]

[1] "George Emil Palade" University of Medicine, Pharmacy, Science, and Technology of Targu Mures, Gh. Marinescu str. 38, 540142 Targu Mures, Romania
cristian.dumitru@umfst.ro

[2] D.E.E.R. Mures Branch Distributie Energie Electrica Romania S.A., Targu Mures, Romania

Abstract. Many evolutions in current areas of daily life depend on major events occurrence. A more sensitive domain, due its cascade effects on other interconnected fields, is represented by the electricity sector. The analysis of the major COVID-19 influence on the electricity usage and its repercussion on related economic domains is highlighted in this paper. The results and conclusions are based on records from a local electricity distribution branch. The obtained results represent important key points in establishing proper energy management strategies for developing the future flexible electricity distribution systems. This unprecedented perturbation caused pandemic situation in the referenced period reflects possible challenges in designing the management solutions for the future that must be provisioned for secured and high-quality electricity distribution systems. Thus, the available models must be reconsidered for a proper power losses estimation and evaluation, electricity consumptions forecasting, technical resources provisioning and for the path to establish the tariff of delivered electricity. The electricity price is controlled also by the fluctuations in electricity demands both for industrial and household consumers which intend to reduce the amount of energy consumed by local production of energy needed for daily activities.

Keywords: Power losses · Distribution grid · COVID-19 · Electricity consumers profiles

1 Introduction

1.1 Electricity in the Current Context

The use of electricity became more and more diversified with the evolution of nowadays society. From fields of activity such as industry, agriculture, transportation, mining, commercial [1, 2], to medical domains and education the electricity or residential became indispensable in any daily activity. As an example, the electricity plays an important role in the medical sector from ensuring optimal operation of the entire medical equipment necessary to monitor and to preserve the patient live, to the diagnosis and treatment of

L. Moldovan and A. Gligor (Eds.): Inter-Eng 2021, LNNS 386, pp. 616–628, 2022.
https://doi.org/10.1007/978-3-030-93817-8_55

various chronic diseases [3, 4]. Unfortunately, the electricity and the oxygen became in the last period a major factor in the recovery process and in the treatment of various diseases, playing and important role in the current pandemic situation.

The reduction of activity in many energy-consuming industries has led especially in European countries to a decrease in industrial consumption, maintaining or decreasing residential consumption even if an increasing number of electrical appliances following the trend of comfort was adopted in contemporary society due to higher energy efficiency policies [5–7].

1.2 Electricity Sources and Consumers

The electricity sector is divided in four main categories: production, transport, distribution, and consumption of electricity. According to Transelectrica (the Transmission Operator of Power in the Romanian Electricity System), in a regular summer day, the main producers of electricity remain the fuel-fired power plants as it can be seen in Fig. 1.

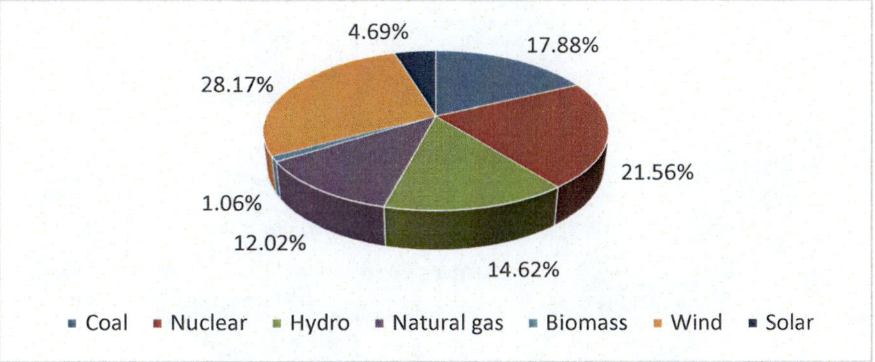

Fig. 1. Electricity production in Romania (15 July 2021) [8]

Nowadays, the transition to renewable resources is promoted and supported by different governmental financial schemes and incentives with an important role of a new type of actor in the energy market namely the prosumer.

The concept of prosumer became more and more frequently used these days. Many consumers choose to locally produce the electricity, the surplus being transferred to electrical distribution grids. The number of prosumers is constantly growing both in Europe and around the world. For example, an important number of prosumers can be found in Germany where it has reached one million [9].

1.3 Regulations Impact in the Evolution of Electricity Sector

The electricity distribution grid has spectacularly evolved in the first decade of the 21st century mainly due to the diversified requirements of consumers that imposed higher standards in terms power quality delivered to consumers. In this regard, the availability of real-time data is very important both for the producer and consumer, but also for the distributor and electricity supplier to reduce incidents in distribution grids and ensure continuous supply of consumers, mainly vulnerable consumers, such as hospitals, public institutions, household consumers with special needs, etc. [10]. In nowadays context, this type of service can be provided by the smart grids.

Thus, it can be concluded that the automation and modern power converter devices, as well as the implementation of smart meters with the possibility of remote reading is a first step in building a smart grid [11]. This trend is primarily given by national and international regulations that require the increasing of the power quality, of energy efficiency and transparency of electricity consumption [12]. A smart grid, as depicted in Fig. 2, is mainly composed of many components such as supervisory control and data acquisition (SCADA) systems, digital protection and control devices, smart metering systems, all interconnected via communication networks that allows data transfer between the key-players involved in production, transport, distribution and consumption of electricity.

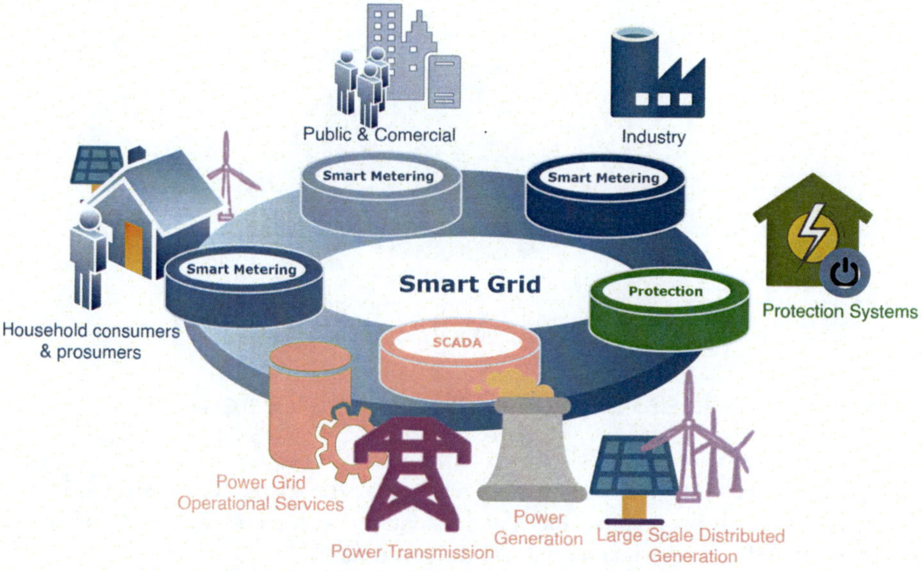

Fig. 2. Smart grid components and services.

This sector is one of the most dynamic one due to rapid of evolution and new technologies development in the field of communications, data storage and processing, monitoring and control.

1.4 COVID-19 Pandemic Situation and Electricity Market

The operation in quality, efficient and safe conditions of the electrical grids can be achieved when the evolution of these systems is predictable. Currently, however, there are several factors that introduce a higher degree of uncertainty, such as renewable sources, but also several external factors such as weather conditions or fluctuations in the economy in general. In addition to these, the social aspect cannot be neglected being derived from the consumption behavior of household consumers and in the last two years as a new aspect by the pandemic situation, unprecedented at this level. Numerous scientific papers have been published lately on this topic [13–15].

The knowledge of the factors that can affect the operation of electricity grids is thus essential by first reducing nondeterminism and achieving adequate operating plans. In addition to this part of operational management, however, we must deal with other aspects, such as energy market forecasting but also energy loss control. In this paper, this last aspect is intended to be determined in terms of the major changes in the consumption profile of this period. This approach is useful for calibrating the parameters of energy demand forecasting algorithms, but also for prediction of energy costs and for the most accurate assessment of losses that allow their identification and reduction with the benefit of both electricity distribution network operators and consumers.

2 Electricity Consumptions in Power Distribution Networks

2.1 Electricity Consumption Map

Electricity consumers, if the Romanian regulations are considered, can be divided into three main categories: industrial consumers corresponding to economical agents, usually considered small consumers with installed power under 100 kW and large industrial consumers over mentioned installed power limit, household consumers and a third category not comprising the previous two mentioned classes, namely tertiary consumers including public institutions, hospitals, schools, etc. The analysis of electricity consumption by different types of consumers is essential mainly for the Electricity Distribution Operator (DO) to establish the optimal operational management decisions and to reduce its own technological consumption (OTC).

Usually, the electricity consumption is not uniform, it varies from one type of consumer to another, depending on: season, geographical area, weather conditions, special conditions such as the occurrence of unpredictable events (such as COVID-19 pandemic) as well as the period of the day for which it is considered.

An important role in detecting energy losses, mainly non-technical ones, is the analysis of energy consumption on the above-mentioned groups of consumers [16], a series of algorithms for detecting electricity consumption being thus implemented. Such algorithms are based on the anomalies appeared in the load curves recorded in data acquisition systems (smart metering) at an imposed time-sampling rate (nowadays established at 15 min) [17].

2.2 System Overview on the Electricity Distribution

In Romania, according to the legislation imposed by the Romanian Energy Regulatory Authority (RERA) by law no. 123/2012 - on the Law on Electricity and Natural Gas, the determination of the energy entered in the contour of any Electricity Distribution Branch can be calculated with the following equation:

$$E_i = E_{Ttr} + E_{export} + E_{tOD} + E_D + E_{OTC} \quad (1)$$

where:

E_i – monthly electricity entered in the contour determined by measurements;
E_{Ttr} – electricity transferred in transport, also determined by measurements;
E_{export} – electricity transferred to other countries;
E_{tOD} – electricity transferred to other grid operators;
E_D - electricity distributed to consumers determined by measurements or estimations;
E_{OTC} – billed electricity corresponding to OTC.

In Eq. 1 it is highlighted that the distributed electricity represents an essential element in establishing the OTC of the distribution operator. So, as the distributed amount of electricity increases, the energy losses from the distribution grids also increase proportionally. Mainly, a high value of the OTC is determined by a large amount of energy that supply the residential consumers.

Figure 3 highlights the energy outline at the power distribution grid level and its most relevant disturbing factors.

The energy losses in electricity transmission and distribution power grids represent the difference between the amount of electricity entering the power grid and the amount of electricity exiting the respective power grid as electricity distributed to consumers or energy delivered to neighboring operators.

These losses together with measurement errors and unauthorized theft of electricity represent the OTC of the DO [18]. To cover these losses, the distribution operator must raise the distribution tariffs, which will be found in the price per kWh paid by the end consumer.

2.3 Economic Trends Influence on Electricity Usage

The evolution of electricity consumption is an important issue for the entire power system contributing to its overall configuration and stability. For the electricity transmission and distribution operators, the importance of an accurate estimation of electricity consumption, to cover the costs due to energy losses from electricity grids is also very important. However, the most important benefits of an accurate energy forecast are for suppliers because the price of electricity is strongly influenced by the supplier's ability to predict energy consumption, to purchase energy in time, without having to buy it from the day-ahead market (DAM), where prices are quite high [19].

The main factors that influence the price of electricity are:

- the price of gas and fossil fuels;
- economic growth indicator (GDP) and salaries evolution;
- climatic factors;
- losses in electrical grids;

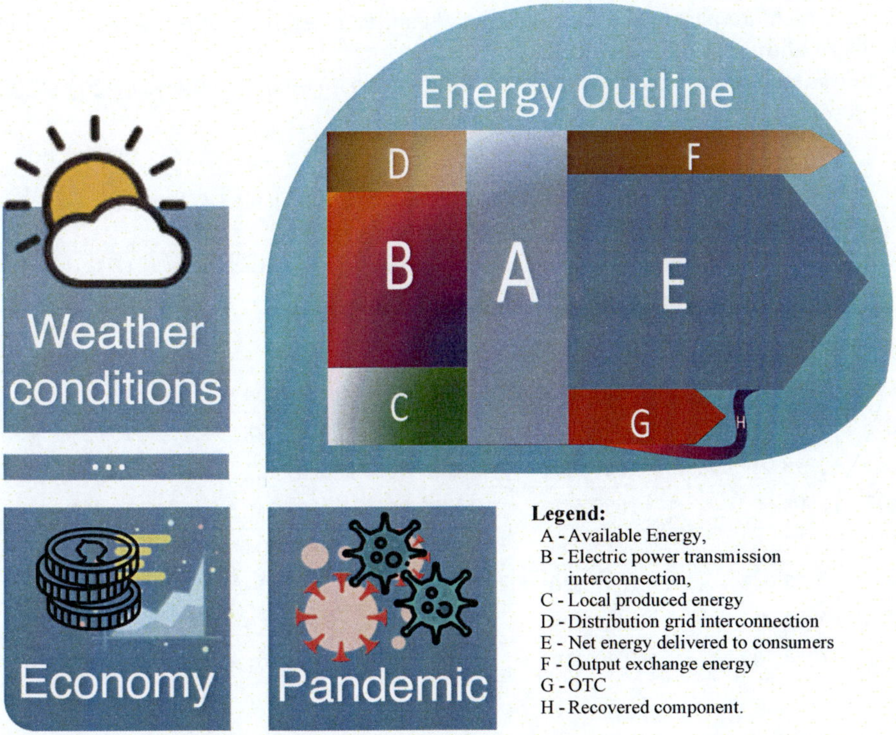

Fig. 3. Energy outline at the power distribution grid level and disturbing factors.

- extreme factors such as the ones related to weather conditions as well as the appearance of a pandemic (COVID 19).

The mentioned factors are specific to each distribution power grid, but it must be considered that the price of electricity largely depends on the ratio between market demand and electricity production, without additional costs.

3 Research Methodology on Electricity Consumptions in Power Distribution Grids

The electricity entered in the power distribution grid, as highlighted in Fig. 3, represents the sum of the measured quantities of electricity received by the DO from: producers/prosumers of electricity, the neighboring DOs, as well as by the Electricity Transmission Operator (TO). The electricity output of the power grid represents the sum of the measured quantities of electrical energy: distributed to consumers, ceded to neighboring DOs, as well as to the interconnection lines.

The distributed electricity represents the amount of energy transferred to the final consumers and which represents the sum of the quantities of electricity transferred by DO registered or estimated, on all electricity meters mounted in all delimitation/measurement points, which are part of the same category (captive or eligible consumers), for which the distribution service is invoiced.

In this study are considered data of a Romanian DO related to the years 2014–2021 are presented in Table 1.

Table 1. The circulation of the electricity amounts in the DO contour

Year	Electricity input [GWh]	Electricity within contour [GWh]	Distributed electricity [GWh]	OTC Electricity [GWh]	OTC (%) – reported to distributed electricity
2014	2069.730	1501.285	1321.801	179.484	13.58%
2015	2315.550	1554.060	1374.631	179.428	13.05%
2016	2278.911	1569.066	1399.903	169.162	12.08%
2017	2508.602	1650.572	1482.326	168.246	11.35%
2018	2322.631	1713.064	1545.160	167.904	10.87%
2019	2137.641	1647.536	1391.506	153.429	10.03%
2020	2109.077	1667.429	1413.579	156.849	10.73%
2021	1254.793	1017.217	922.726	94.491	10.24%

The main goal is to analyze the existing records for evaluating the evolution of energy input and output, from the household and industrial perspective, and for each category OTC were intended to be highlighted.

Based on existing data statistical calculations and non-technical loses have been determined to evaluate the particular feature exhibited in evolution of these terms in the pandemic period.

Results represents the base for formulating the conclusions on how pandemic period compared to other type of unpredictable events influenced consumptions and DO performance indices.

4 COVID-19 Influence Analysis on Electricity Consumptions

4.1 Distributed Electricity Evolution Analysis

As mentioned in the previous section for this study ware considered data available for last 7 years.

Firstly, was analyzed the distributed energy comparatively for 2019, 2020 and 2021 to highlight the evolution from pre until current situation of pandemic period. As shown in Table 2.

As it can be seen in Table 2, the electricity distributed to consumers within the analyzed DO contour during the COVID 19 pandemic increased significantly in the case of

Table 2. Distributed electricity in years 2019–2021.

Month	DE 2019 [MWh]		DE 2020 [MWh]		DE 2021 [MWh]	
	Household	Industrial	Household	Industrial	Household	Industrial
January	41,265.488	92,952.805	38,925.044	90,825.104	44,465.903	90,279.259
February	39,315.490	85,250.608	38,240.374	89,227.538	41,504.079	88,196.167
March	40,601.292	92,736.349	44,967.449	76,566.196	44,015.093	97,969.078
April	39,295.559	86,173.986	43,199.726	67,568.802	39,158.405	91,369.611
May	38,161.034	88,375.746	51,249.690	67,935.635	41,397.716	92,143.304
June	28,828.780	66,433.821	52,446.474	69,522.070	41,415.639	76,914.757
July	34,081.921	79,457.817	53,415.191	73,763.835	45,525.016	88,372.091
August	40,779.569	94,985.662	58,160.686	71,085.283	N/A	N/A
September	38,218.588	88,843.371	59,673.155	70,051.094	N/A	N/A
October	40,009.604	91,855.742	59,025.530	66,560.704	N/A	N/A
November	38,454.419	87,060.312	60,885.495	74,415.605	N/A	N/A
December	36,710.629	81,658.134	62,499.780	73,369.306	N/A	N/A

household consumers, with over 10 GWh, but decreased in the case of industrial consumers. The total amount of electricity distributed in the pandemic year 2020 increased by about 2% compared to the reference year 2019.

Mainly, this conclusion highlights the necessity of a special attention given to the household consumer by enhancing the quality of the electricity metering service, however planned for sustained development by regulations and development plans. This leads to one of known ways to decrease the unregistered consumptions, either by long period between metering readings, low precision readings or unauthorized electricity usage.

Performed calculation was synthetically presented in Fig. 4 and illustrate the level of increasing rate in electricity consumption on the household profile that directly influences the increase in OTC in the distribution grid.

According to the situation presented in Fig. 4, the OTC corresponding to the analyzed DO increased significantly during the pandemic period by over one percent in each month of 2020 compared to the reference year 2019. Overall, the losses related to 2020 increased by 3 GWh: if in year 2019 the energy losses were about 153 GWh, in 2020 the energy losses reached 156 GWh. As it can be seen, the trend of increasing losses is maintained at the beginning of 2021, but from April, with the resumption of activities in the normal parameters of economic agencies and large consumers, they began to decline significantly.

An important factor that keeps the OTC related to each DO within normal limits is the regular and periodic reading of energy meters related to consumers, a reading that was completely missing during the pandemic period due to the health rules with a negative impact on OTC.

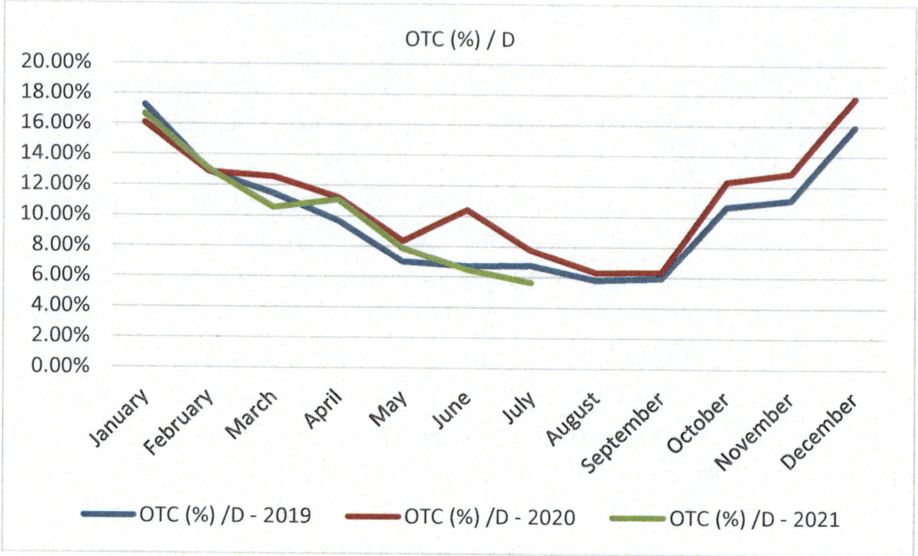

Fig. 4. OTC reported to the electricity distribution in 2019–2021 period.

Any DO forecasts in advance the amount of energy needed to cover technical and non-technical losses in electricity grids, to buy energy at a relatively low price, compared to the day-ahead market prices. An unpredictable event such as COVID-19, upset all energy estimations, the considered DO, according to the evolution of the energy losses in the last 3 years, forecasted an even smaller amount of energy to cover OTC for 2020, but according to Table 3, the values were totally different.

Table 4 shows the difference in energy distributed to household consumers corresponding to the years 2021 - 2014, compared to the average of the reference years 2014 - 2019. As it can be seen in the table, the largest standard deviation from the considered mean value is recorded in 2020 with a value of +218.872 GWh (pandemic year) illustrating a significant increase in household consumption, which led to increased losses in electricity distribution grids.

Table 5 shows the difference in energy distributed to industrial consumers corresponding to the years 2021–2014, compared to the average of the reference years 2014–2019. As it can be seen in this table, the largest standard deviation from the considered mean value is recorded in 2020, with a value of -141.181 GWh (pandemic year) related to a decreasing in activity of the industrial area, most of the activities being carried out online from home.

As it can be seen in Tables 4 and 5, in recent years there has been a slight increase in distributed energy, compared to years 2014 and 2015. Even if 2020 was a year full of unpredictable events, the energy distributed to consumers from DO increased compared to 2014–2016 by over 2 GWh and compared to 2017–2019 by almost 1 GWh.

Table 3. The amount of forecasted and actual energy required for OTC coverage for 2019–2020

Month	OTC [GWh] - 2019			OTC [GWh] - 2020		
	Real	Estimated	Difference	Real	Estimated	Difference
January	22.408	23.238	0.83	21.567	22.597	1.03
February	16.427	17.057	0.63	16.036	16.836	0.8
March	13.930	13.21	−0.72	15.239	15.769	0.53
April	10.724	10.954	0.23	10.586	11.216	0.63
May	8.319	8.069	−0.25	10.576	10.826	0.25
June	8.168	7.848	−0.32	9.929	10.249	0.32
July	8.586	9.306	0.72	8.822	8.942	0.12
August	7.499	7.189	−0.31	7.343	7.553	0.21
September	7.698	7.558	−0.14	8.053	8.293	0.24
October	13.392	14.382	0.99	12.944	14.064	1.12
November	15.060	16.11	1.05	13.817	15.067	1.25
December	21.633	22.563	0.93	21.112	21.942	0.83
Total	153.849	157.484	3.64	156.029	163.354	7.33

Table 4. Difference of energy distributed to household consumers for the years 2021–2014, compared to the average of 2014–2019.

Month	STD_{DE} 2021 [GWh]	STD_{DE} 2020 [GWh]	STD_{DE} 2019 [GWh]	STD_{DE} 2018 [GWh]	STD_{DE} 2017 [GWh]	STD_{DE} 2016 [GWh]	STD_{DE} 2015 [GWh]	STD_{DE} 2014 [GWh]
January	6.323	0.783	3.123	2.443	0.200	−1.609	−1.498	−2.658
February	5.105	1.841	2.916	2.711	−0.74	−0.632	−1.475	−3.445
March	5.903	6.856	2.489	1.930	0.939	−1.183	−1.145	−3.030
April	2.433	6.475	2.570	1.789	1.282	−1.769	−2.044	−1.828
May	5.343	15.195	2.106	3.511	2.564	−3.819	−1.524	−2.839
June	8.195	19.226	−4.391	5.032	0.031	−2.582	2.348	−0.438
July	10.499	18.390	−0.943	4.515	-3.490	0.69	1.075	−1.227
August	N/A	21.476	4.095	3.143	2.086	−0.891	−1.176	−7.257
September	N/A	24.278	2.823	2.238	2.894	0.583	−2.609	−5.930
October	N/A	21.663	2.647	0.891	2.864	−0.031	−4.218	−2.153
November	N/A	23.339	0.908	2.249	2.413	0.124	−3.195	−2.500
December	N/A	24.302	−1.486	2.469	2.257	0.870	−3.025	−1.086
Total	N/A	218.872	16.685	32.781	13.931	−10.79	−18.377	−34.226

STD_{DE} – Standard deviation of distributed energy.

4.2 Discussions and Findings

The obtained results for the considered period and scenario show that a significant electricity consumption decreasing in case of industrial customers was recorded. The amount of this decreasing is illustrated in Fig. 5.

Table 5. Difference of energy distributed to industrial consumers for the years 2021–2014, compared to the average of 2014–2019.

Month	STD_{DE} 2021 [GWh]	STD_{DE} 2020 [GWh]	STD_{DE} 2019 [GWh]	STD_{DE} 2018 [GWh]	STD_{DE} 2017 [GWh]	STD_{DE} 2016 [GWh]	STD_{DE} 2015 [GWh]	STD_{DE} 2014 [GWh]
January	4.614	5.160	7.288	5.700	0.467	−3.756	−3.496	−6.203
February	9.438	10.469	6.492	6.034	0.166	−1.408	−3.283	−7.669
March	11.042	−10.360	5.809	4.504	2.192	−2.761	−2.672	−7.071
April	10.918	−12.882	5.722	3.982	2.854	−3.938	−4.551	−4.069
May	8.683	−15.524	4.915	8.194	5.984	−8.912	−3.557	−6.624
June	0.235	−7.157	−10.245	11.742	0.072	−6.024	5.479	−1.023
July	6.713	−7.894	−2.200	10.536	−8.143	0.163	2.508	−2.863
August	N/A	−14.344	9.556	7.334	4.869	−2.080	−2.745	−16.934
September	N/A	−12.204	6.588	5.224	6.753	1.361	−6.089	−13.836
October	N/A	− 9.117	6.177	2.080	6.683	−0.073	−9.843	−5.024
November	N/A	−10.524	2.120	5.248	5.631	0.289	−7.455	−5.834
December	N/A	−11.756	−3.467	5.762	5.268	2.030	−7.058	−2.534
Total	N/A	−141.181	31.862	76.490	32.506	−25.189	−42.879	−79.860

STD_{DE} – Standard deviation of distributed energy.

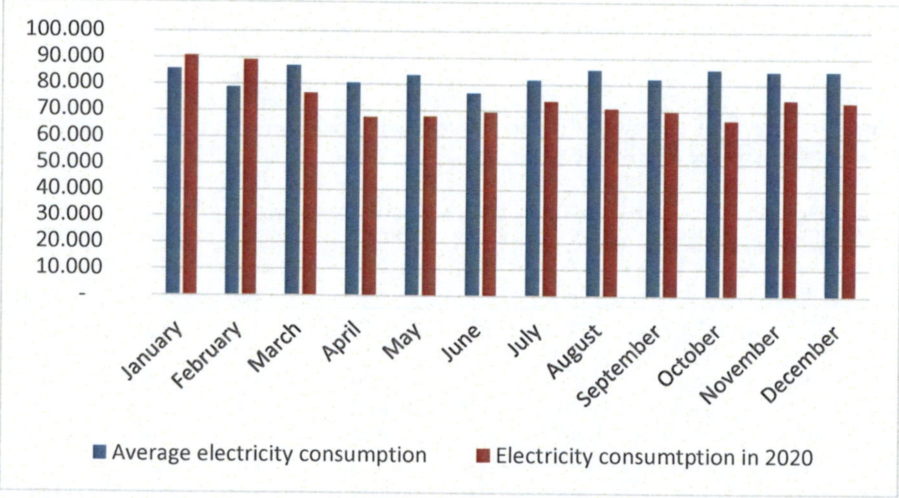

Fig. 5. Industrial average consumption for 2014–2019 period and the consumption for 2020.

The obtained results for most of the considered period and scenario show that a significant electricity consumption increasing in case of household customers was recorded, excepting the winter period where the trend was almost similar regardless of the analyzed period. Graphically this situation can be depicted from Fig. 6.

These differences obtained for the two categories of analyzed consumers are reflected in the significant variation of OTC related to DO, illustrated above in Fig. 4.

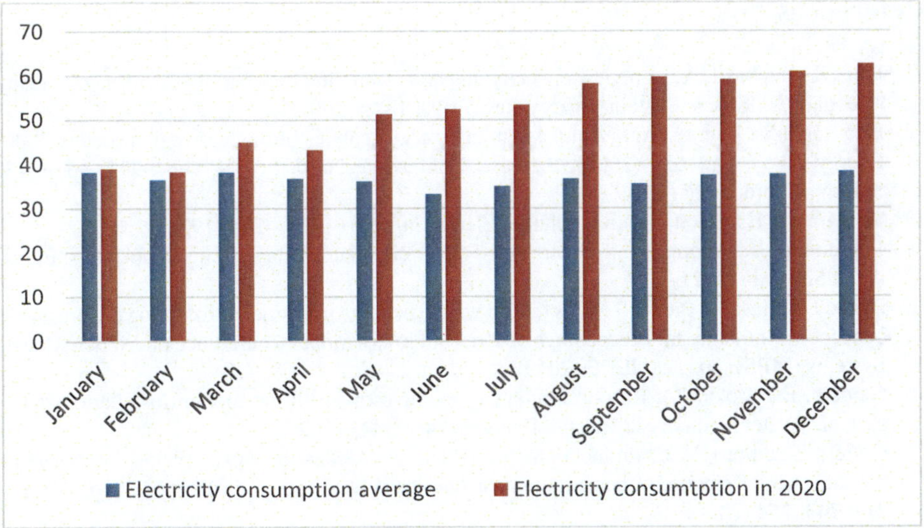

Fig. 6. Household average consumption for 2014–2019 period and the consumption for 2020.

5 Conclusion

In this work the proposed research was focused on assessment of the pandemic situation influence on electricity balance of a DO from Romania and its important performance indices represented by OTC.

The main results were obtained by considering recordings available from a period of almost 8 years (2014- mid of 2021). For an accurate image of the situation a wider period could be considered. However as smart grid evolved in general and smart metering technologies improved in particular during the last period of time, we can consider that no relevant results might be obtained.

Research performed show that household consumption has increased and given the nowadays level of offered metering service in general and smart metering system implementation in particular and overall, the OTC increased. This requires the parameters of energy forecast methods updating or even the adoption of new algorithms.

In future, if other major event might arise, in the nowadays condition of high level of penetration of less predictable and controllable renewable energy sources, a real support from the new technologies might solve the problem of maintaining the OTC under a desired threshold with multiple benefits for DO and consumers.

The mentioned situation can be solved if new forecast algorithms are developed, able to adapt faster to changes in consumption behaviors. The nowadays technologies from the field of data analytics or cognitive computing could constitute a stating point. Realtime data analytics with automated classification using artificial intelligence were developed and successfully tested in other scientific domains and can be transferred and adopted for future researches and developments in the field of electricity distribution.

References

1. Utlu, Z., Hepbasli, A.: A review on analyzing and evaluating the energy utilization efficiency of countries. Renew. Sustain. Energy Rev. **11**(1), 1–29 (2007)
2. Ertesvåg, I.S.: Energy, exergy, and extended-exergy analysis of the Norwegian society 2000. Energy **30**(5), 649–675 (2005). S. (eds.) Conference 2016, LNCS, vol. 9999, pp. 1–13. Springer, Heidelberg (2016)
3. Williams, R.E.: Electricity in medicine. Electr. Eng. **57**(6), 237–244 (1938)
4. Vourdoubas, J.: Applications of distributed electricity generation systems in hospitals. Eur. J. Appl. Sci. **9**(1) (2021)
5. Bălan, D., Vlasa, I., Bălan, S.M.: Is there a way to become prosumer? Promoting the prosumer concept in Romania. In: Proceedings of 2019 8th International Conference on Modern Power Systems (MPS), pp. 1–8. IEEE (2019)
6. Gajdzik, B., Sroka, W.: Resource intensity vs. investment in production installations—the case of the steel industry in Poland. Energies **14**(2), 443 (2021)
7. Gutiérrez-Pedrero, M.J., Tarancón, M.Á., del Río, P., Alcántara, V.: Analysing the drivers of the intensity of electricity consumption of non-residential sectors in Europe. Appl. Energy **211**, 743–754 (2018)
8. Cialani, C., Mortazavi, R.: Household and industrial electricity demand in Europe. Energy Policy **122**, 592–600 (2018)
9. Transelectrica, Electricity production, consumption and exchange. https://www.transelectrica.ro/widget/web/tel/sen-grafic/-/SENGrafic_WAR_SENGraficportlet. Accessed 15 July 2021
10. Mendiola, J.E., Pedrasa, M.A.A.: Detection of pilferage in an ami-enabled low-voltage network using energy reading anomalies. In: Proceedings of 2019 International Conference on Smart Grid Synchronized Measurements and Analytics (SGSMA), pp. 1–6 (2019)
11. González-Sotres, L., Mateo, C., Frías, P., Rodríguez-Morcillo, C., Matanza, J.: Replicability analysis of PLC PRIME networks for smart metering applications. IEEE Trans. Smart Grid **9**(2), 827–835 (2016)
12. Kuc-Czarnecka, M.E., Olczyk, M., Zinecker, M.: Improvements and spatial dependencies in energy transition measures. Energies **14**(13), 3802 (2021)
13. Werth, A., Gravino, P., Prevedello, G.: Impact analysis of COVID-19 responses on energy grid dynamics in Europe. Appl. Energy **281**, 116045 (2021)
14. Krarti, M., Aldubyan, M.: Review analysis of COVID-19 impact on electricity demand for residential buildings. Renew. Sustain. Energy Rev. 110888 (2021)
15. Kirli, D., Parzen, M., Kiprakis, A.: Impact of the COVID-19 lockdown on the electricity system of Great Britain: a study on energy demand, generation Pricing and Grid Stability. Energies **14**(3), 635 (2021)
16. Su, C.L., Lee, W.H., Wen, C.K.: Electricity theft detection in low voltage networks with smart meters using state estimation. In: 2016 IEEE International Conference on Industrial Technology (ICIT). IEEE (2016)
17. Otuoze, A.O., et al.: Electricity theft detection framework based on universal prediction algorithm. Indones. J. Electr. Eng. Comput. Sci **15**, 758–768 (2019)
18. Vlasa, I., Gligor, A., Dumitru, C.D., Iantovics, L.B.: Smart metering systems optimization for non-technical losses reduction and consumption recording operation improvement in electricity sector. Sensors **20**(10), 2947 (2020)
19. Maciejowska, K., Nitka, W., Weron, T.: Day-ahead vs. Intraday—Forecasting the price spread to maximize economic benefits. Energies **12**(4), 631 (2019)

Automation, Robotics, Biomedical Technologies and Intelligent Systems

Improvement of the Vector Control for DFIG Integrated into a Wind System by Artificial Neural Networks Accompanied by a Reliability Study of the Control System

Aicha Bouzem[1(✉)], Othmane Bendaou[1], and Bousselham Samoudi[2]

[1] Department of Physics, Faculty of Sciences, Abdelmalek Essaadi University, Sebta Avenue, 93002 Tetouan, Morocco
aicha.bouzem@etu.uae.ac.ma

[2] Civil Engineering, Energetic and Environment Department, National School of Applied Sciences, Abdelmalek Essaadi University, Sidi Bouafif Ajdir, 32003 Al-Hoceima, Morocco

Abstract. The aim of the current work presented in this paper is to integrate artificial intelligence (AI) techniques into field-oriented command to control the active and reactive powers of doubly fed induction generator (DFIG) integrated into a wind system, in order to improve the performances of the conventional vector control.

In the first step we are particularly interested in the application of indirect vector control by stator field orientation of DFIG, using two types of controllers: conventional regulators (PI) and Neural Networks (NN) controllers, and compare their simulation results using Matlab/Simulink software. And in the second step we performed an uncertainty analysis to verify the reliability of our designed model by applying a stochastic analysis and a sensitivity study based on the Monte Carlo simulation.

The obtained results are satisfactory and consistent with those of the references, which confirm that the intelligent controller's presents best performances in term of the response time as compared to that of the PI regulators. Moreover, the Monte Carlo test confirms the effectiveness and the reliability of the NN controllers against machine parameters variations.

Keywords: Wind energy · Doubly Fed Induction Generator (DFIG) · Field oriented control (FOC) · PI (proportional integral) controller · Artificial Neural Networks (ANN) · Monte Carlo · Stochastic · Uncertainty · Sensitivity · Reliability

1 Introduction

Wind energy is among the largest and most promising sources of renewable energy in the world in terms of development, with the aim of improving the efficiency of energy conversion and the quality of the energy provided. In this sense, different control strategies for wind power systems are constantly evolving. Artificial intelligence technology

L. Moldovan and A. Gligor (Eds.): Inter-Eng 2021, LNNS 386, pp. 631–640, 2022.
https://doi.org/10.1007/978-3-030-93817-8_56

and especially the artificial neural network are currently widely used in industrial control applications [1], and they are increasingly used in the realization of many highly sensitive control mechanisms and high security devices due to its many advantages, such as, their ability to solve problems related to industrial processes, and to cope with changing environmental requirements, and also they can be less complicated and less costly compared with some other proposed schemes of regulators [1, 2].

This paper presents the vector control of a DFIG integrated in wind power system using Artificial Intelligence (AI) techniques, in order to improve the behavior of the classical vector control. However, our system has risks and uncertainties that need to be addressed. The current work suggests a stochastic analysis based on Monte Carlo simulation, which can be utilized to apply a reliability study in order to have a more precise understanding about the behavior and quality of our control against system parameters variations.

This paper is organized as follows: Sect. 2 presents a mathematical model of DFIG in the d-q reference frame, and explains the Field Oriented Control (FOC) strategy.

Section 3 is reserved to improve the performances of the FOC by integrating the techniques of artificial intelligence to replace the PI regulators by neural regulators, and presents the simulation results using Matlab/Simulink software, the obtained results show that the NN controller has a faster time response than the PI controller.

In order to perform the reliability test of the NN controllers, a stochastic analysis based on the Monte Carlo simulation is established in Sect. 4, this study provides as results a statistical description of the controlled powers, a classification of the parameters that have the most impact on the control, and the validation of the reliability of our control. Finally, conclusions and suggestions for further research are presented in Sect. 5.

1.1 A Brief State-Of-The-Art

During the last decade, the concept of the variable speed wind turbine equipped with a doubly fed induction generator (DFIG) has received increasing attention, because this combination offers many advantages than other wind turbine concepts [3]. In fact, DFIG are commonly selected as one of the suitable wind energy conversion systems due to many reasons, such us, the ability to operate in wide range of speed variation approximately the synchronous speed $\pm 30\%$, and the accessibility to the stator and rotor that allow to control the generator powers injected into the grid as well as power factor [2].

To control DFIG, many techniques have been proposed in the literature, traditionally, the control of DFIG is commonly based on either stator flux oriented control (FOC) [4], or stator voltage oriented control (VOC) [5] using PI regulators. Therefore, the conventional PI are known by their limitations according to the disturbances and parameter variation effects, and the difficulty to tune the PI gain correctly due to the nonlinearity and high complexity of the system [6]. To solve this problem, many techniques have been developed recently by integrating advanced techniques based on the artificial intelligence algorithms, such as, genetic algorithms [7], neural networks [2–6], fuzzy logic,... etc. These techniques are currently known for their great potential to be able to solve these types of problems.

In this context, this work contributes to validate the performance and evaluate the reliability of the FOC using one of the artificial intelligence techniques.

2 Modeling and Indirect Field Oriented Control of DFIG

The model of the DFIG on the Park referential linked to the rotating field is represented as follows [1–3]:

- Stator and rotor voltage:

$$\begin{cases} V_{sd} = R_s I_{sd} + \frac{d}{dt}\varnothing_{sd} - \omega_s \varnothing_{sd} \\ V_{sq} = R_s I_{sq} + \frac{d}{dt}\varnothing_{sq} + \omega_s \varnothing_{sq} \\ V_{rd} = R_r I_{rd} + \frac{d}{dt}\varnothing_{rd} - (\omega_s - \omega_r)\varnothing_{rd} \\ V_{rq} = R_r I_{rq} + \frac{d}{dt}\varnothing_{rq} + (\omega_s - \omega_r)\varnothing_{rq} \end{cases} \quad (1)$$

- The active and reactive stator powers:

$$\begin{cases} P_s = V_{sd} I_{sd} + V_{sq} I_{sq} \\ Q_s = V_{sq} I_{sd} - V_{sd} I_{sq} \end{cases} \quad (2)$$

In order to control the power generation of the wind turbine, we will control the exchange of the active and reactive power between the stator of the DFIG and the grid, based on the Indirect Field Oriented Control, by aligning the stator flux with d-axis, [2–6], and taking into account the coupling terms, and compensating for them with a two-loop system.

We have: $\varnothing_{sd} = \varnothing_s$, $\varnothing_{sq} = 0$, and $\frac{d}{dt}\varnothing_{sd} = 0$.

By combining different equations, the voltage equations of DFIG can be simplified as:

$$\begin{aligned} V_{sd}^* &= 0 \\ V_{sq}^* &= Vs = \omega_s \varnothing_s \\ V_{rd}^* &= R_r I_{rd}^* + Lr\sigma \frac{d}{dt} I_{rd}^* - g\omega_s \sigma I_{rq}^* \\ V_{rq}^* &= R_r I_{rq}^* + Lr\sigma \frac{d}{dt} I_{rq}^* + g\omega_s \sigma I_{rd}^* + g\frac{L_m Vs}{L_s} \end{aligned} \quad (3)$$

with: $\sigma = 1 - \frac{L_m^2}{L_s L_r}$.

Based on these simplifications, the stator powers can express by:

$$\begin{cases} P_s^* = -\frac{L_m Vs}{L_s} I_{rq}^* \\ Q_s^* = -\frac{L_m Vs}{L_s} I_{rd}^* + \frac{L_m \varnothing_s}{L_s} \end{cases} \quad (4)$$

By establishing the IFOC strategy, and based of the Eqs. (3) and (4), the global block diagram of the controlled system using PI controllers can be established as shown in the Fig. 1. The input of PI is the error $E_i(k)$ between the desired value of i = (Ps, Qs, I_{rd}, I_{rq}) and the measured value at the instant k, while the regulated value present its output. Therefore, its transfer function is $K_p + \frac{K_i}{P}$.

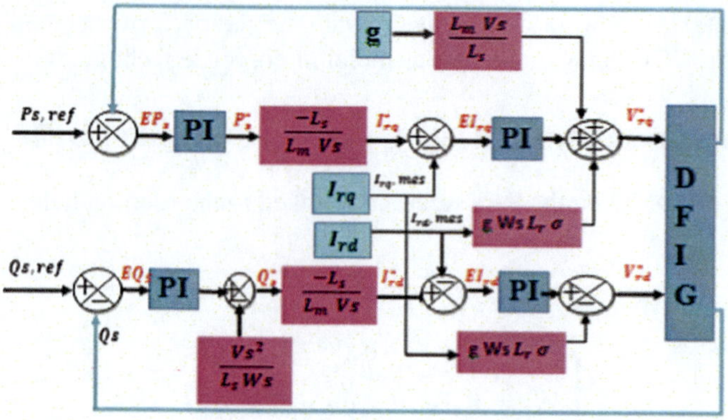

Fig. 1. Global block diagram of indirect field-oriented control technique.

3 Power Control of DFIG Using Artificial Neural Networks and the Simulation Results

3.1 Power Control of DFIG Using Artificial Neural Networks

In this part we train the ANN (supervised learning) [8] to the PI regulators to design the NN controllers witch replace all PI controllers used in the regulation of DFIG Fig. 2, in order to improve the performances of the control and solve the problems related to the conventional PI (Response time, Sensitivity to parametric variations of the system) [2–6]:

Generally, each neuron has an input x, an output y, an input-output function, and learning algorithm (Levenberg-Marquardt algorithm in our case), the properties of the neuron is decided by the input-output function, while the adjustment of the weights is performed by the learning algorithm by minimizing the mean square error (MSE) between the target and the ANN output. [8, 9].

The equations represent the output of a neuron k can be written as:

$$v_k = \sum_{x=1}^{x=n} (W_{ki}.x_i + b) \tag{5}$$

$$y_k = f(v_k) \tag{6}$$

Where, for neuron k, x_i is the ith input and y_k is the output, W_{ki} is a weight from input i to neuron k, b is the bias, f is the activation function. The training data are gained from the conventional IFOC which the inputs is the error and the target is the regulated power/current.

The performances of the NN controller are guaranteed by choosing the number of hidden layers and the number of neurons by performing several tests. In our case we obtained the optimal architecture by taking one hidden layer containing 7 neurons.

Fig. 2. Replacing the PI controller by NN controller.

3.2 Simulation Results

The simulations are investigated with a DFIG of 10KW connected to a 230V/50 Hz grid. The obtained results shown in Fig. 3 and Fig. 4 confirm that with the IFOC strategy, the powers of the controlled system track their optimal values during all the simulation time, and show that the NN controller presents best performances in term of time response as compared to that of the PI regulators.

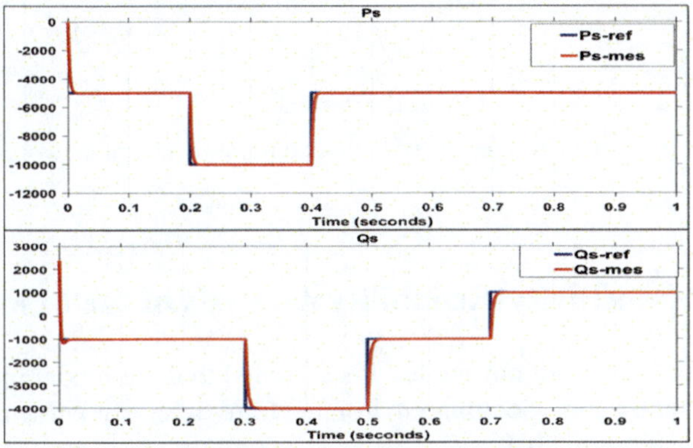

Fig. 3. PI controller responses.

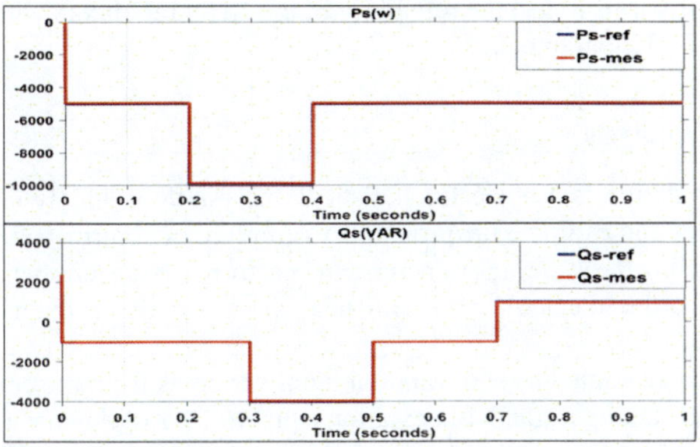

Fig. 4. ANN controller responses.

In order to investigate the performances very well, and test the learning quality of the NN controller, the system is subjected to another reference signal profile as illustrated in Fig. 5. We show that despite the fast fluctuations of the signal profile, the ANN controller reacts rapidly to track the optimal values of the DFIG powers. Who makes it more suitable for industrial control applications because an industrial control system also has time-varying uncertainties and effects.

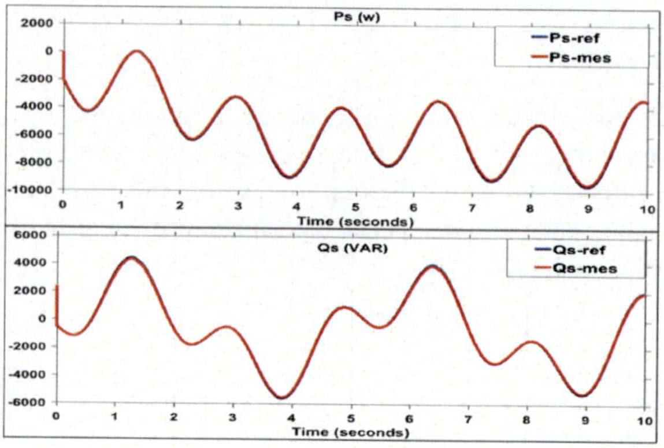

Fig. 5. Simulation results of indirect control by neural networks.

4 Stochastic-Sensitivity-Reliability Analysis of the Model

When creating and analyzing our model, it's assumed that the model inputs, definition parameters (constants) and operating environment are precisely known. But in reality, there is always uncertainty [10].

The behavior of the PI controllers is sensitive to variations caused by system parameters which influence the reliability of the system during real operation [2]. Therefore, the purpose of this part is to test the reliability of the ANN regulators with respect to the variations of the DFIG parameters.

4.1 Stochastic Analysis

Monte Carlo simulation is a powerful statistical analysis tool widely used to perform stochastic analyses and estimate the probability of failure of control systems. It consists in randomly sampling the system parameters, and perform a large number of experiments on a computer, in order to calculate the statistical characteristics of the model outputs to their distributions.

In this test, we generated a set of random input variables from the controlled model with 1000 samples, taking a standard deviation equal to 10% of the mean value for each parameter, and plotting in parallel the change of outputs against the variations of the inputs (Fig. 6) to visualize the behavior of the model with the affected change, and then calculate the statistical characteristics of the model outputs.

The mean values of the active and reactive power obtained under the changes of parameters is equal successively: −5743,38 W, −1740,62 Var, while their standard deviation equal: Ps_std = 74,12 (1.29% of Ps_mean), Qs_Std = 47,56 (2.73% of Qs_mean).

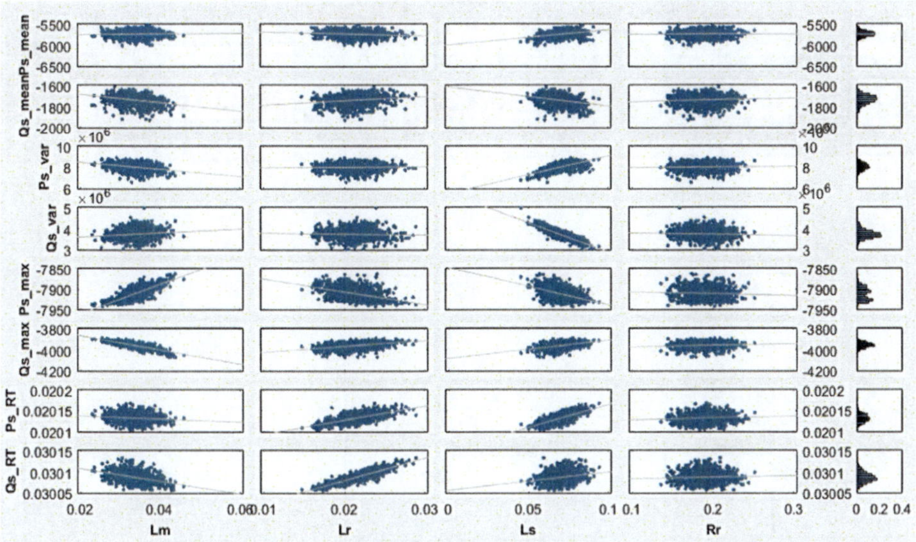

Fig. 6. Results from the Monte-Carlo simulation for Ps and QS.

4.2 Sensitivity Analysis

This section is a continuation of the previous part, where the results of the Monte-Carlo simulation will be used for further analysis by statistical methods in order to identify the parameters of the model that have the most impact on controller requirements and behavior.

To realize this study, we applied a correlation test by measuring Pearson's correlation coefficient and plotting the statistical relationship between the system parameters and the statistical characteristics of the DFIG powers.

The Pearson's correlation coefficient returns a value between -1 and 1, and more than the absolute value of this coefficient is important (close to 1) more the correlation is considered strong and significant. Whereas, +1 indicate a strong correlation which mean that both variables move in the same direction. Conversely, a value of -1 represents a perfect negative relationship.

According to the results presented in the Table 1 and the Fig. 7, we conclude that the correlation degree of most variable doesn't have significant correlations, except to the case of the Rise Time of Qs that have a positive correlation with Lr (Qs_RT, Lr = 0.89), and between the reactive power variance and Ls (Qs_var, Ls = −0.87).

But generally the result of this test confirm that the sensitivity of the NN controllers against the variations of the system parameters is not critical.

Table 1. Pearson's correlation coefficient.

	Ps_mean	Qs_mean	Ps_var	Qs_var	Ps_max	Qs_max	Ps_RT	Qs_RT
Lm	0,038	-0,29	-0,31	0,15	0,78	-0,8	-0,11	-0,4
Lr	-0,14	0,18	0,03	-0,041	-0,43	0,41	0,68	0,89
Ls	0,35	-0,27	0,7	-0,87	-0,42	0,43	0,74	0,24
Rr	0,0076	0,066	0,03	-0,053	-0,052	0,087	0,077	0,045

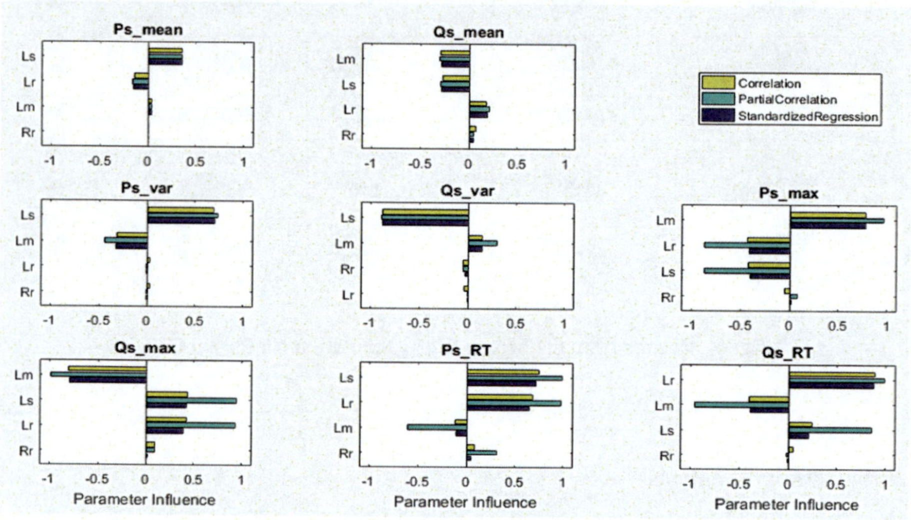

Fig. 7. Statistical analysis for Ps and Qs.

4.3 Reliability Analysis

Because the reliability is an important step in the indeterminist analysis phase, and especially for critical systems that operate under risky conditions, this part is dedicated to evaluate the reliability of our control.

Firstly, we specify a critical power threshold (−9000 W for Ps and −5000VAR for Qs) which must not be exceeded during the operation of the system to prevent damaging the equipment, and we subject the model to random changes to test whether the power will exceed this threshold.

The Fig. 8 shows that the variations of the active power do not diverge too much from the nominal value (-8000 W), and do not reach its critical threshold (-9000 W). And the same thing for the reactive power Fig. 9, which allowed us to say that the probability of failure for our controller is 0% under these changes, and also the NN controllers are well designed and it is able to track the variations of the parameters.

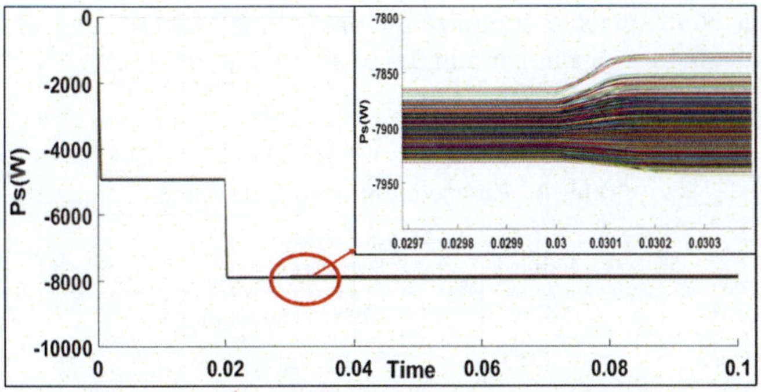

Fig. 8. The response of Ps control for each sample.

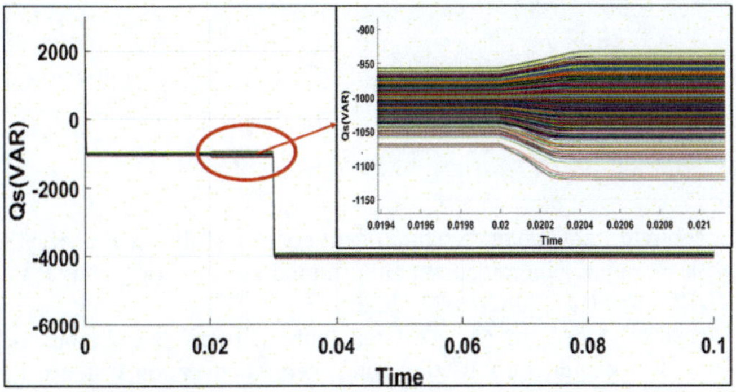

Fig. 9. The response of Qs control for each sample.

5 Conclusions and Future Directions

The work presented in this paper had two main objectives:

- Improve the indirect vector control of DFIG by using the Artificial Intelligence Techniques.
- Perform a stochastic analysis based on Monte Carlo simulation to determine the reliability and the sensitivity of NN controllers.

The results obtained show, on the one hand, the effectiveness of neural controllers in improving the response time compared to conventional PI-type controllers, and that NN regulators provide a more appropriate control of sudden changes in target values. On the other hand, the confirmation of the reliability and the ability of the NN controller to track parametric changes that are a challenge when operating real systems.

Depending on the results achieved we can specify some perspectives such as:

- Based on the results of the sensitivity study, optimize the Rise Time of the control in the face of the expected changes in system parameters and operating conditions.

- Apply advanced methods to minimize the time of the Monte-Carlo calculation to be able to test the model on a larger number of samples (Table 2).

Table 2. Parameters of the tested system.

Parameter	Value	Parameter	Value
Stator resistance Rs	0.455	Mutual inductance Lm	0.034
Rotor inductance Lr	0.0213	Stator inductance Ls	0.07
Viscous coefficient	0.0016	Pair poles p	2
Rotor resistance Rr	0.19	Gearbox ratio G	5.4
Moment of inertia	0.031		

References

1. Bensaid, A., Zebirate, S., Chaker, A.: Intelligent control of flywheel energy storage system associated with the wind generator for uninterrupted power supply. Int. J. Power Electron. Drive Syst. (IJPEDS) **11**(4), 2062–2072 (2020)
2. Djeriri, Y., Meroufel, A., Allam, M.: Artificial neural network-based robust tracking control for doubly fed induction generator used in wind energy conversion systems. J. Adv. Res. Sci. Technol. **2**(1), 173–181 (2015)
3. Yasmine, E., Chakib, B., Badre, B.: Improved performance of DFIG-generators for wind turbines variable-speed. Int. J. Power Electron. Drive Syst. (IJPEDS) **9**(4), 1875–1890 (2018)
4. Meroufel, A., Djeriri, Y.: Commande vectorielle par les réseaux de neurones artificiels de l'énergie d'une MADA intégrée à un système éolien. Revue des Energies Renouvelables **13**(4), 669–682 (2010)
5. Zhou, Y., Bauer, P., Ferreira, J.A., Pierik, J.: Control of DFIG under unsymmetrical voltage dip. In: IEEE Power Electronics Specialists Conference, pp. 933–938 (2007)
6. Ahmed, H.M., Djeriri, Y., Bentaallah, A.: Robust power control of DFIG using artificial neural networks for a wind energy conversion system based energy storage unit. In: 2nd International Symposium Mechatronics Renewable Energy, ISMRE'2018, 2018, pp. 1–6 (2018)
7. Ameur, F., Kouzi, K.: Genetic algorithm optimized PI and fuzzy logic speed vector control of dual stator induction generator in wind energy conversion system. In: 3rd International Conference on Systems and Control, pp. 1036–1042 (2013)
8. Shu, H.: Decoupled temperature control system based on PID neural network. In: ACSE 05 Conference, pp. 19–21. CICC, Cairo (2005)
9. Benbouhenni, H., Boudjema, Z., Belaidi, A.: Power ripple reduction of DPC DFIG drive using ANN controller. Acta Electrotechnica et Informatica **20**(1), 15–22 (2020)
10. Aoues, Y.: Optimisation fiabiliste de la conception et de la maintenance des stuctures. Génie des procédés. Université Blaise Pascal - Clermont-Ferrand II, 2008. Français. ffNNT: 2009CLF21908ff. fftel-00726003f

The Perturbation Method for Dynamic Analysis of Pole Vaulting

Ouadie El Mrimar[1(✉)], Othmane Bendaou[1], and Bousselham Samoudi[2]

[1] Department of Physics, Faculty of Sciences, Abdelmalek Essaadi University, Sebta Avenue, 93002 Tetouan, Morocco
{elmrimar.ouadie-etu,o.bendaou}@uae.ac.ma

[2] Civil Engineering, Energetic and Environment Department, National School of Applied Sciences, Abdelmalek Essaadi University, Sidi Bouafif Ajdir, 32003 Al-Hoceima, Morocco
b.samoudi@uae.ac.ma

Abstract. Pole vaulting has progressed slowly in terms of performance since the 1960's. The world record varies by a few centimeters due to the use of flexible poles. In this research, we propose a new stochastic optimization approach to take into account the uncertainties. A simplified mass-spring model has been implemented to model the pole vault. This model is considered as a tool to explore the possibilities of trajectories and performances, in which the parameters associated with this system are uncertain. This approach is based on the association of the perturbation method with the fourth order Runge-Kutta method (RK4). The new stochastic approach is used to optimize the dynamic response of the system and the computation time. The results of the perturbation method (PM) are compared to those of the reference Monte Carlo method (MC).

Keywords: Monte Carlo · Perturbation method · Pole vaulting · Mass-spring · Stochastic parameters

1 Introduction

Pole Vaulting is an Olympic discipline in which the athlete tries to clear the bar as high as possible through a flexible pole transforming kinetic energy into potential energy, therefore, the performance obtained changes considerably with the combination of two determining factors such as the properties of the pole (stiffness, weight,) and the physical ability of the athlete (speed, force…).

The pole vault has been the interest of many scientific works for the last thirty years. Researchers seek to identify performance factors either through experimental studies [1–4] or through numerical simulations based on mechanical or mathematical models [5–12].

Motivated by the mechanics of the pole vault, the aim of this research is to perform a stochastic study on a simplified model of the pole vault consisting of a point mass and a linear spring, then the determination of the nonlinear dynamic response of a vault using the differential equations of motion. In general, assumes that the physical parameters

L. Moldovan and A. Gligor (Eds.): Inter-Eng 2021, LNNS 386, pp. 641–650, 2022.
https://doi.org/10.1007/978-3-030-93817-8_57

of the model are deterministic (material properties, initial and boundary conditions...) but in reality, these parameters are random. To evaluate the variability of the dynamic response with respect to the variability of the uncertain parameters of the model, a Monte Carlo simulation is used [13, 14]. This method is often used as a reference, even if the prohibitive calculation time limits the use of this method. therefore, we use the perturbation method [15–17] associated with the fourth order Runge-Kutta method [18]. This method based on the Taylor series development of the response around the average values of the random variables, and allows to calculate directly the averages and standard deviations of the solutions.

2 Dynamic Modeling

The dynamic modeling of the pole vault is not an easy problem due to the nonlinearity of the system. Therefore, we decided to model the dynamics using a linear spring-mass system which we base on a non-dimensional study thanks to the assumption of the behavior of the vault. The point mass attach to the spring with an initial velocity V_0 and detach when the spring has no energy store and the mass has no horizontal velocity.

2.1 Linear Mass-Spring System

We consider the system composed of a point mass m and a spring of stiffness k and length R which can rotate freely around the x axis and θ denotes its angle with the y axis, measured counterclockwise.

It is assumed that the length of the spring in its initial state is R_0, its initial angle of rotation is θ_0. The mass has an initial velocity V_0 along the negative y axis when it contacts the free end of the spring can detach from the spring at any time.

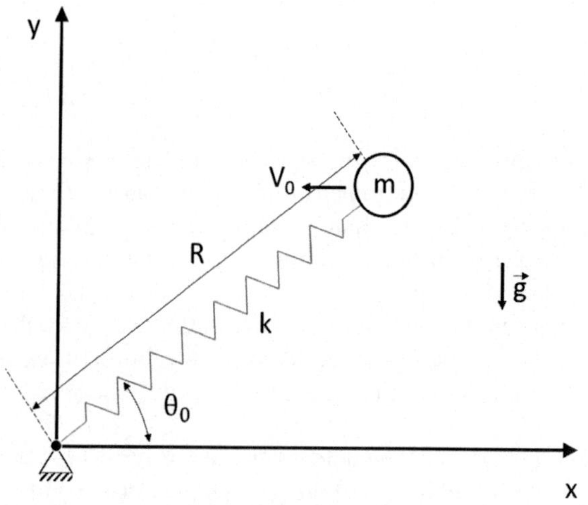

Fig. 1. Point mass system and linear spring in its initial state

2.2 The Equation of Motion

The equation of motion of the mass-spring system in Fig. 1 is determined by the Lagrange method in the polar coordinate base, where R and θ are the generalized coordinates, the equations of motion are as follows:

$$\begin{bmatrix} \ddot{R} \\ \ddot{\theta} \end{bmatrix} = \begin{bmatrix} R\theta^2 + \frac{k}{m}(R_0 - R) - g\sin\theta \\ -\frac{2}{R}\dot{\theta}\dot{R} - \frac{g}{R}\sin\theta \end{bmatrix} \tag{1}$$

And the initial conditions are:

$$\begin{array}{ll} R(0) = R_0 & \theta(0) = \theta_0 \\ \dot{R}(0) = -V_0\cos\theta & \dot{\theta}(0) = \frac{V_0}{R_0}\sin\theta \end{array} \tag{2}$$

We non-dimensionalize Eq. (1) by introducing the following non-dimensional variables of time, position and velocity:

$$\tau = t\sqrt{\frac{k}{m}}, \quad r = \frac{R}{R_0}, \quad v_0 = \frac{V_0}{R_0}\sin\theta \tag{3}$$

The non-dimensional equation is:

$$\begin{bmatrix} \ddot{r} \\ \ddot{\theta} \end{bmatrix} = \begin{bmatrix} r\dot{\theta}^2 + (1 - r) - \varepsilon\sin\theta \\ -\frac{2}{r}\dot{r}\dot{\theta} - \frac{\varepsilon}{r}\sin\theta \end{bmatrix} \tag{4}$$

Due to the weight of the mass, the non-dimensional static deflection of the spring is ε, where:

$$\varepsilon = \frac{\delta}{L_0}\delta = \frac{\text{mg}}{\text{k}} \tag{5}$$

The non-dimensional initial condition is:

$$\begin{array}{ll} r(0) = 1 & \theta(0) = \theta_0 \\ \dot{r}(0) = -\vartheta_0\cos\theta_0, & \dot{\theta}(0) = \vartheta_0\sin\theta_0 \end{array} \tag{6}$$

The mass must be released from the spring at a certain time, provided that the kinetic energy of motion of the mass converted into potential energy. Then these two conditions that must be satisfied:

$$\dot{r}(\tau_f) = 1, \qquad \dot{r}(\tau_f)\cos\theta(\tau_f) = r(\tau_f)\dot{\theta}(\tau_f)\sin\theta(\tau_f) \tag{7}$$

Finally, the point mass moves freely under the influence of gravity only, then its dynamics will be given by the relation:

$$\ddot{Z} = -g \rightarrow \ddot{z} = -\varepsilon \tag{8}$$

The dynamics of the mass-spring system is described by four state variables $r, \dot{r}, \theta, \dot{\theta}$. Since the final time t_f is also unknown, we have to solve a two-point boundary value problem (TPBVP) where (7) provides two conditions that must be satisfied at the final

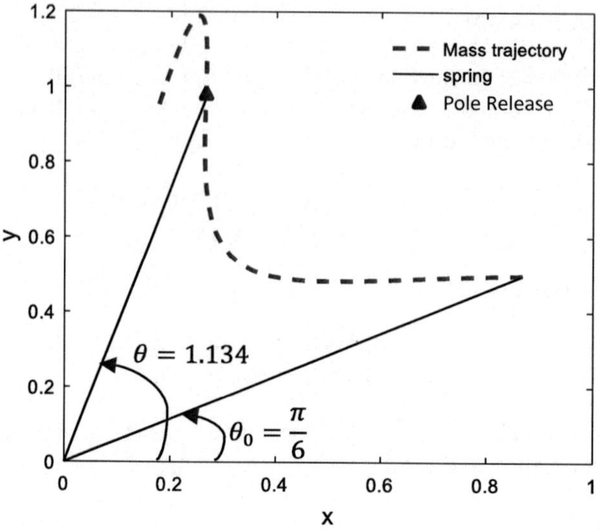

Fig. 2. Trajectory of the mass of Fig. 1 with $\varepsilon = 0.1$ for the initial speed $\vartheta_0 = 0.432$ and the initial angle $\theta_0 = \frac{\pi}{6}$

time. We first solve the TPBVP assuming that ε and θ are constants. A solution was obtained through trial and error for the specific case where $\varepsilon_0 = 0.1$ and $\theta_0 = \frac{\pi}{6}$ rad, Eq. (4) was integrated in time using the initial conditions in (6) for different values of the velocity ϑ_0.

The results are presented in Fig. 2, by transforming the non-dimensional polar coordinate system to the non-dimensional Cartesian coordinate system:

$$\begin{aligned} x &= r\cos\theta \\ y &= r\sin\theta \end{aligned} \quad (9)$$

Only $\vartheta_0 = 0.432$ satisfies the five boundary conditions at $t_f = 3.337$ which provides the solution of TPBVP. At this moment the mass detaches from the spring, the maximum height reached by the mass is h = 1.191. The equations of motion of the mass integrated with respect to time using the method of (RK4). The model developed in this section has been implemented in MATLAB 2020.

3 Stochastic Study

The classical method followed when studying mechanical systems is based on the assumption that the model is deterministic i.e., that its parameters are constant. However, if we conduct experiments, we will realize the limits of deterministic modeling. Because there is always a difference between the calculated and measured results, this is due to the uncertainty of (the material properties, and the initial conditions….). These uncertainties have an impact on the dynamic displacement behavior of the mass-spring system. It is therefore necessary to use stochastic methods, in particular the Monte Carlo method and the perturbation method.

3.1 Monte Carlo Method

The estimation of the moment (mean and variance) of the dynamic response function of the mechanical system can be obtained by Monte Carlo simulation. Although its computational cost is high, this method has been widely used by dedicated software (such as MATLAB, ANSYS...) and provides a reference for approximate calculations. The Monte Carlo method has the advantage of taking into account all types of uncertainties on the parameters of a mechanical system. However, one of its main drawbacks is the computational time required due to its iterative nature. The dynamic response function is considered as a random variable image of the base random variable. The simulation involves constructing a sample of random variable $Y_1, Y_2, \ldots, Y_n$ and processing the sample using standard statistical techniques. The n simulations are performed independently according to the distribution law of the basic random variables.

The mean of Y is given by:

$$E[Y] = \frac{1}{n}\sum_{i=1}^{n} Y_i \tag{10}$$

The variance of Y is given by:

$$Var[Y] = \frac{1}{n-1}\sum_{i=1}^{n} [Y_i - E(Y)]^2 \tag{11}$$

3.2 Perturbation Method

The perturbation method is widely used in the field of stochastic finite elements, and is based on the expansion of the Taylor series, which is a function of the basic random physical variables, mechanical properties, geometric characteristics, (the random parameters must clearly appear in the dynamic matrix). The perturbation method allows to calculate the mean and the standard deviation of the displacement of a dynamic system with uncertain parameters. This method has been used in many fields to solve linear and nonlinear problems, for static or dynamic modes.

The perturbation method can be used for mechanical systems with independent random parameters. It is based on the expansion of the first order Taylor series.

In this section we will apply the perturbation method. This method consists of approximating the dynamic function of movements Eq. (4) of random variables by their Taylor expansion around their mean values, depending on the order considered of the Taylor expansion. The method is said to be first order, second order or higher.

For the mechanical system of Eq. (4) with uncertain parameters, we suppose that the static deviation parameters and the initial velocity are functions of the random variables, $\{\beta_p\}_{(p=1,\ldots,P)}$.

The vector of the average parameters is defined by $\{\overline{\beta}\}$, and the quantity:

$$\{d\beta\} = \{\beta\} - \{\overline{\beta}\}$$

The following notation is used to simplify the writing of derivatives:

$$[A]^0 = [A](\beta)|_{\beta=\overline{\beta}}$$
$$[A]^n = \left.\frac{\partial[A](\beta)}{\partial\beta_n}\right|_{\beta=\overline{\beta}} \quad (12)$$

$[A]^0$, $[A]^n$ are deterministic corresponding to the derivatives, the repetition of the index " n " twice implies a summation.

The unknown position, velocity and acceleration vectors are also developed by Taylor series as follows:

$$\begin{aligned}
\{r\} &= \{r\}^0 + \{r\}^n d\alpha_n \\
\{\dot{r}\} &= \{\dot{r}\}^0 + \{\dot{r}\}^n d\alpha_n \\
\{\ddot{r}\} &= \{\ddot{r}\}^0 + \{\ddot{r}\}^n d\alpha_n \\
\{\theta\} &= \{\theta\}^0 + \{\theta\}^n d\alpha_n \\
\{\dot{\theta}\} &= \{\dot{\theta}\}^0 + \{\dot{\theta}\}^n d\alpha_n \\
\{\ddot{\theta}\} &= \{\ddot{\theta}\}^0 + \{\ddot{\theta}\}^n d\alpha_n
\end{aligned} \quad (13)$$

Substituting these developments in the equation of motion, writing the terms of the same order, we obtain the following differential systems:

- Equation of order zero:

$$\begin{aligned}
\{x\}^0 &= \{r\}^0 \cos\{\theta\}^0 \\
\{y\}^0 &= \{r\}^0 \sin\{\theta\}^0
\end{aligned} \quad (14)$$

- First order equation:

$$\{x\}^n = \{r\}^n \cos\{\theta\}^0 + \{r\}^0\{\theta\}^n \sin\{\theta\}^0 \quad (15)$$

$$\{y\}^n = \{r\}^n \sin\{\theta\}^0 + \{r\}^0\{\theta\}^n \cos\{\theta\}^0 \quad (16)$$

The mean is given by:

$$\begin{aligned}
E[x(t)] &= \{x(t)\}^0 \\
E[y(t)] &= \{y(t)\}^0
\end{aligned} \quad (17)$$

The variance is given by:

$$\begin{aligned}
Var[x(t)] &= \{x(t)\}^{n2} Var(\beta_n) \\
Var[y(t)] &= \{y(t)\}^{n2} Var(\beta_n)
\end{aligned} \quad (18)$$

4 Simulation and Results

To evaluate the variability of the nonlinear dynamic motion response with random parameters, we calculated the two moments (mean, standard deviations) of displacement of the point mass.

In this section, numerical results are presented after the formulation derived in Sect. 3, using MATLAB 2020 computer language scripts. The results of the perturbation method are compared with the Monte Carlo results for 3000 simulations. Some parameters are considered as random variables, the first random variable is the static spring deflection ε, and the second random variable is the initial velocity ϑ_0.The stochastic behavior is described using the normal random variables so these parameters are described by the following relations:

$$\varepsilon = \varepsilon_0 + \sigma_\varepsilon \aleph_1, \quad \vartheta = \vartheta_0 + \sigma_\vartheta \aleph_1 \tag{19}$$

ε_0 et ϑ_0 designate the average values, $\aleph_1$,$\aleph_2$ are the normal random variables, σ_ε, σ_ϑ, are the associated standard deviation.

Table 1. Non-dimensional initial parameters and conditions of the system.

ε_0	ϑ_0	θ_0	r_0
0.1	0.432	30°	1

To see the influence of multiple uncertain parameters on the dynamic response of the mass-spring system, it is assumed that the static deflection ε the velocity ϑ_0 are all uncertain parameters and the angle θ_0 and the initial spring length r_0 are constants Table 1.

For $\sigma = \sigma_\varepsilon = \sigma_\vartheta$ the mean values and standard deviations of the dynamic components of the nonlinear displacements along the two directions x and y have been calculated with the method of first order perturbation.

The results obtained are shown in Fig. 3 and Fig. 4 for $\sigma = 5\%$ and in Fig. 5 for $\sigma = 10\%$. These results are compared to those obtained with the Monte Carlo referential technique using 3000 simulations.

The results of the mean value response are very satisfactory, the instantaneous mean values of the displacements and the standard deviations are consistent with the Monte Carlo reference solutions, and the errors are still acceptable.

For the results of the standard deviations of the displacements, we can clearly see that for $\sigma = 5\%$ and $\sigma = 10\%$, the proposed perturbation method provides very similar results with the reference Monte Carlo method as well as the reduced computation time, from other Table 2.

On the other hand, we can see that the error increases when the standard deviation of uncertain parameters increases.

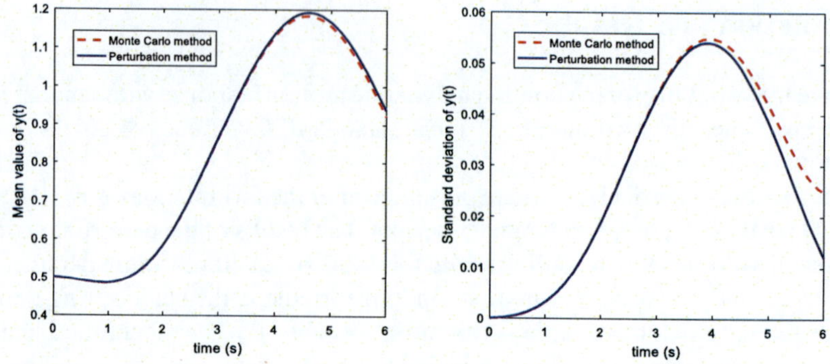

Fig. 3. Instantaneous mean value and standard deviation of x(t) for $\sigma = 5\%$.

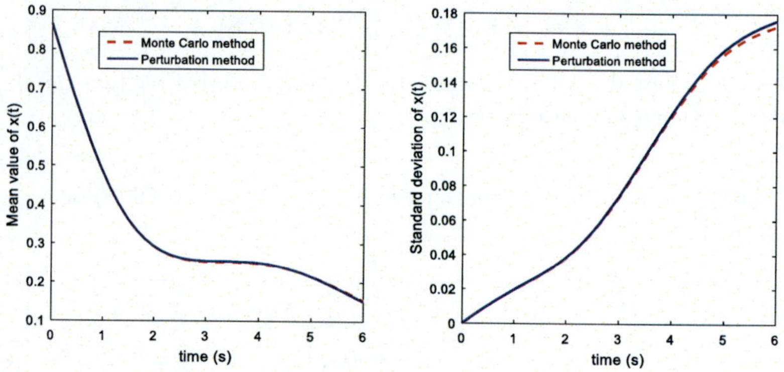

Fig. 4. Instantaneous mean value and standard deviation of y(t) for $\sigma = 5\%$.

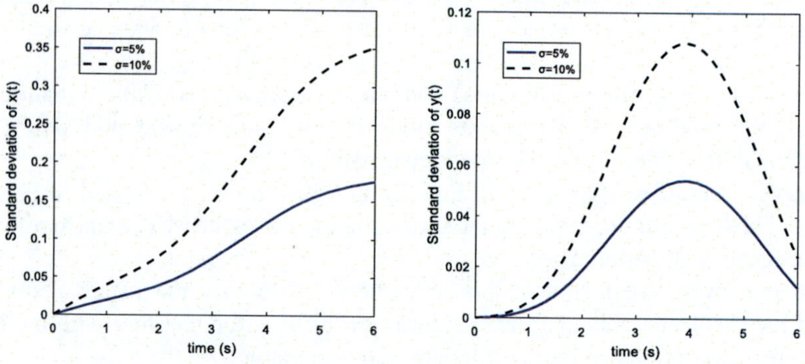

Fig. 5. Standard deviation of x(t) and y(t) for $\sigma = 5\%$ $\sigma = 10\%$

Table 2. Comparison of the CPU time in (s) between the two-simulation methods Monte Carlo method and perturbation method of order 1.

	Monte Carlo Simulation MCS	First order perturbation method
CPU time in (s)	45.45	1.34

5 Conclusion

This paper presents the stochastic method based on the development of the perturbation method by the first order Taylor expansion in combination with the RK4 method.

We have obtained similar results by the perturbation method and the Monte Carlo method. The results showed that these methods are efficient in terms of saving computational time.

This method allowed us to optimize the dynamic response of the mass-spring system and the computation time. We concluded that the two stochastic parameters ε and ϑ_0 significantly affect the performance of the jump.

References

1. Morlier, J., Mesnard, M., Cid, M.: Pole-vaulting: Identification of the pole local bending rigidities by an updating technique. J. Appl. Biomech. **2**(24), 140–148 (2008)
2. McGinnis, P., Bergman, L.: An inverse dynamic analysis of the pole vault. Int. J. Sport Biomech. **2**, 186–201 (1986)
3. Schade, F., Arampatzis, A., Bruggemann, G.-P.: Reproducility of energy parameters in the pole vault. J. Biomech. **8**(39), 1464–1471 (2006)
4. Arampatzis, A., Schade, F., Brüggemann, G.P.: Effect of the pole-human body interaction on pole vaulting performance. J. Biomech. **37**(9), 1353–1360 (2004)
5. Chau, S., Mukherjee, R.: Kinetic to potential energy transformation using a spring as an intermediary: application to the pole vault problem. J. Appl. Mech. Trans. ASME **5**(86) (2019)
6. Hoepffner, J.: Models for an alternative pole vault. Phys. Sports, 1–6 (2012)
7. Hubbard, M.: Dynamics of the pole vault. J. Biomech. **11**(13), 965–976 (1980)
8. Ekevad, M., Laundberg, B.: Simulation of smart pole vaulting. J. Biomech. **9**(28), 1079–1090 (1995)
9. Drucker, S., Schneider, K., Ghotra, N.K., Bargmann, S.: Finite element simulation of pole vaulting. Sport Eng. **2**(21), 85–93 (2018)
10. Ohshima, S., Nashida, Y., Ohtsuki, A.: Optimization of pole characteristic in pole vaulting. Procedia Eng. **2**(2), 3191–3196 (2010)
11. Jahromi, A.F., Atia, A., Bhat, R.B., Tie, W.F.: Optimizing the pole properties in pole vaulting by using genetic algorithm based on frequency analysis. Int. J. Sports Sci. Eng. **1**(6), 41–53 (2012)
12. Choi, J., Jeong, K., Seo, T.: Comparison of linear and torsion-based dynamic modeling of a jumping robot via energy conversion. Int. J. Precis. Eng. Manuf. **18**(11), 1529–1535 (2017). https://doi.org/10.1007/s12541-017-0181-6
13. Shinozuka, M.: Monte Carlo solution of structural dynamics. Comput. Struct. **5–6**(2), 855–874 (1972)
14. Bendaou, O., Bendaou, O., Bourzeix, F., Agouzoul, M., El Hami, A.: Measurements and stochastic F.E.A. with application in thermomechanical characterization of electronic packages. J. Eng. Technol. Sci. **6**(48), 700–714 (2016)
15. Bendaou, O., Rojas, J.E., El Hami, A., Aannaque, A., Agouzoul, M.: Stochastic and reliability analysis of a propeller with model reduction. Eur. J. Comput. Mec. **2**(18), 195–215 (2009)
16. Ding, C., Hu, X., Cui, X., Li, G., Cai, Y., Tamma, K.: Isogeometric generalized n th order perturbation-based stochastic method for exact geometric modeling of (composite) structures. Comput. Methods Appl. Mech. Eng. **346**, 1002–1024 (2019)

17. Kleiber, M., Hein, T.D.: The Stochastic Finite Element Method (Basic Perturbation Technique and Computer Implementation). Wiley, Chichester (1992)
18. Boyce, W., Diprima, R.: Elementary Differential Equations and Boundary Value Problems, Textbook and Student Solutions Manual Set. 9nd edn. Laurie Rosatone, Department. Wiley (2009)

Reducing Traffic Congestion Through Optimal Planning

Eugenia Alina Roman(✉) and Vasile Dragu

University Politehnica of Bucharest, Bucharest, Romania

Abstract. The increasing mobility has led to unbearable levels of congestion and a deterioration in the quality of life due to its negative externalities. Today, Bucharest city has reached a degree of motorization of about 700 vehicles per 1000 inhabitants and it is constantly growing. Under these conditions, correlated and simultaneous measures are needed to reduce traffic congestion. Among them, it is rigorous planning of trips distribution, modal split and route assignment, and large-scale use of public passenger transport. Rigorous transport studies need to be conducted to determine the current transport demand and to formulate some empirical laws for estimating its evolution. The paper discusses the models for estimating transport demand, as they are of great importance in establishing the capacity of the transport infrastructure, the size of the investments to be made as well as the amortization duration of the invested funds. The paper presents an original case study conducted for a city using a synthetic gravitational model to identify the transport demand and the parameters of travel as well as their influence on transport potentials. This highlights the differences between planning models and their implications in conducting traffic studies.

Keywords: Congestion · Passenger transport · Optimal planning

1 Introduction

Rapidly increasing population induces growing cities and increasing car ownership. Mobility is a fundamental component in all models of spatial structure. Mobility, expressed in circulation/movement, plays a role of particular importance in all the functions of society, being both a condition and a consequence. [1] Spatial mobility is correlated and depends on traffic infrastructure, means of transport, technologies, but also on the connection between urbanism and transport. Consequently, transportation and land use problems become significant issues due to their economical effects. Interaction between land use and transportation is the basic factor for the trip generation. This interaction is also strengthened through planning models. The conventional planning paradigm primarily builds the environment and afterwards tries to overcome the existing transportation problems. Through transport planning models the users' travel needs are identified, the transport demand is estimated, trips are sorted by destinations, arranged by modal choice and after that assigned to links on a transport network. Each of these stages has a decisive role in achieving a quality transport in conditions of maximum

L. Moldovan and A. Gligor (Eds.): Inter-Eng 2021, LNNS 386, pp. 651–660, 2022.
https://doi.org/10.1007/978-3-030-93817-8_58

efficiency and effectiveness. Among these conditions, the travel time must fall within a time frame assumed by the transport operator.

The sharp increase in the degree of motorization has led to congestion in large cities with major implications on the quality of life of residents.

The overarching role of mobility and transportation in modern societies has generated a fast growing field of social-science-based mobilities research. This rising field focuses on large-scale as well as regional movements of people, goods, capital, and information. The cross-fertilization of disciplines and academic traditions in this field brings about strong and innovative approaches concerning the future of cities that integrate the human as well as the systematic and the global scale of current transformations [2].

The current main challenge of any urban system deals with the externalities produced by the road transportation system. It is well-known that about 50% of pollution steams from road transport and, in particular, from internal combustion engines [3]. Within this context, different actions have been deployed in the last few decades to reduce pollution, such as acting on the vehicle technology or different types of fuels, and through different and sophisticated mobility/travel demand management policies or traffic flow control strategies [3].

Although the term congestion is used extremely frequently, it has different interpretations from stakeholders who, with often divergent interests and points of view, interfere on the transportation market. We can name here: *the person in charge of the transport infrastructure development strategy, the infrastructure user, the traffic engineer, the transport beneficiary and the transport economist*. For each of them the notion of congestion has different meanings depending on the degree of involvement in the transportation process [4].

Environmental sustainability entails improvement in the quality of urban environment and reduction of emissions and energy consumption (greenhouse gasses emission variation; pollutant emission variation; impact variation in other sectors). By contrast, social sustainability entails improvement in the quality of life and social equity (e.g., easy access to transportation) and improved safety (e.g., reduction in the frequency of accidents). Finally, economic sustainability entails making mobility of people and goods more efficient and effective and ensuring that the economic benefits produced by the project are greater than the costs [5].

Bucharest ranks 4th in the top of the most congested cities in Europe compiled by the car navigation company TomTom. The company has compiled a report (*TomTom Traffic Index*) showing traffic congestion in 416 cities in 57 countries. In the global ranking, on the first 5 places is the city of Bengaluru (in India), with a congestion index of 71% (the percentage of additional time spent traveling compared to the hours spent outside peak hours) followed by the Philippines capital Manila (71%), the capital of Colombia, Bogota (68%), Mumbai (65%) and the city of Pune (59%) also in India [6].

In the European top, on the first 5 places are Moscow (59%), Istanbul (55%), Kiev (53%), Bucharest (52% compared to 41% in 2015) and St. Petersburg (49%) and on the 14th places, 15th and 17th respectively are Paris (39%), Rome (38%) and London (38%). The level of congestion increased between 2018 and 2019 (in the last report an additional 239 cities were included in the list of congested cities), only a number of 63 cities registered measurable decreases [6].

From the data supplied by the Driving Licenses and Vehicle Registration Directorate of the Ministry of Internal Affairs – DRPCIV, information was obtained regarding the number of vehicles registered in Bucharest. [7] From the analysis of the number of vehicles registered in the period 1990–2019, it is concluded that the evolution was a rapidly increasing one, after a polynomial type function of degree 2, this having the highest correlation coefficient [8].

2 Transport Planning Model

Classic congestion reduction measures focus exclusively on the development of road infrastructure leading to land usage, degradation of the natural landscape, increased risk of accidents (due to increased road width) and attracting new traffic, which will again lead to congestion and the process is cyclical until the expansion of road infrastructure is no longer possible. Thus is described a vicious circle of possibilities to eliminate congestion [9, 10].

Since 2013, the European Commission has developed a guideline for the Development and Implementation of a Sustainable Urban Mobility Plan (SUMP), according to which each urban locality must achieve a SUMP through which to achieve a sustainable urban planning in transport [11].

In the SUMP for Bucharest-Ilfov Region, 2016–2030 it is presented the four stage transport model through which the travel planning in the analyzed region was made (SUMP Bucharest Ilfov, p. 241–278). [12] After the development of the transport model, measures were proposed to reduce congestion and to promote a sustainable urban mobility (good accessibility, safety and security, environment, economic efficiency, quality of the urban environment) but which have not yet shown their usefulness and especially efficiency.

A more recent treatment of urban transport planning is the one based on TOD policies - Transit-Oriented Developments [13]. Today, many cities have developed large financial investments to use, mainly public transport to achieve sustainable mobility. The TOD areas are compact, mixed-use developments that make it easier to walk, bike and use public transport through their urban design. Consequently, they are seen as a way to environmental sustainability by conserving resources and energy, using better use of urban space, reducing the number of kilometers traveled by vehicles and encouraging the use of greener modes of transport [14].

TOD offers many social, environmental, economic benefits but also for health. It is associated with high-density, mixed-use urban development that brings many opportunities closer to residential locations and facilitates the choice of sustainable modes of transport (walking, cycling or using public transport). Therefore, TOD affects travel behavior mainly by increasing choice options and encouraging the use of non-motorized modes of transport. People living in TOD are expected to have a more sustainable, active lifestyle and less dependence on their personal car [15].

The classic transport planning model comprises four main stages (trip generation, trip distribution, modal split and traffic assignment) by which the transport demand is estimated and on its base the public transport supply is designed [16].

Transport modeling and planning are the main tools that can be used to achieve efficient transport in cities [17].

The origin - destination matrix is the result of the second stage of the model. In a transport study this matrix must be rigorously dimensioned, because it represents a transposition of human behavior in the transport process [18]. The matrix provides the total number of trips having the origin zone i (g_i) or the destination zone j (a_j), and the model used seeks to determine the g_i trip distribution by destinations and a_j trip distribution by origins on a certain network [19].

Basically, two methods are used in trip distribution stage [20, 21]:

Growth factor methods (constant factor method, average factor method, Detroit factor method, FRATAR method, FURNESS method).

Synthetic methods using gravity type models or opportunity models.

The model presented below is called *the composed gravity model*, because it starts from estimating the number of internal trips (trips made inside the origin zone) and then the remaining trips (external trips) are distributed to the adjacent zones depending on the distance. So, it is a model that uses as a function of travel impedance the distance between zones, but the travel time or cost of travel also can be used.

The transport flows of the destination zones are determined by the number of jobs, commercial, administrative and other elements that can attract travelers, and the transport potentials of the origin zones by the number of the population that moves.

In finding out the elements of the distribution matrix, it is assumed that the transport flows from the origin and destination zones are known, which define the generated trips, as follows:

$$\begin{aligned} g_1 &= n_{11} + n_{12} + \cdots + n_{1n} \\ g_2 &= n_{21} + n_{22} + \cdots + n_{2n} \\ g_n &= n_{n1} + n_{n2} + \cdots + n_{nn} \end{aligned} \tag{1}$$

respectively the attracted trips:

$$\begin{aligned} a_1 &= n_{11} + n_{21} + \cdots + n_{n1} \\ a_2 &= n_{12} + n_{22} + \cdots + n_{n2} \\ a_n &= n_{1n} + n_{2n} + \cdots + n_{nn} \end{aligned} \tag{2}$$

Graphically, the situation can be represented as in Fig. 1, from where the trips made between all the areas in which the study locality was divided can be identified.

The n_{ij} values can be determined graphically in Fig. 2.

The meanings of the notations in Fig. 1 are as follows:

H_i represents the travel density at the level of the destination zone a_j, measured in trips/km;

h_i - travel density for origin zone's internal trips, trips/km;
$d_{ij\,max}$ - maximum travel distance in the studied area;
d_{ij} - travel distance of interal trips (it is considered that i = j).

Considering that a_j it is determined by the surface of the triangle ABC and n_{ij} (for i = j) by the surface of the trapezoid AEDC, the following relations can be written:

$$H_i = \frac{2 \cdot a_j}{d_{ij\,max}} \tag{3}$$

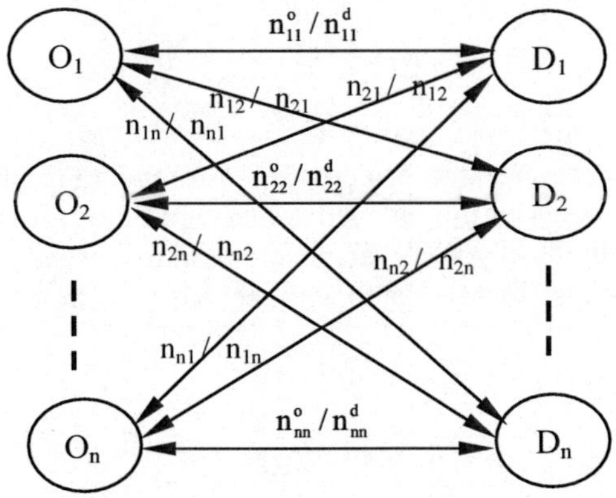

Fig. 1. Geometric interpretation of trips.

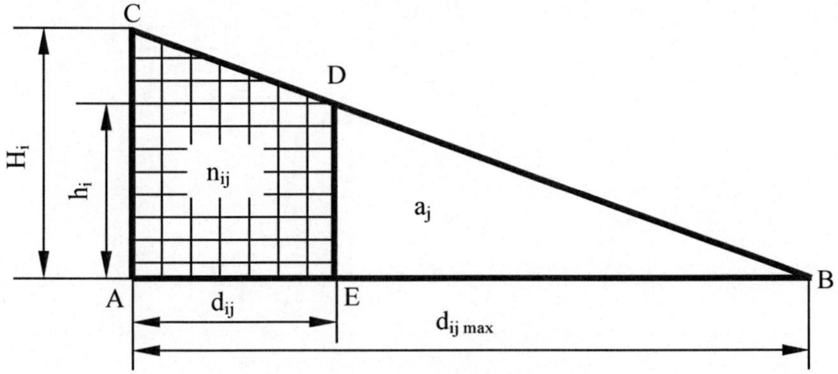

Fig. 2. Geometric interpretation of internal trips

$$h_i = \frac{H_i\left(d_{ij\ max} - d_{ij}\right)}{d_{ij\ max}} \tag{4}$$

$$n_{ij} = \frac{(H_i + h_i)\, d_{ij}}{2} \tag{5}$$

The relation (5) is valid when $i = j$. In order to determine the values of n_{ij}, with the condition $i \neq j$ the following relations are used:

$$\frac{n_{12}}{a_1'} = \frac{g_2'}{d_{12}} \cdot \frac{1}{\frac{g_2'}{d_{12}} + \frac{g_3'}{d_{13}} + \cdots + \frac{g_n'}{d_{1n}}} = \frac{g_2'}{d_{12}} \cdot \frac{1}{A} \tag{6}$$

$$\frac{n_{13}}{a_1'} = \frac{g_3'}{d_{13}} \cdot \frac{1}{\frac{g_2'}{d_{12}} + \frac{g_3'}{d_{13}} + \cdots + \frac{g_n'}{d_{1n}}} = \frac{g_3'}{d_{13}} \cdot \frac{1}{A} \tag{7}$$

$$\frac{n_{1n}}{a_1'} = \frac{g_n'}{d_{1n}} \cdot \frac{1}{A} \tag{8}$$

$$\frac{n_{n\ n-1}}{a_n'} = \frac{g_{n-1}'}{d_{n\ n-1}} \cdot \frac{1}{A_n} \quad (9)$$

where g_i' andi a_j' represents the corrected values of the transport flows.

If the differences between the planned values and those resulting from the matrix are large, then the methods of growth factors are used to correct the values [21].

The indicators can also be calculated with the values from the O-D matrix:

- average travel distance in each zone, $\overline{d}_i$

$$\overline{d}_i = \sqrt{\frac{d_{max}^2 g_i}{\sum_{i=1}^{n} g_i}}\ (km) \quad (10)$$

- the average distance traveled in studied area, $\overline{D}$,

$$\overline{D} = \frac{\sum_{i=1}^{n} g_i \overline{d}_i}{\sum_{i=1}^{n} g_i}\ (km) \quad (11)$$

- moment of transport, M,

$$M = \sum_{i=1}^{n} \sum_{j=1}^{n} n_{ij} d_{ij} (cal\ km) \quad (12)$$

3 Case Study

For a city divided into five zones, the transport flows of the origin and destination zones (Table 1), the matrix of distances (Table 2), as well as the maximum distance traveled in the studied area ($d_{ij\ max} = 7$ km) were determined by specific procedures.

It is mentioned that the locality used for the case study is a hypothetical one, the numerical values used being similar to a locality with mixed type areas, considering that trips are made inside the residential areas as well.

The number of trips made between city zones and specific travel indicators must be determined. The values of H_i, h_i şi n_{ij} for $i = j$ are determined using the relations (3), (4) and (5). The results are presented in Table 3.

The transport flow matrix is corrected with the n_{ij} values calculated for $i = j$, using relations:

$$g_i' = g_i - n_{ii} \text{ and } a_j' = a_j - n_{ii} \text{ for } i = j = 1, 2, \ldots, 5. \quad (13)$$

It is obtained the transport flow matrix presented in the Table 4.

To determine the number of trips from i to j, with $i \neq j$, we use the relations (6),…, (10) customized for $i = 1, 2, \ldots, 5$ and $j = 1, 2, \ldots, 5$. The results are shown in Table 5.

Table 1. Transport flows

Area	Origin (g_i)	Destination (a_j)
1	900	1100
2	800	700
3	1100	1000
4	700	800
5	1000	900

Table 2. The matrix of distances

d_{ij}	1	2	3	4	5
1	2	3	4	6	3
2	3	1	4	5	6
3	4	4	2	3	5
4	6	5	3	1	2
5	3	6	5	2	2

Table 3. Trips made inside the areas

i	1	2	3	4	5
H_i	314.29	200.00	285.71	228.57	257.14
h_i	224.49	171.43	204.08	195.92	183.67
n_{ij} $(i = j)$	539	186	490	212	441

Table 4. The corrected transport flow matrix with internal trips

i	1	2	3	4	5
$g_i^{\prime}$	361	614	610	488	559
$a_j^{\prime}$	561	514	510	588	459

Table 5. Primary trips distribution matrix

Dest.	Orig.						
	1	2	3	4	5	Total	g_i
1	539	184	137	73	167	1100	900
2	133	186	170	108	103	700	800
3	89	151	490	160	110	1000	1100
4	53	108	180	212	247	800	700
5	94	80	95	190	441	900	1000
Total	908	709	1072	743	1068	4500	–
a_j	1100	700	1000	800	900	–	4500

It is observed that the sum of the distributed values differs from the values of the generation and attraction flows (exceeding is greater than 20%) which requires corrective iterations for balancing. After each iteration, the correction coefficients C_i (i = 1,2,..., 5) were calculated for each zone and iteration (Table 6). The coefficients were used to multiply the rows/columns of the O-D distribution matrix to obtain the convergence of the solution.

Table 6. Correction coefficients used in iterations

Coefficient	1	2	3	4	5
C_1	1.2114	0.9873	0.9328	1.0767	0.8427
C_2	0.7608	1.1331	1.1247	0.9044	1.1641
C_3	1.1088	0.9971	0.9560	0.9650	0.9667
C_4	0.9534	0.9963	1.0185	1.0204	1.0132
C_5	1.0195	1.0014	0.9921	0.9913	0.9923

The distribution matrix obtained after iteration 4 is shown in Table 7.

Table 7. Distribution matrix obtained after four iterations

Dest.	Orig.						
	1	2	3	4	5	Total	g_i
1	526	132	89	55	98	900	900
2	201	204	172	126	97	800	800
3	137	171	499	189	104	1100	1100
4	65	99	148	203	185	700	700
5	150	93	100	234	423	1000	1000
Total	1079	699	1008	807	907	4500	–
a_j	1100	700	1000	800	900	–	4500

The average travel distance, calculated with relation (10) is shown in Table 8.

The average distance traveled by the entire area, calculated with the relation (11) is 3.1596 km, and the moment of transport is 13176 km.

Table 8. Average travel distance for each zone

	Average travel distance (km)				
Area	1	2	3	4	5
	3.130	2.951	3.461	2.761	3.300

4 Conclusions

In reducing traffic congestion, transport planning models play an important role due to the way in which the generating and attracting potentials of the zones, in the case of the most numerous trips, trips to work, are correlated by the O-D matrix.

A determination and then a correct calibration of the OD matrix are decisive for the next stages (modal split and traffic assignment) because the transport demand will be correctly estimated and in this way the supply, especially the public transport supply will be designed so as to respond to the travel needs of the inhabitants this being a method of reducing traffic congestion.

The paper presents a distribution model that initially determines the number of internal trips (these are important because they can be made on foot or by non-motorized means) and then the remaining trips (external trips) are distributed to other zones depending on the distance. It is mentioned that the value of the distance between the zones is not always the Euclidean distance but the influence of congestion can be included by adding an additional distance between the zones. It is mentioned that the accuracy of the model is consistent with the accuracy of the data entered (distances between zones and generating and attracting potentials of the zones, elements that can not always be known rigorously or changes in socio-economic activities of the city can occur - the most eloquent example is the emergence of the COVID 19 pandemic which radically changed the size and structure of transport demand).

The case study highlighted that the composed gravitational model works well for the distribution of zones' generated trips, and convergence to the generated and attracted traffic values is achieved relatively quickly. For the case study considered, the convergence was achieved in a percentage of 5% after only three iterations and after the fourth convergence was 2%, which attests to the ease of application of the model even for large cities.

References

1. Raicu, Ş., Costescu, D.: Mobilitate. Infrastructuri de trafic [Mobility. Traffic Infrastructure]. AGIR Publishing house, Bucharest (2020)
2. Freudendal-Pedersen, M., Kesselring, S., Servou, E.: What is smart for the future city? Mobilities Autom. Sustain. **11**, 221 (2019). https://doi.org/10.3390/su11010221
3. De Luca, S., Di Pace, R., Memoli, S., Pariota, L.: Sustainable traffic management in an urban area: an integrated framework for real-time traffic control and route guidance design. Sustainability **12**, 726 (2020). https://doi.org/10.3390/su12020726
4. Raicu, Ş.: Sisteme de Transport [Transport systems]. Publishing House AGIR, Bucharest, Romania (2007)
5. Henke, I., Cartenì, A., Molitierno, C., Assunta Errico, A.: Decision-making in the transport sector: a sustainable evaluation method for road infrastructure. Sustainability **12**, 764 (2020). https://doi.org/10.3390/su12030764
6. Newspaper: Univers ingineresc [Engineering Universe] no.3 (697) Bucharest, Romania, 1–15 February 2020. https://www.agir.ro/univers-ingineresc/numar-3-2020/
7. https://www.drpciv.ro/

8. Dragu, V., Burciu, Ş., Roman, E.A.: Dezvoltarea transportului public urban de mare capacitate – Soluţie pentru un oraş inteligent [Development of high capacity urban public transport - Solution for a smart city], AGIR Bulletin no. 1/2017, Romanian (2017). https://www.buletinulagir.agir.ro/articol.php?id=2818
9. Banister, D.: The sustainable mobility paradigm. Transp. Policy **15**, 73–80 (2008). https://www.sciencedirect.com/science/article/abs/pii/S0967070X07000820?via%3Dihub
10. Holden, E., Linnerud, K., Banister, D.: Sustanable passenger transport. Back Brundtland Transp. Res. Part A **54**, 67–77 (2013)
11. European Commission Guidelines for Developing and Implementing a Sustainable Urban Mobility Plan (2013). http://www.eltis.org/sites/eltis/files/guidelinesdeveloping-and-implementing-a-sump_final_web_jan2014b.pdf
12. SUMP-Bucharest - Sustainable Urban Mobility Plan Bucharest Ilfov, 2016–2030. https://tpbi.ro/files/proiect_pmud.pdf
13. Olaru, D., et al.: Place vs node transit: planning policies revisited. Sustainability **11**, 477 (2019). https://doi.org/10.3390/su11020477
14. Attard, M., Shiftan, Y.: Sustainable urban transport–an introduction. In: (eds.) Attard, M., Shiftan, Y., Sustainable Urban Transport, pp. 14–24. Emerald Group Publishing Limited, Bingley, UK (2015)
15. Nasri, A., Zhang, L.: How urban form characteristics at both trip ends influence mode choice: evidence from TOD vs. Non-TOD Zones of the Washington, D.C. Metropolitan Area. Sustainability **11**, 3403 (2019). https://doi.org/10.3390/su11123403
16. Ortuzar, J.D., Willumsen, L.: Modelling Transport 4th edn. John Wiley and Sons, Ltd Publication, Hoboken (2011). ISBN: 978-0-470-76039-0
17. Daubayev, K., et al.: The transport model as a necessary condition for the construction of an efficient transport system. Espacios **38**(54), 22–31 (2017)
18. Rasouli, A., Chegenizadeh, A., Nikraz, H.: A Cricical Review of Curent Transport Models, EJGE, vol. 22, Bund 14, pp. 5453–5464 (2017)
19. Profillidis, V.A., Botzoris, G.N.: Modelling of Transport Demand – Analyzing, Calculating and Forecasting Transport Demand, 1st edn. Elsevier, Amsterdam (2018)
20. Oppenheim, N.: Urban Travel Demand Modeling. Wiley, New York (1994)
21. Dragu, V.: Trafic urban şi suburban de călători – Îndrumar de laborator [Passengers Urban and Sub-urban Traffic – Student activities guide]. BREN Publishing house, Bucharest (2001)

Considerations on Monitoring the Drowsiness of Drivers Through Video Detection and Real-Time Warning

Maria Claudia Surugiu(✉) and Ion Nicolae Stăncel

Department of Telematics and Electronics in Transportation, Faculty of Transportation, University "Polytechnica" of Bucharest, Bucharest, Romania
claudia.surugiu@upb.ro

Abstract. Microsomnia, decreased concentration and fatigue at the wheel are particularly dangerous and are the cause of many accidents. However, the initial signs can be detected in advance: tired, low-attention drivers perform less precise steering maneuvers and have to make minor path corrections more often.

The willingness to take over the vehicle control in driving scenarios, in autopilot mode, is an important factor for road safety. This paper presents a low-cost system for automatic recognition of driver activity by eye monitoring. Thus, an architecture based on eye movement and blink tracking data is introduced in this system, thus analyzing several features. It is estimated that this technology will help prevent acci-dents caused by drivers who become drowsy. Various studies have suggested that about 20% of all road accidents are related to fatigue.

Keywords: Video sensors · Driving attention · Fatigue detection · Algorithm

1 Introduction

Automotive electronic systems are experiencing an unprecedented age of development. For luxury vehicles, it is estimated that electronic systems account for about 23% of the total cost of the vehicle [1]. Experts also estimate that about 80% of automotive innovations are mostly in the electronic system field. The main purpose of automotive electronic systems is for driver assistance in the vehicle control, by using traction (engine), steering (electric power steering), braking system (ABS, ESP) or suspension (active suspension). In early experiments conducted by Google with the autopilot vehicle division, the company reported test drivers falling asleep while driving at a speed of 55 km/h on the highway. The possibility of vehicle control taking in certain driving scenarios, in autopilot mode, is an important factor for road users' safety. Drowsiness is one of the most common causes of road accidents [1, 2]. The effects of fatigue can be compared to those of driving under the influence of alcohol. Like alcohol, drowsiness slows down the reaction time, which leads to an increased risk of being involved in a car accident [2].

Causes of drowsiness or fatigue include.

L. Moldovan and A. Gligor (Eds.): Inter-Eng 2021, LNNS 386, pp. 661–671, 2022.
https://doi.org/10.1007/978-3-030-93817-8_59

- Insufficient sleep;
- Fatigue accumulated over time;
- Untreated sleep disorders;
- Use of sedative drugs.

Drowsiness or fatigue can cause the following problems:

- Slowing the reaction time, visual deficiencies, judgment deficiencies;
- Difficulties in processing information;
- Decreased performance, motivation and alertness;
- May cause irritable and aggressive behavior.

The incidence of accidents caused by fatigue usually occurs between 2:00–6:00 and 14:00–16:00. There is a 20 times higher probability that a driver will fall asleep behind the wheel at 6:00 compared to 10:00. Motorways are roads where most accidents are common due to fatigue driving, because of the lack of incentives for drivers. A study conducted on professional drivers, indicated that the risk of having an accident due to fatigue begins to increase after 9–10 h of driving [1]. This risk doubles after this period. According to a study by the AAA (Foundation for Traffic Safety) on 3500 drivers in the United States, monitored over a period of six months on board mounted cameras, it was shown that they were involved in 700 road accidents. The drowsiness/fatigue condition caused 9.5% of those accidents. Depending on the country and region, statistics indicate that fatigue and drowsiness while driving can cause between 10% and 30% of all road accidents. According to the Police Headquarters in Romania, in 2017, 190 accidents resulting in 227 deaths were caused by drivers who fell asleep at the wheel. Most of the accidents occurred on weekends, on national roads. Unfortunately, Romania ranks 10th in the top of road accidents caused by drowsiness. More than a million Romanians have sleep disorders, but only 5000 are diagnosed.

2 State of the Art

In the last 5 years, new technologies are used to manage the amount of information, obtained from video tools, in the form of objects and images. Different hybrid systems in terms of how to interpret this image and object date are concatenated in homogeneous applications, such as detection systems that use the Convolutional Neural Networks (CNN) with the Scale Invariant Feature Transform (SIFT) algorithm integration [3]. New methods of implementing visual objects recognition based on deep learning with CNN are adapted to both demanding applications in terms of hardware, which has driven the development and research of low-cost systems, which consists in the use of open-source platforms, such as be Rasberry Pi [4]. Also, in terms of latency and accuracy, CNN networks are studied, from the perspective of eye indicators status which controls the detection system. The main types of networks used are Fully Designed Neural Network (FD-NN), Transfer Learning in VGG16 and VGG19 with extra designed layers (TL-VGG) [5].

Vehicle manufacturers use video cameras, sensors to track head position, steering wheel monitoring and audible alerts to ensure drivers pay attention when using advanced driver assistance systems, such as autopilot. Several car manufacturers offer drowsiness detection systems [6]. In various studies, [7] the most used techniques for detecting fatigue are based on monitoring physiological parameters: brain waves, heart rate and respiration. These methods are intrusive and require the use of sensors attached to the driver, which is inconvenient. Other techniques have also been studied to detect the fatigue of drivers by moving their eyes and blinking, [8] thus proving that if the blink rate increases, it indicates a state of fatigue.

For example, BMW uses a video camera mounted in the car to solve a critical challenge with autopilot systems, ensuring that drivers pay attention to the road. The driver-monitoring system proposed by Ford consists of a small traffic-oriented camera connected to an on-board computer. The system offered by Mercedes draws up an individual profile of the driver's driving style, which is constantly compared with the latest sensor feedback. At the center of the system is an extremely sensitive sensor, which records direction and speed of the vehicle [9]. The warning system used by Nissan presents the following solution: drivers usually make small adjustments and corrections to the steering wheel while driving. It is common for tired drivers to stop or slow down their steering wheel movements. The system used by Subaru uses a dedicated video camera and facial recognition software that tracks the driver's activity to calculate two stages of fatigue - sleepy and extremely sleepy. The system offered by VW closely monitors the driver's behavior, noticing any irregular movement of the steering wheel, the pedal use and any deviations from the lane (Lane Assist), so it can judge when the driver begins to feel drowsy and must stop and take a break [10]. If the system detects that the driver is starting to lose focus, he will be warned with a visual display on the instrument panel and a warning sound.

3 Description of Work

The proposed system includes a video camera aimed at generating images of the driver, including the eyes. It also includes a processor for processing images generated by the camcorder. The processor monitors the acquired image and determines whether the eye is in an open or closed position. The processor also determines a proportion of the closing time of the eyes as the proportion of a time interval in which the eye is in the closed position and determines a state of drowsiness when the proportion of time exceeds a certain threshold value. The sleep detector system will use a single-color video camera to capture images from the passenger compartment. The camera is mounted firmly on the driver's side to see his upper body. Image analysis will be performed using the eye tracking algorithm and motion detection.

Eye Detection

Algorithms commonly used for eye detection include the Hough transform, matching predefined patterns, principal component analysis (PCA), and the Adaboost algorithm.

In this paper, the eye *is monitored by facial cues detection* to locate important regions of the face [10].

Blinking Detection
Blinking is detected using the Eye Appearance Report (EAR), introduced by [11].

Extraction of Imaging Features
The pre-trained facial landmark detector, iBUG 300 W, from the Dlib library is used to estimate the location of 68 coordinates (x, y) that match the facial structures (see Fig. 1) [11–13].

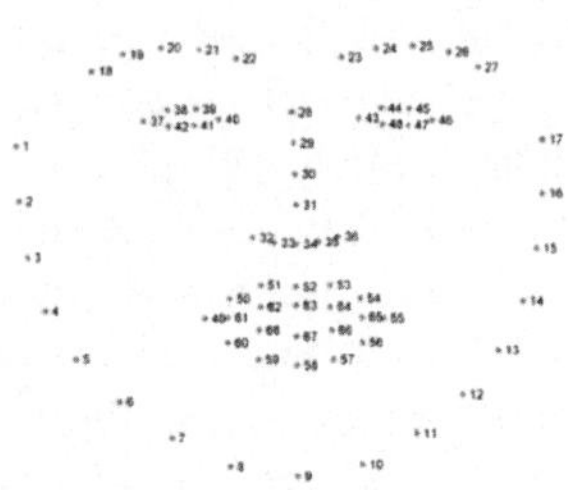

Fig. 1. View over the 68 coordinates of the face marker in the iBUG 300 W data set [11].

The IBUG database was released as part of the first 300-W version. It consists of 135 images downloaded from the web, with large variations in expression, lighting conditions and positions [11, 12]. Face Marker Detection uses an input image and a shape detector that tries locating key points of interest along the face.

Therefore, facial detection cues are a two-step process [11, 12]: face location in the image and facial structures detection.

Face Location in the Image
For this first stage, an object detector will be applied: Oriented Gradient Histogram (HOG) and SVM (Support Vector Vectors) classification algorithm pre-trained for the face detection task.

Oriented Gradient Histogram (HOG)
HOG descriptors are feature descriptors used in image processing for object detection purposes. The technique is based on monitoring the number of occurrences of the orientation of a gradient (variation per unit length of a scalar quantity) in a certain region of the image. Below, a generated face model by HOG is presented (see Fig. 2). The main idea is for the local appearance and shape of an object in an image to be described by the gradient intensity distribution and the edge orientation distribution [11].

The calculation of the gradient values is done by applying one-dimensional bypass filters horizontally and vertically. The filtering is done with two convolution nuclei:

$$D_x = [-1\ 0\ 1]$$
$$D_y = [-1\ 0\ 1]^T$$

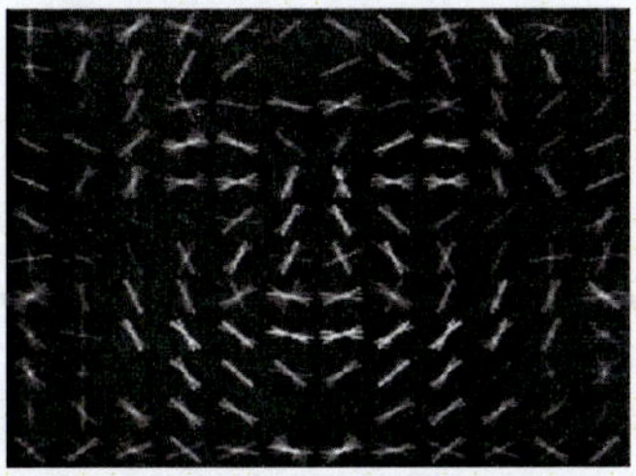

Fig. 2. Face model generated by HOG

Using a convolution operator $I_x = I \cdot D_x$ și $I_y = I \cdot D_y$ we will obtain horizontal and vertical derivatives. The magnitude of the gradient is $|G| = \sqrt{I_x^2 + I_y^2}$ and the gradient orientation is $\arctan\left(\frac{I_y}{I_x}\right)$. The second step is to create the histogram for each cell. Each pixel that forms this cell will be represented in the resulting histogram.

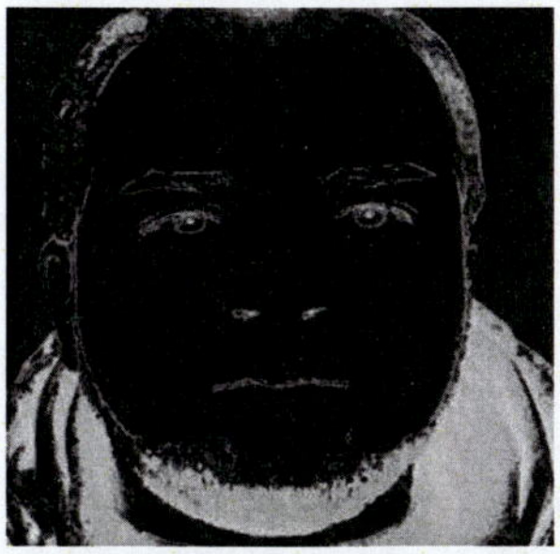

Fig. 3. The magnitude of the gradient

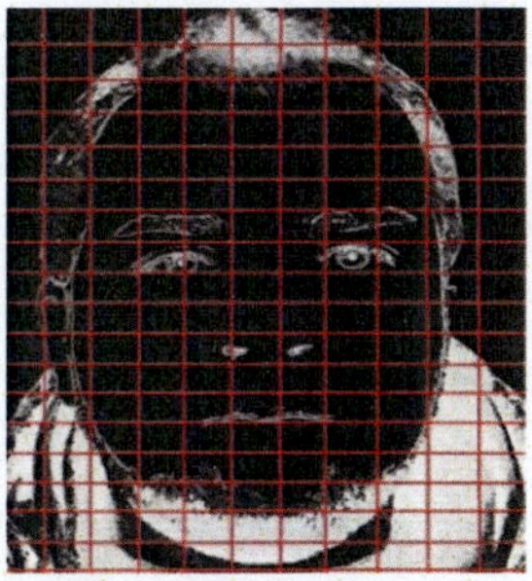

Fig. 4. Division into cells

To consider, the variations in brightness and contrast, the grouping of cells in larger structures (blocks) (see Fig. 3 and Fig. 4) will take place to locally normalize the magnitude of the gradients. The HOG descriptor will result from the normalized components of the histogram. The blocks partially overlap so that each cell contributes several times to the final descriptor [11].

Normalization of blocks

The normalization factor is calculated with one of the Eqs. (1), (2) below, [12]:

$$L_2 - norm : f = \frac{v}{\sqrt{\|v\|_2^2 + e^2}} \tag{1}$$

$$L_1 - norm : f = \frac{v}{\sqrt{\|v\|_1 + e}} \tag{2}$$

where

$v =$ the non-normalized vector containing all the histograms of a block;

$\|v_k\| =$ is the norm k for $k = 1, 2$; $e =$ constant of very low value.

Euclidean distance between two vectors $P = (p_1 p_2, \ldots p_n)$ and $Q = (q_1 q_2, \ldots q_n)$ is:

$$d = \sqrt{(p_1 - q_1)^2 + (p_2 - q_2)^2 + \ldots + (p_n - q_n)^2} \quad (3)$$

Support vector machine (SVM)
Mainly, the studied topic can be classified into two classes (sleepy or non-sleepy). SVM was chosen because it is used generically for binary classification issues and has attributes that make it fit perfectly [10, 11]. SVM is a set of supervised learning methods used to classify, regress, and detect external values. It uses a subset of decision- making training points called support vector. Different Kernel functions can be specified for the decision function. An SVM creates a hyper-plane or set of hyper-planes in a large or infinite dimensional space that can be used for classification, regression, or other objectives [9, 11]. A good separation is achieved by the hyper-plane that has the greatest distance to the nearest data points formation of any class (the so-called functional margin), because, in general, the higher the margin the lower the classifier generalization error gets.

The separation surface is described by the following Eq. (4):

$$(w \cdot x) + b = 0 \quad (4)$$

where vectors are represented by w and x; b represents a scalar.

The SVM method assumes that the separation surfaces equations are normalized. The two classes will have the limits that pass-through support vectors in Fig. 5, and Eqs. (5), (6):

$$(w \cdot x) + b = -1 \quad (5)$$

$$(w \cdot x) + b = 1 \quad (6)$$

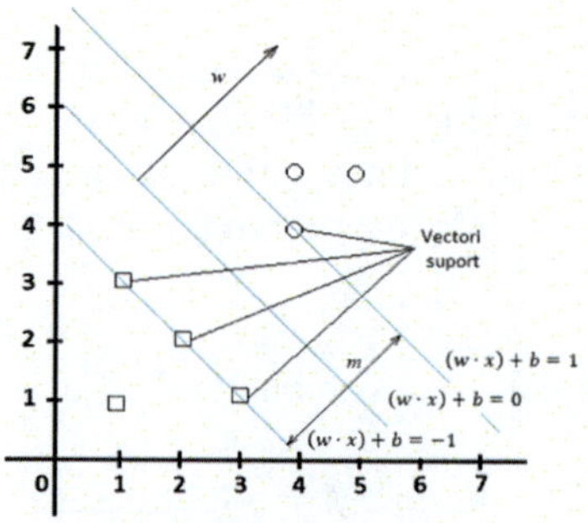

Fig. 5. Support vectors and characteristic problem notions illustration.

The actual classification involves determining w and b parameters that maximize the separation margin between classes.

Facial Structures Detection
Facial mark detectors try locating and label the following facial regions: Mouth, Nose,

Jaw, Left and Right Eye, Left and Right Eyebrow [11]. The detection method begins with a training data set with facial markers manually labeled on an image. These images are part of a library that specifies the coordinates (x, y) of the regions surrounding each facial structure [11]. Using this dataset, a set of regression trees is "trained" to estimate the positions of facial marks using pixels [12, 13].

3.1 Real-Time Blink Detection Using Facial Cues

The proposed algorithm estimates the mark positions, extracts a single scalar quantity, the Eye Aspect Ratio (EAR), [11, 13] characterizing the opening of the eye in each frame. Finally, an Support vector machine (SVM) [13] classifier detects blinking as a pattern of EAR values in a short time window. Existing methods are active or passive for Eye aspect ratio (EAR), [11–13]:

- Active methods are reliable, but they use special hardware, often expensive and intrusive, such as infrared cameras and illuminators, wearable devices, glasses with a special video camera that observes the eyes;
- Passive methods are based on a standard remote camera only.

In this paper, the passive system was studied. Regarding the detection of blinking; only two sets of facial structures, namely, the eyes, were of interest. Each eye is represented by 6 coordinates (x, y), starting from the left corner of the eye and then clockwise around the rest of the region (see Fig. 6).

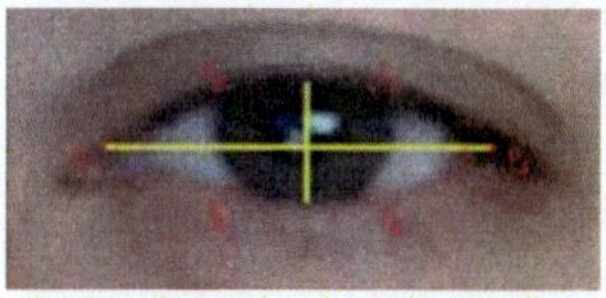

Fig. 6. The 6 facial features associated with the eye.

In these coordinates, there is a relationship between the width and height. Based on the real-time detection of the clip using facial cues, an equation can be derived that reflects this relationship called the Eye Aspect Ratio in (7), [11]:

$$EAR = \frac{\|p_2 - p_6\| + \|p_3 - p_5\|}{2\|p_1 - p_4\|} \tag{7}$$

where $p_1, \ldots, p_6$ are 2D facial landmarks.

The numerator of this equation calculates the distance between the vertical landmarks of the eye, while the denominator calculates the distance between the horizontal landmarks of the eye. Using this simple equation, we can avoid image processing techniques and simply rely on the eye reference distance ratio to determine whether a person is blinking. To make this clearer, we will demonstrate the results [11–13] in the images below (see Fig. 7.a), b)). In the Fig. 7 a) a completely open eye is presented, so the aspect ratio of the eye here is large and relatively constant over time. In the Fig. 7 b), you can see that once the person blinks, the aspect ratio of the eye decreases dramatically, approaching zero [11]:

$$EAR = \begin{cases} X > 0, & eyes\,opened \\ 0, & eyes\,closed \end{cases} \tag{8}$$

While the eyes are closing, the EAR result will be about 0, while during open eyes, the EAR can be any integer x greater than 0, [12, 13, 15].

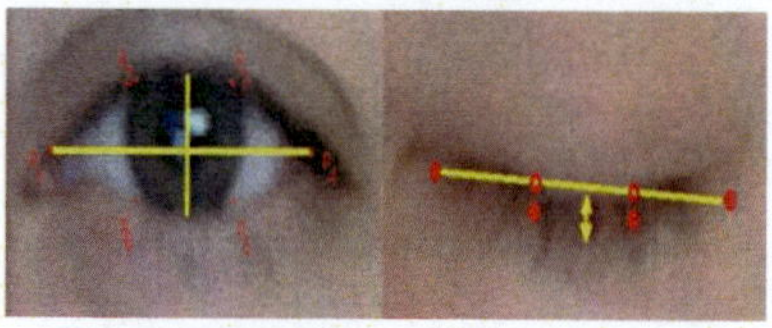

Fig. 7. Eye marks when the eye is open and closed.

3.2 Vehicle Equipment

The paper presents a low-cost system for automatic recognition of driver activity by eye monitoring. Therefore, an architecture based on eye movement and blink tracking data is introduced in this system, thus analyzing several features. The camera used in this project is an HP HD-3110. It has a video resolution of 720p, 1280 * 720 pixels, and autofocus. The camera is mounted on top of the dashboard and is connected to the Raspberry Pi control unit [14, 15]. After the assembly was completed, the drowsiness detector was built using computer vision techniques (see Fig. 8).

Fig. 8. The position of the camera in the vehicle.

OpenCV was used, an open-source library (which provides certain finished products, allowing users to modify and improve it without any obligation) for the field of image processing and not only, originally developed by Intel. The library has interfaces for C/C++, Python, Java, Matlab and runs on Windows, Linux and Mac OS X. This technology is especially based on continuous image processing. At this time the OpenCV library supports continuous time captures, object detection, application of simple filters on images.

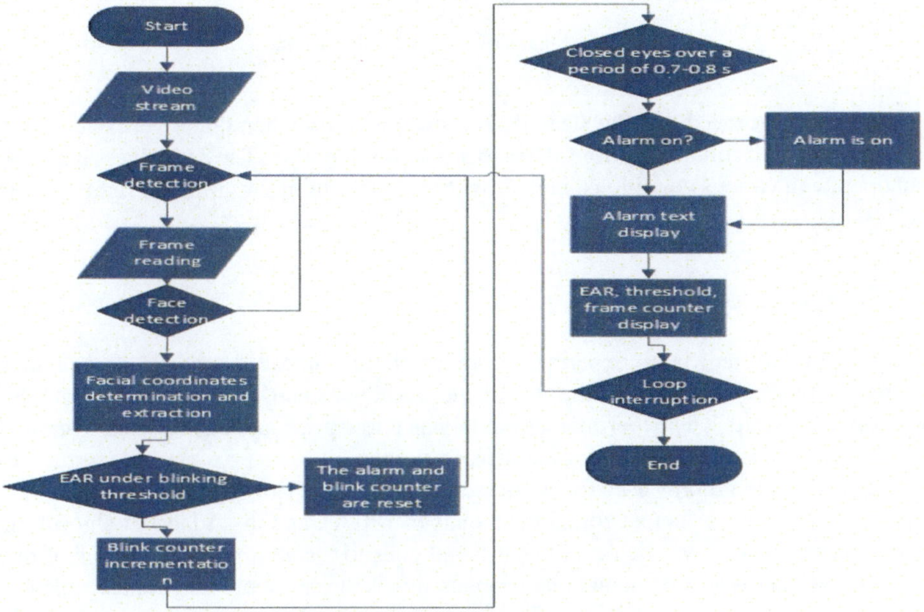

Fig. 9. Proposed system algorithm.

3.3 Drowsiness Detector Test

The general flow of the proposed algorithm for detecting drowsiness is simple. First, the camera will be configured to monitor the driver of the vehicle. When a face is identified, face detection is applied, and the eye regions are extracted. After determining the eye regions, the Eye Aspect Ratio (EAR) is calculated to determine whether the eyes are closed and if they are closed for a period, an alarm is activated [11, 12, 15] (see Fig. 10 a) and b)).

Once the software was running, the video stream began to be processed, followed by careful testing of the drowsiness detector. After completing the initial tests in a closed perimeter, the aim was to perform the test on a slightly crowded road because driving with closed eyes, even for a second, can be dangerous.

a) Face identification in the input video stream b) Locating facial features to extract eye regions c) Alarm status

Fig. 10. Stages of sleepiness detection.

The results showed that the sleep detector can identify when the driver is at risk of falling asleep and emits an audible alarm to attract his attention. The drowsiness detector can operate in various conditions, including direct sunlight or low artificial lighting (see Fig. 10 c)).

4 Conclusions

The proposed system is an experimental model based on the use of a video camera for face detection mounted on the top of the dashboard and a Raspberry Pi Zero W development board. The experimental model has a low price and can be used to emit a sleep signal in case of detection of the driver's fatigue.

Algorithms commonly used for eye detection include the Hough transform, matching predefined patterns, principal component analysis (PCA), and the Adaboost algorithm. In this paper, the eye was detected using facial cues to locate important regions of the face. Drowsiness is a relevant and real problem that is the cause of many road accidents. The number of victims of these accidents is high and has been constant recently, with an associated economic impact. All these determine a strong motivation for developing a measure to solve this problem. The implementation of this system on autonomous vehicles will lead to increased traffic safety for both drivers and pedestrians.

References

1. Joly, A., Zheng, R., Kaizuka, T., Nakano, K.: Efect of drowsiness on mechanical arm admittance and driving performances. IET Intell. Transp. Syst. **3**(12), 220–226 (2018)
2. Tripathi, A., Kumar, T.V., Dhansetty, T., Kumar, J.: Real time object detection using CNN. Int. J. Eng. Technol. (UAE). **7**, 33–36 (2018). https://doi.org/10.14419/ijet.v7i2.24.11994
3. Shehab, M.A., Al-Gizi, A., Swadi, S.M.: Efficient real-time object detection based on convolutional neural network. In: International Conference on Applied and Theoretical Electricity (ICATE) 2021, pp. 1–5 (2021). https://doi.org/10.1109/ICATE49685.2021.9465015
4. Hashemi, M., Mirrashid, A., Beheshti Shirazi, A.: Driver safety development: real-time driver drowsiness detection system based on convolutional neural network. SN Comput. Sci. **1**(5), 1 (2020). https://doi.org/10.1007/s42979-020-00306-9
5. Vural, E.: Video based detection of driver fatigue, Graduate School of Engineering and Natural Sciences, Sabanci University, Spring (2009)

6. Dong, Y., Hu, Z., Uchimura, K., Murayama, N.: Driver inattention monitoring system for intelligent vehicles: a review. IEEE Transp. Intell. Transp. Syst. **12**, 596–614 (2011)
7. Meireles, T., Dantas, F.: A low-cost prototype for driver fatigue detection. Multimodal Technol. Interact. **3**(5), 1–11 (2019)
8. https://www.ebay.com/itm/Mercedes-ME9-7-ME-9-7-ECU-ECM-Engine-Computer-Programming-Cloning-Unlocking-x2-/262689159375
9. Barea, R., Boquete, L., Mazo, M., Lopez, E.: System for assisted mobility using eye movements based on electrooculography. IEEE Trans. Neural Syst. Rehabil. Eng. **10**(4), 209–218 (2002)
10. Cech, J., Soukupova, T.: Real-time eye blink detection using facial landmarks. In: Cehovin, L., Mandeljc, R., Struc, V. (eds.) 21st Computer Vision Winter Workshop Luka, Rimske Toplice, Slovenia, 3–5 February 2016
11. Chen, M.-C., Chen, J.-L., Chang, T.-W.: Android/OSGi-based vehicular network management system. Comput. Commun. **34**(2), 169–183 (2011)
12. Tzimiropoulos, G.,: Project-out cascaded regression with an application to face alignment. In: Proceedings of the IEEE Conference on Computer Vision and Pattern Recognition (CVPR), Boston, MA, USA, June 2015
13. Fischer, I., Hennecke, F., Bannes, C., Zell, A.: Java neural network simulator web site (2001). http://www.ra.cs.uni-tuebingen.de/software/JavaNNS/welcome.html
14. Mouser Electronics. https://pt.mouser.com
15. Leon, F.: Artificial Intelligence: Cars with Support Vectors. Tehnopress, Iasi (2014)

Model for Bus Line Planning in an Intermodal Urban Transport Network

Dorinela Costescu(✉), Sergiu Olteanu, and Eugenia Alina Roman

Polytechnic University of Bucharest, Spl. Independentei 313, 060042 Bucharest, Romania
dorinela.costescu@upb.ro

Abstract. Reducing car usage in modal share is a continual goal, with more and more ambitious targets in European strategies for sustainable urban mobility. Consequently, the efficacy and the efficiency of public transport represents a major theme in urban planning and the development of sustainable solutions for urban mobility needs. Benefits as reducing road traffic congestion, increasing the attractiveness of public transport, reducing operating costs due to economy of scale are arguments for financing rapid public transport projects (e.g., rail, subway, light rail). However, to ensure their performance, rapid public transport projects need to be integrated into the urban public transport system and, in particular, into the urban mobility system. This paper presents a model for bus line planning in an intermodal transport network developed to increase the efficiency of a high-capacity public transport line. A case study is presented to exemplify the bus line network design to feed and to increase the level of transport capacity usage for a new operating subway line in Bucharest.

Keywords: Sustainable urban mobility · Public transport · Intermodal network

1 Introduction

Urban mobility is an important incentive for the growth and welfare of the population. But it causes negative effects like congestion, air and noise pollution, and traffic risk, which are common problems in many cities. The question of how to enhance mobility while at the same time reducing congestion and pollution is a common challenge to all major cities [1].

According to the objectives of sustainable urban mobility [2], public transport systems must be developed to become an attractive alternative to travel by personal car. Improving the efficiency of public transport can be achieved by using several modes of transport. Relative to the morphology and structure of the urban area, but also the particularities of the behavioural pattern of the urban population, the most appropriate modes of travel and transport must be identified in terms of capacity and operating costs [3]. Therefore, services for different modes of transport and travel need to be correlated in an intermodal system, corresponding to the economic, social, and environmental missions of urban public transport.

L. Moldovan and A. Gligor (Eds.): Inter-Eng 2021, LNNS 386, pp. 672–683, 2022.
https://doi.org/10.1007/978-3-030-93817-8_60

Public transport planning is an ample topic, investigated in numerous studies in the field of operational research [4–8]. In general, the planning process is considered hierarchically, in which, sequentially, the next steps are solved [8]:

- network design
- line planning
- timetabling
- vehicle scheduling, and crew scheduling.

The public transport network design is a complex problem, framed in the strategic planning of public transport, as it is based on the existing transport infrastructure. In the case of major modifications of the urban structure and socio-economic activities, the phase of public transport network design may lead to investment measures for new components of transport infrastructure (which must be well-substantiated). In addition, the change in the high-capacity transport network needs to redesign the public surface transport network.

This paper aims to develop a model for the redesign of the public surface transport network in an area where a new subway line has been introduced. The problem is to design bus lines as feeder lines for the subway line. The goals are to expand the coverage area of the rapid public transport and to increase passenger volumes on the rapid public transport line. These measures will contribute to the enhance of the usage of the subway transport capacity and, finally, to improve the efficiency of urban public transport.

The remainder of the paper is structured as follows. After a brief literature review on transport network design problems, the next section discusses the characteristics of the problems proposed to design an intermodal network to enhance the access to rapid public transport provided by a subway line. Then, it introduces the basic notation and presents the model developed to integrate the bus lines and the subway line into an intermodal system. Section 3 discusses the problem of the connection of the bus lines and a new operating subway line in Bucharest. Finally, Sect. 4 concludes with the research major findings and suggestions for future work.

2 Intermodal Transport Line Planning

2.1 Background

The line planning phase consists in the determination of a set of proposed public transport lines for operation, as well as their transit unit frequencies. Note that transit route or transit line is a designated set of streets or separated rights-of-way that transit units regularly serve [3]. An overview of the line planning problem is presented by Guihaire and Hao [7], Kepaptsoglou and Karlaftis [9], and Schöbel [10].

In general, in the line planning problem, generation of the route is included in the optimization problem, A two-stage model is developed [9]:

- In the first stage, a conceptual model is developed, in which the objectives and the main restrictions are defined. Linear programming, heuristic and metaheuristic algorithms are applied to solve the model [9–12]. The result consists in a network of potential lines.

- In the second stage, a methodological model is used to establish a set of lines for operating and their parameters.

E.g., a two-step solution is developed by Guihaire and Hao [7], in which a genetic algorithm is applied to optimize travel time and operating costs.

In most studies, the objectives in line planning problem are defined as minimizing operating costs and maximization of the user benefits [4, 9]. A distinct category of models optimizes passenger convenience, in terms of eliminating transfers for a satisfactory distance [13] or reducing a generalized cost defined function of travel time [14]. Recently, models have been developed to minimize energy consumption and protect the environment [4, 15].

The next section presents a model for the development of a surface public transport network that will feed the rapid public transport network. The particularities of the model consist in the union of the elements of two modes of transport. The goal is to design an intermodal transport network that will contribute to increasing public transport efficiency. The model objectives are minimizing operating costs and minimizing travel time.

2.2 Problem Definition

Rapid public transport lines (rail, subway, light rail with grade-separated or exclusive right-of-way) are considered solutions to increase the attractiveness of public transport. The financing of rapid public transport projects is supported by benefits associated with reducing road traffic congestion, reducing travel times, reducing operating costs due to economy of scale.

The implementation of urban rapid transport projects requires significant investments and have important implications on the transport capacity (more than 30,000 passengers per hour per direction [16]). Therefore, the projects have to be integrated into the whole urban public transport system and, in particular, into the urban mobility system, to efficiently use the provided capacity.

In this frame, the proposed problem is to design bus lines in an area where a subway line is implemented. The role of the bus line network is to feed the subway line and to establish the basis of an intermodal network with increased efficiency of the subway line.

Let consider an urban area served by bus lines in which a project for a subway line is implemented (see Fig. 1). In the initial state, bus lines converge to one destination located at a point of major interest (see Fig. 1.a). The new operating subway line provides access to the same destination (see Fig. 1.b). The problem is to redesign the bus lines to eliminate duplication of bus and subway services on certain sections, increase the accessibility of the subway network and feed the subway line with additional passenger volumes than those in the coverage area (see Fig. 1.c).

The integration of bus and subway lines can reduce costs at the intermodal transport system level by eliminating sections doubled by the two modes and increasing the capacity utilization on each section of the system.

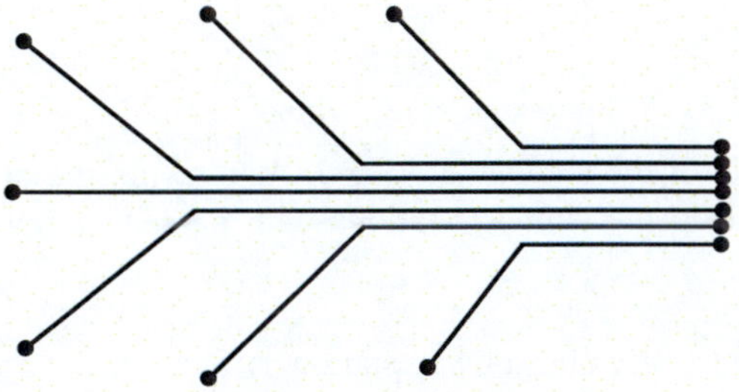

a) trunk-and-branch bus lines (surface public transport)

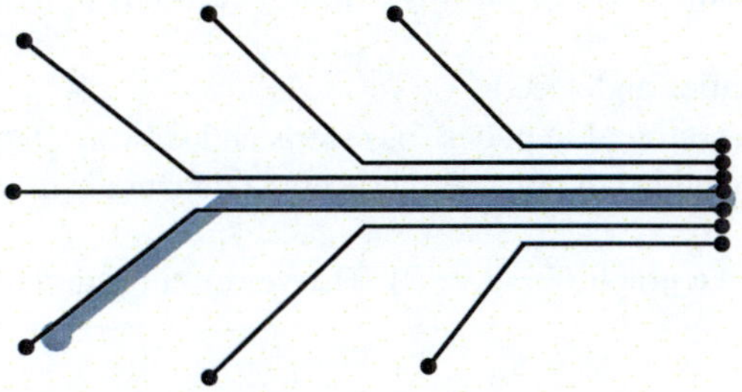

b) subway line implementation in bus served area

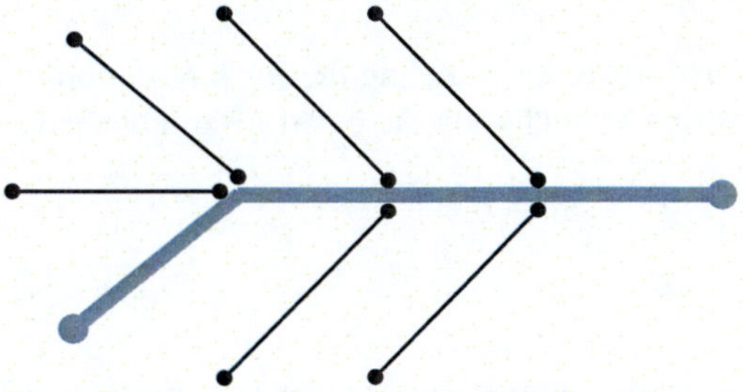

c) trunk-and-feeder intermodal network

Legend

Surface public transport (bus) which operate on road infrastructure with mixed traffic

Subway line (rapid transit, on fully controlled right-of-

Fig. 1. Steps in network design.

2.3 Model Formulation

The public transport network design is a complex problem that involves a multitude of parameters such as attributes of the line interstations, frequencies, transport capacities of vehicles, the particularities of demand and user behaviour.

In this case, the main components of the system are:

- Subway line stations

- Elements of the subway line
- Bus line stations.

The stations and the route of the subway are fixed. As a result, the decisions concern the bus line design to feed the subway stations with passenger flows. The next assumptions are considered:

- Each bus line is connected to only one subway station
- A bus stop is located next to each subway station to ensure the connection to the rapid public transport
- Except for the bus stops located next to subway stations, each bus station serves only one bus line
- Buses have standard capacities and speeds
- Demand is considered concentrated in nodes (bus stops and subway stations)
- Temporal variation of demand is not considered (average demand is used).

The problem is defined on a graph $G = (N, R)$. The vertex set is defined as

$$N = S_M \cup S_B \tag{1}$$

where

$S_M = \{s_1, s_2, ..., s_i, ..., s_m\}$ is the vertex set including the subway stations;
$S_B = \{1, 2, ..., b, ..., n\}$ - the vertex set including the bus stops established as

$$S_B = B \cup A \tag{2}$$

where

B is the vertex subset including the bus stops located at distances more than 250 m from any subway station;
A - the vertex subset including the bus stops located next to a subway station, ensuring access to the subway network.

The edge set is defined as

$$R = T \cup R \tag{3}$$

where

$T = \{(s_i, s_{i+1})|s_i \in S_M, i \leq m-1\}$ is the edge set including the subway interstations;
$R = \{(i, j)|i, j \in B, i \neq j\}$ - the edge set including the road sections between each pair of bus stations.

The proposed model aims to minimize the operating bus costs and the travel time by bus. The next notations for parameters and decision variables are used $i \in B$:

q_i – Average passenger demand per hour in bus station i (passengers/h); $q_i \geq 0$ if $i \in B$ and $q_i = 0$ if $i \in A$.
c_B – Unit bus operating cost (euro/veh. km)
c_t – Value of passenger riding time ((euro/pass. hour)
f_k – Bus operating frequency on the bus line k (veh./h)
l_{ij} – Distance from bus stop i to bus stop j (km)
l_{is}– Distance from bus stop i to subway station s (km)
V – Average bus operating speed (km/h)
Q_B – Bus transport capacity (passengers/veh.)
α – Load capacity coefficient applied to ensure passenger comfort in periods with high demand.
X_i^s – Binary decision variable indicating that bus stop i is allocated to subway station s; $X_i^s = 1$ if stop i is allocated to feed station s, $X_i^s = 0$ otherwise.
Y_i^k – Binary decision variable indicating that bus stop i is assigned to bus line k; $Y_i^k = 1$ if stop i is assigned to line k, $Y_i^k = 0$ otherwise.
W_k^s – Binary decision variable indicating that bus line k is assigned to subway station s; $W_k^s = 1$ if bus line k feeds subway station s, $W_k^s = 0$ otherwise.
Z_{ij}^{ks} – Binary decision variable indicating that bus stops i and j are assigned to subway station s and bus line k.
K– Set of the proposed feeder bus lines.

The objective function is formulated as follows:

$$\min\left(\sum_s\sum_k\sum_i\sum_j c_B f_k l_{ij} Z_{ij}^{ks} + \sum_s\sum_i c_t \frac{l_{is}}{V} q_i X_i^s\right) \tag{4}$$

subject to

$$\sum_{i\in B} X_i^s = 1, \forall s \in M \tag{5}$$

$$\sum_{i\in S_B} Y_i^k = 1, \forall k \in K, \tag{6}$$

$$\sum_k W_k^s = 1, \forall s \in M \tag{7}$$

The objective function (4) seeks to minimize the operating bus cost (first term) and the passenger travel time by bus (second term). In the term of the operating bus cost, the frequency is computed function of the passenger demand assigned to the bus stops on the line k:

$$f_k = \frac{1}{\alpha Q_B}\sum_s\sum_i q_i X_i^m Y_i^k W_k^s, \tag{8}$$

and the decision variable Z_{ij}^{km} is computed as:

$$Z_{ij}^{km} = X_i^m X_j^m Y_i^k X_j^k W_k^m. \tag{9}$$

Constraint (5) ensures that each bus stop located not next to the subway station (meaning $i \in B$) is assigned to only one feeder bus line. Several lines can converge in the bus stops that provide access to the subway (meaning $i \in A$). Constraint (6) ensures that each stop is assigned to one subway station. Constraint (7) ensures that each feeder bus line is assigned to one subway station.

Due to the combinatorial nature of the problem and large number of variants, the problem can be solved only by heuristic algorithms. The next section exemplifies the model application for a subway line in Bucharest. At this stage of the research, the problem is solved only for the second term in Eq. (4), corresponding to the minimization of the travel time by bus.

3 Case Study

Gradually built (since 1979) and equipped, the subway network in Bucharest has a length of around 78 km and it is designed on five routes, denoted by *M1*, …, *M5* [17]. The fifth route (*M5*) has operated since 2020. The new line has 10 stations on 6.8 km in length and is connected to the *M1* and *M3* lines (see Fig. 2). In the peak periods, the *M5* line provides a capacity of 12,000 passengers per hour per direction.

The line coverage indicators are defined as the percentage of residents and jobs located within 800 m of walking distance around subway stations [18]. For *M5*, we computed these indicators based on data and traffic analysis zones (TAZ) given in Bucharest - Ilfov SUMP [19]. It results that 7.85% of 1.88 mills. Residents and 5.5% of 1.02 mills. Employees have satisfactory access to the new subway line (see Fig. 3.a).

The entire coverage area of the new subway line is served by existing bus lines for which the coverage contour is computed for 400 m of walking distance around bus stops (see Fig. 3.b).

Consequently, the objective is to redesign the bus lines as a feeder network to achieve increased accessibility to the subway line and reduced operating costs on the intermodal transport network. To solve the problem, based on the model presented in the previous section, we used procedures available in *Spatial Analysis Toolbox* and *Network Analyst Extension* included in the *ArcGIS software package* [20].

The algorithm designed to solve the minimization of the passenger travel time by bus follows the next steps (see Fig. 4).

1. Create the set of bus stops used as input for the line planning problem:

   - The bus stops located in the coverage area of the subway line (CovS) are selected.
   - The existing bus lines with stops in CovS are selected. All the distinct stops of the selected bus lines are included in the set of the line.
   - Proximity areas for input bus stops are created. To each bus stop located at more than 250 m, average passenger demand is assigned based on TAZ attributes [19].

2. Create the sets of bus stops that feed each subway station

   - The bus stops are allocated to the closest subway stations. Attributes are edited to allow the selection of the subset of the bus stops allocated to each subway station.

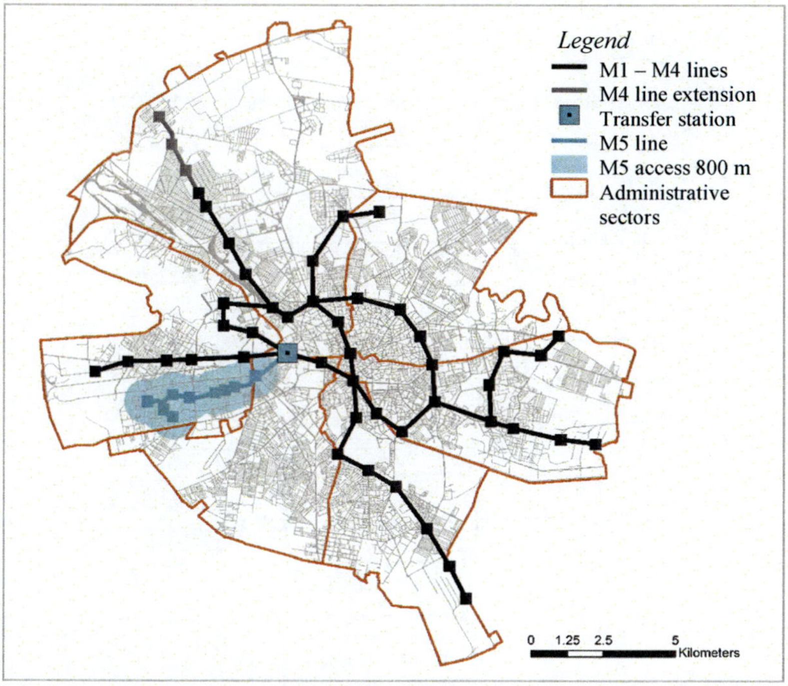

Fig. 2. Location of the new subway route (M5) of the subway network in Bucharest.

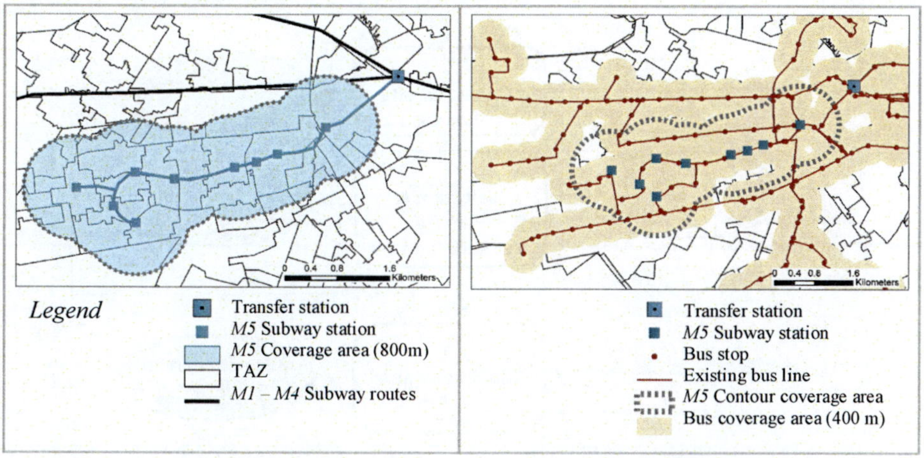

a) Coverage area (800 m) of the new subway line

b) Coverage area (400 m) of the existing bus lines

Fig. 3. Coverage study area

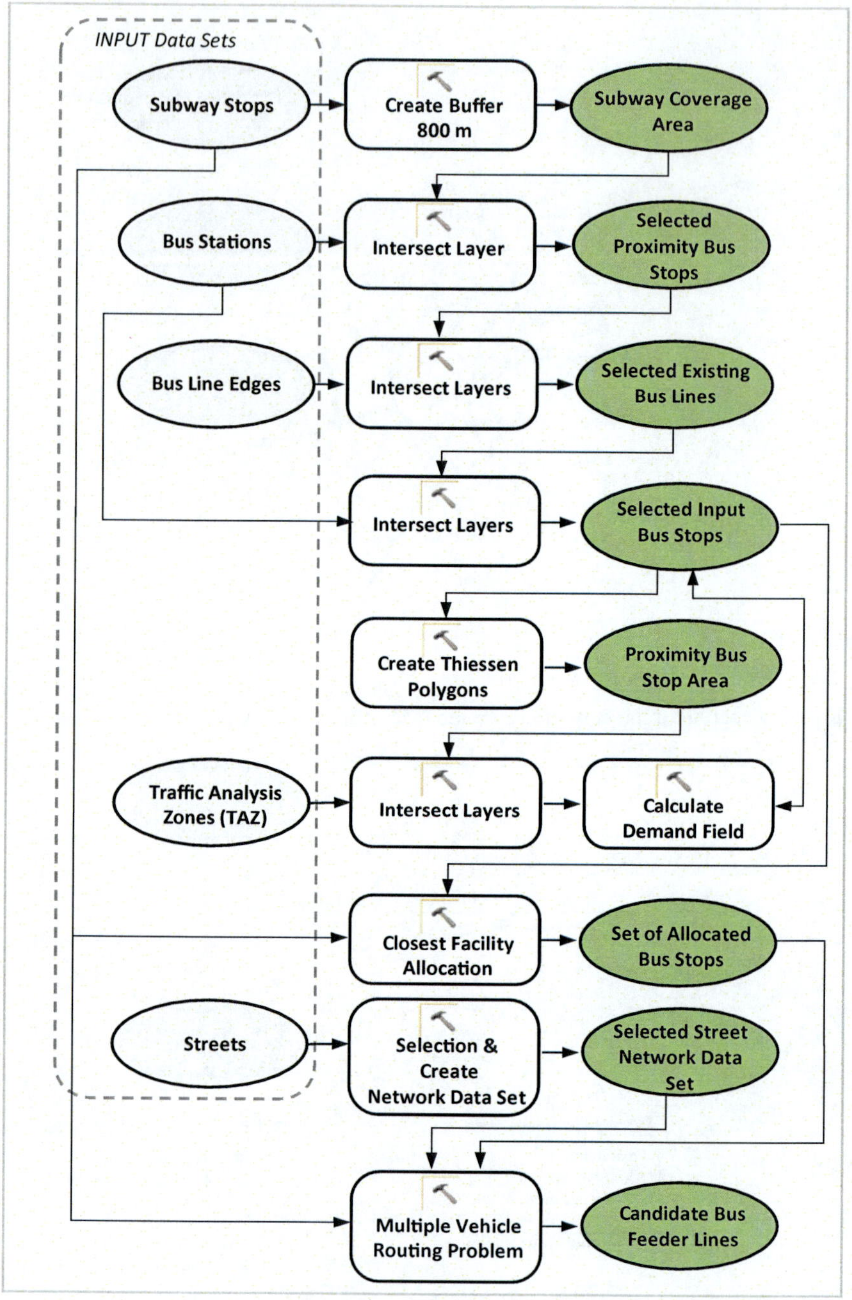

Fig. 4. Flowchart of the main procedures applied for feeder bus line design.

3. Design the bus feeder lines

   - The streets with more than two lines and appropriate characteristics for bus circulation are selected.
   - Average operating bus speed is used to compute travel time on each selected street feature.
   - The network data set is created for the selected street features.
   - For each subset of stops allocated to a subway station, vehicle routing procedures are applied to determine the bus lines.

The solution of the sequential algorithm is depicted in Fig. 5. Future work aims to develop procedures to minimize operating bus costs and establish the bus operating parameters.

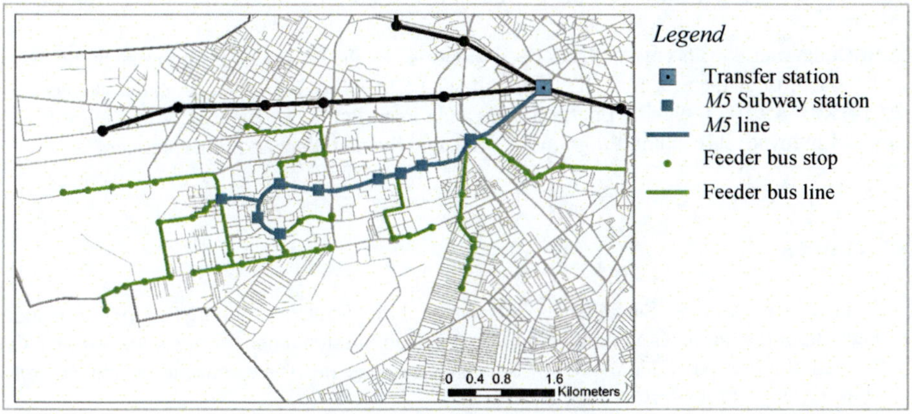

Fig. 5. Feeder bus line network

4 Conclusion

To achieve sustainable mobility goals, cities with a high proportion of motorized individual transport need to redesign their mobility systems (defined by the set of infrastructures, travel pattern and related services). Rapid public transport (rail, subway, light rail with grade-separated or exclusive right-of-way) are considered solutions to increase the attractiveness of public transport. But financing rapid public transport projects requires significant investments. Further, for the efficient use of the investments, the high transport capacities provided by these lines must be used appropriately. The development of intermodal public transport network can enlarge the area of accessibility to rapid transport lines and can contribute to increasing the weight of utilization of the provided transport capacities. Thus, the efficiency of public transport can be enhanced.

The presented study for an intermodal transport network design aims to enhance the access to rapid public transport provided by a subway line. The coverage urban area of the new implemented subway line is considered. The objective is to redesign existing bus lines in the area as feeder lines to the subway. Thus, sections doubled by bus and subway lines are eliminated and the coverage subway area is extended. More efficient use of subway transport capacity can be achieved.

The proposed model seeks to minimize bus operating costs and travel times. It is assumed that each feeder bus line serves only one subway station. The locations of the current bus stations (which are familiar to passengers as access points to public transport) are preserved.

The model is exemplified for the *M5* subway line in Bucharest, which opened for operation in 2020. At this stage of the research, the model component regarding the minimization of travel times has been solved.

Further work is required to develop the algorithm to solve the bus operating cost component. Also, to identify appropriate intermodal transport planning, future research aims to evaluate the effects on the efficiency of public transport for different scenarios for travel demand allocation on the components of the intermodal transport system.

Acknowledgements. This research has been supported by the Erasmus+ Programme of the European Union, through the Erasmus+ KA203 Strategic Partnerships for higher education, Project no. 2020-1-TR01-KA203-094242, "Energy Usage and Green Public Transportation in Future Smart Cities: An Innovative Teaching Program for Students, Stakeholders and Entrepreneurs".

References

1. European Commission: Sustainable and Smart Mobility Strategy – putting European transport on track for the future. Communication from the Commission to the European Parliament, the Council, the European Economic and Social Committee and The Committee of the Regions. COM (2020) 789 final, Brussels (2020)
2. Consult, R. (ed.): Guidelines for Developing and Implementing a Sustainable Urban Mobility Plan. 2nd edn. (2019)
3. Vuchic, V.: Urban Transit Systems and Technology. Wiley, Hoboken (2007)
4. Pternea, M., Kepaptsoglou, K., Karlaftis, M.: Sustainable urban transit network design. Transp. Res. Part A **77**, 276–291 (2015)
5. Borndörfer, R., Grötschel, M., Jäger, U.: Planning problems in public transit. In: Grötschel, M., Lucas, K., Mehrmann, V. (eds.) Production Factor Mathematics, pp. 95–121. Springer, Berlin, Heidelberg (2010). https://doi.org/10.1007/978-3-642-11248-5_6
6. Bunte, S., Kliewer, N.: An overview on vehicle scheduling models. Public Transp. **1**(4), 299–317 (2009)
7. Guihaire, V., Hao, J.: Transit network design and scheduling: a global review. Transp. Res. Part A **42**(10), 1251–1273 (2008)
8. Desaulniers, G., Hickman, M.: Public transit. Handb. Oper. Res. Manag. Sci. Transp. **14**, 69–127 (2007)
9. Kepaptsoglou, K., Karlaftis, M.: Transit route network design problem: review. J. Transp. Eng. **135**(8), 491–505 (2009)
10. Schöbel, A.: Line planning in public transportation: models and methods. OR Spectr. **34**(3), 491–510 (2012)
11. Shafahi, Y., Khani, A.: A practical model for transfer optimization in a transit network: model formulations and solutions. Transp. Res. Part A **44**(6), 377–389 (2010)

12. Roca-Riu, M., Estrada, M., Trapote, C.: The design of interurban bus networks in city centers. Transp. Res. Part A **46**(8), 1153–1165 (2012)
13. Bussieck, M., Kreuzer, P., Zimmermann, U.: Optimal lines for railway systems. Eur. J. Oper. Res. **96**(1), 54–63 (1997)
14. Schöbel, A., Scholl, S.: Line planning with minimal transfers. In: 5th Workshop on Algorithmic Methods and Models for Optimization of Railways, vol. 6901 (2006)
15. Jovanovic, A.D., Pamucar, D.S., Pejcic-Tarle, S.: Green vehicle routing in urban zones – a neuro-fuzzy approach. Expert Syst. Appl. **41**(7), 3189–3203 (2014)
16. UN-Habitat: Planning and design for sustainable urban mobility. Global report on human settlements 2013. United Nations Human Settlements Programme, Routledge (2013)
17. Metrorex: Activity report. METROREX S.A., Bucureşti (2019). (In Romanian)
18. Curtis, C., Scheurer, J.: Performance measures for public transport accessibility: learning from international practice. J. Transp. Land Use **10**(1), 93–118 (2017)
19. Bucharest - Ilfov SUMP: Sustainable Urban Mobility Plan 2016–2030. Bucharest - Ilfov Region. Rom Engineering Ltd., AVENSA Consulting SRL. Bucuresti (2016). (In Romanian)
20. ESRI ArcGIS Pro. Projects in ArcGIS Pro. https://pro.arcgis.com/en/pro-app/latest/help/projects/what-is-a-project.htm. Accessed 31 May 2021

Robust Control Design of MIMO Systems

Mircea Dulau and Stelian-Emilian Oltean(✉)

"George Emil Palade" University of Medicine, Pharmacy, Sciences and Technology of Targu-Mures, no. 38 Gh. Marinescu Street, 540139 Targu Mures, Romania
{mircea.dulau,stelian.oltean}@umfst.ro

Abstract. A solution for designing the controllers in the case of systems with uncertainty is offered by the robust control theory, which can be generalized to MIMO (multiple-input multiple-output) systems. Some principles of the design methodologies of the robust controllers based on H2 and H-infinity synthesis are presented in this paper and applied on a multiple-input multiple-output system to achieve good performances and stability in the presence of modeling errors. For this purpose, the SISO (single-input single-output) closed-loop control diagrams are first adapted to MIMO systems. The study case used in the paper consists of the typical example of two-coupled tanks process which has two inputs and two outputs. The plant mathematical model includes also the possible additive uncertainty of the hydraulic resistances of the outlet and cross-coupling valves. The design of the H2 and H-infinity controllers and the study of the behaviors of the global systems were made in the Matlab environment using the Robust Control Toolbox.

Keywords: Robust control · H-infinity synthesis · H2 synthesis · MIMO systems

1 Introduction

Conventional controllers (e.g. PID proportional-integrative-derivative controllers) are often used in the industrial control of the SISO plants. These controllers have a known structure and setup, are flexible, and easier to be implemented. Several issues should be considered in the case of MIMO plants in comparison to SISO systems because of the additional inputs and outputs: the performances and evolution for each output, the interaction between the variables, the compensations of the disturbances, and uncertainty. All the design methodologies based on the closed-loop system from Fig. 1 include the transfer matrix of the plant $\mathbf{H}_P(s)$ and the transfer matrix of the controller $\mathbf{H}_C(s)$.

There is a continuous concern among researchers and industrial applications when control structures $\mathbf{H}_C(s)$ should achieve the desired performances for the MIMO systems.

The QFT (Quantitative Feedback Theory) is a technique presented in [1] and could be used first to decompose the MIMO plant in several MISO (Multiple-Input-Single-Output) systems and design the controllers and filters which assures the robust performances for the systems containing uncertainty.

L. Moldovan and A. Gligor (Eds.): Inter-Eng 2021, LNNS 386, pp. 684–695, 2022.
https://doi.org/10.1007/978-3-030-93817-8_61

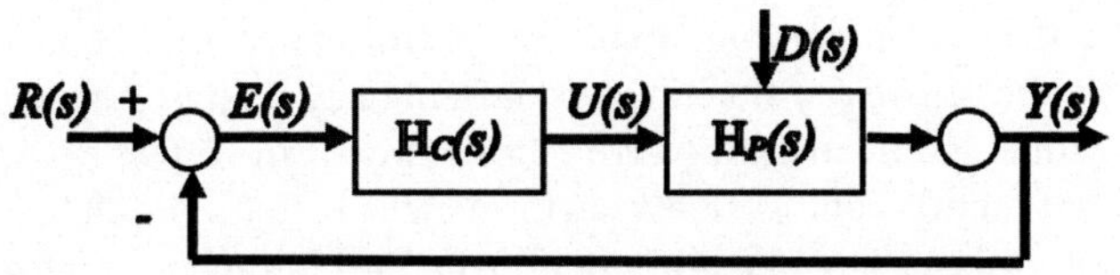

Fig. 1. The general closed-loop control diagram for MIMO systems.

Aspects of control law synthesis for a linear MIMO process are described in [2] considering the quantized output and external disturbances. The determined controller assures a reference tracking accuracy which depends on the quantization.

The control problem of MIMO systems with different time delays often found in the industrial applications is addressed in [3]. The authors propose a solution with multiple-loops control to reinforce the tracking performances and controller robustness in case of uncertainty and disturbances. The disturbance rejection and robust stability are assessed with frequency performance indicators.

Several robust control methods are tested on the quadruple tank process in [4]. The MIMO controller is designed based on H2 and H-infinity synthesis for the linearized system with uncertainty.

The robust control strategy is also applied in [5] to achieve the desired response with required performances and disturbance rejection. The synthesis method is based on multi-objective optimization used for systems with uncertainty. The authors tested the method using two processes, including the two-coupled tanks process.

A robust controller that satisfies the H-infinity criterion in the frequency domain is developed in [6] for LTI (linear time-invariant) single-input single-output systems. The design method uses fixed-order controllers and the necessary and sufficient conditions for the existence of such controllers are described by a set of convex constraints.

In [7] the identification of the MIMO active magnetic bearing (AMB) system shows an unstable open-loop behavior because the system has two poles in the right-half of the complex plane and the authors use the H2 and H-infinity approaches to obtain the MIMO controller. The cross-interaction channels have negligible gains in the low-frequency region, and in this way, the system is almost decoupled (diagonalized).

The case of nonlinear MIMO systems is treated in the presence of unknown disturbances in [8]. The backstepping technique and the dynamic surface method are used to design the robust controller for a UAV (unmanned aerial vehicle).

The robust control problem of MIMO systems, H-infinity norm, sensitivity functions, performance requirements, mixed sensitivity problem, and robustness analysis are well described in [9]. Authors of [10] include also in their study the robustness of the systems with uncertainty, the choice of the weighting functions, and the H2, H-infinity and loop-shaping synthesis.

The mathematical modeling of the two-coupled tanks process which has two inputs and two outputs is presented in [11]. For the MIMO system was designed and tested a robust control algorithm based on H-infinity loop-shaping method.

Considering the sensitivity functions, the performance requirements include the desired shape for the sensitivity function in the frequency domain, the maximum amplitude, the bandwidth, and the tracking error. The study of the effect of these parameters on the weighting functions and systems performances is realized in [12].

The textbook [13] is a good introduction to the automatic control theory, contains aspects of mathematical modeling, conventional control, LQG (Linear Quadratic Gaussian) control, and also robust control design using H2 and H-infinity technique.

In [14] different types of conventional or robust control structures are determined and the behavior of the control schemes in time or frequency domain is simulated using the Matlab environment and its toolboxes.

The present paper proposes a short analysis of a MIMO system based on simulation of the behavior of the two-coupled tanks control system. In the design stage, robust control methods are used. In this sense, Sect. 2 presents the general diagram of the system with uncertainty and control diagrams used for determining the robust controllers. In Sect. 3 are detailed the control design aspects using the H2 and H-infinity synthesis. These aspects and the Matlab environment are used in Sect. 4 to obtain the robust controllers for the MIMO process. Section 5 emphasizes some conclusions regarding the robust control design of the two-coupled tanks process.

2 MIMO Systems with Uncertainty. The Weighting Matrix

The mathematical models should contain the dynamic behavior of the systems, but also information about the disturbances. The modeling procedure of the physical systems involves a compromise between simplicity and accuracy. So, some of the components are neglected to obtain a usable model, and differences from the real system are included in the modeling errors. These modeling errors should be considered somehow in the analysis and design of the automatic control as uncertainty based on available information and reasonable assumptions.

Figure 2 shows the closed-loop control diagram of the system with additive/multiplicative (unstructured) uncertainty [9, 10, 13].

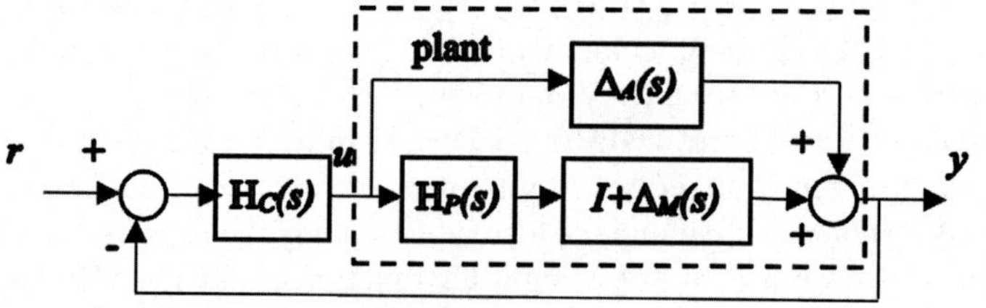

Fig. 2. The closed-loop control diagram of a system with unstructured uncertainty.

The transfer function of the system with uncertainty is

$$H_G(s) = \Delta_A(s) + H_P(s)[I + \Delta_M(s)]. \tag{1}$$

In (1) if $\Delta_A(s) = 0$ the relation emphasizes the multiplicative uncertainty and for $\Delta_M(s) = 0$ relation contains the additive uncertainty.

Determination of the control structure based on robust synthesis techniques depends on the weighting functions for the SISO case and the weighting matrix for the MIMO case (Fig. 3).

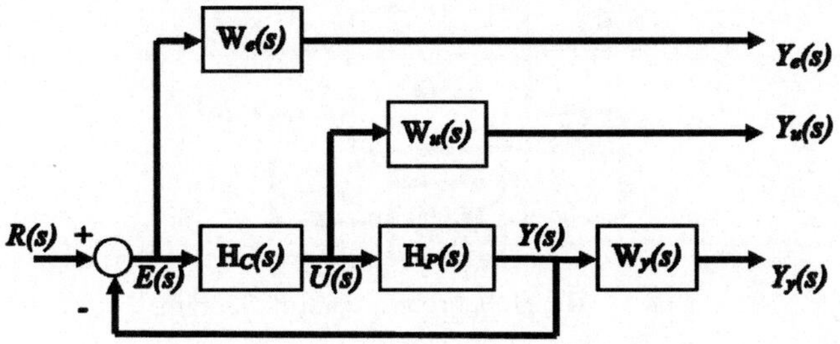

Fig. 3. The robust control diagram for the MIMO system including the weighting matrix.

Once the weighting matrix is known, then the H-infinity design problem can be addressed for the augmented system and the output vector is [9, 10, 13].

$$Y(s) = \left[\, Y_e(s)\ Y_u(s)\ Y_y(s) \,\right]^T. \tag{2}$$

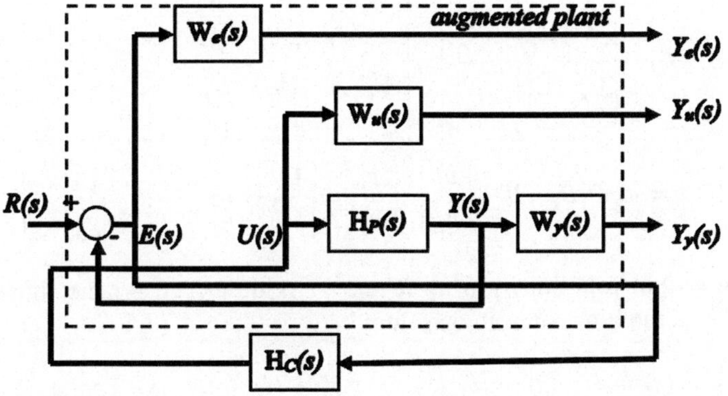

Fig. 4. Robust control diagram with the augmented plant model

The robust control design consists of:

- modeling the plant transfer matrix, $\mathbf{H}_P(s)$;
- choosing the weighting matrix, $\mathbf{W}_e(s)$, $\mathbf{W}_u(s)$, $\mathbf{W}_y(s)$ penalizing the error signal, control signal, and output signal, respectively;
- developing the augmented system, $\mathbf{H}_G(s)$ (see Fig. 4);
- designing the transfer matrix of the controller, $\mathbf{H}_C(s)$.

3 Robust Control Design Techniques

The general problem of robust design consists of finding a controller which assures the stability and robustness performances of the global system in the presence of disturbances or uncertainty.

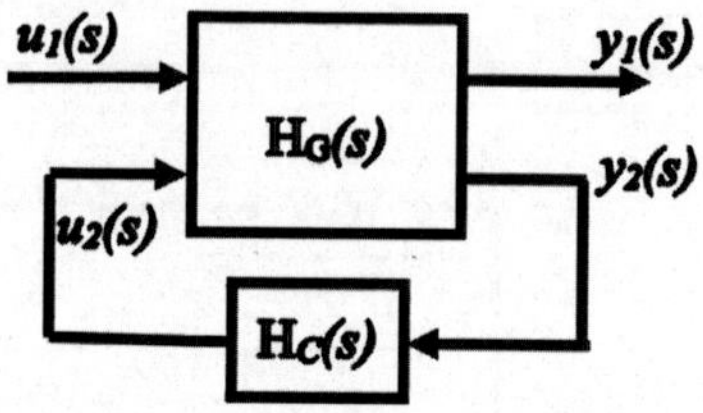

Fig. 5. The global robust control diagram.

The control diagram from Fig. 4 is modified in Fig. 5 to emphasize [9, 10, 13]:

- the augmented model of the system $\mathbf{H}_G(s)$ which may contain the nominal plant model and the disturbances;
- the controller matrix $\mathbf{H}_C(s)$;
- the input signal (which could include also the disturbances) $u_1(s)$;
- the output signal $y_1(s)$.

$$H_G = \begin{bmatrix} H_{11} & | & H_{12} \\ -- & | & -- \\ H_{21} & | & H_{22} \end{bmatrix} = \begin{bmatrix} A & | & B_1 & B_2 \\ -- & | & -- & -- \\ C_1 & | & D_{11} & D_{12} \\ C_2 & | & D_{21} & D_{22} \end{bmatrix} \tag{3}$$

$$\dot{x} = Ax + \begin{bmatrix} B_1 & B_2 \end{bmatrix} \cdot \begin{bmatrix} u_1 \\ u_2 \end{bmatrix}; \begin{bmatrix} y_1 \\ y_2 \end{bmatrix} = \begin{bmatrix} C_1 \\ C_2 \end{bmatrix} x + \begin{bmatrix} D_{11} & D_{12} \\ D_{21} & D_{22} \end{bmatrix} \begin{bmatrix} u_1 \\ u_2 \end{bmatrix}$$

The input-output relation from u_1 to the y_1 is described by the linear fractional transformation of the interconnected system:

$$T_{y1u1}(s) = H_{11}(s) + H_{12}(s)H_C(s)[I - H_{22}(s)H_C(s)]^{-1}H_{21}(s). \tag{4}$$

3.1 H-infinity Technique

In the H-infinity technique the control law including $\mathbf{H}_C(s)$ is [10, 13]

$$u_2(s) = H_C(s) \cdot y_2(s). \tag{5}$$

The global system with disturbances and uncertainty is stable and the H-infinity norm of the closed-loop system is bounded by a positive value γ.

$$\left\| T_{y1u1} \right\|_{\infty} < \gamma \tag{6}$$

Using the terms of the augmented model (3) the ARE (algebraic Riccati equations) are:

$$\begin{aligned} A^T X + XA + X\left(\gamma^{-2} B_1 B_1^T - B_2 B_2^T\right)X + C_1 C_1^T = 0 \\ AY + YA^T + Y\left(\gamma^{-2} C_1^T C_1 - C_2^T C_2\right)Y + B_1^T B_1 = 0 \end{aligned} \tag{7}$$

and the solutions X, Y are used then by the controller represented by (8).

$$H_C(s) = \left[\begin{array}{c|c} A + \gamma^{-2} B_1 B_1^T X - B_2 B_2^T X - \left(I - \gamma^{-2} YX\right)^{-1} Y C_2^T C_2 & \left(I - \gamma^{-2} YX\right)^{-1} Y C_2^T \\ \hline - B_2^T X; & 0 \end{array} \right] \tag{8}$$

The existing conditions of the optimal controller are:

- D_{11} should be small enough so that: $D_{11} < \gamma$;
- the solutions X, Y of the ARE should be positive definite;
- $\lambda_{\max}(XY) < \gamma^2$ shows that eigenvalues are smaller than γ^2($\lambda_{\max}$- maximum value).

If all the weighting matrix are imposed than the general mixed sensitivity problem and the linear fractional transformation (4) becomes:

$$T_{y1u1}(s) = \begin{bmatrix} W_e(s)S(s) \\ W_u F(s)S(s) \\ W_y T(s) \end{bmatrix}. \tag{9}$$

The sensitivity function *S(s)* and the complementary sensitivity function *T(s)* are given in relation (10).

$$\begin{aligned} S(s) &= E(s)R^{-1}(s) = [I - H_C(s)H_P(s)]^{-1} \\ T(s) &= I - S(s) = H_C(s)H_P(s)[I - H_C(s)H_P(s)]^{-1} \end{aligned} \tag{10}$$

3.2 H2 Technique

In the H2 technique the matrix $\mathbf{H}_C(s)$ results from the optimization of the linear fractional transformation norm [10, 13]:

$$\left\| T_{y1u1}(s) \right\|_2 = \left\| H_{11}(s) + H_{12}(s)H_C(s)[I - H_{22}(s)H_C(s)]^{-1} H_{21}(s) \right\|_2 < 1. \tag{11}$$

The symmetric matrix V, N satisfies the existing conditions and the H2 controller is given by (12).

$$H_C(s) = \left[\begin{array}{c|c} H_{C11} & H_{C12} \\ \hline H_{C21} & H_{C22} \end{array}\right] = \left[\begin{array}{c|c} A - H_{C12}C_2 - B_2H_{C21} + H_{C12}D_{22}H_{C21} & H_{C12} \\ \hline H_{C21} & 0 \end{array}\right]$$

$$H_{C12} = \left(VC_2^T + B_1D_{21}^T\right)\left(D_{21}D_{21}^T\right)^{-1}; H_{C21} = \left(D_{12}^TD_{12}\right)^{-1}\left(B_2^TN + D_{12}^TC_1\right) \tag{12}$$

If $D_{21} \rightarrow 0$ then the H2 optimal control problem is similar to the LQG design.

4 Designing the MIMO Controller for the Two-Coupled Tanks

The robust control design is applied to the two-coupled tanks from Fig. 6 which is a MIMO system with two inputs and two outputs. According to [11] and [15]: LT_1, LT_2 are the level transducers; LC is the level control; FV_1, FV_2 are the flow valves, R_1, R_{12}, R_2 are the hydraulic resistances of the outlet and cross-coupling valves, F_{a1}, F_{a2} are the inlet flows; F_{e1}, F_{e12}, F_{e2} are the outlet flows; A_1, A_2 are the surfaces at the base of the tanks; L_1, L_2 are the levels of the fluid inside the tanks.

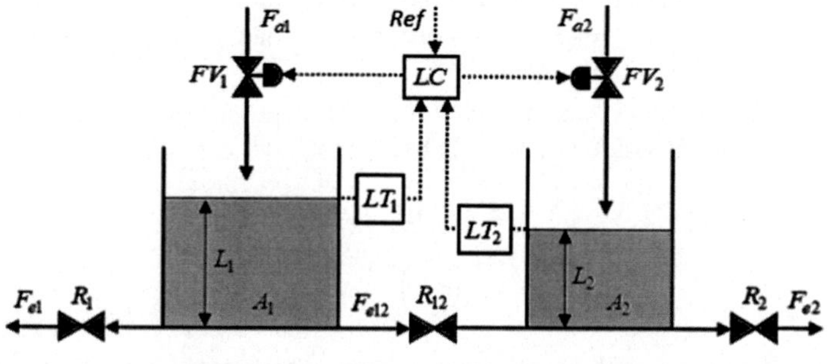

Fig. 6. Two-coupled tanks MIMO plant.

The ISO (input-state-output) mathematical model which describes the dynamics of the system and time evolution of the levels of fluid in the tanks regarding the inlet and outlet flows of the fluid is deducted in (13) and (14) [11].

$$\begin{bmatrix} \frac{d}{dt}L_1 \\ \frac{d}{dt}L_2 \end{bmatrix} = \begin{bmatrix} -\frac{R_1+R_{12}}{A_1R_1R_{12}} & \frac{1}{A_1R_{12}} \\ \frac{1}{A_2R_{12}} & -\frac{R_2+R_{12}}{A_2R_2R_{12}} \end{bmatrix} \begin{bmatrix} L_1 \\ L_2 \end{bmatrix} + \begin{bmatrix} \frac{1}{A_1} & 0 \\ 0 & \frac{1}{A_2} \end{bmatrix} \begin{bmatrix} F_{a1} \\ F_{a2} \end{bmatrix}$$

$$\begin{bmatrix} L_1 \\ L_2 \end{bmatrix} = \begin{bmatrix} 1 & 0 \\ 0 & 1 \end{bmatrix} \begin{bmatrix} L_1 \\ L_2 \end{bmatrix} + \begin{bmatrix} 0 & 0 \\ 0 & 0 \end{bmatrix} \begin{bmatrix} F_{a1} \\ F_{a2} \end{bmatrix} \tag{13}$$

$$\begin{bmatrix} L_1(s) \\ L_2(s) \end{bmatrix} = \begin{bmatrix} H_{P11}(s) \; H_{P12}(s) \\ H_{P21}(s) \; H_{P22}(s) \end{bmatrix} \begin{bmatrix} F_{a1}(s) \\ F_{a2}(s) \end{bmatrix} \equiv \mathbf{Y}(s) = \mathbf{H}_P(s) \cdot \mathbf{U}(s) \qquad (14)$$

with $H_{P11}(s)$, $H_{P22}(s)$ the transfer functions on the direct channels of the MIMO system; $H_{P12}(s)$, $H_{P21}(s)$ the transfer functions on the cross-interaction channels.

In the case of MIMO systems is necessary to obtain the desired performances on each direct channel and compensate the cross-interaction effect, disturbances, and uncertainty. The solution for such issues is given by the robust control synthesis because its theory has the methodologies to design robust controllers.

The values $L_1 = 0.4$; $L_2 = 0.25$; $A_1 = 0.028$; $A_2 = 0.02$ are the main parameters used in simulations and $R_1 = 0.8 \pm 25\%$; $R_2 = 1.2 \pm 30\%$; $R_{12} = 1 \pm 20\%$ contain the additive uncertainty of the hydraulic resistances. So, the transfer matrix of the nominal plant given by (14) is:

$$H_P(s) = \begin{bmatrix} \frac{35.71s+3274}{s^2+172s+5580} & \frac{1786}{s^2+172s+5580} \\ \frac{1786}{s^2+172s+5580} & \frac{50s+4018}{s^2+172s+5580} \end{bmatrix}.$$

The step response behavior from Fig. 7 shows the effect of the uncertainty on each channel from the MIMO system and the weighting matrix for designing the robust controller is chosen according to [10, 12, 13].

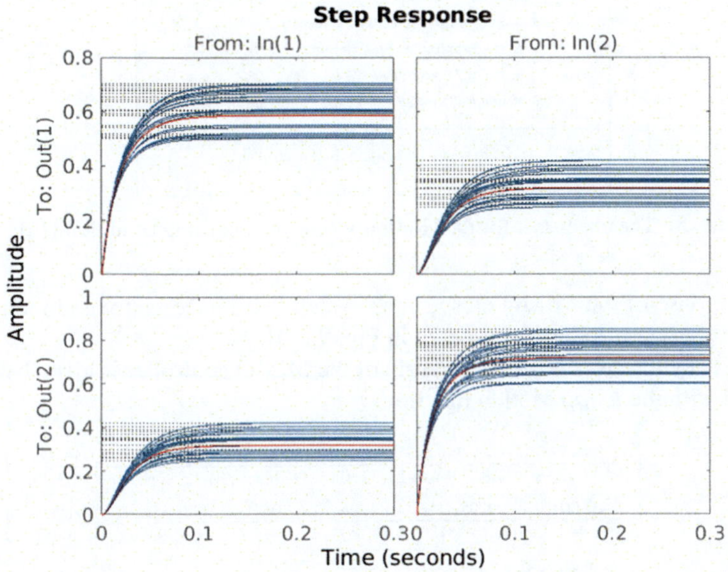

Fig. 7. The step response of the system with additive uncertainty (red is the nominal plant) [14].

$$W_e(s) = \begin{bmatrix} \frac{100}{s+0.5} & 0 \\ 0 & \frac{100}{s+1} \end{bmatrix}; \quad W_e(s) = \begin{bmatrix} 10^{-3} & 0 \\ 0 & 10^{-3} \end{bmatrix}; \quad W_y(s) = \begin{bmatrix} \frac{3}{300} & 0 \\ 0 & \frac{s}{200} \end{bmatrix}.$$

Before the determination of the transfer matrix (8) of the robust controller, the *hinf* Matlab function verifies the existing conditions (Fig. 8) [14].

$$H_C(s) = \begin{bmatrix} \frac{1253s^2+5.298e05s+3.448e07}{s^3+677.2s^2+1.148e05s+5.724e04} & \frac{-0.001296s^2-4.475e04s-1.532e07}{s^3+677.2s^2+1.148e05s+5.724e04} \\ \frac{-0.007265s^2-3.932e04s-1.314e07}{s^3+677.7s^2+1.152e05s+1.145e05} & \frac{786.3s^2+3.349e05s+2.409e07}{s^3+677.7s^2+1.152e05s+1.145e05} \end{bmatrix}$$

Computing the 4-block H-inf optimal controller
using the S-L-C loop-shifting/descriptor formulae

Solving for the H-inf controller F(s) using U(s) = 0 (default)

Solving Riccati equations and performing H-infinity existence tests:

1. Is D11 small enough? OK
2. Solving state-feedback (P) Riccati ...
   a. No Hamiltonian jw-axis roots? OK
   b. A-B2*F stable (P >= 0)? OK
3. Solving output-injection (S) Riccati ...
   a. No Hamiltonian jw-axis roots? OK
   b. A-G*C2 stable (S >= 0)? OK
4. max eig(P*S) < 1 ? OK

----

all tests passed -- computing H-inf controller ...
DONE!!!

Fig. 8. The check test for the H-infinity controller existing condition [14].

The response of the MIMO closed-loop system with uncertainty and including the H-infinity robust controller is presented in Fig. 9 [14].

In the H2 synthesis, the transfer matrix of the MIMO controller is given by (12) and is obtained with the *h2lqg* Matlab function.

$$H_C(s) = \begin{bmatrix} \frac{836.5s^2+2.337e05s+1.338e07}{s^3+443.9s^2+4.887e04s+2.432e04} & \frac{0.00063s^2-2.988e04s-5.946e06}{s^3+443.9s^2+4.887e04s+2.432e04} \\ \frac{-0.00128s^2-1.98e04s-4.84e06}{s^3+444.4s^2+4.909e04s+4.865e04} & \frac{396s^2+1.331e05s+8.874e06}{s^3+444.4s^2+4.909e04s+4.865e04} \end{bmatrix}$$

The response of the MIMO closed-loop system with uncertainty and including the H2 robust controller is presented in Fig. 10 [14]. Figure 11 shows a comparison of step input responses of the nominal closed-loop systems with both robust controllers, designed using the H2 and H-infinity synthesis.

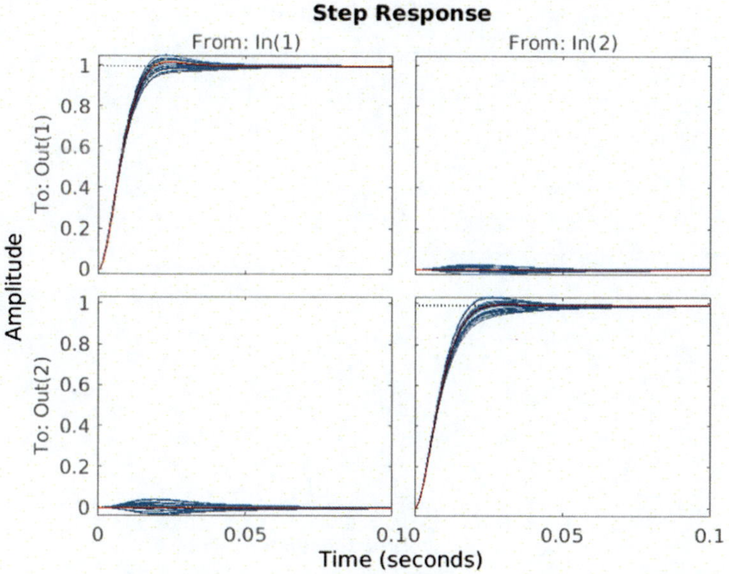

Fig. 9. The step input response of closed-loop MIMO system including the H-infinity robust controller (red is the nominal response) [14].

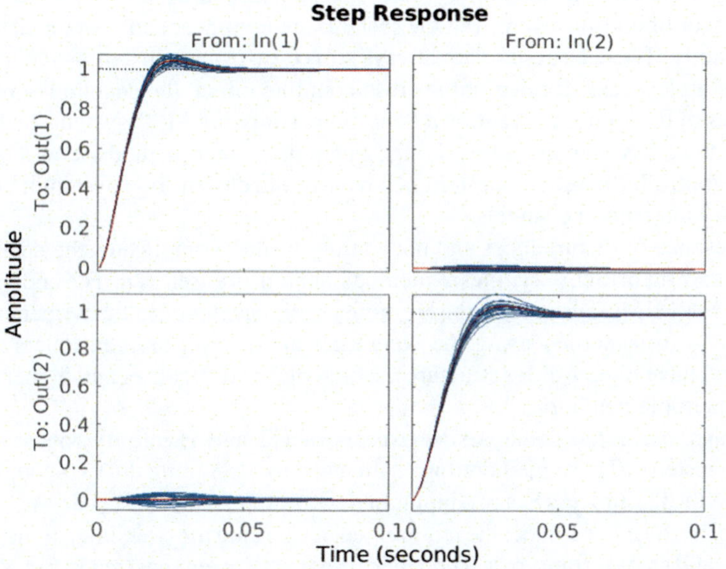

Fig. 10. The step input response of the closed-loop MIMO system including the H2 robust controller (red is the nominal response) [14].

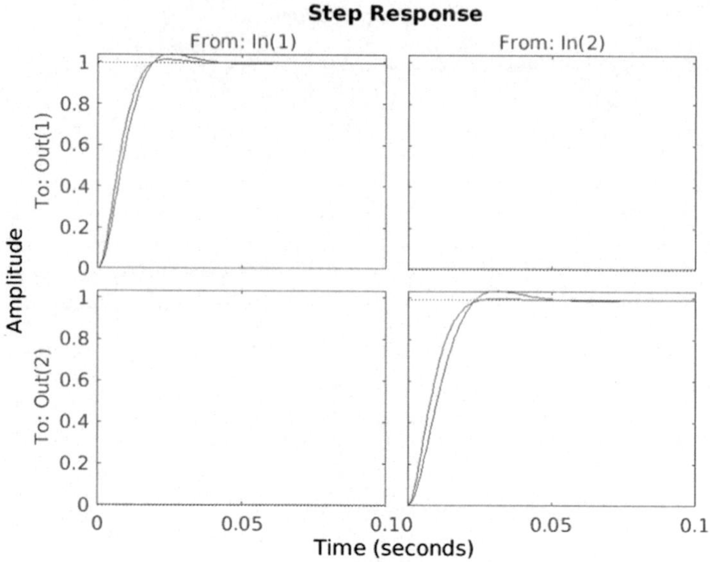

Fig. 11. The step input responses of the closed-loop MIMO systems with both robust controllers, H2 controller (with blue) or H-infinity controller (with red) [14]

5 Conclusions

In a lot of applications, e.g. industrial control, the processes are cross-connected resulting in a MIMO configuration. Moreover, the systems are nonlinear and contain disturbances and uncertainty. These issues are not always solved by using conventional control.

In the simplest case, for negligible cross-coupling effect, the decoupling methodology consists of designing conventional controllers similar to SISO systems on the direct channel. If the cross-coupling effect is significant then other methods should be used to obtain the controller laws, e.g. desired behavior of the closed-loop system including or not the cross-interaction channels.

The presence of disturbances and uncertainty further complicates the design of the controller and then robust synthesis methods should be used (i.e. H2 and H-infinity synthesis). For MIMO systems, in the H-infinity design technique, the frequency domain performances are specified using the weighting matrix. So, the controllers are more complex and have higher-order structures, which means difficulties and higher costs for the practical implementation.

The paper presented some design aspects of H2 and H-infinity robust synthesis for MIMO systems. These methodologies are used for designing the two-coupled tanks control system. The process has two inputs and two outputs. In both scenarios, the closed-loop control performances meet the requirements regarding the overshoot, transient time, rising time, and steady-state error. These performances are emphasized in Fig. 9, Fig. 10, Fig. 11.

The control diagrams were tested including the uncertainty and the simulation results show stable responses of the closed-loop systems. The cross-influences (from F_{a1} to L_2

and from F_{a2} to L_1) are very small, thus being minimized. Robust Control Toolbox was used for designing the H2 and H-infinity robust controllers and finally, the closed-loop MIMO systems were simulated in the Matlab environment.

Further, in the study of the MIMO synthesis, some limitations of the control signals and possibilities to reduce the higher-order of the robust controllers will be considered, of course without affecting the overall performances of the dynamic regime.

References

1. Yousfi, N., Almalki, H., Derbel, N.: Robust control of industrial MIMO systems based on fractional order approaches. In: 2020 Industrial & Systems Engineering Conference (ISEC) Proceedings, pp. 1–6 (2020)
2. Margun, A., Furtat, I.: Robust control of linear MIMO systems in conditions of parametric uncertainties, external disturbances and signal quantization. In: 2015 20th International Conference on Methods and Models in Automation and Robotics (MMAR) Proceedings, pp. 341–346 (2015)
3. Tong, B., Chen, J., Wang, H., Yu, Y.: A control method for MIMO systems with multiple time delays. In: 2017 IEEE International Conference on Systems, Man, and Cybernetics (SMC) Proceedings, pp. 2203–2208 (2017)
4. Hypiusová, M., Rosinová, D.: Robust control of quadruple-tank process via LMI. In: Cybernetics & Informatics (K&I), pp. 1–6 (2016)
5. Gonçalves, E.N., Bachur, W.E.G., Palhares, R.M., Takahashi, R.H.C.: Robust H2/H∞ reference model dynamic output-feedback control synthesis. Int. J. Control **84**(12), 2067–2080 (2011)
6. Karimi, A., Nicoletti, A., Zhu, Y.: Robust H∞ controller design using frequency-domain data via convex optimization. Int. J. Robust. Nonlinear Control **00**, 1–19 (2015)
7. Noshadi, A., Shi, J., Lee, W.S., Shi, P., Kalam, A.: System identification and robust control of multi-input multi-output active magnetic bearing systems. IEEE Trans. Control Syst. Technol. **24**(4), 1227–1239 (2015)
8. Zhou, Y., Chen, M., Jiang, C.: Robust tracking control of uncertain MIMO nonlinear systems with application to UAVs. IEEE/CAA J. Autom. Sin. **2**(1), 25–32 (2015)
9. Sename, O., Fergani, S.: Robustness and H∞ control of MIMO systems. Gipsa-lab Grenoble and LAAS-CNRS Toulouse, France (2017)
10. Zhou, K.: Essentials of Robust Control. Upper Saddle River, NJ, USA (1999)
11. Dulău, M., Oltean, S.E., Duka, A.V.: Robust control of a multivariable system. In: 2016 10th International Conference Interdisciplinarity in Engineering Proceeding. Procedia Engineering, p. 181 (2017)
12. Dulău, M., Oltean, S.E.: The effects of the weighting functions on the performances of the robust control systems. In: 2020 14th International Conference Interdisciplinarity in Engineering INTER-ENG. MDPI Proceedings, vol. 63, no. 1, pp. 1–9 (2020)
13. Xue, D., Chen, Y., Atherton, D.P.: Linear Feedback Control: Analysis and Design with Matlab. Society for Industrial and Applied Mathematics (SIAM), Philadelphia (2007)
14. Mathworks. https://matlab.mathworks.com/?s_tid=tah_po_start. Accessed 21 Mar 2021
15. ANSI/ISA-S5.1. Instrumentation symbols and identification. American National Standard (1984, R1992)

Comments on the Solutions Set of Equilibrium Problems Governed by Topological Pseudomonotone Bifunctions

Marcel Bogdan(✉)

University of Medicine, Pharmacy, Science and Technology "George Emil Palade", Târgu Mureş, Romania
marcel.bogdan@umfst.ro

Abstract. A recent assertion given by Sadeqi, I., Salehi Paydar, M., in: A comparative study of Ky Fan hemicontinuity and Brezis pseudomonotonicity of mappings and existence results, J. Optim. Theory Appl. (2015), that the set of solutions of the variational inequality problem governed by a pseudomonotone operator is closed, is obtained as a particular case of a result from Bogdan, M., Kolumbán, J.: Some regularities for parametric equilibrium problems, J. Global Optim. (2009). An example of a set in a Hilbert space where $\nabla\|\cdot\|$ is not Fan-hemicontinuous is given. A different approach to show that $\nabla\|\cdot\|$ is indeed topologically pseudomonotone, is expressed. From the corresponding definitions, Fan-hemicontinuity implies topological pseudomonotonicity, but the reverse implication does not hold in general. This strict relationship is strengthened by $\nabla\|\cdot\|$. Moreover, another counterexample is given in a Lebesgue space, instead of the Sobolev space $W_0^{1,3}$ used in Steck, D.: Brezis pseudomonotonicity is strictly weaker than Ky Fan hemicontinuity. J. Optim. Theory Appl. (2019). Stability of pseudomonotonicity with respect to the composition with a linear operator is formulated as open question.

Keywords: Fan-hemicontinuity · Topological pseudomonotonicity · Solutions set · Equilibrium problems · Positively oriented set · Simple pendulum

1 Introduction

Equilibrium problems unify many mathematical models among them variational inequalities and optimization problems (see [1]). These class of problems have been found useful in the study of many problems from the fields of economics, mechanics, and engineering sciences [2–4]. The mathematical tools involved are mainly algebraic ones such as convexity and topological ones such as continuity ([5]), in particular topological pseudomonotonicity for a special class [6,8]. The notion of pseudomonotonicity is related to the class of semi-linear operators (see

L. Moldovan and A. Gligor (Eds.): Inter-Eng 2021, LNNS 386, pp. 696–703, 2022.
https://doi.org/10.1007/978-3-030-93817-8_62

[3,4,7,8]). Existence result for the case of topological pseudomonotone bifunction is given in [9] (see also [10], Theorem 2.3).

It was noted in [11] that the notion of pseudomonotonicity is not equivalent to Fan-hemicontinuity, as it was affirmed in [12]. It is worth to be noted that in [12], sufficient conditions were imposed on a function to obtain pseudomonotonicity for its gradient, and the weak closedness of the solution set of a variational inequality governed by pseudomonotone operators was proved.

In this paper are added some comments and remarks on the two papers mentioned above. Some restrictions on the domain are expressed in such a way the gradient of the norm defined on a Hilbert space to be Fan-hemicontinuous.

The setting is the usual one, $(X, \|\cdot\|)$ a real Banach space with dual X^*, the duality pairing between X^* and X being denoted by $\langle\cdot,\cdot\rangle$; the strong and the weak convergence is indicated by $\longrightarrow$ and $\rightharpoonup$, respectively. Let $K \subseteq X$ be nonempty and convex. The following two notions are discussed.

Definition 1. *$F : K \to X^*$ is said to be* Fan-hemicontinuous *if for every $y \in X$, $x \longmapsto \langle F(x), x - y\rangle$ is weakly sequentially lower semicontinuous on K, that is $\forall (x_k) \subset K$ with $x_k \rightharpoonup x \in K$, one has*

$$\liminf_k \langle F(x_k), x_k - y\rangle \geq \langle F(x), x - y\rangle.$$

This notion was used in the existence results for problems in equilibrium theory ([1]). Since the norm of a Hilbert space is weakly sequentially lower semicontinuous, it is straightforward that the identity operator on a Hilbert space is Fan-hemicontinuous.

Definition 2. *$F : K \to X^*$ is said to be* pseudomonotone *if whenever*

$$(x_k) \subset K \text{ with } x_k \rightharpoonup x \in K, \text{ and } \limsup_k \langle F(x_k), x_k - x\rangle \leq 0, \tag{1}$$

then

$$\liminf_k \langle F(x_k), x_k - y\rangle \geq \langle F(x), x - y\rangle, \text{ for all } y \in K. \tag{2}$$

Recently, an example of nonlinear operator, defined on a Sobolev space which is pseudomonotone but not Fan-hemicontinuous was given in [11]. A result from [12] on the solution set of a variational inequality governed by topological pseudomonotone operators is remarked.

This present content is organized as follows. Subsection 1.1 contains an example of a closed, convex but not positively oriented set (see Definition 3) where $\nabla\|\cdot\|$ is not Fan-hemicontinuous. Also, an alternative proof that $\nabla\|\cdot\|$ is Fan-hemicontinuous on a restrictive set is given. Subsection 1.2 contains a remark about the closedness of the solution set of a variational inequality problem governed by a pseudomonotone operator, stated in [12]. It is a particular case of a result given in [6]. Section 2 is concerned with another example of a pseudomonotone operator that it is not Fan-hemicontinuous. The content ends with an open question related to the composition of a pseudomonotone operator with a linear one.

1.1 Fan-hemicontinuity of the Gradient of the Norm

In this section, some properties of Fan-hemicontinuity and pseudomonotonicity, respectively are emphasized for $\nabla\|\cdot\|$ (in particular, if $\nabla\|\cdot\|$ is Fan-hemicontinuous, then it is pseudomonotone). These led to the property for its domain to ensure that it is Fan-hemicontinuous.

In [12] it was claimed that the gradient of a convex Gateaux differentiable is Fan-hemicontinuous. It is proved that this does not hold in general.

Definition 3. ([13]) *The set $K \subset X$ is said to be positively oriented if $\forall x, y \in K, \langle x, y\rangle \geq 0$.*

Imposing this condition on the set K, one obtains the Fan-hemicontinuity of $\nabla\|\cdot\|$.

Proposition 1. ([13]) *Let X be a Hilbert space, $K \subset X$ closed, convex, with $0_X \notin K$. If K is positively oriented, then $\nabla\|\cdot\|$ is Fan-hemicontinuous on K.*

The positive orientation property of the set K is sufficient in order to obtain the Fan-hemicontinuity of $\nabla\|\cdot\|$. However, there are sets that are not positively oriented.

Example 1. The set

$$K = \{x = (x^j) \in \ell^2 \,|\, x^1 \in \mathrm{IR}, x^2 = 1/2, (-1)^j \cdot x^j \geq 0, \text{for } j \geq 3\}$$

is closed, convex but not positively oriented. Moreover, $\nabla\|\cdot\|$ is not Fan-hemicontinuous.

Consider $\bar{x} = (-1, 1/2, 0, 0, ..., 0, ...)$ and $\bar{y} = (1, 1/2, 0, 0, ..., 0, ...)$ that belong to K but $\langle \bar{x}, \bar{y}\rangle = -3/4 < 0$.

Now, it is shown that $\nabla\|\cdot\|$ is not Fan-hemicontinuous on K. For this, it is needed $(x_n) \subset K$, $x_n \rightharpoonup x \in K$, and $y \in K$ such that

$$\liminf_n \left[\|x_n\| - \frac{1}{\|x_n\|} \cdot \langle x_n, y\rangle\right] < \|x\| - \frac{1}{\|x\|}\langle x, y\rangle. \tag{3}$$

Let be defined $x = (-1, 1/2, 0, 0, ..., 0, ...) \in K$, $(x_n) \subset \ell^2$, for $n \in \mathrm{IN}$,

$$x_n^j := \begin{cases} -1, & j = 1 \\ \frac{1}{2}, & j = 2 \\ (-1)^j \frac{1}{2}, & j = n+2 \\ 0, & \text{otherwise,} \end{cases}$$

and $y = (\frac{13}{4}, \frac{1}{2}, 0, 0, ..., 0, ...) = (y^j) \in K$. Observe that $\|x\| = \frac{\sqrt{5}}{2} =: b$, $\|x_n\| = \frac{\sqrt{6}}{2} = a > b$, and $\langle x_n, y\rangle = \langle x, y\rangle = -3$. One has $(x_n) \subset K$, $x_n \rightharpoonup x \in K$. For the values indicated, (3) is checked, that becomes

$$a - b < \left(\frac{1}{a} - \frac{1}{b}\right) \cdot (-3),$$

or equivalently $\frac{\sqrt{30}}{4} = ab < 3$, that is true.

It is true that for f convex and Gateaux differentiable, ∇f is pseudomonotone ([12]). Just for considering one different approach, it is remarked below that $\nabla\|\cdot\|$ is indeed pseudomonotone, in particular, if X is a Hilbert space, $K \subset X, 0_X \notin K, F : K \to X, F(x) = \nabla\|x\|$.

Remark 1. $\nabla\|\cdot\|$ is pseudomonotone if whenever

$$(x_k) \subset K \text{ with } x_k \rightharpoonup x, \text{ and } \limsup_k \frac{1}{\|x_k\|} \cdot \langle x_k, x_k - x\rangle \leq 0, \tag{4}$$

then (2) holds.

It is sufficient to show that

$$\liminf_k \|x_k\| + \liminf_k \left(- \frac{1}{\|x_k\|} \cdot \langle x_k, y\rangle \right) \geq \|x\| - \frac{1}{\|x\|}\langle x, y\rangle. \tag{5}$$

Let be denoted $a_k = \langle x_k, x_k - x\rangle$ and $b_k = \|x_k\|$. The sequence (x_k) is bounded since it is weakly convergent. Let $b > 0$ be such that $0 < b_k \leq b$ for all $k \in \mathrm{IN}$. For an $\varepsilon > 0$ arbitrary, there exists $k_\varepsilon \in \mathrm{IN}$ such that

$$\frac{a_k}{b_k} < \varepsilon \Longleftrightarrow a_k < \varepsilon b_k \leq \varepsilon b,$$

for all $k \geq k_\varepsilon$. It results $\limsup_k a_k \leq \varepsilon b$. Since $\varepsilon > 0$ was arbitrary this implies $\limsup_k a_k \leq 0$. By the inequality (4) one has

$$0 \geq \limsup_k \frac{1}{\|x_k\|} \cdot \langle x_k, x_k - x\rangle = \limsup_k \frac{\|x_k\|^2 - \langle x_k, x\rangle}{\|x_k\|},$$

thus

$$\begin{aligned} \limsup_k \|x_k\|^2 - \|x\|^2 &= \limsup_k \|x_k\|^2 - \lim_k \langle x_k, x\rangle \\ &= \limsup_k (\|x_k\|^2 - \langle x_k, x\rangle) = \limsup_k a_k \leq 0. \end{aligned}$$

Hence $\limsup_k \|x_k\| \leq \|x\|$, therefore $\lim_k \|x_k\| = \|x\|$, that together with $x_k \rightharpoonup x$ give $x_k \longrightarrow x$, since the Kadec-Klee property stands in the Hilbert space setting. Remark that, if (x_n) strongly converges to $x \neq 0$, then obviously $\frac{1}{\|x_n\|} \cdot \langle x_n, y\rangle \to \frac{1}{\|x\|}\langle x, y\rangle$, and consequently (5) is verified. Consequently (2) holds, meaning that $\nabla\|\cdot\|$ is pseudomonotone.

1.2 The Solution Set for the Variational Inequality

Let $K \subseteq X$ be a nonempty convex set and $F : K \to X^*$ be a given mapping. The variational inequality problem is the following:

$$VI(F, K) \text{ find } x \in K \text{ such that } \langle F(x), y - x\rangle \geq 0, \forall y \in K.$$

Denote by $S_{VI} = S_{VI(F,K)}$ the set of its solutions.

The result in [12], Proposition 4.1, is the following: "Let K be a closed and convex subset of X and let $F : K \to X^*$ be pseudomonotone such that S_{VI} is nonempty. Then, S_{VI} is weakly closed." The mentioned assertion is a particular case of Lemma 1 from [6].

Remark 2. Theorem 1 from [6] states the closedness of the solution map for some equilibrium problems, in particular for $n \in \mathrm{IN} \longmapsto S_{VI(F_n,K_n)}$, on a general setting of a Hausdorff topological space X, endowed with two comparable topologies.

Let $f_n, f : X \times X \to \mathrm{IR}$ be the bifunctions given by $f_n(x,y) = \langle F_n(x), x-y\rangle$ and $f(x,y) = \langle F(x), x-y\rangle$, respectively. The hypotheses of the theorem mentioned above include the convergence for (K_n) to K in some sense, a condition that relates F_n and F, and the pseudodomotonicity of F. If these hold, then S_{VI} is closed at ∞, i.e. for each sequence (x_n) of solutions to $VI(F_n,K_n), (x_n)$ convergent with respect to the less strong topology to x in X, imply $x \in S(\infty) = S_{VI(F,K)}$. If $(F(x_n))$ is bounded, it can be applied for the variational form of the bifunction, a constant sequence of operators, $F_n = F$, and for the constant domains $K_n = K$. If one takes the less strong topology be induced by $\rightharpoonup$ on X, then obtains the following result.

Corollary 1. ([6], Lemma 1) *If K is weakly closed and F is pseudomonotone, then $S_{VI(F,K)}$ is weakly closed.*

2 Another Example of a Pseudomonotone Operator That It Is Not Fan-hemicontinuous

The subject of [11] was to prove that there exists a pseudomonotone operator that is not Fan-hemicontinuous. Recall that the error pointed out in the nonequivalence between these two notions was in the proof of Proposition 3.5 from [12], where the implication $(x_k) \subset X, x_k - x \rightharpoonup 0$, in $X, F(x_k) \rightharpoonup f \in X^*$, then $\langle F(x_k), x_k - x\rangle \to 0$ is not correct.

In this section, another example of a pseudomonotone operator that it is not Fan-hemicontinuous is given. Let $X = L^3(0,1)$ be the Lebesgue space and let $F : X \to X^*$ be defined by

$$\langle F(u), v\rangle = \int_0^1 |u(t)| \cdot u(t) \cdot v(t)\, dt, \; u, v \in X. \tag{6}$$

By a direct consequence of Proposition 2.3 from [3] (p. 41), F is pseudomonotone since it is monotone and continuous. Indeed, let $S : \mathrm{IR} \to \mathrm{IR}, S(\xi) = |\xi| \cdot \xi$. Since $S'(\xi) = 2|\xi| \geq 0, S$ is increasing. Actually since $[S(\xi) - S(\eta)] \cdot (\xi - \eta) \geq \frac{1}{2}(|\xi| + |\eta|) \cdot (\xi - \eta)^2, \forall \xi, \eta \in \mathrm{IR}$, it follows,

$$\begin{aligned}\langle F(u) - F(v), u - v\rangle &= \int_0^1 [S(u(t)) - S(v(t))] \cdot [u(t) - v(t)]\, dt \\ &\geq \int_0^1 \frac{1}{2}(|u(t)| + |v(t)|) \cdot [u(t) - v(t)]^2\, dt.\end{aligned}$$

The continuity holds by the following inequalities. One has $|S(\xi) - S(\eta)| \leq 3(|\xi| + |\eta|) \cdot |\xi - \eta|$, $\forall \xi, \eta \in \mathrm{IR}$, so that, by Hölder's inequality

$$\begin{aligned}
\big|\langle F(u) - F(u_k), v\rangle\big| &\leq \int_0^1 \big|S(u(t)) - S(u_k(t))\big| \cdot |v(t)|\, dt \\
&\leq 3\int_0^1 \big[|u(t)| + |u_k(t)|\big] \cdot |u(t) - u_k(t)| \cdot |v(t)|\, dt \\
&\leq 3\big[\|u\|_{L^3} + \|u_k\|_{L^3}\big] \cdot \|u - u_k\|_{L^3} \cdot \|v\|_{L^3}.
\end{aligned}$$

Hence $F(u_k) \to F(u)$ as $u_k \to u$, in $L^3(0,1)$.

To prove that F is not Fan-hemicontinuous, one needs to show that there exist $v \in X$ and $(u_k) \subset X$ with $u_k \rightharpoonup u$ in X, such that

$$\liminf_k \langle F(u_k), u_k - v\rangle < \langle F(u), u - v\rangle.$$

To achieve this, it is defined

$$u_k(t) = \begin{cases} l \cdot a, \text{ if } t \in \left(\frac{i}{k}, \frac{i+\theta}{k}\right), i = 0, \ldots, k-1 \\ -a, \quad \text{if } t \in \left(\frac{i+\theta-1}{k}, \frac{i}{k}\right), i = 1, \ldots, k, \end{cases}$$

where $a > 0, l > 1$, and $\theta \in (0,1)$ are such that $m := \theta \cdot la + (1-\theta) \cdot (-a) = 0$, hence $\theta = \frac{1}{l+1}$. For this,

$$\begin{aligned}
\int_0^1 u_k(t)\, dt &= \sum_{i=0}^{k-1} \int_{i/k}^{\frac{i+\theta}{k}} u_k(t)\, dt + \sum_{i=1}^{k} \int_{\frac{i+\theta-1}{k}}^{i/k} u_k(t)\, dt \\
&= la \cdot \frac{\theta}{k} \cdot k + (-a) \cdot \frac{1-\theta}{k} \cdot k = 0.
\end{aligned}$$

One has $u_k \rightharpoonup u := 0 = m$, in $L^3(0,1)$, since there exists $M > 0$ with $\|u_k\|_{L^3} \leq M$ and $\lim_{k\to\infty} \int_D u_k(t)\, dt = 0, \forall D = [c,d] \subset (0,1)$ (see [14], Lemma 1.4). Indeed,

$$\begin{aligned}
\|u_k\|_{L^3}^3 &= \sum_{i=0}^{k-1} \int_{i/k}^{\frac{i+\theta}{k}} |u_k(t)|^3\, dt + \sum_{i=1}^{k} \int_{\frac{i+\theta-1}{k}}^{i/k} |u_k(t)|^3\, dt \\
&= \ell^3 a^3 \cdot \frac{\theta}{k} \cdot k + a^3 \cdot \frac{1-\theta}{k} \cdot k = a^3 \cdot (\ell^3 \cdot \theta + 1 - \theta) \\
&= a^3 \cdot l \cdot (\ell^2 + 1)/(l+1) =: A,
\end{aligned}$$

thus $M = \sqrt[3]{A}$, can be taken. For the second condition, the computations are the following. If $c < d$, there exists $k, i = [c \cdot k] + 1 \in \mathrm{IN}$, such that $c < \frac{i}{k} < \frac{i+1}{k} < d$. Denote $i_d = \max\{p \in \mathrm{IN} \mid \frac{i+p}{k} < d\}$. On the interval $\left[\frac{i}{k}, \frac{i+1}{k}\right]$, one has $\int_{i/k}^{\frac{i+1}{k}} u_k(t)\, dt = \int_{i/k}^{\frac{i+\theta}{k}} u_k(t)\, dt + \int_{\frac{i+\theta}{k}}^{\frac{i+1}{k}} u_k(t)\, dt = \frac{m}{k} = 0$. Similarly on $\left[\frac{i+j}{k}, \frac{i+j+1}{k}\right], 1 \leq j \leq p-1$. On the interval $[c,d] \setminus \left[\frac{i}{k}, \frac{i_d+1}{k}\right]$, one has $\left|\int_c^{i/k} u_k(t)\, dt\right| \leq la \cdot (i/k) \to 0$, as $k \to \infty$; similarly on $\left[\frac{i_d+1}{k}, d\right]$.

For $v(t) = \alpha > 0, t \in (0,1)$, so $v \in X$, one has $\langle F(u), u - v\rangle = 0$. Compute

$$\begin{aligned}\langle F(u_k), u_k - v\rangle &= \int_0^1 |u_k(t)| \cdot u_k(t) \cdot [u_k(t) - v(t)]\, dt \\ &= \int_0^1 |u_k(t)| \cdot u_k^2(t)\, dt - \int_0^1 |u_k(t)| \cdot u_k(t) \cdot v(t)\, dt \\ &= \|u_k\|_{L^3}^3 - \alpha \cdot \int_0^1 |u_k(t)| \cdot u_k(t)\, dt \\ &= A - \alpha \cdot B, \end{aligned} \quad (7)$$

where the value B from (7) is given by

$$\begin{aligned} B = \int_0^1 |u_k(t)| \cdot u_k(t)\, dt &= \sum_{i=0}^{k-1} \int_{i/k}^{\frac{i+\theta}{k}} \ell^2 a^2\, dt + \sum_{i=1}^{k} \int_{\frac{i+\theta-1}{k}}^{i/k} a \cdot (-a)\, dt \\ &= \ell^2 a^2 \cdot \theta + (-a^2) \cdot (1-\theta) = a^2 \cdot l \cdot (l-1)/(l+1). \end{aligned}$$

Now, chose α such that $A - \alpha \cdot B < 0$, in (7), thus $\liminf_k \langle F(u_k), u_k - v\rangle < 0$. This means that F is not Fan-hemicontinuous.

3 Open Question

In this section, stability of pseudomonotonicity with respect to the composition to a linear operator is questioned. Let us consider two examples of pseudomonotone operator that are not Fan-hemicontinuous.

Let $F : X \to X^*$ be given in (6). Let $\tilde{F} : Y \to Y^*$ be defined on the Sobolev space $Y = W_0^{1,3}(0,1)$, given by

$$\langle \tilde{F}(u), v\rangle = \int_0^1 |\nabla u(t)| \cdot \nabla u(t) \cdot \nabla v(t)\, dt,\ u, v \in Y,$$

that it is known as being pseudomonotone [11]. Let $S : \mathrm{IR} \to \mathrm{IR}, S(\xi) = |\xi| \cdot \xi$. Since

$$\langle F(\nabla u), \nabla v\rangle = \langle -div\, S(\nabla u), v\rangle = \langle \tilde{F}(u), v\rangle,\quad v \in Y,$$

one can ask about the pseudomonotonicity of $F \circ \nabla$. In this matter, for an arbitrary F, it is formulated the following question.

Q: What (sufficient) conditions should be imposed on $L : Y \to X$ such that for $F : X \to X^*$ pseudomonotone, then $F \circ L$ to be pseudomonotone as well?

References

1. Kassay, G., Rădulescu, V.D.: Equilibrium Problems and Applications, Mathematics in Science and Engineering. Academic Press, Cambridge (2019)
2. Panagiotopoulos, P.D., Fundo, M., Rădulescu, V.: Mathematical Theory of Hemivariational Inequalities, Dekker, New-York (1999)

3. Showalter, R.E.: Monotone Operators in Banach Space and Nonlinear Partial Differential Equations, vol. 49. American Math. Society, USA (1997)
4. Zeidler, E.: Nonlinear Functional Analysis and its Applications, vol. III. Springer Verlag, Berlin, Germany (1990)
5. Bigi, G., Castellani, M., Pappalardo, M., Passacantando, M.: Existence and solution methods for equilibria. Eur. J. Oper. Res. **227**, 1–22 (2013)
6. Bogdan, M., Kolumbán, J.: Some regularities for parametric equilibrium problems. J. Glob. Optim. **44**, 481–492 (2009). https://doi.org/10.1007/s10898-008-9345-3
7. Hartman, P., Stampacchia, G.: On some non linear elliptic differential functional equations. Acta Math. **115**, 271–310 (1966). https://doi.org/10.1007/BF02392210
8. Leray, J., Lions, J.L.: Quelques résultats de Visik sur les problémes elliptiques non linéaires par les méthodes de Minty-Browder. Bull. Soc. Math. Fr. **93**, 97–107 (1965)
9. Gwinner, J.: On fixed points and variational inequalities-a circular tour. Nonlinear Anal. **5**, 565–583 (1997)
10. Gwinner, J., Ovcharova, N.: From solvability and approximation of variational inequalities to solution of nondifferentiable optimization problems in contact mechanics. Optimization **64**, 1–20 (2015)
11. Steck, D.: Brezis pseudomonotonicity is strictly weaker than Ky Fan hemicontinuity. J. Optim. Theory Appl. **181**(1), 318–323 (2019). https://doi.org/10.1007/s10957-018-1435-x
12. Sadeqi, I., Salehi Paydar, M.: A comparative study of Ky Fan hemicontinuity and Brezis pseudomonotonicity of mappings and existence results. J. Optim. Theory Appl. **165**(2), 344–358 (2015). https://doi.org/10.1007/s10957-014-0618-3
13. Bogdan: Fan-hemicontinuity for the gradient of the norm in Hilbert space (under review)
14. Dacorogna, B.: Direct Methods in the Calculus of Variations, Applied Mathematical Sciences 78, vol. III. Springer Verlag, Berlin, Germany (1989)

A New Ultra Wide Band Antenna Design with Dual Band for WLAN and WiMAX Applications

Nisrin Sabbar[1], Khalid Hati[2](✉), Hassan Asselman[1], and Abdellah Elhajjaji[2]

[1] Optics and Photonics Group, FS Tetuan, Abdelmalek Essaadi University, Tétouan, Morocco
[2] Systems of Communications and Detection Laboratory, FS Tetuan, Abdelmalek Essaadi University, Tétouan, Morocco

Abstract. In this paper, a compact microstrip-fed planar ultra-wideband (UWB) monopole antenna with dual band for Wireless Local Area Network (WLAN) and worldwide interoperability for microwave access (WiMAX) application is presented. The proposed design consists of a circular radiating patch and a partial ground plane. A T-shaped slot resonator and four rectangular shaped slot resonators are respectively etched in the radiating patch and partial ground plane for WiMAX band (3.3–3.8 GHz) and WLAN band (5.1–6 GHz). The proposed antenna is printed on the FR4 substrates and is optimized by ANSOFT High Frequency Structure Simulator (HFSS).

Keywords: UWB antenna · WLAN · WiMAX · T-shaped slots · Slotted ground pane

1 Introduction

Since the Federal Communication Commission (FCC) allocates rules for the commercial use of ultra-wideband (UWB) in 2002 [1], the feasible design and the implementation of UWB system becomes a very competitive topic for both academic and industrial communities of telecommunications.

In particular, as a key component of the UWB system, an extremely broadband antenna will be launched in the frequency range from 3.1–10.6 GHz, which has attracted significant research power in the recent years [2]. The major Challenges of a UWB antenna design include the UWB performances of the impedance matching, radiation stability, the compact appearance of the antenna size, and the low manufacturing cost for consumer electronics applications [3].

However, the frequency spectrum of UWB systems will cause interference to the existing WLAN (Wireless Local Area Network) and WiMAX (Worldwide interoperability for Microwave Access) networks operating, respectively, at 5.1–6 GHz and 3.3–3.8 GHz. Thus, the best antenna for dual band applications should have dual bands at (3.3–3.8 GHz) and (5.1–6 GHz) to avoid interference. Some antennas with this characteristic have been reported in the published literature [4]. A recent report [5] describes an antenna designed by the use of the slots to obtain dual band characteristic.

L. Moldovan and A. Gligor (Eds.): Inter-Eng 2021, LNNS 386, pp. 704–717, 2022.
https://doi.org/10.1007/978-3-030-93817-8_63

The multi-band antennas have aroused high interest in recent years for its application to multimode communication systems [6, 7]. Printed monopole antennas are very popular candidates for these applications, because of their low cost and process simplicity. The currently popular designs suitable for WLAN operation in the 5.2/5.8 GHz (5.15–5.35 GHz/5.725–5.825 GHz) bands and the WiMAX applications (3.3–3.8 GHz) have been reported in [8, 9].

In this paper, a novel dual-band circular antenna is proposed in the first part. In the second part, a dual-band is also achieved by embedding T-shaped inserted in the radiation patch and modified four rectangular slots in the ground plane.

2 Antenna Configuration

The geometry and the configuration of the proposed antenna with partial plane is illustrated in Fig. 1. The antenna is fed with a 50 Ω micro-strip line and fabricated on the FR4 substrate with a thickness of 1.6 mm and a relative permittivity of 4.4. The shape of the radiating element is circular with a radius of "R" and a modified ground plane with four rectangular slots. The radiator fed by a micro-strip line is printed on the top side of the substrate, while the partial ground plane of size W * G is printed on the bottom side of the substrate. The radiating element is fed by 50 Ω micro-strip transmission line and has a center width w = 3 mm, which is terminated with a sub miniature A (SMA) connector for measurement purposes.

On the bottom of the antenna, four rectangular slots are embedded on the partial ground plane. The gap between the radiating patch and ground plane is h. T-shaped slot is etched on the element radiation.

3 Antenna Design and Parametric Study

3.1 UWB Monopole Antenna Configuration

The T-slot is embedded in the radiating patch in order to get an UWB can yield an ultra wideband. A circular monopole is used as the base antenna [10] and to obtain better

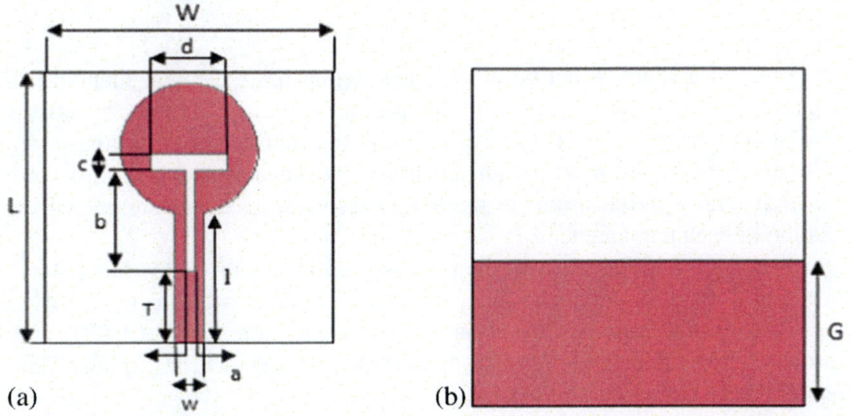

Fig. 1. Geometry of the proposed antenna UWB: (a). front view; (b). back view.

VSWR performance over the whole UWB band. Recently, it was demonstrated in [11] that by etching T slot in the radiation patch, the total bandwidth of the monopole antenna can be significantly increased up to 128%. As a result, by inserting a T-shaped slot at the radiating patch and carefully adjusting their parameters (a, b, c, T and d) as shown in Fig. 1, additional resonance can be excited and hence much enhanced impedance bandwidth may be achieved.

The optimum dimensions of designed antenna are shown in Table 1 below.

Table 1. Design specification of the proposed antenna.

Parameters	Values
h	1.6
W	30
L	30
w	3
l	14.5
a	1
b	11
c	2
d	8
G	12.8
e	2
f	9
m	4.8
n	2
g	1
T	8

The proposed antenna structure is simulated using finite element method (FEM) software, HFSS. The return losses S11 for different values of 'T' and 'b' are shown in Figs. 2 and 3, respectively. It can be seen that the return loss curves have similar shapes for different values of the parameters at low frequencies, but the high-frequency impedance matching and positions of the higher resonances change significantly with the variation of each parameter.

Figure 4 shows the return loss of the proposed antenna with and without slots. We note that this is an ultra-wideband antenna with a bandwidth between 2.7 GHz and 13 GHz. Two resonance modes are observed, one centered around 3.61 GHz and the other around 11.35 GHz. The presence of these resonances can be explained by the presence of the T-shaped slot.

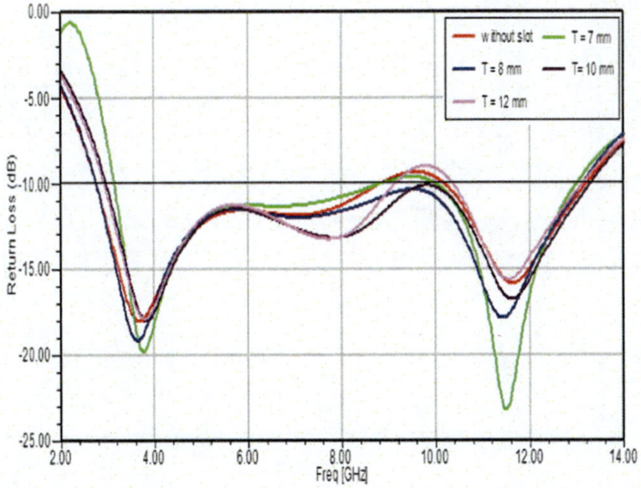

Fig. 2. Simulated return losses for different values of T with fixed values of a = 1 mm, b = 11 mm, c = 2 mm, and d = 8 mm.

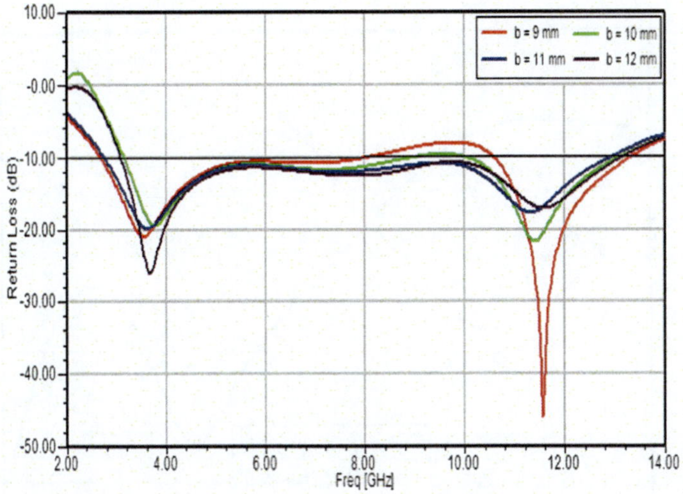

Fig. 3. S11 parameter for different values of b with fixed values of a = 1 mm, T = 8 mm, c = 2 mm, and d = 8 mm.

Figure 5 shows the antenna's peak gain in the frequency range from 2–14 GHz for the proposed UWB antenna. We can observe that the significant antenna gain increase at 3.61 GHz and 11.35 GHz indicate the effect of band function clearly.

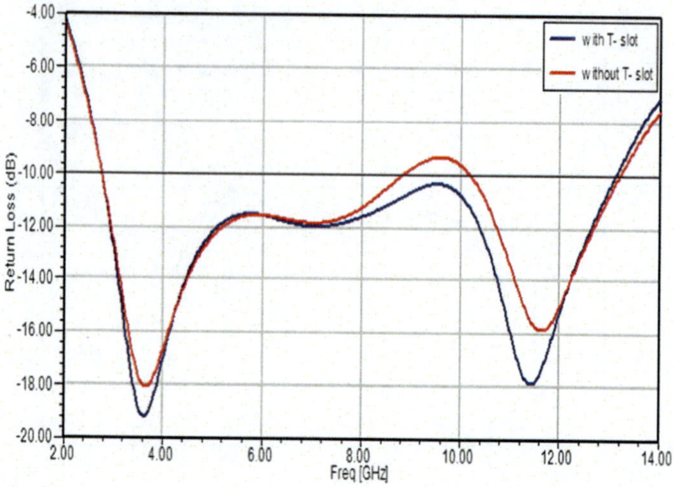

Fig. 4. Simulated return losses of the proposed antenna with and without T-shaped slot. The optimized parameters of the T-slot are: a = 1 mm, b = 11 mm, c = 2 mm, d = 8 mm and T = 8 mm.

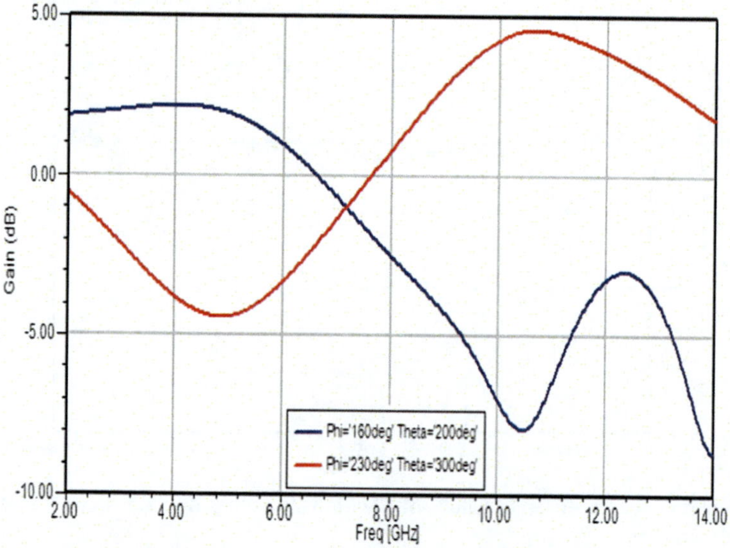

Fig. 5. Simulated gain of the proposed UWB antenna.

Two principal planes are selected to present the radiation pattern of the proposed antenna (Fig. 6). These are called E plane ($\varphi = 0$) and the H plane ($\varphi = \pi/2$). Figure 6 shows the radiation patterns in the H-plane and E-plane of the proposed antenna at frequency 3.61 GHz, 7 GHz and 11.35 GHz. We can see that the radiation patterns in the E plane are nearly almost omni-directional for the two frequencies 3.61 GHz, 7 GHz and bidirectional for 11.35 GHz. H-plane, the radiation pattern is bidirectional and as a dipole antenna.

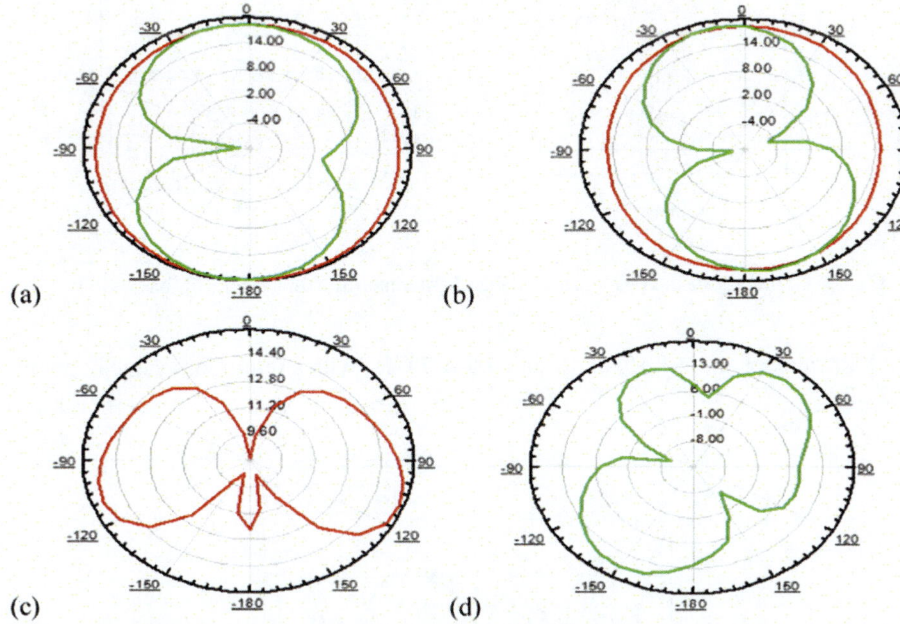

Fig. 6. Simulated radiation patterns of the proposed UWB antenna in the E plane and the H plane at (a). f = 3.61 GHz, (b). f = 7 GHz, (c, d). f = 11.35 GHz.

3.2 Single and Dual Band Antenna

The rapid development of wireless communication urges the need of antennas covering multiple bands with good radiation characteristics. Thus, many researchers have been paying much attention to design this kind of antennas. Wide-band antenna and multi-band antenna designs have become very important for wireless communications. In [12], by the presence of an L-shaped parasitic strips, three resonant modes of the antenna for the 2.6/3.5/5.5 GHz-bands can be excited to meet the WiMAX system. By introducing dual U-shaped strips, multi-resonant modes for WiMAX applications are proposed in [13]. In [14], with the inclusion of an additional small radiation patch, a dual-band antenna designed from 3.1 to 10.6 GHz out of the band 5.0–6.0 GHz can be achieved.

In this section, a novel printed monopole antenna with dual-band is presented to satisfy WLAN (5.1–6 GHz) and WiMAX (3.3–3.8 GHz) applications (Fig. 7). The antenna structure consists of a circular monopole with a micro-strip feed-line and a partial ground plane.

Figures 8, 9 and 10 show the simulated return loss of the proposed antenna with different parameters which affect the dual band characteristics. We can observe that the proposed antenna provides the dual band in the frequency range of 2.6–3.8 GHz with a

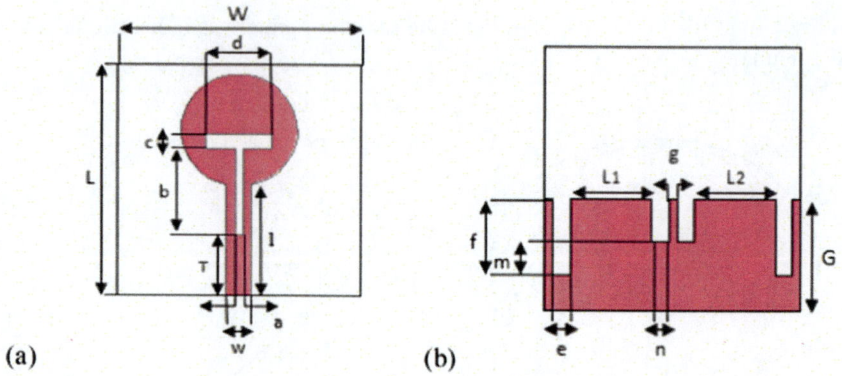

Fig. 7. Geometry of the proposed dual-band antenna: (a). front view; (b). back view

return loss value of about −15.43 dB and 4.5–8.7 GHz with a return loss value of about −30.2 dB.

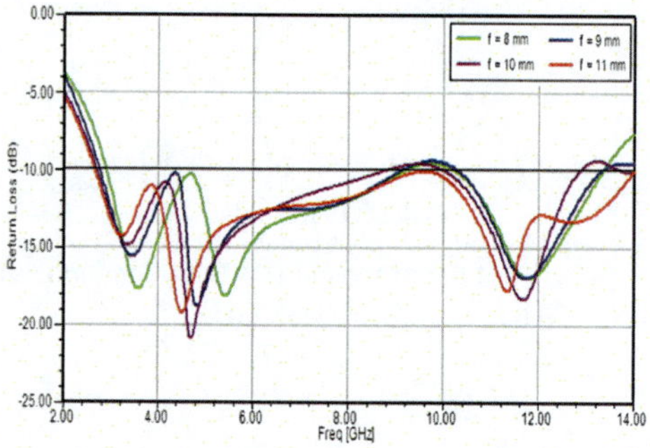

Fig. 8. Simulated return losses for different values of f with fixed values of e = 2 mm.

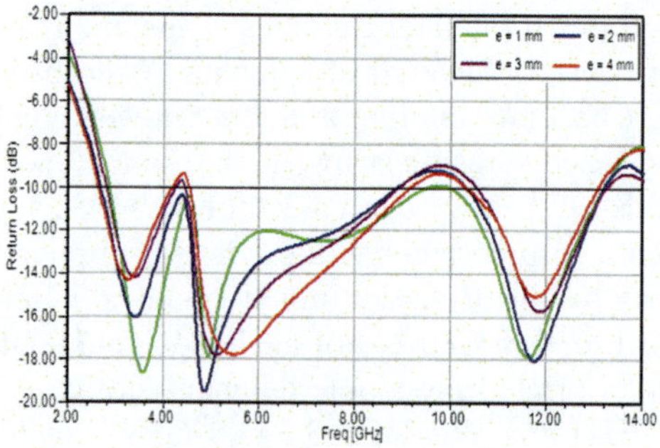

Fig. 9. Simulated return losses for different values of e with fixed values of f = 9 mm.

Figures 11, 12, 13 and 14 show the simulated return loss of the proposed antenna with different parameters which affect the dual band characteristics.

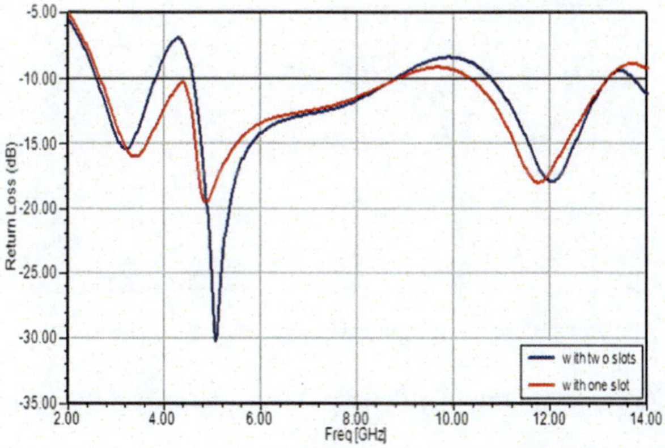

Fig. 10. Simulated return losses of the proposed antenna with one and two slots.

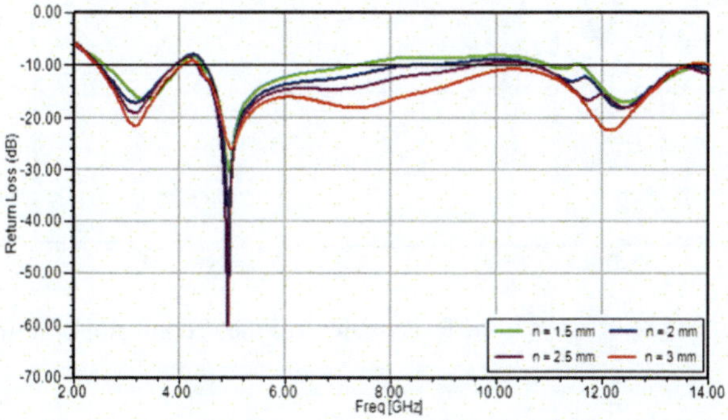

Fig. 11. Simulated return losses of different values for the first slot of n, m = 4.8 mm.

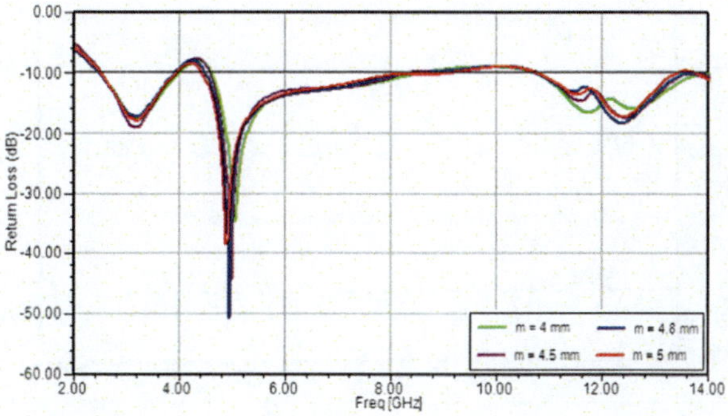

Fig. 12. Simulated return losses of different values for the first slot of m, n = 2 mm.

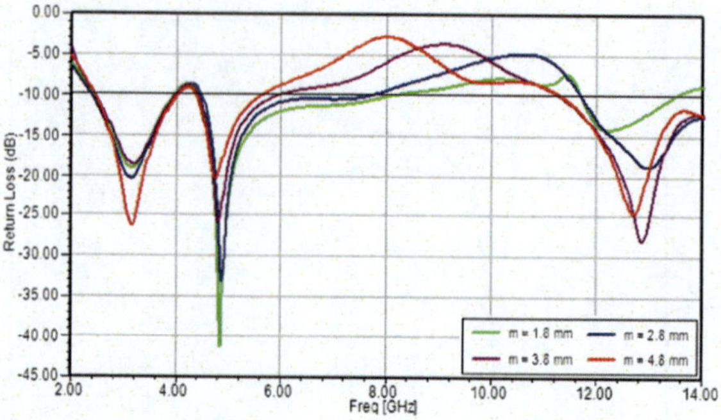

Fig. 13. Simulated return loss of different values to the second slot of m, n = 2 mm.

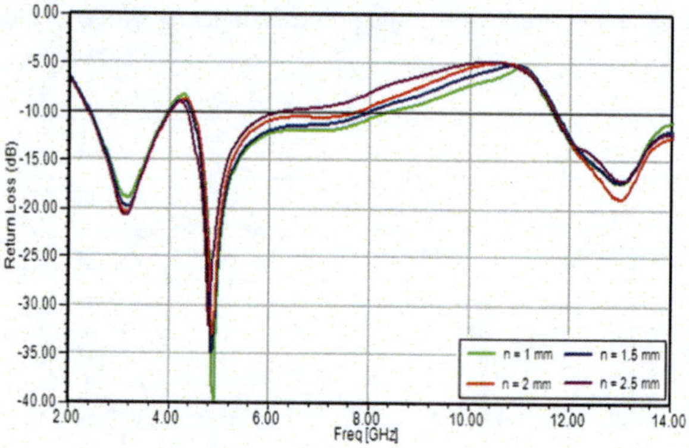

Fig. 14. Simulated return loss of different values to the second slot of n, m = 2.8 mm.

From Fig. 15 we observe the simulated VSWR of the dual band antenna with slots. The VSWR ≤ 2 for the proposed antenna is from 3 to 3.4 GHz and 4.7 to 5.1 GHz.

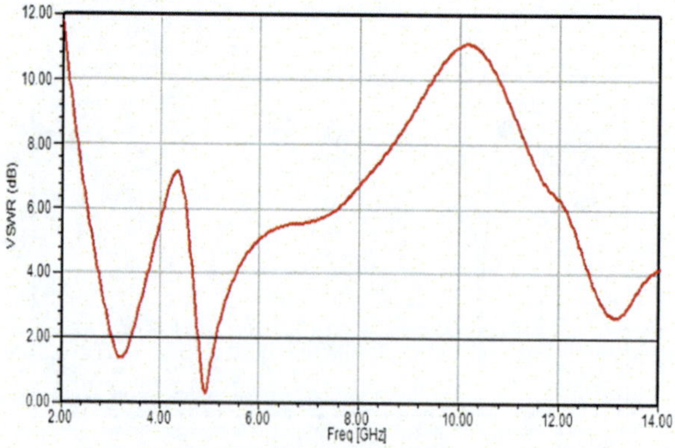

Fig. 15. Simulated VSWR of the proposed antenna.

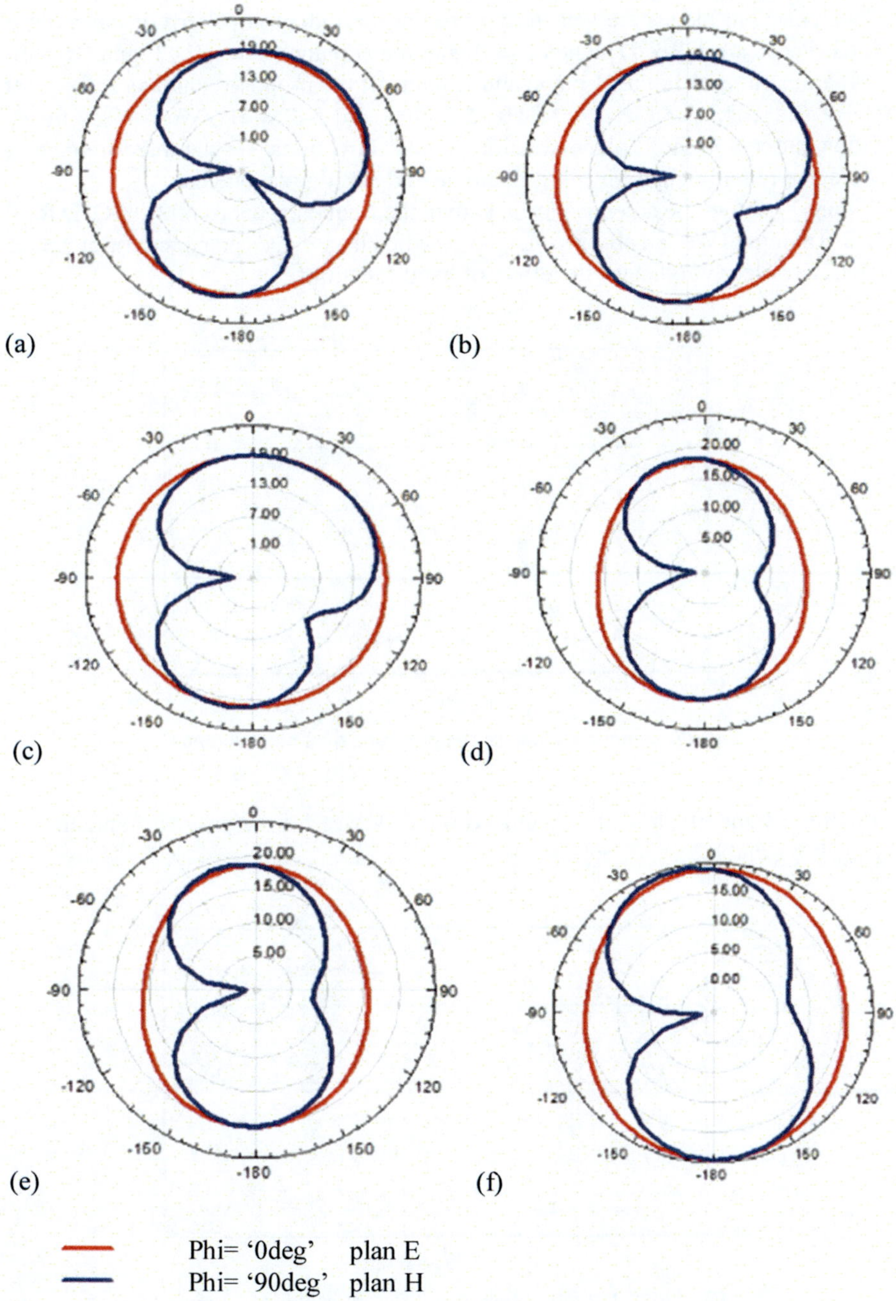

Fig. 16. Simulated radiation patterns of the proposed antenna in the E plane and the H plane at (a). f = 2.5 GHz, (b). f = 3.2 GHz, (c). f = 3.3 GHz, (d). f = 4.9 GHz, (e). f = 5.1 GHz, (f). f = 5.82 GHz.

Two principal planes are selected to present the radiation pattern of the proposed antenna. These are called E plane (φ = 0) and the H plane (φ = π/2). Figure 16 shows the radiation patterns in the H-plane and E-plane of the proposed antenna at frequency 2.5 GHz, 3.2 GHz, 3.3 GHz, 4.9 GHz, 5.1 GHz and 5.82 GHz. We can see that the radiation patterns in the E plane are almost omni-directional for the three frequencies. H-plane, the radiation pattern is bidirectional and as a dipole antenna.

Figure 17 shows the antenna's peak gain in the frequency bands of 2–14 GHz for the proposed antenna. We can observe that the significant antenna's gain increase at 3.3 GHz and 5.1 GHz clearly indicates the effect of band function.

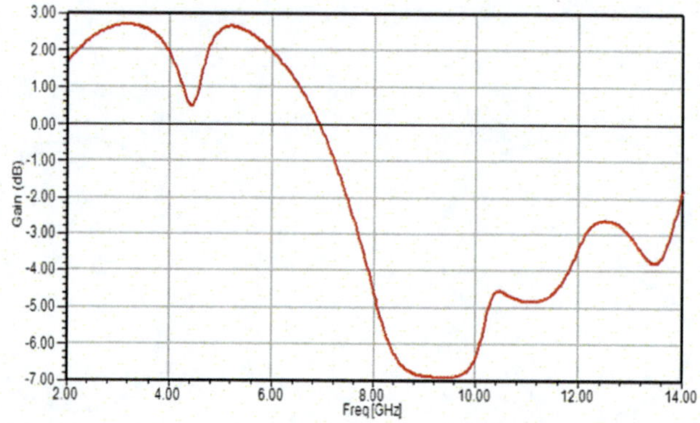

Fig. 17. Simulated gain of the proposed dual band antenna.

Figures 18 and 19 show the simulated return losses for the proposed antenna with varying T, L1 and L2.

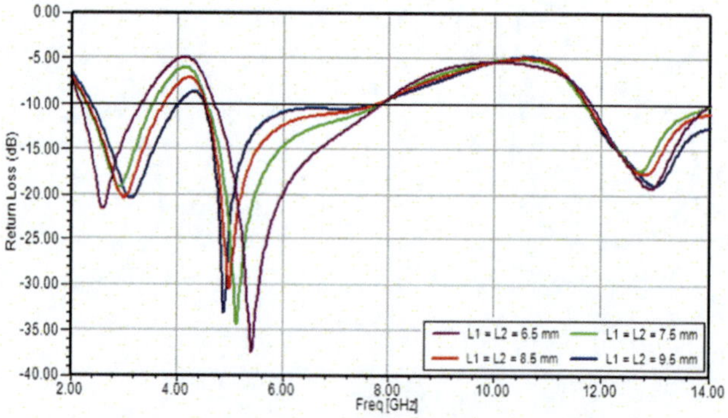

Fig. 18. Simulated return losses of different values of L1 and L2.

Figure 19 shows that the center frequency of the dual band at 3.2 GHz and 4.9 GHz is increased as T decreases. Furthermore, we can observe in Fig. 18 that the length of T obviously affects the impedance matching. The simulated impedance of the proposed

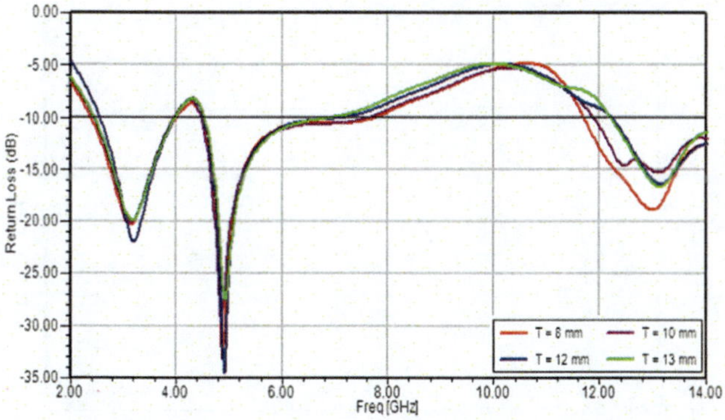

Fig. 19. Simulated return losses of different values of T.

antenna shows that the return loss decreased in the WiMAX and WLAN bands. By properly tuning the dimensions L1, L2 and spacing T. It is empirically found that band broadening can be optimized with choosing the dimensions: T = 12 mm, L1 = 9.5 mm and L2 = 9.5 mm.

Figure 20 shows the simulated return loss of the proposed antenna with the optimal parameters which affect the dual band characteristics. We can observe that the proposed antenna provides the dual band in the frequency range of 2.5–3.99 GHz with a return loss value of about −22 dB and 4.5–6.6 GHz with a return loss value of about −34.44 dB.

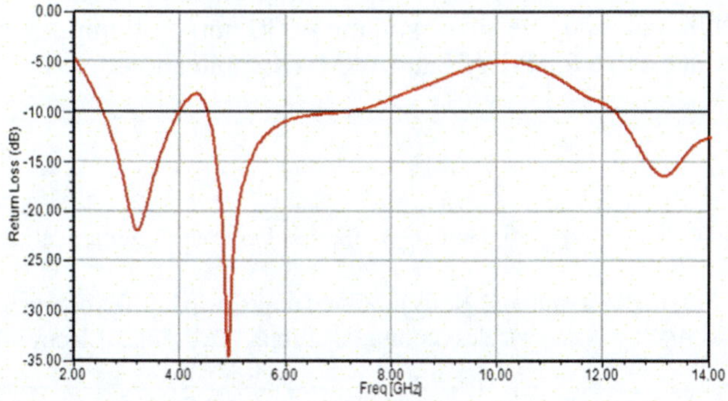

Fig. 20. Simulated return losses of the final proposed antenna.

After completing the parametric study of the proposed antenna, we present in Fig. 20, the final optimization of S11 for the base and the proposed antenna. The antenna in the circular patch with slots allows getting a dual band antenna. For the obtained antenna, we get two modes of resonance centered on the 3.2 GHz frequency and 4.9 GHz.

In Fig. 21, we represent the variations of the reflection coefficient S11 by using both the software HFSS and CST to validate our results.

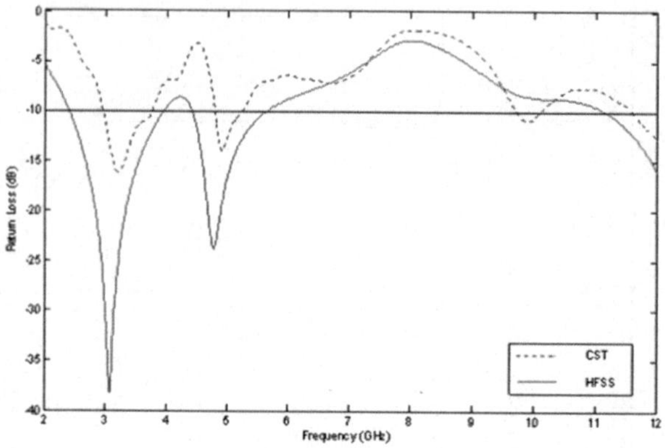

Fig. 21. Return loss for the antenna design, from 2 to 12 GHz.

4 Conclusions

A new design of a compact planar circular antenna's ultra wide band antenna with dual band for WLAN and WiMAX has been proposed and discussed. The circular patch antenna ring ultrawide-bandwidth radiating between 2.7 GHz and 13 GHz, a wide impedance bandwidth 131%. Dual bands antenna is achieved by etching four rectangular slots on the partial ground plan and a T-shaped slot in the radiation element. It has a dual bands frequency with 2.5–3.99 GHz and 4.5–6.6 GHz (WLAN and WiMAX technology). The proposed antenna indicates not only a dual band frequency but also a good radiation performance, while retaining the small volume of 30 mm * 30 mm * 1.6 mm^3. These features are very attractive for WLAN and WiMAX applications.

References

1. Ultra-Wideband Transmission System FCC 02-48: Federal Communications Commission (2002)
2. Tiwari, R.N., Singh, P., Kanaujia, B.K.: A modified microstrip line fed compact UWB antenna for WiMAX/ISM/WLAN and wireless communications. AEU. Int. J. Electron. Commun. **104**, 58–65 (2019)
3. Varkiani, S.M.H., Afsahi, M.: Compact and ultra-wideband CPW-fed square slot antenna for wearable applications. AEU Int. J. Electron. Commun. **106**, 108–115 (2019)
4. Yadav, D., Abegaonkar, M.P., Koul, K.S., Tiwari, V., Bhantnagar, D.: A compact dual band notched UWB circular monopole antenna with paracsitic resonators. AEU Int. J. Electron. Commun. **84**, 313–320 (2018)
5. Cheng, D., Wang, J., Wei, Y., Jin, C., Li, M.: Design of dual-band and high gain waveguide slot antenna. AEU Int. J. Electron. Commun. **98**, 208–212 (2019)
6. Srivastava, K., Kumar, A., Kanaujia, B.K., Dwari, S., Kumar, S.: Multiband integrated wide-band antenna for Bluetooth/WLAN applications. AEU Int. J. Electron. Commun. **89**, 77–84 (2018)
7. Aanweer Ali, T., Khaleeq, M.M., Biradar, R.C.: A multiband reconfigurable slot antenna for wireless applications. AEU Int. J. Electron. Commun. **84**, 273–280 (2018)

8. Kumar, S.A., Shanmuganantham, T., Dileepan, D.: Design and development of CPW fed monopole antenna at 2.45 GHz and 5.5 GHz for wireless applications. Alexandaria Eng. J. **56**(2), 231–234 (2017)
9. Chitra, R.J., Nagarajan, V.: Double L-slot microstrip patch antenna array for WiMAW and WLAN application. Comput. Electr. Eng. **39**, 1026–1041 (2013)
10. Wang, Z., Lim, E.G., Chen, X.: Study the ground plane effect of a multiband antenna for multiple wireless communication systems. Procedia Eng. **15**, 2521–2526 (2011)
11. Luoa, S., Wangab, D., Chenc, Y., Lia, E., Jianga, C.: A compact dual-port UWB-MIMO antenna with quadruple band-notched characteristics. AEU Int. J. Electron. Commun. **136**, 153770 (2021)
12. Abou Al-Alaa, M., Elsadek, H.A., Abdallah, E.A.: Compact multi-band frequency reconfigurable planar monopole antenna for several wireless communication applications. J. Electr. Syst. Inf. Technol. **1**(1), 17–25 (2014)
13. Lu, J.H., Chou, W.-C.: Planar dual U-shaped monopole antenna with multiband operation for IEEE 802.16e. IEEE Antennas Wirel. Propag. Lett. **9**, 1006–1009 (2010)
14. Thomas, K.G., Sreenivasan, M.: A simple ultrawideband planar rectangular printed antenna with band dispensation. IEEE Trans. Antennas Propag. **58**(1), 27–34 (2010)

Review on Microbial Bioinformatics: Novel and Promoting Trend for Microbiomics Research and Applications

Ben Amar Cheba[1,2](✉)

[1] Biology Department, College of Science, Jouf University, P.O. Box: 2014, Sakaka, Saudi Arabia

[2] Department of Biotechnology, Faculty of Nature and Life Sciences, University of Sciences and Technology of Oran-Mohamed Boudiaf (USTOMB), BP 1505 Al Mnaouar, 31000 Oran, Algeria

Abstract. The recent biotechnological advances and the rapid development of next-generation sequencing technologies accompanied by efficient computational facilities and tools have led to explosive and extensive data generation from finished complete genomes and draft genomes, because of this development, an urgent need arises for fast computing and automated approaches to analyze these bio big-data issued from microbial genomes and metagenomes in effective and a comparative way. Bioinformatics has come to play this major role via microbial bioinformatics and microbiomics which fill in the gap between just data accumulation and theoretical speculations to solution discovery and ready for use applications. In this perspective, our review gives an overview on microbial bioinformatics disciplines and subdisciplines and surveys their items, roles, influencing technologies and challenges, as well as lists the most microbial bioinformatics and metagenomics online resources, assemblers, and software, Furthermore, summarizes all the possible biotechnological applications of microbial bioinformatics and microbiomics in agriculture, food, medicine, industry, energy, and environment.

Keywords: Microbial bioinformatics · Microbiomics · Metagenomics · Biotechnological applications

1 Introduction

The sequencing technological revolution generated huge bio big data that require the harnessing of bioinformatics to archive, organize, analyze and use it to solve thorny biological problems [1]. Furthermore, the continuous development of high-throughput sequencing technologies during the last two decades, especially whole-genome sequencing (WGS), led to cumulative and extensive data issued from finished microbial genomes, draft genomes and metagenomes have irreversibly changed the way of microbial research and paved the way for microbial bioinformatics.

Since the sequencing of the first bacterial genome of *Hemophilus influenzae*, several microbial genomes, draft genomes and metagenomes, were finished and need analysis and exploitation, microbial bioinformatics via cloud computing, artificial intelligence

L. Moldovan and A. Gligor (Eds.): Inter-Eng 2021, LNNS 386, pp. 718–729, 2022.
https://doi.org/10.1007/978-3-030-93817-8_64

and machine learning provides an alternative approach that facilitates the fast treatment of this large genome dataset [2].

Table 1. Microbial bioinformatics disciplines and subdisciplines.

	No	Disciplines	Subdisciplines
Fundamental Applications	1	Informatics	Data science, cloud computing, machine learning and artificial intelligence, physical and computational infrastructures
	2	Mathematics	Numerical analysis, algorithms
	3	Statistics	Probability, statistical analysis, statistical modeling
	4	Physics	Material, laser, optics, quantum, nano, bio-radiation, spectroscopy
	5	Chemistry	Physical, material, inorganic, organic, biochemistry, analytical, quantum, photo, nano, chemistry, cheminformatics
	6	Microbiology	Agricultural, food, pharmaceutical, medical and clinical microbiology
	7	Microbial genetics	Viral, bacterial, fungal, microalgal, and protozoal genetics, microbial population genetics, microbial phylogenetics
	8	Microbial genomic	Ceramics, bacteriomics, fungomics, microbiomics, microbial population genomics, comparative microbial genomics
	9	Genomics	Microbiomics, plant, animal, rumen and human microbiomics, oncogenomic, epigenomics, pharmacogenomics, phylogenomic, microbial paleogenomics, structural, functional and comparative genomics, phylogenomic, population genomics
	10	Omics	genomic, transcriptomic, proteomic, secretomics, metabolomics, physiolomics, phenomics,
	11	Meta-Omics	Metagenomic, metatranscriptomics, metaproteomic
	12	Biology	Cellular, molecular, structural, synthetic, integrated, systems, evolution and population biology
	13	Biotechnology	Microbial, plant, animal and human biotechnology
			Environmental, industrial, agricultural, food, pharmaceutical, medical and nano biotechnology
	14	Medicine	Clinical microbiology, microbial epidemiology, health microbiomics, preventive, molecular medicine
	15	Pharmacology	Biologics, pharmacogenomics, antimicrobial drugs, pharmaceutical microbiology
	16	Veterinary sciences	Veterinary microbiology, animal and rumen microbiomics, animal breeding, animal health microbiomics
	17	Agriculture	Soil Microbiology, plant microbiology, phytopathology, terragenomics, soil microbiomics, rhizosphere microbiomics, plant microbiomics
	18	Geology	Geomicrobiology, geobiology, geomicrobiomics, micropaleontology, glacier geomicrobiomics
	19	Environment	Terrestrial, aquatic ecology, marine metagenomics, microbial diversity, marine and environmental microbiology, ecogenomics, environmental genomics and metagenomics
	20	Engineering	Electronics, electrotechnics, robotics, biorobotics, nano technology, biomimetics, bionics or biologically inspired engineering

In this review, we define microbial bioinformatics and list their disciplines and sub disciplines, roles, influencing technologies and challenges in the post-genomic era, as well as discuss the concepts of microbiomics, enumerate their computational tools, softwares and possible biotechnological applications.

2 Microbial Bioinformatics Multi-interdisciplinarity

While bioinformatics is an interdisciplinary research field that applies methodologies from computer science, mathematics and statistics to the study of biological phenomena,

Table 2. Microbial bioinformatics items, roles, influencing technologies and challenges.

Microbial bioinformatics Items	Roles	Influencing sciences and technologies	Challenges
Computational hardwar and software	Create, organize, develop, optimize and standardize, improve	High-Throughput Sequencing (HTS) Technologies	Data sharing,
Data bank	Creation, organization, classification	Next-Generation Sequencing (NGS) Technologies	Data scalability,
Data/metadata	Storage, access, organization, mining, processing, integration, visualization, binning, interpretation, management, shift the focus from data to discovery	Whole Genome Sequencing (WGS)	Data security,
Gene/multigenes	Finding, location prediction, abundance, profiling, annotation,	Whole Shotgun Metagenomic (WSM) sequencing	Data backup,
Genome/metagenome	Preprocessing, processing, binning, analysis, taxonomy, structural annotation, functional annotation, assembly, interpretation, gene prediction, novel genes discovery, recovery of more complete genomes from the metagenome	Mass Spectrometry (MS)	Integrated analytics,
Microbiome/meta microbiome	Archive, assembly, de novo assembly, visualize, binning, taxonomy, interpretation, gene prediction, novel genes discovery	Meta microbiomics/metagenomics	Accuracy and performance,
Phylogenomic/phylomicrbiomics	Analysis, phylogenomic reconstruction, phylogeny inference	Cloud computing, artificial intelligence, and machine learning	Focus on what matters

microbial bioinformatics is extremely multi-interdisciplinary, covering fundamental disciplines such as mathematics, statistics, informatics, physics, chemistry biology, microbiology and computational science, as well as applied disciplines such as agriculture, food, medicine, industry, energy and environment (Table 1).

In this review we define microbial bioinformatics and lists their disciplines, roles, influencing technologies and challenges in the post-genomic era, as well as discuss the concepts of microbiomics and their possible biotechnological applications.

3 Microbial Bioinformatics and Microbiomics

Microbial bioinformatics intersecting microbiology and bioinformatics and become to bridges the gap between data and discovery via analyzing the wealthy data gleaned from the fully sequenced microbial genomes and metagenomes to solve the complex biological problems. Microbial bioinformatics items, roles, influencing technologies, and challenges were summarized in Table 2 as well as online resources, assemblers, and software was listed in Table 3. These data banks and computational tools were very useful for promoting various kinds of analyses and meta-analysis which facilitate the discovery of hidden secrets of microbial ecosystem and microbiomic research in various fields.

The microbiome defined as the total genomes of the resident microorganisms of a particular organism [3]. Microbiomics harnessing bioinformatics to the study the microbial genes, genomes, and metagenomes via high throughput technologies accompanied by efficient computing tools which drastically changing the way of microbial DNA analysis, and revolutionized our understanding of microbial diversity, their functional roles in their environment, and association with plant, animal and human hosts. The various microbial bioinformatics applications spanning from environment, energy and agri-foods to bio-pharma- medicine fields were recapitulated in Table 4.

Table 3. Most microbial bioinformatics and metagenomics online resources, assemblers, and software

No.	Metagenomics		Computing tools or software [4]	
	Online resources	Assemblers	Major	Minor
1	Integrated Microbial Genomes and Microbiomes (IMG) [5]	IDBA_UD [6]	QIIME2(Qualitative Insights into Microbial Ecology) is an open-source bioinformatics tool for performing microbiome analysis	Explicit, Qiita, otupipe, mockrobiota, NGS preprocessor, LOTUS (Less OTUs)
2	EBI metagenomics [7]	MetaVelvet.[9]	MOTHUR: an open-source software package, a command-line computer program for analyzing sequence data from microbial communities	
3	MG-RAST: Metagenomics rapid annotation using subsystems technology [8]	MEGAHIT [10] and Ray Meta [11]	METAREP: a software tool for comparative metagenomics	

Table 4. Microbial bioinformatics and microbiomics applications

Application field	Microbiomics type	Examples/details	Ref.
Agriculture	Plant microbiomics	Microorganisms associated with plants, may exacerbate the dissemination of antibiotic resistance via the food chain, direct contact and globalization	[12]
	Plant growth-promoting microbiomics	Microbiome engineering enhance plant growth-promoting and biocontrol activities	[13]
	phylloplane microbiomics	Improve plant growth, stress resilience and health	[14]
	Rhizosphere microbiomics	Rhizosphere microbiome studies have provided tools for manipulating the environment, improving the health of crops and enabling them to reach their maximum genetic potential	[15]
	Soil microbiomics	Soil health conservation under changing climate	[16]
	Terragenomics	Soil Terragenomics (the study of the complete sequencing of all microbial genomes that inhabits the soil environment)	[17]
	Forest microbiomics	Exploring the functional diversity of fungal microbes in their various types of symbionts, decomposers, or saprophyte	[18]
Medicine	Wound microbiome	Precise and rapid identification of wound-associated microbial communities (microbiomes) for fast clinical management treatments and accelerate healing	[19]
	Personalized medicine	Human gut microbiome and xenobiotic degradation	[20]
	Respiratory microbiome	Aberrant change in the airway microbiome promote risk for the development of pulmonary allergic inflammation	[21]
	Air ways allergy and microbiomics	Study the allergic airway dysregulation in the absence of bacterial settlement	[22, 23]
	Skin microbiome	Studying skin microbial diversity for inflammatory diseases prevention and treatment	[24]
	Human probiotics and microbiomics	Next-generation therapeutic bacteria	[25]
	Human microbiomic	Understanding changes in the human microbiome in health and disease	[26]
		Characterization of the gut microbiome and studying their drug resistance genes	[27]
		Analysis of the genetic and phylogenetic diversity of gastrointestinal bacteria by metagenomic approaches	[28]
		Discovery of emerging evidence corelated the gut microbiome to neurological disorders	[29]

(*continued*)

Table 4. (*continued*)

Application field	Microbiomics type	Examples/details	Ref.
	Human milk microbiome	Studying milk microbiome contribution to the healthy metabolism and functioning of the immune system in the infant's intestines	[30]
	Oncogenomics	Bioinformatics and oncogenes functional analysis enhanced diagnosis and treatment	[31]
	Pathogenomics and microbiomics	using meta-genomic approaches in studying host-microbe interactions involved in disease lead in rapid infectious disease diagnosis	[32]
	Cancer microbiomics	Oral cancer, colorectal cancer, head and neck squamous cell carcinomas	[33–35]
Veterinary science	Cow microbiomics	Recovery of more than 900 bacterial and archaeal genomes from cow rumen predicted to have around 69000 genes related to carbohydrate metabolism, shows great potential for biomass-degrading enzymes prospection	[36]
	Fish gut microbiomics	Improve nutrition, growth, reproduction and feeding strategies	[37]
	Animal population genomics and microbiomics	Cattle breeding to increase traits such as protein production, carcass quality, disease resistance, or even feeding efficiency, and genetic resources conservation	[38]
	Animal gut microbiomics	Metagenomic methods by analyzing animal genome sequences have greatly helped in accelerating the discovery of animal diseases and the development of appropriate vaccines for them	[25]
	Feedomics	Improve the quantity, quality, safety and functional properties of food animal products	[39]
Pharmacology	Pharmaceutical Bioinformatics	The use of bioinformatics methods in the analysis of biological and chemical-pharmaceutical processes; Enable to understand the mechanisms of xenobiotics interaction with the human body and accelerate the discovery of smart drugs	[40]
	Pharmacogenomics and metamicrobiomics	Discovering antibiotics through soil metagenomics	[41]
		Malacidin antibiotics discovery via metagenomic approaches	[42]
	Antibiotic resistome	Sequencing analysis of bacterial genomes for antibiotic resistance genes detection (resistome)	[43]
Cosmetics	Human skin microbiome	Studying the role of everyday cosmetics in altering the skin microbiome to develops personalized skin products	[44]
Food	Food quality and safety	Characterization and molecular diagnosis of foodborne bacterial pathogens via microarrays technique	[45, 46]

(*continued*)

Table 4. (*continued*)

Application field	Microbiomics type	Examples/details	Ref.
		Allergy assessment to genetically modified foods	[47]
	Fermented foods improvement	Analyzing the microbial metabolism of desirable food microbes to develop optimal sensory properties via bioinformatics and metabolomics tools	[48]
	Milk microbiomics	Improve starter cultures and cheese making	[49]
	Personalized nutrition	Personalized nutrition by prediction of glycemic responses	[2]
	Foodomics	The application of advanced and integrated omics technologies greatly facilitated food and nutrition studies, which were reflected in improving the health and quality of life of the consumer	[50]
	Food microbial informatics	The use of comprehensive analysis of integrated omics technologies and their databases accelerated predictions and assessment of the desirable and undesirable effects of microorganisms on foods	[51]
Environment/ecology	Microbiome metagenomics	Discovery of novel enzymes and biocatalysts	[52]
	Metagenomic-based bioremediation	Petroleum hydrocarbons treatment	[53]
		Adoption of microbial metagenomic techniques has improved control and biological treatment strategies for waste and pollutants	
		Reducing the harmful impact of pollutants on ecosystems and restoring contaminated sites with metagenomic bioremediation	
	Climate change research	Mitigating climate change by microbiome engineering and synthetic biotechnology	[54–56]
	Marine microbial ecogenomics	Understanding the diversity and ecological processes involving marine archaea, bacteria and their viruses	[57]
	Water microbiomics	Detecting herbicides, pesticides and fertilizers in polluted water and devising ways to dispose of them	[58]
	Eco-microbiomics	Analysis of microbial diversity and their environment through metagenomic approaches	[59]
	Forest microbiomics	Studying forest ecosystems diversity, complexity and dynamics	[60]
	Environmental viromics	Exploration and analysis of viral communities and their diversity in the salty desert and dry valleys of Antarctica	[61]
	Biodiversity informatics	Create and design computerized taxonomic databases	[62]

(*continued*)

Table 4. (*continued*)

Application field	Microbiomics type	Examples/details	Ref.
Energy	Bio ethanol	Potential application of omics tools comprising transcriptomics, genomics, proteomics and metabolomics in bioenergy and ethanol industry	[63]
	Biofuel	Analysis and understanding of symbiotic or cooperating microbial communities in the production of biodiesel and the decomposition of cellulose into sugars and then fermentation into ethanol	[64, 65]
	Bioenergy	Studying microbes that produce methane and hydrogen	
	Biogas	Understanding and analyzing the microbial communities involved in biogas production	[66]
Industrial	Microbial metagenomic	Bioactive compounds and antibiotics exploring and production	[67]
	Industrial enzymes	Amylase, cellulase, lipase, protease, xylanase and other enzymes	
Archaeology	Archaeogenomics and phylogenomics	Studying and exploring the completely sequenced archaeal genomes using comparative genomics analyses and functional annotation	[68]
	Microbial paleogenomics	Analysis of ancient materials of human, animal and microbial origin have provided important insights into the microbiome, resistome, dietary habits and health and disease states of ancient times and civilizations	[69, 70]
	Paleopathogen genomics	Study and analysis of DNA sequences of pathogens and parasites from archived and archaeological remains	[71]

4 Conclusion

Network-based analytical approaches in general and microbial bioinformatics specially become to transform the raw data generated by sequencers to final outputs, furthermore, have proven useful to study systems with complex interactions, such as microbial genomes, metagenomes, and ecomicrobial systems. Development of new techniques in multi-omics approaches a long with sophisticated platforms will open new future directions for the field and facilitate the microbiomic research from diverse perspectives.

Based on the recent advances in sequencing technologies accompanied with latest research findings, we conclude clearly that microbial bioinformatics will become in the 21st century, the main creative and promoting discipline for microbiome research expansion and will provide a deeper understanding for microbiomes structurally and functionally which may fill in the gaps between pure science and practice for solving the medical, agricultural and environmental challenges.

Acknowledgment. Thanks, are due to Jouf University and their Deanship of Library Affairs for providing necessary documentations for the completion of this work. The author also, wishes to thank anonymous referees for their helpful comments during the review process.

References

1. Sousa, S.A., Leitão, J.H., Martins, R.C., Sanches, J.M., Suri, J.S., Giorgetti, A.: Bioinformatics applications in life sciences and technologies. Biomed. Res. Int. **2016**, 3603827 (2016)
2. Zeevi, D., Korem, T., Zmora, N., Halpern, Z., Elinav, E., Segal, E.: Personalized nutrition by prediction of glycemic responses. Cell **163**, 1079–1094 (2015)
3. Xie, K., Guo, L., Bai, Y., Liu, W., Yan, J., Bucher, M.: Microbiomics and plant health: an interdisciplinary and international workshop on the plant microbiome. Mol. Plant **12**(1), 1–3 (2019)
4. Carriço, J.A., Rossi, M., Moran-Gilad, J., Van Domselaar, G., Ramirez, M.: A primer on microbial bioinformatics for nonbioinformaticians. Clin. Microbiol. Infect. **24**(4), 342–349 (2018)
5. Chen, I.M.A., et al.: IMG/M: integrated genome and metagenome comparative data analysis system. Nucleic Acids Res. **45**(D1), D507–D516 (2017)
6. Peng, Y., Leung, H.C., Yiu, S.M., Chin, F.Y.: IDBA-UD: a de novo assembler for single-cell and metagenomic sequencing data with highly uneven depth. Bioinformatics **28**(11), 1420–1428 (2012)
7. Mitchell, A.L., et al.: EBI Metagenomics in 2017: enriching the analysis of microbial communities, from sequence reads to assemblies. Nucleic Acids Res. **46**(D1), D726–D735 (2018)
8. MG-RAST: (n.d.) http://metagenomics.anl.gov/ Accessed 28 March 2018
9. Namiki, T., Hachiya, T., Tanaka, H., Sakakibara, Y.: MetaVelvet: an extension of velvet assembler to de novo metagenome assembly from short sequence reads. Nucleic Acids Res. **40**(20), e155–e155 (2012)
10. Li, D., Liu, C.M., Luo, R., Sadakane, K., Lam, T.W.: MEGAHIT: an ultra-fast single-node solution for large and complex metagenomics assembly via succinct de Bruijn graph. Bioinformatics **31**(10), 1674–1676 (2015)
11. Boisvert, S., Raymond, F., Godzaridis, É., Laviolette, F., Corbeil, J.: Ray meta: scalable de novo metagenome assembly and profiling. Genome Biol. **13**(12), 1–13 (2012)
12. Chen, Q.L., Cui, H.L., Su, J.Q., Penuelas, J., Zhu, Y.G.: Antibiotic resistomes in plant microbiomes. Trends Plant Sci. **24**(6), 530–541 (2019)
13. del Carmen Orozco-Mosqueda, M., del Carmen Rocha-Granados, M., Glick, B.R., Santoyo, G.: Microbiome engineering to improve biocontrol and plant growth-promoting mechanisms. Microbiol. Res. **208**, 25–31 (2018)
14. Compant, S., Samad, A., Faist, H., Sessitsch, A.: A review on the plant microbiome: ecology, functions, and emerging trends in microbial application. J. Adv. Res. **19**, 29–37 (2019)
15. Bakker, P.A., Berendsen, R.L., Doornbos, R.F., Wintermans, P.C., Pieterse, C.M.: The rhizosphere revisited: root microbiomics. Front. Plant Sci. **4**, 165 (2013)
16. Dubey, A., et al.: Soil microbiome: a key player for conservation of soil health under changing climate. Biodivers. Conserv. **28**(8–9), 2405–2429 (2019)
17. Simonet, P.: Terragenome, an international consortium for the sequencing of a reference soil metagenome. In: The Soil Metagenome: Prospects on basic and applied research A workshop to promote the Italian Soil Metagenomic Project(2010)
18. Vogel, T.M., et al.: TerraGenome: a consortium for the sequencing of a soil metagenome. Nat. Rev. Microbiol. **7**(4), 252–252 (2009)

19. Dominati, E., Patterson, M., Mackay, A.: A framework for classifying and quantifying the natural capital and ecosystem services of soils. Ecol. Econ. **69**(9), 1858–1868 (2010)
20. Hodkinson, B.P., Grice, E.A.: Next-generation sequencing: a review of technologies and tools for wound microbiome research. Adv. Wound Care **4**(1), 50–58 (2015)
21. Rampelli, S., et al.: Shotgun metagenomics of gut microbiota in humans with up to extreme longevity and the increasing role of xenobiotic degradation. mSystems (2020). https://doi.org/10.1128/mSystems.00124-20
22. Morris, A., et al.: Comparison of the respiratory microbiome in healthy nonsmokers and smokers. Am. J. Respir. Crit. Care Med. **187**(10), 1067–1075 (2013)
23. Herbst, T., et al.: Dysregulation of allergic airway inflammation in the absence of microbial colonization. Am. J. Respir. Crit. Care Med. **184**(2), 198–205 (2011)
24. Hilty, M., et al.: Disordered microbial communities in asthmatic airways. PLoS One **5**(1), e8578 (2010)
25. Jiménez-Sánchez, G., Philp, J.: Omics and the bioeconomy: applications of genomics hold great potential for a future bio-based economy and sustainable development. EMBO Rep. **16**(1), 17–20 (2015)
26. Rosa, B.A., Hallsworth-Pepin, K., Martin, J., Wollam, A., Mitreva, M.: Genome sequence of Christensenella minuta DSM 22607T. Genome Announce. **5**, e01451-e1516 (2017)
27. Bäckhed, F., et al.: Dynamics and stabilization of the human gut microbiome during the first year of life. Cell Host Microbe **17**(5), 690–703 (2015)
28. van Schaik, W.: The human gut resistome. Philos. Trans. R. Soc. B. Biol. Sci. **370**(1670), 20140087 (2015)
29. Suen, G., et al.: An insect herbivore microbiome with high plant biomass-degrading capacity. PLoS Genet. **6**(9), e1001129 (2010)
30. Griffiths, J.A., Mazmanian, S.K.: Emerging evidence linking the gut microbiome to neurologic disorders. Genome Med. **10**(1), 1–3 (2018)
31. Zivkovic, A.M., German, J.B., Lebrilla, C.B., Mills, D.A.: Human milk glycobiome and its impact on the infant gastrointestinal microbiota. Proc. Natl. Acad. Sci. **108**(Supplement 1), 4653–4658 (2011)
32. Gonzalez-Perez, A., et al.: Computational approaches to identify functional genetic variants in cancer genomes. Nat. Methods **10**(8), 723 (2013)
33. Pallen, M.J., Wren, B.W.: Bacterial pathogenomics. Nature **449**(7164), 835–842 (2007)
34. Gallimidi, A.B., et al.: Periodontal pathogens *Porphyromonas gingivalis* and *Fusobacterium nucleatum* promote tumor progression in an oral-specific chemical carcinogenesis model. Oncotarget **6**(26), 22613–22623 (2015)
35. Vipperla, K., O'Keefe, S.J.: Diet, microbiota, and dysbiosis: a 'recipe' for colorectal cancer. Food Funct. **7**(4), 1731–1740 (2016)
36. Wang, H., et al.: Microbiomic differences in tumor and paired-normal tissue in head and neck squamous cell carcinomas. Genome Med. (2017). https://doi.org/10.1186/s13073-017-0405-5
37. Stewart, R.D., et al.: Assembly of 913 microbial genomes from metagenomic sequencing of the cow rumen. Nat. Commun. (2018). https://doi.org/10.1038/s41467-018-03317-6
38. Ghanbari, M., Kneifel, W., Domig, K.J.: A new view of the fish gut microbiome: advances from next-generation sequencing. Aquaculture **448**, 464–475 (2015)
39. Soetan, K.O., Awosanya, E.A.: Bioinformatics and its application in animal health: a review. Trop. Vet. **33**(1–2), 3–22 (2015)
40. Sun, H.Z.: Feedomics: promises for food security with sustainable food animal production. TrAC, Trends Anal. Chem. **107**, 130–141 (2018)
41. Wikberg, J., et al.: Introduction to pharmaceutical bioinformatics. Oakleaf Academic (2010)
42. Wrighton, K.H.: Discovering antibiotics through soil metagenomics. Nat. Rev. Drug Discovery **17**(4), 241 (2018)

43. Hover, B.M., et al.: Culture-independent discovery of the malacidins as calcium-dependent antibiotics with activity against multidrug-resistant Gram-positive pathogens. Nat. Microbiol. **3**(4), 415–422 (2018)
44. Li, L., et al.: Colistin and amoxicillin combinatorial exposure alters the human intestinal microbiota and antibiotic resistome in the simulated human intestinal microbiota. Sci. Total Environ. **750**, 141415 (2021)
45. Wallen-Russell, C.: The role of every-day cosmetics in altering the skin microbiome: a study using biodiversity. Cosmetics **6**(1), 2 (2018)
46. Fang, H., et al.: An FDA bioinformatics tool for microbial genomics research on molecular characterization of bacterial foodborne pathogens using microarrays. BMC Bioinf. **11**(6), 1–11 (2010)
47. Brul, S., Schuren, F., Montijn, R., Keijser, B.J.F., Van der Spek, H., Oomes, S.J.C.M.: The impact of functional genomics on microbiological food quality and safety. Int. J. Food Microbiol. **112**(3), 195–199 (2006)
48. Mari, A., Scala, E., Palazzo, P., Ridolfi, S., Zennaro, D., Carabella, G.: Bioinformatics applied to allergy: allergen databases, from collecting sequence information to data integration. The Allergome platform as a model. Cell. Immunol. **244**(2), 97–100 (2006)
49. Waidha, K.M., Jabalia, N., Singh, D., Jha, A., Kaur, R.: Bioinformatics approaches in food industry: an overview. In: National Conference on Recent Trends in Biomedical Engineering, Cancer Biology, Bioinformatics and Applied Biotechnology (BECBAB-2015) vol. 1, pp. 1–4 (2015)
50. Tilocca, B., et al.: Milk microbiota: characterization methods and role in cheese production. J. Proteomics **210**, 103534 (2020)
51. He, M., Tan, C.P., Liu, Y., Xu, Y.J.: Foodomics: a new perspective on gut probiotics nutrition and health research. Curr. Opin. Food Sci. **41**, 146–151 (2021)
52. Alkema, W., Boekhorst, J., Wels, M., van Hijum, S.A.: Microbial bioinformatics for food safety and production. Brief. Bioinform. **17**(2), 283–292 (2016)
53. Zhang, H., et al.: dbCAN2: a meta server for automated carbohydrate-active enzyme annotation. Nucleic Acids Res. **46**(W1), W95–W101 (2018)
54. Paul, D., Pandey, G., Pandey, J., Jain, R.K.: Accessing microbial diversity for bioremediation and environmental restoration. Trends Biotechnol. **23**(3), 135–142 (2005)
55. Sinha, S.: Role of bioinformatics in climate change studies. J. Sci **1**, 1–8 (2015)
56. Gupta, C., Prakash, D.G., Gupta, S.: Role of microbes in combating global warming. Int. J. Pharm. Sci. Lett. **4**, 359–363 (2014)
57. Dubey, A., et al.: Soil microbiome: a key player for conservation of soil health under changing climate. Biodivers. Conserv. **28**(8–9), 2405–2429 (2019)
58. Coutinho, F.H., Gregoracci, G.B., Walter, J.M., Thompson, C.C., Thompson, F.L.: Metagenomics sheds light on the ecology of marine microbes and their viruses. Trends Microbiol. **26**(11), 955–965 (2018)
59. McLeod, M.P., Eltis, L.D.: Genomic Insights into the Aerobic Pathways for Degradation of Organic Pollutants, pp. 1–23. Caister Academic Press, Norfolk (2008)
60. Vogel, T.M., et al.: TerraGenome: a consortium for the sequencing of a soil metagenome. Nat. Rev. Microbiol. **7**(4), 252–252 (2009)
61. Baldrian, P.: Forest microbiome: diversity, complexity and dynamics. FEMS Microbiol. Rev. **41**(2), 109–130 (2017)
62. Kerepesi, C., Grolmusz, V.: The "Giant Virus Finder" discovers an abundance of giant viruses in the Antarctic dry valleys. Adv. Virol. **162**(6), 1671–1676 (2017)
63. OECD Megascience Forum Working Group on Biological Informatics: Final Report of the OECD Megascience Forum Working Group on Biological Informatics, January, 1–74 (1999)

64. Verma, S., Pandey, A.K.: Omics tools: approaches for microbiomes analysis to enhance bioenergy production. In: Kashyap, B.K., Solanki, M.K., Kamboj, D.V., Pandey, A.K. (eds.) Waste to Energy: Prospects and Applications, pp. 207–234. Springer, Singapore (2020). https://doi.org/10.1007/978-981-33-4347-4_9
65. Li, L.L., McCorkle, S.R., Monchy, S., Taghavi, S., van der Lelie, D.: Bioprospecting metagenomes: glycosyl hydrolases for converting biomass. Biotechnol. Biofuels **2**(1), 1–11 (2009)
66. de Carvalho, L.M., et al.: Bioinformatics applied to biotechnology: A review towards bioenergy research. Biomass Bioenergy **123**, 195–224 (2019)
67. Gaspari, M., et al.: Microbial dynamics in biogas digesters treating lipid-rich substrates via genome-centric metagenomics. Sci. Total Environ. **778**, 146296 (2021)
68. Dhanjal, D.S., Sharma, D.: Microbial metagenomics for industrial and environmental bioprospecting: the unknown envoy. In: Singh, J., Sharma, D., Kumar, G., Sharma, N.R. (eds.) Microbial Bioprospecting for Sustainable Development, pp. 327–352. Springer, Singapore (2018). https://doi.org/10.1007/978-981-13-0053-0_18
69. Ettema, T.J., de Vos, W.M., van der Oost, J.: Discovering novel biology by in silico archaeology. Nat. Rev. Microbiol. **3**(11), 859–869 (2005)
70. Cano, R.J., Toranzos, G.A., Santiago-Rodriguez, T.M.: Special Issue "Microbial Paleogenomics: The Role of Microbes on Evolutionary Trends and Future Discoveries". A special issue of Genes (ISSN 2073-4425) (2018)
71. Santiago-Rodriguez, T.M., et al.: Gut microbiome and putative resistome of Inca and Italian nobility mummies. Genes **8**(11), 310 (2017)
72. Sereno, D., Dorkeld, F., Akhoundi, M., Perrin, P.: Pathogen species identification from metagenomes in ancient remains: the challenge of identifying human pathogenic species of Trypanosomatidae via bioinformatic tools. Genes **9**(8), 418 (2018)

Virtual Modeling of an Electro-mechanical Powertrain and Steering System with Optical Proximity Sensors for Driverless Ambulance Vehicles in Unity 5

Doru-Laurean Băldean[1](✉), Lavinia Andrei[2], Tudor Oniga[1], Viorel Chindea[1], and Adela-Ioana Borzan[1]

[1] Automotive Engineering and Transportation Department, Faculty of Automotive Engineering, Mechatronics and Mechanics, Technical University of Cluj-Napoca, Muncii 103-105, Cluj-Napoca, Cluj, Romania
{doru.baldean,adela.borzan}@auto.utcluj.ro

[2] Public Health and Management, Faculty of Medicine, University of Medicine and Pharmacology, Victor Babeş, Cluj-Napoca, Cluj, Romania

Abstract. Since automated driving and vehicles are gaining popularity, the optical parking systems, electric power steering (EPS) solutions, as well as the electronically controlled powertrain, represent important technological landmarks. They facilitate driving now by improving the ergonomic features and through their specific function of lowering the physical stress and specific efforts that are required from the driver or user in the real-road operating conditions, such as steering, parking, cornering, forward and reverse maneuvers. Also, electric powertrain and steering systems give the possibility and all basic components for high precision and advanced control of the most important systems (such as steering, braking, propulsion, and suspension), thus improving the infrastructure and making the way to autonomous operation and automated driving capabilities. In these conditions, the present scientific paper shows a virtual modelling process of an electric powertrain and steering system designed with optical proximity sensors for a battery-electric four-wheel-drive smart prototype ambulance as important component in Emergency Medical Unit (EMU) fleets. Specifically, the vehicle control system is working based on a linear actuation model of the steering and acceleration. The ambulance is tested in virtual reality with Unity 5 application. Modeling and testing techniques for this automated ambulance are outlined. Thus, virtual model is tuned through the training sessions. Finally, the performances of the EMU vehicle are tested virtually on the Unity platform with artificial intelligence. The powertrain and propulsion systems performance (especially steering, but also braking and suspension) are measured in terms of the percentage or capability for following the reference track, thus showing important outcomes in the case of the proposed architecture.

Keywords: Ambulance · Automation · Intelligent systems · Powertrain · Steering · Virtual reality

L. Moldovan and A. Gligor (Eds.): Inter-Eng 2021, LNNS 386, pp. 730–745, 2022.
https://doi.org/10.1007/978-3-030-93817-8_65

1 Introduction

Driverless vehicles seem to be both a necessity and utility in the future of transport and for a segment of the road traffic. It is an both a practical idea and an ideal scenario to use as much as it is possible driverless robotic vehicles for specific tasks such as highly predictable traffic and transport tasks. There are multiple companies and developers of automated vehicles that compete in the field. One such company is Alphabet Inc which has indicated the necessity for autonomous or automated driving technologies since 2009. This company came into the playground of automated driving industry with the Google vehicle, known as Self Driving Car Project, and today has its own independent automated vehicle (AV) tech project, popularly indicated as Waymo.

Another player, such as Intel Corporation's project known as Mobileye, made an important step in taking up the challenge and task of making complex experimental tests with the AV in the City of New York. The semiconductor companies are rallying for the AV development and testing. Luminar company develops the specialized lidar technology for automated vehicles, such as passenger cars and cargo trucks. Practically, this company, designs, develops, and finally sells multiple long-range lidar components (Hydra and Iris sensors) and systems for AV. Ambarella, the popular computer vision chip designer, is a different important player in the AV technology. It makes the design and development for high definition (HD), ultra-HD video codding/compression, specific processing of images, and digital solutions for computer vision, offering important support in blind-spots visual detection. It offers components, assemblies and products that are frequently applied in multiple computer vision applications. Latter one includes automated vehicles with autonomous driving capabilities, reconning vehicles, advanced driver assistance systems (ADAS), and specialized robots. In the first part of the year 2021 CV2FS CVflow AI vision processor has been created to be installed on the Arrival Buses and Vans due to its capability to enhance ADAS features, and autonomous driving. These systems provide the necessary artificial intelligence based on neural network processing power, environmental perception, stereovision capability, and HD image acquisition. Aptiv PLC designs and develops electric, and electronic components for active safety systems in automotive industry. The company has recently made a project with Hyundai, known as Motional, to test the AVs' capabilities for monitoring the infrastructure in real-time.

Qualcomm Inc created a platform project, known as Snapdragon Ride, that allows the automotive designers and producers to custom program their AVs for specific tasks and requirements through a programmable framework. MicroVision company has developed a scanning technology with laser beams, light detection and ranging (LiDAR) to create high precision maps of AV surroundings and objects. The minimal elevations of the detected objects and obstacles is at an accuracy around 10 cm. The onboard digital control units use maps of the surrounding environment to navigate and drive the AVs.

General Motors, an important and well-known automaker, supports the AV technologies, through Cruise, a subsidiary company. It is testing AVs for passenger rides in California, with the support from other companies like Microsoft and Walmart. Tesla, NVIDIA and XPeng Motors are also very important players in the field of AVs design and development. Designing, researching, and developing program models for controlling with artificial intelligence the AVs both in virtual reality (VR) and in actual road traffic is still a big challenge, because some of the over-the-air updates may pose as security weaknesses in some cases [1, 2]. Defining the system-on-a-chip electronic control units, the interfaces and the programs used for the mobile platforms and automotive industry, facilitates the digitization and computing process in AV technology [3]. Artificial intelligence applications are to be used both in electric vehicles (EV) and AVs, to improve safety and accessibility [4–7].

The new crowdsourcing program for micro-mobility drive-sharing platforms outlines the challenges and weak spots of the AVs in road traffic and transportation system, operated and exploited based on integrated demand-supply relations. Different scenarios are simulated to gain the proper data [8]. Energetic efficiency and operation safety may be improved through the implementation of the artificial intelligence [9]. To avoid accidents, road traffic events, both AVs, and connected cars, must scan the surrounding environment to properly detect the other road traffic participants and the objects in their proximity [10]. In hazard scenarios, data transmission between different vehicles and road users faces multiple challenges such as the message flooding or contention, and broadcast storm [11].

The Fig. 1 presents some of the scenarios in which the communicating vehicles may reside during the road use and when communicating to each other. The [11] research work showed us that it is possible to create a topological sequence recognition mechanism to generate a connected car-based subgroup. This leads to a recognizable geolocation definition in real time. It facilitates the position characterization in relation to the dynamic topological variation of the vehicles in the subgroup. They applied this mechanism in the Connected Mobile Virtual Fence (CMVF) and have tested its capabilities in real traffic. To avoid road traffic events, both AVs and cars with advanced driving assistant system (ADAS) will be considered safe when relevant data are available (obtained, process and eventually displayed) about vehicles in proximity [11, 12].

Through the appearance of vehicular ad hoc networks (for short known as VANETs) and the so-called Internet of vehicles (IoVs), there is produced a large amount of digital data. These are available both for vehicle users and traffic control systems. The amount of available data is proportionally related to the density of road traffic and participants. It is higher than vehicle's processing and storage capacity. To handle this challenge, VANET is supported by combined cloud computing systems. The later one is also known as vehicular cloud computing (VCC). It manages vehicle's data and offers support to the drivers [13]. Proposing a secure key agreement with authentication protocol to obtain a data message confirmation through VCC the [13] protocol is proved as safe during attacks. Meanwhile it offers privacy to the users and mutual authentication.

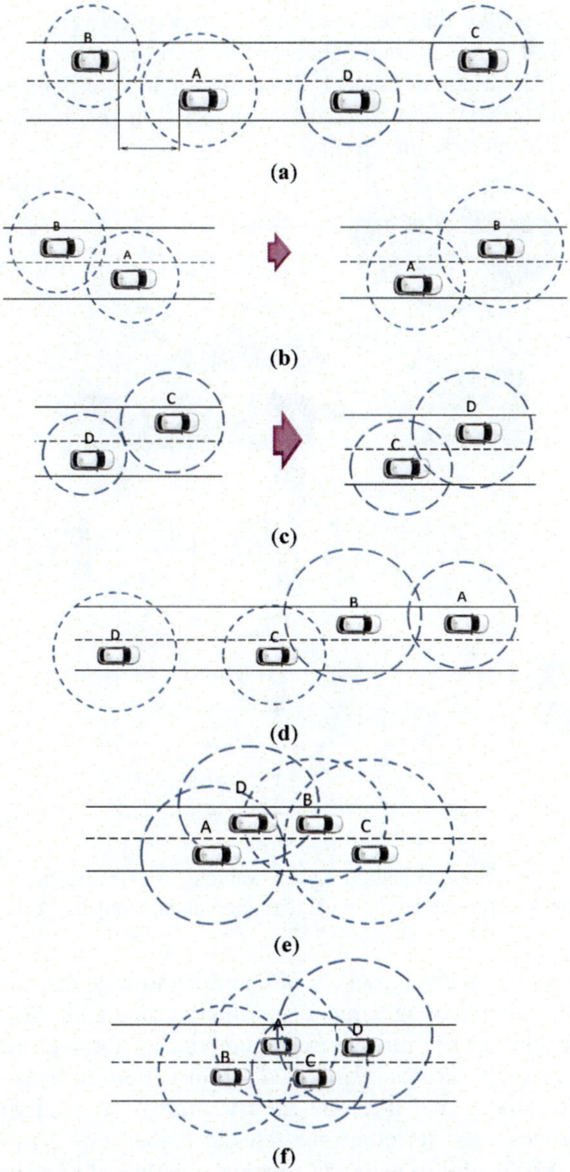

Fig. 1. Basic scenario models for interacting vehicles in road traffic communications system to each other [11]: (a) independent vehicles in road traffic; (b) forming a group and overtaking step; (c) group initiation and overpassing; (d) reformation of the group and overpassing; (e) full connected group and its topologic phase; (f) complex all-to-all overpassing.

In the Fig. 2 is shown the communication protocol between the vehicle and the infrastructure, as well as the internal data transfer for autonomous driving and steering, supported by [13, 14]. Vehicular cloud 1 allows information exchange 2 with the server of the infrastructure 3. Mutual authentication and session key agreement 4, as well as registration 5 of the vehicle to the server 3.

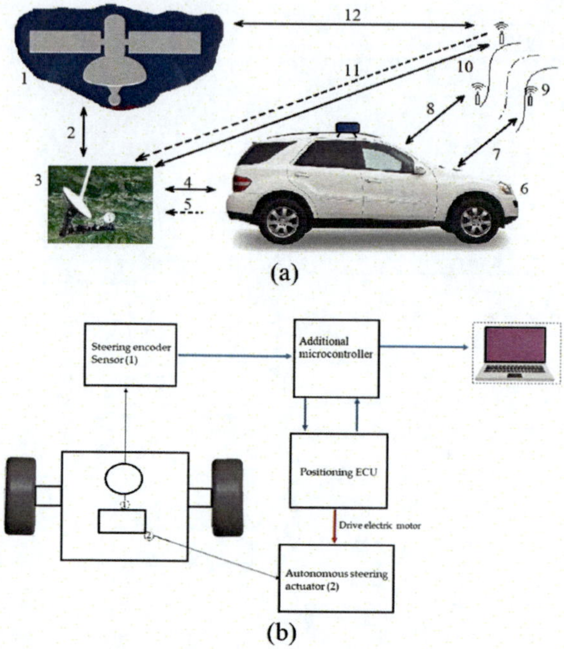

Fig. 2. Schematic of the communication model vehicle to infrastructure and onboard data management: (a) external communication [13]; (b) internal data transfer [14].

Vehicle 6 sends and receives packets of data for message confirmation 7, mutual authentication and session key agreement 8 to the roadside units (RSU) 9. Like these transfers is data exchange of infrastructure components for registration 10, mutual authentication and session key agreement 11. Information exchange 12 takes place between vehicular cloud 1 and the road infrastructure 9. Considering the technical aspects, the challenges with detecting obstacles at higher speeds and long distances may be the greatest difficulty to solve. In relation to road traffic control, most studies support the idea that vehicles should exchange information. General vehicle to vehicle (V2V), vehicle to infrastructure (V2I), vehicle to everything (V2X) cooperation, communication, vehicular ad-hoc networks, and platooning are solutions being studied, with multiple variables [15]. Experimental research of the management system and data exchange should allow the optimization of the electronic control for the vehicles [16]. Integration of the mechanical and electronic components in the design and modelling of mobile robotic applications [17], as well as AVs, is the basis for development of smart cars. Virtual modeling and testing in real road traffic conditions are the way to develop and design electric and smart vehicles [18]. Simulation approach as vehicle-in-the-Loop (VIL) is an innovative way for analyzing ADAS vehicles that is the basis for the safety in the critical testing scenarios. In real-road traffic tests some variables are specific but some of them may be recreated based on virtual environments [19]. By development of the automated systems the mobile tracking, vehicle location and remote accessibility are closer than they appear to be [20]. Lidar technology and cameras installed on the vehicles are two solutions that are considered in the automated driving [21]. Light detection and ranging (LiDAR) system improves the features of monitoring the surrounding traffic and supports the advanced driver assistance systems. With the 360-degrees capability of monitoring the proximity it should improve the quality of safety alerts and their accuracy also. Some of its critics support the idea that Lidar is inferior to the cameras, as well as more expensive.

In Fig. 3 is presented the compared analyze for both systems of proximity monitoring.

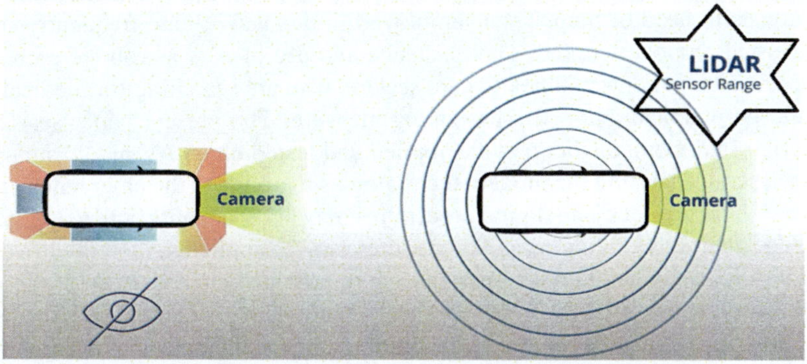

(a) (b)

Fig. 3. Simplified models of proximity safety control: (a) camera-based system with visual image analyze; (b) LiDAR sensor range with 360-degree vision that solves the problem of blind spots.

Virtual scenarios simulations are the proper method for studying the context and performance in given conditions for the considered system or vehicle. Autonomous driving simulators allow the researchers to recreate some of the road traffic and transport conditions to find solutions and solve the problems [22]. The main objective of the present paper is to show the steps and challenges with the virtual modelling of an electric powertrain and steering system for an automated ambulance in virtual simulation and with ADAS in real road traffic. The virtual model of control unit has been designed, studied, and tested in Unity 5 for an automated ambulance. Other objectives are: using the optical proximity sensors; making the SWOT analyze and interpretation of the results.

2 Materials and Methods

The method applied in this study is mainly based upon virtual modeling, simulation, and validation experimental tests. In the first step, there is created a virtual model and after that the practical set up validates the simulation results on a real size model. In the last two years, with Covid pandemic, the robotization, automation, and digital modeling were speed up. Anyway, the obtained results did not satisfy all the proposed expectations. Communications were certainly gaining profit, but multiple other fields have not recorded significant benefits. Technology advancement toward robotization and automation are missing still from the field of emergency rooms and ambulances. Increasing digitization process with "online school" and other "online procedures" in public institutions and companies made an upgrade in software, hardware, and methods of interaction. Although the things were progressing in some parts, other systems and their related activities are less suited for upgrading with highly automatic, robotic, and digital equipment. The emergency rooms and healthcare system in general has been strongly stressed and used during the COVID-19 pandemic crisis, but for example it was not profiting fully from the existence of the AVs. Ambulances, during COVID-19 crisis, have been stressed and challenged by emergency events and hazardous scenarios which haven't been managed always as expected. The ambulances and emergency personnel must be very responsive and act rapidly to save lives and protect the health of individuals in need of help. Those employed in this activity are frequently stressed out, exhausted and overworked. The existence and use of AVs as robotic ambulances to transport some patients and health care seekers who are searching for medical attention could be practical and a life-protecting measure. The present paper searches all the materials and methods that may be applied and tested on an AV model suitable for ambulance services. Using the modern technologies to optimize the emergency services with AVs, robotic devices, and digital procedures may bring benefits both to the staff and society. The alternative solutions and procedures may be applied to increase life quality, to mitigate stresses in health risk situations and to provide better services. If there is no need for complex resuscitation procedures on the patient the AV ambulance seems to be a proper solution. If there is required transportation between home and emergency unit, or between different health care units, the use of AVs would be the proper solution to ease up the process. These additional solutions are somehow adequate and realistic now because they lead to an improvement of emergency units. The powertrain is one of the few most important systems of a mobile robot or AV. Its operation must be either defined through calculation or measured experimentally and practically to be considered when starting the main stages of the AV virtual modeling and programming process. This is so because of the wide variation in the kinematic and dynamic parameters defining the operation. The appropriate methodology for the research, design, and development of a robotic ambulance (as AV) is the virtual modeling and simulation, followed by the experimental test and optimization according to the real-world scenarios. Unity 5 is the proper application to be used for creating the virtual model and to simulate the AV. Different scenarios are studied to gain the realistic data relative to the kinematics and dynamics of AV in virtual and actual conditions.

Equipment used in the research is the Unity 5 software and a real vehicle platform with ADAS for testing and inspecting capabilities of the optical proximity sensors, if suited for automated driving. Technical data brief overview is shown in Table 1.

Table 1. Technical specifications regarding the AVs used for the present study.

Applied protocol	Research 1	Research 2
Method	VR simulation	ADAS test
Vehicle model	Digital	Standard
Destination	Ambulance	Emergency services
Detection system	Combo (LiDAR + cameras)	Conventional OPS – optical proximity sensors
Transmission	Electromechanical	Mostly mechanical with electronic controls[a]

[a] Supporting advanced driver-assistance systems features.

3 Virtual Modeling and Results

Results of the virtual modeling both in CAD apps and in Unity 5 environment are compared with the actual data. Important findings of the virtual modeling and research through simulation, followed by testing, outline some kinematic and dynamic limitations for the studied AV model. Safety operation has been reached at speed ranging between 20–45 km/h (5.5–12.5 m/s). Data available regarding tires status, road surface friction coefficient variation in the virtual modeling is still needed to be defined in higher level of detail.

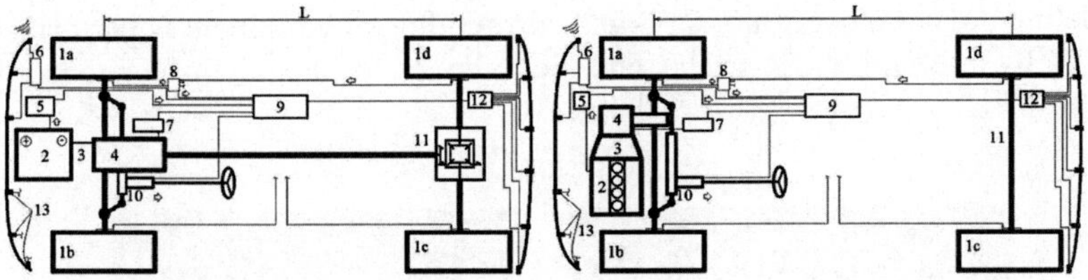

(a) (b)

Fig. 4. Simplified CAD models of vehicles with electro-mechanical powertrain and OPS for study: (a) virtual model used in simulation; 1-wheels, 2-batery, 3-electric connections, 4-electric engine, 5-battery control unit, 6-front OPS hub, 7-engine control unit, 8-antilock braking system unit, 9-onboard central processing unit, 10-steering unit, 11-mechanical transmission, 12-rear OPS hub, 13-OPS electrical connectors; (b) schematic of the physical powertrain with OPS capabilities; 1-wheels, 2-engine, 3-clutch, 4-gerbox, 5-engine control unit, 6-front OPS hub, 7-transmission control unit, 8-antilock braking system unit, 9-onboard central processing unit, 10-steering unit, 11-rear axle, 12-rear OPS hub, 13-OPS electrical connectors.

Unity 5 environment, known as Virtual Reality platform, offers the programing possibility with artificial intelligence to control the AV model. Figure 4 shows the simplified schematics of CAD models for the vehicles equipped with OPS and used in the present research.

Robotic ambulance virtual model in Unity 5 has advanced automated driving capabilities supported by AI. Selecting and defining the equipment and materials in the Unity program is shown in Fig. 5.

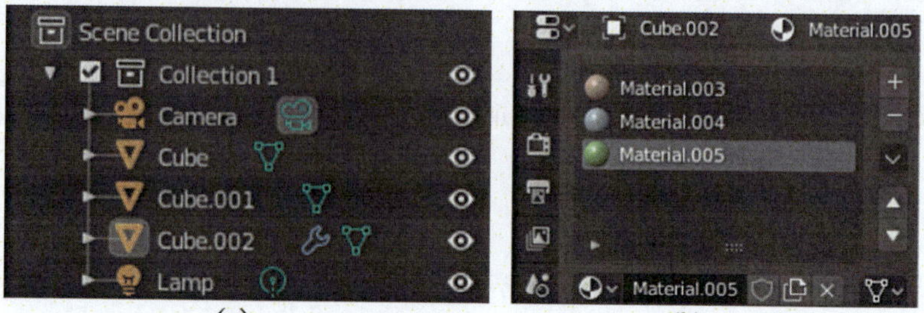

(a) (b)

Fig. 5. Unity 5 panels for scene collection and material definition regarding the virtual model of the AV: (a) equipment and scene collection menu; (b) materials definition board in VR application.

The actual car model than facilitates the validation of the program, with its advanced driver assistance system (ADAS). In the first place, it is quite important to consider the fact that, nowadays, autonomous, AVs, or self-driven cars do not appear for sale in the market to everyday customers. Platforms like those popular cars made at Tesla factory, or the Super Cruise-equipped Cadillacs are featuring the capability to ride and use the hands-free technology for limited (but significant) time intervals. They do this for a precise and limited set of conditions (meaning highway and interstates).

Modeling the car powertrain, OPS and surrounding environment is performed with in Unity 5, by applying those scripts and menu boxes for data definition, according to Fig. 6.

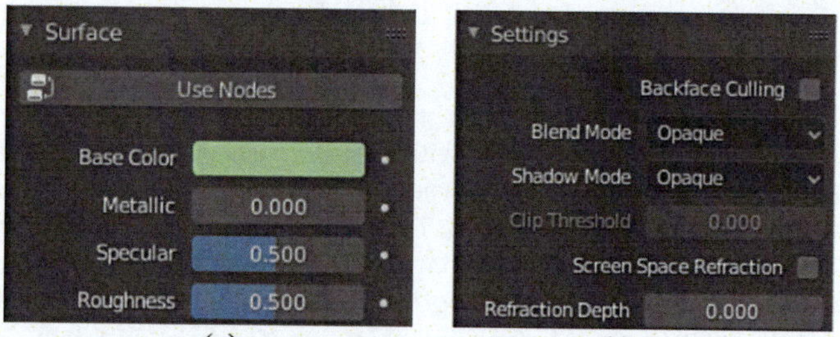

(a) (b)

Fig. 6. Basic software features for defining the surfaces with Blender script for Unity 5 program: (a) Surface virtual modeling editor; (b) Setting menu for Blender application in VR environment.

To model the virtual driverless ambulance (as an AV) in Unity 5 is necessary to apply the general and specific editing features, shown in Fig. 7.

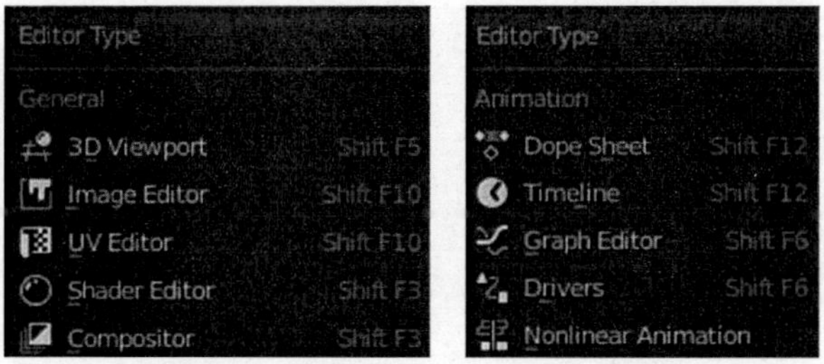

(a) (b)

Fig. 7. Editing features for compositor, shader, images, and animation in Unity 5 digital platform: (a) General features in Editor type section; (b) Animation features menu useful to run the VR model.

Other specific editing features regarding scripting and data, are shown in Fig. 8.

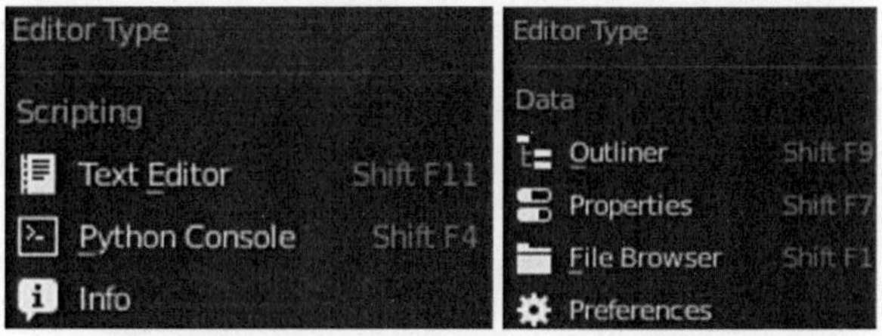

(a) (b)

Fig. 8. Editor type in Unity 5 application for scripting and data: (a) Scripting for the following options: Text Editor; Python Console; Other Info; (b) Data menu for the following: Preferences, File Browser; Properties; Outliner.

Results obtained through virtual modeling are based on artificial intelligence protocol which is using vehicle-to-everything and context awareness data systems. Automated robotic vehicle is thus generated and tested in Unity 5. Object perception and situation awareness capability are important requirements to perform avoidance maneuvers around the obstacles. The AV with OPS is designed to drive around the objects for completing the transport task. Dynamic parameters such as the following: resistance air force (Raf), gravitational force (Fg), road vertical reactions to weight force on each axle of the vehicle (front axle reaction = Rf, rear axle reaction = Rr) and vehicle speed (as kinematic parameter) are most important to be considered in virtual modeling, along with engine's output performances (such as torque and power). In Fig. 9 are presented the AV's resulted forces.

The developed tests are consisting in straight road segment and follow the track, sine-sweep steering maneuver, and steering maneuvers created by the AI program which controls the driverless AV dynamics. The performed sine-sweep procedures are followed with both virtual model and physical vehicle. It aims to the system characterization and dynamic behavior definition. Anyway, the AV's tracking performance is evaluated when

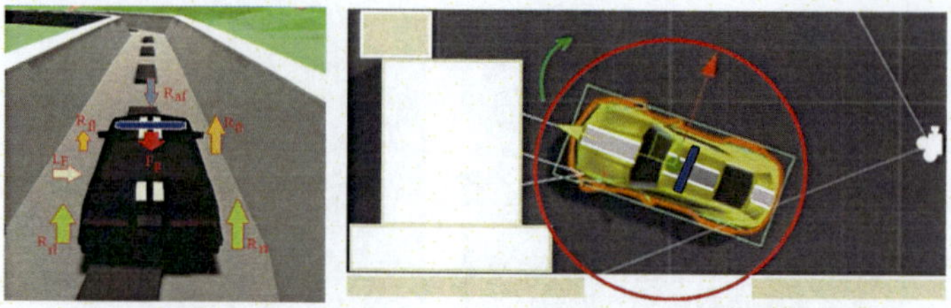

(a) (b)

Fig. 9. AV robotic ambulance model in Unity 5, used as Virtual Reality drive-test program: (a) dynamic parameters definition as forces; (b) Torque and longitudinal velocity in object's proximity.

the reference road profile generated by the program for the autonomous task is also offered. It is important to consider the integration of new technologies, such as Connected Mobile Virtual Fence. If AVs are equipped with AEB (Autonomous Emergency Braking system), the vehicle will almost instantly and automatically operate the emergency brakes. This is done when road traffic is moving around the AV, when they overtake and intervene in car's trajectory. Virtual Fence Technology makes the idea of continuous communication possible to avoid events and accidents. It is the basis for mutual communication of AVs to increase their awareness or perception factor. Vehicle-to-vehicle (V2V), vehicle-to-infrastructure (V2I), and vehicle-to-everything (V2X) protocols must be implemented to provide the connected car feature. When AV do participate in the road traffic, the volume of information exchange needed and the frequency at which it must be acquired and processed is tremendous. To have a decision-making system similar in complexity with a human brain, the AV must be provided with a real-time image of the context around them. This is quite accurate in road traffic urban environments, where human drivers encounter and perceive in road traffic other individuals, different animals, as well as a multitude of vehicles in a small timeframe. Vehicle-to-vehicle communication is the basis for the creating the mutual awareness system. The performance of the studied AV with electromechanical powertrain and steering system was investigated based on the driving profile based on an autonomous tracking mission. The AV is travelling at a predefined longitudinal speed of 10 m/s while the steering wheel rotation is modified by the electronically controlled actuator (2) for a time of 60 s (see Fig. 2b). In automated driving scenario the steering angle value is determined through the encoder sensor (10) provided in the virtual model (see Fig. 4). The present research is based on multiple protocols which allow the optimization of the Connected Mobile Virtual Fence (CMVF) system using LiDAR, smartphones, cameras and/or the connected vehicle technology. A connected car is the one that is using wireless networks to communicate with nearby devices. Connected cars are a big step forward toward the IoT. The AI program distinguishes the direction to which vehicles are traveling and allows individual cars from the connected group to communicate and mutually recognize each other's position. It gives a precise topological sequence. That is one very important task, and when the preceding car gets closer to an event, the accident may be avoided by warning the following traffic in advance about the hazardous situation of the preceding vehicles. As vehicles travel in the other direction, they don't need to read data of those driving forward, thus the risk of message contention or congestion to send and receive false emergency encoded messages may appear rarely. Unity 5 is useful both for environment development and for testing whole vehicle as shown in Fig. 10.

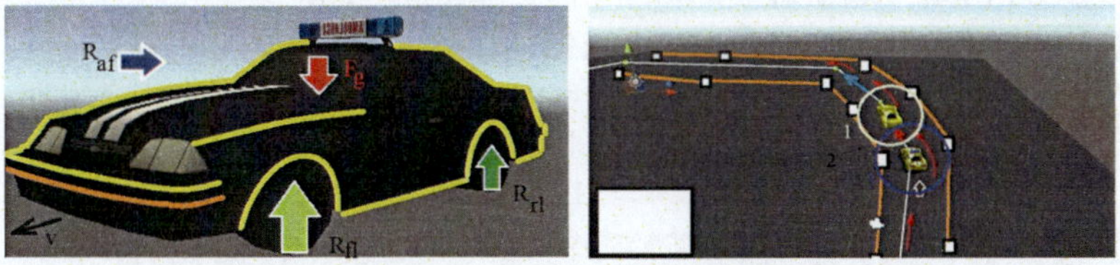

(a) (b)

Fig. 10. Virtual model of a driverless ambulance vehicle in Unity 5: (a) driverless ambulance; (b) Connected Mobile Virtual Fence and tracking in Unity 5 software application.

The "Fitness" is determined based on the distance traveled by the vehicle, the average speed at which the car travels, as well as the "Sensor Multiplier" value that considers how close the vehicle is to the road's margins during driving. The distance traveled by the car is in fact also the decision-making component in the calculation of "fitness". This component of calculus has an important role in determining the "fitness" value. Other factors are also important, namely" Avg Speed Multiplier" and "Sensor Multiplier", as shown in Fig. 11. They are referring to the average speed of the vehicle and the position of the vehicle on the road. They are considered, but the role played is less important compared to the distance traveled by the car. When lacking one of these components which allows the assessment of how the vehicle travels the route, it is quite hard to arrange them in such manner that the AI of AVs may learn. Context Awareness Mechanism based upon Mobile Virtual Fence System helps the artificial intelligence program to control driving actions, with secure maneuvers and direction change, to avoid the physical objects.

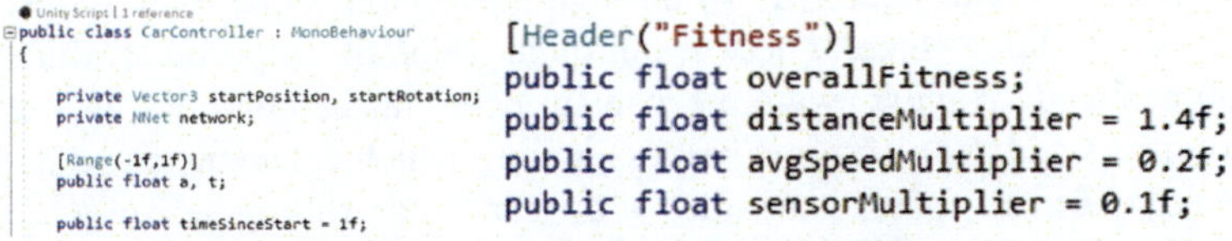

(a) (b)

Fig. 11. Unity code script: (a) The code which makes possible the calculation of the "Fitness"; (b) "Fitness" component code presentation.

Using the Mobile Virtual Fence System "(MVFS) in simulation and in real vehicle (with ADAS) is very important for automated driving process optimization. MVFS is generated via OPS. Other technologies may be also used (LiDAR, smartphones, cameras etc.). Each obstacle is triggering the OPS and subsequently leads to signals and actions related to context aware mechanism. To avoid accidents, AVs or conventional vehicles equipped with advanced driving assistant systems can be labeled as safe if specific data can be obtained on the scenario of other object in the proximity. Real world traffic is the ultimate basis for the detection of proximity obstacles and for vehicle mutual awareness in Connected Mobile Networks through contactless and wireless information exchange.

With Context Awareness Mechanism (CAM), supported by OPS, the real vehicle in obstacle approaching situation is signaling and acts (visually, acoustically and emergency braking if needed) if the probability of an accident is increasing, as shown in Fig. 12.

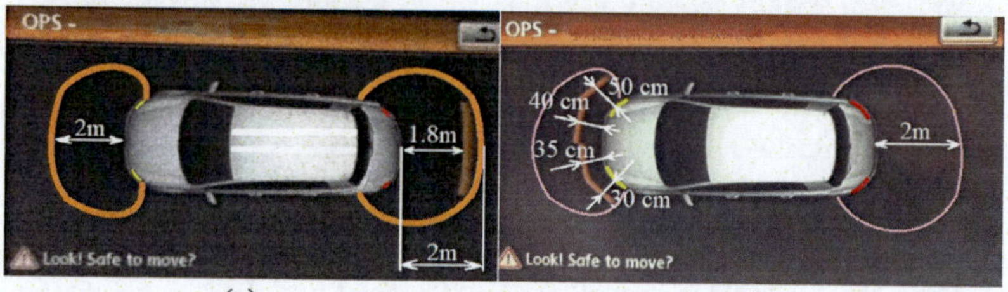

(a) (b)

Fig. 12. Real vehicle is equipped with OPS both in front and rear side to detect physical objects and avoid events: (a) with ADAS field of awareness is 2 m long and 2 m wide; (b) Different states of contactless perception of the obstacles in the front of the vehicle.

Virtual modeling of an electro-mechanical powertrain and steering system with optical proximity sensors for a driverless AV, designed to operate as an ambulance, with the artificial intelligence support, both in the Virtual Reality and for real road traffic, encounters problems and challenges, which are presented in Table 2. OPS support the design and development of partial or complete virtual fence. The later one signals the presence of objects in the proximity of the vehicle. Wireless capabilities and/or hot spot option on smartphones facilitates the implementation of a mobile virtual fence.

This one may be used to exchange information with the surrounding environment. Using the mobile virtual fence (MVF) through smartphones one may almost directly join or initiate a mobile vehicular network through which the important traffic data may be sent and received. Context aware mechanism based instead on OPS, is made by the proximity sensors and the electronic processing unit which calculates the data received simultaneously from all channels or sides of the car, is controlling behavior of the vehicle in hazardous conditions. Evaluation and centralizing of the strengths, weaknesses, opportunities, and threats highlights a package of the most important observations and hotspots, as shown in Table 2.

Table 2. Observations of AVs' Strengths, Weaknesses, Opportunities, and Threats (SWOT) research.

1 Strengths	2 Weaknesses	3 Opportunities	4 Threats
Stored data	Complex program	More programing	Man-in-the-middle attack
Self-driven	Limited perception	Increased productivity	Tracking via open networks
IoT (Internet of things)	Malicious attacks[a]	Precise AV service	Masquerade as rightful user[a]

[a]Malicious attack consist in one adversary entity that is capable of masquerading as a legitimate user and thus tricking the authority entities for accessing and wrongful use the resources; Anyway, malicious adversaries may steal or take up a legitimate user's account or device and make side-channel intrusions and attacks to take key data stored in the robotic ambulance; A particular adversary party may take an server (authority entity)'s secret keys. Thus, the attacker may compute other previous or alternate session keys to trick rightful users and/or authority entities [13].

Writing the AI program based on Bayes Theorems is using the probability factors. The first or primary probability is considered by the AI and defined using a training group as it is shown below:

$$Q_I(x|w) = [Q_I(d|w) \cdot Q_I(x)]/Q_I(w) = \left[Q_I(x) \cdot \Pi_{i-1}^{N} \cdot Q_I(w_i|x)\right]/Q_I(w), \quad (1)$$

where $Q_I(x)$ is the first probability; $Q_I(w)$ – final state probability; $Q_I(w|x)$ – probability of training group.

4 Conclusions

Designing, virtual modeling, testing, and experimenting with digital control programs in Unity 5 for AVs and ADAS equipped cars has been a step forward and supported us to understand the challenges which are important to be studied in further research, both through simulations and road traffic measurements. Testing in road like scenarios on actual vehicles is still a challenge because there are no series produced AVs yet for daily customers. There are only concept cars and limited series models with different levels of automation. The conventional advanced driver assistance systems, on the other hand, are quite useful for the vehicle's kinematic and dynamic control in hazardous situations, but still needs the presence and actions of the human driver.

Some of the significant and challenging topics that were studied in the present work are as follows: virtual modeling of an electro-mechanical powertrain; designing the power steering system on the same virtual model; indicating the architecture of the optical proximity sensors installed on the passenger car with ADAS. Primary phase of research consists in simulations with Unity 5 program and then comes the second step which makes practical measurements of distances between the vehicle and the surrounding objects with ADAS and OPS in real road traffic. To avoid traffic events and crashes, ADAS that are nowadays installed on conventional cars are a step forward but neither perfect systems nor complete for the automated driving scenarios. They are lacking both the feature of connected car technology and self-driving capabilities.

References

1. Andrei, L., Baldean, D.-L., Borzan, A.: Designing an artificial intelligence control program model to be tested and implemented in virtual reality for automated chevrolet camaro. Proceedings **63**(1), 44 (2020). https://doi.org/10.3390/proceedings2020063044
2. Bec, P., et al.: Study of vulnerabilities in designing and using automated vehicles based on SWOT method for chevrolet camaro. In: Grebenisan, G. (ed.) The IOP Conference Series: Materials Science and Engineering, Proceedings of the Annual Session of Scientific Papers "IMT ORADEA 2020", Oradea, Felix SPA, Romania, 28–29 May 2020, pp 1–6. IOP Publishing, Philadelphia, PA 19106, USA, Accessed 06 June 2020
3. Borzan, A.I., et al.: The development of a new interface for intelligent control of energy supply in dynamic environment with process digitization. Procedia Manuf. **46**, 914–921 (2020)
4. Covaciu, F., Băldean, D.-L.: Contribution to research the applied engineering protocol to implement a fuzzy regulator for autonomous driving of an automotive model implemented in virtual reality. In: Dumitru, I., Covaciu, D., Racila, L., Rosca, A. (eds.) SMAT 2019, pp. 468–476. Springer, Cham (2020). https://doi.org/10.1007/978-3-030-32564-0_54
5. Covaciu, F.A., et al.: Developing the communication of autonomous vehicles controlled with the aid of artificial intelligence for person and capital safety. In: Nechita Iancu, E.A. (ed.) Safety of the person and building the social capital, vol. 1, 1st edn., pp. 478–84. Universul Juridic, Arad, Romania (2020)
6. Covaciu, F., Bec, P., Băldean, D.-L.: Design and development of a low-cost automated all-terrain intelligent robotic vehicle for detection to study its faults and vulnerabilities from SWOT perspective. Proceedings **63**(1), 38 (2020). https://doi.org/10.3390/proceedings2020063038
7. Covaciu, F.A., et al.: Design and control of a mobile robot of exploration, using virtual reality as a simulation environment. S I, **34**(43), eISSN 2359 – 828X (2018).http://stiintasiinginerie.ro/34-43
8. Elhenawy, M., et al.: A novel crowdsourcing model for micro-mobility ride-sharing systems. Sensors **21**, 4636 (2021). https://doi.org/10.3390/s21144636
9. Ferenti, I., et al.: Artificial intelligence implemented in rally vehicles for increasing energetic efficiency in competitions. S I, **34**(18), eISSN 2359 – 828X (2018).http://stiintasiinginerie.ro/34-18
10. Borzan, A.I., et al.: Application of advanced engineering methods in studying a road traffic event between a 12-wheeler truck and a small tourism in a local junction from Cluj-Napoca. J. Automot. Eng. **24**, 97–106 (2018)
11. Eom, Y.H., et al.: Topological sequence recognition mechanism of dynamic connected cars using the connected mobile virtual fence (CMVF) system for the connected car technology. Appl. Sci. **10**(12), 4347 (2020). https://doi.org/10.3390/app10124347
12. Jovrea, S., et al.: Researching on-board display of essential information concerning technical conditions in operation and fuel-economy of a motor-vehicle in operation. S I, **31**(67), eISSN 2359 – 828X (2017). http://stiintasiinginerie.ro/31-67
13. Lee, J.Y., et al.: Secure key agreement and authentication protocol for message confirmation in vehicular cloud computing. Appl. Sci. **10**(18), 6268 (2020). https://doi.org/10.3390/app10186268

14. Manca, R., et al.: Performance assessment of an electric power steering system for driverless formula student vehicles. Actuators **10**, 165 (2021). https://doi.org/10.3390/act10070165
15. Martinez-Diaz, M., et al.: Autonomous vehicles: theoretical and practical challenges. Transp. Res. Procedia **33**, 272–282 (2018). https://doi.org/10.1016/j.trpro.2018.10.103
16. Moldovan, A., et al.: Experimental research of the management system from the Peugeot 4007 Sport Utility Vehicle. Sci. Eng. **31**(67), eISSN 2359 – 828X (2018).http://stiintasiinginerie.ro/31-71
17. Ollero, A., et al.: Integrated mechanical design and modelling of a new mobile robot. In: IFAC Intelligent Components and Instruments for Control Apps, pp 461–466. Malaga, Spain (1992)
18. Pappalardo, C.M., Lombardi, N., Dašić, P.V., Guida, D.: Design and development of a virtual model of an electric vehicle of category L7. IOP Conf. Ser.: Mater. Sci. Eng. **568**(1), 012114 (2019). https://doi.org/10.1088/1757-899X/568/1/012114
19. Park, C., Chung, S., Lee, H.: Vehicle-in-the-loop in global coordinates for advanced driver assistance system. Appl. Sci. **10**(8), 2645 (2020). https://doi.org/10.3390/app10082645
20. Popescu, D., Borzan, A., Băldean, D.-L.: Development of an automated system for fuel tank level checking and machinery location management to optimize remote accessibility and mobile tracking. Proceedings **63**(1), 17 (2020). https://doi.org/10.3390/proceedings2020063017
21. Teague, C.: What Lidar Is and Why It's Important for Autonomous Vehicles. Autoweek, **4**, 27 (2021). https://www.autoweek.com/news/a36190274/what-lidar-is/
22. Wen, M., et al.: Virtual scenario simulation and modeling framework in autonomous driving simulators. Electronics **10**, 694 (2021). https://doi.org/10.3390/electronics10060694

Check for updates

A Fuzzy Logic-Based LabVIEW Implementation Aimed for the Detection of the Eye-Blinking Strength Used as a Control Signal in a Brain-Computer Interface Application

Oana Andreea Rușanu(✉)

Product Design, Mechatronics and Environment Department, Transilvania University of Brașov, Brașov, Romania
oana.rusanu@unitbv.ro

Abstract. The paper proposes a Fuzzy Logic-based LabVIEW application to determine the strength of the voluntary eye-blinks used as commands in a brain-computer interface to control a mobile robot. Relevant statistical features (standard deviation, root mean square, Kurtosis coefficient, and maximum value of amplitude) of the raw electroencephalographic signal acquired from the biosensor provided by a portable NeuroSky headset determine the input linguistic variables. A customized counting algorithm of the voluntary eye-blinks generates the various movement commands (move forward, move backward, turn left, turn right, stop). By implementing LabVIEW graphical custom code sequences, it resulted in developing this algorithm. The Bluetooth-based communication between the LabVIEW application and Arduino allows the sending of commands to the mobile robot. The proposed BCI experimental system is necessary to provide an efficient working principle employed by robust mechatronic systems that support people with neuromotor disabilities to regain their confidence and independence in performing simple everyday activities.

Keywords: Brain-computer interface · Fuzzy logic · EEG headset · LabVIEW

1 Introduction

The brain-computer interface (BCI) is a multidisciplinary application involving broad knowledge and advanced technical abilities from different research areas: biomedical engineering, mechatronics, computer science, neuroscience, and psychology. The most efficient non-invasive BCI systems involve electroencephalographic (EEG) signals, which can be acquired even with portable commercial headsets.

Processing methods and artificial intelligence techniques are applying to the EEG signal to enable the detection of particular patterns associated with the task executed by the user, as a command, for example: focusing the attention on something [1], keeping a relaxing state of mind [1], executing the voluntary eye-blinks, counting the number of times a specific element is flashing, eliciting P300 evoked potential [2] or imagining something, especially a specific movement [3], triggering slow cortical potentials

L. Moldovan and A. Gligor (Eds.): Inter-Eng 2021, LNNS 386, pp. 746–756, 2022.
https://doi.org/10.1007/978-3-030-93817-8_66

[4]. The accomplishment of the previously mentioned tasks will determine the real-time control of mechatronic devices, such as a robotic arm [5], a robotic hand [6], an automated wheelchair [7], necessary in the everyday assistance of disabled people, who suffered from neuromotor illnesses, for example, locked-in syndrome, amyotrophic lateral sclerosis, cerebral stroke, spinal cord injuries, and tetraplegia.

The voluntary eye-blinking is an artifact across the EEG signal. The eye-blink pattern is easily recognizable and precisely identifiable among the other variations of the EEG biopotentials. A voluntary eye-blink lasts for 100–400 ms and is frequently chosen as a control signal in BCI applications [8] when neuromotor-disabled patients usually can execute such a simple gesture. The researchers used either the thresholding approach or the raw EEG data analysis to detect voluntary eye-blinks, according to the previous work. Tran D.-K et al. [9] implemented and tested an offline thresholding-based algorithm for intentional eye-blinks detection by identifying peaks across the raw EEG signal acquired from the Emotiv headset. Prem S. et al. [10] presented preliminary stage research using thresholding values measured by the voluntary eye-blinks and the attention level acquired from a NeuroSky headset connected to a smartphone. Miranda M. et al. [11] performed the analysis of the EEG signal acquired from the NeuroSky headset by designing an ad-hoc mother wavelet to precisely detect the temporal location and measure the duration of each eye blinking occurrence also captured with a camera.

The current paper presents a Fuzzy Logic-based LabVIEW application to determine the strength of the voluntary eye-blinks used as BCI commands to control a mobile robot. The paper novelty consists of the processing methods applied to the raw EEG signal to identify the most relevant features used in the configuration of a Fuzzy Logic System that can run in real-time and classify the voluntary eye-blinks used as movement commands sent to Arduino based mobile robot. The raw EEG signal processing for detecting the eye-blinking pattern has benefited from getting a quicker response and the possibility of identifying other significant EEG patterns used simultaneously. Across scientific literature, it is tough to identify explicit evidence of applying fuzzy logic methods on features determined by EEG signals to recognize the voluntary eye-blinks. This paper reveals a novel approach to analyzing the raw EEG signal to measure the eye-blink strength efficiently. In contrast to using the predefined LabVIEW function aimed at accessing the functionality of the ThinkGear chip, the proposed Fuzzy Logic-based LabVIEW application removes the time delays or any other interruption. Therefore, other mental processes (for example, attention or meditation) get measured in parallel with the detection of the voluntary eye-blink.

The structure of the paper is as follows: Sect. 2 describes the hardware system of the proposed BCI solution, Sect. 3 comprises information about the implemented software application, Sect. 4 includes the obtained results and related discussions, and Sect. 5 focuses on conclusion and future research directions.

2 Hardware System – Arduino Based Mobile Robot Controlled by NeuroSky Mindwave Mobile Headset

The brain-computer interface application involved designing a mobile robot (Fig. 1) based on a chassis connecting two wheels and two DC motors to an Arduino Uno board using an L298N motor driver. The mobile robot can change its movement directions according to the received commands based on the voluntary eye-blinks deciphered from the raw EEG signal acquired from the biosensor embedded by a portable NeuroSky headset. NeuroSky is paired by Bluetooth protocol to the computer running a LabVIEW application, receiving the EEG signal, and sending commands to the Arduino Uno board. An HC-06 Bluetooth device enables the communication between the Arduino and the computer.

Fig. 1. The hardware system (NeuroSky headset and mobile robot) of the BCI application.

The NeuroSky Mindwave Mobile headset is one of the most popular portable monitoring devices used in BCI research. It provides the developers a toolkit of the pre-defined functions enabling the reading of specific values characterizing: the meditation level, the attention degree, and the eye-blinking strength. Moreover, the samples of the raw EEG signal can be acquired and stored in data structures for further processing of the extraction of different EEG frequencies: delta, theta, low alpha, high alpha, low beta, high beta, low gamma, and high gamma. One of the most convenient ways of accessing the full functionality provided by the ThinkGear chip of the NeuroSky portable headset is the use of the toolkit offered by NI LabVIEW [12].

3 Software System – LabVIEW Application and Arduino IDE

The software system consists of two programming environments, LabVIEW used to acquire and analyze the EEG signal, and Arduino IDE used to control the mobile robot.

3.1 The LabVIEW Based EEG Signal Processing

Achieving the communication between the NeuroSky headset and LabVIEW application (Fig. 2) resulted in using the NI driver [9], which enabled both basic and advanced functionality of the ThinkGear embedded chip. The principal used function/virtual instrument is ThinkGear Read, and it provides the benefit of acquiring/reading an array of 512 samples of the raw EEG signal during a time interval of 1 s. Further processing methods targeted the raw EEG signal. For example, a Fast Fourier Transform (FFT) function resulted in getting the frequency domain of the raw EEG signal. Likewise, different types of digital filters (high pass, low pass, bandpass) were necessary to extract certain frequency rhythms, for example, delta (0.5–4 Hz), theta (4–8 Hz), alpha (8–12 Hz), beta (12–30 Hz) and gamma (>50 Hz). Thus, it has resulted in a series of arrays, each of them containing 512 samples of various types of EEG signals, both in the time and frequency domain. Ten statistical features (mean, standard deviation, sum, range – the difference between the maximum and the minimum value of amplitude, median, Kurtosis coefficient, skewness, root mean square – RMS, mode, and maximum value of the amplitude) got computed across the 512 samples of each array defining a specific type of signal. As previously mentioned, 1 s of acquisition is equivalent to getting 512 samples. An experiment was performed, consisting of two stages, each involving a time interval of EEG data acquisition equal to 30 s.

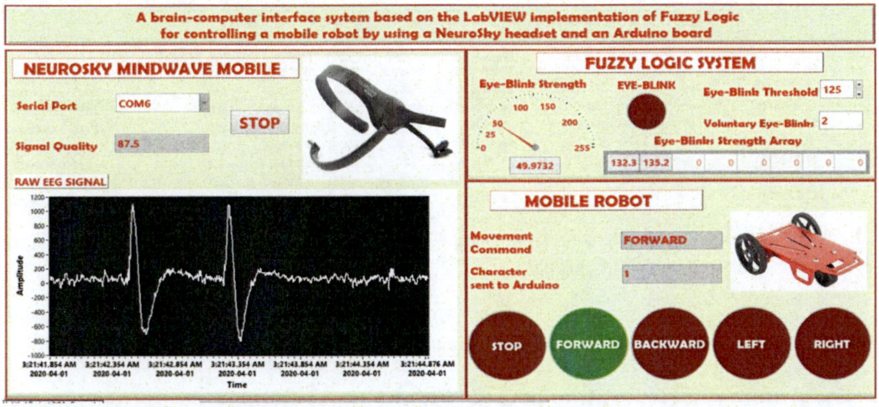

Fig. 2. The LabVIEW GUI of the BCI application based on Fuzzy Logic System.

During the first stage (meaning the first 30 s), the user had to execute voluntary eye-blinks, according to an audio signal (a beep) transmitted at each time interval of one second. Therefore, 30 sequences have resulted, each consisting of a series of 512 samples of different types of the EEG signal (raw, delta, theta, alpha, beta, gamma), both in time and frequency domain, showing the pattern of a voluntary eye-blink. Ten statistical features got computed for each of the 30 sequences: mean, standard deviation, sum, range – the difference between the maximum and minimum value of amplitude, median, Kurtosis coefficient, skewness, root mean square, mode, and maximum value of amplitude. Each of those 30 sequences was also graphically displayed so that the pattern of the voluntary eye-blink got observed. Thus, irrelevant sequences got removed if they could not offer a correct representation of the eye-blink pattern. This situation happened when the user mistakenly executed a voluntary eye-blink at an erroneous time interval, incorrectly recorded by the sequence indicated by the audio signal (beep). In the test, two such sequences got removed. Therefore, it resulted in 28 sequences associated with the detection of a voluntary eye-blink.

During the second stage (meaning the last 30 s), the user had to avoid voluntary eye-blinking and keep a neutral state of mind. This way, other 30 sequences, each of them containing 512 samples of EEG signal, were obtained and taken as a reference for distinguishing between the moment when the user has executed a voluntary eye-blink, intending to send a command, and any other situation when he/she has avoided the eye-blinking. Analog to the analysis applied in the first stage, there were ten statistical features for every 30 sequences. Each of those 30 sequences was also graphically displayed to observe any possible anomaly. In the test, a single irrelevant sequence got removed. Therefore, it resulted in 29 sequences corresponding to an EEG signal, which did not include any eye-blink.

Finally, the two stages got integrated so that 57 sequences resulted, classified by two categories (1 – one voluntary eye-blink got executed and 0 – no eye-blink got performed). The obtained numerical data were exported to a .csv file using the Write Delimited Spreadsheet function from the File I/O palette offered by LabVIEW. Figure 3 shows an example of a .csv file related to the raw EEG signal from the current paper.

Sequence	Mean_Raw	Median_Raw	RMS_Raw	Standard	Coeficient K_Raw	Mode_Raw	Sum_Raw	Skewness	Maximum	Range (Max-Min)_Raw	Class
24	54.792969	65	213.368	206.314	15.450372	63.085	56108	0.48054	1287	2057	1.000000
25	51.62793	60	227.196	221.361	12.437232	73.37	52867	0.31887	1148	1902	1.000000
26	54.875	64.5	245.809	239.722	11.760207	56.435	56192	0.4099	1220	1989	1.000000
27	52.481445	61	241.808	236.159	13.260499	65.575	53741	0.46076	1284	2119	1.000000
28	53.807617	55	61.3143	29.4114	3.169657	55.465	55099	-0.32635	129	191	0.000000
29	44.030273	43	47.0974	16.7264	3.483667	37.55	45087	-0.13346	92	110	0.000000
30	36.875977	37.5	47.5949	30.1052	3.481755	33.05	37761	-0.01104	137	210	0.000000
31	37.876953	40	57.9334	43.8577	3.330285	65.775	38786	-0.17838	161	293	0.000000

Fig. 3. A portion of the .csv file displaying eight sequences (assigned to one of the two different classes) of the analyzed raw EEG signal.

3.2 The Automatic Generation of a Series of Charts Displaying the Variations of Statistical Features Applied on the Raw EEG Dataset

The next objective was to investigate the influence/impact each of the ten statistical features has on each of the 57 sequences of different types of the EEG signal. The accomplishment of this objective resulted in an intuitive solution given by the implementation of a LabVIEW original application enabling the automatic generation of charts displaying the variations of statistical features applied on a dataset consisting of the 57 sequences defining two different classes: 1 – one voluntary eye-blink executed and 0 – no eye-blink executed. Figure 4 shows the LabVIEW application's graphical user interface (GUI) running the automated charts of statistical features applied to the .csv file related to the raw EEG signal (Fig. 3).

Similar charts were also obtained by uploading the other .csv files related to different EEG signals (delta, theta, alpha, beta, and gamma). By visually comparing all the resulted charts, it resulted in the conclusion that the most significant differences between the two classes (1 – a voluntary eye-blink detected and 0 – no eye-blink detected) were given by the computing of four statistical features (standard deviation, Kurtosis coefficient, root mean square and maximum value of the amplitude) applied to the raw EEG signal. Moreover, it resulted in the raw EEG signal being the most significant type because it includes all the possible variations to offer an accurate and detailed description of the pattern of voluntary eye-blinking. Otherwise, different EEG signal processing methods

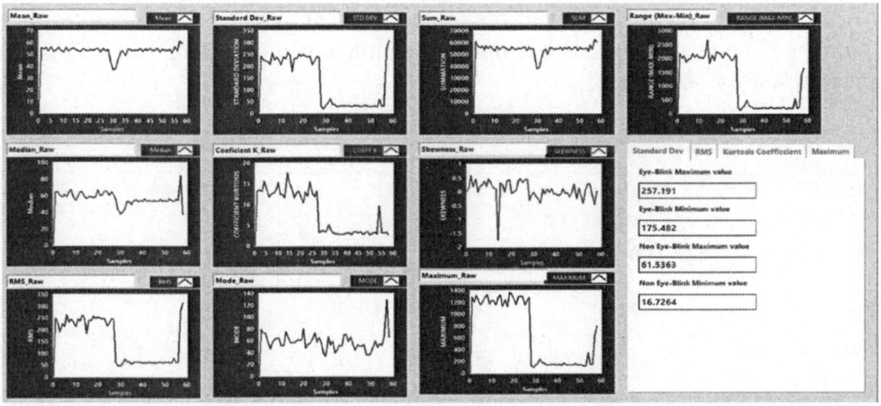

Fig. 4. The LabVIEW GUI application aimed for the generation of the charts based on the statistical features determined across the sequences of raw EEG signals stored in a .csv file.

(FFT and filtering) are unnecessary by choosing the raw EEG signal, resulting in a higher degree of simplicity and convenience regarding the LabVIEW code design.

As seen in Fig. 4, there is a Tab graphical window that displays two maximum values and two minimum values of the statistical features calculated on the array consisting of the first 28 sequences related to the class 1 – voluntary eye-blink detected and the last 29 sequences related to the class 0 – no eye-blink executed. For example, it resulted in a maximum/minimum value across the first 28 sequences of the standard deviation measured on the raw EEG signal and a maximum/minimum value across the last 29 sequences of the same statistical feature measured on the raw signal. Determining the minimum and maximum thresholds is necessary for the configuration of the Fuzzy Logic system.

3.3 LabVIEW Based Fuzzy Logic System Aimed to Measure the Strength of the Eye-Blinking

Designing a Fuzzy Logic system resulted in measuring the strength of the eye-blink to compare the resulted value to an established threshold. Thus, a voluntary eye-blink gets detected if its strength is higher than the given threshold. In contrast to Boolean Logic based on total membership, meaning that an input variable can be included either in the first class or in the second class, Fuzzy Logic enables the partial membership [13], assigning the input variable to some degree of membership to both classes.

The most significant statistical features give the input linguistic variables (standard deviation, root mean square, Kurtosis coefficient, and the maximum amplitude value) computed for the raw EEG signal in the time domain. The Fuzzy System Designer provided by NI LabVIEW [13] was necessary to implement the algorithm to generate the value of the output variable represented by the strength of the eye-blink. According to Fig. 5, the shape of the membership functions, called Low, Medium, High, and their intermediate points were set after the experimentation with different mixtures of functions types and partial ranges to obtain the relevant values of the strength of the eye-blinking. The resulted value was updated continuously according to the real-time execution of the eye-blinking.

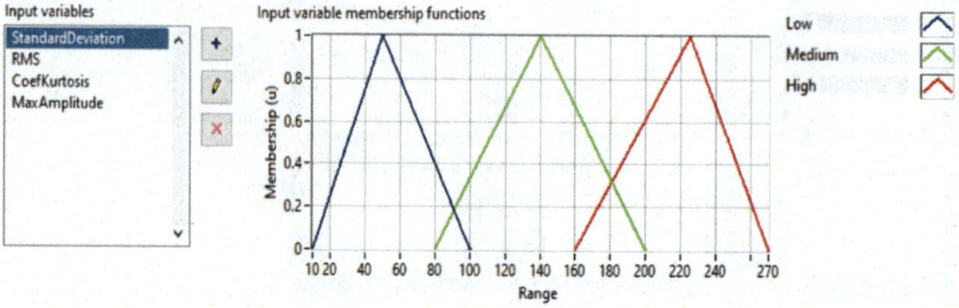

Fig. 5. The Input Variables configuration panel of the Fuzzy Logic System LabVIEW Designer.

According to Fig. 6, the range of the output variable, the eye-blinking strength, has a minimum value equal to 0 (zero) and a maximum value equal to 255. Also, the membership functions of the output variable have the characteristics shown in Fig. 6.

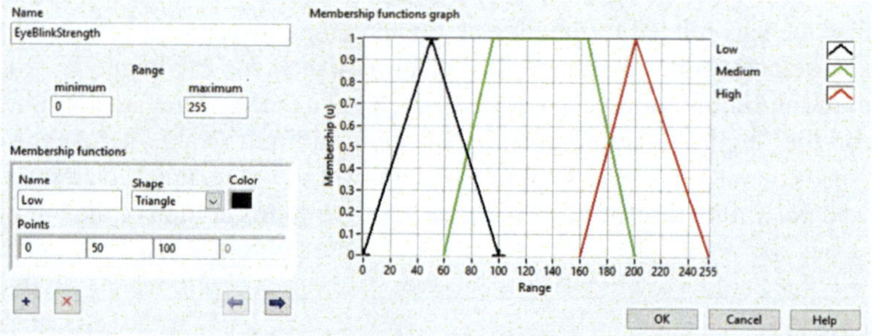

Fig. 6. The Edit Variable configuration panel of the Fuzzy Logic System LabVIEW Designer.

According to Table 1, out of 12 possibilities, three Fuzzy Logic-based rules got tested.

Table 1. The three tested rules of the proposed LabVIEW-based Fuzzy Logic System.

Standard deviation		RMS		Kurtosis coefficient		Maximum amplitude	Eye-blink strength
Low	Or	Low	Or	Low	Or	Low	Low
Medium	Or	Medium	Or	Medium	Or	Medium	Medium
High	Or	High	Or	High	Or	High	High

3.4 The Integration Between the Raw EEG Data Acquisition and Fuzzy Logic

The eye-blinking strength, previously calculated by using the Fuzzy Logic System, is further compared with a threshold established in the real-time execution of the LabVIEW application. Designing a state-machine paradigm integrated the algorithm of computing the number of voluntary eye-blinks with the real-time running of the Fuzzy Logic system and the real-time detection of the commands aimed to control the movement direction of the mobile robot. The state-machine paradigm is consisting of a case structure (known as if-else or the switch statement in procedural programming) including seven sub-diagrams aimed for the achievement of the following seven tasks: initialization, stop command, move forward, move backward, turn left, turn right and compute the number of voluntary eye-blinks. At every moment, only the instructions code of a single sub-diagram can be executed based on the value of a variable (enumerated type) assigned to the case selector. As described below, these sub-diagrams have a similar instructions code:

- The acquisition of an array containing 512 samples of raw EEG signal;
- The computing of statistical features (standard deviation, root mean square, Kurtosis coefficient, the maximum value of amplitude) for that array;
- The real-time running of the Fuzzy Logic System Controller;
- The generation of the eye-blink strength, calculated as an output variable of the Fuzzy Logic System;
- The comparison of the resulted strength of the eye-blink with a given threshold so that two alternative sequences of instructions code will get executed;
- The first sequence is related to a favorable condition (it resulted in detecting a voluntary eye-blink, whose amplitude exceeds the given threshold), and it consists of inserting the current value of the eye-blinking amplitude in a numerical array, followed by the transition to the next state.
- The second sequence is related to a false condition (it did not detect an eye-blink), and the transition to the last state is enabled.
- In the Move Forward and Move Backward states, a Flag Boolean variable is set or unset, to indicate the previously selected movement direction, when the robot should turn to the left or the right;
- The last state detected the commands changing the movement directions of the mobile robot, according to the size of the array containing the strength of the voluntary eye-blinks. Therefore, there were deciphered the following commands, based on the array size, consisting of:

  o One element – stop (one voluntary eye-blink was detected);
  o Two elements – move forward (two voluntary eye-blinks got achieved);
  o Three elements – move backward (three voluntary eye-blinks got executed);
  o Four elements – move forward left or move backward left, depending on the value of the Boolean Flag variable (four voluntary eye-blinks got executed);
  o Five elements – go forward right or go backward right, depending on the value of the Boolean Flag variable (five voluntary eye-blinks got executed).

3.5 The Commands Sent by LabVIEW Application to the Arduino Program Necessary to Change the Movement Directions of the Mobile Robot

A string variable is assigned to a particular character, depending on the identified command. The Arduino board received this particular character and executed the previous commands by changing the movement directions of the mobile robot. Using the VISA LabVIEW-based toolkit resulted in implementing the Bluetooth protocol-based communication between the LabVIEW application and the Arduino program.

4 Results and Discussions

A video demonstration showing the results from running the proposed software and hardware system is available at this YouTube unlisted link: https://youtu.be/Mh9ibydqd5w. By testing the brain-computer interface application based on the Fuzzy Logic LabVIEW algorithm, the values returned by the eye-blinking strength can successfully determine the voluntary eye-blinks after setting a suitable threshold customizable for each user. Nevertheless, the proposed BCI system is just a prototype for training, simulation, and educational purposes. Disabled people could use it to experiment with controlling a mobile robot without involving muscles and peripheral nerves. The Fuzzy Logic approach can be considered a starting point based on the raw EEG signal processing to identify the eye-blinking amplitude. Currently, the presented BCI application constitutes a proof of concept. Conducting various testing scenarios should validate the accuracy and response time of the implemented algorithm. A deeper analysis of the feature extraction methods and testing of all possible Fuzzy Logic rules are also necessary to improve the detection accuracy of the voluntary eye-blinks.

5 Conclusion

This research work aims to prove the possibility of using the Fuzzy Logic system to detect voluntary eye-blinks, also used to determine the movement commands of a mobile robot. It resulted in developing a LabVIEW application and designing a BCI prototype by the acquisition of the raw EEG signal from the biosensor of a NeuroSky portable headset, the computing of the most significant statistical features (the standard deviation, the root mean square, the Kurtosis coefficient and the maximum value of amplitude) and the calculation of the eye-blinking strength compared to a given threshold. The strength of a voluntary eye-blink should exceed this threshold. Implementing a state-machine-based algorithm was necessary to compute the number of voluntary eye-blinks and generate the movement commands sent to the Arduino board, which controls the mobile robot.

The proposed Fuzzy Logic System provides good precision and accuracy regarding measuring the strength of eye-blinking. It resulted in a quicker response and a simple working principle. It also resulted in an efficient integration between the acquisition of the raw EEG signal, the Fuzzy Logic System controller, and the counting algorithm of the voluntary eye-blinks. Considering the current status of the scientific literature, implementing a LabVIEW-based Fuzzy Logic system to classify the eye-blinks is an underexplored research field. Therefore, this paper provided the foundation of knowledge necessary to develop advanced systems. Regarding the future research directions, the proposed BCI experimental system needs testing by different categories of users in various psychological conditions and environments to accomplish suitable adjustments and enhance the Fuzzy Logic-based detection of voluntary eye-blinks.

References

1. Galíndez-Floréz, I., Coral-Flores, A., Moncayo-Torres, E., Mayorca-Torres, D., Guerrero-Chapal, H.: Biopotential signals acquisition from the brain through the mindwave device: preliminary results. In: Botto-Tobar, M., Zambrano Vizuete, M., Torres-Carrión, P., Montes León, S., Pizarro Vásquez, G., Durakovic, B. (eds.) ICAT 2019. CCIS, vol. 1193, pp. 139–152. Springer, Cham (2020). https://doi.org/10.1007/978-3-030-42517-3_11
2. Wang, Y., Wang, Z., Clifford, W., Markham, C., Ward, T.E., Deegan, C.: Validation of low-cost wireless EEG system for measuring event-related potentials. In: Proceedings of the 2018 29th Irish Signals and Systems Conference (ISSC), Belfast, pp. 1-6 (2018). https://doi.org/10.1109/ISSC.2018.8585297
3. Rosca, S., Leba, M., Ionica, A., Gamulescu, O.: Quadcopter control using a BCI. IOP Conf. Ser.: Mater. Sci. Eng. **294**, 012048 (2018). https://doi.org/10.1088/1757-899X/294/1/012048
4. Harsono, M., Liang, L., Zheng, X., Jesse, F.F., Cen, Y., Jin, W.: Classification of imagined digits via brain-computer interface based on electroencephalogram. In: Wang, Y., Huang, Q., Peng, Y. (eds.) Image and Graphics Technologies and Applications: 14th Conference on Image and Graphics Technologies and Applications, IGTA 2019, Beijing, China, April 19–20, 2019, Revised Selected Papers, pp. 459–471. Springer Singapore, Singapore (2019). https://doi.org/10.1007/978-981-13-9917-6_44
5. Kubacki, A., Milecki, A.: Control of the 6-axis robot using a brain-computer interface based on steady state visually evoked potential (SSVEP). In: Trojanowska, J., Ciszak, O., Machado, J.M., Pavlenko, I. (eds.) Advances in Manufacturing II: Volume 1 – Solutions for Industry 4.0, pp. 213–222. Springer International Publishing, Cham (2019). https://doi.org/10.1007/978-3-030-18715-6_18
6. Reust, A., Desai, D., Gomez, L.: Extracting motor imagery features to control two robotic hands. In: Proceedings of the 2018 IEEE International Symposium on Signal Processing and Information Technology (ISSPIT), pp. 118–122. Louisville, KY, USA (2018). https://doi.org/10.1109/ISSPIT.2018.8642627
7. Mistry, K.S., Pelayo, P., Anil, D.G., George, K.: An SSVEP based brain computer interface system to control electric wheelchairs. In: Proceedings of the 2018 IEEE International Instrumentation and Measurement Technology Conference (I2MTC), pp. 1–6. Houston, Texas (2018). https://doi.org/10.1109/I2MTC.2018.8409632
8. Zhi-Hao, W., Hendrick, K., Yu-Fan, C., Chuan-Te, L., Shi-Hao,, Gwo-Jia, J.: Controlling DC motor using eye blink signals based on LabVIEW. In: Proceedings of the 2017 5th International Conference on Electrical, Electronics and Information Engineering (ICEEIE), pp. 61–65. Malang (2017). https://doi.org/10.1109/ICEEIE.2017.8328763
9. Tran, D.-K., Nguyen, T.-H., Nguyen, T.-N.: Detection of EEG-based eye-blinks using a thresholding algorithm. Eur. J. Eng. Technol. Res. **6**(4), 6–12 (2021). https://doi.org/10.24018/ejers.2021.6.4.2438
10. Prem, S., Wilson, J., Varghese, S.M., Pradeep, M.: BCI integrated wheelchair controlled via eye blinks and brain waves. In: Pawar, P.M., Balasubramaniam, R., Ronge, B.P., Salunkhe, S.B., Vibhute, A.S., Melinamath, B. (eds.) Techno-Societal 2020: Proceedings of the 3rd International Conference on Advanced Technologies for Societal Applications—Volume 1, pp. 321–331. Springer International Publishing, Cham (2021). https://doi.org/10.1007/978-3-030-69921-5_32

11. Miranda, M., Salinas, R., Raff, U., Magna, O.: Wavelet Design for Automatic Real-Time Eye Blink Detection and Recognition in EEG Signals. Int. J. Comput. Commun. Control **14**, 375–387 (2019). https://doi.org/10.15837/ijccc.2019.3.3516
12. NI LabVIEW Toolkit – NeuroSky Brain Computer Interface, https://forums.ni.com/t5/NI-Labs-Toolkits/NeuroSky-LabVIEW-Driver/ta-p/3520085?profile.language=en. Accessed 09 Aug 2021
13. LabVIEW PID and Fuzzy Logic Toolkit User Manual, http://www.ni.com/pdf/manuals/372192d.pdf. Accessed 09 Aug 2021

SCADA System for Monitoring and Reconfiguring an Electrical Distribution Network After a Fault

Traian Turc[1(✉)] and Claudiu Damian[2]

[1] Department of Electric Engineering and Information Technology, University of Medicine, Pharmacy, Sciences and Technology of "G.E. Palade" of Târgu-Mureş, 540139 Târgu-Mureş, Romania
traian.turc@umfst.ro

[2] Electrical Company "Transilvania Sud SA", "SDEE Mureş", 540320 Târgu-Mureş, Romania

Abstract. The main goals of any power system are represented by the continuity of electricity delivery and by maintaining voltage and frequency parameters in an acceptable disturbances range. If, during operation, the parameters of the power lines are different from normal due to defects, then the power line is in an abnormal operating mode, in which malfunctions or damage to electrical installations may occur. Then the problem of defect analysis appears in order to manage them quickly. The present paper aims to present a SCADA system that includes the defect analysis in a distribution network by analyzing the signals in several different points of the electrical network.

Keywords: Current control · Power system faults · Power system protection · SCADA systems

1 Introduction

In case of defects, the power lines have an abnormal operation regime. The effect of abnormal power lines regimes is more unfavorable as the operating time in these regimes is higher.

Determination and elimination of defects depends on the structure of the electrical networks in the respective area, on the functioning scheme at the moment of fault occurrence of and on the existing automation in the stations involved in the fault regime [1, 2]. In an electric power distribution system, detecting High Impedance Faults (HIFs) are generally a difficult problem [3]. Each fault type needs to be analyzed individually [4].

2 Aim of Paper

The main objective of the paper is the research and implementation of new facilities in SCADA (Supervisory Control and Data Acquisition) systems for power distribution networks, facilities which will provide to the operators of SCADA systems operational

L. Moldovan and A. Gligor (Eds.): Inter-Eng 2021, LNNS 386, pp. 757–762, 2022.
https://doi.org/10.1007/978-3-030-93817-8_67

safety, time reducing between received information about an incident which took place in the energy system monitored by the SCADA system and its signaling and the possibility of taking the best decisions in supplying consumers after a damage in the energy system.

Different studies conducted on various SCADA systems, showed that in frequent cases when either HMI (Human Machine Interface) displayed incomplete or too much information, the SCADA operator was determined to take wrong decisions when consumer's path of power supply had to be restored. It was also noticed that an inappropriate decision of the SCADA system operator could lead to worsening of the event that took place and, instead of feeding as quickly as possible the consumers, generated new damages. Thus, instead of supplying electricity to the consumer, the operator could operate or supply damaged equipment's through good equipment's, which may lead to an increased fault area and damages to other equipment's. It is thus necessary to develop SCADA systems as safe as possible to provide clear and precise information.

Traditionally, measurements used for power system monitoring are collected by low updating rate SCADA systems via Remote Terminal Units (RTUs). Estimations based on those measurements could not be accurate enough in capturing how the system states evolve in time [5].

An efficient way to increase information accuracy is to interconnect SCADA systems with specialized devices and equipment's that monitor power equipment's in an electrical transformer station. Interconnection with these devices requires new facilities in SCADA systems, facilities that can provide to SCADA system operators the possibility of reducing consumers feed times and operational safety.

By introducing these new facilities, SCADA systems will become much more reliable and safer in operation, reducing the risk of a fault spreading if the operator takes the wrong decisions.

Control Center operators possess knowledge about the network's dynamic behavior, namely its physical laws, its protection devices and their operation and how to combine their knowledge with incoming SCADA information (alarms) to assess the Electrical Network's state, in order to perform fault diagnosis and to plan and execute power supply restoration [6–8].

3 Case Study in Different Fault Situations

The case study refers to an 110 kV electric power grid consisting of the power lines and power stations to which these lines are connected.

The analysis was performed on two operating cases, which are monitored by the same SCADA system.

The studied electrical network consists of 110 kV overhead power lines, which interconnect five power stations A, B, C, D, E.

Stations C, D, E are system stations and can be considered sources for the analyzed network, each of them being capable of delivering electricity to the national power system, and therefore across the network on which the study is conducted.

The network on which the case study is performed is illustrated in Fig. 1 where a configuration of 110 kV power lines and power stations is presented, with the possibility of receiving the voltage from three power sources, namely from station E, from station C and station D. High voltage lines have the same construction type and the same electrical characteristics, depending on the length of each line.

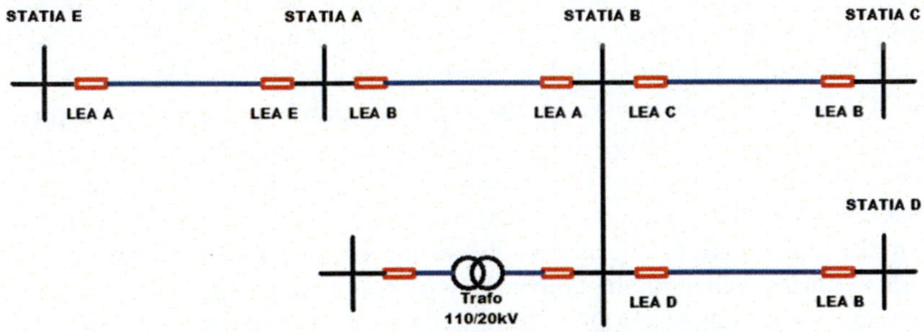

Fig. 1. Network configuration for the study case.

On all overhead power lines, independent protection and automation devices are installed, including five-stage distance protection, four of them directed and one not, as well as two-stage directed homopolar protection.

On these lines were implemented a series of automations, included in the SCADA system for LRP (Line Reconnecting Process) automation on all power lines, and in power station B, on LEA A, the AROL (Automatic Open Loop Reconnecting) automation was implemented, so it could function with an open loop.

The first case study was made when the loop is open, the breaker in power station B on LEA A is disconnected and the AROL automation is in operation.

The role of this automation is to reconfigure the normal scheme when one of the energy sources becomes unavailable.

In this case, the defect is initially fed from one end, but during the event, the power flow changes, as well as the power supply of the fault.

The second case study was made on the same network, but when all the circuit breakers are connected, the network is running loose and the fault is fed by several parts. The SCADA system will receive information from the main power stations that are supplying the fault and the information regarding the proper or improper operation of the protection and automation will be analyzed and transmitted to the operator together with the state of consumers (properly/improperly fed) correlated with the network situation at the end of the event.

The SCADA system will receive information from distance protection, homopolar directed protection, maximum current reserve protection and LRP automation. The HMI will display this information based on special integration features able to synthesize the displayed information.

4 SCADA System

The SCADA application receives digital and analogue information from five power stations, compares them to each other and informs the SCADA system operator by displaying on the screen if the operations of the protection, automation and circuit breakers positions are appropriate at the end of an event.

The SCADA system verifies each time the confirmation of the connected position of the high voltage circuit breakers after an automation operation, LRP. If, after the operation of these automations, the disconnected position of the circuit breaker is not displayed, the fault is signaled and the system operator is informed about the event.

The SCADA system integrates the startup information from the numerical protection terminals. Based on this information, the functioning of the system is assessed and it is possible to check the startup currents on the phases at which the fault occurs.

The SCADA system compares the startup current signals on defect phases from the source to the fault location. The system checks the startup of the same phase in at least two or three places. When on the fault supply path, it cannot be found the same startup phase or zero currents, the SCADA system will signal an error, and the system operator will be informed of the event.

The SCADA system performs a comparison of defect distances resulting from calculations with the defect distances determined by protection relays and displays them on HMI (Human Machine Interface) (Fig. 2).

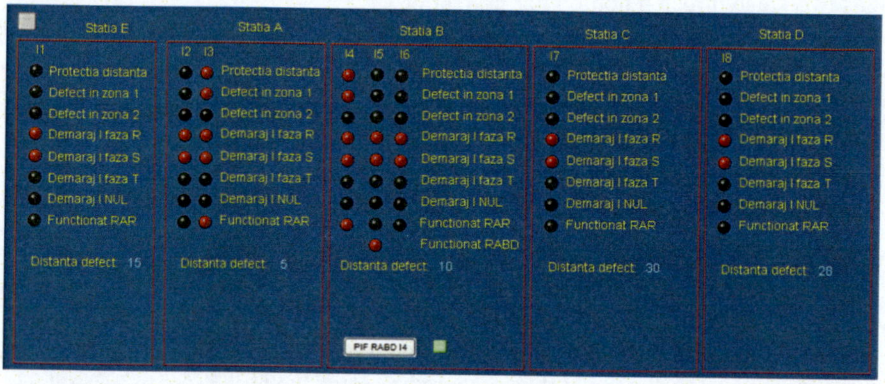

Fig. 2. HMI SCADA.

The distance to the fault site calculation is performed by the localization defect tool of each numerical protection terminal which is a component of each power line cell and the length to the defect is transmitted by the numerical protection terminal to the SCADA system.

The fault locator function is included in the line protection terminal and provides high precision measurements and indications with an error of $<\pm 3\%$ of the distance to the fault location.

The used algorithm eliminates the influence of the load current on the overcharging from the opposite end and reduces the influence of the pass-through resistance at the fault site. The distance to the fault location can be indicated as a percentage of the line length or in kilometers.

The fault locator algorithm is based on the values of measured voltages and currents at the mounting location of the numeric protection terminal.

The length to the fault location is calculated at each station that is on the fault current supply path by summing the section length to the fault of the affected line and the lines interposed on the fault current circuit lengths.

The SCADA system displays the length to the fault location and reconfigures the network to isolate the fault.

Figure 3 shows the distance to the fault and isolation of the affected network by opening the circuit breakers I3 and I4.

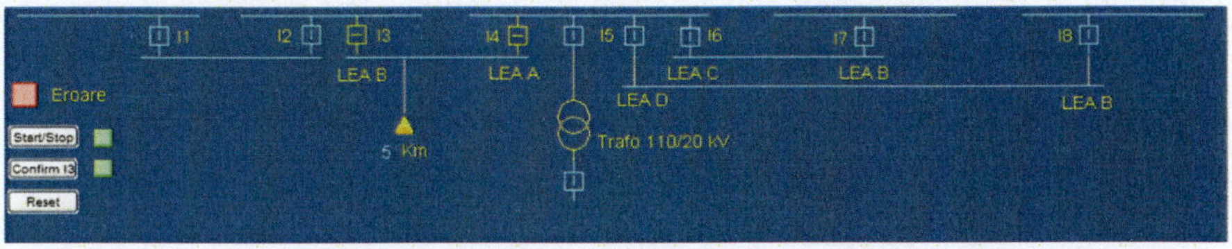

Fig. 3. Network reconfiguration after fault.

The comparison lengths checking to the point of defect can only be done on one section of the network. The lengths checking can't be done on two sections because countless defects occurring in the power networks are not net short-circuits and the fault circuit often interferes with the internal resistance of the electric arc produced at place of defect. In this situation, the length calculated by the fault locator to the defect location is greater than the true length of the line.

A new comparison of the occurred signals is performed by the SCADA system by comparing the missing voltage signal from a line with the triggered position of all the circuit breakers on that line. If, following the event that occurred, the line breakers remain triggered, with an unsuccessful LRP occurring, the SCADA system waits for line voltage signaling at both stations and if it fails to function, an error will be signaled, and the system operator will be informed of the event.

The SCADA system uses information from the 5 power stations, information that are gathered from the numerical protection terminals mounted on the lines in the power stations.

5 Conclusions

The paper proposed a SCADA system development for displaying signaling and protection monitoring in five power stations of a power system. In addition to monitoring, command and data acquisition, the system also analyzes the information received from independent protection and automation systems. The SCADA system analyzes the signals occurred starting from a trigger on one of the lines, compares the onset of all lines along the fault line, verifies the display of the circuit breakers connected position after a LRP cycle or after an AROL (Automatic Reconnection in Open Loop) operation, and, if identifies wrong signaling, displays the error signal and communicates the incorrect signals to the SCADA system operator.

In the proposed SCADA system new features were introduced that allow the calculation of the parameters required for the distance protection settings by taking into consideration the line specific values. Thus, by considering the values of resistances and reactance, the control impedances for each stage of the distance protection are calculated.

Based on studies and analyzes performed, it can be concluded that there is a strong dependence between the power distribution network and the data transmission networks used by SCADA systems. The two cases presented lead to the conclusion that the existence of several distribution power lines leads to the need of developing a complex data transmission network for SCADA systems.

Through studies based on the simulation of the interconnected operation different power systems, optimal operation can be achieved, with as few as possible damage to the interconnected systems, a low degree of danger and smaller fault times.

If a fault situation occurs, by using the new facilities included in the SCADA systems, the weaknesses in the electrical power systems can be identified.

Based on the risk assessment models, short-term forecasts can be performed, regarding the quality of the services provided by a SCADA operator compared to the quality of the electricity supplied to the customers.

References

1. Penescu, C., Cain, S.: Protectia prin Relee electronice. Editura Tehnică, Bucharest (1969)
2. Cristea, P., Mihai, I., Leca, M.: Protectia Sistemelor Electroenergetice. Editura "Crenguta Gâldau", Iaşi (1996)
3. Kavaskar, S., Mohanty, N.K.: Detection of high impedance fault in distribution networks. Ain Shams Eng. J. **10**(1), 5–13 (2019). https://doi.org/10.1016/j.asej.2018.04.006
4. Aithal, A., Li, G., Jianzhong, W., James, Y.: Performance of an electrical distribution network with Soft Open Point during a grid side AC fault. Appl. Energy **227**, 262–272 (2018). https://doi.org/10.1016/j.apenergy.2017.08.152
5. Alcaide-Moreno, B.A., Fuerte-Esquivel, C.R., Glavic, M., Van Cutsem, T.: Electric power network state tracking from multirate measurements. IEEE Trans. Instrum. Meas. **67**(1), 33–44 (2018). https://doi.org/10.1109/TIM.2017.2754838
6. Malheiro, N., Ramos, Z., Couto, V.: Decision support for power system control centers – a model based reasoning component. Int. J. Eng. Intell. Syst. (2005)
7. Meng, J., Wang, W., Tang, X., Xu, X.: Zero-sequence voltage trajectory analysis for unbalanced distribution networks on single-line-to-ground fault condition. Electr. Power Syst. Res. **161**, 17–25 (2018)
8. Vlasa, I., Gligor, A., Dumitru, C.D., Moldovan, C.E.: Real-time power supply service monitoring in industrial electric power distribution grid. In: 2019 International Conference on Electromechanical and Energy Systems (SIELMEN), pp. 1–6 (2019)

Dynamic Calibration of Tyre-Road Contact Patch Stress Tri-Axial Transducer

Alexandra-Raluca Moisescu and Gabriel Anghelache(✉)

University Politehnica of Bucharest, Splaiul Independentei 313, 060042 Bucharest, Romania
gabriel.anghelache@upb.ro

Abstract. The paper presents the dynamic behaviour verification for the sensing elements of the tyre-road contact stress measuring system developed at the Tyre Research Laboratory in University POLITEHNICA of Bucharest. The complex transducer for investigating distributions of tri-axial stress in the tyre-road contact patch consists of a transversal array of strain gauged pins covering the entire contact patch width. A verification of sensing elements behaviour in transient conditions was required, taking into consideration the highly variable character of tyre contact patch stress phenomena. The dynamic calibration of sensing elements has been performed by applying transient force successively on each measuring direction, using an impact hammer with piezoelectric force transducer. Tests have been performed in laboratory conditions, by measuring force simultaneously using the strain gauged sensing elements and the piezoelectric transducer of the impact hammer. Very good correlation has been obtained between shapes and magnitudes of forces measured on each direction using the impact hammer force transducer and the strain gauged sensing element. Thus, the dynamic calibration of sensing elements has been verified for tyre speed ranges corresponding to real rolling conditions.

Keywords: Dynamic calibration · Sensing element · Complex transducer

1 Introduction

In the tyre-road contact patch, distributions of stresses appear on three orthogonal directions as consequence of tyre deformations due to simultaneous actions of internal pressure, wheel load and road contact [1]. These tyre-road stresses represent the source of all forces in the contact patch which control the vehicle dynamics, consequently influencing safety, rolling tyre vibration and noise, tyre wear and road damage [2]. The main challenges concerning measurements of tyre contact stress pertain to the requirement of maintaining contact of the tyre with the road unaltered by insertion of sensors.

The distributions of tyre-road stresses are influenced by various factors, some related to tyre design and construction (radial or bias, tread design, tyre dimensions, etc.), and others to exploitation and rolling conditions (inflation pressure, vehicle speed, tyre vertical load, slip angle, etc.) and therefore cannot be precisely determined by using simple physical models. Finite element models allow obtaining distributions of contact

L. Moldovan and A. Gligor (Eds.): Inter-Eng 2021, LNNS 386, pp. 763–771, 2022.
https://doi.org/10.1007/978-3-030-93817-8_68

patch stresses [3, 4], but also require measurements of tyre-road stress distributions, which are necessary for evaluation and verification of models.

An overview of the main equipment for measurement of tyre-road stress distributions developed worldwide has been presented in [5] and has ascertained that most of the existing test rigs are designed for indoor use, and many include sensing elements consisting of strain gauged pins for measuring tri-axial stresses [6–18].

For experimental investigation of the tyre-road contact stress distributions, outdoor (road-embedded) as well as indoor (laboratory-use) test rigs have been designed and produced in recent years at the Tyre Research Laboratory in University POLITEHNICA of Bucharest [5, 19, 20].

Results of experimental research performed in real rolling conditions using the road-embedded test rig have been presented in [20] and in [21], for truck tyres in free rolling and braking conditions. Alongside, the study of contact stress distributions continued using finite element models of truck tyres [22], and some comparisons between results obtained through finite element modelling and experimental data have also been shown in [21].

The calibration of sensing elements in static conditions has been presented in [5]. In addition to the calibration with forces applied statically, the dynamic calibration of sensing elements has also been considered necessary, taking into consideration that the stress distributions in the tyre contact patch represent dynamic phenomena. The current paper objective is to investigate the dynamic behaviour of sensing elements, representing an important phase in test rig calibration.

2 Overview of Tyre Contact Stress Measuring System

Details about technical specifications of the road-embedded test rig with complex transducer for measuring stresses distributions have been presented in [5, 20, 21]. The transducer includes a transversal array of 30 sensing elements with strain gauges, each sensing element having three half-bridges that correspond to the three orthogonal directions, resulting 90 data acquisition channels for measuring tri-axial stress distributions in the entire contact patch, mainly for truck tyres but also for other tyre types.

The transducer design provides minimum change in contact surface, as well as the option of performing either road or laboratory measurements; the measuring system can be used for free rolling/traction/braking tyres, and allows different rolling speed ranges.

The simplified shape of a sensing element is represented in Fig. 1, together with force components applied schematically on three orthogonal directions, and with the strain gauges constituting half bridges that correspond to each of the measuring directions [23]. Foil strain gauges for steel, each with one measuring grid, have been used.

The measuring system also includes signal conditioning modules and data acquisition system with embedded computer manufactured by National Instruments, and software applications developed in LabVIEW for measuring strain channels, converting and decoupling force components, computing and displaying stress distributions.

Longitudinal displacement of the tested tyre is also monitored and stored using a mobile data acquisition system with dedicated software applications developed in LabVIEW for correlation with the stress measuring system in view of spatial representation of tyre-road stress distributions over the contact area.

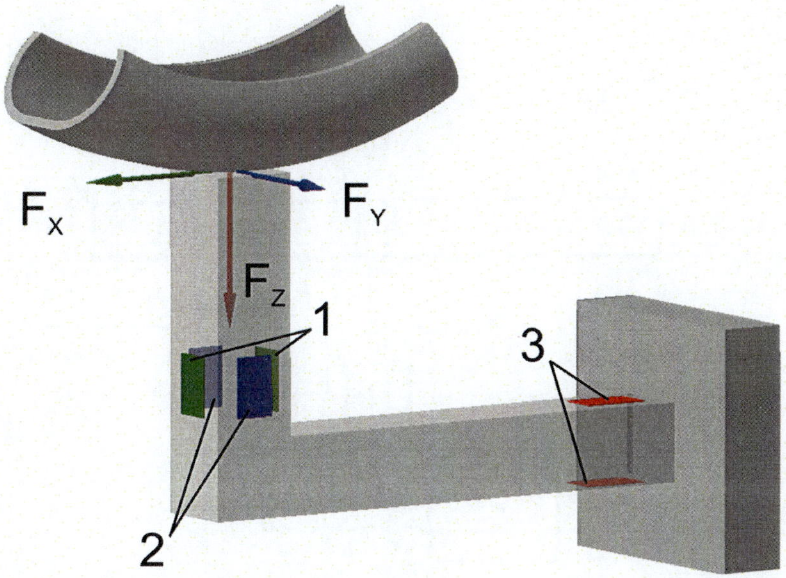

Fig. 1. Representation of a sensing element under the effect of three orthogonal forces: 1, 2, 3 strain gauges [23]

3 Dynamic Behaviour of Sensing Elements

Taking into account the dynamic character of tyre contact patch stress distributions measurement, dynamic calibration of sensing elements has also been needed. Impact duration was taken into account to be smaller than the duration of contact between tyre tread and sensing element of measuring system, in real rolling conditions. For verification of sensing elements dynamic behaviour, tests have been performed in laboratory conditions using an impact hammer with piezoelectric force transducer.

The layout for dynamic calibration is presented in Fig. 2 and an overall image of the equipment used is shown in Fig. 3 where the complex transducer, originally developed at the Tyre Research Laboratory in University POLITEHNICA of Bucharest, is displayed without the layer of coarse material on the sensing elements upper surface and on the top plates. The signal conditioning modules for strain gauges and the data acquisition system with embedded computer are manufactured by National Instruments Corporation; the impact hammer with piezoelectric force transducer and corresponding signal conditioning module are manufactured by PCB Piezotronics.

Dynamic force on each orthogonal direction has been applied successively on sensing elements; due to the configuration of complex transducer, an adapter has been used to facilitate impact application on longitudinal and lateral directions.

Software applications, such as the example shown in Fig. 4, have been specially developed by the authors in LabVIEW graphical programming environment [24] for

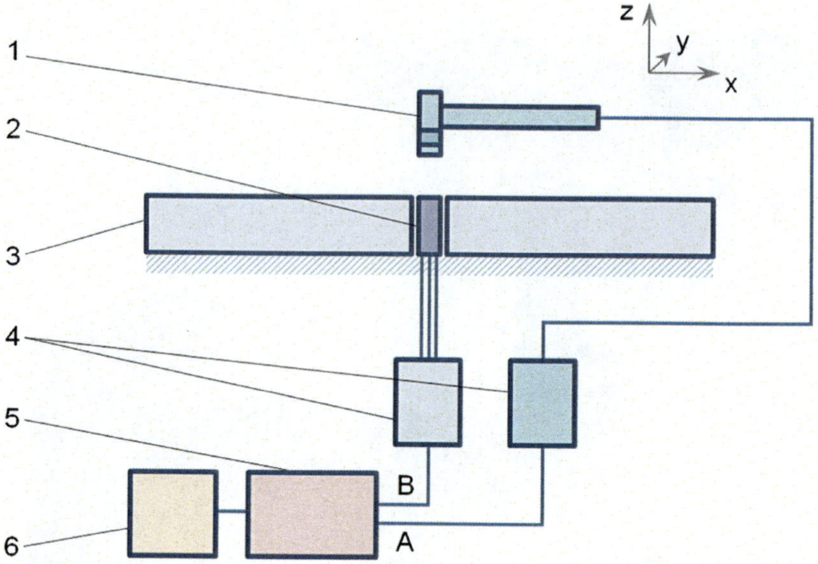

Fig. 2. Layout for the verification of sensing elements dynamic behaviour: 1. impact hammer with piezoelectric force transducer; 2. strain gauged pin; 3. complex transducer with array of strain gauged pins; 4. signal conditioning modules; 5. data acquisition system with embedded computer; 6. monitor; A. signal from force transducer of the impact hammer; B. signal from strain gauged pin

Fig. 3. Overall image of the equipment for verification of sensing elements dynamic behaviour

measuring force simultaneously using the strain gauged sensing elements and the piezoelectric transducer of the impact hammer. For the strain gauges signals, a lowpass Butterworth filter has been used. Details about software applications and connection of data acquisition channels to the measuring system have been presented in [5].

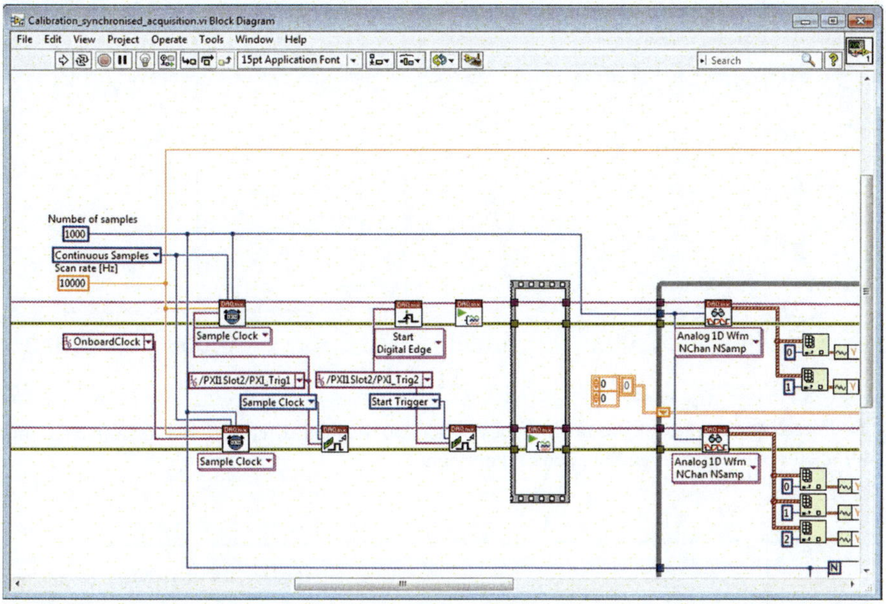

Fig. 4. Example of LabVIEW software application for simultaneously measuring force

Vertical force measured by one of the strain gauged sensing elements during an impact applied on vertical direction is shown in Fig. 5 in comparison to impact force measured simultaneously by the piezoelectric force transducer of impact hammer. Impact duration, shape and magnitude can be noticed, and good correlation between shapes and magnitudes of the two measured forces can be verified, with insignificant delay in the strain gauges results with respect to the piezoelectric force transducer of the impact hammer. Lateral force measured by one of the strain gauged sensing elements during an impact applied on lateral direction is shown in Fig. 6 in comparison with the impact force measured simultaneously by the piezoelectric force transducer of the impact hammer. The lateral impact has lower magnitude and slightly longer duration than the vertical impact. Also in this case, the good correlation between the shapes and magnitudes of the two measured forces can be remarked, with negligible delay in the strain gauges results with respect to the piezoelectric force transducer.

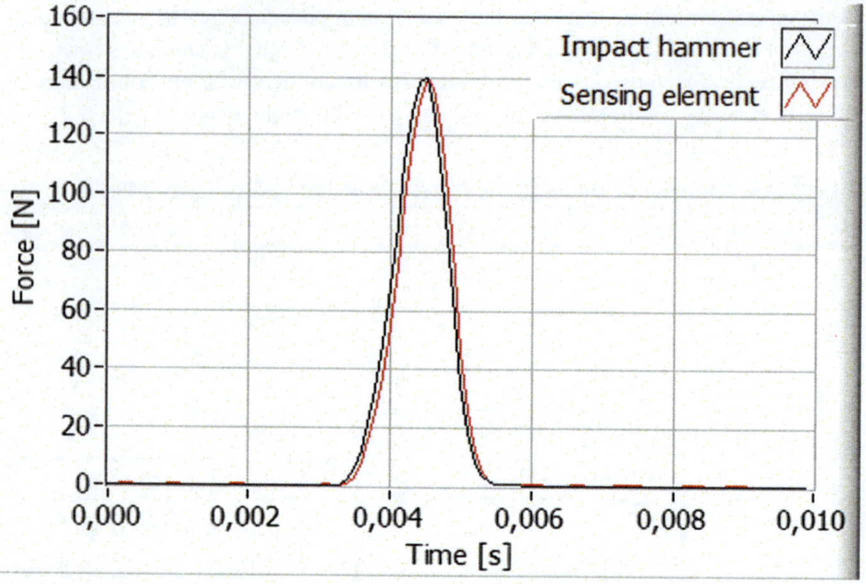

Fig. 5. Forces measured during an impact applied on vertical direction, using impact hammer force transducer and strain gauged sensing element

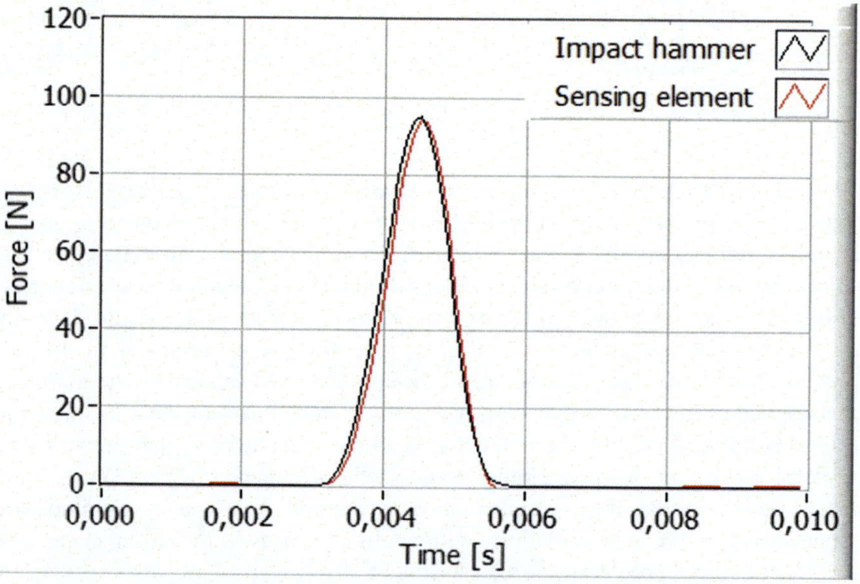

Fig. 6. Forces measured during an impact applied on lateral direction, using impact hammer force transducer and strain gauged sensing element

Longitudinal force measured by one of the strain gauged sensing elements during an impact applied on longitudinal direction is shown in Fig. 7 in comparison with the impact force measured simultaneously by the piezoelectric force transducer of the impact hammer. The longitudinal impact has lower magnitude and longer duration than the vertical and lateral impacts. However, the 50 N force applied on the sensing element with a contact area of 10 mm × 10 mm would be equivalent to a 500 kPa contact stress, therefore above the estimated range for the magnitude of longitudinal stress in the contact patch of a passenger car tyre.

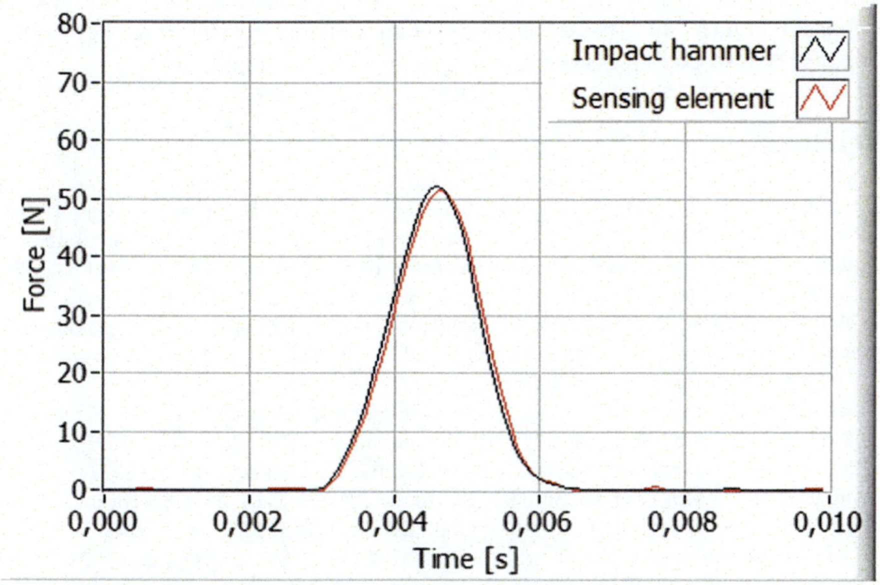

Fig. 7. Forces measured during an impact applied on longitudinal direction, using impact hammer force transducer and strain gauged sensing element

For the longitudinal impact also a good correlation between the shapes and magnitudes of the two measured forces can be noticed, with insignificant delay in the strain gauges results with respect to the piezoelectric force transducer. Additionally, in this case, some minor vibrations can be detected on the curve corresponding to the strain gauged sensing element immediately after the impact.

The abovementioned comparisons have proven very good correlation between the forces measured by strain gauged sensing elements during impacts applied on each orthogonal direction and the impact forces measured simultaneously by the piezoelectric force transducer of the impact hammer. Therefore, the calibration of sensing elements has been verified in dynamic conditions, thus allowing measurement of contact patch stress distributions for tyre speed ranges corresponding to real rolling conditions.

4 Conclusions

The experimental layout including an impact hammer with piezoelectric force transducer has allowed performing dynamic calibration on the tyre-road stress measuring system. Transient force has been applied successively on each orthogonal direction for verification of dynamic behaviour of sensing elements in laboratory conditions. Impact duration has proven to be smaller than the duration of contact between tyre tread and sensing elements in real rolling conditions.

Very good correlation exists between the impact forces measured by sensing elements and the impact forces measured simultaneously by the piezoelectric force transducer of the impact hammer. Thus, the dynamic calibration of sensing elements has also been verified for tyre speed ranges corresponding to real rolling conditions.

References

1. Pacejka, H.: Tire and Vehicle Dynamics, 3rd edn. Butterworth-Heinemann, Oxford (2012)
2. Pottinger, M.G., McIntyre, J.E.: Effect of suspension alignment and modest cornering on the footprint behavior of performance tires and heavy duty radial tires. Tire Sci. Technol. **27**(3), 128–160 (1999). https://doi.org/10.2346/1.2135981. Accessed 01 Mar 2021
3. Olatunbosun, O.A., Bolarinwa, O.: FE simulation of the effect of tire design parameters on lateral forces and moments. Tire Sci. Technol. **32**(3), 146–163 (2004). https://doi.org/10.2346/1.2186779. Accessed 01 Mar 2021
4. Wei, C., Olatunbosun, O.A., Yang, X.: A finite-element-based approach to characterising FTire model for extended range of operation conditions. Veh. Syst. Dyn. **55**(3), 295–312 (2017). https://doi.org/10.1080/00423114.2016.1264613. Accessed 01 Mar 2021
5. Anghelache, G., Moisescu, R., Sorohan, S., Buretea, D.: Measuring system for investigation of tri-axial stress distribution across the tyre–road contact patch. Measurement **44**(3), 559–568 (2011). https://doi.org/10.1016/j.measurement.2010.11.021. Accessed 01 Oct 2021
6. Koehne, S.H., Matute, B., Mundl, R.: Evaluation of tire tread and body interactions in the contact patch. Tire Sci. Technol. **31**(3), 159–172 (2003). https://doi.org/10.2346/1.2135266. Accessed 01 Mar 2021
7. Hall, W., Mottram, J.T., Dennehy, D.J., Jones, R.P.: Characterisation of the contact patch behaviour of an automobile tyre by physical testing. Int. J. Veh. Des. **31**(3), 354–376 (2003). https://doi.org/10.1504/IJVD.2003.003365. Accessed 01 Mar 2021
8. Douglas, R.A.: Tyre/road contact stresses measured and modelled in three coordinate directions. NZ Transport Agency research report 384, Wellington, New Zealand (2009). https://www.nzta.govt.nz/assets/resources/research/reports/384/docs/384.pdf. Accessed 01 Mar 2021
9. Tuononen, A.J.: Vehicle lateral state estimation based on measured tyre forces. Sensors **9**(11), 8761–8775 (2009). https://doi.org/10.3390/s91108761. Accessed 01 Mar 2021
10. De Beer, M., Fisher, C.: Stress-In-Motion (SIM) system for capturing tri-axial tyre-road interaction in the contact patch. Measurement **46**(7), 2155–2173 (2013). https://doi.org/10.1016/j.measurement.2013.03.012. Accessed 01 Mar 2021
11. Hernandez, J.A., Al-Qadi, I., De Beer, M.: Impact of tire loading and tire pressure on measured 3D contact stresses. In: Proceedings of the 2013 Airfield & Highway Pavement Conference: Sustainable and efficient pavements, pp. 551–560. ASCE (2013). https://doi.org/10.1061/9780784413005.04. Accessed 01 Mar 2021

12. Turc, T., Gligor, A., Dumitru, C.D.: Web-based wireless sensor system for SCADA environment. Procedia Engineering **181**, 546–551 (2017). https://doi.org/10.1016/j.proeng.2017.02.432. Accessed 01 Apr 2021
13. Garcia-Pozuelo, D., Yunta, J., Olatunbosun, O., Yang, X., Diaz, V.: A strain-based method to estimate slip angle and tire working conditions for intelligent tires using fuzzy logic. Sensors **17**(4), 874 (2017). https://doi.org/10.3390/s17040874. Accessed 01 Mar 2021
14. Garcia-Pozuelo, D., Olatunbosun, O., Strano, S., Terzo, M.: A real-time physical model for strain-based intelligent tires. Sens. Actuat. A **288**, 1–9 (2019). https://doi.org/10.1016/j.sna.2018.12.010. Accessed 01 Mar 2021
15. Stan, V.A., Buretea, D., Gheorghiu, R.A., Iordache, V.: Simple method to increase precision and stability for frequency counters. In: Proceedings of the 2020 12th International Conference on Electronics, Computers and Artificial Intelligence (ECAI-2020), pp. 1–4. IEEE (2020). https://doi.org/10.1109/ECAI50035.2020.9223216. Accessed 01 Apr 2021
16. Zhao, W., Zhang, C., Zhang, J.: Continuous measurement of tire deformation using long-gauge strain sensors. Mech. Syst. Signal Process. **142**, 106782 (2020). https://doi.org/10.1016/j.ymssp.2020.106782. Accessed 01 Mar 2021
17. Leiva-Villacorta, F., Vargas-Nordcbeck, A., Aguiar-Moya, J.P., Loría-Salazar, L.: Influence of tire footprint area and pressure distribution on pavement responses. In: Aguiar-Moya, J., Vargas-Nordcbeck, A., Leiva-Villacorta, F., Loría-Salazar, L. (eds) The Roles of Accelerated Pavement Testing in Pavement Sustainability, pp. 685–700. Springer, Cham (2016). https://doi.org/10.1007/978-3-319-42797-3. Accessed 01 Mar 2021
18. Costanzi, M., Rouillard, V., Cebon, D.: Effects of tyre contact pressure distribution on the deformation rates of pavements. In: 9th International Symposium on Heavy Vehicle Weights and Dimensions, International Forum for Road Transport Technology, pp. 892–903. Pennsylvania State University, Pennsylvania (2006). https://hvttforum.org/conference-papers/2006-hvwd9. Accessed 01 Mar 2021
19. Anghelache, G., Negrus, E.M., Ciubotaru, O.: Investigation of shear stresses in the tire-road contact patch. SAE 2003 World Congress & Exhibition, Detroit, USA, SAE Technical Paper 2003-01-1273 (2003). https://doi.org/10.4271/2003-01-1273. Accessed 01 Oct 2020
20. Anghelache, G., Moisescu, R.: Measurement of stress distributions in truck tyre contact patch in real rolling conditions. Veh. Syst. Dyn. **50**(12), 1747–1760 (2012). https://doi.org/10.1080/00423114.2012.674143. Accessed 01 Oct 2020
21. Anghelache, G., Moisescu, R.: Truck tyre-road contact stress measurement and modeling. In: Gruber, P., Sharp, R.S. (eds.) Proceedings of the 4th International Tyre Colloquium: Tyre Models for Vehicle Dynamics Analysis, WOS 000375226400026, pp. 218–227. University of Surrey, Guildford (2015). http://epubs.surrey.ac.uk/807823/13/Proceedingsofthe4thInternationalTyreColloquium_Gruber_Sharp_2015.pdf. Accessed 01 Oct 2020
22. Moisescu, A.R., Anghelache, G., Cristea, G.: Investigation of radial modal behaviour using finite element analysis for truck tyres without road contact. In: The 11th International Conference Interdisciplinarity in Engineering INTER-ENG 2017. Procedia Manufacturing, vol. 22, pp. 99–106 (2018). https://doi.org/10.1088/1757-899X/227/1/012006. Accessed 01 Oct 2020
23. Anghelache, G., Moisescu, R.: Measuring Tri-axial stress distribution across the tire contact patch width. Tire Technology International 2010 Annual Review, pp. 46–49 (2010). http://viewer.zmags.com/publication/c4b7c87c#/c4b7c87c/1. Accessed 01 Oct 2020
24. LabVIEW Function and VI Reference Manual, National Instruments Corporation, Austin, Texas, USA (2014)

Dimensional Optimization in Screw Fixation for Personalized Treatment of the Tibial Plateau Fracture

Flaviu Moldovan[1](✉), Adrian Gligor[2], and Tiberiu Bataga[3]

[1] IOSUD Doctoral School, "George Emil Palade" University of Medicine, Pharmacy, Science, and Technology of Targu Mures, 540139 Targu Mures, Romania
moldovan.flaviu95@yahoo.com

[2] Biomedical Research Center, "George Emil Palade" University of Medicine, Pharmacy, Science, and Technology of Targu Mures, 540139 Targu Mures, Romania

[3] Department of Orthopedics—Traumatology, "George Emil Palade" University of Medicine, Pharmacy, Science, and Technology of Targu Mures, 540139 Targu Mures, Romania

Abstract. **Background:** Tibial plateau fractures represent 1% of all fractures in adults, and the aim of the fracture treatment is stable fixation. There is a lack of consensus regarding the optimal treatment method for complex fractures. Closed reduction and percutaneous screw fixation for tibial plateau fractures is minimally invasive and has many advantages. **Objective:** The aim of this study was to develop a method for the optimal sizing of cannulated cancellous screws for fixing bone fragments of tibial plateau fractures according to the patient's characteristics, followed by a calculation of the crushing strength of bone fragments. **Method:** The dimensional parameters of the cannulated cancellous screws for fixing fractured bone fragments in the basal bone are the length and diameter. These parameters are determined in the surgical planning stage using virtual models and resistance calculations. Bone fragments fixed with cannulated screws are checked for crushing and the number of screws required to meet the crushing strength condition is calculated. **Results:** Validation in practice of the methodology for fixing tibial plateau fractures with cannulated cancellous fixation screws was performed in a patient who had a Schatzker type IV fracture. The patient required the fixation of the tibial fracture fragment using a cannulated screw sized with 3 mm diameter and 54 mm length. The crushing verification of the fractured bone fragment was performed according to the methodology developed in this paper. **Conclusion:** The results indicate that the present methodology is recommended in the preoperative planning stage of the tibial plateau fractures reductions and screw fixations, offering the guarantee of a biomechanical quality fixation which ensures the good health of the patient. It opens new perspectives for fixing optimization and calculating the strength of other types of fractures.

Keywords: Orthopedic surgery · Tibial fracture · Screw · Personalized treatment

1 Introduction

Tibial plateau fractures represent 1% of all fractures in adults. These may be related to high-energy trauma or, in older people with osteoporosis, to mild traumatic injuries.

L. Moldovan and A. Gligor (Eds.): Inter-Eng 2021, LNNS 386, pp. 772–783, 2022.
https://doi.org/10.1007/978-3-030-93817-8_69

From the beginning the aim of the fracture treatment is stable fixation [1] which is based on the four basic principles formulated by AO in 1985, that become guidelines for internal fixation (see Fig. 1) [2]:

- Principle 1 - Anatomic reduction aims to restore the anatomical relationships by reduction and fixation of bones;
- Principle 2 - Stable fixation aims to offer relative or absolute stability according to the requirements of the patient, type of fracture and surgical intervention;
- Principle 3 - Continue to irrigate with blood the soft tissues and bones applying gentle techniques of precise reduction and careful manipulation;
- Principle 4 - Active mobilization in the shortest possible time and restoration of the fractured area and the patient.

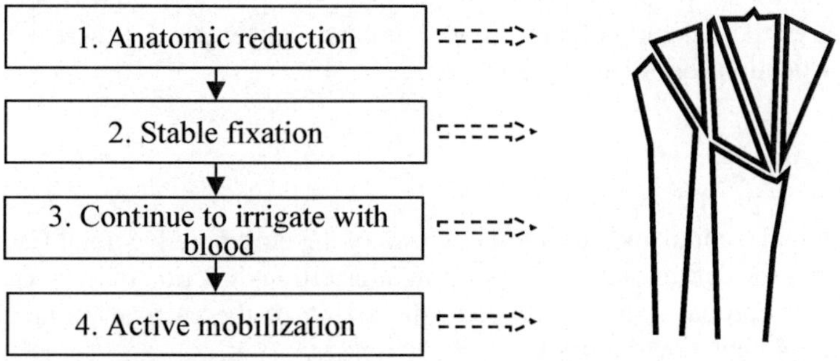

Fig. 1. Basic principles for internal fixation applied in management of tibial plateau fractures.

There is a lack of consensus regarding the optimal treatment method for complex fractures [3], but before undertaking surgical repair patient factors, surgeon experience, and fracture pattern must be carefully considered [4].

Even if it appears that Fiberwire fixation of eminence fractures provides biomechanical advantages over cannulated screw fixation [5], no significant difference was found between the two fixations in terms of stiffness and displacement after the cyclic testing [6].

Invasive internal fixation of tibial plateau fractures can be done with screws, plate and screws or with "T" plate and screws. New approaches and trends in health care [7], the rapid advances in trauma knowledge in the past decade have oriented the orthopedic surgery to minimally invasive techniques, soft-tissue handling, osteoporosis and biotechnology [8].

Arthroscopy-assisted percutaneous fixation was first described in the 1980s by Caspari et al. [9] and Jennings [10] and subsequently proved to be predominantly effective in Schatzker type I, II and III fractures [11]. Compared to open reduction and internal fixation, the low invasiveness of arthroscopy-assisted percutaneous fixation results in decreased morbidity rates.

In a cadaveric study Bowles et al. [12] demonstrated that for the treatment of Schatzker type II tibial plateau fractures percutaneous screw fixation conferred comparable fixation strength with plate/screw constructs. The biomechanical stability of the double-screw fixation is similar with screw-plate fixation for the treatment of fractures of the lateral tibial plateau [13].

Closed reduction and percutaneous screw fixation for tibial plateau fractures is minimally invasive reducing the length of hospitalization. It enables early mobilization with minimal instrumentation, and achieves satisfactory outcomes [14], avoids extensive soft tissue dissection and gives excellent results [15].

In internal osteosynthesis, percutaneous screw fixation of tibial plateau fractures requires adequate sizing of cannulated cancellous screws in order to increase the efficiency of fracture fragments fixation.

The objectives of this study are to develop a method for the optimal sizing of cannulated cancellous screws for fixing the bone fragments of the tibial plateau fractures, which depend on the patient's characteristics. The sizing of the cannulated cancellous screw requires the calculus of the screw diameter and length, followed by crushing strength verification for the bone fragments.

2 Methodology

The fixation with cannulated cancellous screws of the comminutive tibial fracture fragments requires the determination of the optimal insertion direction of the fixing screws, the number of cortical fixing screws and the sizing of the cortical fixing screws by choosing their length and diameter.

2.1 Anatomy

Tibial plateau fractures are joint injuries that threaten short and long term knee function. Tibial plateau fractures are severe joint fractures that are difficult to treat. Surgical treatment is superior to orthopedic treatment because it can restore the joint surface of the tibia.

Indications for surgical treatment are open fractures, fractures associated with compartment syndrome, fractures with associated vascular or nerve damage and bills with

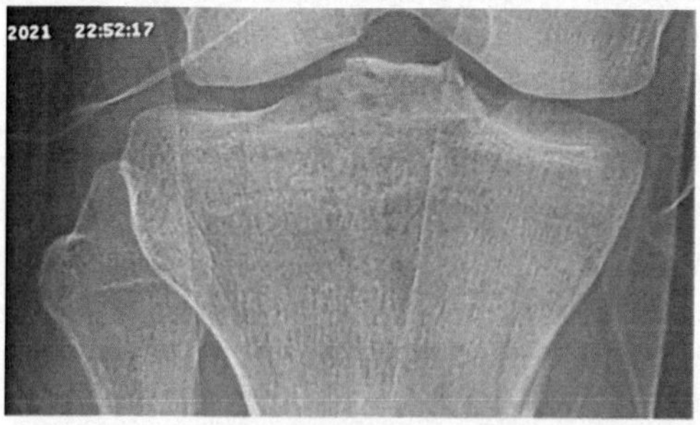

Fig. 2. Anatomy of the tibial plateau

displacement over 2 mm. Of the two articular surfaces, the largest is the medial plateau, which is concave in both transverse planes. A little higher than the surface of the medial joint is the lateral plateau, which is smaller and convex, and allows its identification on the lateral X-rays (see Fig. 2).

2.2 Sizing of Fixing Screws

The dimensional parameters of the cannulated cancellous screws for fixing fractured bone fragments in the basal bone are the diameter and the length. These parameters can be determined in the surgical planning stage with the support of virtual and physical models developed in this research.

On the virtual model built with the support of the three-dimensional technologies [16], the length of the screw can be determined by direct measurement, so that it can penetrate the opposite cortical wall without piercing it (see Fig. 3).

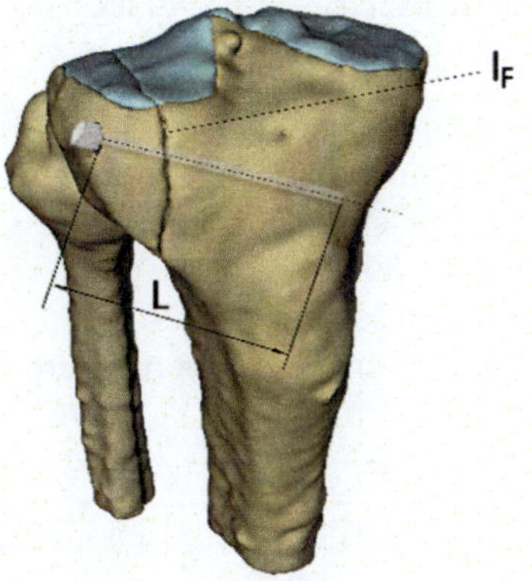

Fig. 3. Determination of cannulated cancellous screw length (L – length of screw, l_F - fracture line).

The sizing of the cortical fractures fragments fixation with cannulated cancellous screws, requires the determination of the screw diameter, a calculation that is performed according to the number of screws that are inserted.

Under the action of the F forces, the bone fragments fixed with cannulated screw tend to slide towards the tibia. The screw prevents this slip and takes over through the bone fragments the forces acting on the body (see Fig. 4).

If the fixing is done with more than one cannulated screw, two equal and opposite forces act on each screw, which are called forces on the screw (see Fig. 4):

$$F_1 = \frac{F}{n} \tag{1}$$

where, n represents the number of screws that take over the force F.

In order to determine the stresses in the screw rod, it is sectioned at the fracture surface and is removed the bottom. The forces transmitted through the section of the

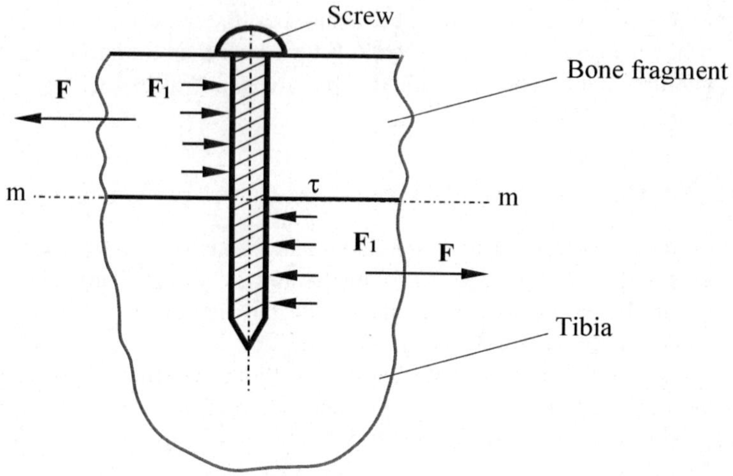

Fig. 4. Cannulated cancellous screw fixation of the bone fragment on the tibia.

screw rod from the lower to the upper part, balance the F_1 force, i.e. act parallel to it in the plane of the section and by summation give a result equal to F_1.

In the plane of the section m-m acts the tangential stresses τ. For cannulated screws belonging to the category of small thickness parts, the tangential stress τ is considered to be uniformly distributed over the section.

The tangential stress τ, from the cross section of a screw in a joint with n screws, is calculated with the relation:

$$\tau = \frac{F_1}{\frac{\pi \cdot d^2}{4}} = \frac{F}{n \cdot \frac{\pi \cdot d^2}{4}} \leq \tau_a \tag{2}$$

In the literature it is shown that cannulated screws provide better fixation for tibial plateau fractures, but also are counterintuitive in that smaller screws provid greater fixation than larger screws of the same type. For this reason, it is necessary to precisely dimension the screws for fixing the bone fragments [17] so that they can hold the weight of the whole body, following proximal tibial fractures osteosynthesis [18], which can also ensure a good control of the tightening moments [19].

From relation (2) results the expression for the required diameter of a screw, when the number of the fixing screws is known:

$$d = \sqrt{\frac{4F}{n \cdot \pi \cdot \tau_a}} \tag{3}$$

where F has the maximum value of the body weight estimated at 100 kg·9,8 mm/s^2 = 980 N, $\tau_a = 150\ \text{N/mm}^2$.

In the case of fixing with a screw, its diameter is:

$$d = \sqrt{\frac{4.980}{1.3, 14.150}} = \sqrt{\frac{3920}{471}} = \sqrt{8, 32} = 2, 88 \approx 3\ \text{mm} \tag{4}$$

Knowing the diameter of the cannulated cancellous screw, further from the relation (2) it can be determined the capable force of a joint with cannulated screws:

$$F_{cap} = n \cdot \frac{\pi \cdot d^2}{4} \cdot \tau_a \quad (5)$$

2.3 Resistance Calculation of Bone Fragments

Bone fragments fixed with cannulated cancellous screws should be checked for crushing. Since the transmission of forces to the screw is done by pressing between the walls of the hole and the screw rod, it is necessary to check if there is no crushing of the hole walls from the bone fragment. In this way it is checked the pressure crushing on the hole.

The way it is transmitted the pressure (tension) on the walls of the hole is represented in Fig. 5. The pressure distribution diagram on the semi-cylindrical surface of the hole is not known. It depends on the irregularities of the shape of the hole and the screw rod and their particularities (see Fig. 5a).

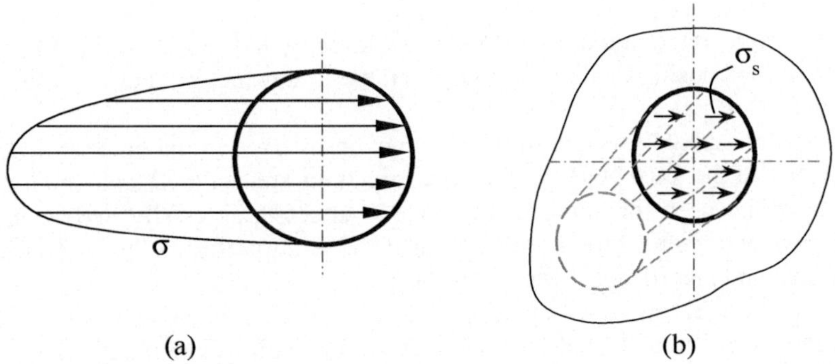

Fig. 5. Crushing stress of the fracture fragment: (a) Pressure distribution on the semi-cylindrical surface of the hole; (b) Uniform conventional pressure distribution.

For this reason, the crushing calculation is conventional assuming that the uneven pressure transmitted from the screw to the surface of the hole is evenly distributed on the diametrical plane of the hole section (see Fig. 5b).

The crushing stress σ_s on the diametrical plane is approximately equal to the maximum stress at the pressure on the hole σ_g.

To calculate this conventional crushing stress σ_s, it is necessary to divide the force returning to the cannulated screw F_l by the area of the diametrical section (see Fig. 5b).

The diametrical surface is a rectangle, which has as sides the diameter of the cannulated cancellous screw d and the thickness of the fracture fragment t, which transmits the pressure on the fracture fragment.

For a joint with n screws, the crushing check relationship is:

$$\sigma_{max,s} = \sigma_s = \frac{F_l}{d \cdot t} = \frac{F}{n \cdot d \cdot t} = \leq \sigma_{as} \quad (6)$$

where in the admissible crushing stress is $\sigma_{as} = (2\ldots2{,}5)\sigma_a$, for bone with values of $\sigma_{as} = 50\,\mathrm{N}\cdot\mathrm{mm}$.

For the numerical values employed, it can be calculated:

$$\sigma_{max,s} = \frac{9{,}8\cdot 100}{1\cdot 3\cdot 10} = 32{,}66 \le \sigma_{as} \tag{7}$$

Based on relation (6), it is calculated the number of screws required to meet the crushing strength condition:

$$n \ge \frac{F}{d\cdot t\cdot \sigma_{as}} \tag{8}$$

which numerically becomes:

$$n \ge \frac{9{,}8\cdot 100}{3\cdot 10\cdot 50} = 0,65 \approx 1 \text{ cannulated screw} \tag{9}$$

2.4 The Screw Fixation

The screws act as internal fixations. This is achieved by fixing the fractured fragments in the determined position without the screws loosening, which could lead to secondary displacement.

The type of screw by which the bone fragment is fixed in the base bone depends on the quality of the fixed bones. If the bone structures are healthy, stable monocortical angular bolts are used for fixing. If the bone is affected, especially by osteoporosis, stable bicortical angular screws are used, or if the fixation is done by several screws, monocortical and bicortical screws can be used.

By using fixing screws without fixing plates, firm contact is avoided and no pressure is created between the plate and the bone surface for stability, and blood flow from the periosteum is less affected than in other fixation pockets, such as conventional open ones.

If the screw is inserted from the lateral to the medial, the surgical maneuver must be performed carefully to avoid penetration of the screw into the concave surface of the medial joint.

3 Results

The validation in practice of the fixation methodology for tibial plateau fractures with cannulated cancellous fixation screw was performed on a 38-year-old female patient who presented at the Emergency Reception Unit of the Mobile Emergency Service for Resuscitation and Extrication Mureș, where was diagnosed with Schatzker type IV tibial plateau fracture. Tibia scans were performed with a CT OPTIMA 580 W.

Based on the initial clinical examination, diagnosis and imaging examination performed in the emergency department, she was sent to the traumatology orthopedics department for a fracture reduction and fixation surgery.

3.1 Installation, Clinical Examination and Evaluation of Lesions

The patient diagnosed with tibial plateau fracture that requires surgery for reduction and fixation lies on his back, with a support or stirrup at the base of the limb, so that limb mobility is unlimited and the knee can be positioned in the varus or valgus during the procedure (see Fig. 6).

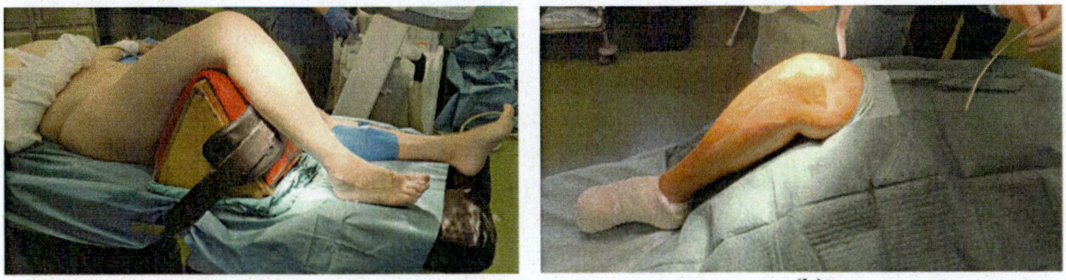

(a) (b)

Fig. 6. Installing the patient for fixation, while leaving the leg free: (a) Installation in supine position with a support at the base of the thigh; (b) Installation of sterile fields.

Depending on the surgeon's preferences, the foot can be placed on the operating table in flexion or left free. A gauze is applied to the base of the lower limb. If bone graft collection is required, a sterile tourniquet and sterile U-shaped fields are used to facilitate access to the iliac crest. On the side of the operated knee is placed a Rontgen TV image amplifier that allows the collection of images in the antero-posterior and latero-lateral incidence (see Fig. 7). The image amplifier is placed in the optimal position before isolation.

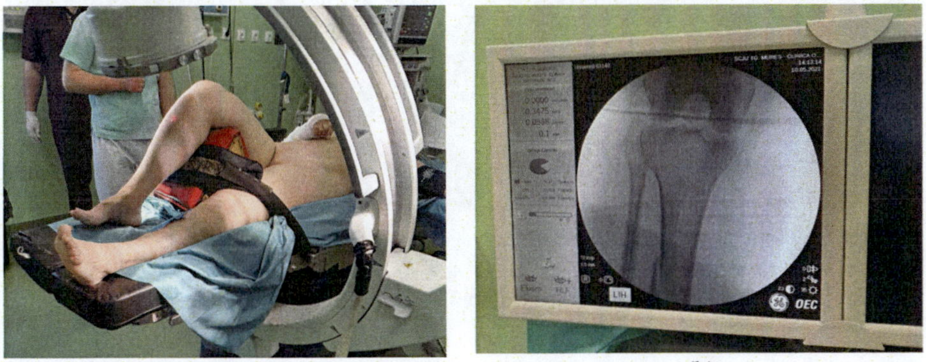

(a) (b)

Fig. 7. Collection of images in the antero-posterior and latero-lateral incidence: (a) Rontgen TV image amplifier; (b) Radiographic image.

3.2 Reduction and Fixation

By manipulating the fractured bone with traction, rotation or bending forces, the indirect reduction of the fracture is achieved. To facilitate the reduction, it is possible to apply direct forces to the fracture site by means of soft tools such as a hammer over which a towel is passed, but the surgical maneuver must be performed carefully so as not to cause additional soft tissue damage. If the indirect reduction does not give the expected results, a limited open reduction can be performed by proximal and distal incisions through which a clamp is placed on the bone fragment. Checking the reduction quality can be done by using three-dimensional technnologies that allow the simulation of the reduction degree of bone fragments [20, 21].

In the reconstruction of the proximal tibial extremity, some important features were taken into account: the antero-posterior slope of the tibial plateau is 10°, the medial plateau has a larger surface and is concave, unlike the lateral one located above, it has a smaller surface and it is convex in shape. That is why the screws under the joint surface that pass from the side to the medial can protrude into the joint. The 'screw-joint distance' may help to guide screw placement intra-operatively [22].

The treated patient required fixation of the Schatzker type IV tibial fracture fragment with a cannulated cancellous screw of 3 mm in diameter and 54 mm long whose sizing and crushing verification of the fractured bone fragment was performed according to the methodology described in this paper.

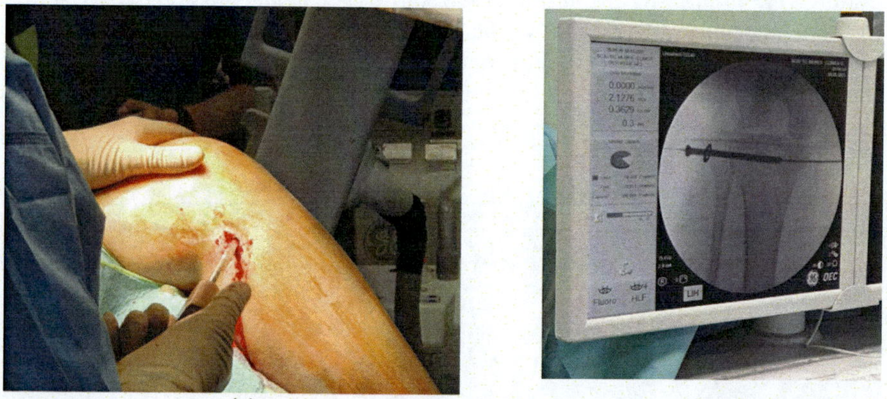

(a) (b)

Fig. 8. Fixing the medial tibial plateau with the cannulated cancellous screw: (a) Surgical maneuver; (b) Image on Rontgen TV image amplifier.

Treatment of tibial plateau depressed fracture with a net-fixation of Kirschner wires is a biological fixation and is a reliably method [23]. For this reason, in order to guide the cannulated cancellous screw, its trajectory materialized in a slightly inclined direction to the medial with the help of a Kirschner brooch, after which the screw fixing was performed (see Fig. 8).

3.3 Aftercare

In the first 48 h after the operation, a careful monitoring of the tibial compartments was performed in order to avoid the compartment syndrome. The neorovascular condition of the foot was also closely monitored for blood deficiency and neurological loss.

The mobilization was performed from the first postoperative day because there were no injuries or complications. For recovery, static quadriceps executions were performed with the passive movement of the knee and the recommendation of subsequent active movements of the knee and ankle in order to perform a full range of movements in an interval of 4–6 weeks [24, 25]. Wound healing was performed in the first two weeks with follow-up at 6 and 12 weeks.

4 Conclusion

The conclusion of this study is formulated in an interdisciplinary context of clinical, biomedical and technological considerations, starting from the finding that during orthopedic surgeries surgeons are fixing the fracture fragments with cannulated cancellous screws whose sizing is mainly based on surgical experience without a proper dimensioning.

In order to avoid an improper surgery for the screw fixing, this paper presents a calculation algorithm that allows a sizing of the fixing screw. The diameter of the screw is determined according to the mass of the patient, which at the time of recovery, by its weight will load the fixation. The capable force of a screw joint is also determined. Next, the bone fragments fixed with cannulated cancellous screws are checked for crushing, and the number of screws required for a proper fixation is also calculated.

The present methodology is recommended in the preoperative planning stage of the tibial plateau fractures surgeries for reductions and screw fixations, offering the guarantee of a biomechanical quality fixation which ensures the good health of the patient. It opens new perspectives for calculating the strength of other types of fractures.

The limitations of the study arise because in vivo the biological conditions are slightly different, the resistance of the bone is not uniform in its entire volume and the blood circulation can influence the local resistance.

Institutional Review Board Statement. The study was conducted according to the guidelines of the Declaration of Helsinki, and approved by the Ethics Committee of Emergency County Hospital Targu Mures (protocol code 2Ad.6108/02.03.2021).

Informed Consent Statement. Informed consent was obtained from the patient involved in the study.

Conflicts of Interest. The authors declare no conflict of interest.

References

1. Claes, L.: Dynamisierung der Osteosynthese : Zeitpunkt und Methoden [Dynamization of fracture fixation: timing and methods]. Unfallchirurg **121**(1), 3–9 (2018)

2. Hodgson, S.: AO principles of fracture management. Ann. R. Coll. Surg. Engl. **91**(5), 448–449 (2009)
3. Lowe, J.A., Tejwani, N., Yoo, B.J., Wolinsky, P.R.: Surgical techniques for complex proximal tibial fractures. Instr. Course Lect. **61**, 39–51 (2012)
4. Lafrance, R.M., Giordano, B., Goldblatt, J., Voloshin, I., Maloney, M.: Pediatric tibial eminence fractures: evaluation and management. J. Am. Acad. Orthop. Surg. **18**(7), 395–405 (2010)
5. Bong, M.R., et al.: Suture versus screw fixation of displaced tibial eminence fractures: a biomechanical comparison. Arthroscopy **21**(10), 1172–1176 (2015)
6. Ye, M., Chen, J., Hu, F., Liu, Y., Tan, Y.: Suture versus screw fixation technique for tibial eminence fracture: a meta-analysis of laboratory studies. J. Orthop. Surg. (Hong Kong) **28**(1), 2309499020907977 (2020)
7. Moldovan, F.: New approaches and trends in health care. Proc. Manuf. **22**, 947–951 (2018)
8. Moldovan, F.: Recent trends in bioprinting. Proc. Manuf. **32**, 95–101 (2019)
9. Caspari, R.B., Hutton, P.M., Whipple, T.L., Meyers, J.F.: The role of arthroscopy in the management of tibial plateau fractures. Arthroscopy **1**(2), 76–82 (1985)
10. Jennings, J.E.: Arthroscopic management of tibial plateau fractures. Arthroscopy **1**(3), 160–168 (1985)
11. Chen, H.-W., Liu, G.-D., Wu, L.-J.: Clinical and radiological outcomes following arthroscopic-assisted management of tibial plateau fractures: a systematic review. Knee Surg. Sports Traumatol. Arthrosc. **23**(12), 3464–3472 (2014). https://doi.org/10.1007/s00167-014-3256-2
12. Bowles, R.J., Chadayammuri, V., Baldini, T., Brecevich, A., Mauffrey, C.: Split-depressed lateral tibial plateau fractures: a comparison of augmented percutaneous screws versus augmented plate and screw construct in a cadaveric model. J. Orthop. Trauma **32**(7), e270–e275 (2018)
13. Boisrenoult, P., Bricteux, S., Beaufils, P., Hardy, P.: Vis versus plaque vissée dans les fractures séparation enfoncement du plateau tibial latéral. Etude biomécanique cadavérique. Société Française d'Arthroscopie [Screws versus screw-plate fixation of type 2 Schatzker fractures of the lateral tibial plateau. Cadaver biomechanical study. Arthroscopy French Society]. Rev. Chir Orthop. Reparatrice Appar. Mot. **86**(7), 707–711 (2000)
14. Sament, R., Mayanger, J.C., Tripathy, S.K., Sen, R.K.: Closed reduction and percutaneous screw fixation for tibial plateau fractures. J. Orthop. Surg. (Hong Kong) **20**(1), 37–41 (2012)
15. Syed, M.A., Panchbhavi, V.K.: Fixation of tibial pilon fractures with percutaneous cannulated screws. Injury **35**(3), 284–289 (2004)
16. Moldovan, F., Gligor, A., Bataga, T.: Integration of three-dimensional technologies in orthopedics: a tool for preoperative planning of tibial plateau fractures. Acta Inform. Med. **28**(4), 278–282 (2020)
17. Cooper, H.J., Kummer, F.J., Egol, K.A., Koval, K.J.: The effect of screw type on the fixation of depressed fragments in tibial plateau fractures. Bull. Hosp. Joint Dis. **60**(2), 72–75 (2001–2002)
18. Haak, K.T., Palm, H., Holck, K., Krasheninnikoff, M., Gebuhr, P., Troelsen, A.: Immediate weight-bearing after osteosynthesis of proximal tibial fractures may be allowed. Dan Med. J. **59**(10), A4515 (2012)
19. Moldovan, F., Batag, T.: Torque control during bone insertion of cortical screws. Proc. Manuf. **46**, 484–490 (2020)
20. Moldovan, F., Gligor, A., Bataga, T.: Structured integration and alignment algorithm: a tool for personalized surgical treatment of tibial plateau fractures. J. Pers. Med. **11**, 190 (2021)
21. Moldovan, L., Gligor, A., Grif, H.S., Moldovan, F.: Dynamic numerical simulation of the 6-PGK parallel robot manipulator. Proc. Rom. Acad. Ser. A-Math. Phys. Tech. Sci. Inf. Sci. **20**(1), 67–75 (2019)

22. Ye, X., Huang, D., Perriman, D.M., Smith, P.N.: Influence of screw to joint distance on articular subsidence in tibial-plateau fractures. ANZ J Surg. **89**(4), 320–324 (2019)
23. Chen, Z.X., Zhang, W., He, H.H., Liu, M.J., Zeng, D.H.: Biomechanical study on a net-fixation of Kirschner wire in treating depressed tibial plateau fractures. Zhongguo Gu Shang **27**(5), 418–421 (2014)
24. Mikolajczyk, T., Moldovan, F., Ciobanu, I., Chalupczak, A., Marin, A.G.: Brain research using computer test. Proc. Technol. **22**, 1113–1120 (2016)
25. Mikolajczyk, T., Moldovan, L., Chalupczak, A., Moldovan, F.: Computer aided learning process. Proc. Eng. **181**, 1028–1035 (2017)

A Fuzzy Mathematical Model to Estimate the Energy Cost for Heat Pump System

Alexandra Ban(✉) and Constantin Bungău

University of Oradea, Oradea, Romania

Abstract. The heating and cooling of buildings plays a crucial role in achieving the goal of a climate-neutral Europe by 2050. Heat pumps are sustainable technologies that offer an energy-efficient alternative for heating and cooling. This article proposes a fuzzy mathematical model to evaluate the effects that heat demand, secondary agent temperature, insulation level and coefficient of performance of the heat pump have on energy costs.

Keywords: Fuzzy model · Heat pump · Optimization · Cost · Energy · Heating · Cooling · Building

1 Introduction

In the current context of global warming, European Union leaders have endorsed the goal of achieving a climate-neutral Europe by 2050 [1]. This means that the European Union will drastically reduce its greenhouse gas emissions. According to Eurostat, the energy needed for heating and air conditioning have the largest share of total energy consumed in the European Union, around 45% [2–4]. Only 22% of the energy consumed by the European Union in 2019 comes from renewable energy sources [5]. For the residential sector, the energy consumption for space heating is by far the most important end-use with an estimated share of 52% in 2015 [6].

The heating and air conditioning of buildings plays a crucial role in achieving the goal of a transition to neutrality in the European Union. As a result, research into the economic efficiency of heating and air conditioning systems equipped with heat pumps fits perfectly with the current and future needs of sustainability and sustainable development imposed by European directives.

There is a wide range of sustainable heating and cooling technologies available, one of which is the use of heat pumps. With a 14% increase in sales of heat pumps reached in 2018, the European heat pump market increases more and more each year. These systems can use low-grade renewable energy sources (such as air, water, ground, solar) to reduce the demand for fossil fuels and also to reduce greenhouse gas emissions [7]. Heat pump technology only meet 5% of worldwide building heating demand at present [8]. In the long term and at the city scale, a consideration of cost of greenhouse emission must be taken into account which is net in favor of renewable energy systems with heat pumps [9–11]. The heat pump, a very efficient technology, is essential for reducing energy consumption and emission [12].

L. Moldovan and A. Gligor (Eds.): Inter-Eng 2021, LNNS 386, pp. 784–795, 2022.
https://doi.org/10.1007/978-3-030-93817-8_70

The costs identified for heating and air conditioning systems with heat pump having as primary source air, water or soil are shown in the Fig. 1.

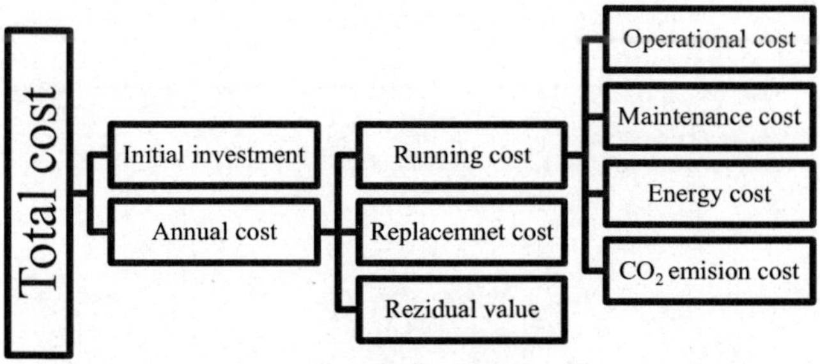

Fig. 1. Classification of costs.

Among these cost categories, the present study addresses to the energy cost. This cost represents the annual costs as well as the fixed and peak energy charges, including national taxes [13]. The energy cost for systems equipped with heat pumps make a significant contribution to the cost of using the whole system.

The key factors influencing the energy cost of a system equipped with a heat pump are the following:

- Heat pump efficiency
- Energy requirements
- Heat source temperature
- The level of insulation of the building

Considering the previous work for developing mathematical models for prediction of running cost, Langdon [14] used a parametric model with a regression analysis. Kirkham et al. [15] used a regression mode, where energy cost can be calculated using only the building area and the number of users.

The goal of this study is to propose a different approach to estimate the monthly energy cost for system using heat pump and to identify how certain parameters influence this energy cost. A study in the UK showed that an air source heat pumps used for underfloor heating could save up to 37% of energy costs compared with a conventional boiler system [16]. Thus, we developed a model based on fuzzy logic that allows the calculation of monthly energy cost of heat pump systems under certain conditions, such as the insulation level of the building, the coefficient of performance of the system and the secondary agent temperature.

2 Materials and Methods

In this research, a fuzzy mathematical model was developed to evaluate the effects that heat demand, secondary agent temperature, insulation level and coefficient of performance of the heat pump have on energy costs. At the same time, fuzzy logic will help to establish the optimal level for energy costs depending on the constraints considered. The model is implemented using Fuzzy Logic Toolbox of MATLAB.

The component parts of a fuzzy controller are fuzzification, interface and defuzzification [17] is shown in Fig. 2.

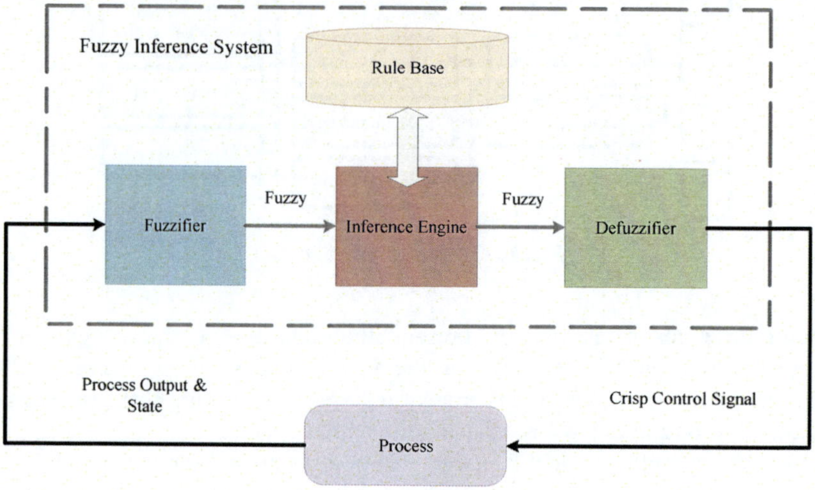

Fig. 2. Conceptual diagram of the fuzzy logic controller.

Fuzzification is the process by which numerical input variables are translated into linguistic variables. The membership functions used in the present research are the triangular and the gaussian functions. The triangular membership function is defined using three parameters {a, b, c} as follows:

$$triang(x; a, b, c) = \begin{cases} \frac{x-a}{b-a,}, a \leq x < b \\ \frac{c-x}{c-b,}, b \leq x < c \\ 0, \ otherwise \end{cases} \quad (1)$$

The Gaussian membership function is defined using two parameters, the standard deviation σ and the mean c.

The role of the interface is to transform the fuzzy input set into a fuzzy output set, using an IF-THEN rules base and an interface mechanism. The interface system is established based on the researcher's experience. Defuzzification is the calculation of a numerical value based on symbolic rules. For the defuzzification process in this research, the center of gravity method was used, and it is defined by the *cog* operator as follows:

$$cog(B) = \frac{\int_Y \mu_B(y) y dy}{\int_Y \mu_B(y) dy} \quad (2)$$

The mathematical model can be defined as $f : \mathbb{R}^4 \rightarrow \mathbb{R}$ and is shown schematically in Fig. 3.

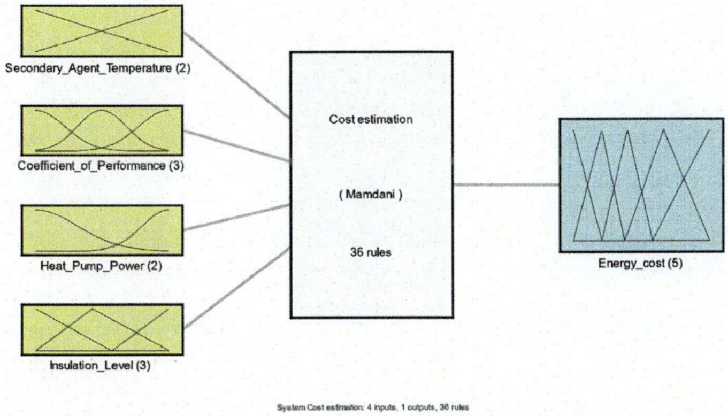

Fig. 3. Logical fuzzy system.

2.1 Input Data

The input variables used in the fuzzy model are the temperature of the secondary agent, the coefficient of performance of the system, the power of the heat pump and the level of insulation. The literature review, the discussions with the experts in the field and the experience facilitated the determination of the relevant points for each function. The relevant data was synthesized in Table 1.

Table 1. Definition of degree of membership for input variable.

Input variables	Fuzzy set	Type	Limiter
Secundary_Agent_Temperature	Low	Triangular	[25, 35, 55]
	High	Triangular	[35, 55, 65]
Coefficient_of_Performance	Low	Gaussian	[0.484, 2.8]
	Medium	Gaussian	[0.396, 3.9]
	High	Gaussian	[0.484, 5]
Heat_Pump_Power	Low	Gaussian	[8, 5]
	High	Gaussian	[5, 30]
Insulation_Level	Low	Triangular	[0.6, 0.9, 1.1]
	Medium	Triangular	[0.2, 0.5, 0.9]
	High	Triangular	[0, 0.2, 0.6]

The universe of discourse for the secondary agent temperature is in the range [35, 55] and is measured in degrees Celsius. The degree of membership is classified as Low for system based on underfloor heating and High for heat distribution systems through fan coils or radiators. The membership functions are shown in Fig. 4.

The universe of discourse for the variable coefficient of performance is in the range [2.8, 5] and is defined as the ratio between the amount of heating and air conditioning

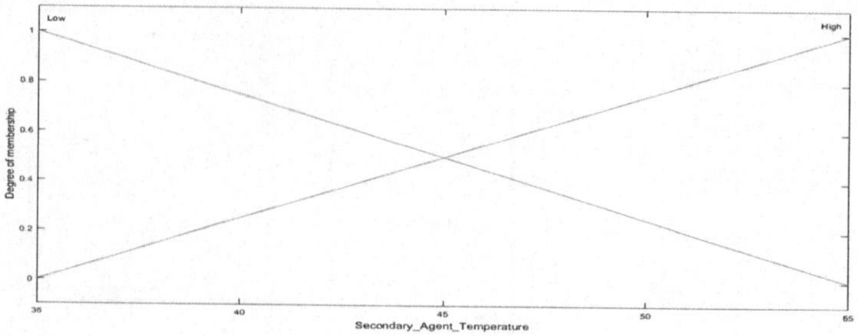

Fig. 4. Degree of membership for fuzzy sets of secondary agent temperature

produced by the system equipped with a heat pump in kilowatts and the consumption of said heat pump in kilowatts.

$$COP = Q/W \tag{3}$$

where:

- Q is the amount of heating and air conditioning produced by the system equipped with a heat pump in kilowatts
- W heat pump consumption in kilowatts.

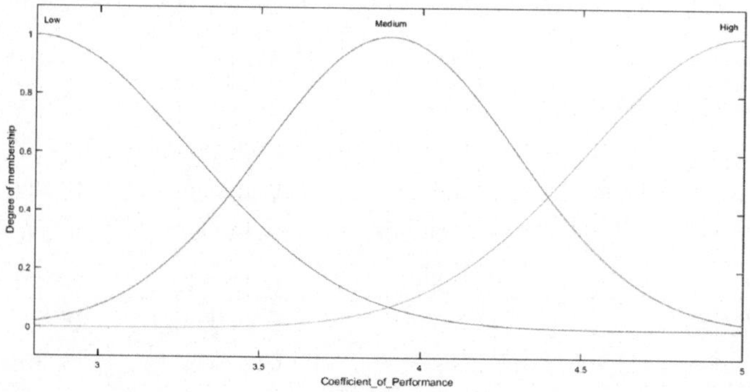

Fig. 5. Degree of membership for fuzzy sets for coefficient of performance

The degree of membership is classified as Low mainly for air source heat pump, Medium mainly for ground source heat pump and High mainly for water source heat pump. The membership functions are shown in Fig. 5.

The universe of discourse for the variable heat pump power is measured in kilowatts and is in the range [5, 30]. The degree of membership is classified as Low and High. The membership functions are shown in Fig. 6.

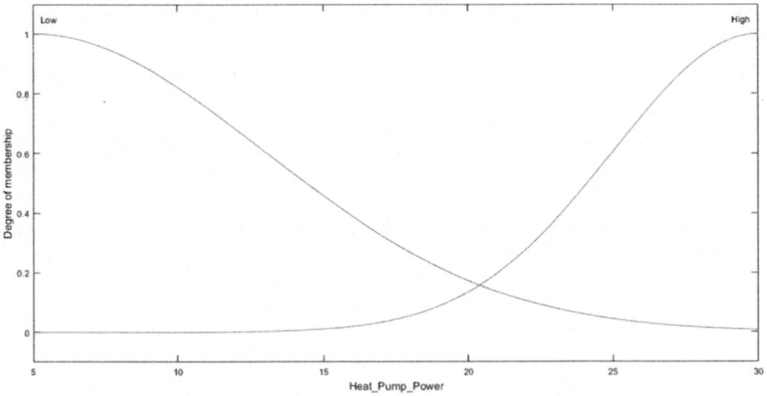

Fig. 6. Degree of membership for fuzzy sets of heat pump power

The universe of discourse for the variable insulation level is in the range [0.2, 0.9] and is measured with the global thermal insulation coefficient of a building (*G*), a thermo-energetic parameter of the building envelop. The global thermal insulation coefficient is calculated with the relation:

$$G = \frac{\sum(L \cdot \tau)}{V} \cdot 0.34 \cdot n \quad \left[\frac{W}{m^3 K}\right] \tag{4}$$

where:

- L the thermal coupling coefficient, calculated with the relation: $L = A/(R'_m)$
- τ outdoor temperature correction factor
- V the interior heated volume of the building
- A area of construction element
- R'_m the corrected average of specific thermal resistance, on the whole building, of a construction element
- n the number of air changes per hour

The degree of membership is classified as Low, Medium and High. The membership functions are shown in Fig. 7.

2.2 Output Data

The universe of discourse for the output variable energy cost is measured in Euro, representing the monthly cost of electricity and is in the range [20, 500]. The membership

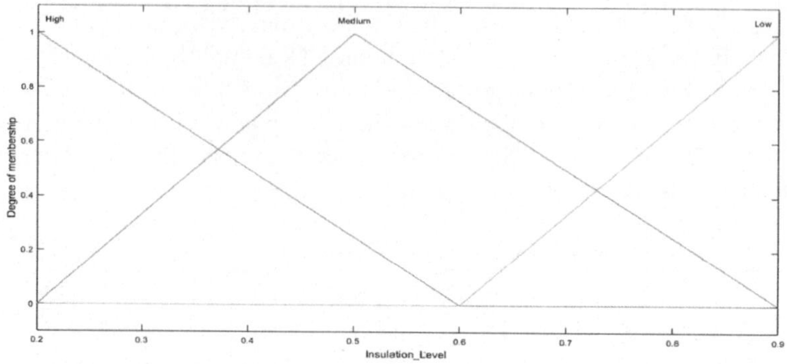

Fig. 7. Degree of membership for fuzzy sets of insulation level

functions are shown in Fig. 8. The relevant points for each function were synthesized in Table 2.

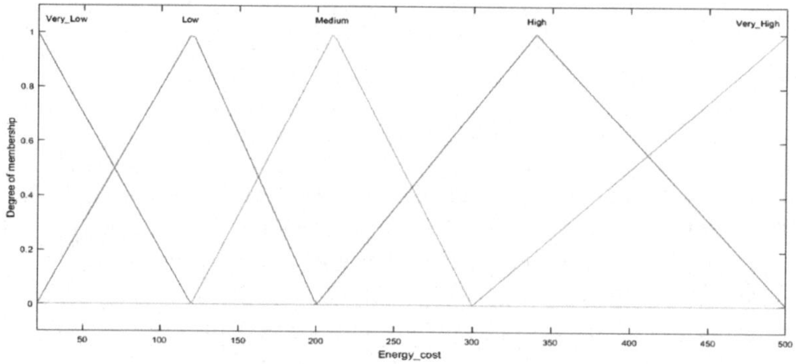

Fig. 8. Degree of membership for fuzzy sets of energy cost

Table 2. Definition of degree of membership for output variable

Input variables	Fuzzy set	Type	Limiter
Energy_Cost	Very low	Triangular	[0, 20, 120]
	Low	Triangular	[20, 120, 200]
	Medium	Triangular	[120, 200, 300]
	High	Triangular	[200, 340, 500]
	Very high	Triangular	[300, 500, 700]

2.3 Validation

The fuzzy mathematical model was validated with a set of five representative buildings, which were monitored from March 2020 until February 2021. The main characteristics of the buildings considered are summarized in Table 3. There were three residential building and two commercial buildings monitored.

Table 3. Profile of buildings

	Building	Sector	Heat pump system	Heat pump power	Heating Type	Insulation level
B1	Single-family building Oradea	Residential 130 m^2	Ground source with borehole	8 kW	Underfloor	0.34
B2	Office Oradea	Commercial 375 m^2	Water source	18 kW	Underfloor	0.29
B3	Single-family building Paleu	Residential 223 m^2	Ground source with borehole	14 kW	Underfloor	0.37
B4	Single-family building Nojorid	Residential 350 m^2	Ground source with horizontal collector	22 kW	Radiator	0.45
B5	Office Oradea	Commercial 320 m^2	Air source	18 kW	Underfloor	0.58

The validation of the model is performed based on the difference between de monitored energy cost for each building and the predicted energy cost based on the fuzzy model. The closer the level of accuracy is to 0, the better the model. As shown in Table 4, the accuracy of predicted energy cost ranges between ±10%.

Table 4. Validation of the developed model

	Monthly energy cost		Difference	Accuracy (%)
	Monitored	Estimated		
B1	€ 51	€ 56	−5	−8.93%
B2	€ 102	€ 109	−7	−6.42%
B3	€ 75	€ 70	5	7.14%
B4	€ 251	€ 269	−18	−6.69%
B5	€ 184	€ 172	12	6.98%

3 Results and Discussion

Between the global thermal insulation coefficient of a building and the cost of energy is a positive relationship. A lower level of insulation implies a higher global thermal insulation coefficient and higher costs of energy. Particularly important is the fact that high insulation, which tends to the standard of passive buildings, implies a drastic decrease in energy costs. There is also a positive relationship between the power of the heat pump and the energy cost. Higher heat demand can be accomplished by using a higher power of the heat pump, which implies higher energy cost. Likewise, the relationship between the temperature of the secondary agent and the cost of energy is positive. A low temperature of the secondary agent due to the use of the underfloor heating system involves lower energy costs. On the other hand, heating systems with radiators or fans, which require high secondary temperatures, result in high energy costs. The relationship between coefficient of performance and energy cost is negative. A small coefficient of performance of the system equipped with a heat pump would imply higher energy costs, and vice versa.

From the developed model it was possible to visualize the surface and the contour maps of the fuzzy system. It graphically shows the dependence of energy cost for systems equipped with a heat pump, to the input variables, namely heat pump power, coefficient of performance, insulation level and temperature of the secondary agent. Given that the heat requirement is determined by the specifics and location of each building, we are interested in the inference of the three input variables, specifically the temperature of the secondary agent, the level of insulation and the coefficient of performance of the system for a pump power. The surface obtained through the Mamdani interface together with the contour maps are presented below.

The graphs in Fig. 9 show the influence of the heat pump performance coefficient and the insulation level on the energy cost, when the temperature of the secondary agent is Low and constant. The optimal variant for the energy cost is for High building insulation with a global thermal insulation coefficient under *0.4*. It is important to note that High insulation of the building plays a decisive role in energy savings. For an insulation level close to the nZEB standard, the coefficient of performance of the heat pump system does not have a significant influence on the energy cost. Low level of insulation of the building implies a significant increase in the monthly cost of energy. As a result, the level of insulation of the building is a key factor in optimizing the cost of energy. The unfavourable variant is Low isolation level and Low coefficient of performance, in which case the costs are maximized.

The graphs in Fig. 10 show the influence of coefficient of performance and the temperature of the secondary agent on the energy cost, when the insulation level is High and constant. High coefficient of performance in correlation with Low secondary agent temperature is a favourable option that results in minimization of energy costs.

It is important to note that for High coefficient of performance, the temperature of the secondary agent does not have a significant influence on the energy cost. For the economic optimization of the system, the option for radiant heating must be considered. On the other hand, Low and Medium coefficient of performance correlates with a High secondary agent temperature has a negative impact on the energy cost.

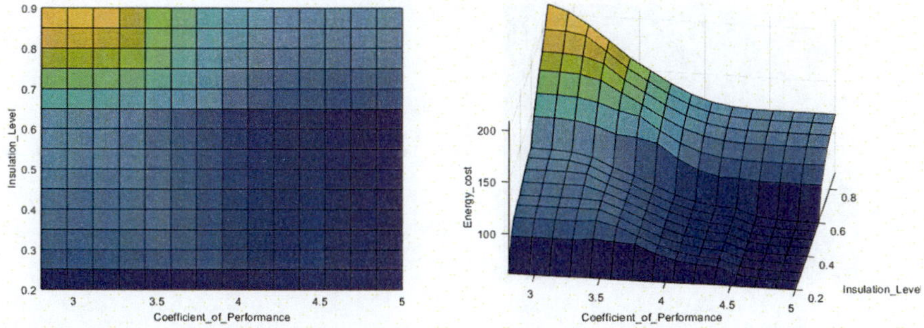

Fig. 9. Energy cost in replay to coefficient of performance and insulation level: surface and contour map

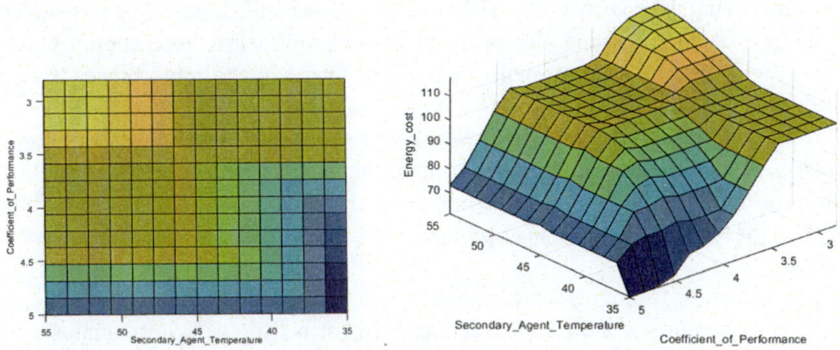

Fig. 10. Energy cost in replay to coefficient of performance and secondary agent temperature: surface and contour map

The graphs in Fig. 11 show the influence of the insulation level and the temperature of the secondary agent on the energy cost, when the coefficient of performance is High and constant. The optimal option to minimize the cost of energy is given by the use of underfloor heating system and medium and High insulation. A global thermal insulation coefficient under 0.35 results in minimization of energy costs, in which case the temperature of the secondary agent has a relatively small influence. For Low insulation level, the energy cost increases with the increase of the secondary agent temperature. The unfavourable variant for which the energy cost is maximized is for Low insulation level of the building and Low and High secondary agent temperature.

Unlike traditional methods based on mathematical programming models which require well-defined coefficients, fuzzy logic is capable to provide the most effective solution to complex issues and help in dealing with the uncertainties.

4 Conclusion

Following the analysis of surface and contour maps, it was found that the level of insulation is the key factor influencing the cost of energy. A High level of insulation,

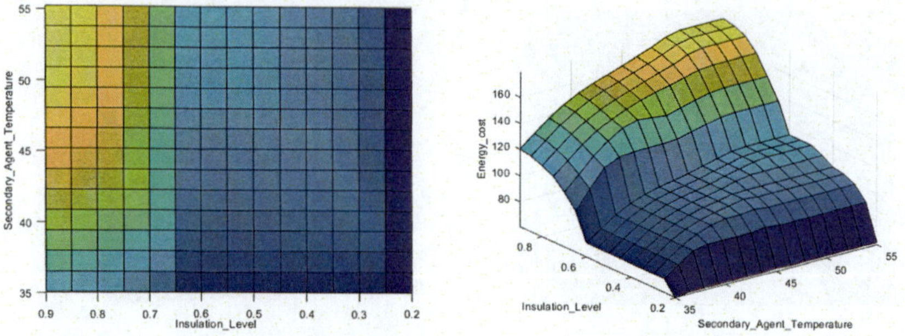

Fig. 11. Energy cost in replay to insulation level and secondary agent temperature: surface and contour map

implies minimizing the energy costs. The European Union rules according to which from 2021 all new constructions must meet the conditions of "nearly zero-energy building" come as a confirmation of the result obtained in terms of the importance of optimal insulation for buildings.

This study highlights the fact that heating and cooling systems with heat pumps are sustainable and economical solutions as long as they are accompanied by a high level of insulation of the building and a low temperature of the secondary agent. Also, there is very important the source of the heat pump, respectively the coefficient of performance of the system.

As a direction of future research, the aim is to identify and quantify the other costs involved in heat pump systems in order to perform a life cycle cost analysis and a sensitivity analysis. In this way we will try to find the most optimal solutions from an economic and ecological point of view for heating and cooling systems.

Acknowledgements. This research benefited from support through the project "SmartDoct—High quality programs for PhDstudents and postdoctoral researchers of the University of Oradea for increasing the relevance of research andinnovation in the context of the regional economy", ID/project code: 123008, co-financed by the EuropeanSocial Fund through The Human Capital Operational Program 2014–2020.

References

1. EU, Directive 2010/31/EU of the European Parliament and of the Council Off. J. Eur. Union (2010)
2. Prada, M., et al.: New solutions to reduce greenhouse gas emissions through energy efficiency of buildings of special importance – hospitals. Sci. Total Environ. **718**, ID 137466 (2020). https://doi.org/10.1016/j.scitotenv.2020.137446
3. Bungau, C.C., Prada, I.F., Prada, M., Bungau, C.: Design and operation of construction: a healthy living environment – parametric studies and new solutions. Sustainability **11**(23), 6824 (2019). https://doi.org/10.3390/su11236824
4. Prada, M., Popescu, D.E., Bungau, C., Pancu, R., Bungau, C.: Parametric studies on European 20-20-20 energy policy targets in university environment. J. Environ. Prot. Ecol. **18**(3), 1146–1157 (2017)

5. European Comission, Heating and cooling (2015). https://ec.europa.eu/energy/topics/energy-efficiency. Accessed 06 May 2021
6. Fleiter, T., et al.: Heat roadmap Europe - profile of heating and cooling demand in 2015, Fraunhofer ISI (2017). http://www.heatroadmap.eu/output.php. Accessed 06 June 2021
7. Averfalk, H., Werner, S.: Essential improvements in future district heating systems. Energy Proc. **116**, 217–225 (2017)
8. Heat Pumps. https://www.iea.org/reports/heat-pumps. Accessed 20 May 2021
9. Popescu, D.E., Prada, M.F., Dodescu, A., Hemanth, J., Bungau, C.: A secure confident cloud computing architecture solution for a smart campus. In: 7th International Conference on Computers Communications and Control (ICCCC), pp. 240–245 (2018)
10. Popescu, D.E., Bungau, C., Prada, M., Domuta, C., Bungau, S., Tit, D.M.: Waste management strategy at a public university in smart city context. J. Environ. Prot. Ecol. **17**(3), 1011–1020 (2016)
11. Ghemis, M.T., Scurt, A.A., Bob, C., Bungau, C.: An analysis of sustainability about rehabilitation and reconversion of an old building. In: The 19th National Technical-Scientific Conference "Modern Technologies for the 3rd Millenium", Romania, p. 273 (2019)
12. Renaldi, R., Kiprakis, A., Friedrich, D.: An optimisation framework for thermal energy storage integration in a residential heat pump heating system. Appl. Energy **186**, 520–529 (2017)
13. ASERO, Performanţa energetică a clădirilor Procedură de evaluare economică a sistemelor energetice din clădiri Partea 1: Proceduri de calcul, Modul M1-14, SR EN 15459-1
14. Langdon, D.: Literature Review of Life Cycle Costing and Life Cycle Assessment. Sage Publication, London (2006)
15. Kirkham, R., Boussabaine, A., Grew, R.: Forecasting the cost of energy in sports centres. In: Proceedings of COBRA Royal Institution of Chartered Surveyros, Salford, UK (1999)
16. Fleuchaus, P., Blum, P.: Damage event analysis of vertical ground source heat pump systems in Germany. Geother. Energy **5**(1), 1–15 (2017). https://doi.org/10.1186/s40517-017-0067-y
17. Cosmescu, A.: Despre fuzzificare, Revista Română de Informatică Şi Automatică, vol. 7, no. 2 (1997)

Cryptographic Key Distribution Protocol with Trusted Platform Module for Securing In-vehicle Communications

Béla Genge(✉) and Piroska Haller

"George Emil Palade" University of Medicine, Pharmacy, Science and Technology of Târgu Mureş, Gh. Marinescu 38, Târgu, Mureş, Romania
bela.genge@umfst.ro, piroska.haller@umfst.ro

Abstract. The modern car comprises dozens of Electronic Control Units (ECUs) running several hundred MBs of code, alongside sophisticated dashboards with integrated wireless communications. While this has brought upon a wide range of advantages and integrated features, it also exposed modern vehicles to significant threats. As a response, we developed a generic scheme for distributing keys between ECUs. The approach leverages the Trusted Platform Module (TPM) 2.0 standard to provide a tamper-proof cryptographic co-processors capable of executing cryptographic operations in a secure manner. The scheme aims at empowering ECUs to leverage state of the art cryptographic techniques by adopting recent technological advances such as the TPM. Implementation details and analysis demonstrate the feasible application of the approach in the context of the modern vehicle.

Keywords: Automotive systems · Cryptographic protocols · Trusted platform module

1 Introduction

The modern car has experienced profound changes in terms of its internal technological ecosystem. Nowadays, we find dozens of Electronic Control Units (ECUs) running several hundred MBs of code, alongside sophisticated dashboards with integrated wireless communications [1].

While this technological advancement has brought upon a wide range of advantages and integrated features, it also exposed modern vehicles to significant threats. As demonstrated by many recent threats [2], vehicle digital communication systems can be exploited in a way that alters the vehicle's behavior in order to gain certain advantages (e.g., financial, performance), or, in more extreme cases, to cause physical damage. Subsequently, changes in the vehicle's parameters, the connection of sensor emulators (e.g., AdBlue emulators), can significantly alter the operation of the vehicle's internal subsystems. Such changes can deactivate critical systems such as the Selective Catalytic Reduction (SCR) dosing system.

L. Moldovan and A. Gligor (Eds.): Inter-Eng 2021, LNNS 386, pp. 796–807, 2022.
https://doi.org/10.1007/978-3-030-93817-8_71

Consequently, it is imperative to ensure that ECUs are able to securely exchange data, but most importantly, it is necessary that ECUs are empowered with the ability to verify the authenticity of critical frames. To this end, we find several techniques to ensure the integrity and authenticity of data transfers within vehicles [3,4]. On the other hand, only few studies addressed the issue of key distribution in these systems. The main rationale is the high computational requirements of modern asymmetric algorithms, which can significantly affect the normal operation of in-vehicle controllers.

To address this issue, this paper documents a generic scheme for distributing keys between ECUs found in modern vehicles. The approach leverages the Trusted Platform Module (TPM) 2.0 standard [5] developed by the Trusted Computing Group (TCG) to provide a tamper-proof cryptographic co-processors capable of executing cryptographic operations in a secure manner. The approach is aimed at empowering ECUs to leverage state of the art cryptographic techniques by adopting recent technological advances such as the TPM. Implementation details and analysis demonstrate the feasible application of the approach in the context of the modern vehicle.

The remainder of this work is structured as follows. Section 2 provides the background terminology and an overview of related studies. Then, Sect. 3 details the developed scheme, while Sect. 4 provides an analysis of concurrency and security. This is followed by Sect. 5, which provides incipient implementation details and an analysis on the possible integration impact. The paper concludes with Sect. 6.

2 Background and Related Work

2.1 Internal Architecture of the Modern Car

From an architectural point of view, the modern car comprises dozens of embedded devices, also known as Electronic Control Units (ECUs), communicating with each other, as well as with digital and/or analog sensors. Here, we encounter a diverse ecosystem of control units (e.g., Engine Control Unit, Communication Control Unit) and sensors (e.g., Nox Sensor), together with communication protocols that range from Controller Area Network (CAN), to SENT (Single Edge Nibble Transmission) or MOST (Media Oriented Systems Transport).

In today's modern vehicles the "backbone" communication is provided by the Controller Area Network (CAN). Initially developed by Robert Bosch GmbH in 1986 for connecting electronic devices in motor vehicles, CAN is no longer restricted to automotive applications. Initially standardized by the International Standardization Organization (ISO) in 1993 as ISO 11898, and later restructured in three parts, the CAN standard specifications cover the data link layer and CAN physical layer for high-speed and low-speed CAN.

The original CAN specifications (Versions 1.0, 1.2 and 2.0A) specify an 11 bit message identifier. This is known as "Standard CAN". Specification V2.0A has been updated (to version 2.0B), to remove the message number limitation and meet the SAE J1939 standard for the use of CAN in heavy duty vehicles.

Version 2.0B CAN, referred to as "Extended CAN", contains a 29-bit identifier in the arbitration field, which allows over 536 Million message identifiers.

Given the limitations of the "classical" CAN protocol in terms of bandwidth (up to 1 Mbit/s) and payload size (each CAN data frame can hold up to 8 data bytes), recently, two main improved communication infrastructures have been proposed. The CAN+ protocol was proposed by Ziermann, *et al.* in 2009 [6], and it exploits the time between transmissions to send additional data. More recently, in 2012, Robert Bosch GmbH developed the CAN with flexible data-rate protocol (CAN-FD version 1.0) [7], which later became part of the ISO 11898-1:2015 standard. This specification allows for increased data lengths and, once the arbitration is decided, a higher bit rate can be configured for the data phase. Up to 8Mbit/s data rates and up to 64 data bytes are now possible with CAN-FD.

The AUTomotive Open System ARchitecture (AUTOSAR) Specification of Secure Onboard Communication standard (AUTOSAR SecOC) [4] provides the principles for achieving authentication and integrity protection of sensitive data exchanged communication peers in an automotive network. The approach is generic, and it supports both symmetric and asymmetric methods. It recommends the use of Message Authentication Code (MAC) for symmetric, and public key-based digital signatures for asymmetric applications. The specification recommends that each Protocol Data unit – PDU (e.g., a CAN frame) is extended with an authentication component (e.g., the MAC) and a freshness value (e.g., freshness counter, timestamp). In addition, for certain protocols, such as the CAN protocol, the size of the MAC tag can be truncated down to 64 bits in order to fit into a single frame.

2.2 Trusted Platform Module

Each component found within the modern car, depending on its hardware configuration, presents different requirements in terms of security objectives. As an example, ECUs require a broader range of features in comparison to a digital sensor, even if the same communication medium is used. Typically, security objectives are implemented within the existing system and are executed on the same processing unit. This affects the computational load on the processing unit. In the case of environmentally restricted units, such an overload can lead to delays that may alter the real-time operation of the CAN ecosystem.

In order to enable security features in environmentally restricted units, the use of security controllers is imperative. To this end, the Trusted Platform Module (TPM) 2.0 standard [5] developed by the Trusted Computing Group (TCG), offers the description for compatible security controllers. TPMs represent tamper-proof cryptographic co-processors capable of executing cryptographic operations in a secure manner, isolated from the main processing unit. In the remainder of this document the TPM abbreviation is used to denote a security controller compatible with the TPM 2.0 specification.

The TPM presents several core capabilities based on which complex security mechanisms can be built. Serving as a Root of Trust (RoT), it offers secure key storage in volatile and non-volatile memory, an Endorsement key (Ek) sealed and only available to the TPM itself, and a set of Platform Configurable Registers (PCRs) capable of performing local integrity measurements by leveraging hash based functions. For persistent storage, usually TPMs contain a non-volatile memory which, in certain cases, given it's modest dimension can't facilitate the storage of a high number of cryptographic keys. As a solution to this problem, the TPM presents two additional methods of processing keys: sealing, which securely stores the cryptographic keys on disk by encrypting the private part of the key pair by leveraging the Ek, and key hierarchies, which allow deterministic key derivation starting from a parent key stored in the TPM. The keys obtained after derivation at run-time follow a tree-like graph structure, where a node can represent a parent or child key. To store the key generated from a key hierarchy, TPM uses the sealing method.

2.3 Related Work

Several techniques have been developed in order to secure CAN communications. Starting with the work of Herrewege, *et al.* [3], it was shown that, while the CAN bus has significant limitations in terms of payload and bandwidth, data authentication is still achievable via Message Authentication Codes (MAC), and more specifically the HMAC standard, alongside a counter in order to ensure freshness and resistance against replay attacks. The approach entitled *CANAuth* presumes the presence of the CAN+ protocol, which permits out-of-band transmissions [6].

With respect to techniques targeting CAN-FD, which is within the main scope of the developed approach, we mention the work of Woo, *et al.* [8], where a secure communication protocol and security architecture were developed for CAN-FD-enabled in-vehicle systems. The approach embraces two distinct layers. A first security layer, where regular ECUs communicate securely by leveraging symmetric cryptographic constructions (e.g., symmetric encryption, and MAC). This is reinforced by a second security layer consisting of an advanced gateway capable of signing and verifying signatures via asymmetric cryptographic computations.

Wang and Liu [9] acknowledged the importance of securing gateway ECUs (i.e., ECUs enabling external communications), which provide various communication interfaces, including the support for Vehicle to Internet communications. These benefit from significant computational power and communication bandwidth and enable the realisation of new infrastructural paradigms such as the Internet of Vehicles. Wang's approach focused on the secure distribution of cryptographic keys in and outside the vehicle. In a similar fashion, we find other approaches, where the emphasis was placed on the distribution of cryptographic keys [10]. Groza and Murvay [10] evaluated several techniques for distributing keys in CAN-FD, including Diffie-Hellman and identity-based cryptography. Compared to this approach, we present a generic set of protocols that distinguish between long-term and short-term keys in order to reduce the number of messages and, overall, the impact on communications. Furthermore, we leverage the capabilities of TPM to reduce the computational overhead.

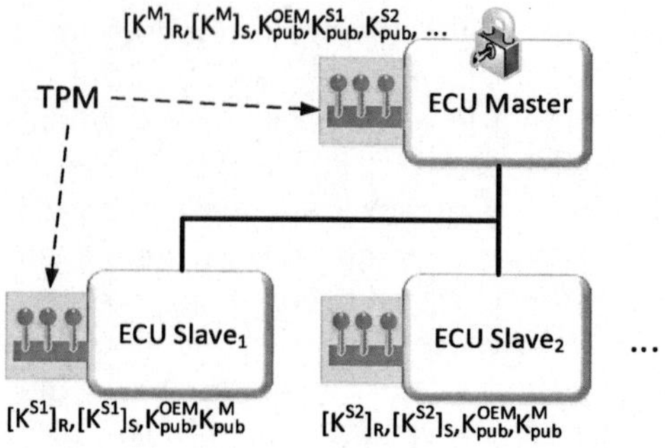

Fig. 1. Bootstrapping.

3 Cryptographic Key Distribution Scheme

Typically, security features are implemented within the existing system and are executed on the same processing unit, which affects the computational load on the processing unit. In the case of environmentally restricted units, such an overload can lead to delays that may alter the real-time operation of the CAN ecosystem. In order to enable security features in environmentally restricted units, the use of security controllers is imperative.

We realize that not all components found within the vehicle can be equipped with a TPM. Furthermore, in the case of data exchanges with digital sensors, complex cryptographic operations (e.g., digital signatures) may not be supported. Therefore, this paper does not focus on the key distribution concerning components not supporting TPM extensions.

3.1 Setup and Bootstrapping

Figure 1 denotes the main entities present in the key distribution scheme. Here, we observe two main entities: "Master" and "Slave". The Master is the one that triggers the generation and distribution of a new cryptographic key, while the Slaves are the recipients of fresh cryptographic keys. The developed key distribution scheme leverages public key cryptography, and several types of symmetric keys.

In terms of cryptographic keys, each ECU, that is, each TPM associated to an ECU, is bootstrapped with several types of keys. More specifically, the following are the relevant keys at bootstrapping in the context of an ECU/TPM:

- Storage Root Key (SRK): A pair of special, public-private keys that never leave the TPM. The SRK is used to encrypt the keys that are stored outside the TPM.
- Key Signing Key (KSK): A pair of public-private keys used in the key distribution procedure, as described later, in order to enforce the non-repudiation property.

- OEM Public key (OemPK): The public key of the manufacturer of the ECU firmware.
- Slave Public Keys (SlavePKs): The public keys of slave ECUs, in the case of the Master ECU.
- Master Public Key (MasterPK): The public key of the master ECU, in the case of Slave ECUs.

To provide a more formal notation of the previous key types, several symbols and notations are introduced. The following notation denotes a public-private key pair:

$$[K^X]_Y = (K^X_{pub}, K^Y_{prv}), X \in \{M, OEM, S_1, S_2, ...\}, Y \in \{R, S\}. \quad (1)$$

In the notation above, $[K^X]_Y$ denotes a pair of public-private key pairs, such that K^X_{pub} is the public component, and K^Y_{prv} is the private component. In the same definition, the value of X is replaced in particular cases with the appropriate symbol from the $\{M, OEM, S_1, S_2, ...\}$ set, while the value of Y is replaced with a symbol from set $\{R, S\}$.

According to this definition, the ECU Master is bootstrapped with the following keys: $[K^M]_R$ – the Master's SRK; $[K^M]_S$ – the Master's KSK; K^{OEM}_{pub} – the Master OEM's public key; $K^{S_1}_{pub}, K^{S_1}_{pub}$ – one or more SlavePKs according to the number of Slave ECUs associated to a Master ECU. Similarly, a Slave ECU is bootstrapped with the following keys: $[K^{S_1}]_R$ – the Slave's SRK; $[K^{S_1}]_S$ – the Slave's KSK; K^{OEM}_{pub} – the Slave OEM's public key; K^M_{pub} – the Master ECU's public key.

3.2 Secure Key Generation and Storage

Once the bootstrapping is completed, Master ECUs are the ones that generate the long-term and the short-term keys. The two key types are distinguished according to their cryptoperiod. The cryptoperiod denotes the time span during which a cryptographic key is authorized for use. According to NIST's "Recommendation For Key Management", long-term keys should be changed once every few weeks, while short-term keys (sometimes referred to as session keys), should be changed once every few days.

Keys need to be generated in a hardware-protected area that cannot be tampered. As it turns out, the TPM addresses this requirement, since it provides secure generation and storage of cryptographic keys. The cryptographic keys generated for this purpose are symmetric keys, that is, the keys used to provide data confidentiality, integrity, and authenticity with the help of encryption algorithms. It is imperative that the generated keys do not leave the TPM unencrypted. For this purpose, TPMs have a particular feature known as sealing. Sealing denotes the procedure of encrypting data (e.g., session keys) with the TPM's SRK. Since the private part of SRK never leaves the TPM, it is practically impossible for someone to decrypt the session key without knowing the SRK's private key. As a result, the TPM can generate a large number of cryptographic keys that can be applied in data authentication, data integrity, and a wide variety of other scenarios. In the case a particular key is needed, the TPM can load, unseal the key, and apply it to enforce certain security properties.

3.3 Long-Term Key Distribution Protocol

In order to reduce the overhead on the CAN-FD bus, the long-term key distribution protocol is defined as a single message protocol. The message is sent by the Master ECU to each Slave ECU. We call this protocol the Long-term Key Distribution Protocol and denote it by `Proto-LTK`:

$$\begin{aligned} ECU_{Master} \rightarrow ECU_Y : pid_{LTK}, kid_{LTK}, N, \{K\}_{K^Y_{pub}}, \\ \{pid_{LTK}, kid_{LTK}, N, \{K\}_{K^Y_{pub}}\}_{K^M_{prv}}, \end{aligned} \quad (2)$$

where $Y \in \{S_1, S_2, ...\}$. In `Proto-LTK`, pid_{LTK} denotes the protocol identifier, which needs to be unique for each type of key. This term is particularly useful in distinguishing between several different use cases, and key distribution scenarios, and, ultimately, to avoid the so-called multi-protocol attacks, where cryptographic terms from one protocol are replayed to other protocols. The second term kid_{LTK} is the key identifier, which can be implemented as a counter. The next term N is a random number used to enforce the freshness of the message, which is in accordance with AUTOSAR SecOc's recommendations regarding message freshness. Next, the term $\{K\}_{K^Y_{pub}}$ is the actual transmission of a freshly generated long-term key K encrypted with each of the target slave's public key. The last term is the digital signature of all prior terms, computed with the Master's private key K^M_{prv}. The purpose of this term is to enforce the authenticity (and integrity via authenticity) and non-repudiation security properties.

In case the message is missed by a target slave, the protocol `Proto-LTK` is extended with a challenge-response sequence, which ensures the explicit interrogation of the currently used long-term key:

$$\begin{aligned} ECU_{S_i} &\rightarrow ECU_M : pid_{LTK}, N_S \\ ECU_M &\rightarrow ECU_{S_i} : pid_{LTK}, kid_{LTK}, N_S, \\ \{K\}_{K^Y_{pub}}, &\{pid_{LTK}, kid_{LTK}, N_S, \{K\}_{K^Y_{pub}}\}_{K^M_{prv}}. \end{aligned} \quad (3)$$

In this case, the slave ECU (ECU_{S_i}, where $i \in \{1, 2, ...\}$) initiates the execution of the protocol by sending a protocol identifier pid_{LTK}, and a freshly generated random number N_S. As a response to this request, the Master issues the encrypted and signed long-term key K.

3.4 Short-Term Key Distribution Protocol

Once the long-term key (LTK) is installed, it is used to periodically distribute a new short-term key (STK). We call this protocol the Short-term Key Distribution Protocol (`Proto-STK`). For this purpose, a protocol similar to the one above is used:

$$ECU_{Master} \rightarrow Broadcast : pid_{STK}, kid_{STK}, N, \{k\}_K, \\ \{pid_{STK}, kid_{STK}, N, \{k\}_K\}_{K^M_{prv}}, \tag{4}$$

where pid_{STK} is the protocol identifier, k_{STK} is the key identifier, N is a freshness counter, and k is the newly generated short-term key. Compared to `Proto-STK`, the protocol exhibits two main differences: (i) `Proto-STK` uses a broadcast message, which means that a single message is issued for distributing the new short-term key k; and (ii) `Proto-STK` digitally signs the newly issued STK, which empowers Slave ECUs with the ability to verify the authenticity of the newly generated key. Once again, the protocol is extended with a challenge-response scheme in order to ensure the subsequent request of the current key:

$$ECU_{S_i} \rightarrow ECU_M : pid_{LTK}, N_S \\ ECU_M \rightarrow ECU_{S_i} : pid_{LTK}, kid_{LTK}, N_S, \\ \{K\}_{K^Y_{pub}}, \{pid_{LTK}, kid_{LTK}, N_S, \{K\}_{K^Y_{pub}}\}_{K^M_{prv}}. \tag{5}$$

4 Concurrency and Security Analysis

4.1 Concurrency Analysis

In order to assess the possible concurrency issues introduced by the developed scheme we use the TLA+ language [12]. TLA+ (Temporal Logic of Actions) is a formal specification language developed by Leslie Lamport. It is a language for constructing specifications, which can then be analyzed by model checking tools. The model checker can then explore all possible behavior, and find violations of desired properties. A language that builds on TLA+ is PlusCal, a formal specification language that can be translated to TLA+ specification. PlusCal was developed by Leslie Lamport to aid the specification of algorithms. While PlusCal resembles more of an imperative language, its expressions use the syntax of TLA+. There is also a translator available that translates PlusCal specifications to TLA+.

PlusCal adopts a wide range of statements from the 'C' programming language, making its learning curve less steep. We modeled the developed key distribution protocol by leveraging PlusCal, and we translated the specification to TLA+ by using the TLA+ toolbox. The specification presumes two entities, a Master and a Slave. The Master periodically issues a new session key, and uses the session key to periodically broadcast authenticated messages. The Slave updates its key, and uses the key with the identifier it receives in order to verify the authentication tag. As it turns out, TLA+ detected miss-synchronization between nodes. According to TLA+ output, certain nodes may miss a specific key update, and would therefore be unable to authenticate subsequent frames. Therefore, it is imperative that nodes have the ability to buffer a sequence of frames and leverage the challenge-response variants of the protocols mentioned earlier in order to synchronize keys.

4.2 Security Analysis

We proceed to the formal analysis of the proposed key distribution scheme with the help of the Scyther [11] tool. Scyther is a model checking tool designed under the perfect encryption assumption, which means that an adversary can only learn an encrypted value if he knows the decryption key. In Scyther, the adversary is assumed to have full control over the underlying communication network. It can eavesdrop messages, it can replay, split and concatenate messages, however, it can correctly encrypt and decrypt a message only if it is in the possession of the right key. Throughout our assessment we used *claim* events to analyze the *aliveness*, *non-injective agreement*, *non-injective synchronization*, and *commitment* security properties:

- *Aliveness*: is a weak form of authentication and it ensures to the protocol initiator I the aliveness of the protocol respondent R, if, after a complete run of the protocol by I, R also executed some protocol events (not necessarily in a recent time).
- *Non-injective agreement (Niagree)*: ensures to the protocol initiator I a non-injective agreement with the respondent R on free variables appearing in the protocol's specification, if whenever I completes a protocol run, then R has also been involved in executing protocol events with I. However, this property does not guarantee a one-to-one mapping of runs. As a result, participant I may complete two runs, while participant R may have taken part only in a single protocol run.
- *Non-injective synchronization (Nisynch)*: is a stronger form of authentication, which means that all messages sent/received by the initiator I have been indeed received/sent by the respondent R.
- *Commitment*: ensures that if whenever I completes a protocol run with a set of terms, then the respondent R also completed the protocol run with the same terms. The property is also useful when verifying a protocol against impersonation attacks.

As part of Scyther's modeling language, protocol participants are defined according to their roles (i.e., initiator, respondent). Message exchanges are modeled with *send* and *recv* events. There are predefined types for generating random numbers (i.e., a *Nonce* - Number once used), while time stamps and session keys are modeled as user defined types. The complete protocol model is illustrated in Fig. 2. Here, the Master was modeled as the protocol's initiator, while a regular node was modeled as the protocol's respondent.

By leveraging the Scyther tool, the developed protocol was proven to be correct within the Dolev-Yao attacker model. More specifically, it was proven that the protocol ensures: (i) the synchronized agreement of roles with respect to runs; and (iii) the secrecy of the LTK K. Similarly to this protocol, the other variants have also been modeled with the help of the Scyther tool, and were found to be correct within the Dolev-Yao attacker model.

```
usertype ProtocolID;
usertype KeyID;
usertype SessionKey;
const pid: ProtocolID;
protocol DIAS-KEYDISTRO(I,R) {
        role I {
                fresh kid: KeyID;
                fresh Ni: Nonce;
                fresh K: SessionKey;
                send_1(I,R, pid, kid, Ni, {K}pk(R),
{pid, kid, Ni, {K}pk(R)}sk(I));
        }
        role R {
                var kid: KeyID;
                var Ni: Nonce;
                var K: SessionKey;
                recv_1(I,R, pid, kid, Ni, {K}pk(R),
{pid, kid, Ni, {K}pk(R)}sk(I));
                claim_R1(R,Secret,K);
                claim_R2(R,Nisynch);
        }
}
```

Fig. 2. The key distribution protocol's description in Scyther's modeling language.

5 Implementation Details

A prototype of the proposed scheme has been implemented in the context of a laboratory testbed. The testbed comprised of an ECU implemented with the help of Raspberry Pi model 3B+ running the Automotive Grade Linux operating system. The ECU was equipped with an An MCP2515 CAN controller, with a TJA1050 CAN transceiver, and an Infineon OPTIGA SLB 9670 Trusted Platform Module. The TPM supports a wide range of cryptographic applications, while providing both symmetric and asymmetric cryptographic primitives. A wide variety of algorithms are supported, including Elliptic Curve Cryptography (ECC), EC Diffie-Helman, 1024 and 2046-bit RSA. The primitives for building the key distribution protocols and for communicating with the TPM were implemented in Python 3.5.

For the following analysis, we used 2048-bit RSA and 256-bit SHA, thus the most resource-intensive algorithms supported by the TPM. Accordingly, we tested the time for generating the message of `Proto-LTK`. As it turns out, the measured time for generating a new symmetric key K of 32 bytes, the sealing of the message, including generating signature was of 2.11 s. Indeed, this is an intensive operation, however, this is also a key advantage of offloading these time-intensive operations to a cryptographic co-processor. An important observation is that, in the case of `Proto-LTK`, a new message needs to be generated for each ECU involved in the communication. However, since this needs to be performed only once every few weeks (as mentioned earlier), we consider that the added value of leveraging state of the art cryptographic constructions outweigh the computational downside. Conversely, `Proto-STK` broadcasts a single message to all involved ECUs, which yields a computational effort for signing only one message. In terms of message dimension, the following implementation variant is proposed for `Proto-LTK`. Since pid_{LTK} could be used to distinguish between different groups of ECUs using the same protocol, a single byte would be sufficient for this element. kid_{LTK} and N can be used together as freshness as well as denoting the identifier of the newly issued key. Accordingly, two bytes for kid_{LTK} would permit releasing a new LTK every week for more than one thousand years. Subsequently, the value of N needs to be sized according to the number of ECUs. Considering the number of ECUs in today's and future vehicles, the size of N can be of one byte. In case more than 256 ECUs are present, we assume there would be certain partitioning of ECUs in several networks/-groups, where pid_{LTK} could be used to distinguish between different groups. The signature with RSA-2048 is of 256 bytes, while the encryption is also of 256 bytes.

Overall, the number of bytes that need to be sent by one run of `Proto-LTK` is of 516 bytes. This value is identical to the number of bytes that are generated by `Proto-STK` (assuming the same ciphers are used). This translates to 9 frames to be sent on CAN-FD. We observe, however, that the maximum supported data rate of CAN-FD is of 8Mbit/s. Therefore, a single run of `Proto-LTK` would consume aprox. 4Kbits, that is aprox. 0.05% of the available bandwidth. Considering that the modern vehicle contains dozens of ECUs, for example in case of 100 ECUs, each ECU receiving a new STK once every day, this would overall consume 5% of the vehicle's bandwidth once every day. Therefore, the overall impact on the vehicle's communication can be seen as negligible, at least from a daily consumption point of view (Table 1).

Table 1. Implementation size for RSA 2048.

	pid	kid	N	$\{K\}_{K_{pub}}$	Signature
Byte count	1	2	1	256	256
TOTAL	516 Bytes				

6 Conclusions

We developed a novel scheme for distributing cryptographic keys in modern vehicles. The scheme includes two protocols, namely: `Proto-LTK` for distributing long-term keys, and `Proto-STK` for distributing short-term keys. Both protocols are supoprted by synchronization constructions such that ECUs may synchronize and actively interrogate the current key. In order to ensure feasibility of integration, the scheme adopted the TPM standard. By using an external cryptographic co-processor, several advantages are provided: (i) the TPM provides a tamper-proof approach for storing and manipulating cryptographic keys; and (ii) the TPM can off-load the computation-intensive operations, and thus empower ECUs with the possibility to use modern cryptographic algorithms. Implementation details and analysis have shown that the impact on the used bandwidth, in the case of CAN-FD is of less than 5%. As future work we intend to develop a full working prototype and assess the performance of the protocol in the context of inter-ECU communications.

Acknowledgement. This work was funded by the European Union's Horizon 2020 Research and Innovation Programme through DIAS project (https://dias-project.com/) under Grant Agreement No. 814951. This document reflects only the author's view and the Agency is not responsible for any use that may be made of the information it contains.

References

1. Coppola, R., Morisio, M.: Connected car: technologies, issues, future trends. ACM Comput. Surv. **49**(3), 46:1–46:36 (2016)
2. Urquhart, C., Bellekens, X., Tachtatzis, C., Atkinson, R., Hindy, H., Seeam, A.: Cyber-security internals of a Skoda Octavia vRS: a hands on approach. IEEE Access **7**, 146057–146069 (2019). https://doi.org/10.1109/ACCESS.2019.2943837
3. Van Herrewege, A., Singelee, D., Verbauwhede, I.: CANAuth - a simple, backward compatible broadcast authentication protocol for CAN bus. In: ECRYPT Workshop on Lightweight Cryptography 2011, pp. 1–7 (2011)
4. AUTOSAR: Specification of Secure Onboard Communication AUTOSAR CP Release 4.3.1. AUTOSAR (2017)
5. Trusted Computing Group: Trusted Platform Module Library Specification, Family "2.0", Level 00, Revision 01.59 (2019)
6. Ziermann, T., Wildermann, S., Teich, J.: CAN+: a new backward-compatible Controller Area Network (CAN) protocol with up to 16"D7 higher data rates. In: Design, Automation Test in Europe Conference Exhibition, pp. 1088–1093 (2009). https://doi.org/10.1109/DATE.2009.5090826
7. Robert Bosch Gmbh: CAN with flexible data-rate. Vector CANtech, Inc., MI, USA, Specification Version 1.0 (2012)
8. Woo, S., Jo, H.J., Kim, I.S., Lee, D.H.: A practical security architecture for in-vehicle CAN-FD. IEEE Trans. Intell. Transp. Syst. **17**(8), 2248–2261 (2016). https://doi.org/10.1109/TITS.2016.2519464
9. Wang, L., Liu, X.: NOTSA: novel OBU With three-level security architecture for internet of vehicles. IEEE Internet Things J. **5**(5), 3548–3558 (2018). https://doi.org/10.1109/JIOT.2018.2800281
10. Groza, B., Murvay, P.S.: Identity-based key exchange on in-vehicle networks: CAN-FD & FlexRay. Sensors **19**(22) (2019). https://doi.org/10.3390/s19224919
11. Cremers, C.J.F.: The Scyther tool: verification, falsification, and analysis of security protocols. In: Computer Aided Verification, 20th International Conference, CAV 2008, Princeton, USA, Proc., pp. 414–418 (2008)
12. Lamport, L.: Specifying Concurrent Systems with TLA+ (1999)

Automated Assessment Generation in Intelligent Tutoring Systems

László Kovács(✉), László Csépányi-Fürjes, and Ghanim Hussein Ali Ahmed

University of Miskolc, 3515 Miskolc, Egyetemváros, Hungary
{kovacs,ghanim}@iit.uni-miskolc.hu,
Laszlo.Csepanyi-Furjes@uni-miskolc.hu

Abstract. Intelligent Tutoring Systems have a huge potential to improve the efficiency of learning and teaching, especially for personalized adaptive self-learning. The investigation presented in this paper focuses on the implementation of a submodule, the Automated Question Generation for Assessment unit, using a combined rule based and machine learning based methods. The proposed module is based on a proposed background domain ontology, and it works in two phases. A generator automaton is used to construct the answer templates in the first phase. In the next phase, an encoder-decoder neural network translator unit was used to generate the question sentence templates for the selected answer templates. According to the tests with the implemented prototype assessment generator application on the SQL domain, the proposed framework is suitable to generate assessment tasks in a very efficient way.

Keywords: Intelligent tutoring system · Assessment generation · Machine learning

1 Introduction

Intelligent Tutoring System (ITS) is a computer-based system that aims to offer direct and customized instruction or feedback to learners with personalized guidelines based on their cognitive skills [1]. ITS systems can be used to simulate the tutoring process of human teachers, including providing instructions, suggestions, decisions regarding the learning path and assessments usually without requiring intervention from a human teacher. It is widely accepted that ITS has a huge potential to improve the efficiency of learning and teaching, especially for personalized adaptive self-learning. According to [2], intelligent tutoring may be viewed as "an effort to capture in computer technology the capabilities and practices of a human instructor who is expert in both the subject matter and one-to-one tutoring". On the market, we can find many successful ITS applications used by thousands of students, including SQLTutor [3], ALEKS [4], Cognitive Tutor [5] and ASSISTments [6].

As most of the current ITS solutions apply different levels of machine intelligence to control the educational processes, the ITS is considered as a subfield of the area of Artificial Intelligence in Education (AIEd). According to the experts [7], the role of

L. Moldovan and A. Gligor (Eds.): Inter-Eng 2021, LNNS 386, pp. 808–819, 2022.
https://doi.org/10.1007/978-3-030-93817-8_72

AI-based educational applications has grown 40% in recent years. Since the appearance of the first AIEd proposal nearly 30 years ago, a wide spectrum of intelligent tools has been developed which can be categorized into three main domains [8]: applications for personal tutors; application for collaborative learning and application for domain knowledge representation (like virtual reality). Regarding the target users, in [9], the following categories are identified: learner-facing, teacher-facing and system-facing intelligent applications. According to [7], the ITS module developments are mainly focused on improving the adaptivity, the assessment processing, student profiling and learning path prediction. In our project works, we were focusing on the teacher-faced applications which is aimed at supporting the teachers with reduction of the workload by automating the key educational activities [7].

ITS applications are designed to behave as intelligent as an expert teacher [10] with intentional application of widely accepted pedagogical strategies [11]. The general architecture of ITS applications involves the following key components to provide the expected level of intelligence [12].

- student model
- teacher model
- domain model
- diagnosis model.

The domain model contains the semantic description of the teaching course content, usually using a semantic network format. Beside the core knowledge model elements, the working of ITS applications is based also on some external components, like fact databases, natural language processing modules, chatbots or web services.

2 Current Issues, Limitations, and Challenges in ITS

The main expectations and properties of an ideal ITS [7] cover the following properties:

- An intelligent tutoring system should be capable of supporting learning
- An intelligent tutoring system would behave as if it genuinely cared about the student's success [13].
- The systems themselves would also learn how to teach [14]
- The systems would study what worked, and when it worked, and they would improve themselves over time [15].

Due to the complexity of the ITS systems, the implemented prototype systems usually cannot meet all the high expectations. Thus, beside the positive evaluations, in the literature, we can find some critical remarks too (see [16]) where the main conclusion of the test experiment was that human tutoring dominates ITS solution both in efficiency and flexibility. As the study in [17] concluded, current AIEd systems can make many mistakes and ITS requires human control. One reason for the weak performance is that the quality of the AI model mainly depends on the quality of the training data, thus the most expensive part in the development of ITS applications is the construction of training data sets having appropriate size and quality.

According to [18], the main barriers of development and adoption of efficient ITS systems can be summarized into the following points:

- limited size of the background knowledge model
- quality of the background knowledge model
- low adaptivity level
- low flexibility in customization.

Based on the experienced difficulties, some works focus on investigation of the shortcomings providing suggestions on the real application options. In [19], a new naming is proposed instead of ITS; the education community can be provided in the near future only with Intelligent Teaching Assistant (ITA) systems. As [20] points out the teachers require such systems which provide.

- more accurate and wider scope assessment of the students
- decision support in selection of efficient learning path for the different students
- better analytical statistics on the students
- providing more explanatory instructions instead of direct commands.

In [21], the author presents a sharp criticism on the shortcomings of the current ITS solutions stating that we need stupid tutoring systems but intelligent teachers: "*Perhaps we do not in fact need intelligent tutoring systems. Perhaps instead what we need, what we are already developing, is stupid tutoring systems. Tutors that do not, themselves, behave very intelligently. But tutors that are designed intelligently, and that leverage human intelligence.*

In other words, perhaps what we need is stupid tutoring *systems, and intelligent humans.*"

These experiments and the outcome of the detailed evaluation in [7] show, that despite the great achievements in AI and in some components of AIEd, much more analysis and development are needed to improve the practical efficiency of ITS applications.

Based on the existing technical difficulties in implementation of high level ITS modules, our investigation focuses on the implementation of sub-module, namely on Automated Question Generation for Assessment (AQGA), using a combined rule based and machine learning based methods. As the proposed method is based on specific knowledge model elements, we also present the supporting knowledge model background of the AQGA unit.

3 Content Model in the Proposed ITS Framework

Considering the functional and data components of current ITS systems, we include the following key components into the proposed ITS system:

- domain module to describe the course topics including the knowledge units and competencies
- student module to model the behavior of the students

- teacher module to model the behavior of the tutor
- training material repository
- assessment management module
- interface module to implement intelligent, human oriented interface to the system
- background supporting units like NLP module or reasoning engine.

This structure can be observed among others in Intelligent Tutoring System Builder (ITSB) [22] or in Cognitive Tutor Authoring Tools (CTAT) [23]. The proposed architecture is presented in Fig. 1.

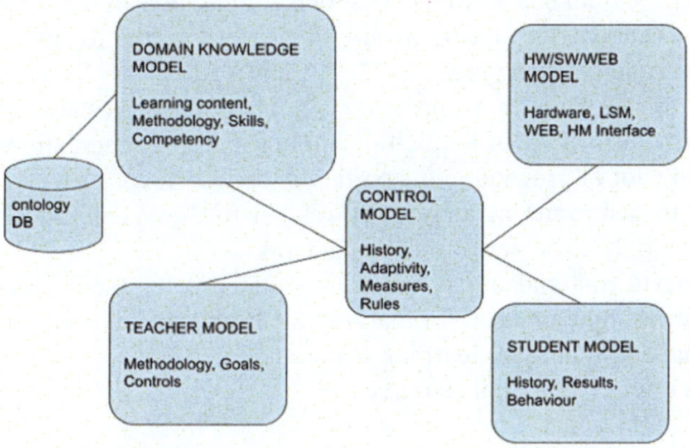

Fig. 1. Proposed e-tutor architecture.

The goal of the domain model is to provide detailed information about the teaching content including all details needed to perform an automated and adaptive implementation and control in the learning process. In [24], the authors investigated the conceptualization of teacher content knowledge from many aspects including pedagogical viewpoints. The work identifies the following dimensions.

- Content knowledge
- Curriculum knowledge
- Pedagogical content knowledge

The Pedagogical content knowledge covers the integration of content knowledge and pedagogy to perform the teaching using a special form of professional understanding. The work highlights the view that teaching a subject is more than knowing the facts, the teachers should understand the organizing principles and structures.

In [25], the following explicit requirements are set on the content model:

- common content knowledge, teachers need to know the material they teach
- competency to perform domain specific operations
- specialized content knowledge with methodological aspects of the domain
- skills in appropriate survey developments.

Beside pedagogical aspects, another important background for domain modelling is the field of knowledge engineering (KE). In KE, the main approaches to describe the knowledge structure cover among others the semantic graph, the framework structures [26], the semantic and the ontology models [27]. The main benefits of ontology are that this model provides explicit conceptualization providing a rich set of concept construction operators and a broad area for integrity validations.

Based on requirement analysis, our proposed model incorporates the following components:

- K-units refers to knowledge units as a unit of education that defines a course, knowledge topics, domains, key concepts, or teaching units of the learning materials.
- K-slots refer to knowledge slots as an atomic property of the K-units. Every slot has a value domain and concept type.
- R-units refer to rule units as a rule or constraints defined on the k-units and the k-slots. The rule unit is a set of explicit or implicit ontology regulations or principles governing behavior or procedure in a particular activity area.
- T-units refer to task units as an activity related to the k-units and k-slots. T-unit describes an activity to be performed by students.
- M-Units refers to material units as teaching material for the k-units. M-units are teaching materials that are used for learning the k-unit. M-units contain any parts of the educational community or learning material.
- M-Resources refers to material resources.

Regarding the main relationships, the ontology model contains the following elements:

- K-unit taxonomy relationship: it describes the specialization among the K-units
- K-unit component relationship: one K-unit consists of other K-units
- K-competency relationship among the K-units
- KT- assignment: every T-unit task is related to a set of K-units.

The basic building blocks and the key relationship components of the proposed domain model are presented in Fig. 2.

Regarding the completeness of the proposed domain knowledge model, we should first consider the main functional expectations from the viewpoint of the ITS systems. The presented model can cover the information requirements of a large set of functionalities, including:

- identification of the relevant knowledge units together with the related training materials, study aids
- automated navigation among the knowledge units during the learning/teaching process
- automated assessment of the student's knowledge level
- providing hints/suggestion for the students.

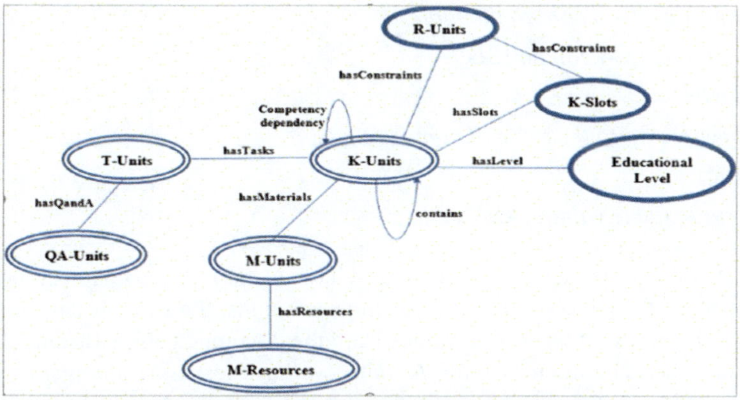

Fig. 2. Schema elements of the domain ontology

On the other hand, the model should be considered as the base framework which can be extended with more sophisticated blocks in the future. Among the possible extensions, we can highlight the following elements:

- compound task structure (current task model uses an atomic approach)
- a more sophisticated grading module for the task units
- direct management of the implicit competency lattice of the tasks

4 Tools for Automated Assessment Generation

According to [7], the automated grading process is the focus area in ITS assessment processing. Due to the large complexity of the assessment module, the existing automated assessment systems require a relatively large human preprocessing and calibration effort. On the other hand, the teachers really need an efficient tool for generation of assessment units, especially in the current epidemic lockdown situation where only the online education mode is possible at each level of the education. Current electronic LMS systems usually do not support automatic assessment generation, thus teachers should devote a lot of time and effort to construction of the assessment units.

In our model, we propose a semi-automatic tool supporting the assessment generation. The description of the question-answer QA-unit is part of the domain model and it has the following structure:

TASK task_name.
WITH variable_declaration.
QUESTION question_schema.
ANSWER answer_schema.

where the variable declaration refers to the variables denoting knowledge K-units or K-slots. For example, in the domain on SQL commands, the variables may denote tables or table columns from the background database. The following sample QA unit uses this SQL domain:

TASK insert a new record to the table
WITH t: Table, c: #t.column[SEQ]
QUESTION
Add a new record to the table #t.name (#c.name)
ANSWER
insert into #t.name values (#c.ANY)

The term Table is defined here as a concept having among others name, column, and PK properties. In the generation of the concrete task units, the variables are substituted randomly with some corresponding values from the domain database instance.

Regarding the generation of the QA-units, we have tested an automatic AI-based solution to provide a more efficient method beside the manual QA-unit construction. Our assumption is that the answer section can be generated with some kind of automaton. For example, in the case of SQL domain, an FSA engine can be used to generate valid SQL commands containing variables. In our prototype system, this automaton also has some parameters to control the generation process like the probability of the different SQL commands.

In the next phase, the question section is generated automatically from the answer section. For this processing step, we have tested two options on our SQL domain:

- A: using an automaton approach
- B: machine learning approach

In case A, the automaton was generated manually. The main benefit of this approach is that the automaton will incorporate the knowledge of the human teacher. Regarding the shortcomings, we can mention the following problems:

- it requires significant programming effort
- it is hard to describe complex cases
- high dependency on the teacher knowledge level

In the case of the machine learning approach, we have implemented an encoder-decoder neural network model that transforms the answer sentence into the question sentence The encoder-decoder network structure [28] contains two GRU-RNN layers, one for encoding and the other for decoding steps. The first RNN unit converts the input sequence into a state vector which contains the key hash description of the input sequence. This state vector will be fed into the decoder unit outputting the translated sequence. The most common use of this structure is automated language translation (Figs. 3 and 4).

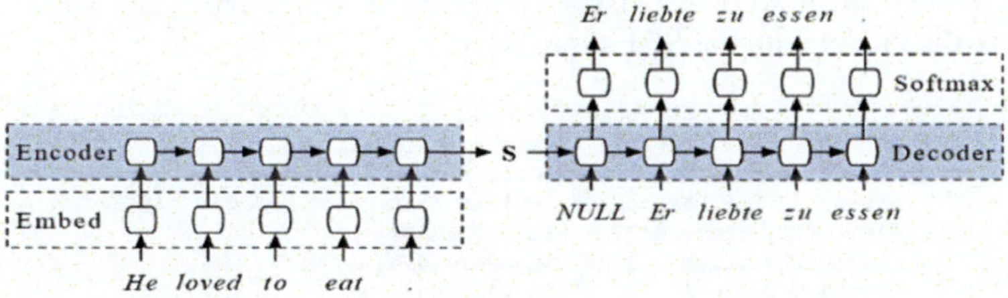

Fig. 3. Encoder-decoder [https://www.guru99.com/seq2seq-model.html]

Layer (type)	Output Shape	Param #
gru_2 (GRU)	(None, 128)	50304
repeat_vector (RepeatVector)	(None, 27, 128)	0
gru_3 (GRU)	(None, 27, 128)	99072
time_distributed_2 (TimeDist	(None, 27, 170)	21930

Fig. 4. Neural network architecture in our test system

Beside the question generation encoding-decoding neural network module, we have developed a domain specific embedding model, too. The NLP embedding models [29] are used to represent the words of a natural language in high dimensional vector space such that words with similar meaning have similar positions in the vector space. The similarity is usually measured with the context similarity of the words, i.e., words occurring in similar contexts should be similar in meaning. The construction of word embeddings is a very expensive operation both in time and computer resources. Also, it requires hundreds of millions of training samples. Thus, usually an existing general embedding model (like the BERT model [30]) is adapted to the specific domain. This general embedding model is called pre-trained model and the domain specific adaptation process is called fine-tuning. During the fine tuning, a domain specific training set is used to update the pre-trained weights so that the model can produce much more accurate results in that certain domain. Two kind of Bert models can be built using a stack of encoder layers. $\text{BERT}_{\text{BASE}}$ consists of 12 encoder layers and produces 768 dimensional vectors and $\text{BERT}_{\text{LARGE}}$ which contains 24 layers and produces 1024 dimensional embedding vectors. The problem with these models is that it is computationally expensive to use them for sentences. For our experiments, we have used a modified variant of the pretrained BERT model called Sentence-BERT [31]. This model is trained to embed whole sentences and to semantically compare them using cosine-similarity. Our aim with this approach was to predict how accurate a free textual answer is. Comparing this experiment with the question generation we used a reverse order here. The QUESTION was an SQL statement (e.g., SELECT City FROM Customers), and the ANSWER was the textual description of that statement. The result of the experiment showed that the cosine similarity of two semantically equal correct answers is higher than the cosine similarity of an incorrect and a correct one, Fig. 5. Though the difference is not significant enough it can give the teacher a hint which answer is incorrect. Our plan is to develop this approach further by fine-tuning the model with the SQL domain.

```
Answer 1: Get the City column from the Customers table
Answer 2: Return the City column from the Customers table
Answer 3: Get the Address column from the Customers table
Similarity score 1-2: 0.8145389556884766
Similarity score 1-3: 0.7955647706985474
Similarity score 2-3: 0.6965268850326538
```

Fig. 5. Cosine similarity of two correct and an incorrect answer

Although the presented approach referred to IT domain, this assessment generation model can be used for a wide variate of knowledge domains. In most cases, a conversion module is needed which will convert the answer activity description into a formal, functional format. For example, in elementary mathematics, the calculation procedure can be given with a function having arguments from the given domain. For example, to calculate the area of a rectangle, we can introduce the function *area_rect(r.a, r.b).* After this preprocessing step, we can apply our approach to generate the parameterized question-answer pairs.

5 Test Results for Automated Assessment Generation

For the generation of valid SQL commands, we have constructed a reduced SQL generator automaton. The automaton was implemented as a graph containing both the SQL syntax and the verbal NLP description (Fig. 6).

```
stree = sql_tree()
n_alter_t = sql_t_node("ALTER TABLE",[" alter the structure of table "," modify the structure of table "])
n_alter_t.add_param("TN_1<Table>"," TN_1<Table> ")
n_add = sql_t_node ("ADD", [" adding column "], True)
n_add.add_param("( MN_1<Any> MT_1<Type> )", " MN_1<Any> of type  MT_1<Type> ")
n_create_t = sql_t_node("CREATE TABLE", [" create a new table  "," construct a new table  "],True)
n_create_t.add_param(" TN_1<Table> ( MN_1*<Any> MT_1*<Type> ) "," TN_1<Table> with columns ( MN_1*<Any> )  ")
n_delete = sql_t_node("DELETE FROM", [" delete all data from "," remove all data from "], True)
n_delete.add_param("TN_1<Table>","  TN_1<Table> ")
n_drop_t =  sql_t_node("DROP TABLE", [" remove the table "],True)
n_drop_t.add_param("TN_1<Table>"," TN_1<Table> ")
n_from1 = sql_t_node ("FROM", [" from table "], True)
n_from1.add_param("TN_1<Table>","  TN_1<Table> ")
n_from2 = sql_t_node ("FROM", [" from matching tables "], True)
```

Fig. 6. A sample definition for generator automata

The next figure (Fig. 7) shows an example for an output item pair:

```
list the  SELECT_0 values  from table  FROM_0  for the records where  column WHERE_0 is equal to WHERE_1
::
SELECT SELECT_0 FROM FROM_0 WHERE WHERE_0 = WHERE_1
```

Fig. 7. An example for generated sentence template pairs

The training set contains both valid exercises collected from the internet, like W3CSchool [https://www.w3schools.com/], and synthetic examples shown in the previous step. Using this hybrid training set generation approach, we could construct a larger size training set with low cost.

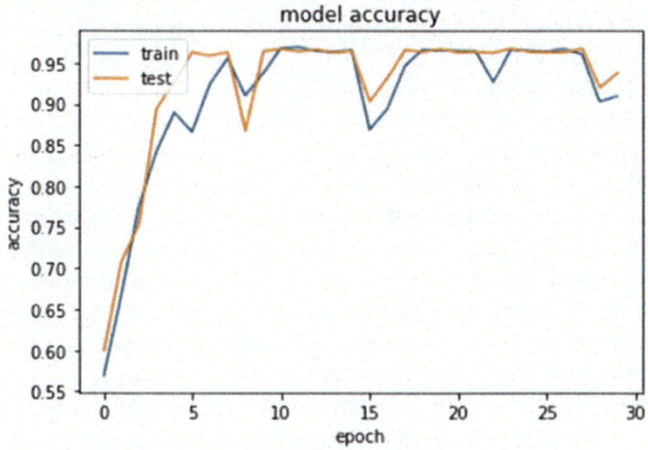

Fig. 8. Accuracy curves for the translation training

Using N = 1000 training size and E = 100 epochs, the validation accuracy reached an acceptable 94.7% value. With a larger dataset (N = 10000), accuracy level reached the 97.7% value. The accuracy values for both train and test are shown in Fig. 8.

The example in Fig. 9, shows a complex syntax, as we can see the prediction from the smaller dataset (N = 1000) was incorrect, especially at near the end of the sentence. In the case of a large dataset (N = 10000), the prediction is correct.

As our experiences show, the encoder-decoder model was able to learn at a high accuracy level also the complex SQL commands generating longer sentences.

```
select select_0 , select_1 from from_0 inner join from_1
  on from_2 = from_3 where where_0 = where_1
::
find the select_0 and select_1 values from matching tables from_0 and from_1
where on matching and
based on the is where for from_2 is where column where_0 is equal to where_1
::
list the select_0 and select_1 from matching tables from_0 and from_1
where the matching is based on from_2 and from_3
for the records where column where_0 is equal to where_1
```

Fig. 9. Examples for the generated assessment templates

6 Conclusion

The automatic assessment management module is one of the main components in ITS systems. In this paper, we presented an automated method to generate assessment question and answer pairs related to a knowledge domain. The proposed module relates to a background domain ontology, where the domain schema is based on the following proposed building blocks: K-units (concepts), K-slots (properties), R-units (constraint rules), M-units (study materials) and T-units (tasks given by question-answer pairs), K-unit taxonomy relationship, K-unit containment relationship and K-competency relationship. The question-and-answer pairs are defined with a template sentence containing placeholders with reference to ontology elements. The proposed method uses a generator automaton to construct the answer templates in the first phase. In the next phase, an encoder-decoder neural network translator unit was used to generate the question sentence templates for the selected answer templates. The paper presents a domain specific embedding model, too. The proposed embedding neural network can be used to compare the semantic similarity of free-text answer sentences. According to the tests with the implemented prototype assessment generator application on the SQL domain, the proposed framework is suitable to generate assessment tasks in a very efficient way.

In the future research, the main goal is to investigate some related open problems like the application of AIEd technology in construction of the generator automaton or the development of efficient crawler engines to collect related examples from the Internet.

References

1. Nkambou, R., Mizoguchi, R., Bourdeau, J. (ed.): Advances in intelligent tutoring systems. Springer Science & Business Media (2010). https://doi.org/10.1007/978-3-642-14363-2
2. Sottilare, R., et al.: Design Recommendations for Intelligent Tutoring Systems. U.S. Army Research Laboratory, Florida (2015)
3. Mitrovic, A.: An intelligent SQL tutor on the web. Int. J. Artif. Intell. Educ. **13**(2–4), 173–197 (2003)
4. Craig, S.D., et al.: The impact of a technology-based mathematics after-school program using ALEKS on student's knowledge and behaviors. Comput. Educ. **68**, 495–504 (2013)
5. Pane, J.F., Griffin, B.A., McCaffrey, D.F., Karam, R.: Effectiveness of cognitive tutor algebra I at scale. Educ. Eval. Policy Anal. **36**(2), 127–144 (2014). https://doi.org/10.3102/0162373713507480
6. Koedinger, K.R., McLaughlin, E.A., Heffernan, N.T.: A quasi-experimental evaluation of an on-line formative assessment and tutoring system. J. Educ. Comput. Res. **43**(4), 489–510 (2010). https://doi.org/10.2190/EC.43.4.d
7. Zawacki-Richter, O., Marín, V.I., Bond, M., Gouverneur, F.: Systematic review of research on artificial intelligence applications in higher education – where are the educators? Int. J. Educ. Technol. Higher Educ. **16**(1), 1–27 (2019). https://doi.org/10.1186/s41239-019-0171-0
8. Luckin, R., et al.: Intelligence unleashed: An argument for AI in education (2016)
9. Baker, T., Smith L., Anissa N.: Educ-AI-tion rebooted? Exploring the future of artificial intelligence in schools and colleges (2020)
10. Graesser, A.C., Conley, M.W., Olney, A.: Intelligent tutoring systems. In: Harris, K.R., Graham, S., Urdan, T., Bus, A.G., Sonya Major, H., Swanson, L. (eds.) APA educational psychology handbook, vol. 3: Application to learning and teaching., pp. 451–473. American Psychological Association, Washington (2012). https://doi.org/10.1037/13275-018

11. Merrill, D.C., Reiser, B.J., Michael Ranney, J., Trafton, G.: Effective tutoring techniques: A comparison of human tutors and intelligent tutoring systems. J. Learn. Sci. **2**(3), 277–305 (1992). https://doi.org/10.1207/s15327809jls0203_2
12. Huang, J., Chen, Z.: The research and design of web-based intelligent tutoring system. Int. J. Multimedia Ubiquitous Eng. **11**(6), 337–348 (2016). https://doi.org/10.14257/ijmue.2016.11.6.30
13. Self, J.A.: Student models: what use are they? In: Ercoli, P., Lewis, R. (eds.) Artificial Intelligence Tools in Education, pp. 73–86 (1988)
14. O'Shea, T.: A self-improving quadratic tutor. In: Sleeman, D.H., Brown, J.S. (eds.) Intelligent Tutoring Systems, pp. 283–307. Academic Press (1982)
15. Beck, J., Woolf, B., Beal, C.R.: ADVISOR: A machine learning architecture for intelligent tutor construction. AAAI/IAAI **2000**(552–557), 1–2 (2000)
16. Cooper, E.: Tutoring center effectiveness: The effect of drop-in tutoring. J. Coll. Read. Learn. **40**(2), 21–34 (2010). https://doi.org/10.1080/10790195.2010.10850328
17. Prinsloo, P.: Fleeing from Frankenstein's monster and meeting Kafka on the way: Algorithmic decision-making in higher education. E-Learn. Digital Media **14**(3), 138–163 (2017). https://doi.org/10.1177/2042753017731355
18. Trausan-Matu, S., Boyer, K.E., Crosby, M., Panourgia, K. (eds.): ITS 2014. LNCS, vol. 8474. Springer, Cham (2014). https://doi.org/10.1007/978-3-319-07221-0
19. Yacef, K.: Intelligent teaching assistant systems. In: International Conference on Computers in Education. Proceedings IEEE (2002)
20. Holstein, K., McLaren, B.M., Aleven, V.: Intelligent tutors as teachers' aides: exploring teacher needs for real-time analytics in blended classrooms. In: Proceedings of the 7th International Learning Analytics, Knowledge Conference (2017)
21. Baker, R.S.: Stupid tutoring systems, intelligent humans. Int. J. Artif. Intell. Educ. **26**(2), 600–614 (2016). https://doi.org/10.1007/s40593-016-0105-0
22. Naser, S.S.A.: ITSB: an intelligent tutoring system authoring tool. J. Sci. Eng. Res. **3**(5), 63–71 (2016)
23. Feng, C., Guo, G., Jin, S., Zhou, C.: Authoring tools for easy creation of adaptive tutoring. In: Spring, BHCI Capstone Project (2018)
24. Shulman, L.S.: Ways of seeing, ways of knowing: Ways of teaching, ways of learning about teaching. J. Curric. Stud. **23**(5), 393–395 (1991). https://doi.org/10.1080/0022027910230501
25. Loewenberg Ball, D., Thames, M.H., Phelps, G.: Content knowledge for teaching: What makes it special. J. Teach. Educ. **59**(5), 389–407 (2008)
26. Tsang, V., Stevenson, S.: A graph-theoretic framework for semantic distance. Comput. Linguistics **36**(1), 31–69 (2010). https://doi.org/10.1162/coli.2010.36.1.36101
27. Maedche, A., Staab, S.: Ontology learning for the semantic web. IEEE Intell. Syst. **16**(2), 72–79 (2001). https://doi.org/10.1109/5254.920602
28. Mastan, I., Shanmuganathan, R.: Multi-level encoder-decoder architectures for image restoration. In: (CVPRW) IEEE (2019)
29. Hao, B., Henghui, Z., Ioannis, P.: Enhancing clinical BERT embedding using a biomedical knowledge base. In: Proceedings of the 28th International Conference on Computational Linguistics (2020)
30. Devlin, J., Chang, M.V., Lee, K., Toutanova, K.: BERT: Pre-training of Deep Bidirectional Transformers for Language Understanding. arXiv:1810.04805. ArXiv: 1810.04805 (2018)
31. Reimers, N., Gurevych, I.: Sentence-BERT: Sentence Embeddings using Siamese BERT-Networks. arXiv:1908.10084. ArXiv: 1908.10084 (2019)

BlockCACert – A Blockchain-Based Novel Concept for Automatic Deployment of X.509 Digital Certificates

Adam Mihai Gergely(✉) and Bogdan Crainicu

University of Medicine, Pharmacy, Science and Technology "G.E. Palade", Târgu-Mureş, Romania
{adam.gergely,bogdan.crainicu}@umfst.ro

Abstract. Digital certificates (also known as X509 public-key certificates) are used to prove the identity of a client or of a server to the communicating partner and to establish an encrypted channel for secure communication. The most prominent use of digital certificates is found in the HTTPS protocol, for securing websites and clients by means of digital signatures and public-key encryption. Because digital certificates are signed by Certificate Authorities that supposedly everyone trusts, it is implicitly said that these certificates provide some degree of trustworthiness. But history shows us that Certificate Authorities can also make mistakes, issue false certificates or be compromised in some form or another. Also, currently there are multiple types of digital certificates that can be issued: DV, OV and EV, which cost differently based on the grade of trust. We believe that real trust should not be quantifiable in degrees of trust: we either trust a certificate or we don't. We propose BlockCACert, a blockchain-based solution for CAs for issuing all types of X509 certificates using the ring signatures concept, which fixes the problem of general trust and multiple-level trust by using a global procedure for issuing all types of digital public-key certificates.

Keywords: Blockchain · X509 · Digital signatures · Digital certificates · Public-key · Private-key · Certification authorities · PKI · Ring signatures

1 Introduction

Within a typical PKI (Public Key Infrastructure) deployment, digital certificates were used to prove the identity of clients and servers by using a tertiary entity known as a Certification Authority (CA).

An entity X that wanted to have a certificate issued for its identity would generate a PKI cryptographic key pair (private key and public key) and a Certificate Signing Request (CSR) using its own identification data bonded to the public-key. In order to obtain a valid signed certificate, this entity would contact a CA and request certification (usually for a fee). The CA would analyze this request, perform a validation on the requesting entity's data, and after the fee is paid, a certificate would be issued. This certificate contains the binding of the CA's digital signature with the CSR resulting in a

L. Moldovan and A. Gligor (Eds.): Inter-Eng 2021, LNNS 386, pp. 820–832, 2022.
https://doi.org/10.1007/978-3-030-93817-8_73

final certificate. The entity would use the certificate throughout the certificate's validity period. After the certificate expires, entity X can apply for recertification by applying for a certificate renewal. If there are no validation or identity proofing problems, the issuing CA reissues a new certificate for the same entity. At any given time, if the entity requests a certificate revocation or the CA finds out that the entity is not eligible for certification (in case of major problems or frauds), the CA can revoke the entity's certificate.

The procedure described above is known as the Standard or Traditional PKI. This form of traditional PKI gives a certain amount of control to the Certification Authorities and assumes that these CAs are honest and do not cheat. However, history shows us that there have been successful attacks on CAs and certificates were issued by fraud.

Another type of issue which we consider a problem is the multiple level trust. We believe that trust exists fully, not partially. Currently, there are multiple main levels of trust: DV (domain validation), OV (organization validation) and EV (extended validation). We believe that having a single type of trust eliminates the need for multiple-level trust mechanisms and helps to avoid confusion regarding trust.

This paper presents a theoretical proof-of-concept and is organized as follows:

In Sect. 2 we briefly discuss about the main components of the PKI ecosystem. Section 3 summarizes the current problems of trust regarding Certificate Authorities. Section 4 argues the problem of multiple-type validations. Section 5 proposes a novel concept called BlockCACert for automatic deployment of X.509 digital certificates and Sect. 6 analyzes a challenging and disruptive approach by which the CA uses ring signatures mechanism in the signing process of the request before publishing the certificate to the blockchain. We conclude the paper with our view regarding the positive impact of our concept on the world of digital certificates.

2 PKI Ecosystem

2.1 Digital Signatures and Digital Certificates

Digital Signatures are small pieces of data that contain authentication information which can be verified using a public key. Digital signatures are used to prove that a message or content belongs to a person or an entity. A digital signature, if implemented properly, is much more secure than a standard hand-written signature and harder to falsify. Of course, the private key used to generate the signature must not be compromised.

Digital Certificates are also known as Public-key certificates, or standard X.509 type documents (the standard which denotes the family of certificates). A digital certificate associates a digital signature to a public key with an identity. The certificate is used to verify if a public key belongs to an entity.

In a general PKI platform, the digital signature from within the digital certificate belongs to the Certification Authority (CA). Regardless of the nature of the digital signature used, the signature proves that the signer trusts the fact that the public key belongs to the certificate owner (information extracted from within the certificate).

2.2 Public PKI Systems

When we talk about the management of digital certificates or about public or private PKI systems, basically we talk about the processes that take place at the Certification

Authority. The CA acts like a third party, which is trusted by all the other communicating parties. The CA certifies that at least one of the communicating parties is genuine and is authorized to execute secure communications in its name.

A typical illustration of a traditional public PKI systems along with the respective workflow are detailed in Fig. 1:

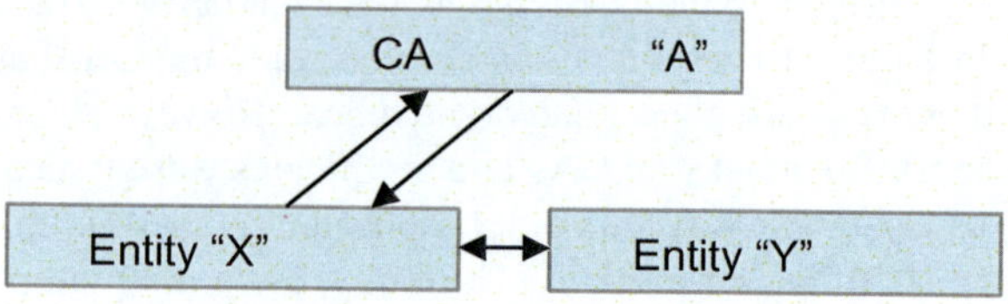

Fig. 1. A typical PKI CA workflow

Step 1: Entity X generates it's keypairs (private and public keys).
Step 2: Entity X creates a CSR binding its public key with its own information.
Step 3: Entity X contacts CA A and proves its identity to the CA. (After paying the certification fee,) The CA signs the CSR with its own private key producing a Certificate (CRT) which is given to Entity X.
Step 4: Entity X gives the certificate to any communicating partner. When entity Y wishes to authenticate Entity X, it validates the certificate against its CA stored public keys. Using the public keys, entity Y can verify the signature of the CA "A" applied to Entity X's certificate.
Step 5: After authentication is performed, secure communication may now commence.

This scenario, however, introduces some security risks which are described later in this paper.

3 Current Problems of Certificate Authorities

CA-based PKI contains numerous single points-of-failure. The centralized trust model of CA-based PKI means that the security of the system is based on single points-of-failure: CAs. Also, the structure of an X509 certificate may rely on a single root certificate for validation. If the Root CA in the chain is compromised, it can compromise the entire chain of trust [4].

There are online services, like Google's Certificate Transparency Project [8] which tries to detect rogue or forged CAs and their certificates. While Certificate Transparency (CT) acts similar to a blockchain, by providing a public logging of digital certificates information, using the blockchain for PKI offers an alternative solution to storing certificates securely and detecting any possible fraudulent activity.

Opsmate Inc. aggregates a list of Certificate Authority failures, in which false CAs or certificates were involved and other attacks which were performed against the PKI system, because of negligence or distrustful practices of CAs [9]. As the list points out, a

CA can suffer from both technical and practice habits problems. While we believe that the technical problems can arise like any other computer software bug, they can also be fixed and a CA cannot be held liable for this if it implemented software guidelines correctly. We can only hope that improved technology, best practices and further regulations in this industry are developed, to try to secure and align CAs to the best practices, highest security and high-quality service grid.

4 Current Shortcomings of Different Validation Types

All X.509 certificates, while having a similar structure and purpose, they differ in the type of validation used to issue them. Any certificate that is issued to a requesting entity is issued by a Certification Authority (CA) after the CA performs some form of validation to ensure that the requesting party proves its identity and its authorization to use the certificate.

In the PKI Digital Certification industry there are multiple types of validations:

- DV – Domain Validation
- OV – Organization Validation
- EV – Extended Validation

Below we present a summary of these validation types:

4.1 Domain Validation (DV)

DV certificates are the most common type of SSL certificates. They are verified using only the domain name as element of verification. Informally the workflow is as follows: the entity that requests a DV certificate sends the request to the CA. The CA then passes on a Challenge of HTTP or DNS type which the client must resolve. After the challenge is resolved and proof of domain control is validated, the certificate can be issued [10].

4.2 Organization Validation (OV)

OV certificates require more validation than DV ones but provide more trust. For this type of certificates, the CA verifies the actual entity or organization that wants to get a certificate. OVs are used by corporations, governments and other entities that want to provide the extra layer of identity information, alongside the technical domain element. OV is also used for code signing, document signing, or client authentication [10].

4.3 Extended Validation (EV)

It is said that EV certificates provide the maximum amount of trust because of the extra verifications being performed, even more than in the case of OV certificates.

The guidelines for issuing EV-type certificates were laid down by the CA/Browser Forum [10] and extra documentation must be provided to issue an EV certificate. As with OV, the EV lists the company/organization name in the certificate itself. EV certificates are the only type of certificates which cannot be locally issued (by creating a local-owend home CA), because it is said that modern browsers hardcode the IDs of the CAs which are authorized to issue EV certificates. While these types of validations offer different grades of trust, obviously for different fees, we believe that real trust must be expressed by a single type of validation, similar to an OV-type validation.

We believe that OV-type validation has enough fields to assure anyone of who the domain/CN is and who operates this certificate (O & OU fields). But, in our opinion, that EV-type validation is nothing more than an extension to OV-type validation, which used to turn the web-browser bar to green color or offer a special marker indicating an EV-type validation – the "O" (Organization) filed is sufficient to denote who is the owner of the certificate in question. A single type of validation is sufficient to ensure trust in a resource (Certificate – "CN" field) ownerd by a Person or an Orgzanization (Certificate - "O" field) [10].

5 BlockCACert Concept

We can use blockchain, as a public ledger, to register any kind of PKI operations, like issuing new certificates, renewing existing certificates or revoking certificates. Our new concept combines the blockchain public ledgering technology with the traditional PKI concept, adding the ring signatures component.

The BlockCACert concept only uses Blockchain technology for securely storing the digital certificates. It does not operate like a cryptocoin's blockchain, hence there is no proof of work involved. There are no miners involved and the transaction is processed automatically without anyone receiving a reward or a monetary value (as opposed to Bitcoin or other cryptocoins).

Implementation details about how BlockCACert is built, how the agents join/leave the system or the protocols involved are not part of the BlockCACert concept.

5.1 Blockchain Technology Basics

Blockchain [2, 3] is an innovative technology which aims to provide a secure, anonymous, based on consensus and decentralized public-ledger solution for transactions between users, without a central authority overseeing any of the transactions and without the possibility of linking the users to their actual identities. Blockchain was created with the concept of security and anonimity in mind [1, 4].

The non-financial aspect of blockchain technology relies in the mechanism used to operate the blockchain. On the blockchain, an easy verification can be made to verify the chain of ownership (Fig. 2).

5.2 Indexes Used by BlockCACert in the Concept Demonstration

C – Client (person, company, institution, etc.);
B – Multilayer Blockchain;
B_e – Blockchain $\rightarrow$ Entities' Layer;
B_c – Blockchain $\rightarrow$ Certificates' Layer;

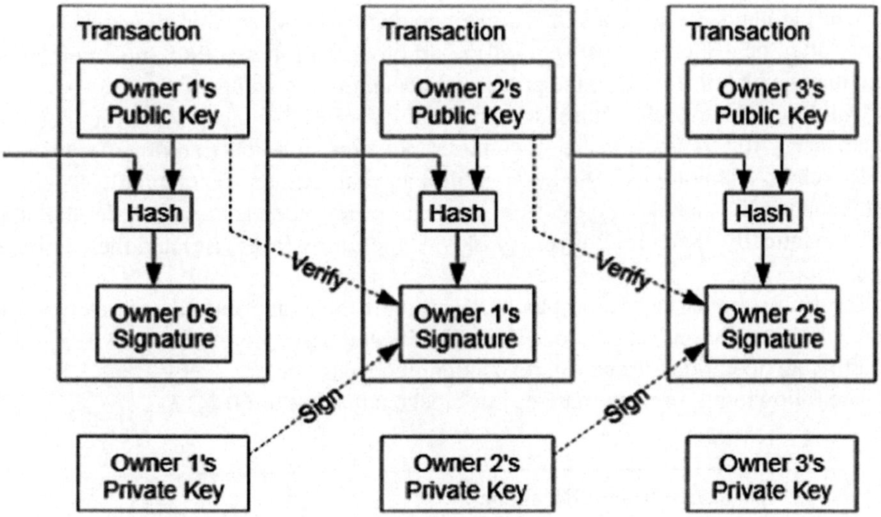

Fig. 2. Original blockchain verification scheme (Bitcoin) [2]

CA_x – Certificate Authorities;
K_{priv} – Private Key;
K_{pub} – Public Key;
CSR – Certificate Signing Request;
CSR_m – Modified Certificate Signing Request (modified standard CSR header);
CRT – Final (signed) Digital Certificate;
ACode – Secret Authorization Code;
EACode – Encrypted Secret Authorization Code;
$\mathrm{K_{priv}CA_x}$ – CAx's Private Key.
$\mathrm{K_{pub}CA_x}$ – CAx's Public Key.

5.3 PKI Component

BlockCACert concept uses the traditional PKI and X.509 standards for issuing, administering and revoking certificates. The only modification of BlockCACert in the PKI area is the introduction of a new type of CSR: CSRm.

The CSRm is the same as a regular CSR, except it contains one field that does the linking between the PKI and the blockchain: The Client Secret Authorization Code (ACode).

5.4 Blockchain Structure Component

Starting from the observation made by Axon in [5] and Axon and Goldsmith in [6], who consider the suitability of a blockchain-based PKI for contexts in which PKI is required, but in which linking of identity with public key is undesirable, our novel concept uses the blockchain technology only for its properties as an unmodifiable public ledger, where no financial elements are present in this implementation. Moreover, according to [7] privacy is very often neglected in most state-of-the-art blockchain-based PKI implementations, topic that we also followed in the process of designing our solution.

Our blockchain architecture does not use proof-of-* concepts and neither does reward users for transactions with some cryptocoins, like the traditional application of blockchain technologies. Our version of blockchain has an abstract multi-layer concept. Technically, there are no multiple layers, as every block comes one after the other. But, semantically, there are 2 layers of blocks: the Entity layer (Be) and the Certificate Layer (Bc).

The Entity Layer stores the blocks regarding to new and retired entities (persons, companies, institutions, and so on.). The Certificate Layer stores the blocks regarding to certificate operations: issuance, revocation, renewals, etc.

The following figure summarizes our blockchain structure (Fig. 3):

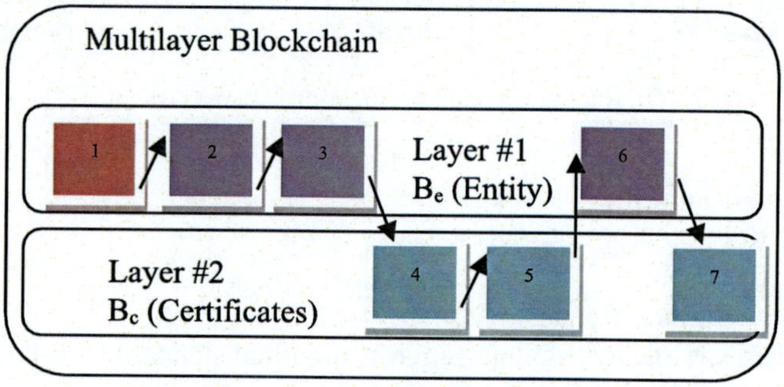

Fig. 3. BlockCACert – the multi-layer blockchain structure

Legend:

1. – Genesis
2. – P.Hash; ID1; Type1
3. – P.Hash; ID2; Type1
4. – P.Hash; ID3; Type2
5. – P.Hash; ID4; Type2
6. – P.Hash; ID5; Type1
7. – P.Hash; ID6; Type2

The contents of the Entity Layer: (Be) (Table 1):

Table 1. Entity layer fields

Field no	Blockchain-registered data for a given ACode(C)
0 × 1	Previous Hash
0 × 2	ID Block
0 × 3	Block Type
0 × 4	Registration Date
0 × 5	Country
0 × 6	State
0 × 7	City
0 × 8	O
0 × 9	OU
0xA	CN
0xB	ACode
0xC	Notes

The contents of the Certificates Layer (Bc) (Table 2):

Table 2. Certificate layer fields

Field no	Blockchain-registered data for a given *ACode(C)*
0 × 1	Previous Hash
0 × 2	ID Block
0 × 3	Block Type
0 × 4	Registration Date
0 × 5	CN
0 × 6	Validity
0 × 7	X509 CRT Encoded (DER/PEM)
0 × 8	Notes

The number of bytes per each field is subject to open discussion, remaining to be established by each implementing entity. There are no fix limits on how much space should the blockchain header occupy. This is a theoretical concept and implementation data may vary based on specific implementation.

5.5 Workflow of PKI-Blockchain Transactions

Let's assume we have a client C which registers on the blockchain as an entity. This entity can be a real person, a company, an institution, and so on. The method by which the client C registers on the blockchain is beyond of this paper's scope. Ideally, it can register on the blockchain via an official authority, via a registering company, etc.

Technically, when registering on the blockchain, C registers on the Entities' Layer, by creating a new block with the following elements:

- (the checksum of the previous block)
- New Block ID
- Block Type: 1
- Registration date (YYYY-MM-DD format)
- Country
- City
- Organization (O)
- Organization Unit (OU)
- Common Name (CN)
- Secret Authentication Code
- Notes

After the registration, C gets a Secret Authentication Code which defined the respective C as being unique and authorizes C to perform blockchain-based PKI operations.

We define the code as follows:

$$ACode(C) = Hash\left(CN + date\left(''U''\right)\right); \quad (1)$$

where,

ACode(C) = The Secret Authorization Code of Client C;
Hash() = A cryptographic hash function like SHA;*
CN = X.509 Standard's Common Name field.
date("U") = a pseudo-code representation of the current date and time, in standard Unix epoch format.

When C decides that it wants to begin cryptographic operations, C will generate its CSR via traditional PKI method:

- It generates its K_{priv}
- It generates its *CSR* (which contains the K_{pub}) with the following fields, actually generating a CSR_m:
- Country
- State
- Cirty
- (Organization)
- OU (Organization Unit)
- CN (Client Name or Server URL)
- EACode – this is the new field in the $\mathrm{CSR_m}$ format;

The following formula describes the *EACode* field:

$$EACode(C) = ENCRYPT\big(ACode(C), K_{pub}CA_x\big); \quad (2)$$

As the formula describes, before inserting the ACode into the EACode field, this code must be encrypted with the Public Key of the Certificate Authority which is going to be contacted in order to perform the certification operation. This encryption assures that no one can steal the client's secret authorization code and only the respective CA can decrypt it using its private key. Modifying the standard CSR to include an encrypted authorization code yields a new CSR fomat: CSRm (CSR modified). The following formula describes the new CSRm format:

$$CSRm = CSR + ACode\ Field; \quad (3)$$

- After generating the CSR_m, C contacts an arbitrary CA (CA_x) and applies for a new certificate by supplying some registration data and the CSR_m.
- CA_x decrypts the *EACode* using its *KprivCA*:

$$ACode(C) = DECRYPT\big(EACode(C), K_{priv}CA_x\big); \quad (4)$$

- The CA contacts the blockchain (B) and begins validating the data against the Decrypted Authorization Code found in the Entity's Layer of the blockchain (B_e) and supplied in the CSR_m (Table 3);

Table 3. Registered vs. supplied fields comparison table

Field no	Blockchain-registered data for a given *ACode(C)*	Validation Condition	Data supplied in the CSRm for a given ACode
1	Country	=/≠	Country
2	State	=/≠	State
3	City	=/≠	City
4	Organization (O)	=/≠	Organization (O)
5	Organization Unit (OU)	=/≠	Organization Unit (OU)
6	Common Name (CN)	=/≠	Common Name (CN)
7	*ACode(C)* (created at registration on B_e)	=/≠	*ACode(C)* (decrypted from CSR_m)

- The condition for data matching is an AND-ing operation which verifies the matching of the values:

$$Field1(B_e)\,AND\,Field1(CSR_m) = 1;$$
$$Field2(B_e)ANDField2(CSR_m) = 1;$$

$$Field3(B_e)ANDField3(CSR_m) = 1;$$
$$Field4(B_e)ANDField4(CSR_m); = 1;$$
$$Field5(B_e)ANDField5(CSR_m) = 1;$$
$$Field6(B_e)ANDField6(CSR_m) = 1;$$
$$Field7(B_e)ANDField7(CSR_m) = 1; \quad (5)$$

- If all the fields match, then the client is validated.
- If, at least one of the fields do not match, the client is invalidated and the transaction is rejected.
- When CA_x validates C on B_e, it proceeds to sign the CSR_m with the $K_{priv}CA_x$, using the ring signatures scheme described below resulting a final signed digital certificate belonging to the C:

$$CRT\ (C) = CASIGN[RINGS]\big(CSRm(C),\ K_{priv}CA_x\big);\ ) \quad (6)$$

- This certificate is issued to the client and is registered on the blockchain (B) as a new transaction on the B_c Layer, by creating a new block with the following elements:

  - (the checksum of the previous block)
  - New Block ID
  - Block Type: 2 (certificate block)
  - Registration date (YYYY-MM-DD format)
  - Common Name (CN)
  - Validity
  - Encoded X.509 Certificate (DER/PEM Format)
  - Notes

Anyone wishing to verity the certificate, can do it by the following means:

- Traditionally, via PKI tools: checking the Signature of the CA within the CRT against a Certificate Root Store;
- Interogating the blockchain (BC of B) by querying the validity of the certificate or the act of certification from CAx who digitally signed the CSRm with its KprivCAx.

6 Ring Signatures in BlockCACert

A ring signature is a type of digital signature that allows any member of a group of users (ring) to sign a message with its own key in an anonymous way. Within the process, no one, but except actual signer, knows which member signed the message - this means it is computationally infeasible to determine which of the group members' keys was used to produce the digital signature. Ring signatures must satisfy two independent notions of security: anonymity and unforgeability.

The notion of ring signature was formalized by Rivest, Shamir and Tauman in [11]. In contrast to group signatures, the authors designed a signature scheme with no managers involved, and which do not require any central coordinator, setup and revocation procedures. Moreover, according to authors, there is no way to distribute specialized keys and to revoke the anonymity of the actual signer, and there are no prearranged groups of users.

A ring signature scheme is defined by two procedures [11]:

- ***ring-sign***(m, P_1, P_2,...,P_n, s, S_s) – produces a ring signature σ for the message m, where P_1, P_2,...,P_n are the public keys of the n ring members, S_s is the secret key of the s-th member, who is the actual signer.
- ***ring-verify***(m, σ) – takes a message m and a signature σ (which includes the public keys of all the possible signers), and outputs either *true* or *false*.

Ring Signatures is fundamental concept used in the CryptoNote/Monero cryptocurrency [13, 14].

In order to preserve the privacy between BlockCACert layer's communication, we propose a formal ring signature mechanism starting from the functional definition of a ring signature presented by Bender, Katz, Morselli in [12]:

- key generation: $Gen(1^k)$, where k is a security parameter, outputs a public key PK and secret key SK. Entity layer (B_e) is responsible for key generation. The keys are not included into the block structure, but each block has an associated pair of keys, which can be changed/regenerated at any given time (due to length restrictions, we are unable to detail the way the keys are regenerated and relinked to the blocks).
- message signature: $Sign_{s,SK}(M, R)$ outputs a signature σ on the message M with respect to the ring $R = (PK_1, ..., PK_s)$, where $(R[s], SK)$ is a valid key-pair output by *Gen*. Here it is important to note that in the signing process we have the secret keys associated with the blocks chained to the current block s at which the signature is applied.
- signature verification: $Vrfy_R(M, \sigma)$ verifies a signature σ on a message M with respect to the ring of public keys R. The verification process takes place only at the Certificate Layer (B_c). If the signature verification fails, that block at the Certificate Layer does not perform any additional analysis to check if a new public key has appeared in the blockchain – the method, although it seems quite rigid, solves quickly two issues: double spending prevention and no more computational resources.

Regarding the implementation details of our proposed ring signature mechanism, one of the viable options would be to define a third layer for BlockCACert architecture, where the entire key management mechanism is modularized and integrated.

7 Conclusions

BlockCACert tries to improve the management of digital certificates by using blockchain technology as a proven security mechanism for CAs trustworthiness. This ensures that no fraudulent certificates can be issued, or even if, somehow, these certificates are issued, the fraud is immediately detectable on the blockchain.

This scheme eliminates the need for multiple types of validation. By using non-modifiable blockchain records, we can ensure that every entity who requests a certificate has a name, and some identification information, thus eliminating the need for multi-level trust. Our solution also introduces a degree of automatic deployment, to ease the task of authentication. As a future direction of research, we will investigate integration of BlockCACert with the privacy data storage protocol proposed in [15], that is based on the ring signature on the elliptic curve.

This concept was implemented in one of our labs, on 10 computers, and the results we got confirmed the desired workings of our concept. There are no numerical results, as this theoretical concept only proves that the workflow of storing digital certificates on blockchains works efficiently when implemented properly.

References

1. Gergely, A., Crainicu, B.: A theoretical study and review on blockchain's privacy mechanisms. In: IFIP SEC 2019. Lisbon, Portugal (2019)
2. Nakamoto, S.: Bitcoin: a peer-to-peer electronic cash system. Cryptography Mailing list at (2008). https://metzdowd.com
3. Xu, M., Chen, X., Kou, G.: A systematic review of blockchain. Financ. Innov. **5**(1), 1–14 (2019)
4. Guegan, D,: Public Blockchain versus Private blockchain, HAL ID: halshs-01524440, version 1 (2017)
5. Axon, L.: Private-awareness in Blockchain-based PKI, Centre for Doctoral Training in Cyber Security, CDT Technical Paper Series 21/15, University of Oxford (2015)
6. Axon, L., Goldsmith, M., PB-PKI: a privacy-aware blockchain-based PKI. In: Proceedings of the 14th International Joint Conference on e-Business and Telecommunications (ICETE 2017) – SECRYPT, vol. 4, pp. 311–318 (2017). ISBN: 978-989-758-259-2
7. Brunner, C., Knirsch, F., Unterweger, A., Engel, D.: A Comparison of blockchain-based PKI implementations. In: Proceedings of the 6th International Conference on Information Systems Security and Privacy – ICISSP, pp. 333–340. Valletta, Malta (2020). ISBN 978-989-758-399-5; ISSN 2184-4356
8. Google Certificate Transparency Project. https://certificate.transparency.dev
9. SSLMate: Timeline of Certificate Authority Failures, Opsmate Inc. https://sslmate.com/certspotter/failures
10. SSL Americas: DV, OV, IV, and EV Certificates. https://www.ssl.com/article/dv-ov-ev-certificates/
11. Rivest, R.L., Shamir, A., Tauman, Y.: How to leak a secret. In: Boyd, C. (eds.) Advances in Cryptology—ASIACRYPT 2001. ASIACRYPT 2001. LNCS, vol. 2248, pp. 552–565. Springer, Berlin, Heidelberg (2001). https://doi.org/10.1007/3-540-45682-1_32
12. Bender, A., Katz, J., Morselli, R.: Ring signatures: stronger definitions, and constructions without random oracles. J. Cryptol. **22**, 114–138 (2009)
13. Shen, N.: Ring signature confidential transactions for monero. Cryptology ePrint Archive, Report 2015/1098 (2015)
14. Shen, N.: Ring confidential transaction. Ledger **1**, 1–18 (2016)
15. Li, X., Mei, Y., Gong, J., Xiang, F., Sun, Z.: A blockchain privacy protection scheme based on ring signature. IEEE Access **8**, 76765–76772 (2020)

Instantaneous Frequency Identification in Nonstationary Signals

Zoltán Germán-Salló(✉)

University of Medicine, Pharmacy, Sciences and Technology "George Emil Palade", Targu-Mures, Romania
zoltan.german-sallo@umfst.ro

Abstract. This paper presents a comparison study of different instantaneous frequency (IF) estimation procedures. These procedures use different types of synthesized test signals, the instantaneous frequencies of these signals are estimated through three procedures, one based on Hilbert Transform (HT) and the other two on time frequency distribution, Short Time Fourier Transform (STFT) provided power spectrum density and Fourier analysis-based spectrogram. The three different signals and the three different estimation procedures gave a comprehensive view over the instantaneous frequency interpretation as the derivative of instantaneous phase. The obtained experimental results from different procedures are compared and analyzed, conclusions that were drawn are presented.

Keywords: Instantaneous frequency · Time–frequency distribution · Analytic signal · Hilbert transformation

1 Introduction

The analysis of nonstationary signals, in order to reveal their hidden properties in time and frequency domains, became a very important task in signal processing. Neither time analysis nor frequency investigation method alone can completely describe or model the nonstationary characteristics of a given signal.

Non-stationary oscillatory signals occur in a wide range of fields, including physics, biology, medicine, electronic engineering and many others. They contain different oscillatory components, which usually reflect properties of the underlying system. In most of situations, in order to extract information from an observed or acquired signal is important to know the properties of the different constituents of the signal. Signal analysis usually means noise elimination from signal (filtering), identification and isolation of individual components in order to contour and evaluate the underlying rules that control the system.

The frequency spectrum-based analysis tools introduced by the Fourier Transform (FT), have been intensely used in systems showing periodic and quasi-periodic behaviors, including oscillations and vibrations. Spectral analysis has been applied to solve various problems in different fields, including mechanical vibrations and fault detection and monitoring, speech and music recognition, telecommunications, power systems and

L. Moldovan and A. Gligor (Eds.): Inter-Eng 2021, LNNS 386, pp. 833–842, 2022.
https://doi.org/10.1007/978-3-030-93817-8_74

medical engineering. The nonstationary nature of the signals and nonlinear systems require the idea of instantaneous frequency (IF). The IF is the basis of the time–frequency distributions (TFD) or time–frequency-energy (TFE) representation and analysis of a signal. The IF is a practically important parameter of a signal which provide explanations for physical phenomenon in many applications.

The spectral methods were improved by D. Gábor [1], who introduced short time, or windowed Fourier transforms in order to help time–frequency analysis and eventually contribute to the development of wavelet theory. Gabor and wavelet transforms provide rich visualizations of time–frequency representations with spectrograms and scalograms respectively [2].

This paper presents instantaneous frequency determination procedures applied to synthetic test signals and a comparison study between them in order to figure out how signal properties and estimation methods are or could be correlated [3].

2 Instantaneous Frequency Estimation Fundamentals

Recently, a variety of approaches to the instantaneous frequency and phase estimation problem, have been developed. These approaches differ by estimation precision, computational resources complexity, and processing duration.

The instantaneous frequency of a nonstationary signal is a time-dependent parameter that reflects the mean value of the frequency components present in the time-varying signal. There are many procedures developed and presented in literature.

Trying to find a valuable definition of instantaneous frequency (IF), Carson and Fry in 1937 [4] arguing that the notion of IF is a generalization of the definition of constant frequency (i.e., it is the rate of change of phase angle at time t), they considered a variable frequency in an electric circuit theory context and then applied the concept to frequency modulated (FM) signals. They defined an FM wave as

$$\omega(t) = exp\left[j\left(\omega_0 t + \lambda \int_0^t m(t)dt\right)\right] \tag{1}$$

where $\omega_0 = 2\pi f_0, f_0$ is a constant carrier frequency, λ is the modulation index and $m(t)$ represents a low-frequency signal to be transmitted ($|m(t) \leq 1|$). They defined the instantaneous angular frequency as

$$\Omega(t) = \omega_0 + \lambda m(t) \tag{2}$$

where m(t) has the dimension of frequency, and the instantaneous cyclic frequency as

$$f_i(t) = f_0 + \frac{\lambda}{2\pi} m(t) \tag{3}$$

The next important step in the study of IF was made by D. Gábor [5] who proposed a method for generating a unique complex signal from a real one. His method for doing so is first to find the FT of the real signal and then to "suppress the amplitudes belonging to negative frequencies and multiply the amplitudes of positive frequencies by two." He also showed that this procedure is equivalent to the following time-domain procedure:

$$z(t) = s(t) + jH[s(t)] = a(t) \cdot e^{j\Phi(t)} \tag{4}$$

where $z(t)$ is a complex signal, introduced by D. Gábor, also named as analytic signal, $s(t)$ is the real signal and H is the Hilbert Transform (HT), defined as a convolution between the real signal $s(t)$ and $\frac{1}{\pi t}$:

$$H[s(t)] = pv \cdot \int_{-\infty}^{+\infty} \frac{s(t-\tau)}{\pi t} d\tau \tag{5}$$

where pv is the Cauchy principle value of the integral [6]. Gabor's motivation to define and work with a complex signal was that only by doing so was he able to define the central moments of frequency of the signal:

$$\langle f^n \rangle = \frac{\int_{-\infty}^{+\infty} f^n \cdot |Z(f)|^2 df}{\int_{-\infty}^{+\infty} |Z(f)|^2 df} \tag{6}$$

where $Z(f)$ is the spectrum of the complex signal [6]. Ville [7] unified the work done by Carson and Fry and Gabor and defined the IF of a signal expressed by $s(t) = a(t) \cdot cos\Phi(t)$

$$f_i(t) = \frac{1}{2\pi} \cdot \frac{d}{dt}\left[arg(z(t))\right] \tag{7}$$

where $z(t)$ is the analytic signal defined previously in Eq. (4). Ville went further and noted that since the IF was time-varying, there should intuitively be some instantaneous spectrum associated with it-with the mean value of the frequencies in this instantaneous spectrum being the IF. Using Gabor's average measures [8], he showed that the average frequency in a signal's spectrum was equal to the time average of the IF [9]:

$$\langle f \rangle = \langle f_i \rangle \tag{8}$$

where

$$\langle f \rangle = \frac{\int_{-\infty}^{+\infty} f \cdot |Z(f)|^2 df}{\int_{-\infty}^{+\infty} |Z(f)|^2 df} \tag{9}$$

$$\langle f_i \rangle = \frac{\int_{-\infty}^{+\infty} f_i \cdot |z(t)|^2 dt}{\int_{-\infty}^{+\infty} |x(t)|^2 dt} \tag{10}$$

Time–frequency moments provide an efficient way to characterize signals whose frequencies change in time (having also a sssnonstationary spectrum).

The Fourier transform (FT) is very good at identifying frequency components present in a signal. However, the FT does not identify when the frequency components occur [10].

Only the time–frequency distributions generated by short-time Fourier transform (STFT) or other time–frequency analysis techniques can observe the time varying behavior [11]. Usually, these techniques require serious computational resources. In contrast, transforming the time–frequency distribution results into low-dimension time–frequency moments provide a method for capturing essential features of the signal [12].

3 Test Signals and the Proposed Procedure

This study uses three types of test signals to compare the instantaneous frequency estimation procedures. All these signals were selected and generated in order to have specific and various time varying frequency components.

The first is a Bessel function-based signal and it consists of a set of pulses of decreasing duration separated by regions of oscillating amplitude and fluctuating frequency with an increasing trend. The second is a quadratic chirp sampled at 1 kHz for 4 s which instantaneous frequency is 0 Hz at t = 0 and crosses 20 Hz at t = 2 s and the third is quadratic chirp modulated by a Gaussian with sample rate of 2 kHz and a duration of 4 s. The three test signals are presented on Fig. 1.

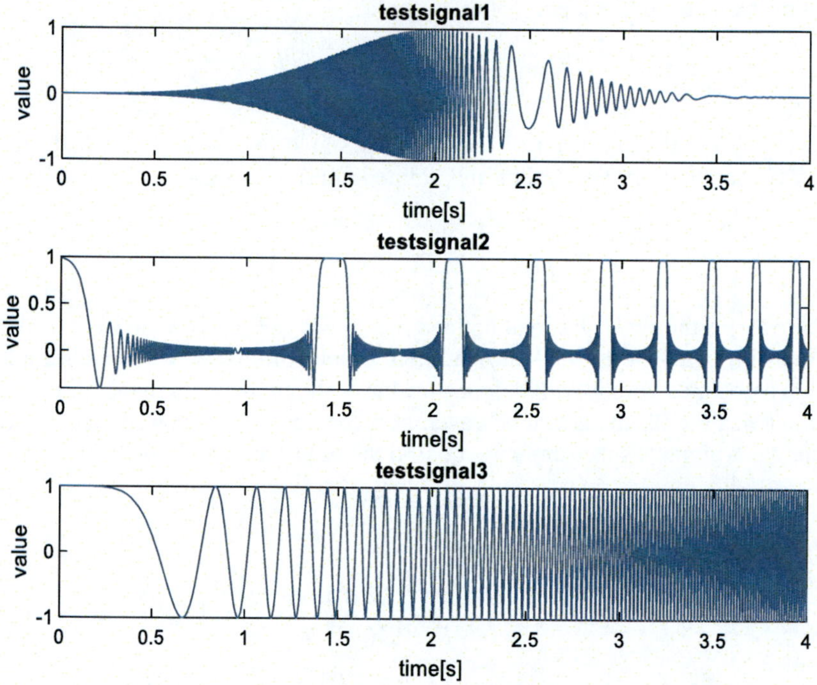

Fig. 1. The used test signals

The signals were resampled in order to have the same length.

Three IF estimation procedures are compared, their schematic representations are illustrated on Fig. 2.

The first estimation computes the instantaneous frequency as the derivative of the phase of the analytic signal using the Hilbert transform. This method accepts only uniformly sampled, real-valued signals.

The second method computes the instantaneous frequency as the first conditional spectral moment of the time–frequency distribution of the signal.

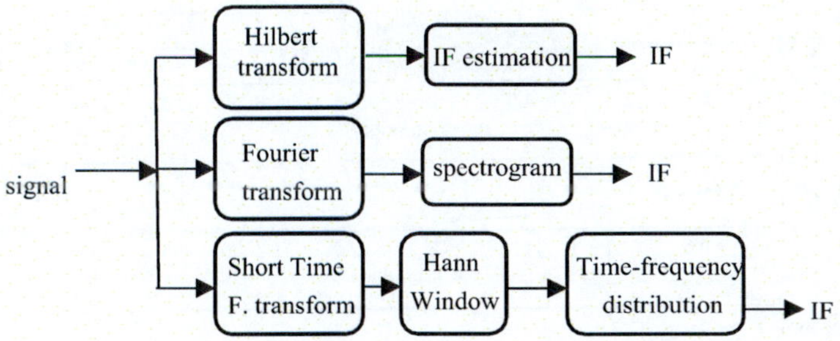

Fig. 2. The proposed IF estimation procedures

The spectrogram which is the visual representation of the spectrum of frequencies of the signal as it changes with time can be generated using several ways, which include Fourier transform, wavelet transform and band-pass filter.

The Short Time Fourier Transform (STFT) serves as time–frequency distribution. By default, the signal is divided into eight segments with 50% overlap, and each segment is windowed with a Hamming window.

The number of frequency points used to calculate the discrete Fourier transforms is 256 or the next power of two greater than the segment length. If the signal cannot be divided exactly into eight segments, it will be truncated.

The third approach uses a the synchrosqueezed Fourier transform, a post-processing method which circumvents the uncertainty relations, inherent to these linear transforms, by reassigning the coefficients in scale or frequency. A 500-sample Hann window is used to divide the signal into segments and window them. From this the spectrogram is estimated and IF computed.

Figure 3,4,5,6 present the obtained instantaneous frequencies for testsignals 1, 2, 3 through the three presented procedures.

Figure 3 allows a good visual verification of the instantaneous frequency both the first image (Hilbert method) and second image (time/frequency distribution from Fourier transform based spectrogram) presents oscillation at the beginning because the signal is truncated to the next power of two greater than the segment length.

All the applied methods produce the same variations in this case.

The estimation of IF for the second signal produces slightly individually results, the fast and irregular changes in the frequency are interpreted in different ways.

As it can be observed, in case of the first test signal, the spectrogram based method produces discontinuities in IF estimation at the final of sequence due to the coefficient truncation.

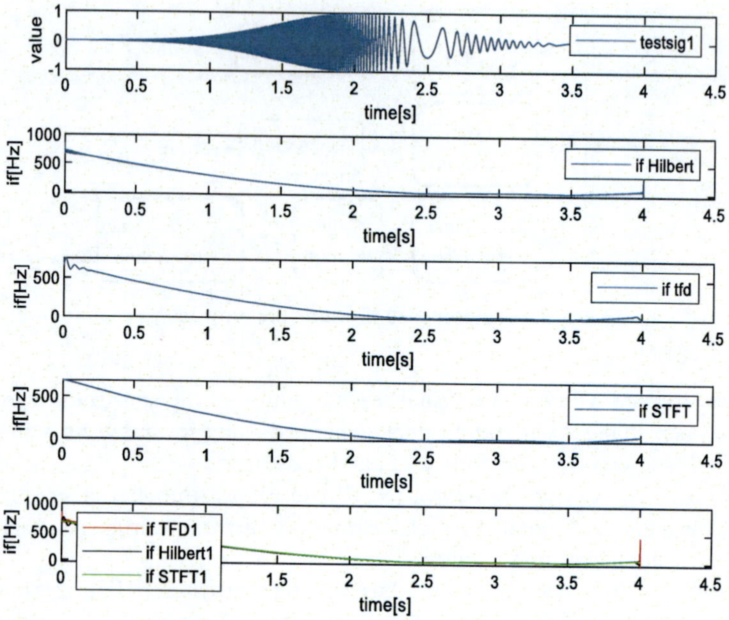

Fig. 3. The obtained Ifs for test signal 1

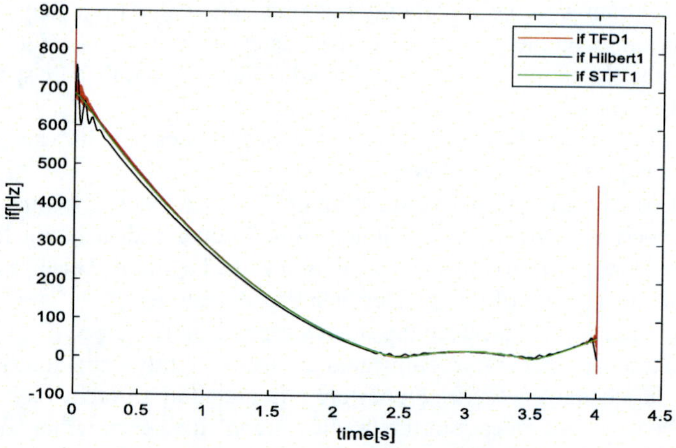

Fig. 4. The discontinuities in IF estimation

The discontinuities at the final of estimated IF are represented more accurately on Fig. 4.

As a comparison study, Fig. 7 presents the three estimations obtained for the first test signal.

In case of the second test signal the time–frequency distribution based method presents slightly differences due to the rescaling in the FFT algorithm (Fig. 8).

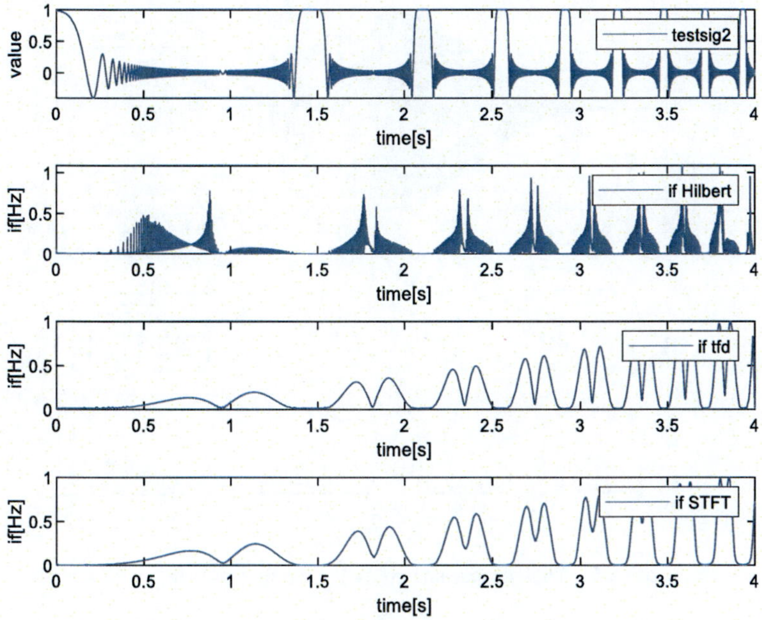

Fig. 5. The obtained Ifs for test signal 2

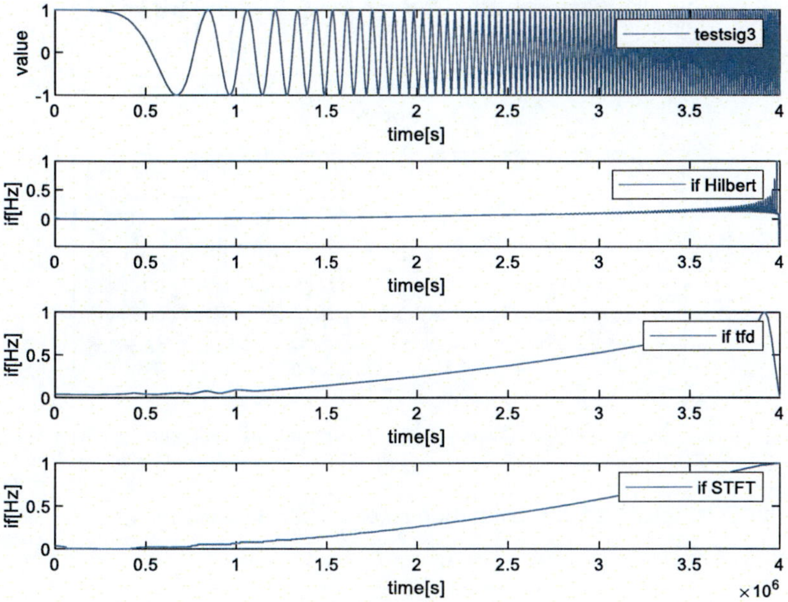

Fig. 6. The obtained Ifs for test signal 3

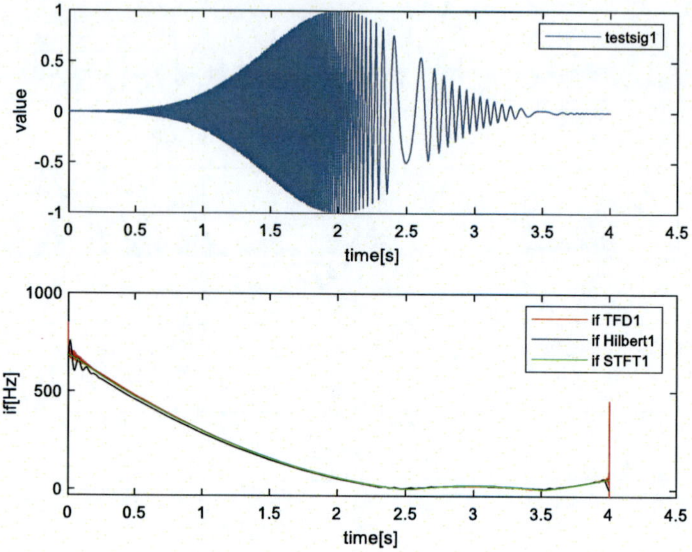

Fig. 7. The comparison study for the first test signal

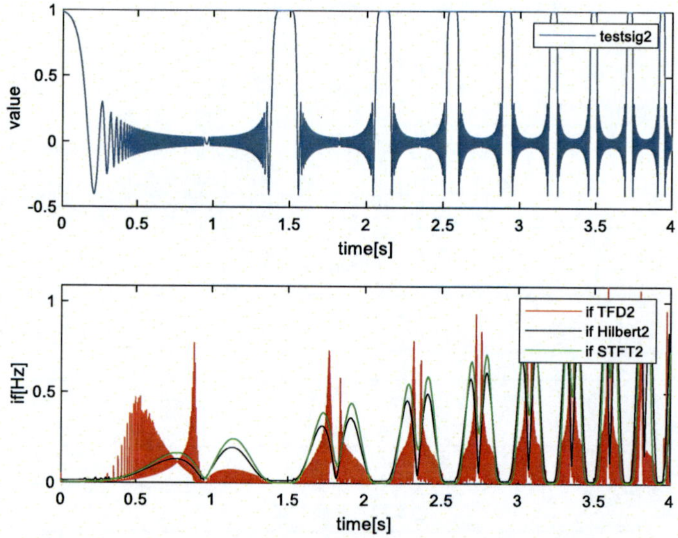

Fig. 8 The comparison study for the second test signal

The third test signal has the estimation results presented on Fig. 9.

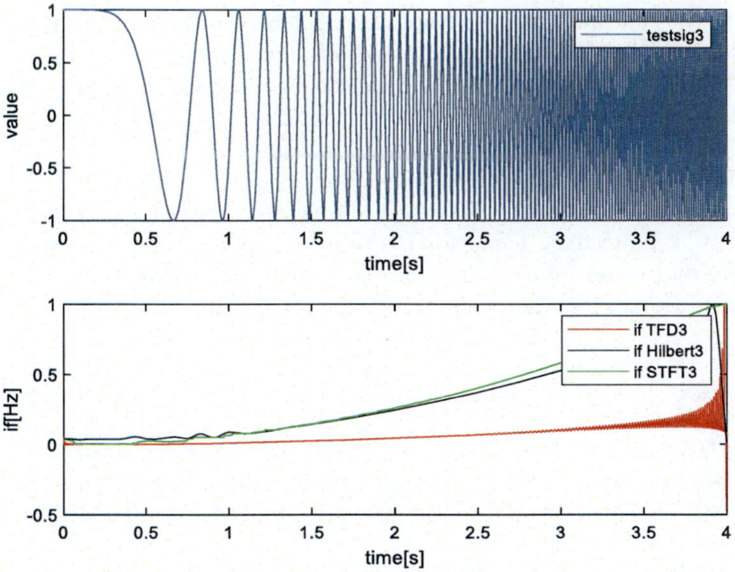

Fig. 9 The instantaneous frequency estimation of the third test signal

4 Conclusions

Instantaneous frequency, taken as the derivative of the phase of the signal, is interpreted in the time–frequency literature as the average frequency of the signal at each time.

In present study three synthesized signals were analyzed, their instantaneous frequencies were estimated in three different ways. From spectral behavior point of view, the synthetic signals have special properties, nonstationary frequency variations and amplitude changes. These test signals were generated to have specific but different properties in time and frequency domains, the applied estimation methods also were different, the results show the advantages and drawbacks of these.

The Hilbert transform estimates the instantaneous frequency of a signal for monocomponent signals only, otherwise is not possible to form the analytic signal.

The Short Time Fourier Transform based spectrogram offers a good time–frequency distribution which depends on sequence length and window type but due to the truncation can appear oscillatory errors. The length of signal is important, because on it depends the number of subbands.

The best results are obtained with the synchrosqueezed Fourier transform, where the reassignment of coefficients in frequency and the windowing smooths the spectrum. provides a sharp, concentrated representation, while remaining invertible. This method received a renewed interest and provides guarantees for the decomposition of a multicomponent signal.

It is obvious from the obtained results that different methods are producing the different time–frequency distributions, but frequencies present in the Fourier spectrum are not always present all the time in a signal. In conclusion, the time–frequency representation depends on the number of bands in which data has been divided. as the signal is decomposed into more narrow bands, true frequencies present in the signal under analysis are revealed, frequency resolution is also increasing.

As a final concluding remark, the IF estimation depends strongly on the analyzed signal, modifying the signal length through truncation to a next power of two, also can have unwanted effects.

As further work, the IF estimation methods could be extended on noisy signals. The relationship between noise level and estimation precision, could be an important task, mostly in biomedical application. The information from a heart rate variability signal spectrum could be extended with the instantaneous frequency information, a better classification of frequency domains would be obtained.

References

1. Gabor, D.: Theory of communication. Proc. IEEE, **93**(III), 429–457 (1946)
2. Robertson, S.: Practical ESM Analysis, pp. 255–315. Artech House, Norwood (2019)
3. Seilmayer, M., Ratajczak, M.: A guide on spectral methods applied to discrete data in one dimension. J. Appl. Math. **1**, 1–27 (2017)
4. Boashash, B.: Time-Frequency Signal Analysis and Processing. A Comprehensive Reference, 2nd edn. Academic Press, New York (2016)
5. Marks, R.J.: Handbook of Fourier Analysis & Its Applications, vol. 800. Oxford University Press, Oxford (2009)
6. Hlawatsch, F., Auger, F.: Time-Frequency Analysis: Concepts and Methods, 1st edn. Wiley, New York (2008)
7. Boashash, B.: Estimating and interpreting the instantaneous frequency of a signal—Part 1: fundamentals. In: Proceedings of the IEEE®, vol. 80, pp. 520–538. (1992)
8. Boashash, B.: Estimating and interpreting the instantaneous frequency of a signal—Part 2: algorithms and applications. In: Proceedings of the IEEE, vol. 80, pp. 540–568 (1992)
9. Feldman, M.: Hilbert transform methods for nonparametric identification of nonlinear time varying vibration systems. Mech. Syst. Signal Process. **47**(1–2), 66–77 (2014)
10. Zeng, Z., Amin, M.G., Shan, T.: Arm motion classification using time-series analysis of the spectrogram frequency envelopes. Remote Sens. **12**, 454 (2020)
11. Tu, G., Dong, X., Chen, S., Zhao, B., Hu, L., Peng, Z.: Iterative nonlinear chirp mode decomposition: a Hilbert-Huang transform-like method in capturing intra-wave modulations of nonlinear responses. J. Sound Vib. **485**, 115571 (2020)
12. Képesi, M., Weruaga, L.: Adaptive chirp-based time–frequency analysis of speech signals. Speech Commun. **48**(5), 474–492 (2006)

Nonlinear Denoising of Nonstationary Signals

Zoltán Germán-Salló(✉)

University of Medicine, Pharmacy, Sciences and Technology "George Emil Palade", Targu-Mures, Romania
zoltan.german-sallo@umfst.ro

Abstract. The paper presents different time-frequency decomposition-based filtering procedures. These procedures are applied to various types of synthesized nonstationary test signals which are contaminated with different levels of Gaussian white noise. Empirical mode decomposition (EMD), discrete wavelet transform (DWT) and wavelet packet transform (WPT) based procedures are implemented in order to perform nonlinear multiscale filtering. The obtained results are compared and analyzed; conclusions summarize the comparison study.

Keywords: Wavelet transform · Time-frequency decomposition · Empirical mode decomposition · Nonlinear filtering

1 Introduction

In signal processing applications, the assumption of stationarity of the signal introduces unwanted limitations, procedures can be applied only in idealized conditions. The signals in acoustics, geophysics, biology, and biomedicine have a nonstationary behavior which demands another kind of approach. The nonstationarity of a signal may be caused by different reasons, as the signal sources, propagation channel or receivers. In all these, noise appears obviously and could have a nonpredictable and undesired effect.

Nowadays, applications of nonlinear filtering range from engineering, machine learning [1], economic science [2] and natural sciences such as geoscience [3]. Valuable applications of nonlinear filtering can be found also in biomedical engineering applications [4].

This paper presents three different nonlinear filtering procedures. A comparison study reflects the different obtained filtering results. The main parameters are the acquired signal to noise ratio and the absolute value of the reconstruction error.

Splitting the signal in components, processing the components separately and reconstruct the signal from these, is the main idea in every multiscale signal analysis and processing task. Usually, the different time-frequency analyzing methods offer various decomposition algorithms, the choice is made depending on followed or required parameters or benefits.

The paper is organized as follows. After the introductive notions, the second paragraph introduces the theoretical background of multiscale analysis, the third paragraph presents the used signals and the proposed procedures. After that, experimental and theoretical results are presented, finally, the concluding remarks end this study.

L. Moldovan and A. Gligor (Eds.): Inter-Eng 2021, LNNS 386, pp. 843–855, 2022.
https://doi.org/10.1007/978-3-030-93817-8_75

2 Time-Frequency Decomposition Techniques

2.1 Empirical Mode Decomposition

The Empirical Mode Decomposition (EMD) was proposed as the fundamental part of the Hilbert–Huang transform (HHT) [5]. The EMD decomposes a signal in the time domain through an iterative sifting process, into a set of functions, known as Intrinsic Mode Functions (IMFs). The IMFs offer a description and a reconstruction of the signal, even though they are not necessarily orthogonal as in case of other transforms.

The decomposition algorithm obtains smooth envelopes defined by local maxima and minima of a given sequence and performs a subsequent subtraction of the mean of these envelopes from the initial sequence. This procedure named sifting process needs the identification of all local extrema (maxim and minim values). These extrema are connected by cubic spline lines to create the upper and the lower envelopes. The sifting process decomposes the signal into IMFs in following way:

For a signal $s(t)$, where m_1 is the mean of its upper and lower envelopes the first component h_1 is computed as follows:

$$h_1 = s(t) - m_1 \tag{1}$$

The first step in procedure when the mean value of the two envelopes is computed, is presented on Fig. 1.

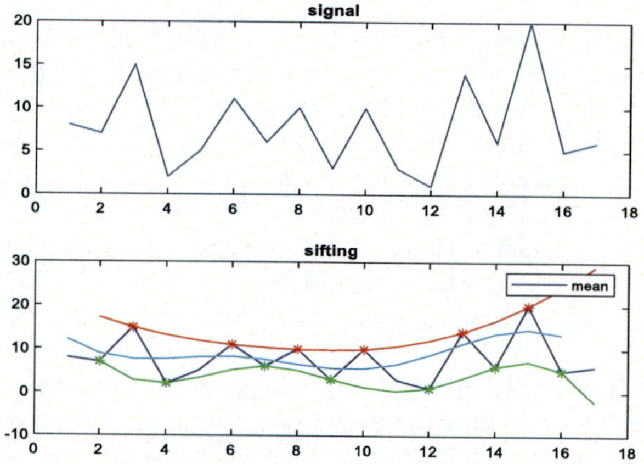

Fig. 1. The first step in sifting process.

In the second sifting process, h_1 is treated as the initial signal $s(t)$, and m_{11} is the mean of h_1's upper and lower envelopes:

$$h_{11} = h_1 - m_{11} \tag{2}$$

This sifting procedure is repeated k times, until h_{1k} is an IMF, that is:

$$h_{1k} = h_{1(\mathrm{k}-1)} - m_{1k} \tag{3}$$

Then $c_1 = h_{1k}$ will be the first IMF component from the analyzed sequence, which contains the shortest period component of the signal. After separation:

$$s(t) - c_1 = r_1 \tag{4}$$

the previously presented procedure is repeated on r_j:

$$r_1 - c_2 = r_2, \ldots r_{n-1} - c_n = r_n. \tag{5}$$

The result is a set of functions; the number of functions in the set depends on the original signal. The decomposition process stops when the residue r(t) becomes a constant or a function without extrema that no longer satisfies the requirements of an IMF.

Finally, the analyzed signal can be expressed as:

$$s(t) = c_1 + c_2 + \cdots c_k + r_k \tag{6}$$

The decomposition is made without external functions, only the signal's properties are used. In conclusion, EMD decomposes any given sequence or signal, whether it is linear, nonlinear or nonstationary, into a set of IMFs [6].

2.2 The Discrete Wavelet Transform (DWT)

In digital signal processing the Fourier Transform is probably the most popular, but many other transforms are available for engineers and mathematicians. Every transformation technique has its applicability with advantages and disadvantages in a certain research area. In Fourier analysis, the Discrete Fourier Transform (DFT) decompose a signal into sinusoidal basis functions of different frequencies and amplitudes. The signal can be completely recovered l from its DFT (FFT) components [7].

In wavelet analysis, the Discrete Wavelet Transform (DWT) decomposes a signal into a set of mutually orthogonal wavelet basis functions, which differ from sinusoidal basis functions in that they are spatially localized. Furthermore, wavelet functions are dilated, translated and scaled versions of a common function φ, known as the mother wavelet. The DWT is equivalent to a tree-structured discrete filter bank where the signal is first filtered by a lowpass and a highpass filter to yield lowpass and highpass subbands, approximation and detail components (A, D) as presented on Fig. 1.

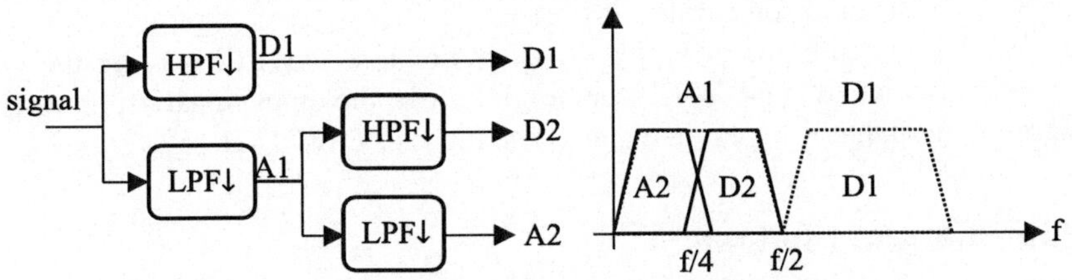

Fig. 2. Illustration of DWT implementation through filter bank.

The lowpass subband is iteratively filtered by the same scheme to yield narrower octave-band lowpass and highpass subbands [8]. In the DWT, the filter outputs are downsampled at each successive stage to eliminate the redundancy, keeping the same length of decomposition, so it returns a data vector of the same length as the input is.

Figure 2 shows that subsequent levels of the DWT operate only on the outputs of the lowpass (scaling) filter. At each level, the DWT divides the signal into octave bands [9].

2.3 Wavelet Packet Transform (WPT)

The wavelet packet transform (WPT) differ from the discrete wavelet transform (DWT) because the transform yields in equal-width subband filtering of signals which is opposed to the coarser octave band filtering obtained in the DWT. WPT is superior at time-frequency analysis because divides the frequency axis into finer intervals (Fig. 3).

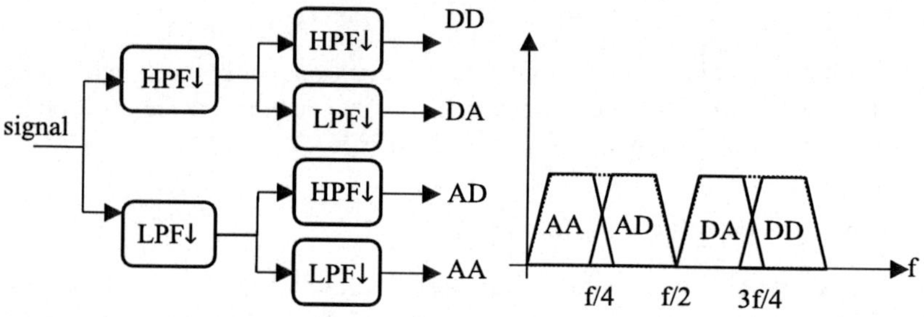

Fig. 3. Illustration of WPT.

Both of lowpass and highpass subbands are iteratively filtered by the same scheme to yield narrower octave-band lowpass and highpass subbands, obtaining detail of approximation and approximation of detail coefficients (AA, AD, DA, DD). In the non-redundant WPT the outputs of the bandpass filters are downsampled by two as in case of DWT [10].

In addition to filtering a signal into equal-width subbands at each level, the WPT partitions the signal's energy among the subbands, providing superior frequency resolution. As a result, the WPT can separate nearby frequency components which usually fall in the same octave band in case of DWT.

The advantage of DWPT is that it is possible to select the optimal representation of the signal with respect to some criterion.

The WPT best representation of a signal is of three types: it may be the DWT representation (included in the WPT scheme); it may be the input signal itself; or it may be a representation that is better than that of the input or DWT [11].

3 Materials and Methods

3.1 The Filtering Procedure

This study uses three types of test signals, three filtering procedures are compared, these can be observed on Fig. 4.

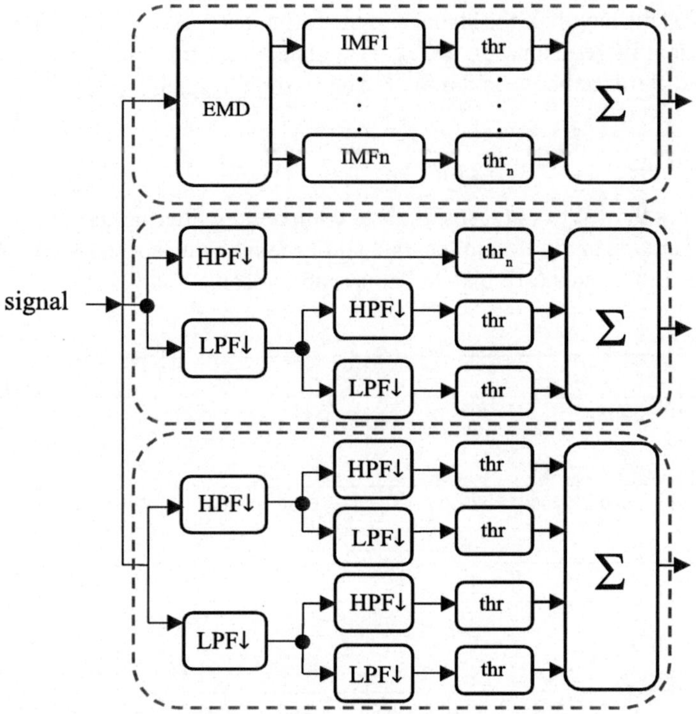

Fig. 4. The proposed filtering methods.

The all three filtering methods use threshold for components, the signal is reconstructed from the truncated components. For every filtering the same thresholds are applied to see the difference between obtained parameters.

The procedures are compared through signal to noise ratio (SNR) and reconstruction error. The power of signal $x = [x_i]$ and noise $n = [n_i]$ are defined as

$$\mathrm{P_S} = \frac{1}{\mathrm{N}}\sum\nolimits_{\mathrm{i=1}}^{\mathrm{N}} \mathrm{x_i^2}, \quad \mathrm{P_n} = \frac{1}{\mathrm{N}}\sum\nolimits_{\mathrm{i=1}}^{\mathrm{N}} \mathrm{n_i^2} \tag{7}$$

The initial signal to noise ratio SNR_i (with known noise) and the obtained signal to noise ratio SNR (where x_f is the filtered signal) are:

$$SNR_i = 10 \cdot lg\left({}^{P_S}/_{P_N}\right) \tag{8}$$

$$P_{Sf} = \frac{1}{N}\sum\nolimits_{i=1}^{N} (x_{fi})^2 \tag{9}$$

$$P_{ne} = \frac{1}{N}\sum\nolimits_{i=1}^{N} (x_{ni} - x_{fi})^2 \tag{10}$$

$$SNR = 10lg\left(\frac{P_{Sf}}{P_{ne}}\right) \tag{11}$$

where N is the length of the signal and the estimated noise is the difference between filtered and nonfiltered signal $n_e = x_n - x_f$. The reconstruction error is the absolute value of the difference between the filtered and original signal.

3.2 Test Signals

This study uses three types of test signals to compare the filtering results. These signals are of different type and different lengths, all of them having special features.

The three test signals (original and noisy) are presented on Fig. 5.

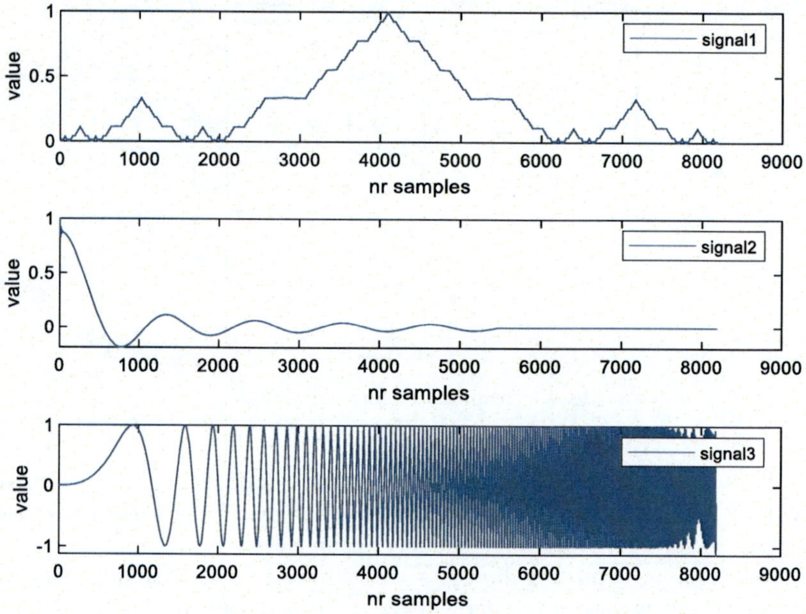

Fig. 5. The test signals.

To estimate the filtering results, different levels of Gaussian white noise were added to the signal, see Fig. 6. The noisy signal is taken as a sum between signal and noise, this study uses six level of added noise. Previously, the signals were preprocessed; resampled to have the same length and normed to have the same level (in this case the added noise level is the percentual the same). The length of the signal is 8192 (a power of two, usually, in this case the numerical procedures product better results).

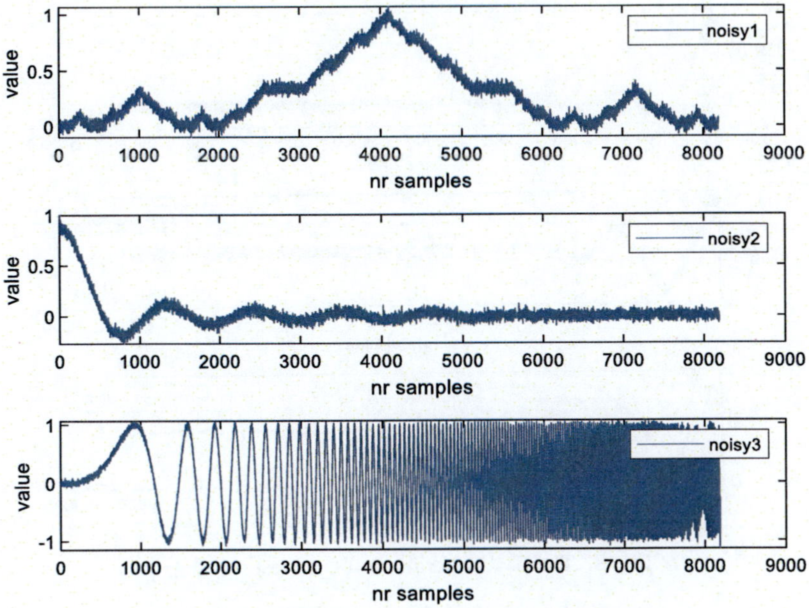

Fig. 6. The test signals with added Gaussian white noise.

4 Experimental Results

This study uses signals of equal lengths, the noise is added in six steps, the level is 1%–6% of signals maximum value (Fig. 7).

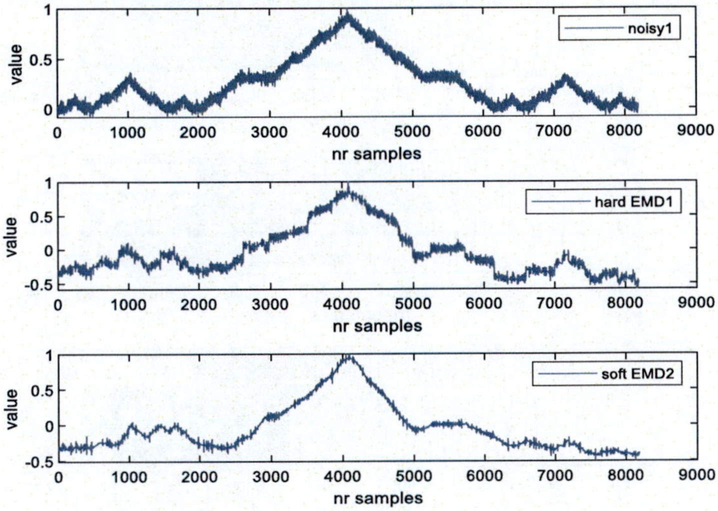

Fig. 7. The EMD filtering results on testsignal1 (level noise 5).

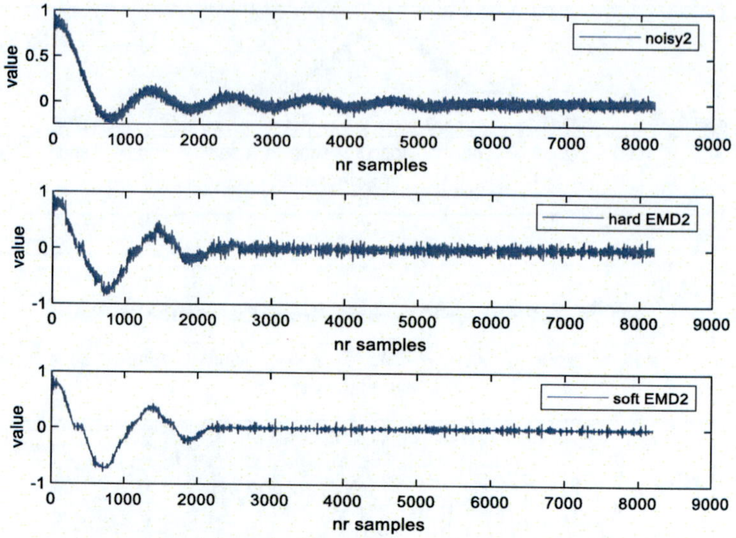

Fig. 8. The EMD filtering results on test signal 2.

The thresholding was made according to signal level in every subband, the threshold values were estimated through the same rules (based on the principle of Stein's Unbiased Risk). Both of soft and hard thresholding procedures were applied, the results can be seen on Fig. 8.

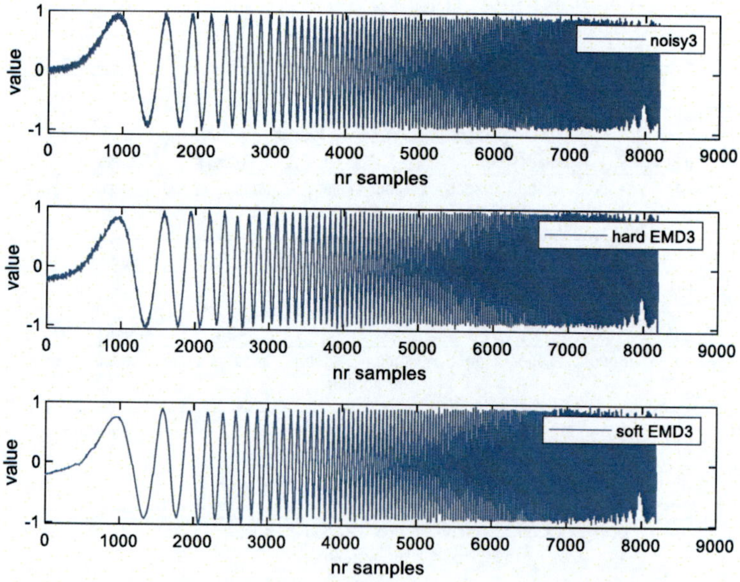

Fig. 9. The EMD filtering results on testsignal3 (noise level 5).

The next table contains the measured parameter obtained through the EMD based filtering procedure (Table 1).

Table 1. The parameters obtained for EMD based filtering results.

n	SNRi1	SNR1	e1	SNRi2	SNR2	e2	SNRi3	SNR3	e3
1	31.510	0.954	0.316	24.009	3.723	0.050	36.535	14.423	0.087
2	25.683	0.765	0.311	18.182	4.130	0.049	30.708	14.691	0.105
3	22.089	0.497	0.339	14.588	5.794	0.054	27.114	13.742	0.093
4	19.511	0.792	0.308	12.010	5.035	0.065	24.536	11.667	0.128
5	17.670	0.678	0.315	10.169	4.117	0.079	22.695	12.634	0.111
6	16.109	0.876	0.297	8.6089	4.839	0.064	21.134	13.448	0.137
7	14.759	0.234	0.368	7.2582	4.121	0.094	19.784	12.940	0.135

The noise was added in seven levels, the obtained signal to noise ratio is strongly dependent on signal features. Figure 9 presents the variation of reconstruction errors, which is taken as the absolute value of the difference between original and filtered signal. The first signal, due to his properties has the greatest reconstruction error. Figure 10 brings the values of obtained signal to noise ratios versus the initial ones.

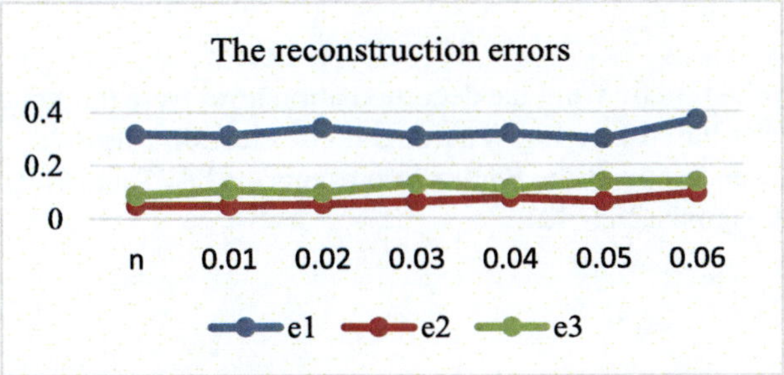

Fig. 10. The reconstruction errors in case of EMD based filtering.

Table 2 contains the initial and the obtained signal to noise ratio values from DWT and WPT filtering and the reconstruction errors of these procedures.

The noise level in this case was set between 1% and 6% of the normed signals value.

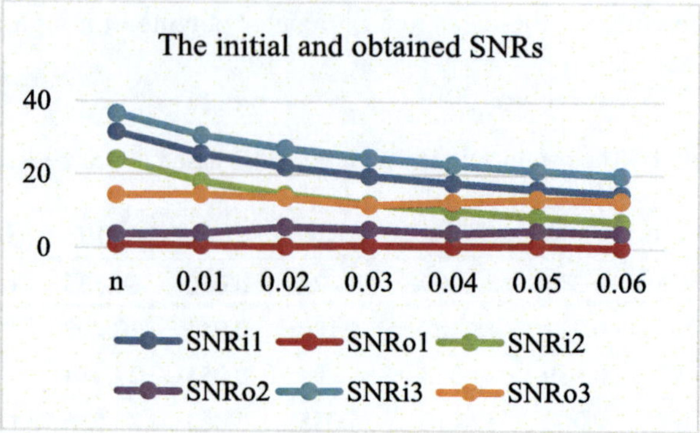

Fig. 11. The initial and obtained signal to ratios in case of EMD based filtering.

Table 2. The obtained parameters in case of DWT and WPT based filtering.

Noise [%]	1	2	3	4	5	6
SNRI1	31.5162	25.6272	21.9741	19.4939	17.6737	16.1093
SNRW1	31.6974	25.8218	22.3774	19.7786	17.751	16.3392
SNRWP1	31.6136	25.8589	22.2189	19.7273	17.9487	16.4154
eW1	0.0025	0.004	0.0055	0.0061	0.0067	0.0084
eWP1	0.0027	0.0044	0.0063	0.008	0.0101	0.012

The used wavelet function and the decomposition level were the same for both DWT and WPT transform based filtering. The threshold selection rules were the same as in case of Ifs in EMD decomposition. Figure 11 presents the DWT and WPT based filtering results for the test signals (Fig. 12).

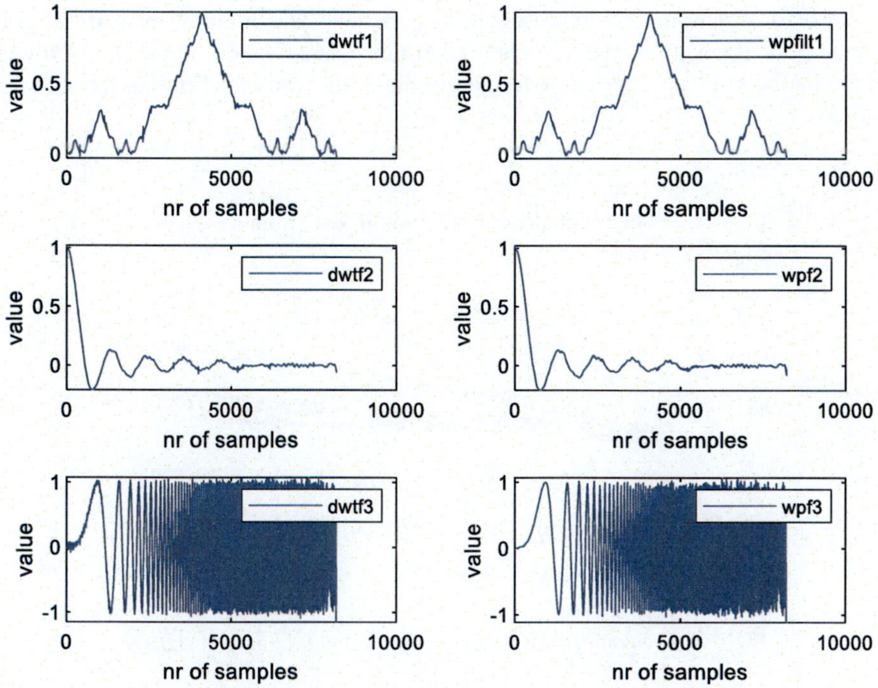

Fig. 12. The DWT and WPT based filtering results.

The resulted signal to noise ratios after the DWT and WPT filtering are illustrated on Fig. 13. The results show that this kind of filtering produces good results due to the decomposition and thresholding in subbands.

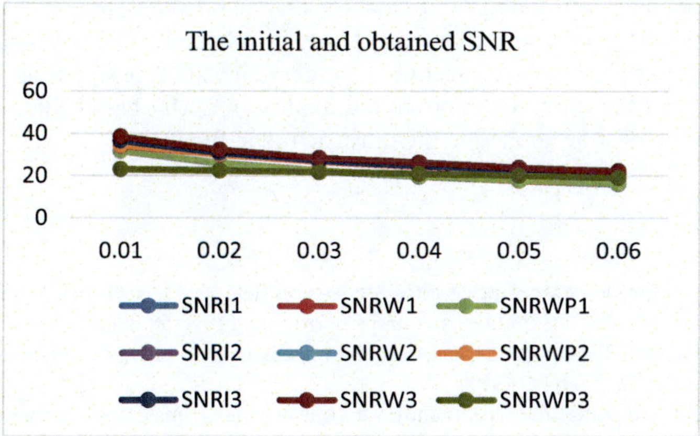

Fig. 13. The initial and obtained SNR in case of DWT and WPT filtering

The reconstruction error representation completes the filtering evaluation, good obtained SNR does not means always a good noise cancellation, the shape of filtered signal must be as close as possible to the original, other way the filtering has no sense (Fig. 14).

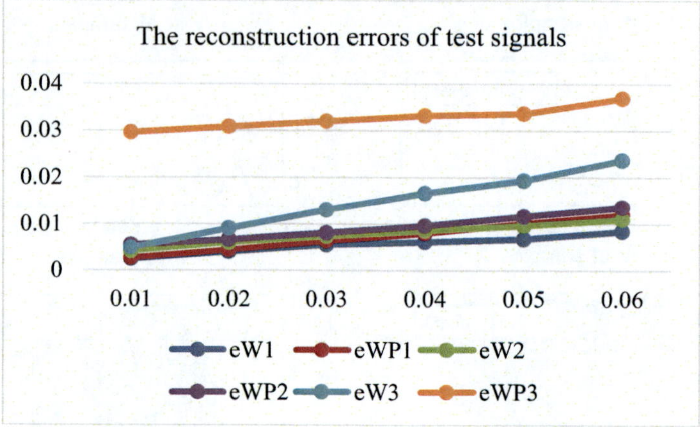

Fig. 14. The comparison between filtered and original test signals

5 Conclusions

The aim of this study is to compare different methods in nonstationary signal filtering, preserving as much as possible the given signal parameters. The main parameters in examining the methods were the obtained signal to noise ratio and the reconstruction error. Experimental results demonstrate that the DWT and WPT subband filtering were more efficient due to the accurate decomposition in subbands. The advantage of EMD is that there is no need of basic functions, the decomposition is signal driven without involving external function. To improve the quality of EMD based filtering, a better thresholding rule must be applied to the intrinsic mode functions.

References

1. Eden, U.T.: Point process adaptive filters for neural data analysis: theory and applications. In: 46th IEEE Conference on Decision and Control, pp. 5818–5825 (2007)
2. Brigo, D., Hanzon, B.: On some filtering problems arising in mathematical finance. Insurance Math. Econom. **22**, 53–64 (1998)
3. Evensen, G.: Sequential data assimilation with a nonlinear quasi-geostrophic model using Monte Carlo methods to forecast error statistics. J. Geophys. Res. **99**, 10143 (1994)
4. Hassani, M., Karami, M.-R.: Improved nonlinear noise estimation in EEG signal based on Volterra filters. Biomed. Eng.: Appl. Basis Commun. **29**(04), 1750026 (2017)

5. Komaty, A., Boudraa, A.O., Augier, B., et al.: EMD-based filtering using similarity measure between probability density functions of IMFs. IEEE Trans. Instrum. Meas. **63**(1), 27–34 (2014)
6. Huang, N.E., et al.: The empirical mode decomposition and the Hilbert spectrum for non-linear and non stationary time series analysis. Proc. R. Soc. London A **454**, 903–995 (1998)
7. Wickerhauser, M.V.: Adapted Wavelet Analysis from Theory to Software. A.K. Peters, Wellesley (1994)
8. Percival, D.B., Walden, A.T.: Wavelet methods for time series analysis. In: Cambridge Series in Statistical and Probabilistic Mathematics. Cambridge University Press, Cambridge (2000)
9. Graps, A.: An introduction to wavelets. IEEE Comput. Sci. Eng. **2**(2), 50–61 (1995)
10. Bultheel, A.: Learning to swim in a sea of wavelets. Bull. Belg. Math. Soc. Simon Stevin **2**, 1–45 (1995). https://doi.org/10.36045/bbms/1103408773
11. Donoho, D.L.: De-noising by soft-thresholding. IEEE Trans. Inf. Theory **42**(3), 613–627 (1995)

Author Index

L. Moldovan and A. Gligor (Eds.): Inter-Eng 2021, LNNS 386, pp. 857–859, 2022.
https://doi.org/10.1007/978-3-030-93817-8